57th International Conference on Experimental Stress Analysis (EAN 2019)

AA004957

Luhacovice, Czech Republic
3-6 June 2019

Editors:

Jindřich Petruška
Tomáš Návrat
Lubomír Houfek
František Šebek

ISBN: 978-1-5108-8969-9

Printed from e-media with permission by:

Curran Associates, Inc.
57 Morehouse Lane
Red Hook, NY 12571

Some format issues inherent in the e-media version may also appear in this print version.

Copyright© (2019) by Czech Society for Mechanics (CSM)
All rights reserved.

Printed by Curran Associates, Inc. (2019)

For permission requests, please contact Czech Society for Mechanics (CSM)
at the address below.

Czech Society for Mechanics (CSM)
Dolejskova 5
Praha 8
182 00
Czech Republic

Phone: +420 266 053 045
Fax: +420 286 587 784

csm@csm.cz

Additional copies of this publication are available from:

Curran Associates, Inc.
57 Morehouse Lane
Red Hook, NY 12571 USA
Phone: 845-758-0400
Fax: 845-758-2633
Email: curran@proceedings.com
Web: www.proceedings.com

EXPERIMENTAL STRESS ANALYSIS

57th International Scientific Conference

June 3 – June 6, 2019, Luhačovice, Czech Republic

EAN 2019
57th conference on experimental stress analysis

CONFERENCE PROCEEDINGS

Edited by
Jindřich Petruška
Tomáš Návrat
Lubomír Houfek
František Šebek

Czech Society for Mechanics
Brno University of Technology – Faculty of Mechanical Engineering

All rights reserved. No parts of this publication may be reproduced, stored in retrieval system or transmitted in any form or by means of: electronic, magnetic tape mechanical, photocopying, recording or otherwise without permission in writing from the publishers.

Title: Experimental Stress Analysis 2019
Conference Proceedings

Editors: Jindřich Petruška
Tomáš Návrat
Lubomír Houfek
František Šebek

Published by: Czech Society for Mechanics

The complete set of EAN 2019 Conference Proceedings consists of two parts:

- Book of Extended Abstracts
- USB Flash Drive (Full Papers)

Organizing committee

prof. Ing. Jindřich Petruška, CSc.	*(Brno University of Technology, Czech Republic)*
doc. Ing. Miloš Vlk, CSc.	*(Brno University of Technology, Czech Republic)*
doc. Ing. Tomáš Návrat, Ph.D.	*(Brno University of Technology, Czech Republic)*
Ing. Lubomír Houfek, Ph.D.	*(Brno University of Technology, Czech Republic)*
Ing. František Šebek, Ph.D.	*(Brno University of Technology, Czech Republic)*

Scientific committee

Ing. Karel Doubrava, Ph.D.	*(Czech Technical University in Prague, Czech Republic)*
prof. Ing. Nikolaj Ganev, CSc.	*(Czech Technical University in Prague, Czech Republic)*
doc. Ing. Radim Halama, Ph.D.	*(VŠB – Technical University of Ostrava, Czech Republic)*
prof. Dr. Ing. František Holešovský	*(J E Purkyně University in Ústí nad Labem, Czech Republic)*
prof. Ing Stanislav Holý, CSc.	*(Czech Technical University in Prague, Czech Republic)*
Ing. Lubomír Houfek, Ph.D.	*(Brno University of Technology, Czech Republic)*
prof. RNDr. Miroslav Hrabovský, DrSc.	*(Palacký University in Olomouc, Czech Republic)*
prof. Ing. Josef Jíra, CSc.	*(Czech Technical University in Prague, Czech Republic)*
Ing. Aleš Lufinka, Ph.D.	*(Technical University of Liberec, Czech Republic)*
doc. Ing. Tomáš Návrat, Ph.D.	*(Brno University of Technology, Czech Republic)*
doc. Ing. Pavel Padevět, Ph.D.	*(Czech Technical University in Prague, Czech Republic)*
doc. Ing. Iva Petríková, Ph.D.	*(Technical University of Liberec, Czech Republic)*
prof. Ing. František Plánička, CSc.	*(University of West Bohemia, Czech Republic)*
prof. Ing. Milan Růžička, CSc.	*(Czech Technical University in Prague, Czech Republic)*
Ing. Jaroslav Václavík	*(Research and Testing Institute Plzeň, Czech Republic)*
doc. Ing. Miloš Vlk, CSc.	*(Brno University of Technology, Czech Republic)*
doc. Ing. Robert Zemčík, Ph.D.	*(University of West Bohemia, Czech Republic)*

All the papers were reviewed by members of the conference scientific committee.

Contents

Comparison comfort of the car seat at a freezing temperature

BARTONÍČEK Pavel[1,a], MARTONKA Rudolf[1,b] and RYVOLOVÁ Martina[2,c]

[1]Technical university of Liberec, Studentska 2, Liberec, Czech Republic

[2]Institute for Nanomaterials, Advanced Technology and Innovation, Bendlova 1407/7, Liberec, Czech Republic

[a]pavel.bartonicek@tul.cz, [b]rudolf.martonka@tul.cz, [c]martina.ryvolova@tul.cz

Keywords: Car seat, Contact pressure field, Thermos-comfort

Abstract. The seat comfort is assessed from several aspects. Static comfort and dynamic comfort are defined. In both cases, the seat is comfortably seated on the seat. The main properties of these comforts are the mechanical, vibro-isolation properties of the PUR car seat. The properties of the PUR filler are mainly temperature-dependent (freezing temperature) and also the PUR hygroscopic for chemical reasons. In this paper, the influence of temperature on the size of the contact pressure distribution and thus on the mechanical properties of the PUR filler and the whole car seat is described.

Introduction

One of the parameters describing the overall characteristics of the seat is the mechanical comfort characteristics. It was not the seated person's seat characteristics. In the short term, the person sitting on the automobile seat mainly considers the feeling of contact with the seat - the size of the contact area, the place with the greater contact pressure. After sitting in the seat and getting the first impressions of static comfort, the seated man starts and uses the vehicle to ride. At this point, the seat behaves differently. The seat is operated in dynamic mode. This state can be described by dynamic comfort. In this case, the sitting person observes the unpleasant feelings of the selections that enter him from the moving vehicle to the seat. One of the characteristics of the incoming vibrations to humans is the transmission characteristics of the car seat Fig.1. This characteristic not only depends on the mechanical characteristics of the seat itself, but also on the size (weight) of the seated person. This is the so-called working point on the hysteresis characteristics of the seat Fig. 2. The PUR padding has the greatest influence on the shape of the seat hysteresis curve. When PUR material is used in non-standard conditions such as frost, the material solidifies, ceases to be flexible and pliable. The seat is hard and no longer has the original comfort features. For this reason, the use of automotive seat heating has started, among other things. In addition to heating a seated person, the task is to return the standard use conditions to the seat in the shortest possible time.

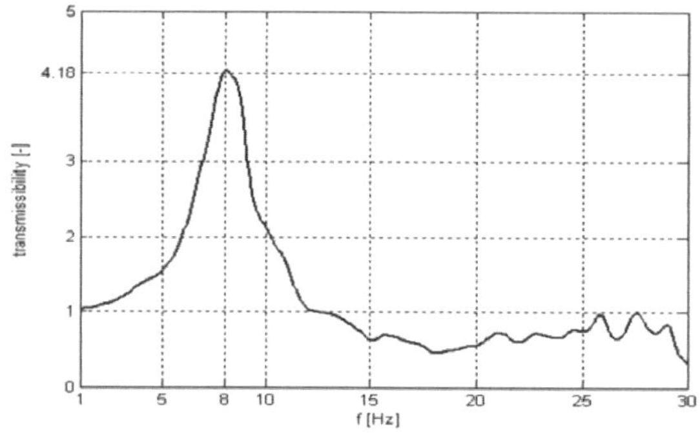

Fig. 1: Transmissibility characteristic of the car seat

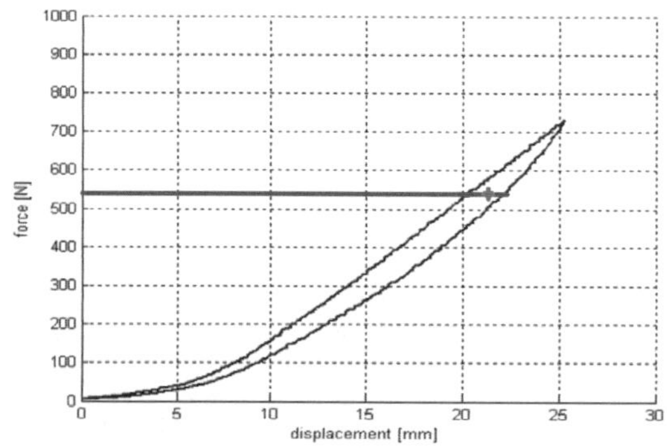

Fig. 2: Hysteresis curve of car seat

Experiment

For the experimental verification, a serial car seat Fig. 3 equipped with electric heating.This seat was placed in a special air-conditioned box. The K-type thermocouple (Ni-NiCr) thermocouples thermocouples with a measuring range of -40 ° C to 200 ° C were placed on the seat. A contact pressure sensor from X-senaor Technology Corporation, Canada was used to determine the shape and size of the contact pressure fields.

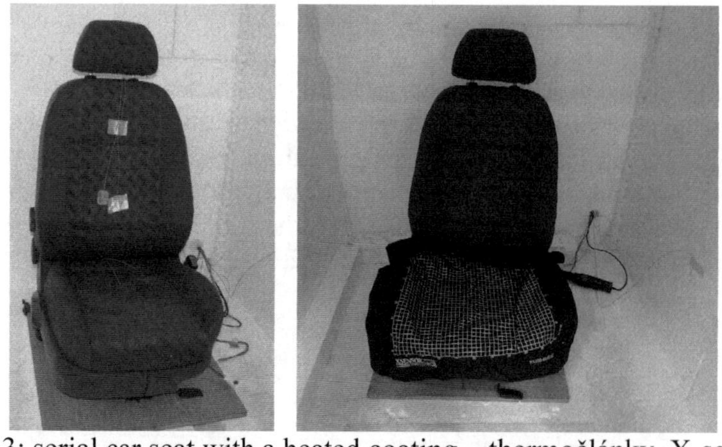

Fig. 3: serial car seat with a heated coating – thermočlánky, X-senzor

The thermocouple set was deployed in the area of the seat and back to specific locations, such as instead of the expected peak contact pressure - below the hip joints and the position of the wire guide and the position beyond that path. A total of 5 temperature characteristics on the seat were measured and 6 were referenced on the inside of the box climate. Experimenting was the simulation of getting into a frozen car, turning on the seat heating to the maximum and "turning on" the heating of the car salon by heating. This warming up of the car salon was simulated by opening one of the box walls and thus the possibility of changing the freezing air with warm air from the outside area by the laboratory. At the same time contact pressure distribution maps were measured over time. Exemplary features are shown in Figs 4 and 5.

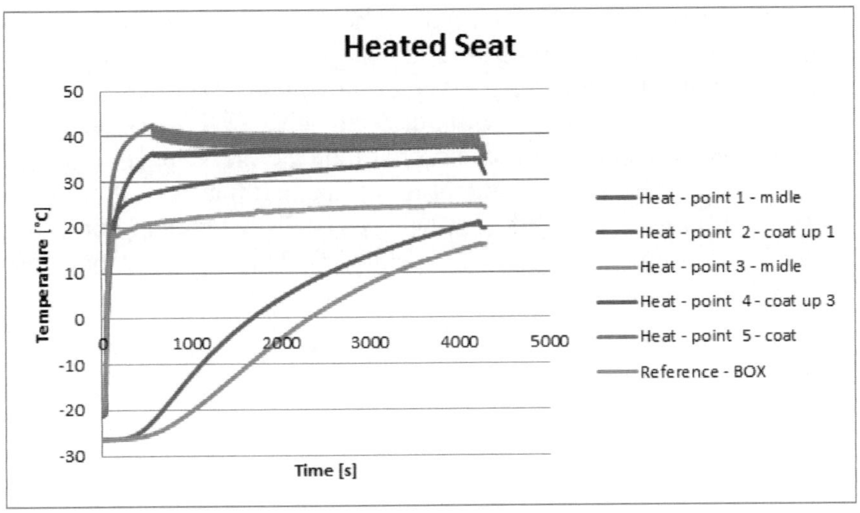

Fig 4: teplotní charakteristiky

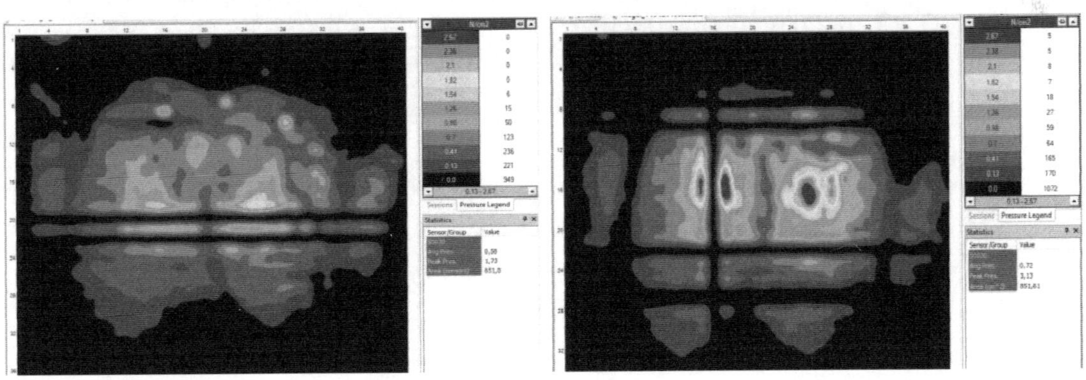

Fig. 5: mapy kontaktního tlaku při teplotě -20°C and 20 °C

Results

As can be seen from the light brown curve, the air in the box changed during the first minute. The remaining curves above this reference curve correspond to the points on the top of the coating, either at or outside the heating wire. The two curves below this reference curve correspond to the locations within the PUR fill. As can be seen from the curves, these sites did not warm up above 0 ° C or 1500 ° C. 2000 sec - 25 and 33 minute. The same applies to pressure maps. Although there was already a temperature of 20 ° C in the contact area after the first minutes, the pressure field showed an incorrect contact pressure distribution of the frozen seat. The return of the standard pressure distribution occurred at the same time as the temperature I in the core of the core fill, ie after 1 hour of warming.

Conclusions

Thus, the following conclusions can be drawn from the measured experimental results. The distribution of contact pressures at different temperatures makes it possible to assess the effect on static and dynamic comfort, the discomfort of the car seat. It is evident from the measured temperature characteristics that even in the case of forced heating of the seat, the car seat does not obtain the original proposed mechanical, vibroisolation properties. Thus, if in winter we only use the vehicle for short, several minute journeys, the utomobil seat behaves like a hard, rigid part without much vibro-isolation properties. Thus, not only cold starts and short journeys do not indicate the vehicle engine but also other components in the vehicle. The worse it is that it is a car seat that is in direct interaction with man. The question would be how such a seat behaves at higher temperatures, for example in hot summer, when the warmth of the car is above 50 ° C. This could be the subject of further research.

Acknowledgement

This publication was written at the Technical University of Liberec as part of the project "Innovation of products and equipment in engineering practice" with the support of the Specific University Research Grant, as provided by the Ministry of Education, Youth and Sports of the Czech Republic in the year 2018. This work was supported by the Ministry of Education, Youth and Sports of the Czech Republic and the European Union - European Structural and Investment Funds in the frames of Operational Programme Research, Development and Education - project Hybrid Materials for Hierarchical Structures (HyHi, Reg. No. CZ.02.1.01/0.0/0.0/16_019/0000843).

References

[1] V. F. Ganong, Přehled lékařské fyziologie. Praha: Galen, 2005. 890 s. ISBN: 80-7262-3111-7.

[2] F.B. Mazari, A Study On The comfort and Thermo-Physiological Properties of Car seats, Summary of the Thesis, TUL, 2017

[3] M. P. Reed and col.: An Investigation of Driver Discomfort and Related Seat Design Factors in Extended-Duration Driving, Journal of Passenger Cars, Vol. 100, Section 6, pp. 130-159, 1991

[4] C. Sengul, H. Abdollahofakhr: Automatic design of wiring pattern for car seat Heaters, Master Thesis, Jönköping University, 2010

[5] A. Elarusi, A. Attar, H. Lee: Optimal Design of a Thermoelectric Cooling/Heating for Car Seat Comfort, Journal of Electronic Materials, April 2017, Volume 46, Issue 4, pp 1984–1995

EAN 2019
57th conference on experimental stress analysis

The validation of numerical models of potted inserts in sandwich structures based on experimental testing

CABRNOCH Bohuslav[1,a], KRÁL Michal[1,b] and KRUML Josef[1,c]

[1]Czech Aerospace Research Centre, Department of Composite Technologies, Beranových 130,199 05 Praha - Letňany, Czech Republic

[a]cabrnoch@vzlu.cz, [b]kral@vzlu.cz, [c]kruml@vzlu.cz

Keywords: spacecraft, potted insert, finite element method, experimental testing, stiffness

Abstract. The main goal of this paper is to describe the research of an appropriate approach to modelling of potted inserts in sandwich panels used in spacecraft structures. Potted inserts are used for the connection of sandwich panel with other parts of the assembly. In other words, they are used for the introduction of forces into sandwich panel. The main goal of the research was not to find the most accurate manner of modelling but to find an optimal manner within the meaning of compromise between accuracy of prediction and model complexity. The main observed property was the out-of-plane stiffness of the insert. This property affects global behaviour of the structures besides in terms of natural frequencies which is one of the key observed characteristics of spacecraft structures.

Introduction

Sandwich structures are frequently used in many industrial sectors especially in the design of rolling stocks and spacecraft structures. Sandwich structures consist of two thin skins with high in-plane strength and stiffness connected by the lightweight core with high out-of-plane shear stiffness. This combination ensures excellent stiffness-to-weight proportion in bending. To utilize this benefit, it is necessary to ensure a sufficient design solution of introduction of forces into the sandwich panel. This problem is very often solved by potted inserts. Based on requirements of individual structure, the insert is designed as partially potted, fully potted or as through the whole thickness insert. During FE modelling of sandwich structures, it must be ensured that model will predict the stiffness and strength of potted insert with sufficient precision. The main goal of this paper is to describe experimental validation of created FEM models, which are used for modelling of panels used in spacecraft structures. These structures are subjected to static and dynamic environments, where the stiffness of the potted inserts significantly influences the global behaviour of the structure.

The main goal of the presented paper is to find optimal proportion between the size of FE model in terms of number of elements and accuracy of predicted potted insert stiffness in out-of-plane direction. The requirements to potted insert model are different for individual analysis types. Linear static analysis is usually performed to determine stress-strain response of the structure and thanks to high computing capacity it is possible to model the structure in detail. Other analysis types like modal analysis or random response are more demanding to computational capacity and the size of FE model should stay on a satisfactory level.

Review of current state of the art

Design and sizing principles of sandwich structures with potted inserts in terms of structural strength is usually based on methodologies described in design handbooks, see e.g. [1]. These handbooks provide sufficient sizing methods. Numerical modelling of damage and growth of individual damage mechanisms is very often the objective of research projects and can provide a good tool for the understanding of damage mechanisms evolution. The research of damage evolution using different types of FE models is performed e.g. in [2], where the main objective is to compare different approaches to discretization of honeycomb core. Core is usually modelled as 3D continuum using volume elements. Authors have shown that better estimation of response of potted inserts is achieved by modelling honeycomb in detail using 2D elements. Undoubtedly, detailed honeycomb model is better for correct modelling of damage growth. Anyway, in engineering practice, detailed modelling seems to be highly time consuming and it is necessary to find a compromise between precision and time requirements. The accuracy can be satisfactory also using 3D continuum model of the core as shown in [3].

FEM models

To evaluate the effect of the number of elements, more FE models were built and further predicted stiffness was compared with experimental data. Models were created in Femap pre-processor. The example of created FE model is presented in Fig. 1. Two planes of symmetry were taken into account. The distribution of load was performed using RBE3 element. The core was modelled using 3D elements using 3D orthotropic material model. Skins were modelled using laminate type element as 2D orthotropic material model. Insert and resin were modelled as 3D isotropic continuum using 3D elements. Analysis was performed as a linear static using Nastran solver.

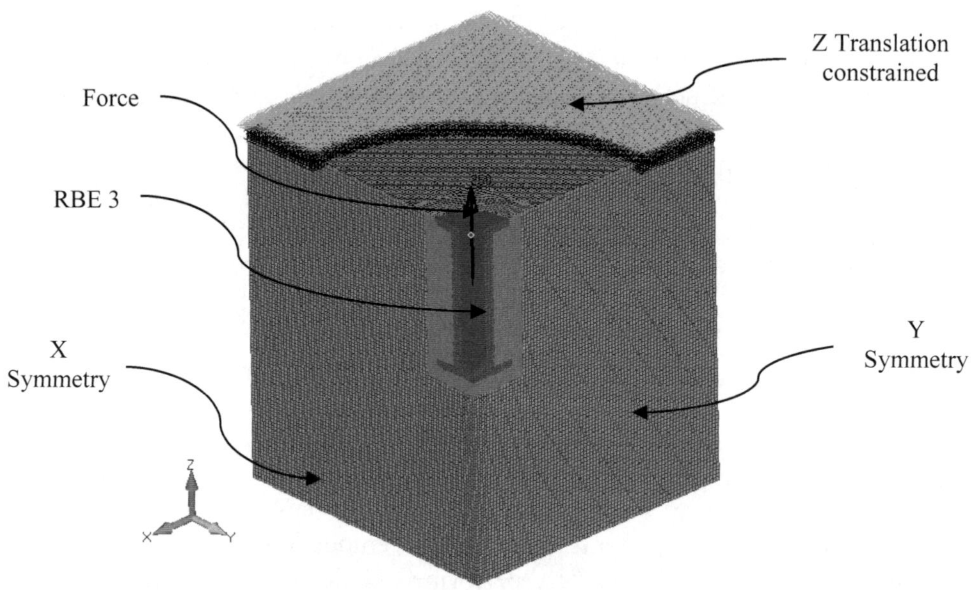

Fig. 1: Example of created FE model with two plane of symmetry

FE model presented in Fig. 1 is highly complex with 675 209 3D volume elements and 13 586 2D laminate elements. Other model with lower complexity was created. In this case 772 3D volume elements and 378 2D laminate elements were used. Both models are compared in Fig. 2.

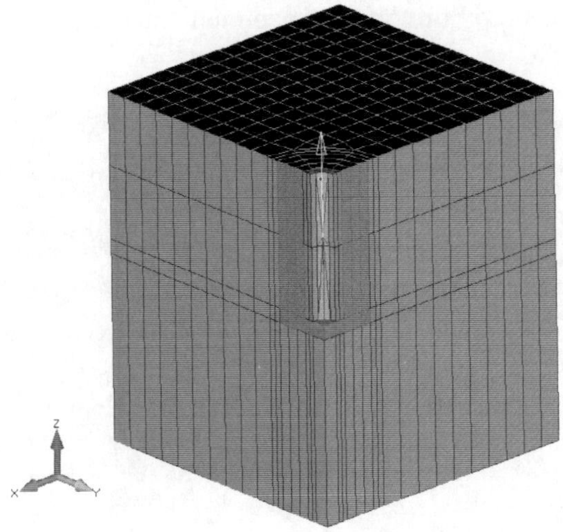

Fig. 2: Detailed FE model of potted insert (left one) and model with the lower number of elements (right one)

Experimental arrangement

The setup of experimental testing is presented in Fig. 3. Testing was performed using Instron test machine. Load rate was 2 mm/min. Loading was stopped when critical load has been reached, i.e. at 2 mm deflection of the skin.

Fig. 3: Detailed FE model of potted insert (left one) and model with the lower number of elements (right one)

Displacements were measured using optical system Q400 DIC - Digital Image Correlation System by Dantec Dynamics, see [4]. Digital image correlation is a full field image analysis method based on grey digital images. This experimental method allows determining displacements of observed object in three dimensions. Speckle pattern was made by spraying. See Fig. 4, where the boxes with recognizable pattern are indicated. Displacements were measured both on test specimen and on the jig to obtain results without affecting by the softness of the jig.

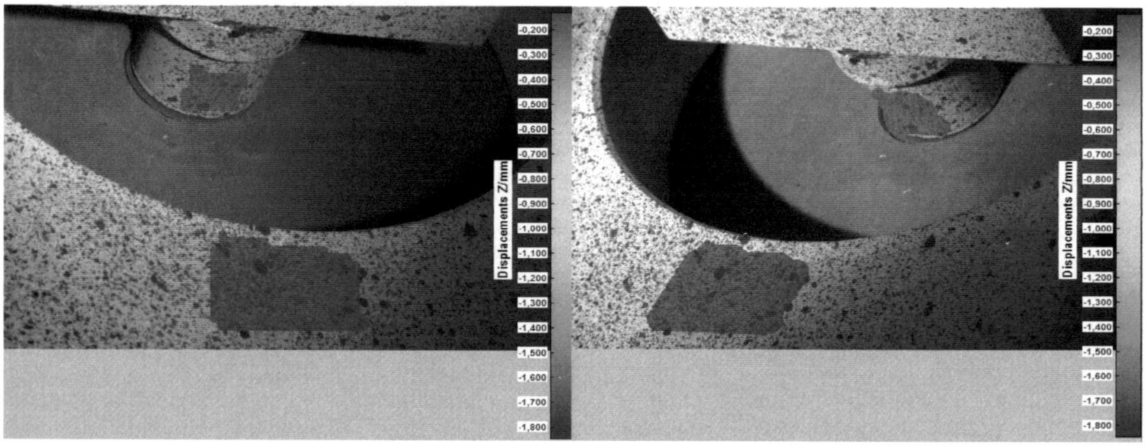

Fig. 4: Designation of boxes with recognizable pattern on test specimen and on the jig

Results

Measured characteristic of insert stiffness are presented as a dependence of displacement and force, see Fig. 5. Stiffness was evaluated as secant stiffness on the force base 0.5 kN – 1.5 kN. Five test specimens were tested for each design solution of sandwich or insert. Example of test specimens is shown in Fig. 6. Evaluated stiffness of inserts with statistical characteristics is shown in Tab. 1. Mean value is 26602 N/mm. It is obvious that the results show a great dispersion. The value of correlation coefficient is 37.5%. This value is caused also by a small statistical sample.

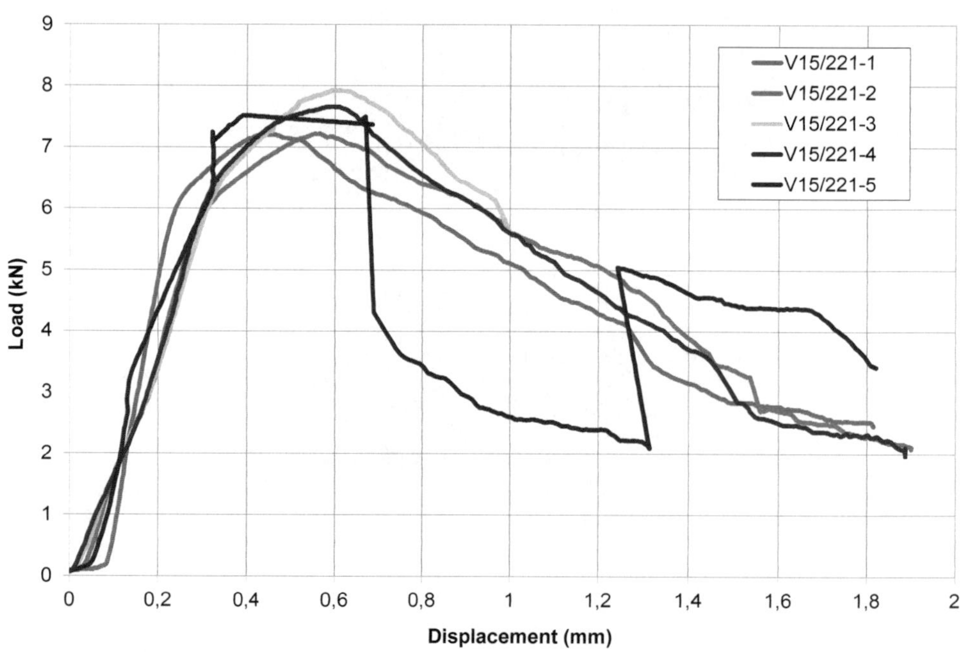

Fig. 5: Results of testing of five specimens V15/221-1 - V15/221-5

Generally it is recommended to use more test specimens. But in order to perform a basic validation of FE model, five specimens is a satisfactory quantity.

Stiffness predicted by FE models is shown in Tab. 2. It is obvious that the complexity of FE model does not have a significant effect on out-of-plane stiffness prediction. This was an important finding because it was proofed that the complexity of FE model can be changed if needed without affecting of results. Also the conformity between experimental data and numerical prediction is satisfactory. The fraction of experiment and simulation is 0.82 for the model with high complexity and 0.79 for model with lower complexity. Numerical prediction can be improved by defining of contact problem between test specimen and jig. Used boundary conditions, defined in Fig. 1, do not allow the creation of the gap between test specimen and jig which is not in the line with reality and causes higher stiffness of the model.

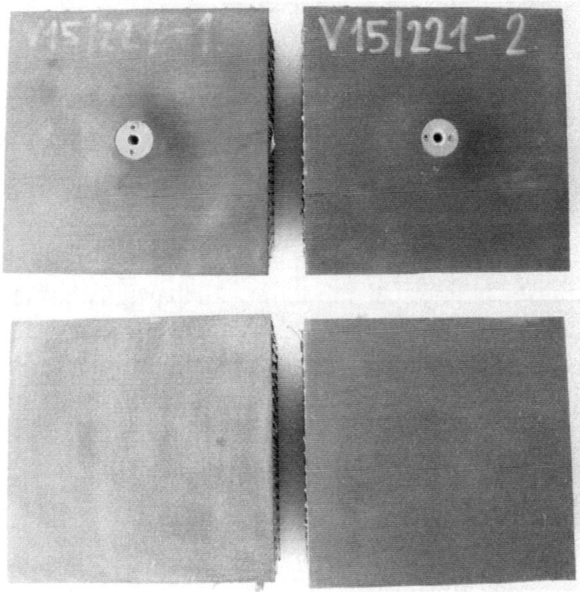

Fig. 6: Two examples of used test specimens after the test - bottom and upper side is shown

Tab. 1: Experimental results

Test specimen	Stiffness [N/mm] 0.5 - 1.5 kN
V15/221 - 1	42853
V15/221 - 2	24639
V15/221 - 3	20067
V15/221 - 4	17451
V15/221 - 5	28000
Mean [N/mm]	26602
St. Deviation [N/mm]	9954,3
Variation Coef. [%]	37,5

Tab. 2: Numerical results

Numerical simulation	Predicted Stiffness [N/mm]
High complexity model	32479
Low complexity model	33412

Conclusions

It is obvious that experimental testing still plays a very important role in the development of aerospace and other structures and in validation of numerical models. The development of spacecraft structures is always associated with detailed testing and verification. The reasons are as follows - high cost of spacecraft structures, very low number of structures already built and therefore resulting low practical experience.

The attention was concentrated to out-of-plane stiffness of inserts. In next step also in-plane stiffness will be verified because in-plane stiffness has also important effect on the behaviour of the sandwich structure. It can be stated that the effect of the number of elements on predicted stiffness is not as significant as expected and the difference is up to 5% which is acceptable.

Acknowledgements

This research was performed with financial institutional support from government budget through the Ministry of Industrial and Trade of the Czech Republic.

References

[1] ESA European Space Agency. Space Engineering – Insert Design Handbook. ESA Requirements and Standards Division, March 2011. ECSS-HB-32-22A.

[2] R. Seeman, D. Krause, Numerical modelling of partially potted inserts in honeycomb sandwich panels under pull-out loading, Composite Structures. 2018 (203) 101-109.

[3] P. Bunyawanichakul, B. Castanié, J. J. Barrau, Non-linear finite element analysis of inserts in composite sandwich structures, Composites Part B: Engineering. 2008, 39(7-8) 1077-1092.

[4] Dantec Dynamics, Q-400 µDIC™ Microscopic Shape, Displacement & Strain Measurement, Publication no. 354_v1, 2015, Dantec DynamicsGmbH.

EAN 2019
57th conference on experimental stress analysis

Design concept of sandwich beams deflection

CÉZOVÁ Eliška

Czech Technical University in Prague, Faculty of Mechanical Engineering, Department of Designing and Machine Components, Czech Republic

eliska.cezova@fs.cvut.cz

Keywords: Beam deflection, Cantilever beam, Sandwich beam, Elastic curve

Abstract. The aim of this paper is to introduce a framework for designing simple sandwich beams. Such sandwiches are assumed to be composed of multiple interconnected individual sub-beams made of various materials, having different cross-sectional profiles and different lengths. The work presented in this paper offers a simple framework for evaluation of the deflection of sandwich beams, with application to solution of cantilever sandwich beam composed of arbitrary number of sub-beams of unequal length.

Introduction

The sandwich beams are not new in principle, but the recent development of material science and technology has brought new ideas and possibilities to their practical implementations. Thus also naturally the attempts to develop a simple methods for their design got an important impulse. The concept itself is inspired by a similar approach previously seen e.g. in leaf springs, multi-chamber beams (aluminum/plastic) or in sandwich composite structural elements.

The individual beams are aligned and connected in parallel to each other in order to assure that they deform together, following the same elastic deflection curve. As an example let's consider a cantilevered beam shown in the Fig. 1.

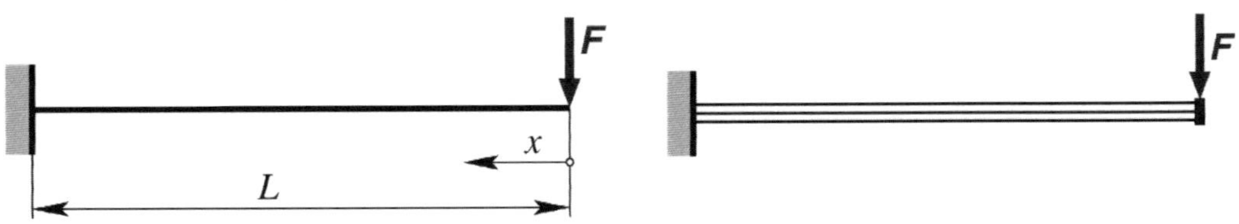

Fig. 1: Cantilever beam scheme (left) and its sandwich version composed of three parallel beams (right).

The beam global load (shear $T(x)$ and moment $M(x)$) distribution is independent to the internal structure (composition) of the whole beam.

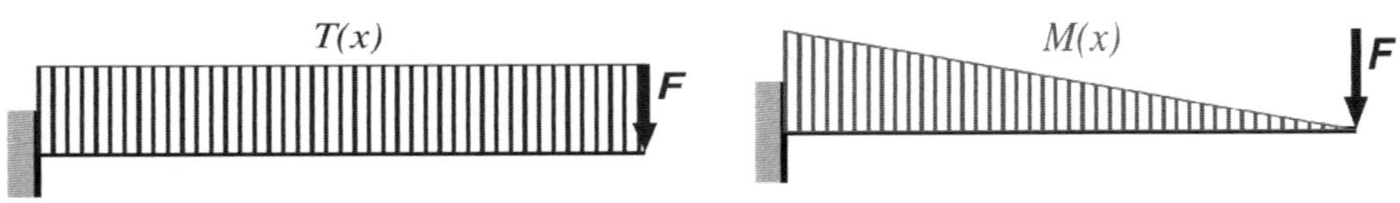

Fig. 2: The shear force (left) and bending moment (right) distribution along the beam.

The (sandwich) internal structure of the beam results into a redistribution of the loading among the individual beams, while keeping their deformation identical for all individual sub-beams of the sandwich group. In the simplest case, assuming that only the shear force is transferred between the individual beams at the point where they are interconnected, shear load is internally redistributed to individual sub-beams depending on their stiffness, while respecting the deformation condition of equal deflection for all sub-beams in the sandwich. The situation is analogical, from the elasticity point of view, to springs attached in parallel and jointly loaded by a force [1,2]. See the illustration in Fig. 3

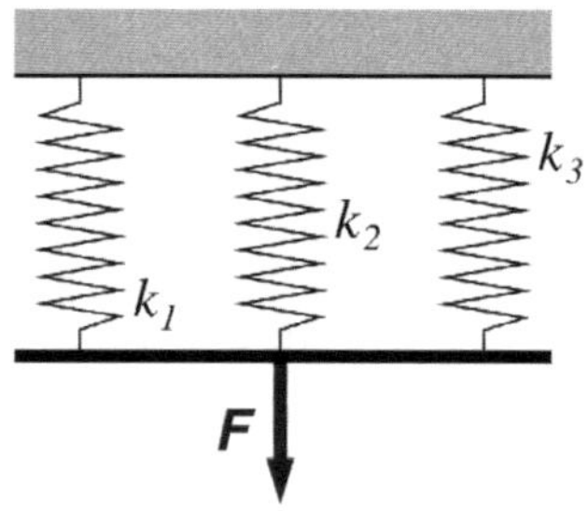

Fig. 3: Springs attached in parallel have a joint equivalent stiffness $k=k_1+k_2+k_3$.

In the same way the total stiffness of the whole sandwich beam can be assumed as an algebraic sum of the stiffnesses of the individual beams. E.g. the sandwich beam depicted in the Fig 1 (right), composed of three individual beams, can be represented by a simple beam with equivalent stiffness (flexural rigidity) $EI=E_1I_1+E_2I_2+E_3I_3$. The material's modulus of elasticity E_i and beam's profile moment of inertia I_i can be different for each individual sub-beam.

Remark: In the case of the mechanical spring analog shown in the Fig. 3 the total force F is redistributed to individual sprigs 1, 2, and 3 proportionally to their stiffness as $F_1:F_2:F_3=k_1:k_2:k_3$, where $F=F_1+F_2+F_3$.

Deflection of a beam

Single sandwich cantilever beam

Assuming that the cantilever beam is composed of a single sandwich, where all the sub-beams have the same length (like in the Fig. 1 - right), the joint equivalent stiffness can be directly used to determine the beam deflection, i.e. the beam elastic curve equation $v(x)$.

$$\frac{d^2v}{dx^2}(x) = \frac{M(x)}{EI} \quad \text{where} \quad M(x) = F \cdot x \tag{1}$$

$$\frac{d^2v}{dx^2}(x) = \frac{F \cdot x}{E_1I_1+E_2I_2+E_3I_3} = \frac{F}{EI} \cdot x \tag{2}$$

From this second order differential equation, the elastic curve can be obtained by double integration [3,4].

$$\frac{dv}{dx}(x) = \frac{F}{EI}\left(\frac{x^2}{2} + C_1\right) \tag{3}$$

$$v(x) = \frac{F}{EI}\left(\frac{x^3}{6} + C_1x + C_2\right) \tag{4}$$

The integration constants C_1, C_2 can be obtained from the boundary conditions at the fixed end of the beam, i.e. at $x=L$, where $v(L) = 0$ and $\frac{dv}{dx}(L) = \varphi(L) = 0$. The displacement and slope of the elastic curve can be finally expressed as:

$$\frac{dv}{dx}(x) = \frac{F}{2EI}(x^2 - L^2) \tag{5}$$

$$v(x) = \frac{F}{6EI}(x^3 - 3L^2x + 2L^3) \tag{6}$$

Remark - on practical realisation of sandwich beams: The sandwich principle can be used to build custom beams with wide range of cross-sectional profiles, allowing to mechanically join together different materials. There exist infinitely many technical realisations of such sandwich beams. As an example, some profiles of sandwich beams composed of simple beams with standard profiles are shown in the Fig. 4.

Fig. 4: Example of several sandwich beam profiles built of three individual sub-beams.

Multiple fields cantilever sandwich beam

Besides of building beams with custom profiles, the sandwich construction can be used to reinforce the beams in parts, where the load is higher. In this way the stress can be better redistributed among the profiles in the sandwich, leading to optimized material use and possibly lower mass of the whole beam for given loading forces. In this context, as an example, considering the moment diagram in the Fig. 2, it seems natural to build a sandwich beam with only one beam at the tip (where the moment is small) and to add reinforcement beams while going towards the wall, where the moment attains its maximum value. A schematic view of such sandwich beam with three individual sub-beams of unequal length is shown in the Fig. 5.

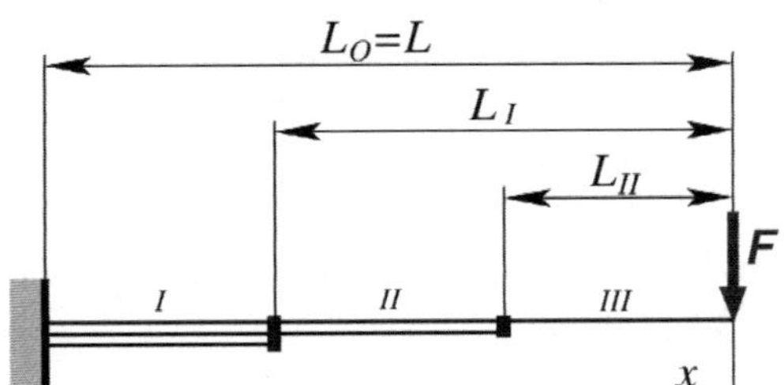

Fig. 5: Sandwich beam with three individual sub-beams and three loading fields.

The deflection of the beam shown in the Fig. 5 can be obtained by applying the above described double integration of the differential equation (1), where *E(x)* and *I(x)* are considered variable (piece-wise constant), with possible jumps in places where the number of sub-beams is changing. E.g. in the field *I*, $(EI)_I= E_1I_1+E_2I_2+E_3I_3$, in the field *II*, $(EI)_{II}= E_1I_1+E_2I_2$, while in the field *III*, $(EI)_{III}= E_1I_1$. The deflection of the beam is solved from the fixed end (where boundary conditions are known), sequentially field by field, by taking the displacement and the slope of the elastic curve at the end of one field as the initial condition for the next field.

Considering the cantilever sandwich beam depicted in the Fig. 5, we can write (according to (3) and (4)) the displacement and slope of the elastic curve in the field I as:

$$\frac{dv_I}{dx}(x) = \frac{F}{(EI)_I}\left(\frac{x^2}{2} + C_{1,I}\right) \tag{7}$$

$$v_I(x) = \frac{F}{(EI)_I}\left(\frac{x^3}{6} + C_{1,I}x + C_{2,I}\right) \tag{8}$$

which is subject to boundary conditions at $x=L$, where $v(L) = 0$ and $\frac{dv_I}{dx}(L) = 0$, leading to integration constants

$$C_{1,I} = \frac{-L^2}{2} \quad \text{and} \quad C_{2,I} = \frac{L^3}{3} \quad . \tag{9}$$

The elastic curve in the field II can be obtained from the equations analogical to (7), (8), i.e.

$$\frac{dv_{II}}{dx}(x) = \frac{F}{(EI)_{II}}\left(\frac{x^2}{2} + C_{1,II}\right) \tag{10}$$

$$v_{II}(x) = \frac{F}{(EI)_{II}}\left(\frac{x^3}{6} + C_{1,II}x + C_{2,II}\right) \ , \tag{11}$$

but using different boundary conditions. The smooth transition of the elastic curve between the fields I and II is guaranteed provided that in the transition point located $x = L_I$ the following conditions on the displacement and slope are met:

$$v_{II}(L_I) = v_I(L_I) \quad \text{and} \quad \frac{dv_{II}}{dx}(L_I) = \frac{dv_I}{dx}(L_I) \quad . \tag{12}$$

These conditions lead to the integration constants for the second field in the form:

$$C_{1,II} = \frac{1}{2}\left(\frac{(EI)_{II}}{(EI)_I}(L_I^2 - L^2) - L_I^2\right) \tag{13}$$

$$C_{2,II} = \frac{1}{6}\left(\frac{(EI)_{II}}{(EI)_I}(L_I^3 - 3L_IL^2 + 2L^3) - L_I^3\right) - C_{1,II}L_I \tag{14}$$

The integration constants $C_{1,II}$ and $C_{2,II}$ can be then plugged in to the (10) and (11) to obtain explicit formulas for the elastic curve displacement and slope. It is good to note here, that setting $(EI)_{II} = (EI)_I$ leads to $C_{1,II} = C_{1,I}$ and $C_{2,II} = C_{2,I}$, meaning that the elastic curve solution in the field II simply keeps following the equation from the previous field I, without any change. The same approach can be used to continue by prolongation of the solution to the field III.

Generalisation to beams with arbitrary number of fields

In fact, the above described procedure can easily be generalized to a sandwich beam with arbitrary number *N* of sub-beams (and fields). In such a case, the first field is solved exactly as described, while for any consequent field with index *K+1,* the solution can be obtained from a recurrent formulas generalizing (12), resp. (13) and (14). Considering the smooth transition from the field *K* to *K+1*, in the transition point located $x = L_K$ the following conditions on the displacement and slope have to be satisfied:

$$v_{K+1}(L_K) = v_K(L_K) \quad \text{and} \quad \frac{dv_{K+1}}{dx}(L_K) = \frac{dv_K}{dx}(L_K) \tag{15}$$

These conditions lead to the integration constants for the *K-th* field in the form:

$$C_{1,K+1} = \frac{1}{2}\left(\frac{(EI)_{K+1}}{(EI)_K}(L_K^2 + 2C_{1,K}) - L_K^2\right) \tag{16}$$

$$C_{2,K+1} = \frac{1}{6}\left(\frac{(EI)_{K+1}}{(EI)_K}(L_K^3 + 6C_{1,K}L_K + 6C_{2,K}) - L_K^3\right) - C_{1,K+1}L_K \tag{17}$$

The recurrent formulas (16) and (17) for integration constants $C_{1,K+1}$ and $C_{2,K+1}$ can be used repeatedly for *K=1,...,N-1* and substituted to the formulas analogical to (10), (11)

$$\frac{dv_{K+1}}{dx}(x) = \frac{F}{(EI)_{K+1}}\left(\frac{x^2}{2} + C_{1,K+1}\right) \tag{18}$$

$$v_{K+1}(x) = \frac{F}{(EI)_{K+1}}\left(\frac{x^3}{6} + C_{1,K+1}x + C_{2,K+1}\right) , \tag{19}$$

providing the solution for elastic curve for $x \in \langle L_{K-1}; L_K \rangle$.

Here we have assumed that each of the *N* sub-beams has constant stiffness (flexural rigidity) along its length, which results into a general expression for the equivalent stiffness of the *K-th* field

$$(EI)_K = \sum_{i=1}^{N-K+1} E_i\, I_i \quad . \tag{20}$$

Remark: Also in this case it can be assumed that the loading shear force *T(x)* redistributes in each field to individual sub-beams, forming the sandwich, proportionally according to their stiffness (E_iI_i). It means that, exactly as in the mechanical spring analog depicted in the Fig. 3, the deformation of all sub-beams is the same, while the loading force is redistributed, with higher load being transferred to the stiffer sub-beams.

Conclusions

The presented method of solution of the deflection of sandwich beams results into simple way of obtaining closed form formulas for elastic curve of the beam. This is in contrast to the recent trends in simulation of the non-trivial beams, where only numerical representation of the solution can be obtained. Despite of this trend, the presented analytical method can be used in providing fast design concepts solutions in the initial stages of the complex mechanical devices.

It should be pointed out and understood that the presented approach is simplified and it heavily depends on the assumption that only the shear forces are transferred from one sub-beam. The practical situation can differ from the described one depending especially on the way how the sub-beams are interconnected in the sandwich. The specific case considered here however represents the worst case scenario, which justifies its use in the preliminary design calculations.

The future extensions of the presented work will focus on the following topics:

- *validation* – The presented analytical approach can (and should) be confronted with experimental results, or at least with more accurate (e.g. finite-element) numerical solutions.
- *extension* – The examples presented here were based on cantilever beam loaded by a single force at its free end. Both the loading and the type (shape) of the beam can be modified and generalized within the presented framework.
- *implementation* – There arised several questions concerning practical implementation possibilities of the sandwich beams. Among them namely the methods of interconnection of sub-beams, treatment of contact between the beams, material and structural properties of sub-beams.
- *applications* – The motivation for the presented study came from solving practical case of a moving arm of a mechanism. There are however many other application areas, including e.g. biomechanics or civil and transport engineering.

Acknowledgment - This work was partly funded by the Czech Ministry of Education, under the Institutional support for the development of research organization, No.:RVO12000 for Faculty of Mechanical Engineering of the Czech Technical University in Prague.

References

[1] D.B. Marghitu, Mechanical enginneer's Handbook, Academic Press, 2001.

[2] J. Bird & C. Ross, Mechanical engineering principles, Routledge, 2015.

[3] P. Puchmajer, J. Řezníčková, Sbírka úloh z pružnosti a pevnosti, Vydavatelství ČVUT, 2002.

[4] L. Šubrt, J. Řezníček, M. Růžička, Příklady z pružnosti a pevnosti, Nakladatelství ČVUT, 2011.

Investigation of residual stresses in high-cycle loaded laser steel welds

ČAPEK Jiří[1,a], ČERNÝ Ivo[2,b], TROJAN Karel[1,c], GANEV Nikolaj[1,d], KEC Jan[2,e] and NĚMEČEK Stanislav[3,f]

[1]Department of Solid State Engineering, Faculty of Nuclear Sciences and Physical Engineering, Czech Technical University in Prague, Trojanova 13, Prague 2, Czech Republic

[2]SVÚM a.s., Tovární 2053, 250 88 Čelákovice, Czech Republic

[3]RAPTECH s.r.o., U Vodárny 473, 330 08 Zruč-Senec, Czech Republic

[a]jiri.capek@fjfi.cvut.cz, [b]ivo.cerny@seznam.cz, [c]karel.trojan@fjfi.cvut.cz, [d]nikolaj.ganev@fjfi.cvut.cz, [e]kec@svum.cz, [f]nemecek@raptech.cz

Keywords: X-ray diffraction, fatigue life, laser welding

Abstract. The paper contains results of a comprehensive experimental programme aimed at an evaluation of residual stresses in and near laser welds of pressure vessels steels using X-ray diffraction technique. Generally, the manufacturing processes of a machine component introduce residual stresses that have an important and may have even essential influence on their behaviour during service life. The goal of this contribution was to find the influence of high-cycle loading on the stresses´ relaxation in and near the welds. It has been found that the surface residual stress state is not the most important factor, either positive or negative, but the fatigue life more or less depends on the specific situation and condition of the materials and welds. On the other hand, the fatigue loading test caused the reduction of microdeformation values. The analysis of microstructure changes by X-ray diffraction could be an indicator of the future cracks initialization and their propagation.

Introduction

The fatigue life of an object with a laser weld or, more generally, its service life depends on numerous structure parameters and can be significantly affected by the state of residual stresses. Even though residual stresses not yet belong to standard parameters for certification of welding processes, they can be considered as an important characteristic of the weld quality [1]. The aim of the laser welding process is not only to produce a joint but also to avoid the generation of undesirable stresses that would significantly promote crack propagation and cause the distortion of the final object. It has been found that distortion and the macroscopic residual stresses due to welding are two mutually affected phenomena and, therefore, upon controlling the residual stresses by means of laser beam energy and welding speed, one can influence the undesirable distortion as well.

Note that the macroscopic residual stresses (RS) can significantly promote fatigue resistance and decelerate life reducing fatigue initiation processes in the weld or bulk material. Moreover, the compressive RS can reduce crack growth from different defects like pores of lack of fusion if they are negative and contrarily if they are positive.

High cycle fatigue damage is a complicated process which can be divided into two categories: micro- and macrostructure changes. The microstructure changes contain two stages. Stage I. is associated with the movement, multiplication of dislocations and development of specific layout inside the material and grains. In case of high cycle fatigue loading with the stress amplitude close to endurance limit, the main effect of these movements is a reduction of microdeformations (microstresses or microscopic stresses). Cyclic softening or hardening may be associated phenomena in some materials. Stage II. corresponds to clustering of dislocations, formation of persistent slip bands, then extrusions and intrusions resulting in an occurrence of microcracks with a crystallographic character and orientation in the first stage. These microcracks result in a smooth reduction of RS and microdeformations in relation to Stage I. Due to macrocracks propagation up to final failure, Stage III. of macrostructure changes show a very high decrease of micro- and macrostresses values [2,3]. The relaxation of the mentioned structure parameters strongly depends on their original values. The main reduction of macroscopic residual stresses is connected with Stage III. It should be pointed out, however, that the reason of this reduction in the Stage III. are not fatigue processes themselves, but simply RS relaxation resulting from the material disruption. The same or similar stress relaxation would occur after any kind of cutting, which is being used, when some of the destructive methods of RS analysis are applied, like e.g. the contour method.

Experiment

The samples for high-cycle fatigue test were prepared from plates made of P355 and P460 steels (marked as "N" and "H") with a ferritic-pearlitic microstructure. The plates of thicknesses 8 mm and 10 mm were double-side laser-welded (the first-welded side is marked as "UP" and the second one as "DOWN", see Fig. 2), for welding parameters see Tab. 1.The X-ray diffraction analysis were performed only in the direction parallel to the axial axis of the samples for high-cycle fatigue test, i.e. in the direction perpendicular to the weld, see Fig. 1.

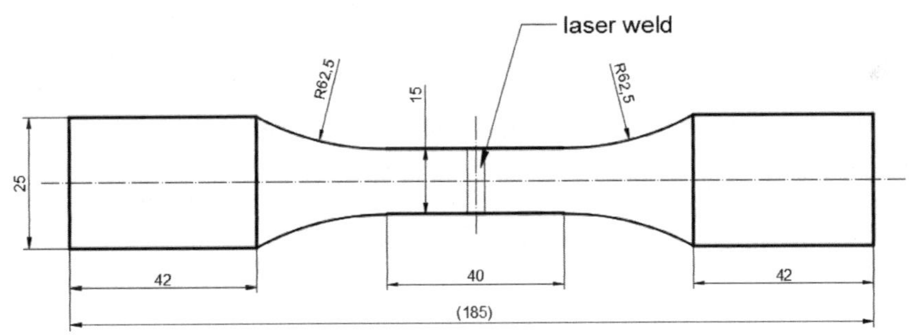

Fig. 1: Shape and dimensions of the fatigue test specimen

Table 1: Welding parameters; P – power of laser beam, v – welding speed, λ – wavelength of laser beam

P [W]	v [mm·s^{-1}]	mode	λ [nm]
3000	5.5	Continuous	900–1080

High-cycle fatigue tests were performed at stress ranges sufficiently low, bellow expected endurance limit, not to cause failures, with the aim to evaluate effects of the cyclic loading on possible relaxation of residual stresses. Load asymmetry was R = 0, frequency 33 Hz, stress range 206 and 185 MPa for the N10 and H8 specimen, respectively. The target number of cycles was 5 M and 17 M for the first and second loading test, respectively.

The *X'Pert PRO MPD* diffractometer was used to measure lattice deformations of the ferrite phase (α-Fe) using the Cr$K\alpha$ radiation. Diffraction angles $2\theta^{hkl}$ were determined from the peaks of the diffraction lines $K\alpha_1$ of the planes *{211}* of ferrite. The *Rachinger's* method was used for the separation of the diffraction lines $K\alpha_1$ and $K\alpha_2$ and diffraction lines $K\alpha_1$ were fitted by the

Pearson VII function. To determine RS, the *Winholtz & Cohen* method [4] and X-ray elastic constants $\frac{1}{2}s_2 = 5.75\ \mathrm{TPa}^{-1}$, $s_1 = -1.25\ \mathrm{TPa}^{-1}$ were used.

The diffraction patterns were obtained using the *X'Pert PRO MPD* diffractometer in the Bragg-Brentano geometry with Co*Kα* radiation. The Rietveld refinement performed on the *MStruct* software [5] was used for calculation of the values of microdeformations. The microstresses can be calculated using Hook's law $\sigma^{micro} = Ee$, where e and E are microdeformation and Young's modulus, respectively.

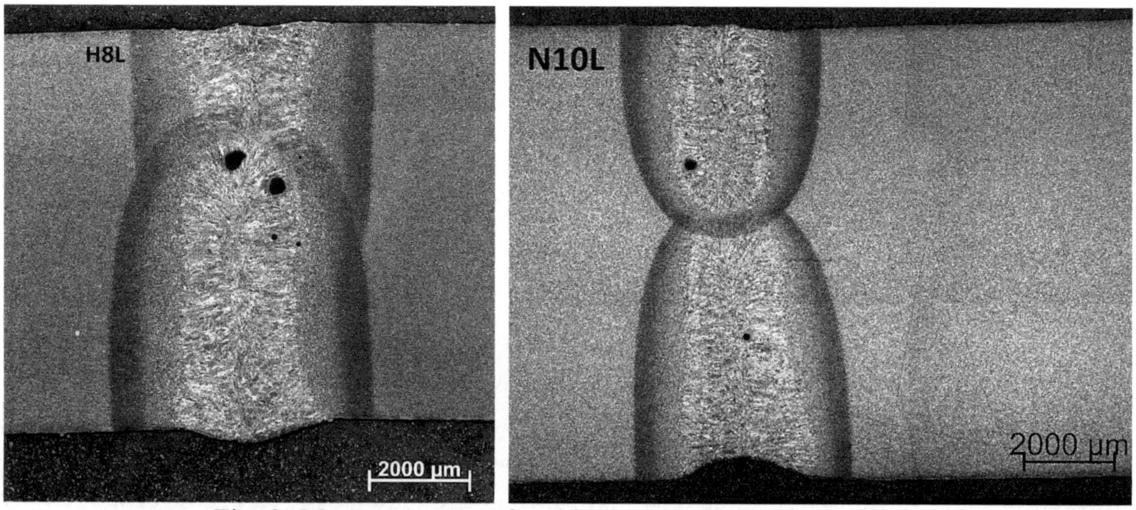

Fig. 2: Macrostructure of weld cross-section after etching

Results and discussions

Surface distributions of RS can be seen in Fig. 3. It is evident, in the case of the H8 sample, that the RS are compressive unlike the N10 sample, where they are tensile on the UP side. After the 1st (or the 2nd) fatigue cycle loading no statistically significant changes in the RS values for the H8 sample were observed. On the contrary, unexpected, premature failure of the N10 sample occurred after approx. 800000 cycles. The crack was initialized on the UP side on the range of the weld fusion and heat affected zones, where the RS were (slightly) tensile. As a result of the cracking, rather than the fatigue loading as mentioned in the Introduction, the RS relaxed to insignificant values.

More significant effect of high-cycle loading is evident from a microscopic point of view. From diffraction patterns, microdeformations and phase contents were determined using Rietveld refinement, see Figs. 4–5. All three stages, see Introduction chapter, is possible to find in Fig. 4. It was found that after the 1st fatigue cycle loading there were almost no changes (correlation with Stage I.) but after the 2nd fatigue cycle loading (correlation with the transition from Stage I. to Stage II.), the values of microdeformations of H8 sample were reduced more than two times. This effect is doubtless, interesting and important. It can be explained by the movement of dislocations, their recombination, and formation of different dislocation microstructure layout, which is a very long-time process occurring for a very high number of cycles, may last several tens of million cycles. Whether this phenomenon was connected with an occurrence of microcracks or arrested physically short cracks, it is not known. Nevertheless, further investigations in this direction would be useful and could provide a more exact explanation. The major decrease/relaxation of microdeformations of the N10 sample was evidently caused by the total specimen disruption as a result of the long main fatigue crack and final failure.

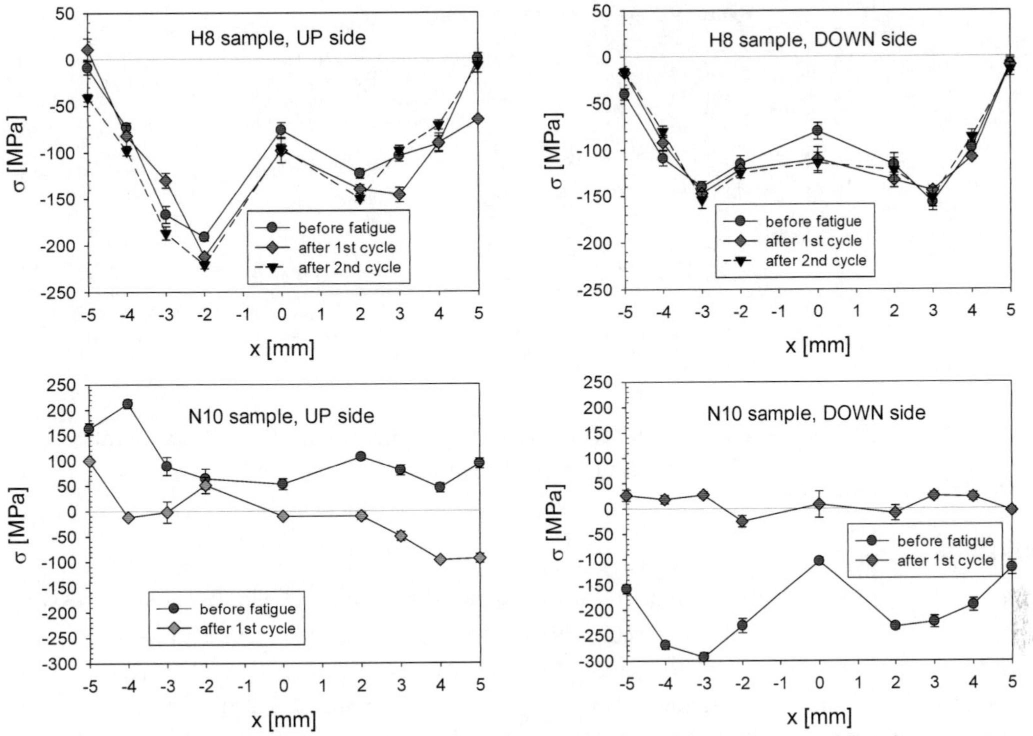

Fig. 3: Residual stresses in respect to the position from weld axis

In the process of initiation or propagation of fatigue cracks, the oxide layer can play in general an important role [6]. The premature fatigue failure of the N10 sample actually was initiated at the fusion zone on the surface with a thin oxide layer of the thickness less than 20 μm. Note that fusion zone represents a microstructural notch and is therefore typically the weakest area of the weld. The tensile RS could slightly accelerate the fatigue initiation process [7] by shifting the means stress to positive values and so increasing the load asymmetry. But such effect was likely small as the RS values were not so high in comparison with the stress amplitude.

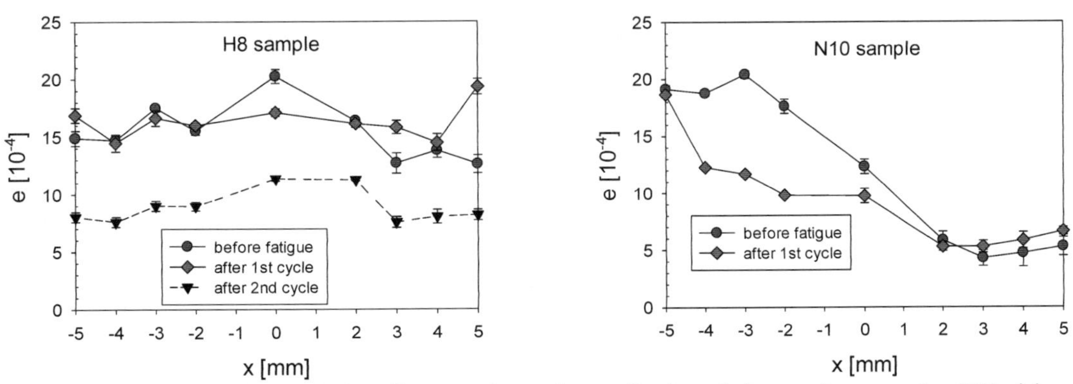

Fig. 4: Influence of high-cycle loading on the values of microdeformations on the UP side of samples, where x denote distance from the weld axis

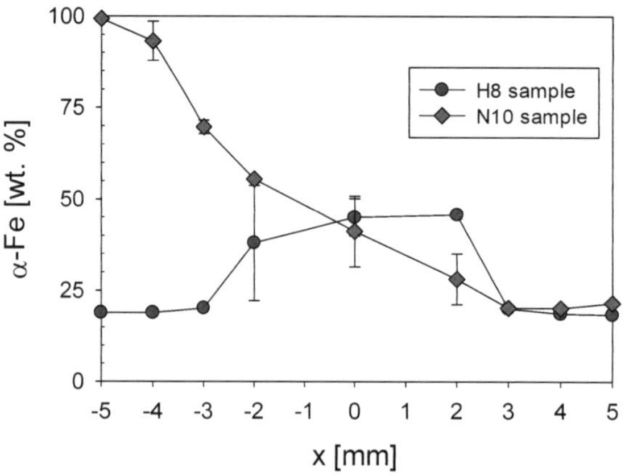

Fig. 5: Ferrite (α-Fe) content in respect to the position from weld axis in the surface layers on the UP side of samples

Conclusions

Results of the study of fatigue loading effects on the macroscopic residual stresses (RS) confirmed negligible effects of high-cycle fatigue loading with the stress amplitude close to the endurance limit on the RS. No statistically significant changes in the RS values for the H8 sample were observed. In the case of cracking of the N10 sample during fatigue loading, the RS values decreased and relaxed almost to zero due to macroscopic disruption. On the other hand, the fatigue loading test of H8 sample caused the significant changes of the microdeformation values after a high number of cycles, namely 17 million. This phenomenon can be explained by changes of dislocation layout and structure inside the material and grains as the first step of the fatigue crack initiation process. For the cracked N10 sample, the reduction of microdeformations values was observed in the area with tensile RS, high microdeformation, the thin oxide layer and within the heat affected zone. The analysis of microstructural changes by non-destructive X-ray diffraction in the surface layers indicated that this method could be a good indicator of the future cracks initialization and propagation. However, further investigations in this field would have to be performed.

Acknowledgement

This work was supported by the project TH02010664 of the Technology Agency of the Czech Republic.

References

[1] Z. Pala, K. Kolařík, N. Ganev, X-ray diffraction analysis of residual stresses in laser welded steel sheets, Fund. Sci. Appl. 19.1 (2013) 319–322.

[2] B. Pinheiro, J. Lesage, I. Pasqualino, E. Bemporad, N. Benseddiq, X-ray diffraction study of microstructural changes during fatigue damage initiation in pipe steels: Role of the initial dislocation structure, Mater. Sci. Eng. A. 580 (2013) 1–12.

[3] W.L. Shan, Y. Yang, K.T. Hillie, W.A. Jordaan, W.O. Soboyejo, Role of oxide thickening in fatigue crack initiation in LIGA nickel MEMS thin films, Mater. Sci. Eng. A. 561 (2013) 434–440.

[4] U. Welzel, J. Ligot, P. Lamparter, A.C. Vermeulen, E.J. Mittemeijer, Stress analysis of polycrystalline thin films and surface regions by X-ray diffraction, J. Appl. Crystallogr. 38 (2005) 1–29.

[5] Z. Matěj, R. Kužel, L. Nichtová, XRD total pattern fitting applied to study of microstructure of TiO2 films, Powder Diffr. 25 (2010) 125–131.

[6] B. Fournier, M. Sauzay, C. Caës, M. Noblecourt, M. Mottot, A. Bougault, V. Rabeau, A. Pineau, Creep-fatigue-oxidation interactions in a 9Cr–1Mo martensitic steel. Part II: Effect of compressive holding period on fatigue lifetime, Int. J. Fatigue. 30.4 (2008) 663–676.

[7] J. Čapek, N. Ganev, K. Trojan, J. Němeček, X-ray diffraction study of residual stress distribution on surface of laser welded steel sheets, in: Proceedings of the International Conference on Technics, Technologies and Education ICTTE 2017, Yambol, Bulgaria, 2017, pp. 40–44.

EAN 2019
57th conference on experimental stress analysis

Modelling of corrosion fatigue for steam turbines by probabilistic Kitagawa-Takahashi diagram

CERNY Miroslav[1,a]

[1] Czech Technical University in Prague; Klokner Institute; Solinova 7, 166 08 Prague 6, Czech Republic

[a] cerny@hppro.klok.cvut.cz

Keywords: Fatigue, Corrosion, Kitagawa-Takahashi Diagram, Probabilistic Modelling, Steam Turbines

Abstract. Probabilistic Kitagawa-Takahashi diagram is proposed for corrosion fatigue modelling of steel alloy in steam turbines.

Introduction

A probabilistic modelling is proposed for corrosion fatigue of steel alloy in steam turbines. Kitagawa-Takahashi diagram represents two distinct fatigue damage mechanisms: one associated with crack initiation (or microstructurally small cracks) and the other with crack propagation (or long cracks). These two fatigue criteria can be combined in order to simultaneously model both damage mechanisms and the transition between them. It results in a probabilistic Kitagawa-Takahashi type diagram [1], which can be used for pitting corrosion fatigue.

Probabilistic Kitagawa-Takahashi type diagram.

Corrosion (pitting) leading to fatigue crack nucleation and crack growth is considered to be among the most significant degradation mechanisms in aging structures [2]. The corrosion pits in steel alloy for steam turbines can be considered as surface (short) microcracks for determining threshold limits for fatigue cracking. The threshold for long crack fatigue can be defined using linear elastic fracture mechanics (LEFM). Then the stress intensity factor should be used for evaluating the influence of corrosion pits on fatigue such materials [3]. Speed of short cracks (surface microcracks) depends not only on the level of stress, but on the *microstructure of material.* The higher the level of stress in critical place, the greater the plastic deformation in the surface grain material and the greater the speed of crack in this grain. An initial gradual decrease in speed is due to the interaction of fatigue crack with the first sub-surface grain border. The surface speed of microcracks is minimal if its length a is equal to the dimension of grains d.

At lower levels of stress, the crack can also stop at the grain border. In this case, the local stress in the grain boundary is not big enough to re-initiation of fatigue cracks in the grain. If, on the other hand, stress is relatively high, the crack depends on the border characteristics of the grain, orientation of microcrack, etc. The crack grows through the grains to further sub-surface grains and starts gradually to behave as a long crack.

It has been observed, that most corroded pits have roughly a semi-elliptical shape with the width at the surface, $2c$, and the pit depth a. Further, investigations show that the pits can be treated as semi-circular surface short cracks.

This choice then makes it possible to define the *probability* of *crack initiation* within a grain. Equation 1 shows the probability density function used to define the threshold stress.

$$f_{01}(\sigma_{th}) = \frac{m_1}{\sigma_{th01}}\left(\frac{\sigma_{th}}{\sigma_{th01}}\right)^{m_1-1} \exp\left[-\left(\frac{\sigma_{th}}{\sigma_{th01}}\right)^{m_1}\right] \tag{1}$$

where σ_{th01} is the scale parameter and m_1 is the shape parameter (the Weibull exponent). A crack will not propagate in cyclic loading if the stress intensity factor ΔK is less than the *crack propagation* threshold ΔK_{th} ,

$$f_{02}(\Delta K_{th}) = \frac{m_2}{\Delta K_{th02}}\left(\frac{\Delta K_{th}}{\Delta K_{th02}}\right)^{m_2-1} \exp\left[-\left(\frac{\Delta K_{th}}{\Delta K_{th02}}\right)^{m_2}\right] \tag{2}$$

After combination of both mechanisms we get the expression for the fatigue strength as a function of the probability of failure, and the crack length, as follows:

$$\sigma_{1,a}(P_F, a) = \left[\frac{\ln\left(\frac{1}{1-P_F}\right)}{\left(\frac{1}{\sigma'_{th01}}\right)^m + \left(\frac{Y2\sqrt{\pi a}}{\Delta K'_{th02}}\right)^m}\right]^{\frac{1}{m}} \tag{3}$$

Equation (3), plotted as a function of the crack size, *a*, for a given probability of failure P_f , results in a probabilistic Kitagawa-Takahashi diagram. After correlation of a half pit width *c* with a crack size *a* (Fig.2) we get a corrosion fatigue strength of a material (Fig.1).

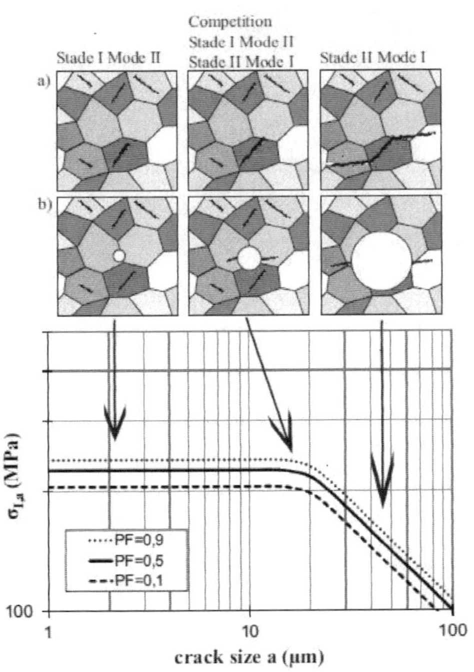

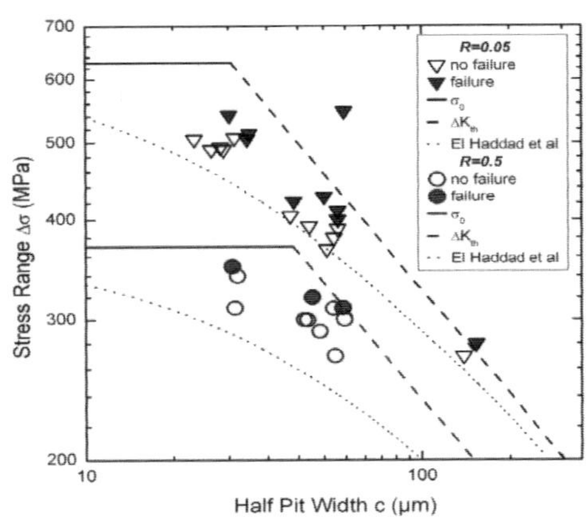

Fig. 1: Kitagawa-Takahashi diagram (Eq.1) for different failure probabilities showing the fatigue damage mechanisms a) without and b) with initial defects ($m = 20, \sigma'_{th01} = 234MPa$, $Y = 2/\pi$, $\_K'_{th02} = 2.3MPa.m^{1/2}$)

Fig. 2: Kitagawa-Takahashi diagram for corroded steel alloy 403/410, c is correlated with crack size a

True pit geometry

The description of true pit geometry has been published by author earlier [2]. The max. and min. depths of pits and their densities have been found. In [3] are shown the mentioned characteristics and Table 1 in [3] comprises the parameters of corrosion evaluated for part of steam turbine at power plant ETU (alloy X22CrMoV12-1, exposed 677 hr in 650°C steam).

Fig.3 True geometry of corrosion

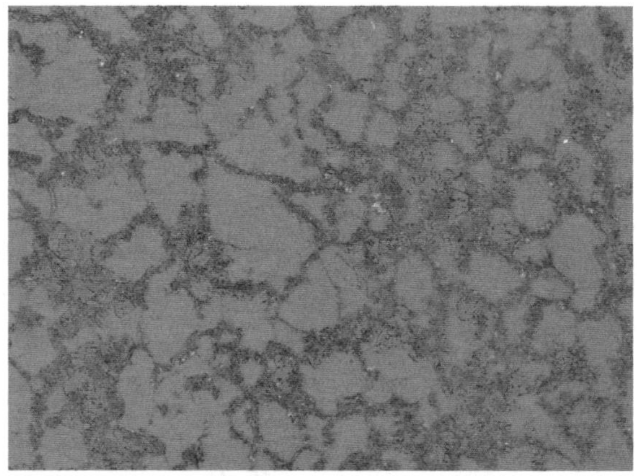

Fig.4 Alloy steel grains after corrosion

Acknowledgement

The presented work has been supported by grant No TE01020068.

Conclusions

A probabilistic modelling is proposed for corrosion fatigue of steel alloy in steam turbines. Crack size a in Kitagawa-Takahashi diagram is correlated with corroded half pit c.

References

[1] Pessard E.,Bellett D.: Probabilistic Kitagawa- Takahashi Diagram, Engineering Fracture Mechanics, 2014.

[2] Černý M.: Localized Corrosion in Alloy Steel of Steam Turbines, Applied Mechanics and Materials, vol.827, Trans Tech Publications, 2016. doi:10.428/www.scientific.net/AMM.827.177

[3] Černý M.J: Evaluation of the Stress Intensity Factor for Corrosion Fatigue on Steel Alloy in Steam Turbines, Conference EAN 2018, Proceedings, 2018.

EAN 2019
57th conference on experimental stress analysis

An Improvement of measurement technique for high speed impact tests analysis

DOUBRAVA Radek[1,a], OBERTHOR Martin[1,b], BĚLSKÝ Petr[1,c] and VÍCH Ondřej[1,d]

[1]VZLÚ (Czech Aerospace Research Centre), Beranových 130, 199 05, Prague – Letňany, Czech Republic

[a]doubrava@vzlu.cz, [b]oberthor@vzlu.cz, [c]belsky@vzlu.cz, [d]ondrej.vich@vzlu.cz

Keywords: impact, hail, high-speed, laser, FEM

Abstract. The paper presents improvement methodology and verification of numerical models by quantitative measurement during high speed impact tests. A comparison between the test and simulation results reveals consistency in predictions of damage initialization on composite material and plastic deformation on metal material. The new technique of displacement measurement by optical sensor during impact provided unique value for verification of numerical models was applicate. The results showed the necessity to increase the measurement range of sensors.

Introduction

Foreign object damage (FOD) and especially high-speed impact such as bird or hailstone strikes are an important phenomenon that must be considered when designing aircraft [1]. As a result, it is more important than ever to study and understand the mechanics of high-speed impact.

An increasing number of aircraft manufacturers, aviation companies, and government authorities have been conducting advanced research and development programmes to reduce the annual cost, number of injuries, and fatalities that result from collisions of aircraft with foreign objects e.g. wildlife [2] or hail-stone [3]. To eliminate the harmful effects of bird or hailstone strikes, two main approaches are typically implemented:

1. Bird or hailstone strike prevention strategies that attempt to reduce the probability of a bird or hailstone strike incident e.g. elimination of food sources, keeping grass mowed short, using sonic cannons, recorded predator calls, and other noise generators, using lasers, flying trained falcons, training dogs, using radar equipment etc.
2. Aircraft certification programmes that employ various measures to maintain the integrity of an aircraft against the high loads resulting from high-velocity impacts in accordance with international certification standards.

The second approach will be further developed in this study. International certification regulations require that all forward-facing aircraft components be proved to withstand bird or hailstone strikes to a certain level before they can be used in an aircraft [4,5]. A bird or hailstone impact test provides a direct method for determining bird strike resistance; however, the design of aircraft structures typically involves many iterations, from design to manufacturing to testing and back, requiring that many bird impact tests be conducted. This is not only time consuming but also costly. Furthermore, experimental data from these tests are often narrowly focused, constituting a barrier for their direct use in refining structural design. Owing to these shortcomings, several numerical methods have been developed to simulate bird or hailstone strikes to reduce the number of intermediate tests required and subsequently shorten the duration of the component design phase [6-8].

With regard to high-speed effect in the course of the impact, it is a major problem to obtain data for the analysis of the mechanism of initialization and propagation of damage especially for composite material [9,10].

Impact tests

High-speed tests on simplified structures represented by flat test composite and metal plates (figure 1) have been carried out in order to verify and optimised numerical analyses.

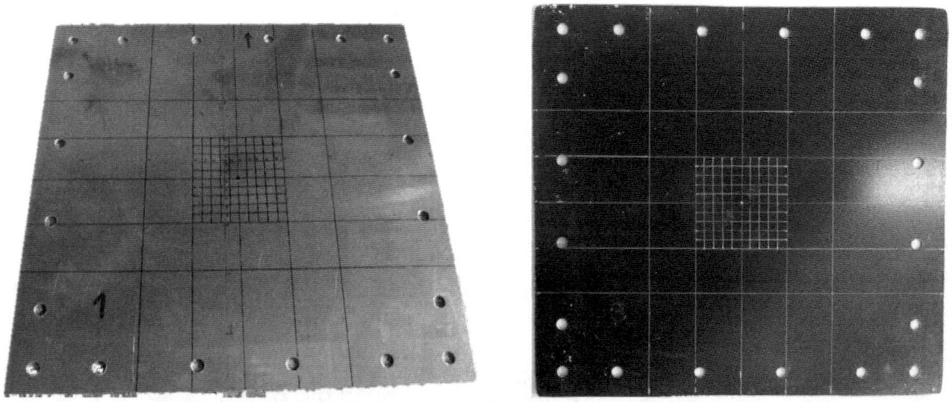

Fig. 1: Metal and composite flat test specimen

The composite flat test panel was manufactured from Hexply 8558/AGP193-PW [11] Carbon Fibre Reinforced Polymer (CFRP) composite laminated shell consisting of 10 layers $(45,0,45,0,45)_s$ impregnated with resin system in the vacuum infusion process. The test specimen was prepared by autoclave curing process.

The metal test panel was manufactured from sheet aluminium alloy 2024-T3 [11] with nominal thickness 2 mm.

The ice ball of 1-inch (2.54 cm) and 2-inch (5.08 cm) diameter [4] were used as hail stone impact projectile (Fig 2).

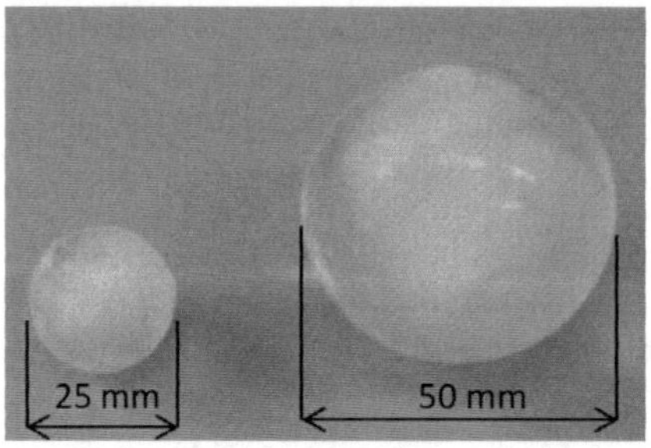

Fig. 2: Hail stone projectiles

The tests were performed in the Strength of Structure Department laboratory of VZLU (Czech Aerospace Research Centre) according to ASTM standard [4]. The air gun of VZLU was used as the test equipment (Fig.3).

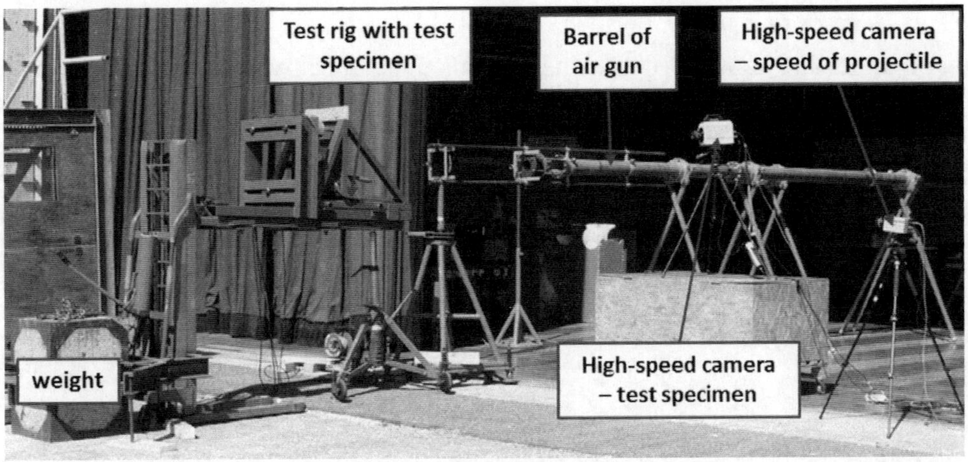

Fig. 3: Test assembly

The standard qualitative technique for high speed impact analysis such as high-speed camera and post-test non-destructive technique (NDT) were used for analysis of test results and verification of numerical model. New technique of displacement measurement by laser triangulation displacement sensor during impact was proposed and used for quantitative analysis and more precision verification of numerical model (Fig. 4).

The test specimen deformation during hail strike was measured using the laser triangulation displacement sensor with controller for analogue signal output. Measuring range of the laser sensor was 20 mm (from 40 to 60 mm off the sensor). Real point of hail impact was then determined by means of high-speed camera record and visual inspection. Measuring rate is adjustable from 1.5 to 49.1 kHz and it was set at value of 20 kHz during the test. Resolution of the sensor is 0.3 μm. Real point of hail impact was then determined by means of high-speed camera record and visual inspection. The video was shot by Photron Fastcam and Dantec NanoSence MkIII high speed cameras. One camera recorded the course of the impact from the front side, second camera was used to measure the hail speed. Velocity measurement by high speed camera was based on measurement of time which is necessary for passing of specific distance.

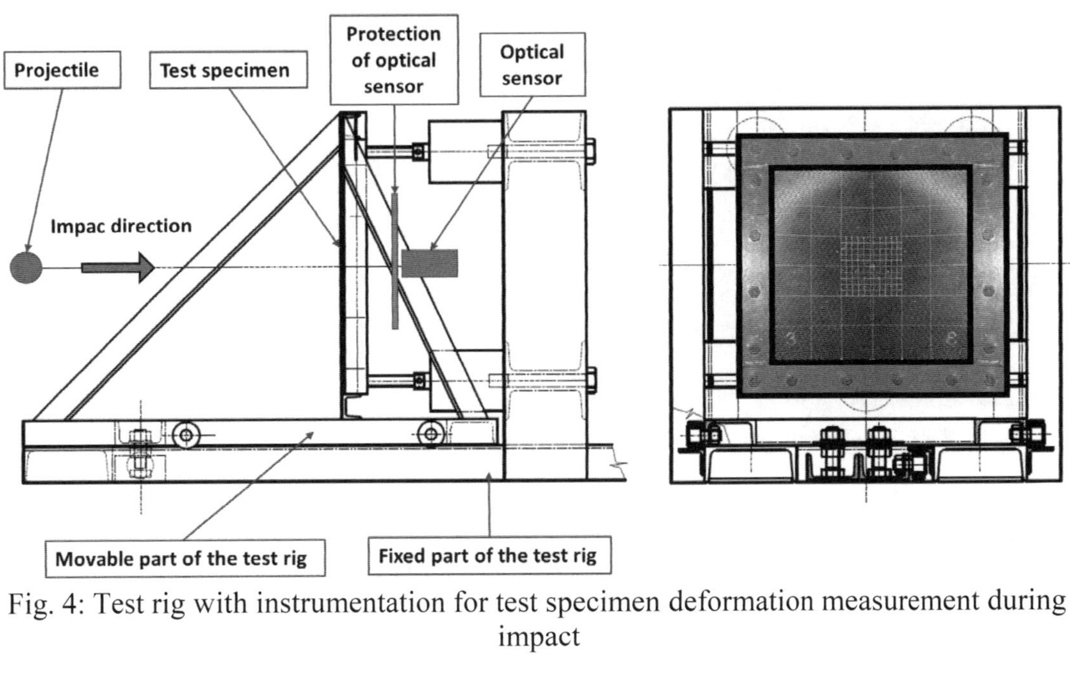

Fig. 4: Test rig with instrumentation for test specimen deformation measurement during impact

NDT of composite panels

Visual and ultrasonic non-destructive inspections of all specimens were performed before and after hail strike tests. For ultrasonic inspections was used Phased Array device (C-Scans) (see Fig.5)

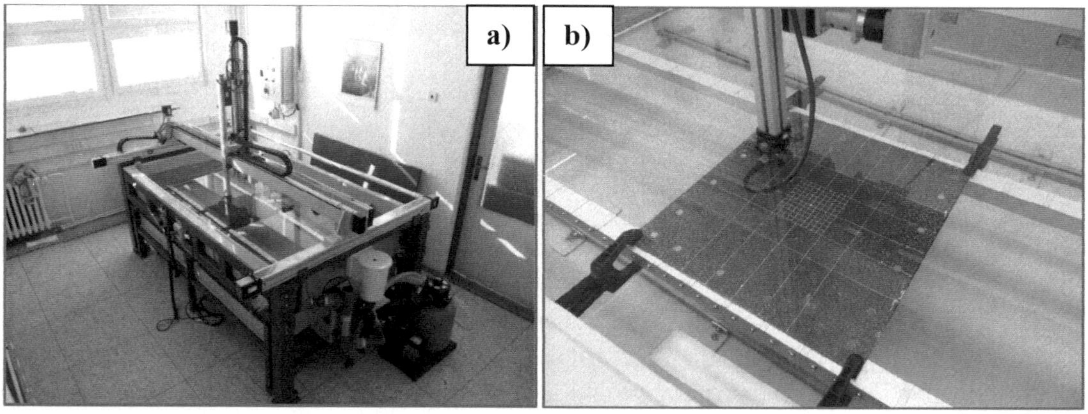

Fig. 5: Ultrasonic test NDI equipment: a) Immersion tank + automated 3-axis scanner system, b) clamping fixture

Amplitude and TOF (Time of Flight) C-scans were generated from measured data during ultrasonic inspections. PEBR (Pulse Echo Back-wall Reflection) and PEFR (Pulse Echo Flaw Reflection) techniques were used for the amplitude C-scans.

For PEBR technique the gate (Gate A) was placed over the back-wall echo and it was synchronized on the interface echo. In case of PEFR technique the start point of the gate (gate B) was behind the interface echo and the end of the gate was before the back-wall echo.

Plastic deformation

Measurement of plastic deformation of the metal test specimens was performed using raster and coordinate recorder after impact test. The plastic deformation was measured in two perpendicular directions in the center of the test specimen (see Fig.6).

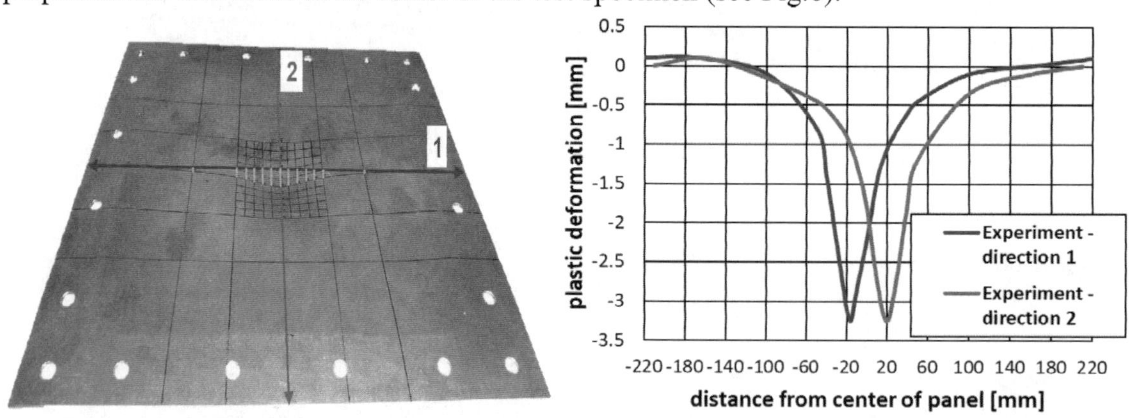

Fig. 6: Measurement of plastic deformation on metal test specimen after impact test

Numerical simulation

The FE simulation was performed using the ABAQUS FE software program [13]. An explicit solver with double precision was used for the analysis. The aim of this analysis was to tune unknown inputs and computational parameters to obtain the same behaviours between the test and simulation, i.e., damage, displacement or reaction force.

In terms of the impactor's (hail) material properties and features, ice is a complex material presenting high degrees of variability. The material density of ice and hail varies and is subject to a range of factors, such as the weather systems through which it was created. Ice may exhibit two forms of non-elastic behaviour under stress. Under low-velocity deformation, ice exhibits ductile behaviour and yet as velocity increases, the material become brittle (Fig.7). The method used to adjust hail strike simulations is certified by the Czech Republic Civil Aviation Authority. The geometry of the hail stone model was meshed by 3,936, C3D8R 8 - node linear brick elements with conversion to particle elements (SPH – Smoothed Particle Hydrodynamics) [14]. The shell elements type S4R were used for simulation of the test specimen. The global mesh size of 10 mm was refined to 3 mm at the impact area.

From the viewpoint of the damage analysis of a composite material, the Hashin's damage material model was used [15]. The material properties used in Hashin's model of energy criterion in ABAQUS were adopted from reference [8]. A Johnson-Cook material model [12] was used for the metal test specimen for simulation of elastoplastic and damage behavior.

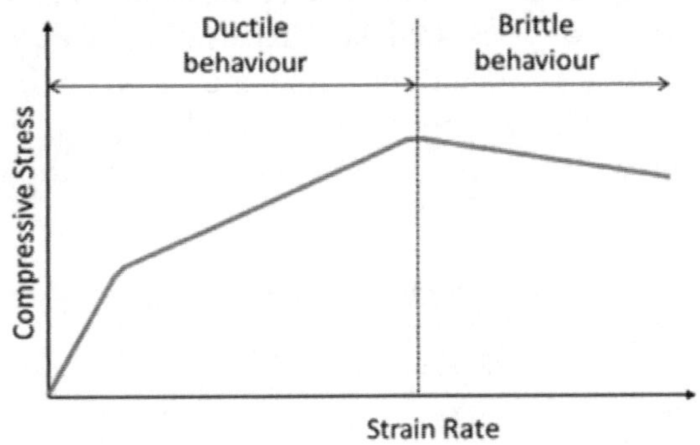

Fig.7: Ice behaviour under stress

Results

The tests results and numerical simulations were compared based on:

- Qualitative measurement – high speed camera pictures, NDT (C-scan)
- Quantitative measurement – displacement during impact and plastic deformation

Analyses of NDT and simulation results

The results from NDT (C-scan) performed after the tests were compared with composite damage criterion from numerical simulation from point of view of size and position of damage. Fig. 8 shows comparison of contour map from C-scan (delamination area) and contour map of Hashin tensile fibre damage initialization criteria. Damage initiation occurs when the index exceeds 1 (red colour in contours maps of FE analysis).

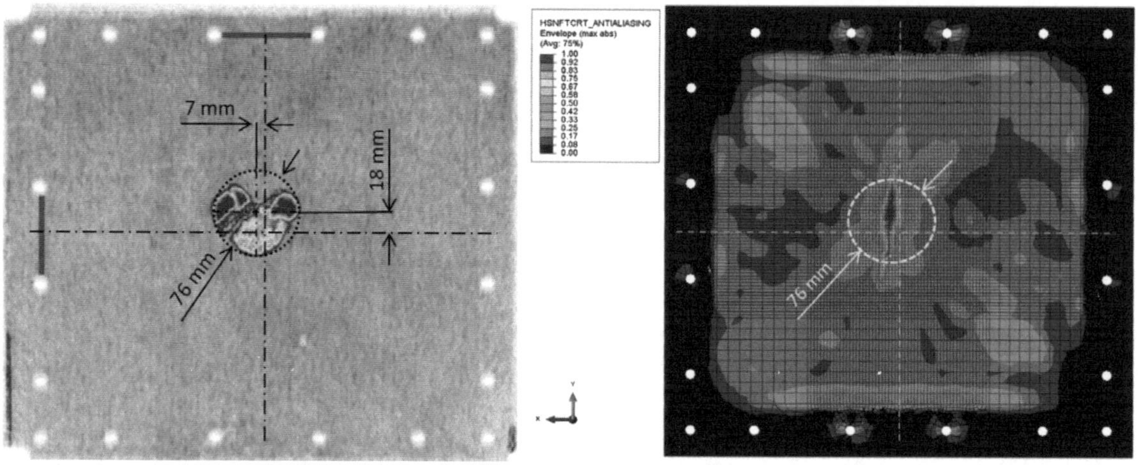

Fig. 8: Comparison between NDT (C-scan) (left) and numerical simulation (tensile fibre damage) (right) for impact test of hail stone 50 mm diameter and speed of impact 362 km/h.

Analyses of high-speed camera pictures with simulation results

The result from high speed camera was analysed from point of view of damage of impactor, analysis of impact point position or uncertainties such as damage of impactor before impact. The second-high speed camera was used for analyses of speed of impactor. The comparison of high-speed camera results with the numerical simulations for the same time of impact (see Fig. 9).

Experiment	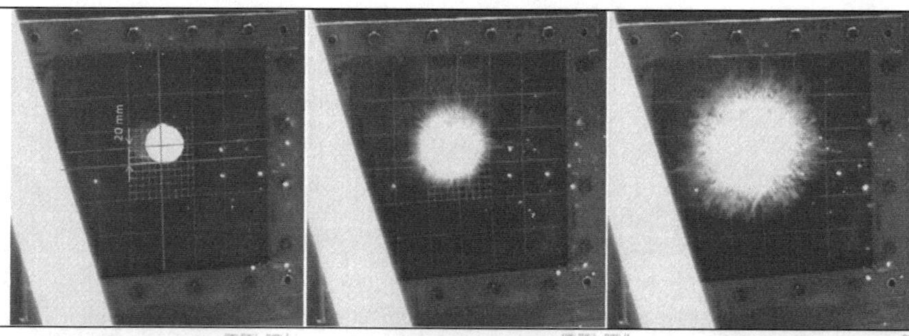		
FEM	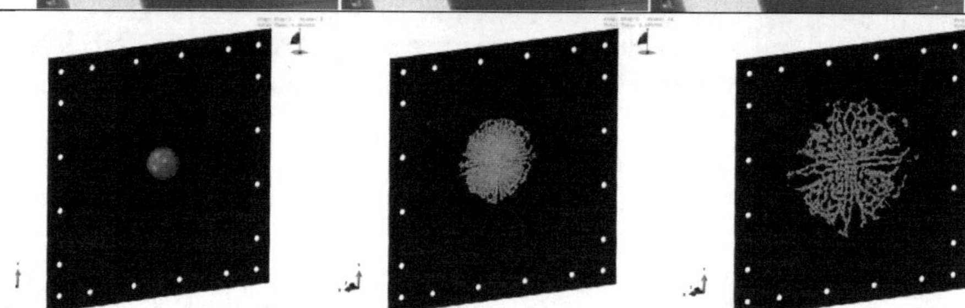		
Time of impact [ms]	0 ms	0.5 ms	1 ms

Fig. 9: Comparison between experiment and simulation (composite test specimen, impact speed 655 km/h, hail stone 50 mm diameter, solid laminate 10 plies)

Analyses of displacement

The displacement in centre of the test panel during the impact was analysed based on the optical sensor data. From point of view of availability of optical sensor with small range measurement, the measurement was realized only for small energy impact (small speed for hail stone of 50 mm and small diameter of hail stone – 25 mm for high speed). Small energy impact also to ensure no failure of panel and decrease of the sensor damage probability.

Figure 10 show comparison between experiment and numerical simulation. Figure 10 also shows the influence of material damping property. The Railigh damping properties (alpha = 29.4, beta = 2.72 E-7) was used from another project measurement performed in VZLU for analyses of acoustic fatigue [16].

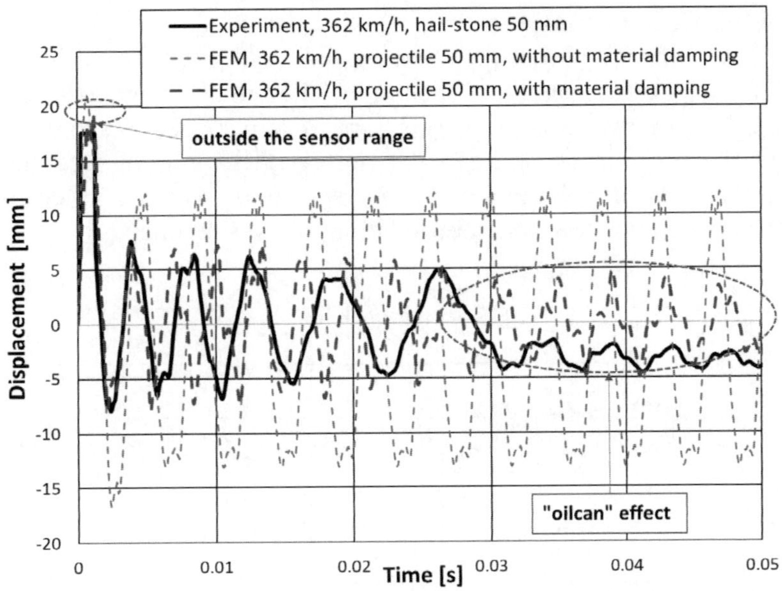

Fig. 10: Result of displacement measurement during impact of hail stone 50 mm diameter on composite panel (10 layers of CFRP) and comparison with simulations

The optical sensor data measured on composite panel shows, that maximal displacement was outside the sensor range (in this case up to 17 mm). The results also show the influence of oilcan effect on the damping and observed negative displacement. The elastic displacement was removed after unmount the test panel from the test rig.

Figure 11 show comparison between experiment and numerical simulation for metal test panel.

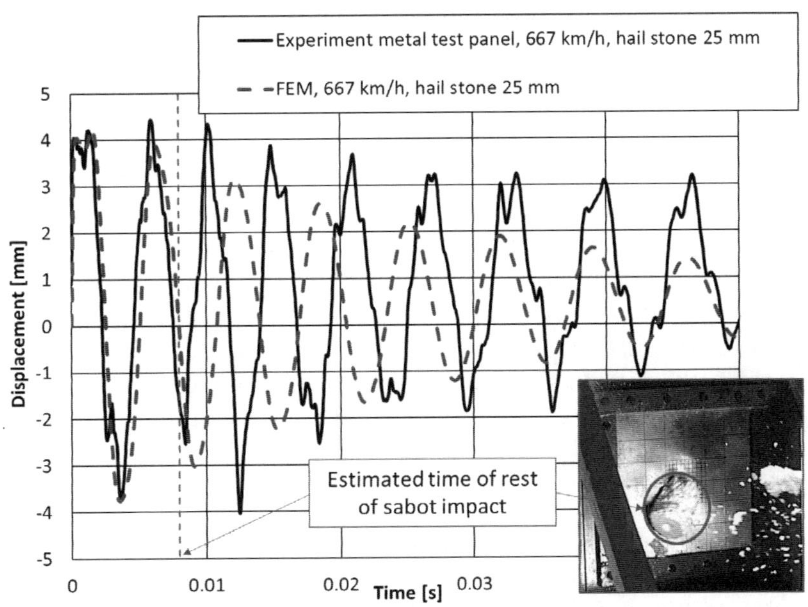

Fig. 11: Result of displacement measurement during impact of hail stone 25mm diameter on metal panel (2024-T3, 2 mm thickness) and comparison with simulation

The result from optical sensor measurement on metal test panel shows the influence of secondary impact of rest of polystyrene sabot on test specimen (Fig. 11). The time of secondary impact was estimated from high speed camera pictures.

Analyses of plastic deformation

Figure 14 shows comparison between experiment and numerical simulation for plastic deformation after impact on metal test specimen. The impact point for numerical simulation was shift on the base of real test results (high speed camera pictures and plastic deformation measurement). The differences between the numerical simulation and the experiment are mainly in the area near to the fixation of the test specimen, wherein reflected the influence of the material spring back due to plastic deformation of test specimen.

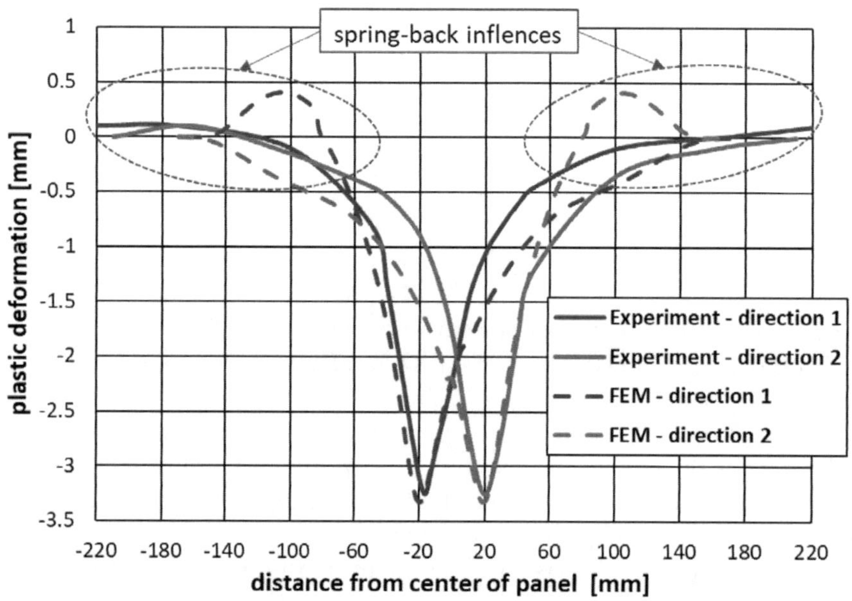

Fig. 14: Comparison between experiment and simulation of plastic deformation after impact on metal test specimen.

Conclusions

A comparison between the test and simulation results reveals consistency in predictions of damage initialization on composite material and plastic deformation on metal material.

The application new technique of displacement measurement by optical sensor during impact provided unique value for verification of numerical models.

These tests provide important information for design and computational analysis. Test and numerical model verification based on different impact speeds also provide important generalizations for the application and optimization of composite and metal structures.

At the same time, the results showed the necessity to increase the sensor measurement range for application to the real application related to high-speed impact analyses.

Acknowledgement

This project result was developed within the institutional support of the Ministry of Industry and Trade of the CR directed to the development of research organizations.

References

[1] J. Wilbeck, Impact Behavior of Low Strength Projectiles. Air Force Materials Lab, Air Force Wright Aeronautical Labs, Report No. AFML-TR-77-134, Wright-Patterson AFB, OH, 1978

[2] E. Cleary, R. Dolbeer, S. Wright, Wildlife Strikes to Civil Aircraft in the United States 1990-2005, Federal Aviation Administration National Wildlife Strike Database, 2005

[3] P.R. Field, W. Hand, G. Cappelluti, A. McMillan, A. Foreman, D. Stubbs and M. Willows, Hail Threat Standardisation, FINAL report for EASA.2008.OP.25, 2008

[4] ASTM F320-05, Standard Test Method for Hail Impact Resistance of Aerospace Transparent Enclosures, 2005

[5] ASTM F330 -10 Standard Test Method for Bird Impact Testing of Aerospace Transparent Enclosures, 2010

[6] J. Liu, Y. Li, X. Gao, Bird strike on a flat plate: Experiments and numerical simulations. Int. J. Imp. Eng., Vol. 70, http://dx.doi.org/10.1016/j.ijimpeng.2014.03006, 2014, pp. 21-37

[7] S. Heimbs (2011). Computational Methods for Bird Strike Simulations: A Review. Computers & Structures, Vol. 89 (23), 2093-2112.

[8] K. Kustron, V. Horak, , R. Doubrava, , Z. J. Goraj, New hail impact simulation models on composite laminated wing leading edge, Aircraft Engineering and Aerospace Technology, https://doi.org/10.1108/AEAT-02-2018-0089, 2018

[9] R. Růžek, K. Tserpes, V. Karachalios, I. Giannopoulos, V. Prentzias, Strain and damage monitoring in CFRP fuselage panels using fiber Bragg grating sensors. Part I: Design, manufacturing and impact testing. Comp. Struct., Vol. 107,doi: 10.1016/j.compstruct.2013.09.053, 2014, pp. 726–736

[10] R. Růžek, M. Kadlec, K. Tserpes, E. Karachalios, Monitoring of compressive behaviour of stiffened composite panels using embedded fibre optic and strain gauge sensors, International Journal of Structural Integrity 8 (1), 2017, 134-150

[11] HexPly®8552– Product Data http://www.aerospares.hu/files/hexcel/hexply_8552.pdf. 2017

[12] G.Kay, Failure Modeling of Titanium 6Al-4V and Aluminum 2024-T3 With the Johnson-Cook Material Model, DOT/FAA/AR-03/57, 2003

[13] ABAQUS 6.14, Theory manual, http://abaqus.software.polimi.it/v6.14/, 2017

[14] R. Doubrava, SPH Technigue Application for Fast Impact Simulation, report VZLU R-5841, http://www.vzlu.cz/en/certified-methodology-n-met-c488.html, 2013

[15] Z. Hashin, Failure criteria for unidirectional fiber composites, ASME Journal of Applied Mechanics, Vol. 47, 1981, pp. 329-334.

[16] J., Běhal, J., P., Zděnek, Effect of interlaminar flaw on composite panel behaviour under acoustic loading, MATEC Web Conf. Volume 188, Art. No. 01005, DOI: https://doi.org/10.1051/matecconf/201818801005, 2018

Experimental and numerical analysis of the head tube joint strength

DOUBRAVA Karel[1,a], PONÍŽIL Tomáš[1,b], KULÍŠEK Viktor[1,2,c], POŠVÁŘ Zdeněk[1,d] and RŮŽIČKA Milan[1,e]

[1] Department of Mechanics, Biomechanics and Mechatronics; Faculty of Mechanical Engineering; Czech Technical University in Prague; Technická 4, 166 07 Praha, Czech Republic

[2] Compo Tech PLUS, spol. s r.o., Nová 1316, Sušice 34201, Czech Republic

[a]Karel.Doubrava@fs.cvut.cz, [b]Tomas.Ponizil@fs.cvut.cz, [c]Viktor.Kulisek@compotech.com, [d]Zdenek.Posvar@fs.cvut.cz, [e]Milan.Ruzicka@fs.cvut.cz

Keywords: Composite Bike Frame, Strain Gauge, Acoustic Emission, Digital Image Correlation, Finite Element Method

Abstract. A composite bicycle frames are produced by a variety of technique and the joint of tubes is weakest link for tube to tube design. The FE model of the joint was created and the joint strength was experimentally verified when the load was monitored by strain gauges the DIC and Acoustic Emission method.

Introduction

The bicycle frame is designed with regard to performance and durability. Composite materials have brought new possibilities to change the frame properties like stiffness and mass. A wide range of composite materials and their manufacturing technologies exists, which might lead to different designs with different frame stiffness, strength and price. The paper is focused on one of possible frame designs, where the frame is built from structural tubes with lamination over the tubes in the joint areas. In this type of the design, the joints of the tubes are the weakest points. The experimental verification of the head tube joint strength under typical loading state with numerical calculation validation is included in this paper. The load of the joint was monitored using a strain gauge and the Digital Image Correlation (DIC) and the damage was monitored by Acoustic Emission (AE) method.

Composite Frame

A composite bike frame was produced by a Czech company Festka using composite tubes by a filament winding technology as the main structural material. The manufacturing technology enables to tailor the frame to the customer's specific requirements. The tube to tube manual lamination method is used to made the frame from the tubes. Carbon fabrics and tapes together with microballoon epoxy are used in the joints manufacturing. The strength of the joint with respect to the manual production had to be verified. For the experimental verification, the head tube joint was selected as the most stressed. For this purposes, simplified frames containing the main triangle of frame were made and tested, see Fig. 1.

Fig. 1: Simplified composite bicycle frame

Experiment

The horizontal force head-tube test [1] was used to obtain the critical failure load of the frame and of the head tube joint. A loading device was designed where the frame is fixed in place of the bottom bracket and loaded with a fork replacement connected to servo hydraulic actuator IST PL63, see Fig. 2. The loading was carried out by position control until specimen failure. All devices were synchronized with the force signal during the test by means DC output from hydraulic actuator control unit.

Fig. 2: Experiment setup

Record of experiments can be seen in Fig. 3, when on y-axes is normalized force and on x axes is displacement.

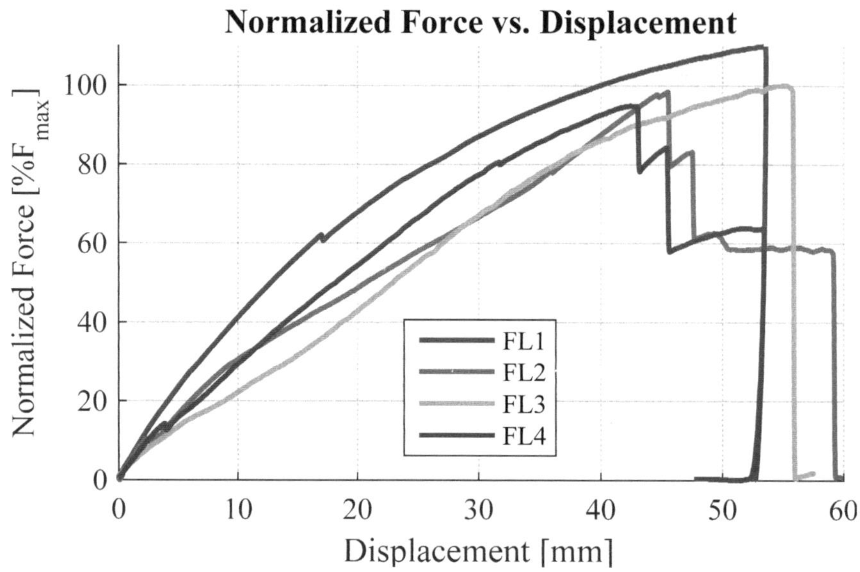

Fig. 3: Tests records normalized force vs. displacement

Strain Gauge measurement. Experiences from the previous tests demonstrated, that the dominant loading is the bending stress in the frame plane. Therefore, strain gauges were only installed in this plane to monitor the stresses around the joint. Two strain gauges were installed on the top tube (TTF) and two on the down tube (DTF). Another strain gauge was installed in the back of the head tube (HT). Standard strain gauges with resistance of 350 Ohms and methylmetacrylate cold curing glue was used. Measurement unit HBM Spider8 was used for data acquisition with quarter bridge connection.

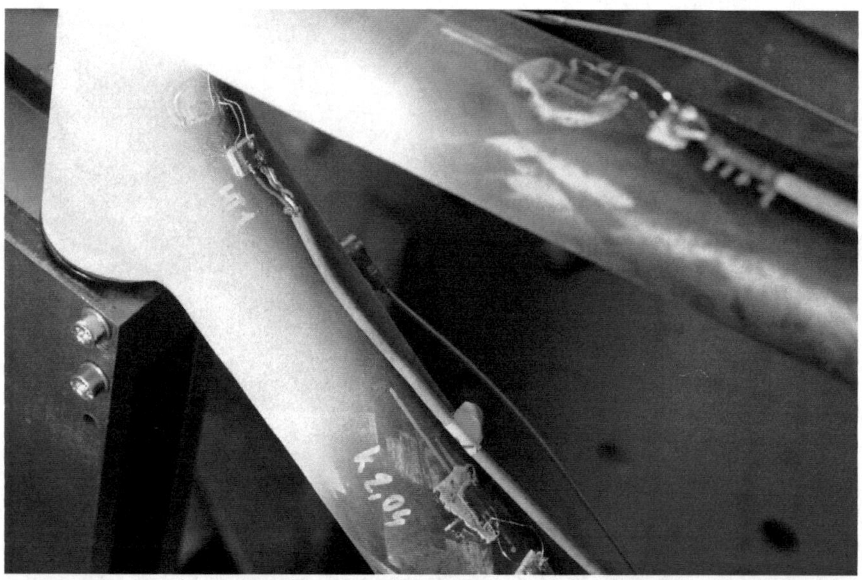

Fig.4: Strain gauge installed on the specimen

An example of evaluated data for one sample can be seen in Fig. 5, where tension and bending signals on down and top tubes were separated. Initialization of the primary damage is evident on the record of strain from HT strain gauge at the level of two thirds of maximum force.

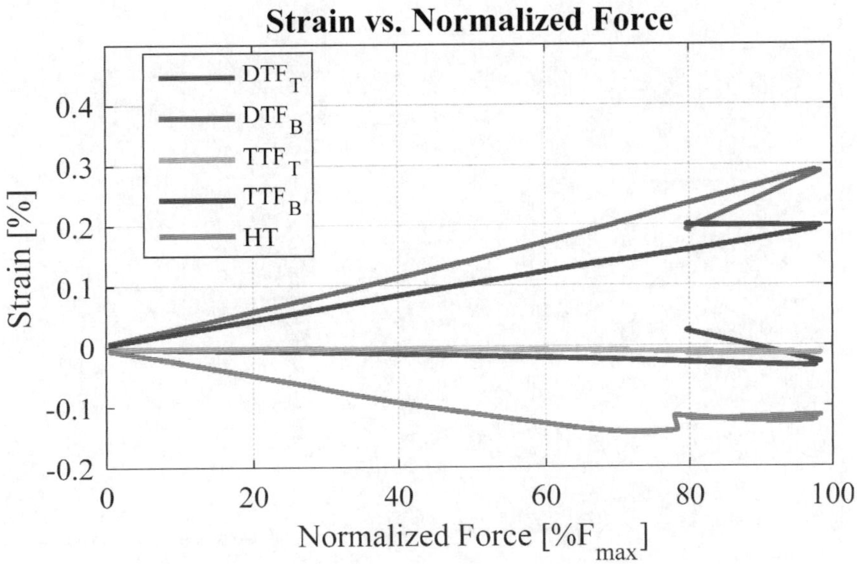

Fig. 5: Tests records strains vs. normalized force

Digital Image Correlation. Area information about displacements and strain was obtained using DIC, where an irregular black and white pattern was made by airbrush on the side of the joint, see Fig 4. System Q400 with 16Mpx firewire camera was used for image recording and software Istra 4D software from Dantec Dynamics for evolution. Principle strains on joint can be seen in Fig. 6, when load level is maximal force.

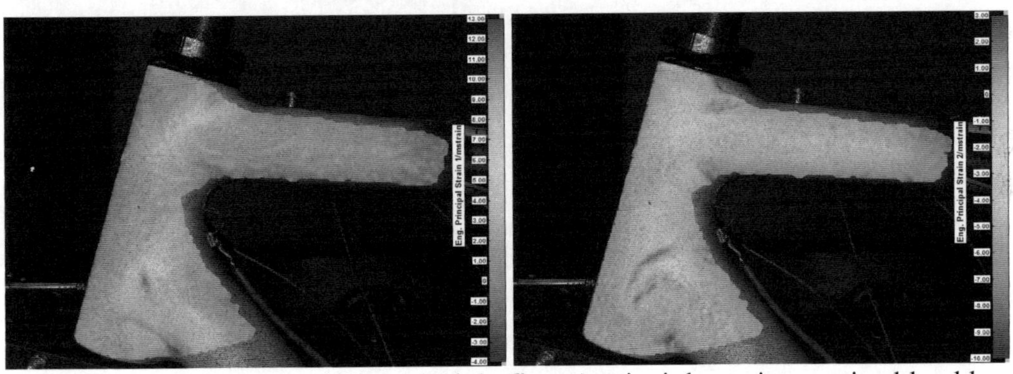

Fig. 6: First (left figure) and Second (right figure) principle strains maximal load level

Acoustic Emission method. AE was used to track damage during strength testing. Three sensors were installed in the vicinity of the head tube joint. One on the side of the top tube, one on the side of the bottom tube and one on the side of the head tube, see Fig. 7. Measurement unit Xedo with software was used for event recording. An example of AE event recording is in Fig. 8. Initialization of the initial damage is evident from the increase of AE events again at the level of two thirds of maximum force.

Fig. 7: Location of AE sensors

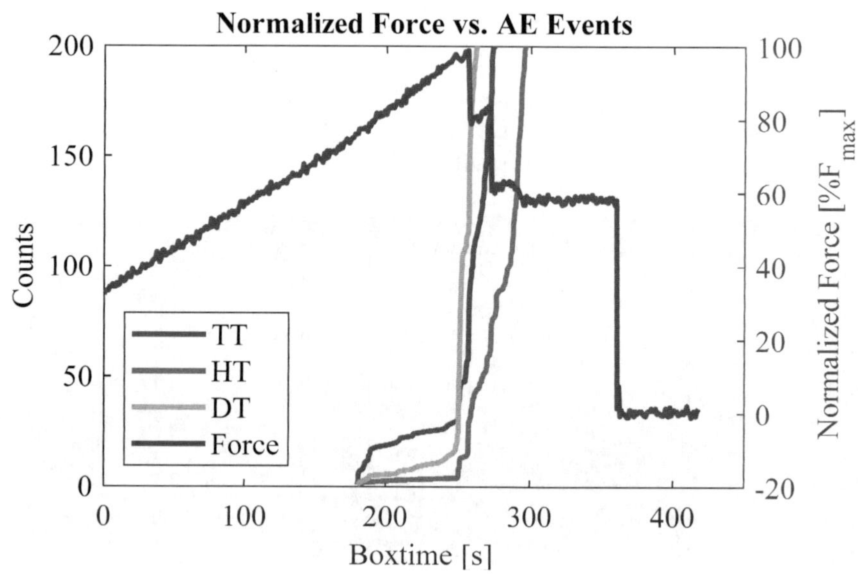

Fig. 8: AE events record

Numerical Simulation

The experimental loading of the composite frame was simulated using finite element analysis. The goal was to predict the frame strength using the finite element analysis and to verify the strength and stiffness behaviour of the model with the experimental results. For the simulations, ANSYS software was used with ANSYS Composite Prepost (ACP) toolkit. The FE model of the frame was made of a head tube; a top tube; a down tube, a seat tube and a bottom bracket tube, see Fig. 9 (left). These tubes were modelled using solid-shell elements (SOLSH190). The head tube joint lamination was modelled using a classic conventional shell element (Shell181), see Fig. 9 (right). The tubes and lamination were connected using "MPC bonded" connections. The fillers in the tube connections were modelled using solid elements.

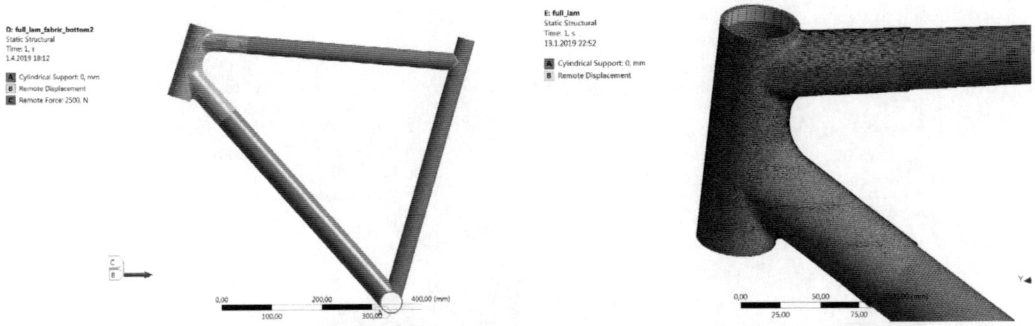

Fig. 9: Geometry of the bike frame model with the detail of the head tube joint shell elements.

The joint lamination, which is critical to the frame strength, was composed of uni-directional carbon fibre plies and woven fabrics combination. The laminate sequence was quite complex, as the joint lamination was made of more than 10 types of layers with a completely different fibre orientation. Two examples of the uni-directional plies of the lamination are given in Fig. 10. Each UD and fabric ply of the lamination was input using ACP, which enabled to have a full laminate specification with one shell through the joint thickness only without any major simplification of the each ply fibre directions.

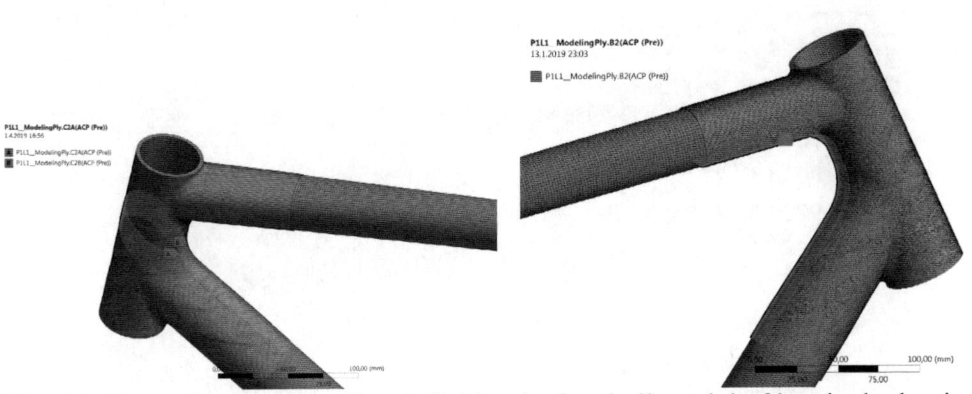

Fig. 10: Examples of joint plies definition in the shell model of head tube lamination.

The goal of the model was to evaluate the frame stiffness and strength. A linear static analysis was used for those purposes. The composite strength was established using a first ply failure analysis. The first ply failure analysis is using the failure index, which is computed as the response to the given loading using the specified composite strength criterion. When the failure index (0 for not-loaded coupon) reaches the critical value (1.0 or higher), the first ply failure criteria indicates the failure of the investigated ply and therefore the failure of the design. However, it must be noted that the first ply failure does not have to indicate the global behaviour change in terms of strength of stiffness. It indicates the end of linear elastic behaviour, however, the indicated failure does not have to be the one, which is causing the major failure.

The failure index was computed for each ply and fabric of the lamination. For the evaluation, following criteria were used: Maximal stress; Tsai-Wu; Tsai-Hill; Puck. The strength data of plies and fabrics were taken from the experiments (tensile strength) or as a nominal data from the literature [2].

The model was loaded by a concentrated force, see Fig. 9 and for this loading the failure indexes through the thickness of the lamination were evaluated. The maximal value was detected at the bottom of the head tube – down tube joint, see Fig. 11. The position of the maximal loading is consistent with the experimental results as the failure during the experiment started at the down-tube – head-tube connection, see Fig. 12, and then continue through the lamination to the top-tube – head-tube connection.

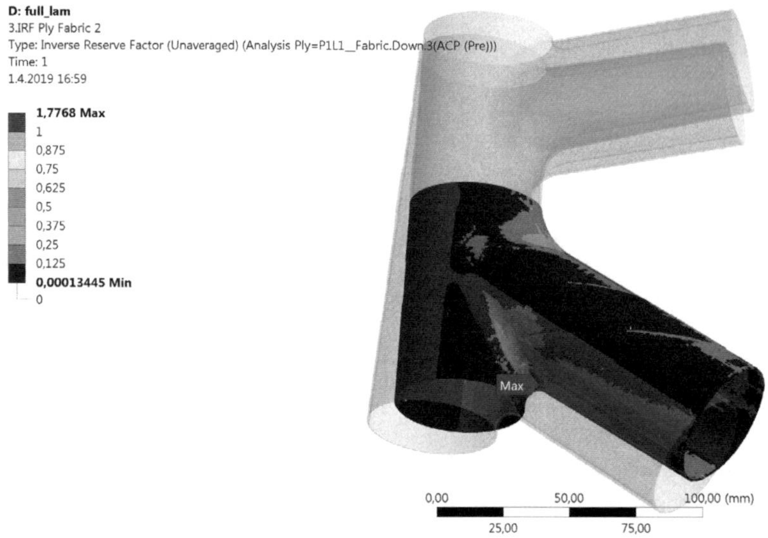

Fig. 11: Maximal failure index in the composite lay-up of the joint lamination.

Fig. 12: The down tube-head tube failure during the frame loading.

Results

The comparison of the numerical model and the experiment can be seen in Fig. 13, where the dependence of the head tube displacement versus normalized force is plotted. The graph for the numerical model ends at a point corresponding to the load when the critical value of failure index was reached. A good match of the FEM model stiffness with the obtained experimental values is evident. The first frame damage detected by the strain gauge signal on the head tube and the increase in the AE event amounted to about two-thirds of the maximum force. The load level at which the failure index value was reached in the FEM model is two-thirds compared to the average force at which the first damage was detected experimentally.

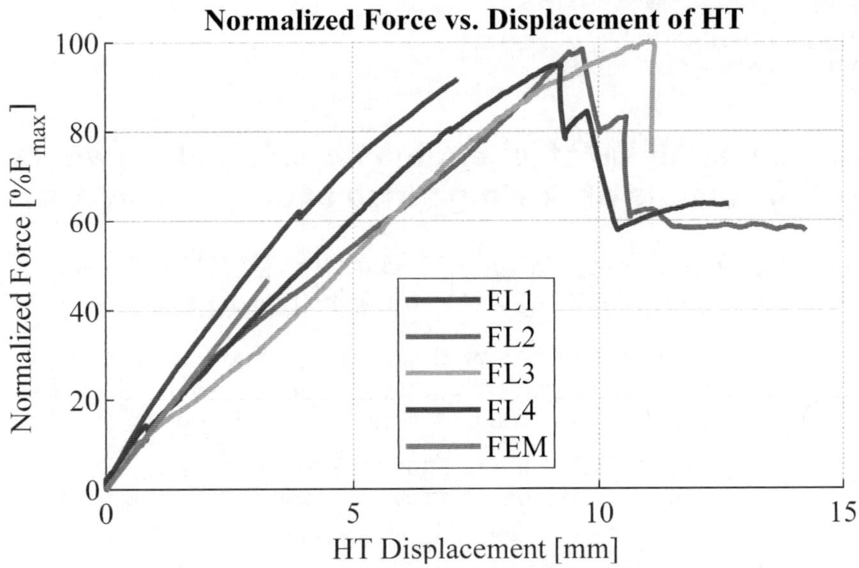

Fig. 13: The down tube-head tube failure during the frame loading.

Conclusions

Suitably positioned strain gauges together with AE sensors are capable of detecting primary joint damage. DIC allows to get a more complex area information of tube joint loading. A good agreement of the experimental results with numerical calculation was confirmed. The numerical model can be used with a high level of credibility for other tube joint variants.

References

[1] ISO 4210-5:2014 Cycles -- Safety requirements for bicycles -- Part 5: Steering test methods. ICS : 43.150 Cycles 2014. International Organization for Standardization.

[2] Ever J. Barbero. Introduction to Composite Materials Design, 2nd edition. CRC Press, 2010, ISBN 978-1420079159.

Experimental analysis of jet engine air channel impact resistance by means of strain gauges and FBG sensors

DOUBRAVA Karel[1,a], DVOŘÁK Milan[1,b], DOUBRAVA Radek[2,c], VÁCLAVÍK Jan[3, d] and RŮŽIČKA Milan[1, e]

[1]Department of Mechanics, Biomechanics and Mechatronics, Faculty of Mechanical Engineering, Czech Technical University in Prague, Technická 4, 166 07 Praha 6, Czech Republic

[2]VZLÚ (Czech Aerospace Research Centre), Beranových 130, 199 05, Prague 9 – Letnany, Czech Republic

[3]AERO Vodochody AEROSPACE a.s., U Letiště 374, 250 70, Odolena Voda, Czech Republic

[a]Karel.Doubrava@fs.cvut.cz, [b]Milan.Dvorak@fs.cvut.cz, [c]doubrava@vzlu.cz, [d]jan.vaclavik@aero.cz, [e]Milan.Ruzicka@fs.cvut.cz

Keywords: Jet Engine Air Chanel; Strain Gauge; FBG Sensor; FEM.

Abstract. New generation of the jet trainer is being developed based on the aerodynamic concept of the original L-39s, but contemporary technology and equipment are utilized. The composite air channel is one of the important parts, which are newly designed. One of the requirements for the air channel is a birds and hail strike resistance. Numerical simulation of impact loading must be compared with experimental results and this article focuses on the description of data acquisition from the impact experiment using a strain gauge and a Fiber Bragg Grating (FBG) optical sensors and comparison of experimental data with numerical solution.

Introduction

The jet trainer L-39 Albatros is legendary aircraft and almost three thousand manufactured planes are still operated by many air forces around the world. The new generation of jet trainer represents deep modernization of the L-39 combining latest engine and aircraft systems technology with traditional advantages of legacy L-39s such as excellent handling qualities, easy maintenance, robust design and low operational costs. One important element of the structure is the air channel. The new air channel design is based on a composite material, with a view to increasing life and using a new jet engine. One of the requirements for aviation structure is resistance to bird and hail impact [2, 3]. The Czech Aerospace Research Centre (VZLÚ) has the equipment and a lot of experience with this type of tests. The prototype of the air channel was attached to the test stand, and tests were carried out in the VZLÚ. The results obtained using a strain gauge and FBG sensors during tests are described in this paper.

Composite Air Channel

A complete air channel (see Fig. 1) is composed of three main components parts. The main material used for the air channel is pre-impregnated carbon fibre reinforced plastic (CFRP), A193-PW Hexply fibre with 8552 epoxy matrix. This particular combination is used due to previous experience of Aero Vodochody Aerospace (AVA) with the material. Carbon fibre is used for its high strength to weight ratio. Woven fabric is desirable for its higher impact resistance and due to the complexity of the part shape. The minimum used number of CFRP layers is 12, while the maximum no. of layers is 24 at the main reinforcement.

Left and right parts of air channel are connected by means the air channel brace, and the brace is fitted to the fuselage by using four rods, which are main load connections between the fuselage and the air channel. Air channel extension was not included in experiment setup.

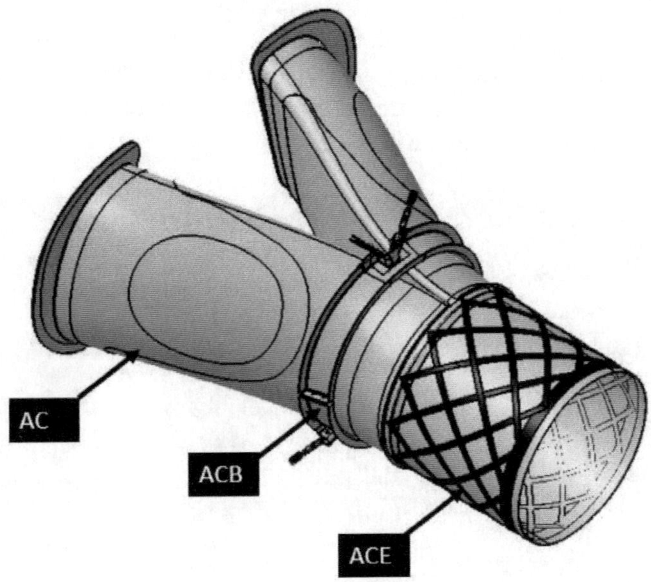

Fig. 1: A figure example (no punctuation here) Composite air channel (AC – Air Cannel, ACB – Air Cannel Brace, ACE – Air Channel Extension) [4]

Impact Experiments

The impact experiments were made at the VZLÚ according airworthiness requirements [1]. The composite air channel without extension (see Fig 2 left) was installed to the test bed simulating real stiffness of fuselage and it was impacted by bird and hails fired by an air gun (see Fig. 2 right).

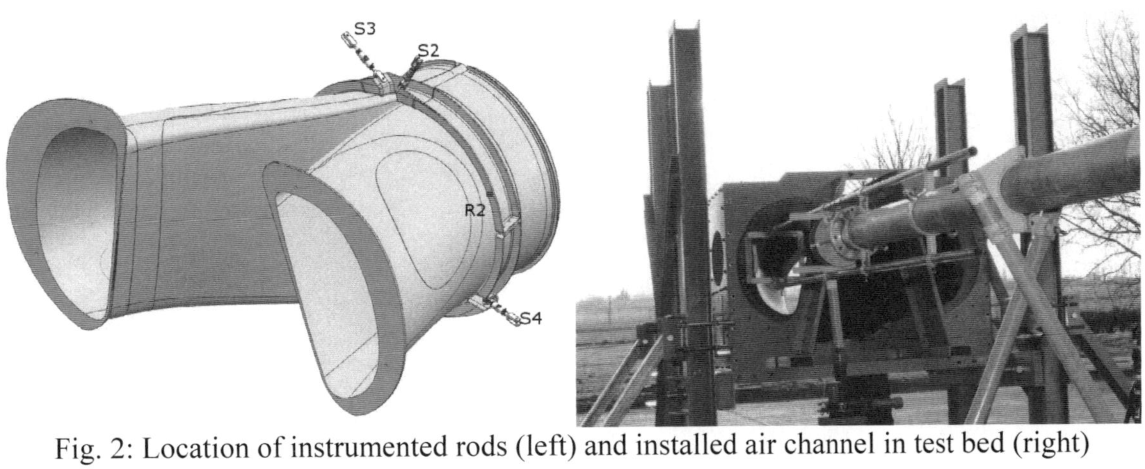

Fig. 2: Location of instrumented rods (left) and installed air channel in test bed (right)

Left inlet was impacted by 1 kg bird (Bird Strike) and right inlet was impacted by series of 50 mm (Hail Strike 1, 2) and 25 mm hails (Hail Strike 3 – 5). Loading of rods and air channel parts was monitored by sensors and displacement was filmed using the high speed cameras.

Strain Gauge Instrumentation

The air channel was connected with the test bed by means of the four instrumented rods. Two strain gauges T rosette HBM XY11-3/350 were installed on each rod and the full bridge connection was used to determine the axial load transmitted by the rod and to eliminate temperature affecting the measured strain, see Fig. 3 left. Calibration of the instrumented rods was done on the material load machine before and after impact tests (see Fig. 3 right) to obtain relationship between measured strain and loaded force.

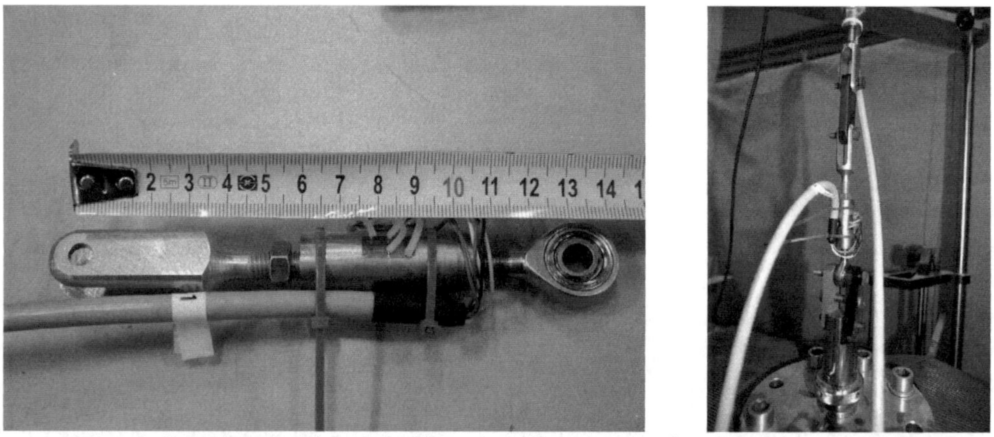

Fig. 3: Instrumented rod (left) and calibration by load machine (right)

Signals from strain gauges were measured by HBM Quantum measurement unit with 19.2 kHz sample frequency, see Fig. 4 where measured data for bird strike experiment are plotted. Reaction forces were computed from signals multiply by obtained calibration constants [5].

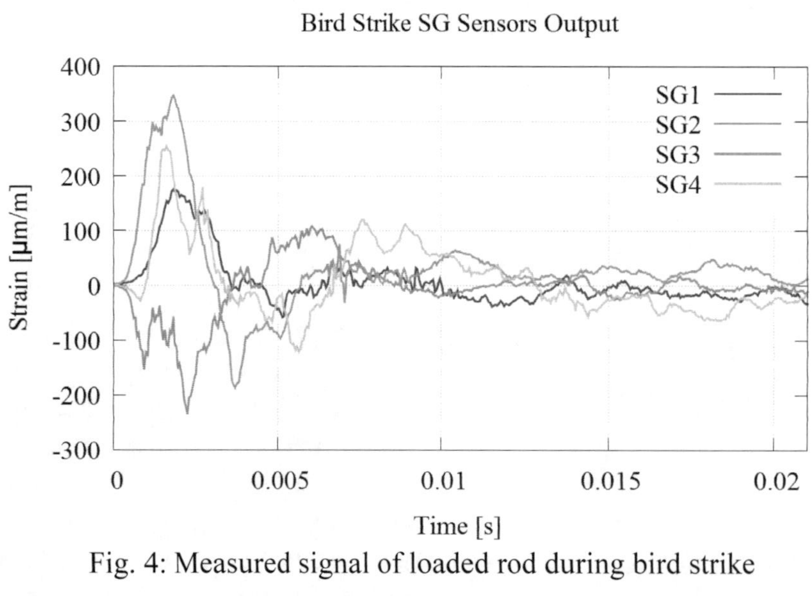

Fig. 4: Measured signal of loaded rod during bird strike

FBG sensors

Two optical fibres with total of nine FBG sensors were installed on the outer surface of air channel. Sensor position on the left part of air channel can be seen in Fig. 5. So-called draw-tower ORMOCER® coated optical fibres were used to ensure reliability during the impact, with the following configuration:

1. Left air channel part: 5 pcs of FBG sensor, outer diameter of optical fibre 0.195 mm, grating length 8 mm, central wavelength: 1505 nm (FBG1), 1510 nm, 1515 nm, 1520 nm, 1525 nm (FBG5), routing of optical fibre and locations of single FBG sensors are pictured in the left part of Fig. 5.
2. Right air channel part: 4 pcs of FBG sensor, outer diameter of optical fibre 0.195 mm, grating length 8 mm, central wavelength 1529 nm (FBG1), 1533.5 nm, 1538 nm, 1542.5 nm (FBG4), routing of optical fibre and locations of single FBG sensors are pictured in the right part of Fig. 5.

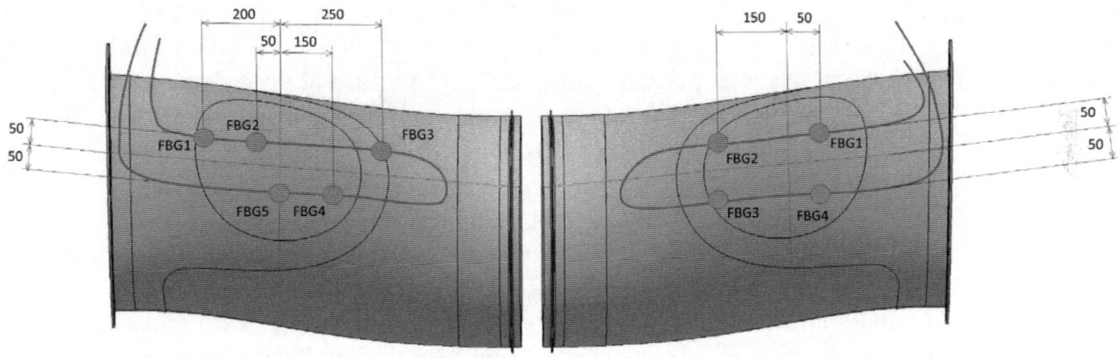

Fig. 5: FBG sensors configuration on the left (left) and right air channel part (right)

Fibre optic sensors were installed onto the cured air channel surface (see Fig. 6), because due to the complicated air channel prepreg composite lay-up process it wasn't possible to embed any sensors inside the structure. In the first step, optical fibres were fixed on the cleaned and degreased surface using the cyanoacrylate adhesive. In the next step, FBG sensors were completely covered using the hand lay-up composite process (combination of thin Glass fabric and L285 epoxy resin system). The remaining parts of optical fibres were protected using the thick layer of epoxy resin with glass micro-balloons. Both ends of the optical fibre were fusion spliced to the protective cables with FC/APC connectors. These splices were also covered and protected by the glass fabric. Careful protection of FBG sensors and optical fibre routing was important because of harsh conditions during the final assembly of the air channel and its preparation for impact test (placing inside the test-bed, other measurement methods instrumentation etc.).

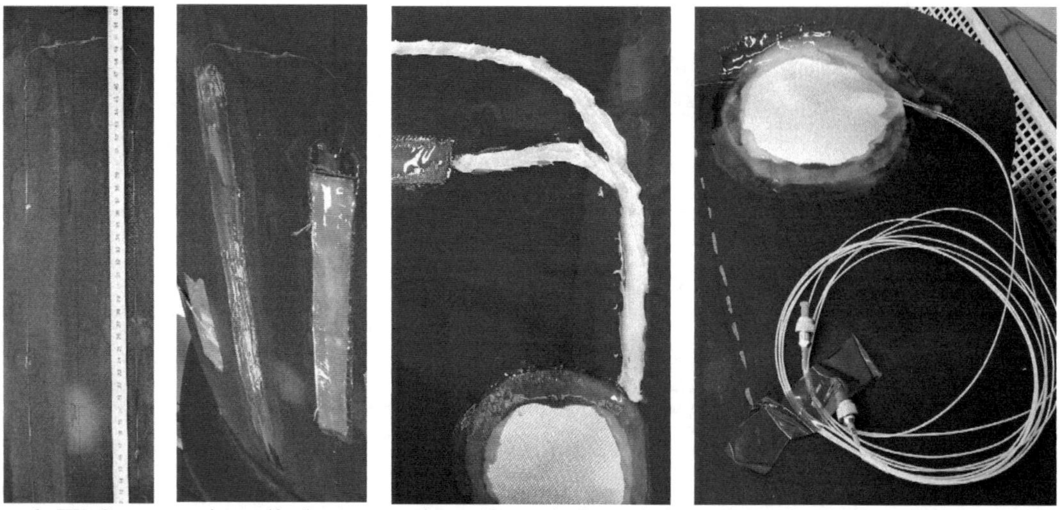

Fig. 6: FBG sensor installation procedure (from left to right): 1. Optical fibre fixed on the air intake surface; 2. Covering layer applied over the FBG sensors area (thin Glass fabric / Epoxy resin); 3. Covering layer to protect optical fibre routing (mixture of epoxy resin and micro-balloons); 4. Glass fabric protection of fusion splices and optical pigtails with FC/APC connectors

Safibra FBGuard 1550 FAST DAQ device was used to capture signal from the FBG sensors. Mechanical strain from FBG sensors versus time was measured with maximum frequency of 11.6 kHz, which is enough to cover impact load itself plus air channel twisting and bending after impact. All FBG sensors were able to measure during the whole test campaign. Signal measured by FBG sensors during bird strike test can be seen in Fig. 7, data from hail strike tests are shown in Fig. 8 and 9.

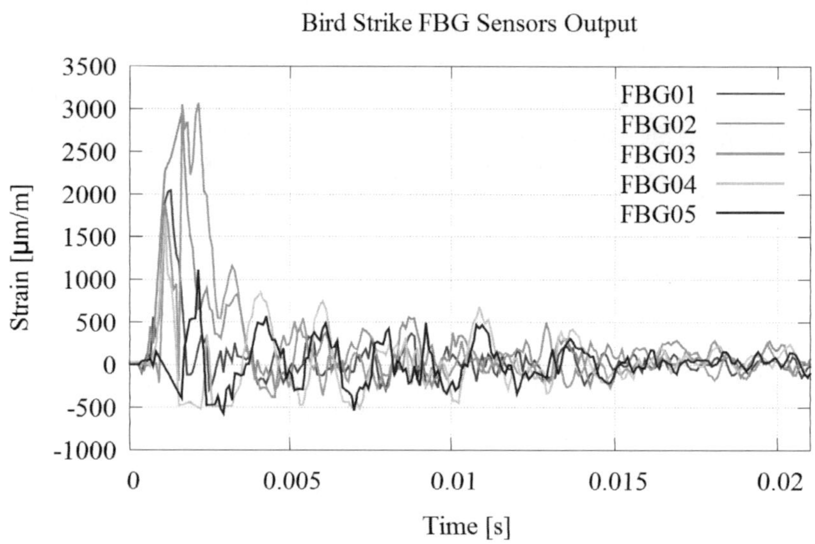

Fig. 7: Strain measured by FBG sensors on left inlet during bird strike

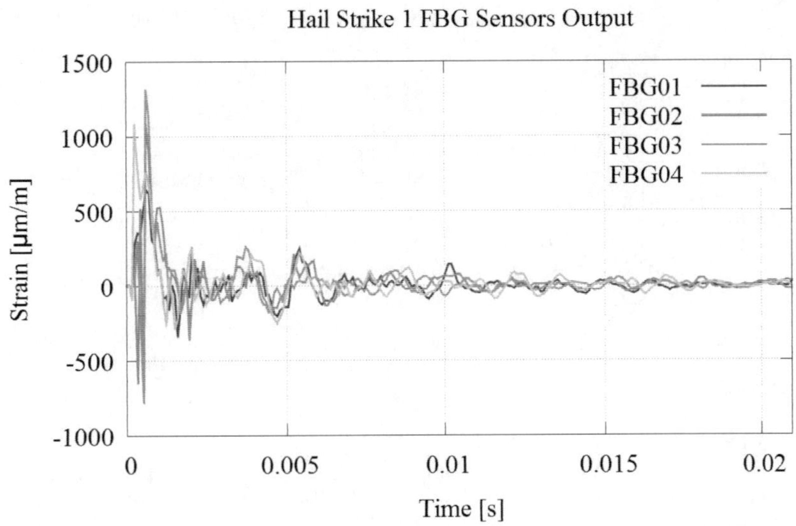

Fig. 8: Strain measured by FBG sensors on right inlet during hail strike 1

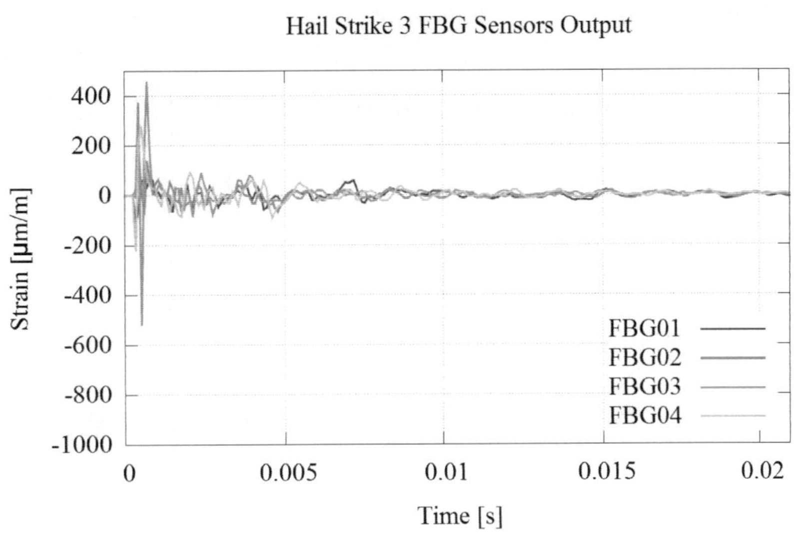

Fig. 9: Strain measured by FBG sensors on right inlet during hail strike 3

Numerical Simulation

Numerical simulation bird strike was made by VZLÚ. A bird strike is a high-velocity impact in which materials with a huge difference in properties come into contact with each other, resulting in nonlinear material behaviour, high strain rates, and extremely large deformations. Nonlinear finite element software has the capability to predict the loads and deformations of both the bird projectile and the complex aircraft component being impacted within acceptable levels of accuracy. In high-velocity impacts, the pressure on the bird tissues severely exceeds their limits, causing the bird material to behave like a fluid. An explicit solver of the ABAQUS FE software was used for numerical simulation, when geometry of bird projectile was simplified and elastic-plastic material model was used [6, 7]. The Smoothed Particle Hydrodynamics technique was used for projectile model damage propagation [8, 9]. Data from numerical simulation of bird strike was compared with experimental data with good agreement, see Fig. 10 where are plotted reaction force measured by means instrumented rods versus computed reactions.

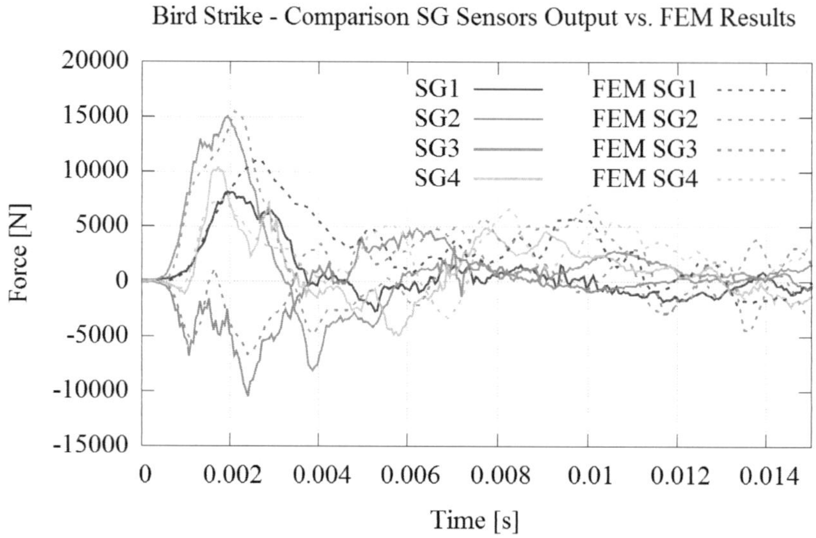

Fig. 10: Comparison measured and computed signal of loaded rod during bird strike

Comparison of strains measured by FBG sensors and computed results is plotted in Fig. 11.

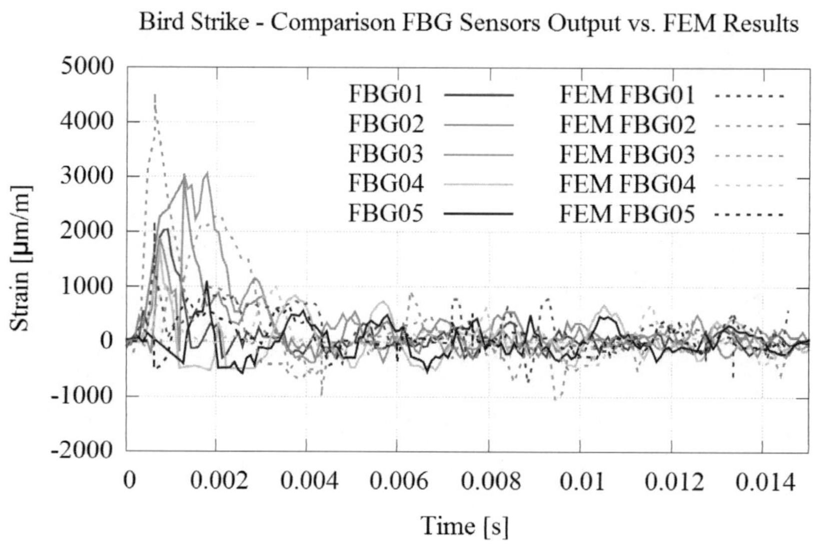

Fig. 11: Comparison measured and computed signals on left inlet during bird strike

Conclusions

Impact resistance of the composite air channel was confirmed. Strain Gauges were used for measurements reaction force during impact tests. Numerical simulation of bird strike was made and obtained experimental data and numerical simulation are in good agreement. It was proven that it is possible to make durable connection between the cured Carbon/Epoxy composite and optical fibre/ FBG sensor using the hand layup manufacturing method. All optical fibres and FBG sensors were functional during the complete impact load test. No sensors were detached from the composite surface. FBG sensors provided good data source for numerical analysis of the structure.

Acknowledgement

This work was supported by the Technological Agency of the Czech Republic in the project TE02000032 - Advanced Aerostructure Research Centre.

References

[1] Defence Standard 00-970 Part 1 Section 3, Design and Airworthiness Requirements for Service Aircraft.

[2] ASTM F330-10 Standard Test Method for Bird Impact Testing of Aerospace Transparent Enclosures.

[3] ASTM F320-05 Standard Test Method for Hail Impact Resistance of Aerospace Transparent Enclosures.

[4] Kučera, J.: Composite air channel design for L-39NG aircraft, report, Aero Vodochody Aerospace a.s., Odolena Voda, TR-AV9954-0011, 2017

[5] Dvořák, M., Doubrava, K.: Test Results Collection (Demonstrator Test) report (in Czech) CTU in Prague 12105/17/29, 2017

[6] Doubrava, R., Strnad,V.: Bird strike analyses on the parts of aircraft structure 27th International Council of the Aeronautical Sciences (ICAS 2010), Nice, France, ISBN 978-0-9565333-0-2

[7] Doubrava, R.: Influence of Impact on the Real Aircraft Structure, NAFEMS World Congress 2009, Crete, Greece, ISBN – 978 1 874376 42 2

[8] Doubrava, R., Oberthor, M., Bělský, P.: Validation of high speed impact numerical simulation on flat test panels for real structure analysis, 55th Conference on Experimental Stress Analysis 2017, ISBN 978-80-553-3167-6, Nový Smokovec, SlovakiaJ. Lemaitre, J. Dufailly, Damage measurements, Eng. Fract. Mech. 28 (1987) 643–661.

[9] Doubrava, R., Raška, J., Oberthor, M.: Verification of numerical analysis for L-39NG air inlet development by bird strike tests on flat composite panels, Transfer 29/2017, ISSN 1801 – 9315 (in Czech)

EAN 2019
57th conference on experimental stress analysis

Monitoring of deep core drilling operational parameters

DUB Martin[1,a], ŠTOČEK Ondřej[1,b], HAAS Filip[1,c], VESELÝ Tomáš[2,d], KUČERA Lukáš[2,e] and DYNYBYL Vojtěch[1,f]

[1]Czech Technical University in Prague, Faculty of Mechanical Engineering, Department of Designing and Machine Components, Technická 4, 166 07, Prague 6, Czech Republic

[2]Czech Technical University in Prague, Faculty of Biomedical Engineering, Department of Information and Communication Technologies in Medicine, nám. Sitná 3105, 272 01, Kladno, Czech Republic

[a]martin.dub@fs.cvut.cz, [b]ondrej.stocek@fs.cvut.cz, [c]filip.haas@fs.cvut.cz, [d]tomas.vesely@fbmi.cvut.cz, [e]lukas.kucera@fbmi.cvut.cz, [f]vojtech.dynybyl@fs.cvut.cz

Keywords: Deep core drilling, Operational parameters, Identification, Sensor, Strain gauge

Abstract. The deep core drilling has currently wide application - construction drills, geological exploration, drills for measuring pendulums in dams etc. It is useful to know the drilling process parameters precisely. The most important parameters for the optimal drilling process are torque, pressure force and rotation speed. These parameters depend on the strength of the drilled material (drilling capacity) and of course on the drilled diameter and applied pressure force. In-time identification of the drilling parameters enables the optimization of the drilling process directly during the drilling process. This contribution deals with the design of the new sensor for in-time identification of above mentioned core drilling operational parameters and its testing. The sensor is based on the strain gauge measurement and Hall probe. The measured data are also stored and could be post-processed in longer time context to obtain also the drilling capacity of the drilled materials. First results from the terrain measurement are also presented.

Introduction

The deep core drilling has currently wide application - construction drills, geological exploration, drills for measuring pendulums in dams etc. It is useful to know the drilling process parameters precisely. The most important parameters for the optimal drilling process are torque, pressure force and rotation speed. These parameters depend on the strength of the drilled material (drilling capacity) and of course on the drilled diameter and applied pressure force. In-time identification of the drilling parameters enables the optimization of the drilling process directly during the drilling process, [1]. This contribution deals with the design of the new sensor for in-time identification of above mentioned core drilling operational parameters and its testing, [2], [3]. The sensor is based on the strain gauge measurement and Hall probe. The measured data are also stored and could be post-processed in longer time context to obtain also the drilling capacity of the drilled materials. First results from the terrain measurement are also presented. The schema of whole system can be seen in Fig. 1.

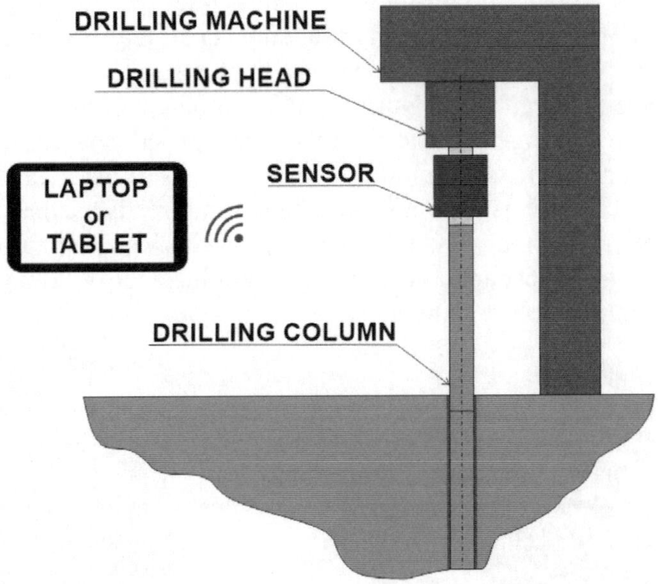

Fig. 1: Scheme of the measurement, sensor installed on the drilling machine

Sensor design

Sensor design is based on the common drilling rod with hexagonal couplings, which is used in short version as a deformation member. Thus, it can be easily connected to the various drilling machines. Above mentioned operational parameters are identified by means of strain gauges (torque and axial force) and Hall probe (rotational speed). For measuring both torque and pressure force, full bridge arrangement with compensation of temperature and other loads is used. Torque is identified by means of strain gauge rosettes HBM 1-XY21-6/120, the pressure load is identified by means of strain gauge rosettes HBM 1-XY91-6/120. The rosettes installed on the deformation member can be seen in Fig. 2.

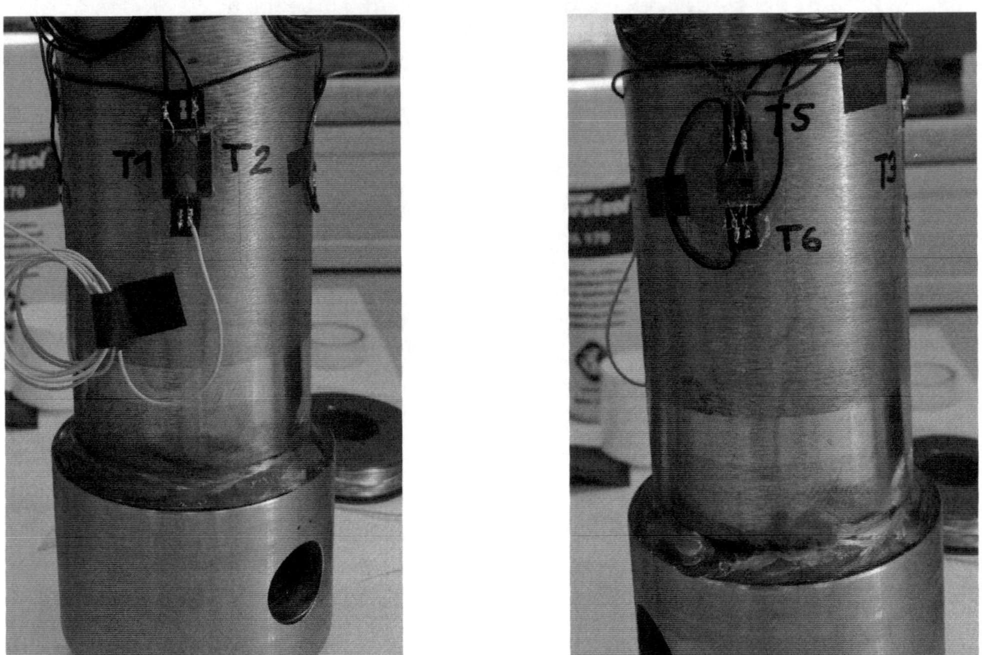

Fig. 2: Strain gauge rosettes installed in the deformation member

Hall probe is used in combination with neodymium magnet placed on the steel console. This console has to be designed for each drilling machine separately because of the drilling head differences. Sensor body is formed by the FDM 3D printed cover, which serves both like a protection and a storage for electronics. During the development of the sensor, two body designs arose. They are illustrated in Fig. 3, in the right there is the final version. It enables much better access to the batteries and micro SD memory card. It consists of the main body, flange and two doors – one is used just for the assemblage and the second one is used for the common access to the batteries and micro SD memory card. The doors with installed batteries and electronics during the testing with the tablet can be seen in Fig. 4. For the energy saving there is furthermore the power switch mounted into the flange.

Fig. 3: Prototypes of the sensor body

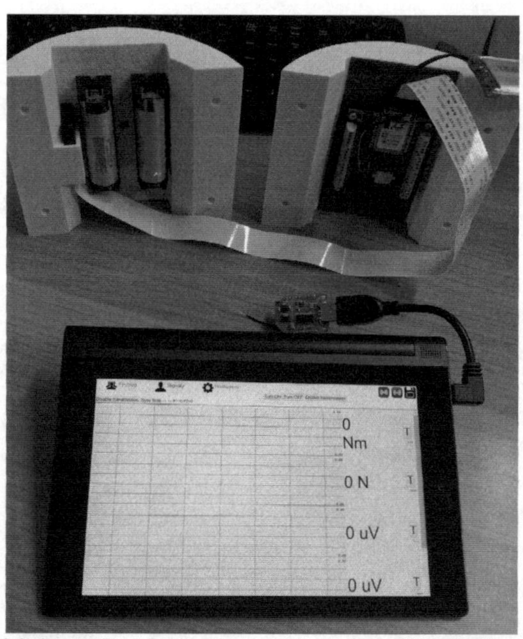

Fig. 4: Sensor doors with installed batteries and electronics during the testing with tablet

Data acquisition and telemetry

Data from the strain gauge Wheatstone bridges and Hall probe are acquired by means of the special datalogger LPDR with 24bit A/D converter and Cortex M3 processor, see Fig. 5 (left). Now it is set to the three-channel acquisition with sampling frequency 500 Hz per channel. The telemetry is realized by the Digi XBee 2.4 GHz transmission. The size of the LPDR is 70 x 55 x 25 mm with weight approximately 165 g. Measured data are processed, transmitted and also stored on the micro SD memory card. The power supply is provided by two Li-ion batteries of type 18650. This datalogger was developed in cooperation with Faculty of Biomedical Engineering.

Fig. 5: Datalogger LPDR with Digi XBee receiver (left), cover for the receiver (right)

Laboratory and terrain testing

Designed sensor was laboratory tested to verify the accuracy of the measurement. First results of laboratory testing were introduced in [1]. During this testing the correct functioning was proved. Further tests were carried out on the test polygon in the area of company Chemcomex, a.s. in Prague – Zbraslav and the area in not closely specified company in Prague - Zizkov. Various drilling machines were used during these measurements so that also the versatility of the sensor was tested. From the measurements, the time courses of torque, pressure force and rotational speed were acquired and post-processed.

In the first destination, Prague – Zbraslav, the RDBS drilling machine was used. It is the machine with tracked chassis and maximum rated torque 15000 Nm. The drilling sensor mounted under the drilling head of the drilling machine can be seen in Fig. 6 (left). There were used the drilling tools with diameter 220 mm and 190 mm. The geological profile of this drill can be seen in Fig. 6 (right). The most of the drill consists of the backfill (approx. 6 m) the rest is the sand (approx. 1.5 m). The measured data are presented in Fig. 7. There can be seen the time course of the torque, pressure force and rotation speed. The data will be correlated with the dept of the drill and with the description of the activity.

Fig. 6: Drilling in Zbraslav – drilling machine RDBS (left), 7.5 m drilled core (right)

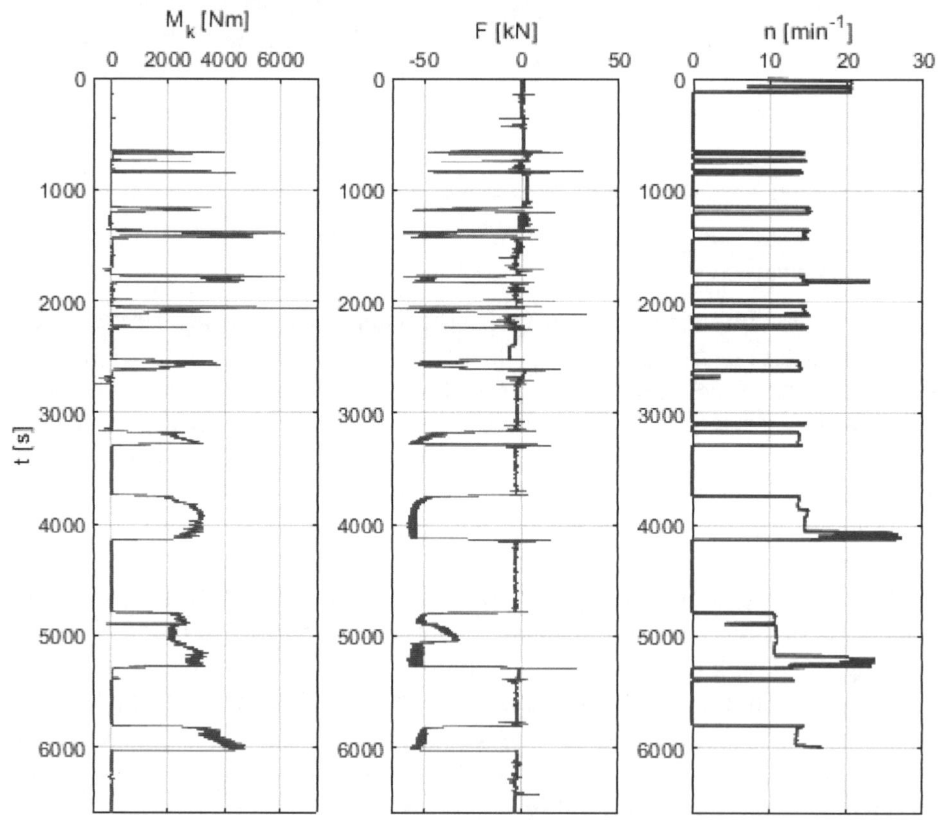

Fig. 7: Drilling in Zbraslav – measured time courses of torque, pressure force and rotation speed

In the second destination, Prague – Zizkov, the SGBO drilling machine was used. It is the drilling machine embedded on the track chassis. The maximum torque of this drilling machine is 10200 Nm. The drilling sensor mounted under the drilling head of the drilling machine can be seen in Fig. 8. There were used the drilling tools with diameter 220 mm and 190 mm. The geological profile of this drill can be seen in Fig. 9. Brief geological profile description follows in Table 1. The water level was approximately in the dept 6.5 m.

Fig. 8: Drilling in Zizkov – drilling machine SGBO

Fig. 9: Drilled core

Table 1: Drilled core brief description

Dept [m]	Material description
0.0 – 0.2	Bitumen
0.2 – 1.1	Backfill
1.1 – 1.7	Clay
1.7 – 3.9	Sandy clay
3.9 – 10.1	Sand
10.1 – 12.0	Slate

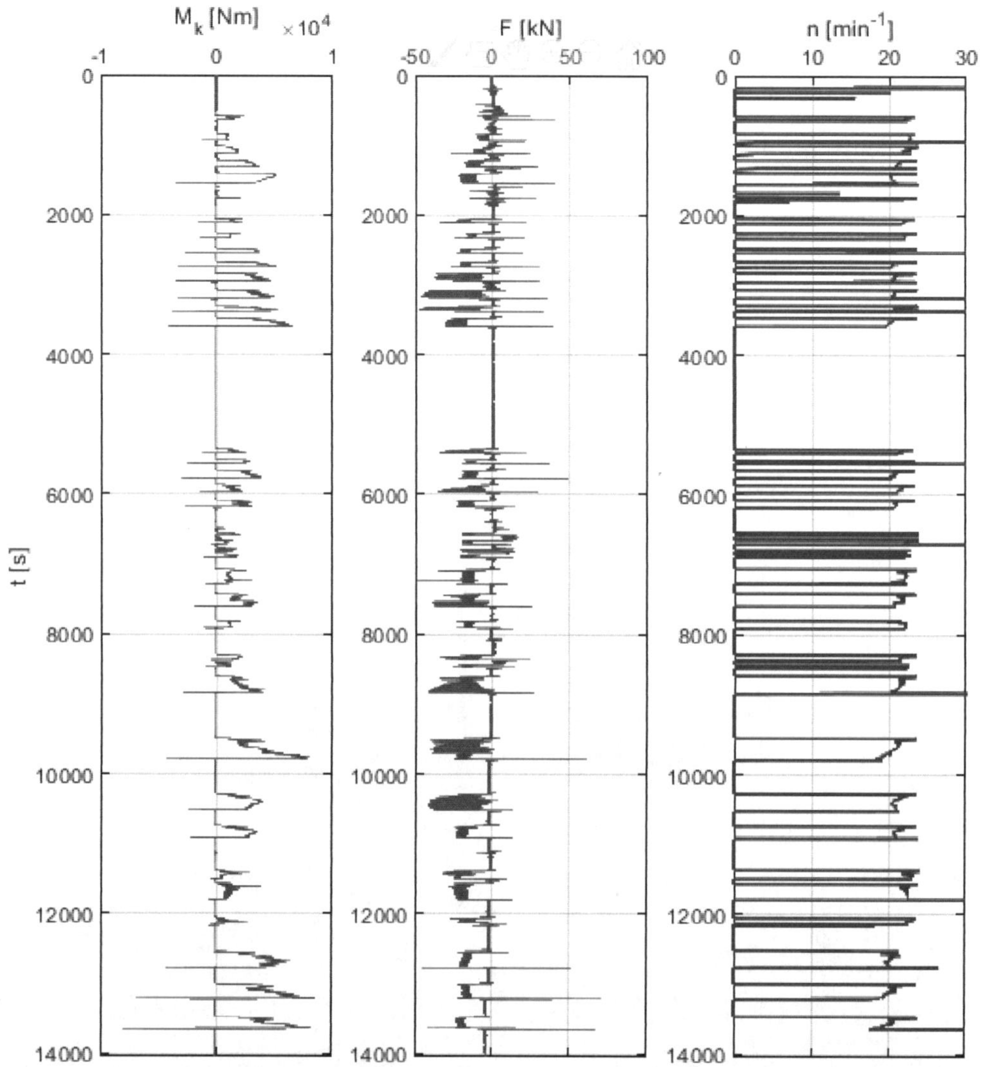

Fig. 10: Drilling in Zizkov – measured time courses of torque, pressure force and rotation speed

Acquired and processed data are presented in Fig. 10. There can be seen again the time course of the torque, pressure force and rotation speed. The data will be correlated with the dept of the drill, geological profile and with the description of the activity. Detailed view on the acquired data can be seen in Fig. 11. The maximal value of the torque reaches approximately 5 kNm, the pressure force is approximately constant 16.5 kN and the rotation speed is also approximately constant 20 rev/min. At the end of the drilling process, there is the significant jump in the torque due to the quick reversation. This is done to enable easier dismounting of the hexagonal rods. There is also significant increase of the axial force. This force is necessary to get the drill bit out of the drill.

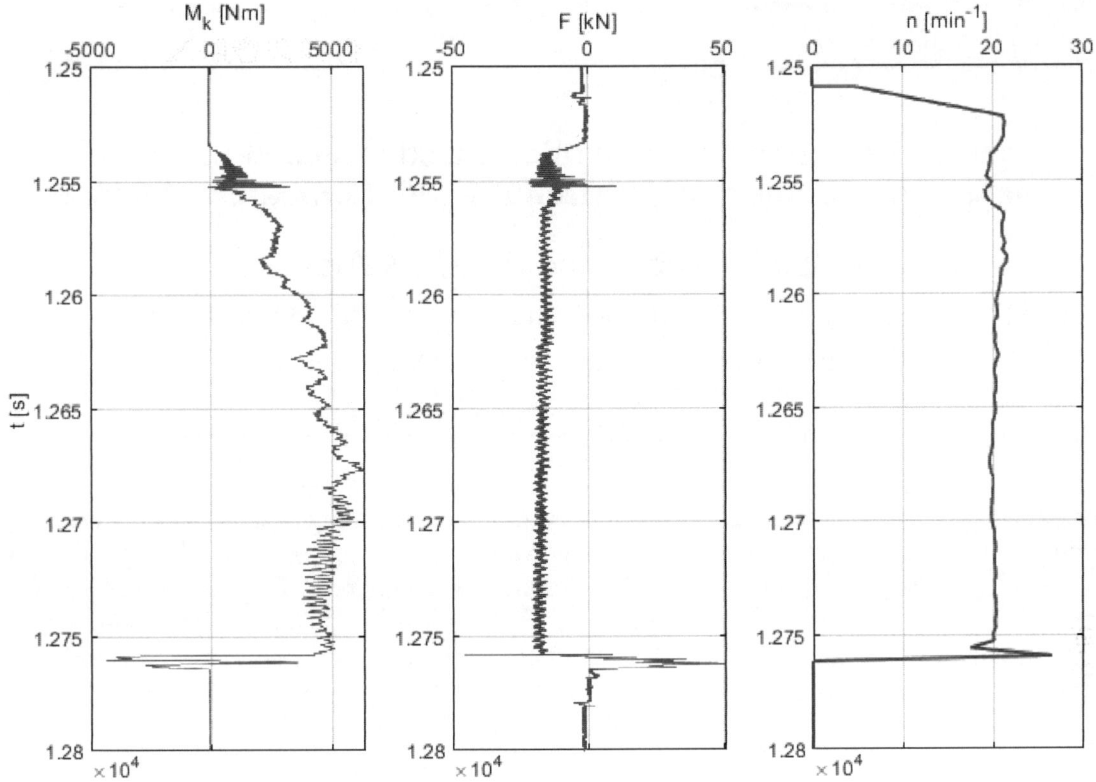

Fig. 11: Drilling in Zizkov – detailed view – drilling of a slate

Conclusions

The sensor for monitoring of deep core drilling operational parameters was introduced in this contribution. It enables acquisition of valuable information about the drilling capacity and also about own drilling process. The sensor is tested both in laboratory and terrain conditions. Special cover for the sensor was manufactured by the FDM 3D printing and will be further improved.

Acknowledgement

This work was supported by The Technology Agency of the Czech Republic project No: TH02010958 “Sensor for measuring of operational parameters of deep core drilling”.

References

[1] M. Dub and V. Dynybyl, Sensor for In-time Identification of Deep Core Drilling Parameters, in: The 59th International Conference of Machine Design Departments, 2018.

[2] P. Bauer, J. Kalivoda, System of axle-box force measurement for experimental railway bogie, in: Experimental Stress Analysis - 56th International Scientific Conference, EAN 2018 - Conference Proceedings, pp. 9-16, 2018.

[3] F. Lopot, M. Dub, M. Janda, O. Berka, Use of notches in the shafts loaded by torque for improvement in accuracy of strain gauge measurements, in: Experimental Stress Analysis - 56th International Scientific Conference, EAN 2018 - Conference Proceedings, pp. 9-16, 2018.

EAN 2019
57th conference on experimental stress analysis

Foamed cement paste with micronized recycled concrete - measuring of compressive strength and thermal conductivity

ĎUREJE Jakub [1,a] and PROŠEK Zdeněk [1,b]

[1]CTU in Prague, Faculty of Civil Engineering, Thákurova 7, 166 29 Praha 6, Czech Republic

[a]jakub.dureje@fsv.cvut.cz, [b]zdenek.prosek@fsv.cvut.cz

Keywords: Lightweight blocks, Concrete recyclate, Foaming agent, Recycled blocks

Abstract. This paper describes the possibility of using fine concrete recycled material to produce lightweight blocks. We focus on mechanical properties of foamed cement matrix. The samples contain 1: 1 cement and finely ground waste concrete. The matrix was lightened with various foam agents and subsequently compressive strength and thermal conductivity were measured. The results obtained from these samples were compared with reference samples.

Introduction

Waste materials from the construction industry have a negative impact on the environment. Some building materials can be partially or completely recycled and reused [1]. Old waste concrete could be recycled. First, it is sorted out and removed foreign materials (such as steel, bricks, wood, etc.) and subsequently milled into concrete recyclate [2]. Recycled concrete is commonly produced in many different fractions, large fractions are most commonly used as foundation under roads, but the fine fraction has almost no used. In previous researches, it has been shown that after grinding fine fractions with a high speed mill, the resulting product - micronized concrete can be used as a filler for cementitious composites [3].

One of the possibilities of using micronized recycled concrete is its application to lightweight blocks intended for peripheral masonry. This application is suitable, for example, for the Middle East, where there are many damaged or destroyed buildings due to frequent war conflicts. Concrete debris could be recycled and reused [4].

To achieve thermal insulation properties, it is advisable to lightweight the blocks. [5] Appropriate additives will be used to increase the strengths and obtain suitable properties for serial production. [6]

Materials and Samples

Six different foaming agents were selected for the tests. Two foaming agents for concrete, two industrial cleaners, fire-fighting foam and natural soap were selected. The foaming agent was selected in consultation with the manufacturers. In the case of building chemistry manufacturers, a foaming agent for cement composites was directly required. Mapeiair LP 800 from Mapei and Lightcrete-400 from Sika were used. Highly foaming and stable products were required at the industrial cleaner manufacturer. Based on experience, product 75-32 SUPER Universal Cleaner (Pink Cleaner) and 79-50 ULTRA Universal Cleaner (White Cleaner) from Metaflux were recommended. These products are intended as degreasing agents for industrial use. Foam used by firefighters was selected based on good results from previous works and natural soap from small-scale production was tested to compare results with conventional industrial products.

To determine the influence of foaming additives on the presence of the recycled concrete, test specimens containing only Portland cement CEM 42.5R (Radotin), micronized recycled concrete, water and foaming agent were produced. For comparison, test specimens with the same composition without micronized recycled concrete were produced too (Table 1). Sika Lightcrete - 400 (S), Mapei Mapeiair LP 800 (M) and Firefighters foam (H) were selected to produce test samples based on foaming and stability results.

Table 1: Composition of the samples

Set/Material	Cement [g]	Recyclate [g]	Foaming agent [ml]	Water [ml]	w/(c+r)
C	1500	0	0	480	0.32
CR	750	750	0.0	480	0.32
CS1.5	1500	0	1.5	480	0.32
CRS1.5	750	750	1.5	480	0.32
CS3	1500	0	3.0	480	0.32
CRS3	750	750	3.0	480	0.32
CM1.5	1500	0	1.5	480	0.32
CRM1.5	750	750	1.5	480	0.32
CM3	1500	0	3.0	480	0.32
CRM3	750	750	3.0	480	0.32
CH1.5	1500	0	1.5	480	0.32
CRH1.5	750	750	1.5	480	0.32
CH3	1500	0	3.0	480	0.32
CRH3	750	750	3.0	480	0.32

Experimental Methods and Results

Different types of foams were selected to produce concrete blocks. The most important properties of foams for a given application are stability and frothiness. Foam measurement was carried out in a graduated cylinder into which air is supplied by means of a compressor. The maximum amount of foam produced at a constant air flow rate was measured. Extruded polystyrene with a lead weight was placed on the foam surface at the beginning of the experiment to obtain flat level of foam to easy reading amount of foam. Moreover, the extruded polystyrene loaded the foam, which simulates the load from the cement matrix.

Stability measurements were made in the same measuring apparatus as the frothiness. In the graduated cylinder was produced 2000 ml of foam and then the time to reduce its amount to 1000 ml was measured.

The preliminary measurement showed insufficient foaming of the natural soap. Based on preliminary measurements, 3 different concentrations of foaming solutions were selected. The foaming solution was made up of 50 ml of water and 30, 60 or 120 μl of foaming agent. The measurement of each solution was repeated six times. The measurement results are shown in Fig. 1. With increasing amounts of foaming agent, frothiness increasing approximately linearly.

Foaming solution for stability measurement contained foaming agent and water in ratio 1:1. The measurement of each solution was repeated four times. The measurement results are shown in Fig. 2.

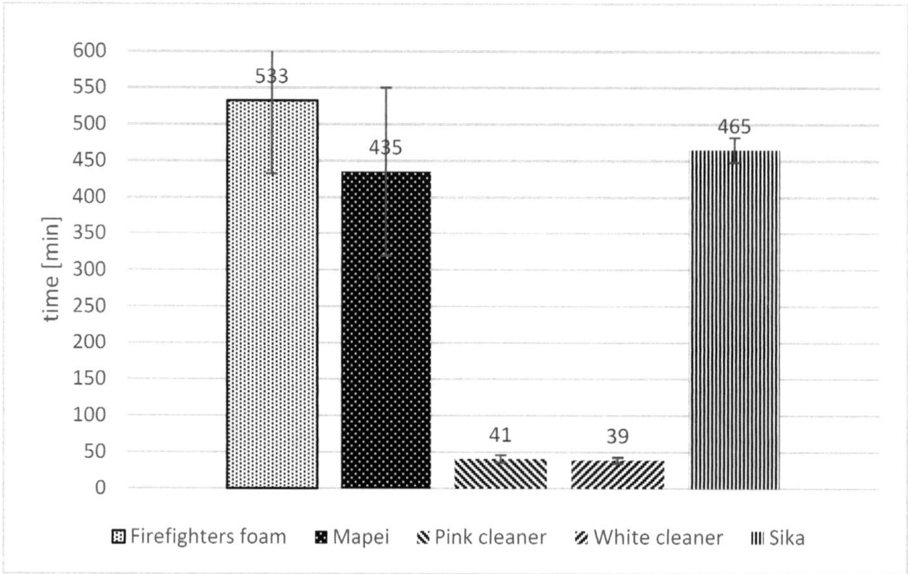

Figure 1: Stability results

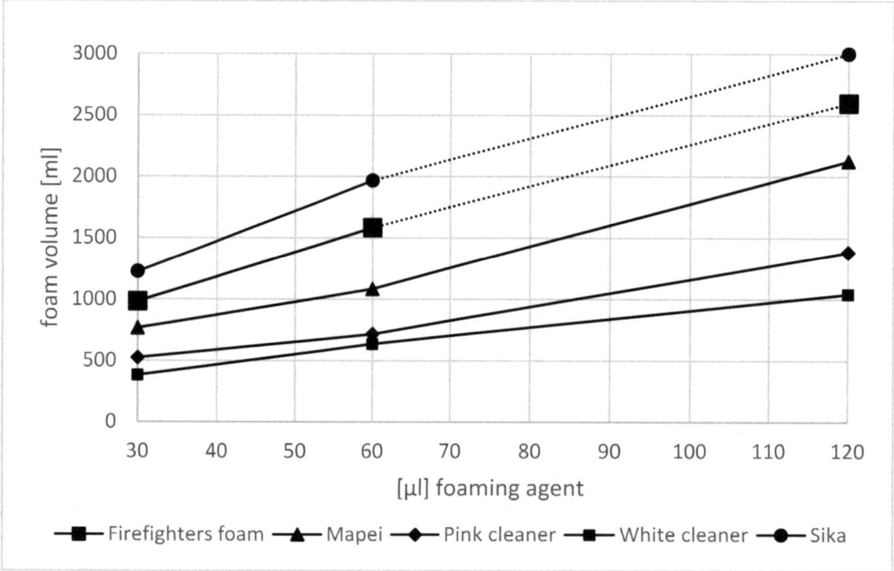

Figure 2: Frothiness results

There were made six sets of samples, each set contains three testing samples. The dimensions of the test samples were 40 x 40 x 160 mm. First, all the dry ingredients were mixed thoroughly at 132 rpm and then the mixing water was added together with the foaming agent and the mixture was stirred for 3 minutes at 421 rpm. Subsequently, all specimens were stored for 28 days under water in a laboratory environmental at 22 ± 1 °C. The specimens were halved during the bending test. The samples were subsequently stored for 28 days in a laboratory environment at 22 ± 1 ° C and a relative humidity of 50 ± 2%. The thermal conductivity coefficient λ was determined by ISOMET 2104 device of all 8 halves of the samples and then a compressive strength test was performed (Fig. 3).

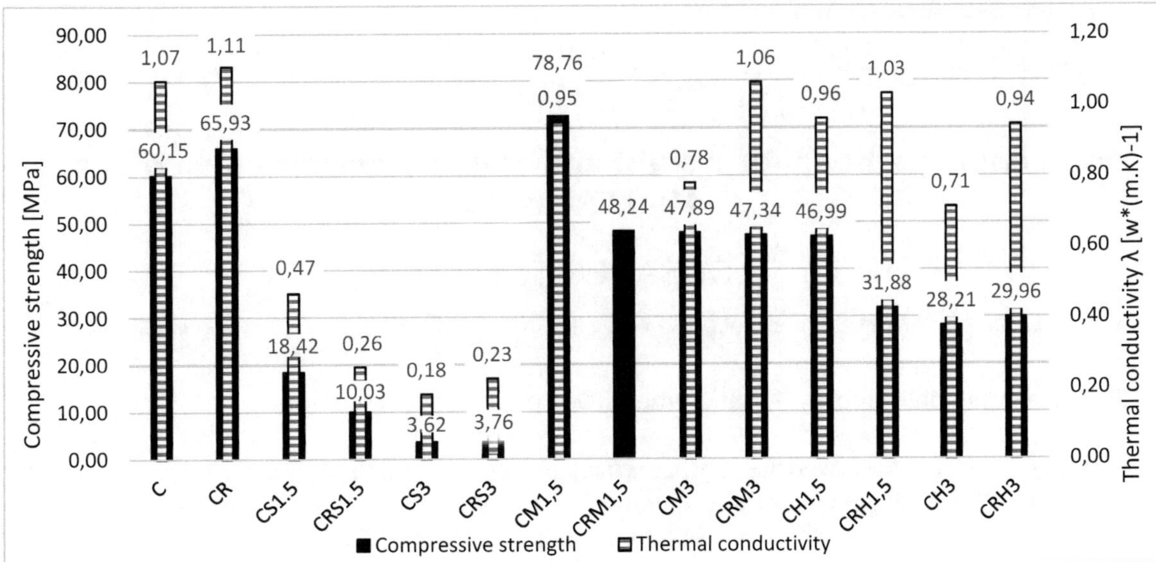

Figure 3: Compressive strength and thermal conductivity

Conclusions

The most suitable foaming agent for this application is SIKA Lightcrete 400, other foaming agents could achieve better results, for example, when using a foam generator. The most foamed samples achieve a sufficiently low thermal conductivity to utilize blocks in Middle East. The compressive strength is relatively high, but it is not yet sufficient for practical use. To produce blocks, it is planned to add suitable additives and dispersed fiber reinforcement to the mixture to provide sufficient compressive strength.

Acknowledgement

This paper was financially supported by the Czech Technical University in Prague under No. SGS project SGS19/148/OHK1/3T/11. The authors also thank company LAVARIS Ltd, where a high-speed mill from was used for micronizing of the recycled old concrete.

References

[1] M.S. Ibrahim, et al., Health Risk Assessment of Marble Dust at Marble Workshops, Nature and Science, 9 (2011) 144-154.

[2] N.D. Oikonomou, Recycled concrete aggregates, Cement and Concrete Composites (2005) 315-318.

[3] Z. Prošek et al., Influence of Recycled Concrete Composition on its Elastic Stiffness, Key Engineering Materials, Vol. 677 (2019) 288-291.

[4] J. Hrůza, Návrh kompozitní směsi zdicího bloku s využitím recyklovaného betonu a přísad, Diplomová práce (2017).

[5] J. Trejbal, V. Hrbek, J. Ďureje, R. Hlůžek, Lightweight masonry block made from recycled concrete, in: 56th International Scientific Conference on Experimental Stress Analysis, EAN 2018. Harrachov, 2018, pp. 435-438.

[6] Z. Prošek, V. Nežerka, R. Hlůžek, J. Trejbal, P. Tesárek, G. Karra'a, Role of lime, fly ash, and slag in cement pastes containing recycled concrete fines, Construction and Building Materials. 201 (2019) 702-714. doi: 10.1016/j.conbuildmat.2018.12.227.

EAN 2019
57th conference on experimental stress analysis

Predicting methods for weight and stability performance of a new Mini Excavator

DUSANEK Tomas[1,a]

[1] Doosan Bobcat EMEA, s.r.o.; U Kodetky 1810; 263 12 Dobříš, Czech Republic

Keywords: Weight, Stability, Measurement, Design, Optimisation

Abstract. This work shows the mathematical and testing methods how to predict stability targets for a newly developed mini excavator.

Introduction

In this work it is shown how to predict weight and stability of a newly developed Mini excavator. The method developed needs to be fast and allow for major changes in the design. The goal is to use this knowledge to drive design changes and to gain confidence that final product will meet requirements.

When developing a new product, the engineers are usually faced with several requirements for the new machine. Often these requirements are in conflict. For the mini excavator there are usually these criterions in order of importance:

1) Reach, Digging depth, max height of working equipment… *(How deep can I dig?)*
2) Total weight *(Can I tow the excavator behind my car?)*
3) Length, width, height *(Will I get through the door?)*
4) Lift capacity *(Can I lift something?)*
5) ….

The engineers need to finely tune the design, to offer the best combination of the required criterions. There are also another sets of requirements that must be complied with and that are the legal (homologation) tests.

To make correct design decision having the information about weight and stability is crucial.

Mathematical model

A simple 2D model of an equilibrium in a plane had been developed. The machine does have rubber tracks on steel undercarriage. Test have been done to find the influence on the results. But with no good results. The compression of the tracks is not taken in to account.

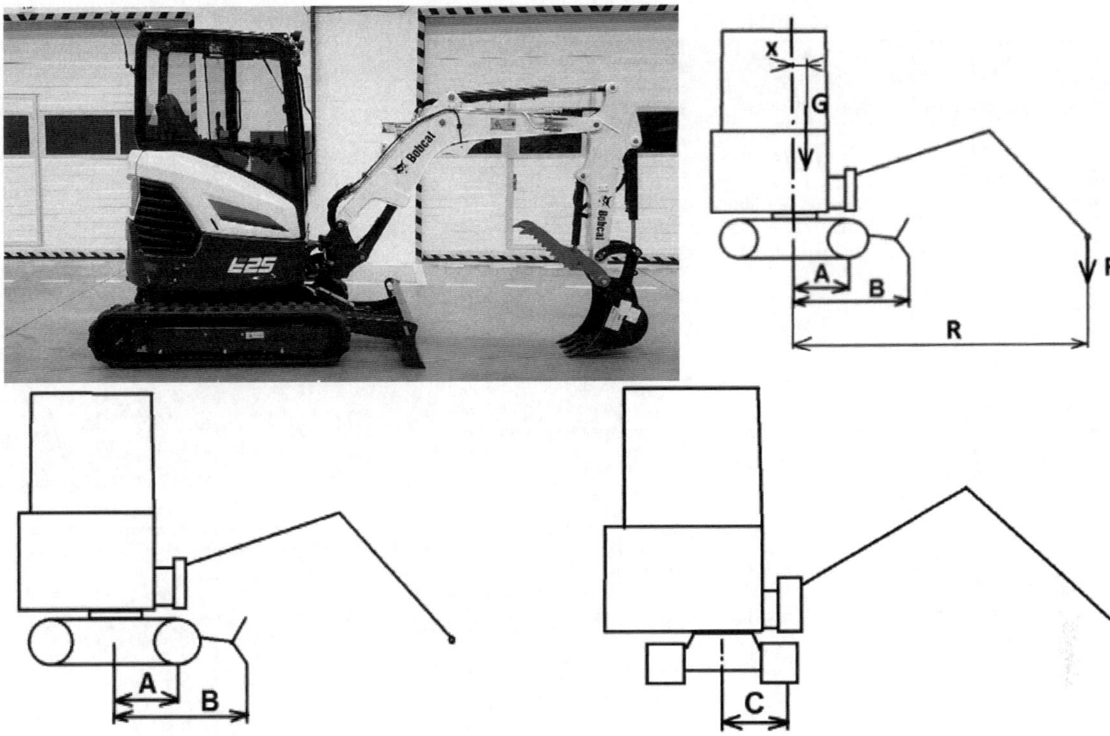

Fig. 1: Real product and simplified mathematical representation

The equation below shows the result for lifting force "F" in front position.

$$F = \frac{A-x}{R-A} \cdot G \cdot g \tag{1}$$

similar equation is governing the stability to side load.

$$F = \frac{A-x}{R-A} \cdot G \cdot g \tag{2}$$

The problem is how to find the weight and Centre Of Gravity (COG) position of a machine that does not exist in reality? The project is using 3D modelling software. The models have been used to find the COG and mass of each individual part in to a excel table.

Several third-party components (Cabin, hydro motors ...) had been unknown. For the first estimate derived values from already known components were used.

The operation of the excavator consists of picking up dirt in front (where the biggest stability is) and dumping it on the side. The top structure including the work group is rotating on the undercarriage. That's why it is important to know stability to front and side. When the excavator turns the "X" (front – aft) value of COG for the rotating structure changes in to "Y" (side to side) value. There fore it is not necessary to get these values from the 3D model and everything can be determined from one position of the 3D model. *(That is important because learning these values is time expensive)*

Table 1: Mass and COG position of individual parts of the machine

	Component	mi [Kg]	Xi [mm]	Yi [mm]	Zi [mm]	Xi*mi	Yi*mi	Zi*mi			
Workgroup	Swing	77.1	740	161	18	57054.0	12413.1	1387.8			
	Boom	142	1920	835	124	272640.0	118570.0	17608.0			
	Arm	97	3092	252	125	299924.0	24444.0	12125.0	Component COG [mm]		
	Bucket		2725.4	-471.8	129.6	0.0	0.0	0.0	X	Y	Z
						0.0	0.0	0.0	**1992**	**492**	**98**
						0.0	0.0	0.0			
						0.0	0.0	0.0			
	Sum	**316.1**				629618.0	155427.1	31120.8	corrected mass [kg]		**311.0**

	Component	mi [Kg]	Xi [mm]	Yi [mm]	Zi [mm]	Xi*mi	Yi*mi	Zi*mi
Upperstructure	UPSTRC W/A	273.2	51.7	165.6	26	14124.4	45241.9	7103.2
	Engine	203	-391	270	-64	-79373.0	54810.0	-12992.0
	Fuel tank	31.3	330	192	-365	10329.0	6009.6	-11424.5

Discussion about the Mathematical model

At this point a discussion is in order about what is so*" great"* about a simple equilibrium on a lever. The design process is a big compromise mainly between weight, stability and size.

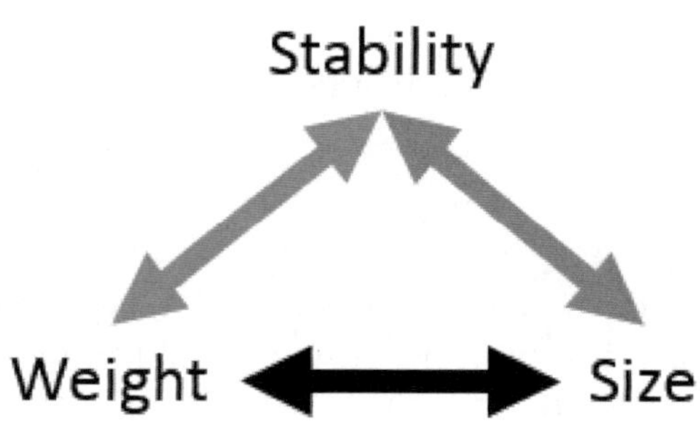

Fig. 2: Design process compromise

Ideally if our 3D models were 100% correct the machine behaviour should be predicted exactly. However, the "reality" of the design is not so simple. The models are not 100% and on top of that they are constantly changing as the design is progressing.

Also, very often the design is facing changes on the requirements. As the design progress the sales department is also working on selling the machine. And very often new or modified requirements are required from the machine. For example. The project here discussed begun as a simple "facelift" of one machine type. But during the design it evolved in to 3 different machines each with different reach, stability and weight. The engineers need to quickly and easily answer questions such as:

- It is needed for the excavator to lift weight M at a radius R. How heavy counterweight is needed?

Using the goal seek function and changing only one cell in the sheet the result is instantly available. Also, a rough estimate can be derived of how much will the new counterweight cost.

- It is needed to increase the lift capacity of the excavator. How much must be the whole upper part moved front or aft to achieve that?

  X values of COG of the Upperstructure, engine... had been linked in such a way that it could be easily modified by a fixed number. That allowed us to get several different results quickly.
- What if we change the size of the undercarriage?

  The length and width are directly responsible for stability response.
- ....

Important for the design process is to quickly show the approximate effect of the proposed change on the final product.

Table 2: Several different solutions of the excavator
The table shows what is the performance of old model compared to the new one in CANopy version, CABin version and CABin version + ADDitional Counterweight

K26 L					over blade-blade up	Over side	over blade-blade up	Over side						
	H [mm]	R [mm]	COG X [mm]	COG Z [mm]	Lift capacity [Kg]		K26 vs Target	K26 vs Target	Mass	MRD mass max 2650±30	Comparable Mass	Slew rad	Overhang	CTWT mass
E26 model CAN	0	3000	191	138	**639**	**496**	100%	99%	2495			770	20	355.0
K26 lift, CAB	0	3000	188	146	**682**	**560**	103%	100%	2651	2696		785	10	370.0
K26 lift, CAN	0	3000	191	147	**650**	**536**	99%	96%	2541	2586	2533	785	10	370.0
K26 lift, CAB + ADD	0	3000	144	104	**769**	**636**	117%	114%	2777	2822		864	89	370.0

The biggest uncertainty of the method is the 3D model which is not 100% precise. This problem is being constantly upgraded both by internal process and by the program development. For example, welds which had been only roughly approximated can be easily modelled as a solid in the new versions. It is believed that the next project will be even better at predicting the COG and mass.

To capture the above-mentioned unknowns, the first batch of prototypes had been used to corelate with mathematical model.

First prototype test

The measurement process had been subjected to Measurement System Analysis (MSA) and it had been found that the uncertainty is +-1kg (with 75% reliability).

Comparing the "real" machine to exact mathematical model has its limitations:

- the prototypes are not built to be used for weight calculations because the protypes need to represent wide variety of different configurations.

- Cabin/ canopy
- Long arm / Short arm….

To help verify the design and manufacturing process.

➢ The prototypes are fitted with variety of testing equipment whose weight can be in the order of 50 kg.

As many real machines as had been available were measured. Than any components that were out of our mathematical specification were subtracted or added. The results are shown below.

Table 3: comparison between test and mathematical model - mass

	Excel	Test	TR vs Excel	
	Weight	**Weight**	**Weight**	
K25	2 437	2 372	-65	-3%

	Excel	Test	TR vs Excel	
	Weight	**Weight**	**Weight**	
K27 H	2 761	2 724	-37	-1%

First iteration had been able to predict the weight with mistake between -65 and -37 kg. *The model did predict smaller weight.*

The table below is comparing our predicted lift capacities with real measurement:

Table 4: comparison between test and mathematical model – Lift capacity

			Excel		Test		Delta Test vs Excel	
			over blade-blade up	Over side	over blade-blade up	Over side	over blade-blade up	Over side
K25	H [mm]	R [mm]	Lift capacity [Kg]		Lift capacity [Kg]		Lift capacity [Kg]	
R27	0	3000	557	509	**507**	**482**	-50	-27
R29	0	3000	542	492	**507**	**482**	-35	-10

			Excel		test		Delta Test vs Excel	
			over blade-blade up	Over side	over blade-blade up	Over side	over blade-blade up	Over side
K27 H	H [mm]	R [mm]	Lift capacity [Kg]		Lift capacity [Kg]		Lift capacity [Kg]	
R27	0	3000	970	821	**910**	**775**	-60	-46
R29	0	3000	957	811	**910**	**775**	-47	-36

The measured lift capacities had been between -10 and -60 kg off from the prediction. *The model did predict smaller lift capacities.*

Discussion about the test and model correlation

Theoretically it should be possible to investigate the model and the test specimens long enough to find all the differences and tie-up the model with reality. How ever in the mean time the design is progressing. It had been decided against such approach. Measured results had been used to fine tune model by modifying some variables.

Updated mathematical model

The test did show several conclusions.

- Adding the measured weight in to mathematical model did show better results. The mass and the mass distribution are the most important input in the calculation.
- For the side lift capacity, the variable "C" needed to be adjusted by -1.6mm to conform with the testing. (This is due to uncertain "tip point". The Tip point is a result of complicated interaction between the rubber tracks and the undercarriage.)
- For the front lift capacity, the variable "A" was adjusted by -8mm. (a value measured in the laboratory) *The variable A is variable in the assembly as well. It depends on a human operator to fill the cylinder to get some approximate tension in the track. Ideally this should be measured during each test. It was not done, and mathematical model used this one adjustment.*

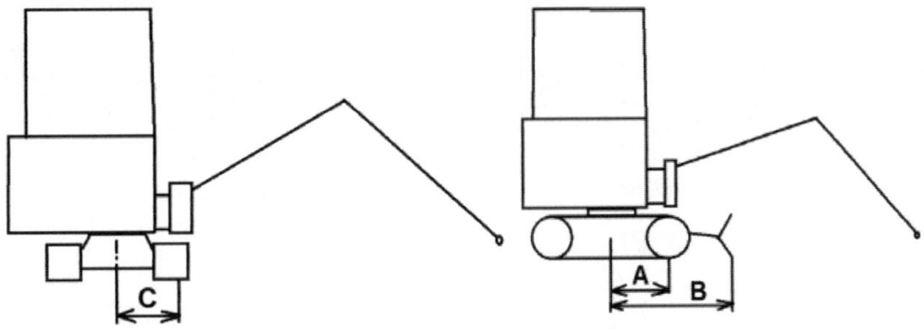

Fig. 3: variable C and A

Using these adjustments, the mathematical model had been tuned and used for the final design of the machine. Weight and stability targets for several variants of the machine had been calculated. A prototype had been built and measured. The table below shows the final comparison between theoretical model and real measurement:

Table 5: comparison between the final calculation and test – weight

	Calculated [Kg]	measured [Kg]		
	R38	-	Delta [Kg]	Delta [%]
E26	2318	2359	41	1.75%
E27z	2458	2453	-5	-0.20%
E27z Long	2472	2467	-5	-0.20%
E27	2530	2578	48	1.86%

Table 6: comparison between of the final calculation and test – Lift capacity

		R38	over blade- blade up			
	H [mm]	R [mm]	Calculated [Kg]	Measured [Kg]	Delta [Kg]	Delta [%]
E26, CAB	0	3000	620	627	6	1.0%
E27z, CAB	0	3000	682	687	5	0.7%
E27zH, CAB	0	3000	755	760	5	0.6%
E27, CAB	0	3000	781	771	-10	-1.4%
E27H, CAB	0	3000	894	884	-10	-1.1%

Conclusions

- Simple tool had been developed which helps to optimise the design.
- The first mathematical iteration is not precise enough. A correlation with test is needed to gain more precise results.
- The updated model can predict the weight of the machine with precision of -5 to +48 kg. (-0,2% to +1,86% of the whole mass.)
- The updated model can predict the stability targets of the machine with maximum mistake of +5 to -10kg. (0,6% to -1,4% of the lift capacity)

EAN 2019
57th conference on experimental stress analysis

Fiber optic strain sensor system for structural analysis of jet engine air intake

DVOŘÁK Milan[1,a], SCHMIDOVÁ Nikola[1,b], DOUBRAVA Karel[1,c], VÁCLAVÍK Jan[2,d] and RŮŽIČKA Milan[1,e]

[1]Czech Technical University in Prague; Faculty of Mechanical Engineering; Technická 4, 160 00 Prague, Czech Republic

[2]AERO Vodochody AEROSPACE a.s.; U Letiště 374, 250 70 Odolena Voda , Czech Republic

[a]milan.dvorak@fs.cvut.cz, [b]nikola.schmidova@fs.cvut.cz, [c]karel.doubrava@fs.cvut.cz, [d]jan.vaclavik@aero.cz, milan.ruzicka@fs.cvut.cz

Keywords: Aircraft structure, Composite, Fiber Bragg Grating (FBG), Structural Health Monitoring (SHM), Fiber optic strain sensor

Abstract. This paper describes experimental verification of a fiber optic strain sensor system for a composite air intake. The new L-39NG jet trainer aircraft from Aero Vodochody is equipped with all-composite air intake. Because there is a need of its acoustic load analysis, both optical and electric strain sensors were installed to its surface. Engine ground test was performed to test fiber optic sensor system at first. It was found that the system is fully functional with capability to provide the required performance. Acquired data were used for further air intake design improvement.

Introduction

Our current work is focused on further development of an optical FBG (Fiber Bragg Grating) sensor system, which was designed to monitor behaviour of a composite air intake (Fig. 1, right) for the L-39NG light jet trainer (Fig. 1, left). Previous works were focused on a performance of the air intake demonstrator during the bird and hail impact tests [1, 2] and on the sensor system detection capability [3, 4]. It was proven that method of installation onto composite structure, which was developed previously, is durable and reliable even during the large energy impact detection testing. Now work has progressed to installation and measurement on the first prototype airplane.

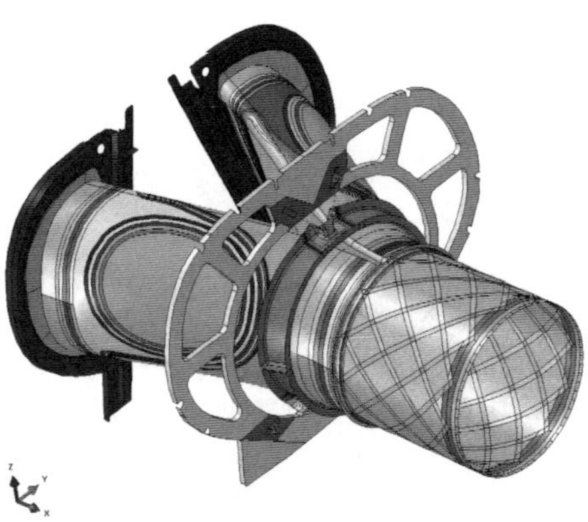

Fig. 1: L-39NG light jet trainer (left) [5]; 3D model of composite ait intake [6]

Sensor System

Four optical fibers with total of forty FBG sensors were installed on the surface of the air intake to monitor its behavior during ground engine test run. Air intake consists of left and righ part, each part is instrumented using two optical fibers with custom-made chains of FBG sensors (6xFBG in the front, 14xFBG in the rear part of the air intake). Optical fibers with outer diameter of 0.150 mm and polyimide primary coating were attached to surface and covered using the thin layer of glass fabric and epoxy resin (see Fig. 2). Individual sensors are placed close to the strain gauges with orientation for the circumferential deformation measurement (in the rear part) or longitudinal deformation measurement (front part). Fiber optic routing on the left side of the air intake can be seen in the Fig. 3 (left). Cured and painted part of the air intake instrumented with strain gauges and FBG sensors (blue cabling) can be seen in Fig. 3 (right).

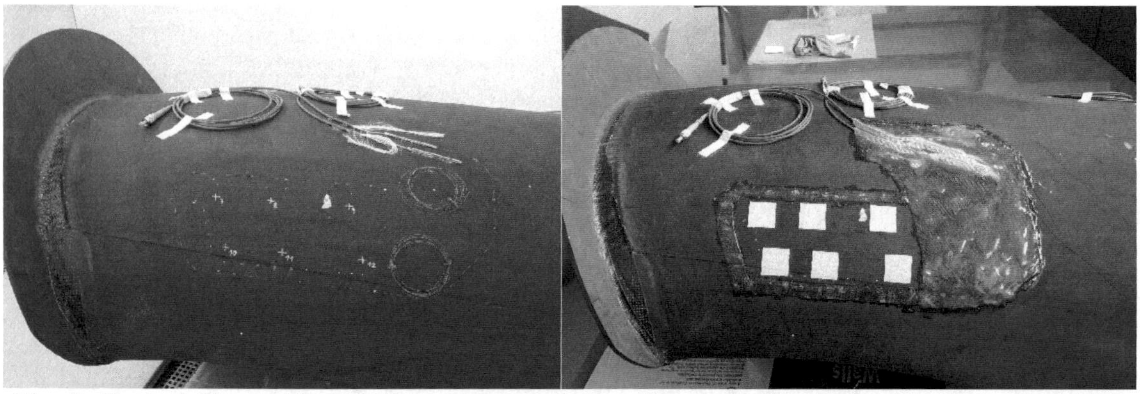

Fig. 2: Optical fiber with 6 FBG sensors attached to the air intake surface (left); optical fiber after lamination, before curing and painting (right)

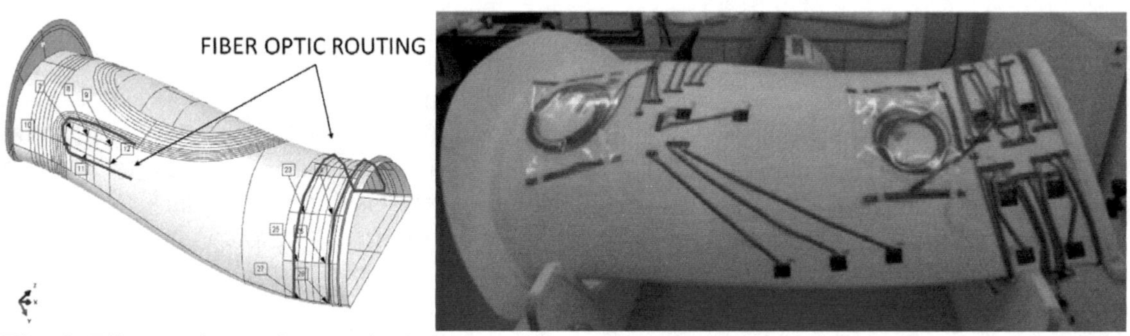

Fig. 3: Fiber optic routing on the left part of the air intake (left); air intake instrumented with strain gauges and FBG sensors (blue cabling)

Engine Ground Run

For the first ground test only FBG sensor system was used for strain measurement, mainly because fiber optic sensors are dielectric. Thus there is no risk of interference with any airplane system, moreover they're immune to interference induced by running engine and related airplane systems. Electric strain gauges will be also used for future testing. Because the main goal of a structural analysis was to monitor acoustic load from the running engine, the sensor system was primarily designed to measure small strain values (estimated strain amplitude about 50 μm/m) in combination with the requirement for a high sampling rate (first blade frequency is about 3.6 kHz). Mechanical strain over time measured during the engine ground test is shown in the Fig. 4. As can be seen, in addition to the expected high-frequency oscillation, the additional quasi-static deformation of the structure was measured, with amplitude ranging from about 90 μm/m (FBG03) on the sides of the structure to about 3100 μm/m in the bottom (FBG14) and top part. Further analysis showed that influence of an air suction during the maximal engine power on the ground (with the zero flight speed) was undervalued. So the air intake structure has deformed excessively in the place where the geometry is complicated (corner between the outer sidewall and the inner straight vertical wall) and where the air intake is connected to the surrounding fuselage structure.

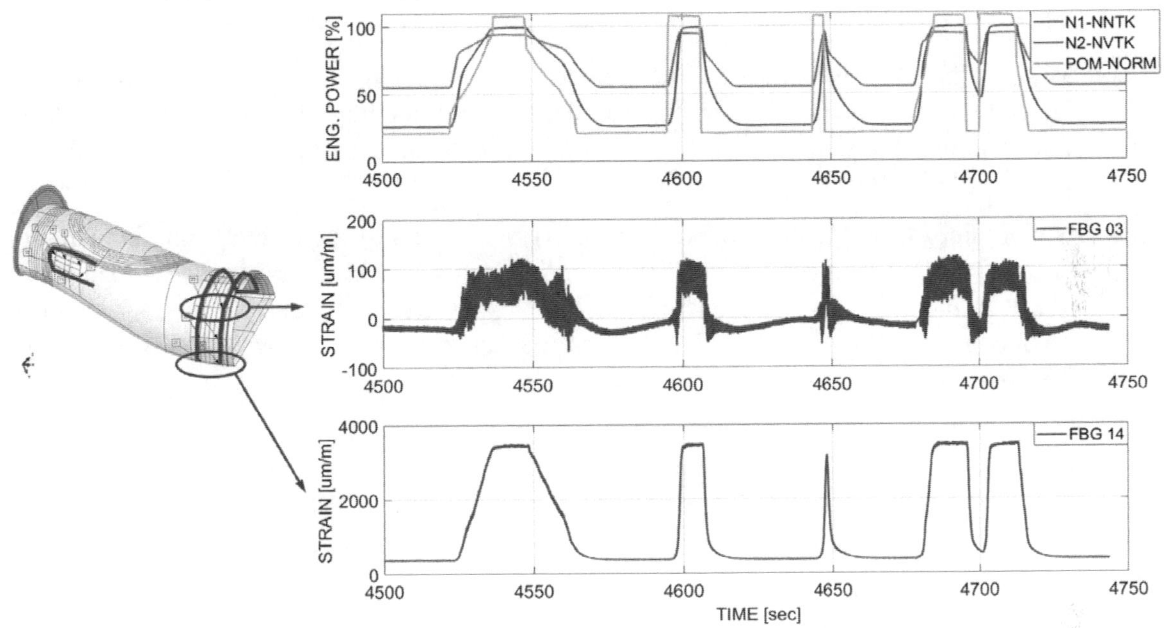

Fig. 4: Mechanical strain measured by the selected FBG sensors over time

Conclusions

First full-scale test of FBG sensor system for the air intake monitoring has shown that it is fully functional with capability to provide the required performance. Both high frequency strain values with small amplitude and quasi-static strain values with large amplitude were measured. No interference induced by the running jet engine or another aircraft system was observed. The measurement data were used to verify the design of the structure and as a basis for modifications.

Moreover, it has been confirmed that the developed procedure of installation and protecting of the optical sensor system allows further steps such as assembly of the air intake, its fitting into the fuselage and installation of the surrounding avionics without damaging the sensors.

The future experimental work will be focused on a comparison of both FBG sensors and electric strain gauges measurement systems during the following engine ground test. The obtained data will be used for further tuning of the finite element model of the air intake structure.

Acknowledgement

The authors would like to thank the Technology Agency of the Czech Republic for supporting this research with the project no. TE02000032.

References

[1] R. Doubrava, M. Oberthor, P. Bělský, M. Dvořák, K. Doubrava, Experimental Verification of Jet Engine Composite Inlet for Bird and Hail Stone Impact Resistance, In: EAN 2018 56th conference on experimental stress analysis, Conference Proceedings. Praha: Česká společnost pro mechaniku, 2018. p. 21-22. ISBN 978-80-270-4061-2.

[2] K. Doubrava, M. Dvořák, R. Doubrava, J. Václavík, M. Růžička, Experimental Analysis of Jet Engine Air Channel Impact Resistance by Means of Strain Gauges and FBG Sensors, In: EAN 2018 56th conference on experimental stress analysis, Conference Proceedings. Praha: Česká společnost pro mechaniku, 2018. p. 19-20. ISBN 978-80-270-4061-2.

[3] M. Dvořák, N. Schmidová, M. Kadlec, M. Růžička, Use of FBG Sensors for Delamination Growth Measurement under Mode I loading, In: 55th Conference on Experimental Stress Analysis 2017. Košice: Technical University of Kosice, 2017. p. 102-107. ISBN 978-80-553-3167-6.

[4] N. Schmidová, M. Dvořák, M. Kadlec, M. Růžička, Monitoring of Delamination Growth on the MMB Specimens Using FBG Sensors and Electrical Resistance Measurement Method, In: EAN 2018 56th conference on experimental stress analysis, Conference Proceedings. Praha: Česká společnost pro mechaniku, 2018. p. 367-373. ISBN 978-80-270-4062-9.

[5] M. Bartošák, J. Jurenka, V. Kulíšek, M. Růžička,Structural-Acoustic Analysis of Air Duct Using FEM, In: 33rd conference with international participation Computational Mechanics 2017 - Extended Abstracts. Pilsen: University of West Bohemia, 2017. p. 3-4. ISBN 978-80-261-0748-4.

[6] Retrieved April 30, 2019, from http://www.aero.cz/

EAN 2019
57th conference on experimental stress analysis

Determination edge strain for fracture prediction of cut edges in sheets made of DP1000 and DC01 steel

FEDORKO Mikulaš[1,a], URBÁNEK Miroslav[1,b] and RUND Martin[1,c]

[1] COMTES FHT a.s. Průmyslová 995, 334 41 Dobřany, Czech Republic

[a]mikulas.fedorko@comtesfht.cz, [b]miroslav.urbanek@comtesfht.cz, [c]martin.rund@comtesfht.cz

Keywords: edge cracks, shear cutting, Diabolo test, finite element simulation

Abstract. Sheared edge quality is a critical variable with respect to the risk of free edge cracking, most notably in advanced high-strength steels (AHSS), which continue to replace conventional steels in automotive industry applications. This paper deals with finding the edge strain limit for edges of different finish qualities. The edge quality is crucial to preventing free edge cracking in deep drawing operations. Therefore, the purpose of this paper was to determine the limit strain which leads to cracking in free edges produced by different sheet parting methods. The variables were the blanking method and the sheet material. Sheared edges and machined edges were produced in conventional DC01 steel and DP1000 AHSS steel. A special fixture was used for measuring the strain level. The strain values were then input into an FE analysis performed using AUTOFORM code for the purpose of their validation. The physical qualities of these edges were compared using a special test procedure, the Diabolo test, which was introduced by Held [1] and Liewald and Gall [2]. In this test, the edge is subjected to loading under plane strain conditions. Another method of testing the susceptibility to edge cracking is the hole expansion test [4], which induces cracking along a circular hole in the centre of a part. The limit edge strain can be derived with the aid of DIC optical-measuring system which monitors strain increments in the course of the test. Factors which affect the susceptibility to edge cracking were identified and validated for DP1000 and DC01 materials

Introduction

In recent years, several approaches have been developed to predict the formation of a free-edge curve by numerical simulation [1]. They share the use of FLC as the basis for calculating reduced forming ability. These models require experimental data obtained from the most commonly used tests for this purpose and hole expansion test according by ISO 16630. In this test, the hole is extended by a conical piston until it breaks in its circumference. The evaluation criterion is defined as the ratio of the extension of the opening between the original diameter and the fracture size. However, this test does not provide the maximum strain value, which is achieved before crack formation. Therefore, several alternative tests [6] have been developed to measure this value. In the present article, the so-called Diabolo test described by Held [1] was used and DC01 steel was analyzed. The susceptibility of the sheared edge was tested on three series of test specimens. Each series was produced by a different cutting method: shearing, chip cutting, and laser cutting. Each specimen has a size of 40×200 mm. The Diabolo test set-up involved optical measurement. Thanks to special punch geometry, the strain was localized into a small region adjacent to the sheared edge. This strain can be measured by GOM Aramis strain measurement system. In this study, the maximum major strain $\varphi 1$ before cracking was evaluated to characterize the edge crack sensitivity. The moment of necking was detected by a time-dependent approach similar to the approach suggested by Volk and Hora for the Nakazima test [5].

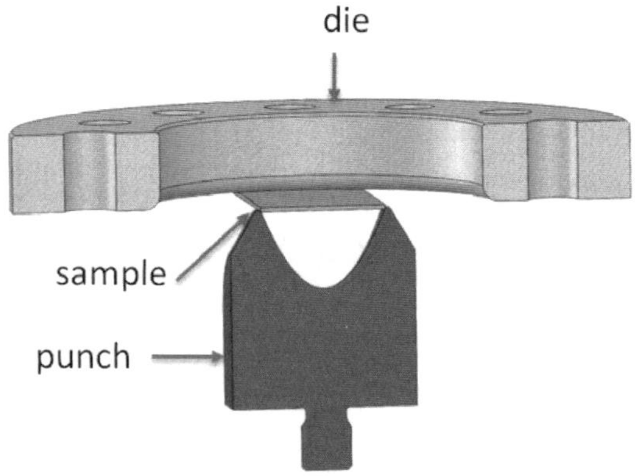

Fig. 1: a) Configuration of tools and sample before forming, b) Sample after measurement

Experimental material and method

Deep-drawing steels exhibit a purely ferritic structure, or consist of a ferritic matrix, in wihich isolated pockets of grain carbide can be embedded, see Fig.2 a). DP steels consist of a ferritic matrix containing a hard martensitic second phase in the form of islands see Fig.2 b). Increasing the volume fraction of hard second phases generally increases the strength. DP (ferrite plus martensite) steels are produced by controlled cooling from the austenite phase (in hot-rolled products) or from the two-phase ferrite plus austenite phase (for continuously annealed cold-rolled and hot-dip coated products) to transform some austenite to ferrite before a rapid cooling transforms the remaining austenite to martensite. Due to the production process, a small amount of other phases (bainite and retained austenite) may be present. A commercial AHSS steel sheet grade DP1000 with the initial thickness of 0,8 mm was used in this work. When these steels deform, strain is concentrated in the lower-strength ferrite phase surrounding the islands of martensite, creating the unique high initial work-hardening rate (n-value) exhibited by these steels. Fig. 1. is an actual micrograph showing the ferrite and martensite constituents. DC 01 is construct low carbon steel. Diabolo test was performed using our Nakazima test machine [7], but the classic spherical punch was replaced by a diabolo punch. All measurement and evaluation methodology was based on ISO12004, which describes the determination of the FLC curve from measurements. ARAMIS DIC system and the time dependent method described by prof. Hora were used for online measurement and evaluation. The measurement procedure consisted of assembling and calibration of the setup, measurement, evaluation and processing of the measurement, and finally application in FEM software.

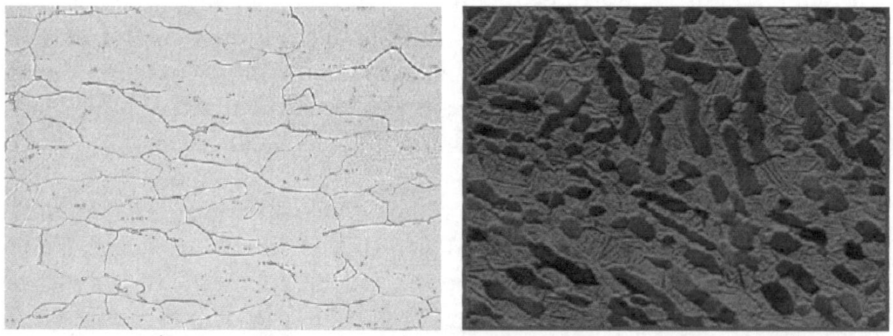

a) b)

Fig. 2 Micrograph of a) deep-drawing steels and b) DP steel

The material model is highly dependent on accurate descriptions of the material´s mechanical behaviour, the selected constitutive model as well as the input data used for its calibration. Therefore, trying to create the appropriate description for the DC01 steel sheet (Tab 1.) and DP1000 steel sheet (Tab 2.), the following material model was selected to describe as accurately as possible the material behaviour based on the hardening law.

Tab. 1: Chemical composition steel DC01

element	C	Mn	P	S	N	Al
%	0.12	0.6	0.045	0.045	0.003	0.04

Tab. 2: Chemical composition steel DP1000

element	C	Si	Mn	Ni	Cr	P	S	N	Al
%	0.16	0,17	2,17	0,01	0,44	0.008	0.003	0.003	0.039

The plastic true stres-strain curves gathered from tensile test are illustrated in Fig. 2.

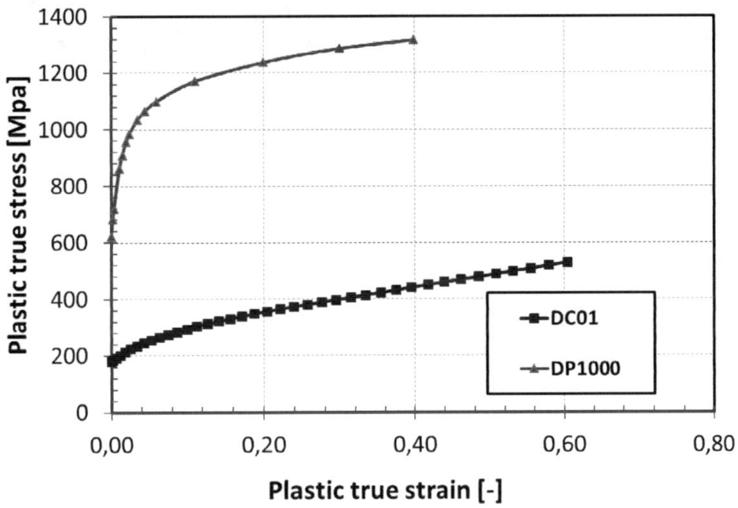

Fig. 3 Plastic true stress-strain curves of steel grade DC01 and DP1000

The shear cutting procedure can be defined as two edges working towards each other to separate material into smaller pieces. The material between the edges is deformed so that fractures appear and eventually the material will separate. Material properties such as strength, Young's modulus and ductility have great impact on the cutting procedure. When the two edges engage in the material, see Fig 3, a fracture will appear on each side and will continue to grow as the edges continue to move through the material. A basic condition that must be met to perform a shear cutting operation is that the material of the cutting tool needs to have larger hardness and strength than the material to be separated. After a shear cut operation an edge with several different zones appear, see Fig. 2, and due to different adjustments of the cutting tool the length of this zones can vary. With optimal adjustments of the cut tool the shear zone represents between 25-35 % of the edge surface. An edge with a large burr zone is a sign that the adjustment of the shearing tool was suboptimal. For an edge with a big burr zone, the adjustments of the cut tool have not been advantageous.

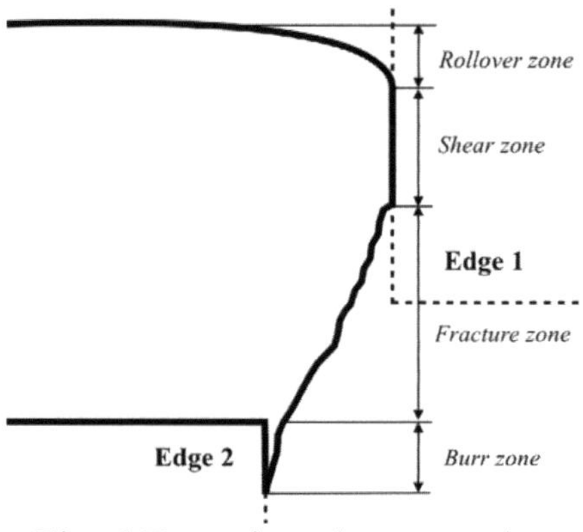

Fig. 4 Zones in a shear cut edge

With the shearing knives appropriately adjusted, the cracks running from both sides meet, resulting in material separation, as in Fig 5. With increasing clearance between or wear in the knife edges, the cracks are less likely to meet. This leads to higher forces needed for separation and poorer quality of the sheared surface. The resultant sheared surface quality can be altered by making changes related to the gap (clearance) between the knife edges, the wear in the knives, the shearing edge geometry and the amount of lubricant.

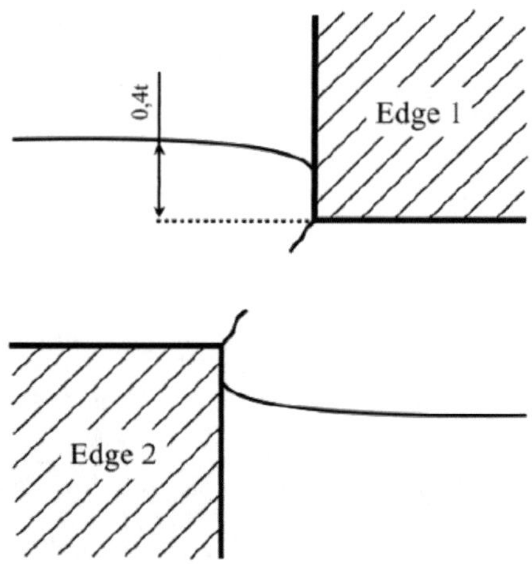

Fig. 5 With optimal adjustments, the fractures will meet and give a fine cut. When the edge advances approx. 40% (0.4t) of the thickness of the material, a fracture will appear.

Sheared edges of specimens

Both specimens were examined under an optical microscope to evaluate the effect of the surface condition on the risk of edge cracking. The edge qualities were analyzed under a microscope and micrographs were taken.

Fig. 6 below shows the quality of the sheared surface in the sheared specimen

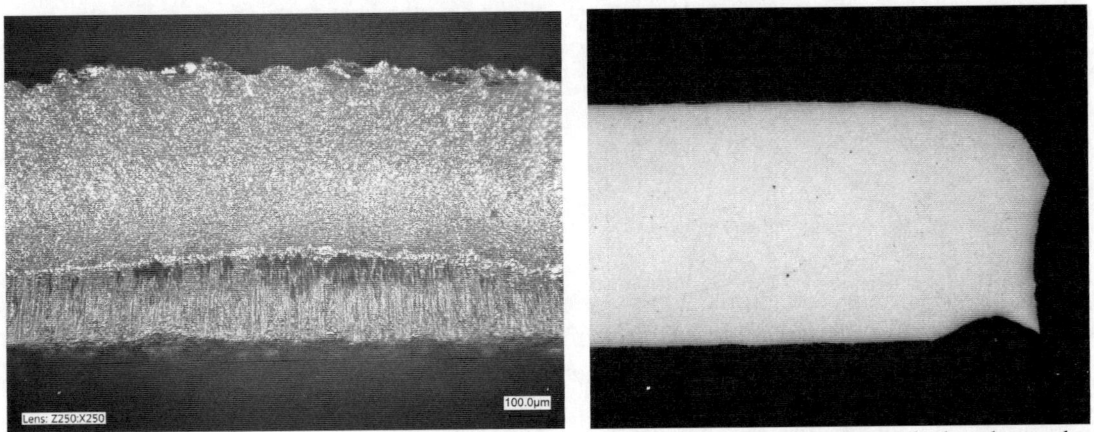

a) Sheared surface b) Cross section through the sheared edge

Fig. 6 Quality of the sheared surface in the sheared specimen of DP1000 steel sheet of 0.8 mm thickness

The figures below document the quality of the edge surface in the machined specimen.

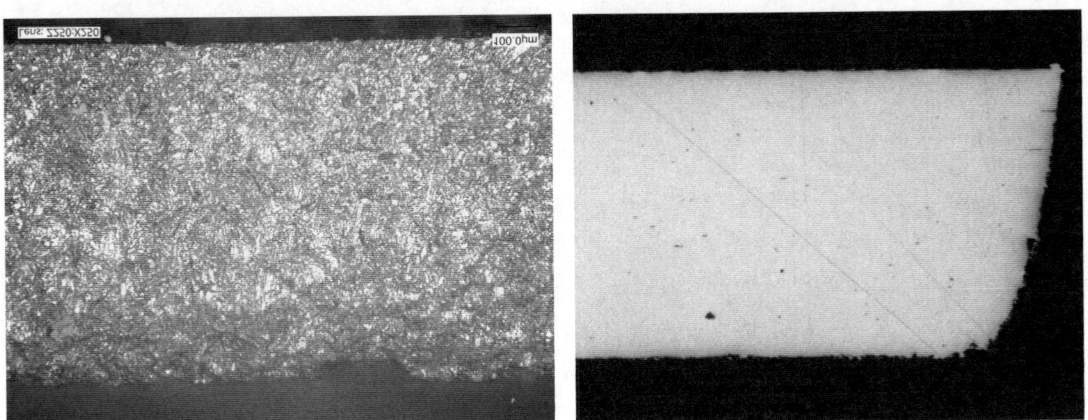

a) Machined surface b) Cross section through the edge

Fig. 7 Quality of the machined surface in the machined specimen of DP1000 steel sheet of 0.8 mm thickness

Measurement results

The measurement revealed considerable differences resulting from the processes that had been used for creating the edge surfaces. The maximum force prior to cracking depended strongly on how the specimen edge had been created. For the specimen with a machined edge, the force was almost 2.5 times higher than for the sheared specimen. The first response recorded early in the graph was the specimen bending. Then, once the entire shaping punch surface had come into contact with the specimen, the force increased steeply, after which a crack initiated and began spreading from the free edge (Fig. 8 and Fig. 9).

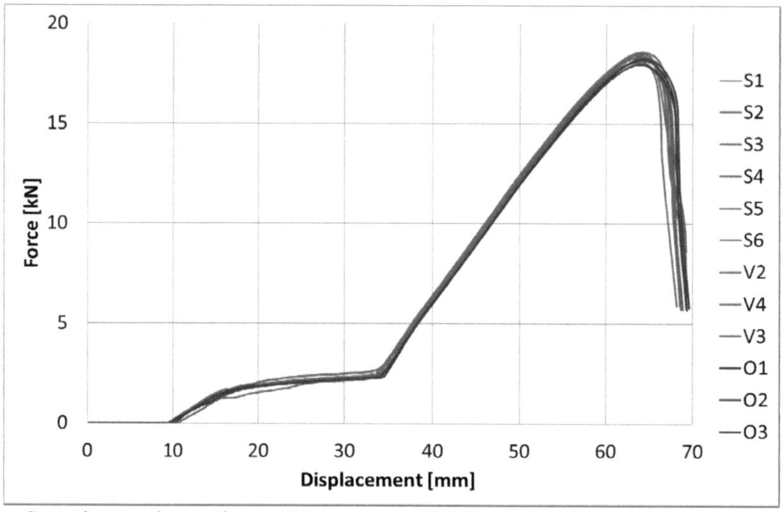

S – sheared specimen V – water cut O – machined specimen
Fig. 8 Force dependent on piston displacement DC01

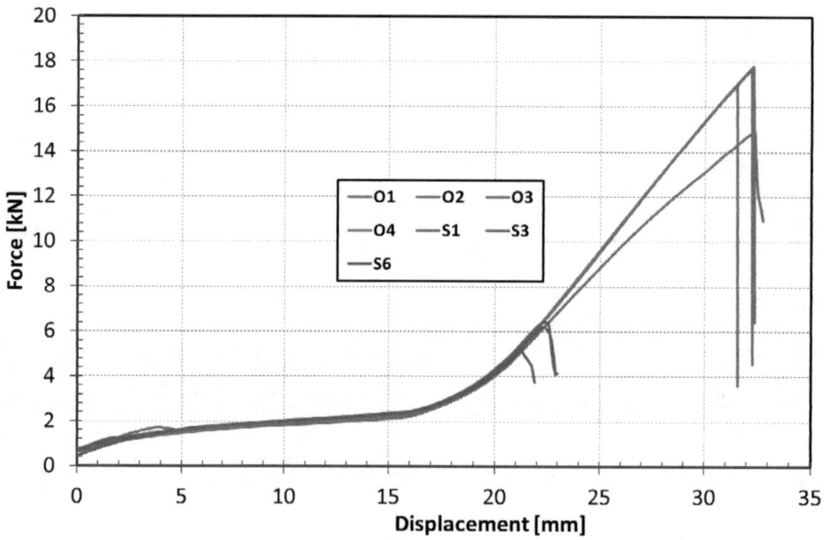

O - machined specimen S – sheared specimen
Fig. 9 Force vs. piston displacement

ARAMIS is a Digital Image Correlation (DIC) system developed to measure displacements, surface strain, velocity and accelerations of a test object. The system creates a 3D measurement and uses digital images to measure changes of the material specimen of just a few millimeters to several meters in size. The measured data is used to determine material properties of the test specimen such as Young's modulus (elastic modulus) and Forming Limit Curves (FLC).
An example of what major strain distribution prior to cracking may look like in the ARAMIS system is shown in Fig. 10.

Fig. 10 The major strain from ARAMIS before a large fracture

Numerical simulation

Numerical modelling of the Diabolo test was performed using the AutoForm software. The longitudinal axis of the specimen was parallel to the rolling direction. The "roll angle" in the software was therefore set at 90°. Only a single-action draw was simulated, where the positions of the punch, the die and the blank holder are precisely defined. In this case, the specimen is held firmly by two high-strength bolts, which is why the blank holder was affixed securely. The friction value in AutoForm was set to different values in order to test which value gives results similar to the actual tests. This was done by exporting force and punch depth data and comparing it with results from the real tests in a diagram.

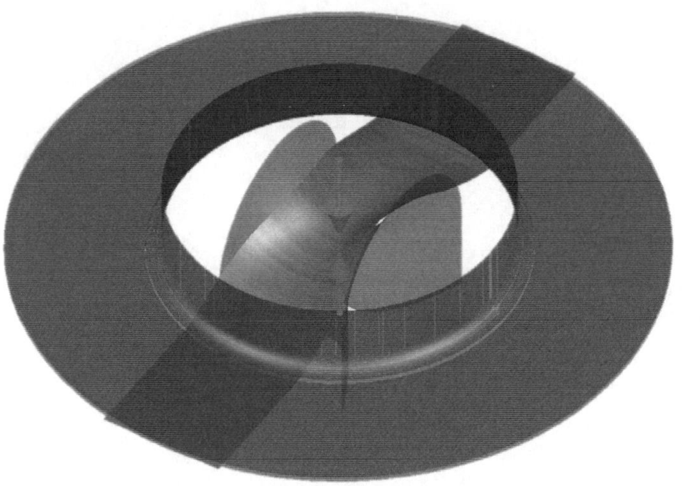

Fig. 11 An overview of the die, blank holder, punch and specimen used for the simulation

The material model embodied in the work-hardening curve has a major impact on the force magnitude computed in numerical simulation. The simulated and actually measured forces were in relatively good agreement. The damage parameters in the AutoForm software were only considered for visual evaluation of the threshold. They had no effect on the force calculation. The main goal was to adjust the damage parameters in the AutoForm software to ensure that the simulated crack forms at the same time and piston position as in the real-world test. Input values were based on the measurement and had to be defined for two manufacturing processes. The threshold strain was determined directly from the DIC ARAMIS data and then calibrated with the AutoForm software. When the threshold value is correct, red regions in the simulation identify the same critical locations for cracking at the same piston positions as in the real-world measurement. The critical region is usually associated with values near unity. This threshold value also depends on the left side of the FLC diagram which depicts the uniaxial stress state generated in the free edge and in its vicinity.

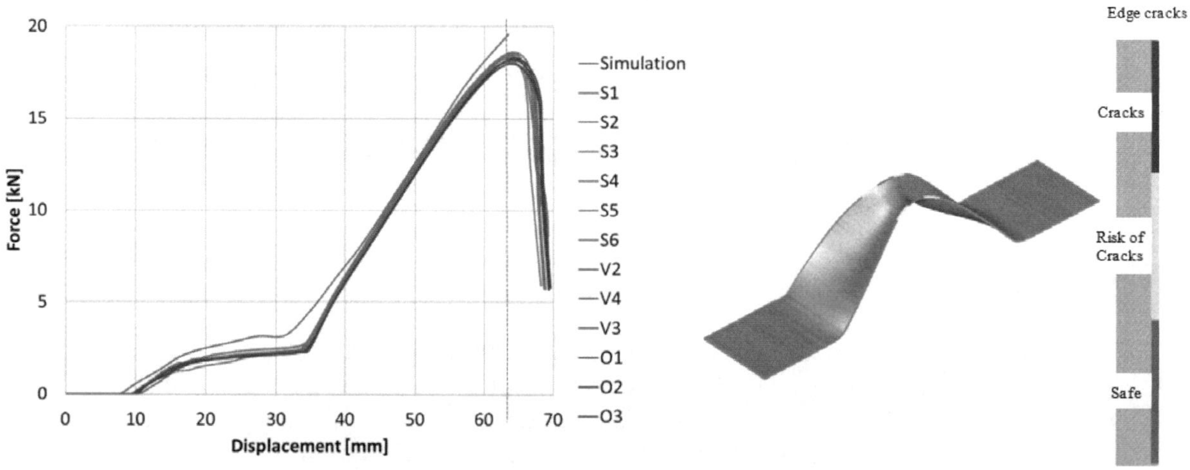

Fig. 12: a) Comparison between values from simulation in AutoForm and diabolo test b) visualization of initial crack on sample for DC01

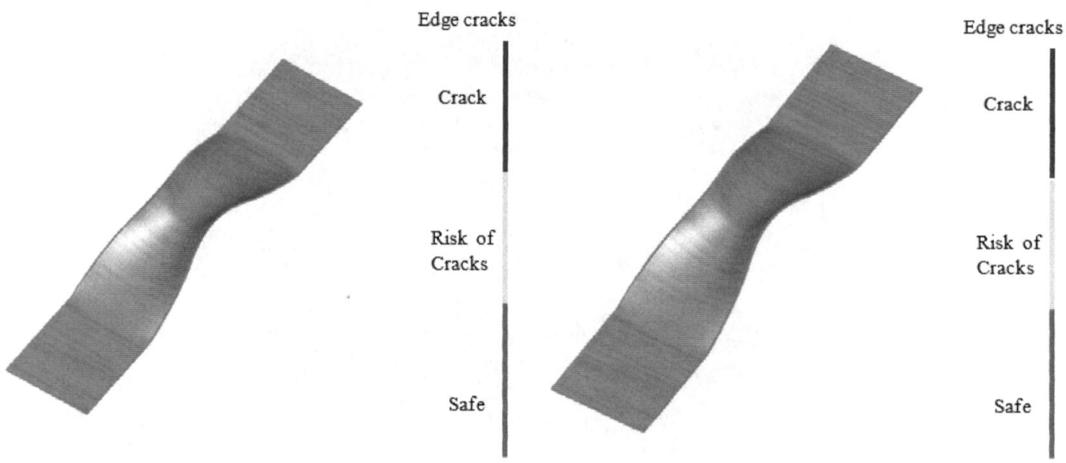

a) Sheared specimen b) Machined specimen

Fig. 13 Evaluation of the risk of cracking by simulation for DP1000

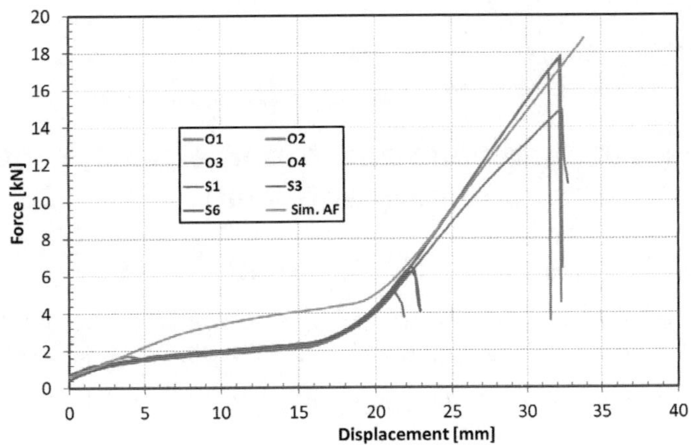

Fig. 14 Comparison between values from simulation in AutoForm and Diabolo test for DP1000

Conclusions

This paper evaluates the effects of various cutting processes on crack formation in the blank edge using the Diabolo test. The test specimens of 0.8 mm thickness were made of DP1000 steel and specimens of 1.5 mm thickness were made of DC01 steel and had two types of sheared edges. Those were produced by shearing and chip cutting. Surface roughness of these edges was measured and the microstructures in their material were examined, which revealed new aspects related to this issue. The measured strain values were fed into numerical simulation of the Diabolo test carried out using AUTOFORM software which offers the capability to use such values for modelling deep drawing processes.

Acknowledgement

This study was created by project Development of West-Bohemian Centre of Materials and Metallurgy No.: LO1412, financed by the MEYS of the Czech Republic.

References

[1] Held, C., Schleich, R., Sindel, M., Liewald, M.: Investigation on advanced forming limit prediction techniques for high strength steels, in: Proceedings of the IDDRG 2009 Conference, 461–470.

[2] Liewald, M., Gall, M.: Experimental investigation of the influence of shear cutting parameters on the edge crack sensitivity of dual phase steels, in: Proceedings of the IDDRG 2013 Conference, Zurich, Switzerland 219–224.

[3] Falk, J.: Fracture prediction of stretched shear cut edges in sheets made of Dual-Phase steel, Master's degree thesis, Blekinge Institute of Technology, Karlskrona, Sweden 2017.

[4] ISO 16630, Metallic materials - Sheet and strip - Hole expanding test, Berlin, Beuth, 2009.

[5] Volk, W., Hora, P.: New algorithm for a robust user independent evaluation of beginning instability for the experimental FLC determination. International Journal of Material Forming, 4, 339–346, 2011.

[6] Bouaziz, O., Douchamps, S., Durrenberger, L., Bui-Van, A.: The double bending test: a promising new way for an optimal characterization of cut-edges ductility, in: Proceedings of the IDDRG 2010 Conference. Graz, 575–582.

[7] Fedorko, M., Urbánek, M., Rund, M., Prediction of thinning of the sheet metal in the program AutoForm and its experimental verification, in: IOP Conference Series: Materials Science and Engineering, ISSN: 17578981, 201

Biomechanics – Probabilistic anthropometry approach for sitting human and seat

FRYDRÝŠEK Karel[1, a], ČEPICA Daniel[1, b] and HALO Tomáš[1, c]

[1]Department of Applied Mechanics, Faculty of Mechanical Engineering, VŠB – Technical University of Ostrava, 17. listopadu 2172/15, 708 00 Ostrava, Czech Republic

[a]karel.frydrysek@vsb.cz, [b]daniel.cepica.st@vsb.cz, [c]tomas.halo.st@vsb.cz

Keywords: Biomechanics, Sitting man, Anthropometry, Stochastics, Segmentations, Loadings

Abstract. The aim of this paper is the calculation of external forces, reactions and internal forces and internal moments acting upon human body in seated position (sedentary lifestyles, travelling, working etc.). According to mechanics/biomechanics, the model of a human is created as a simple (2D truss structure) four segment model with joints (articulationes). Input and output values are given by real anthropometric stochastic parameters of human population. Results are determined and evaluated via the direct Monte Carlo Method (truncated histograms etc).

Introduction

When so many people spend so much time sitting in their lives, at work, at home and in transport between destinations etc., the risks of contracting the different types of disorders that sitting can actually cause increase dramatically. In order to take the right action, we must understand the factors that contribute to the problems (medical treatment, ergonomics, prevention etc.). Hence, there is a conflict between the life humans are adapted to and the sedentary life so many people are now living.

Therefore, the sitting, see Fig. 1, can be sources of medical problems, e.g. causing or influencing injuries, pains or deformities in dorsum etc.

Subsequently, in the engineering design of machines, tools, medical treatment and implants, sports etc., proper databases of anthropometry based on stochastic/probabilistic are needed.

For calculations a simple four segment (feet + legs, thighs, lower part of trunk, upper part of trunk + neck + head) model was made, see Fig. 1. This model has similar traits as truss, therefore can be calculated as one.

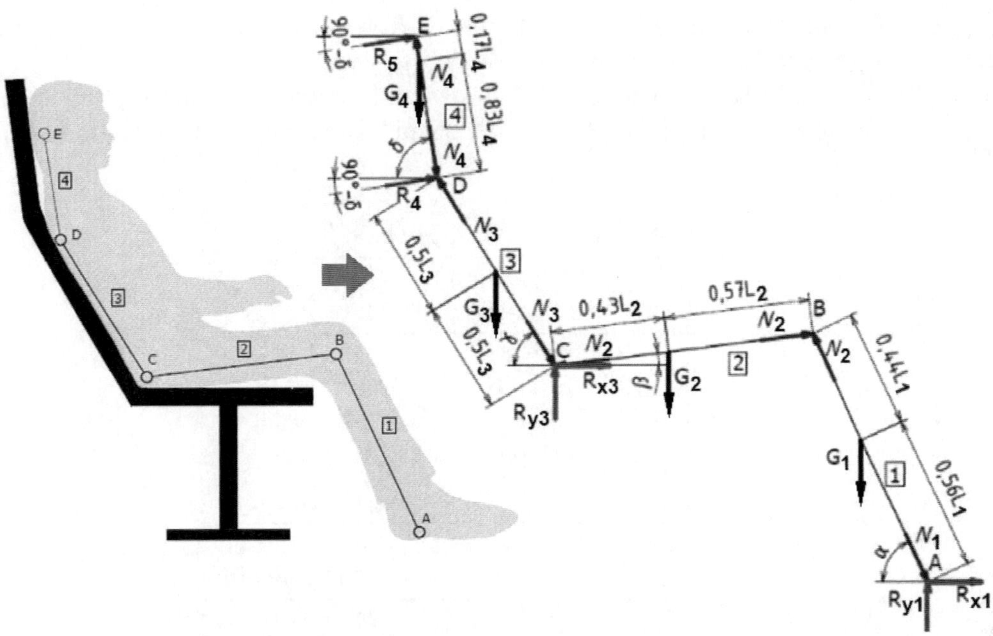

Fig. 1: (a) Sitting human, (b) Model of a sitting human (gravity loadings G_i [N], reactions R_i [N], and internal normal forces N_i [N])

However, human populations and their seats or way of sitting are of stochastic/probabilistic quantities. Therefore, for the interaction between human and seat, the Monte Carlo approach is applied for calculations and evaluations of loading, reactions, normal forces and bending moments.

Reactions and internal forces and moments

To define reactions and normal forces by the Method of Joints, gravity forces acting in centroids must be divided into joints (chosen articulationes), see Fig. 2. Hence, human body can be suitable approximated via 2D truss structure.

Bending moments were calculated considering gravity forces acting outside of joints in (in centroids of body segments).

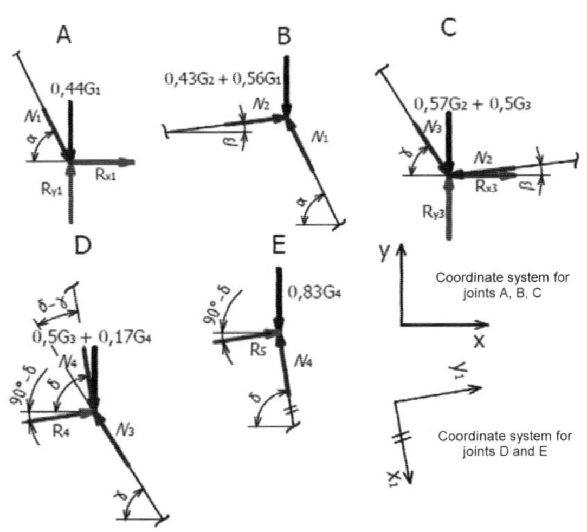

Fig. 2: Free body diagrams of joints and their coordinate systems

Final formulas of reaction and internal normal forces are derived from equilibrium equations, see Fig. 2 and Eq. 1 to 10.

Bending moments can be calculated like 2D truss with loading outside of joints, or, which was used, we can look at the model as "an angled beam". Maximum moments are in Eq. 11 to 14.

$$N_1 = \frac{(0.56 \cdot \mathrm{G}_1 + 0.43 \cdot \mathrm{G}_2) \cdot \cos(\beta)}{\sin(\alpha + \beta)} \tag{1}$$

$$N_2 = \frac{(0.56 \cdot \mathrm{G}_1 + 0.43 \cdot \mathrm{G}_2) \cdot \cos(\alpha)}{\sin(\alpha + \beta)} \tag{2}$$

$$N_3 = \frac{[0.5 \cdot \mathrm{G}_3 + \mathrm{G}_4] \cdot \sin(\delta)}{\cos(\delta - \gamma)} \tag{3}$$

$$N_4 = 0.83 \cdot \mathrm{G}_4 \cdot \sin(\delta) \tag{4}$$

$$\mathrm{R}_{x1} = -\frac{(0.56 \cdot \mathrm{G}_1 + 0.43 \cdot \mathrm{G}_2) \cdot \cos(\beta) \cdot \cos(\alpha)}{\sin(\alpha + \beta)} \tag{5}$$

$$\mathrm{R}_{y1} = 0.43 \cdot \mathrm{G}_1 + \frac{(0.56 \cdot \mathrm{G}_1 + 0.43 \cdot \mathrm{G}_2) \cdot \cos(\beta) \cdot \sin(\alpha)}{\sin(\alpha + \beta)} \tag{6}$$

$$\begin{aligned} \mathrm{R}_{x3} = & -\frac{[0.5 \cdot \mathrm{G}_3 \cdot \sin(\delta) + \mathrm{G}_4 \cdot \sin(\delta)] \cdot \cos(\gamma)}{\cos(\delta - \gamma)} + \\ & + \frac{(0.56 \cdot \mathrm{G}_1 + 0.43 \cdot \mathrm{G}_2) \cdot \cos(\alpha) \cdot \cos(\beta)}{\sin(\alpha + \beta)} \end{aligned} \tag{7}$$

$$\begin{aligned} \mathrm{R}_{y3} = 0.57 \cdot \mathrm{G}_2 + 0.5 \cdot \mathrm{G}_3 + & \frac{(0.56 \cdot \mathrm{G}_1 + 0.43 \cdot \mathrm{G}_2) \cdot \cos(\alpha) \cdot \sin(\beta)}{\sin(\alpha + \beta)} + \\ & + \frac{[0.5 \cdot \mathrm{G}_3 + \mathrm{G}_4] \cdot \sin(\delta) \cdot \sin(\gamma)}{\cos(\delta - \gamma)} \end{aligned} \tag{8}$$

$$\mathrm{R}_4 = 0.5 \cdot \mathrm{G}_3 \cdot \cos(\delta) + 0.17 \cdot \mathrm{G}_4 \cdot \cos(\delta) + \frac{[0.5 \cdot \mathrm{G}_3 + \mathrm{G}_4 \cdot] \cdot \sin(\delta) \cdot \sin(\delta - \gamma)}{\cos(\delta - \gamma)} \tag{9}$$

$$\mathrm{R}_5 = 0.83 \cdot \mathrm{G}_4 \cdot \cos(\delta) \tag{10}$$

$$M_{\mathrm{omax1}} = \mathrm{R}_{y1} \cdot 0.56 \cdot \mathrm{L}_1 \cdot \cos(\alpha) + \mathrm{R}_{x1} \cdot 0.56 \cdot \mathrm{L}_1 \cdot \sin(\alpha) \tag{11}$$

$$\begin{aligned} M_{\mathrm{omax2}} = & \mathrm{R}_{y1} \cdot (\mathrm{L}_1 \cdot \cos(\alpha) + 0.57 \cdot \mathrm{L}_2 \cdot \cos(\beta)) + \\ & + \mathrm{R}_{x1} \cdot (\mathrm{L}_1 \cdot \sin(\alpha) - 0.57 \cdot \mathrm{L}_2 \cdot \sin(\beta)) - \mathrm{G}_1 \cdot (0.44 \cdot \mathrm{L}_1 \cdot \cos(\alpha)) + \\ & + 0.57 \cdot \mathrm{L}_2 \cdot \cos(\beta) \end{aligned} \tag{12}$$

$$\begin{aligned} M_{\mathrm{omax3}} = & -\mathrm{G}_4 \cdot (0.83 \cdot \mathrm{L}_4 \cdot \cos(\delta) + 0.5 \cdot L_3 \cdot \cos(\gamma)) + \\ & + \mathrm{R}_5 \cdot \sin(\delta) \cdot (\mathrm{L}_4 \cdot \sin(\delta) + 0.5 \cdot L_3 \cdot \sin(\gamma)) + \\ & + \mathrm{R}_4 \cdot \cos(\delta) \cdot 0.5 \cdot \mathrm{L}_3 \cdot \cos(\gamma) + \mathrm{R}_4 \cdot \sin(\delta) \cdot 0.5 \cdot \mathrm{L}_3 \cdot \sin(\gamma) + \\ & + \mathrm{R}_5 \cdot \cos(\delta) \cdot (\mathrm{L}_4 \cdot \cos(\delta) + 0.5 \cdot \mathrm{L}_3 \cdot \cos(\gamma)) \end{aligned} \tag{13}$$

$$M_{\mathrm{omax4}} = \mathrm{R}_5 \cdot 0.17 \cdot \mathrm{L}_4 \tag{14}$$

These relationships are used for stochastic/probabilistic evaluation.

Internal (pressure) normal forces and bending moments are presented in Fig. 3.

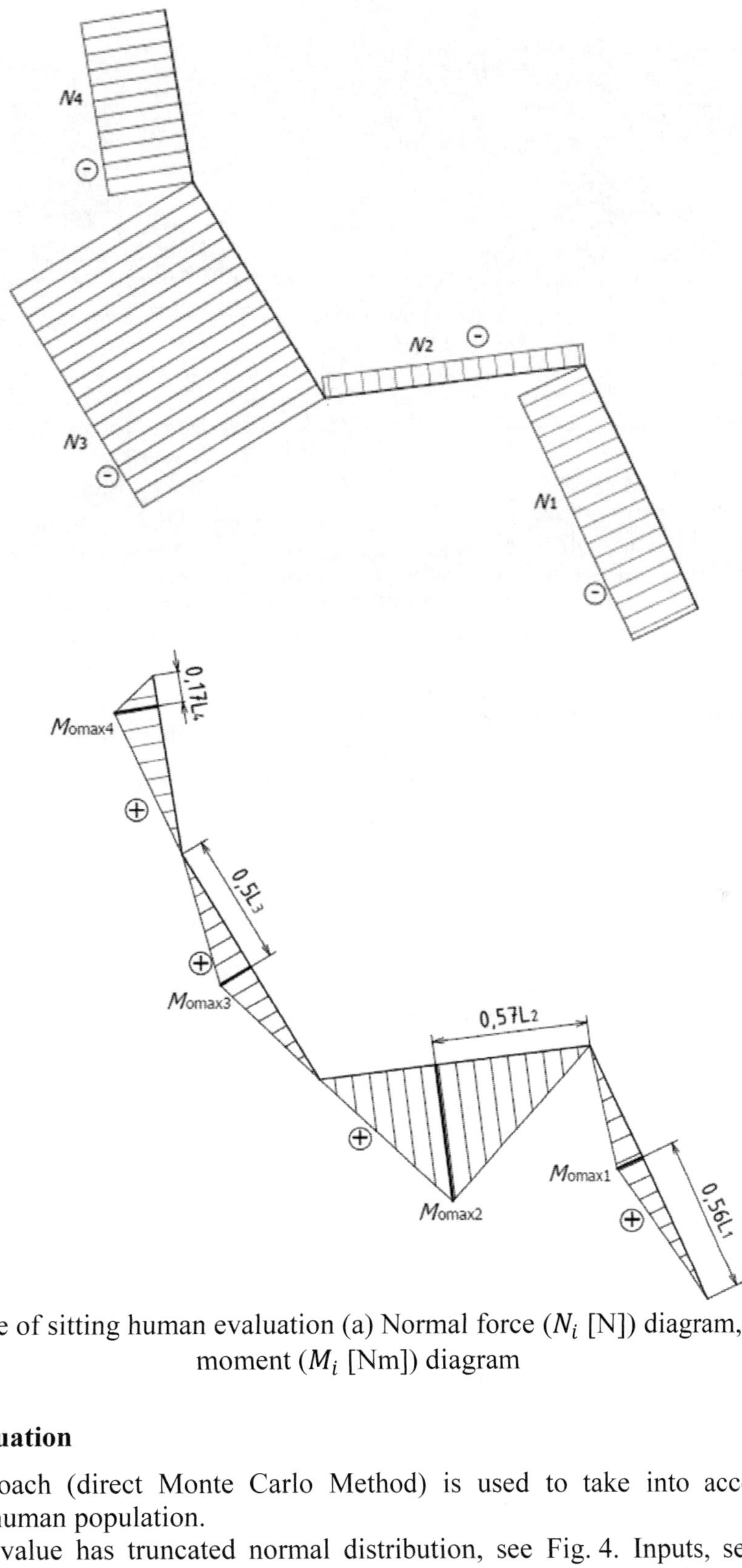

Fig. 3: Example of sitting human evaluation (a) Normal force (N_i [N]) diagram, (b) Bending moment (M_i [Nm]) diagram

Stochastic evaluation

Stochastic approach (direct Monte Carlo Method) is used to take into account the real randomness of human population.

Every input value has truncated normal distribution, see Fig. 4. Inputs, see Tab. 1, are given by real long-time measured anthropometric parameters of human population, angles α, β and δ are depended on seat (chair) design (for more realistic results angle γ is related to δ), and by location on Earth (gravity acceleration g). Outputs are in Tab 2.

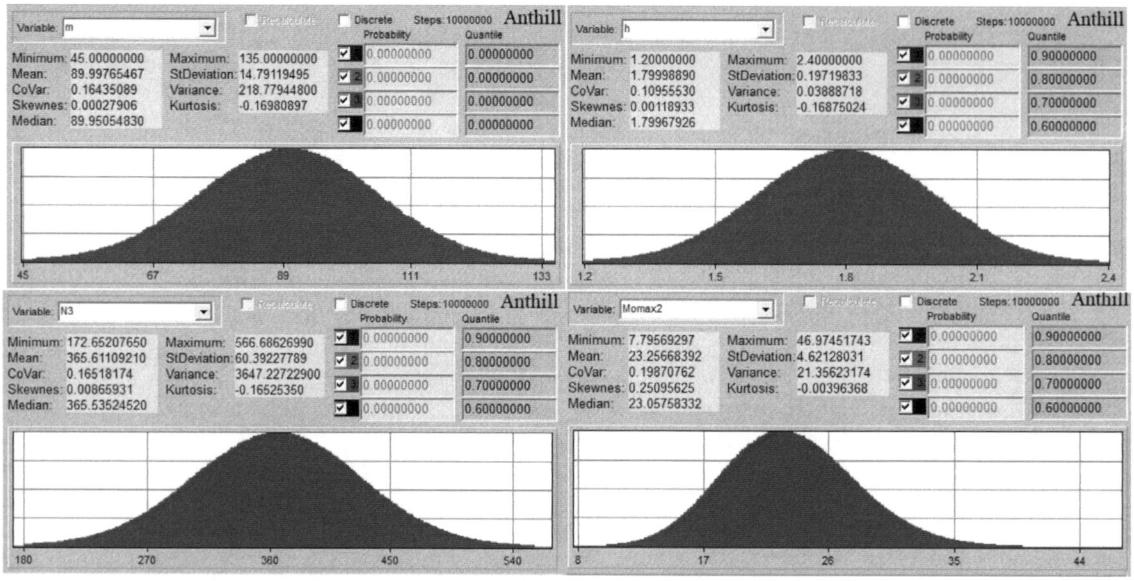

Fig. 4: Examples of inputs and outputs: (a) Total weight, (b) Total height, (c) Maximum normal force, (d) Maximum bending moment

Table 1: Input data (anthropometry, measuring)

Variable name		Symbol	Min. value	Mean value	Median value	Max. value	Histogram
Total weight [kg]		m	45	89.998	89.951	135	
Total height [m]		h	1.2	1.8	1.8	2.4	
Angle of segment 1 [deg]		α	60	74.998	74.991	90	
Angle of segment 2 [deg]		β	0	9.998	9.981	20	
Angle of segment 3 [deg]		$\gamma = \frac{14}{15} \cdot \delta$	60.67	69.999	69.988	79.33	
Angle of segment 4 [deg]		δ	65	74.999	74.987	85	
Gravitational acc. [m/s^2]		g	9.78	9.806	9.806	9.832	
Length [m]	Segment 1	$L_1 = 0.285 \cdot h$	0.342	0.513	0.512	0.684	
	Segment 2	$L_2 = 0.245 \cdot h$	0.294	0.441	0.441	0.588	
	Segment 3	$L_3 = 0.24 \cdot h$	0.288	0.432	0.432	0.576	
	Segment 4	$L_4 = 0.165 \cdot h$	0.198	0.297	0.297	0.396	
Weight [kg]	Segment 1	$m_1 = 0.124 \cdot m$	5.58	11.158	11.138	16.74	
	Segment 2	$m_2 = 0.248 \cdot m$	11.16	22.317	22.275	33.48	
	Segment 3	$m_3 = 0.4 \cdot m$	18	35.996	35.928	54	
	Segment 4	$m_4 = 0.228 \cdot m$	10.26	20.518	20.479	30.78	
Gravit. force [N]	Segment 1	G_1	54.58	109.429	109.435	164.567	
	Segment 2	G_2	109.161	218.859	218.87	329.134	
	Segment 3	G_3	176.066	352.998	353.016	530.862	
	Segment 4	G_4	100.357	201.209	201.219	302.591	

Table 2: Output data (forces, bending moments)

Variable name		Symbol	Min. value	Mean value	Median value	Max. value	Histogram
Internal normal force [N]	Segment 1	N_1	73.367	154.2	154.145	262.296	
	Segment 2	N_2	0	40.535	39.554	128.811	
	Segment 3	N_3	172.652	365.611	365.535	566.686	
	Segment 4	N_4	76.115	161.038	161.006	249.36	
Reaction force [N]	Feet – X direction	R_{x1}	-128.809	-39.856	-38.879	0	
	Feet – Y direction	R_{y1}	91.425	196.529	196.475	305.431	
	Buttock – X direction	R_{x3}	-240.59	-84.724	-83.425	12.204	
	Buttock – Y direction	R_{y3}	305.726	651.419	651.192	1029.748	
	Dorsum	R_4	28.091	86.332	85.601	172.559	
	Head	R_5	7.546	43.15	42.456	105.244	
Max. internal bending moment [Nm]	Segment 1	M_{OMAX1}	0	3.566	3.464	12.567	
	Segment 2	M_{OMAX2}	7.796	23.257	23.058	46.975	
	Segment 3	M_{OMAX3}	3.011	13.02	12.771	34.983	
	Segment 4	M_{OMAX4}	0.324	2.179	2.126	6.63	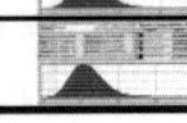

Conclusions

In the engineering design of machines, tools, medical treatment and implants, sports etc., proper databases of anthropometry based on stochastic/probabilistic are needed. The aim of this paper is a new and original biomechanical evaluation of the interaction between load forces to which a sitting man and the seat are exposed; see Fig. 1 and Tab. 1 and 2. All loads and dimensions, which consider actual anthropometry histograms of human population (i.e. measured and applied segmentation of human weight, height, centroids, gravity and shape of seat) are determined using the Monte Carlo Method.

A simple plane model (i.e. calculated probabilistic normal forces and bending moments) shows a sufficient stochastic/probabilistic evaluation connected with biomechanics, ergonomics or industrial design. According to anthropometry, the simple but accurate plane model for the stochastic solution of seat and seating man interaction was applied. The output data show the biggest normal force $N_{MAX} = N_{3MAX} = 566.7$ N in dorsum; see Fig. 2(a) and Fig 4(c). Maximum bending moment $M_{OMAX} = M_{OMAX2} = 46.975$ Nm is in thighs; see Fig. 2(b) and Fig.4(d).

For the further calculations, shear forces and dynamics effects could be added and finally a spatial (3D) model can be applied. This analysis can serve e.g. as an initial part in designing or improving chairs or as a good support for ergonomics, rehabilitation, implant design etc. The acquired results fill the information gap about this problem.

Acknowledgment

This work has been supported by Czech projects SP2019/100, CZ.02.1.01/0.0/17_049/0008407 and CZ.02.1.01/0.0/0.0/17_049/0008441.

References

[1] K. Frydrýšek, D. Čepica, T. Halo, Biomechanics – Simple Stochastic Model of Sitting Human, VSB – Technical University of Ostrava, Faculty of Mechanical engineering, Department of Applied Mechanics, 2019.

[2] D. Čepica. Biomechanics – Stochastic Loading of a Human (unpublished bachelor thesis written in Czech language, VSB - Technical University of Ostrava, 2019, head of thesis Karel FRYDRÝŠEK).

[3] K. Frydrýšek, Biomechanika 1 (Biomechanics 1), monograph written in Czech language, VSB – Technical University of Ostrava, Faculty of Mechanical Engineering, ISBN 978-80-248-4263-92019, (in print).

[4] I. Kingma, J.H. van Dieën, Static and Dynamic Postural Loadings During Computer Work in Females: Sitting on an Office Chair Versus Sitting on an Exercise Ball, Applied Ergonomics, ISSN 0003-6870, Volume 40, Issue 2, 2009, pp. 199-205.

[5] A. Kyvelidou, R.T. Harbourne, J. Haworth, K.K. Schmid, N. Stergiou, Children with Moderate to Severe Cerebral Palsy May not Benefit from Stochastic Vibration when Developing Independent Sitting, Developmental Neurorehabilitation, vol. 21, no. 6, 2018, pp. 362-370, DOI: 10.1080/17518423.2017.1290705.

[6] J. Grepl, K. Frydrýšek, M. Penhaker, A Probabilistic Model of the Interaction between a Sitting Man and a Seat, Applied Mechanics and Materials, Vol. 684, 2014, pp. 413-419.

Biomechanics – Pulling out of infusion hoses

FRYDRÝŠEK Karel[1,a], HALO Tomáš[1,b], ČEPICA Daniel[1,c], FOJTÍK František[1,d], KOLÁŘ Václav[2,e] and PLEVA Leopold[3,f]

[1]Department of Applied Mechanics, Faculty of Mechanical Engineering, VŠB – Technical University of Ostrava, 17. listopadu 2172/15, 708 00 Ostrava, Czech Republic

[2]Department of General Electrical Engineering, Faculty of Electrical Engineering and Computer Science, VŠB – Technical University of Ostrava, 17. listopadu 2172/15, 708 00 Ostrava, Czech Republic

[3]Department of Trauma Surgery, University Hospital Ostrava, 17. listopadu 1790/5, 708 52 Ostrava, Czech Republic

[a]karel.frydrysek@vsb.cz, [b]tomas.halo.st@vsb.cz, [c]daniel.cepica.st@vsb.cz, [d]frantisek.fojtik@vsb.cz, [e]vaclav.kolar@vsb.cz, [f]leopold.pleva@fno.cz

Keywords: Biomechanics, Cannula fixation, Infusion tubes, Safety, Tearing force measurements.

Abstract. The scope of this paper is to determinate critical tearing force. The original measurements were performed for cannula and its securing fixation in patient's body (i.e. applications in medico-chirurgical branches). Data (i.e. pull out or tearing forces) acquired from this experiment may later be used in designing of a device for automatic winding of infusion tubes etc.

Introduction

This article is focused on biomechanics and medical science (i.e. application of infusion hoses for patients in a hospital - cannula and its securing fixation; see Fig. 1). In our case, infusion hoses are placed on patient's forearm. A medical term cannula is a tube that can be inserted into the body, often for the delivery or removal of fluid or for the gathering of data. The anatomical interaction between mechanical circulatory or respiratory support and the patient is achieved by specially designed cannulae. For more information see references [1] to [5].

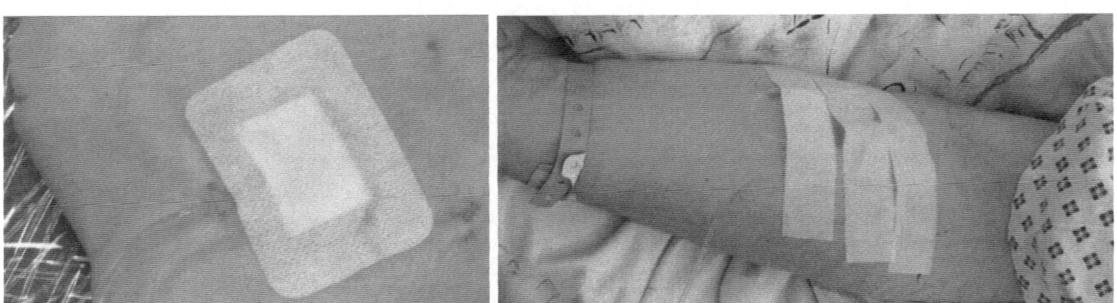

Fig. 1: Intravenous cannulation - (a) Porcine skin, (b) human skin

During a treatment in a hospital etc., if the cannula is pulled (ripped, torn) out from a vein of a patient, by accident or intentionally, it may lead to serious consequences or even death if not checked in time. In order to ensure the safety of patients and make the patients have a good rest during liquid infusion, the cannulae and infusion hoses must be correctly proposed and designed.

The goal of this paper is to determine the minimum force needed for a patient to pull on an infusion tube with cannula at its end to tear it out from vein. And also, to specify force to be pulling on an infusion tube from automatic winding device to keep the tube outstretched, but not to pull too strong.

There is a lack of information about cases of patients tearing cannulas out of their skin. Therefore, it is important to do measurements in this area.

Description of Experiments

First idea for experiment was to use porcine body with skin. However, the problem was that the porcine skin, even after proper preparation was still too oily, therefore the securing tapes did not hold, and cannulas themselves were quite easy to pull out; see Fig. 1(a).

The main element holding cannulas in place are the securing tapes, or in some cases the cannulas can be sewn onto patient's skin, but this was not the case of this experiment.

For more realistic approach, the experimental measurements were done several times in several anatomic directions on right human forearm. For each of seven directions were done two measurements. Directions are shown in Fig. 2. The cannulas were put on human skin and taped in place, see Fig. 1(b). The measurements on humans were performed without inserted cannulae (i.e. infusion hoses were only taped – no intravenous cannulation).

The original measurements of pulling forces were based on general anatomical directions of infusion hoses with cannulae similar to patients with infusions in a hospital; see Fig. 2 and Tab. 1.

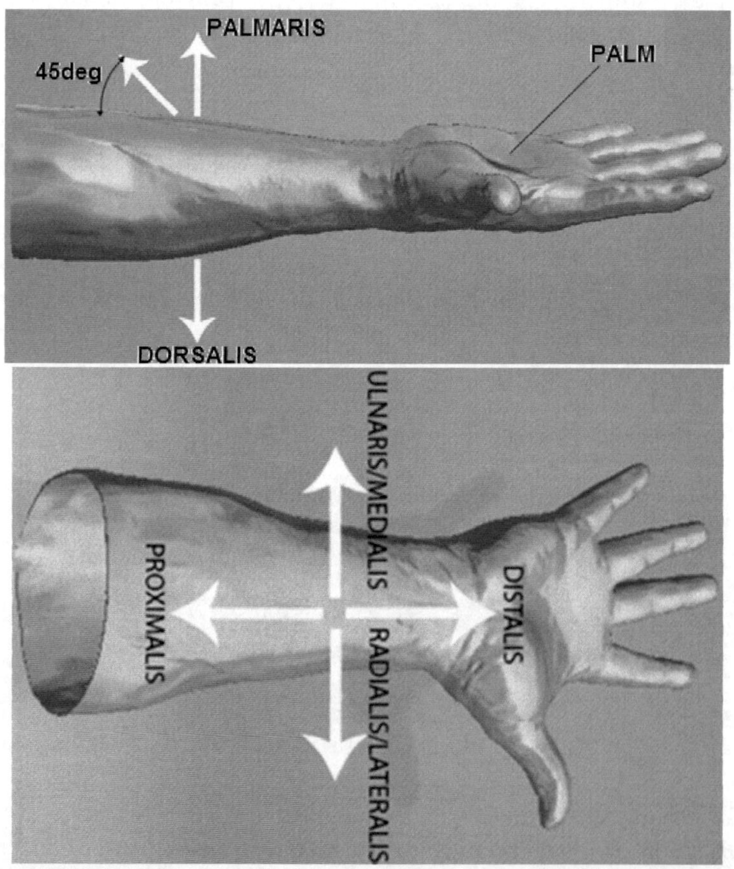

Fig. 2: Anatomical directions applied in measurements of pull out forces (right hand)

Table 1: Anatomical directions and measured data

Pull out Force		
Direction of Tensile Force, see Fig. 2	**Min. & Max. Tensile Force [N]**	**Average [N]**
Palmaris	19.12; 29.50	**24.31**
45 deg	31.87; 34.98	**33.43**
Distalis	21.52; 22.47	**22.00**
Radialis/Lateralis	35.47; 48.46	**41.97**
Ulnaris/Medialis	39.52; 23.51	**31.52**
Proximalis	50.39; 53.30	**51.85**
Dorsalis	34.84; 45.92	**40.38**
Maximum	53.30	**51.85**
Minimum	19.12	**22.00**
Median	34.91	**33.43**
Average	35.06	**35.06**

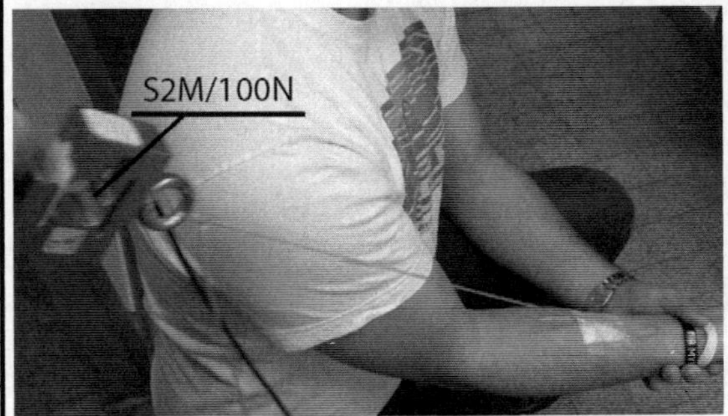

Measured by force transducer (load cell) HBM, type S2M/100N. In this picture, the direction of tensile force is 45 deg)

To measure the tensile force, the S2M/100N load cell (force transducer) developed by HBM was used. This load cell is capable of measuring forces up to 100 N, which was quite enough range for this experiment.

The measurements were performed at VSB – Technical University of Ostrava (Faculty of Mechanical Engineering, Department of Applied mechanics) in cooperation with University Hospital in Ostrava & University of Ostrava.

Results

Some result diagrams; see Fig. 3, 4 and Tab. 1, show dependencies of force on time, until the cannula is torn out or infusion tube is detached.

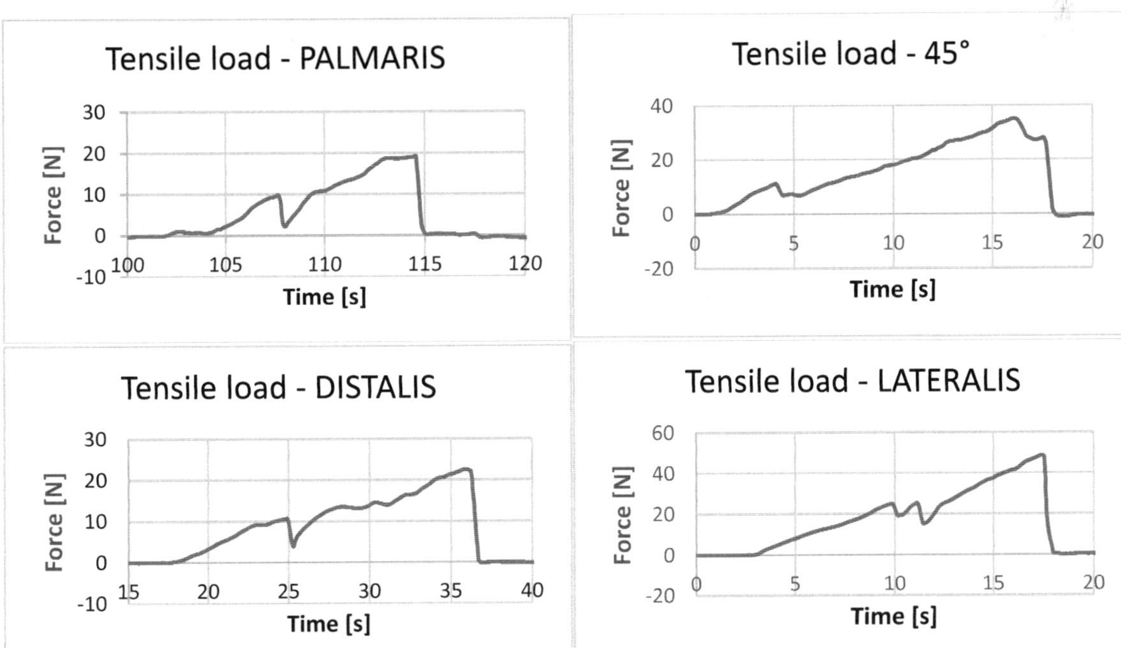

Fig. 3: Examples of pull out force measurement (a) Palmaris direction, (b) Under 45 deg. direction, (c) Distalis direction, (d) Lateralis direction

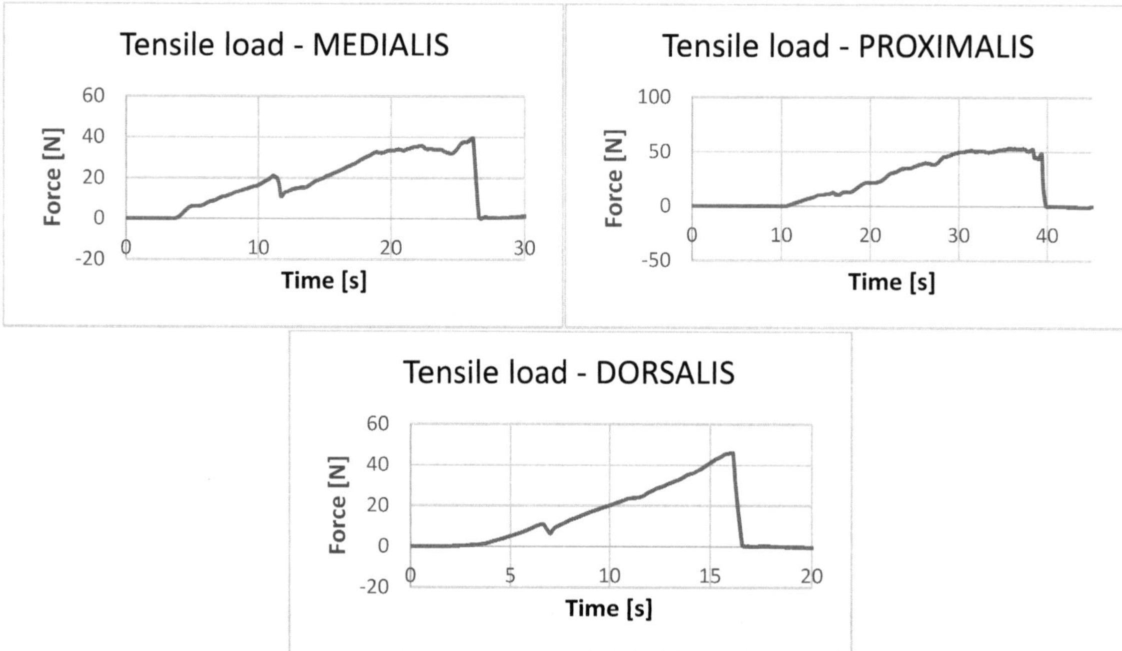

Fig. 4: Examples of pull out force measurement (a) Medialis direction, (f) Proximalis direction, (g) Dorsalis direction

Hence, the safe force (safety factor $k_S = 5$) to be used in auto winding device for hoses is

$$F = \frac{F_{AVERAGE}}{k_S} = \frac{35.06}{5} = 7 \text{ N}. \quad (1)$$

For more information see [1] and [2].

Data acquired from this experiment will be used in designing of a device for automatic winding of infusion hoses.

Conclusions

From measured data; see Tab. 1 and Fig. 3, 4, the minimal & maximal tensile (torn out or pulling out) force can be determined (19.12 N & 51.85 N).

This result can be affected by the fact, that the measurements were done using the same tape without changing it for every measurement. However, the measurements are reliable and sufficient enough. The tapes (i.e. taped cannulae) held well enough in whole experiments.

The acquired results fill the information gap about this problem and can be applied in engineering design and improvement of safe medical treatment.

Acknowledgement

This work has been supported by Czech project SP2019/100 and by Czech project CZ.02.1.01/0.0/0.0/17_049/0008441 under Operational Programme Research, Development and Education supported by European Union.

References

[1] K. Frydrýšek, Halo, T., Čepica, D., Fojtík, F. Pulling out of Infusion Hoses (measurement report), VSB – Technical University of Ostrava, Faculty of Mechanical engineering, Department of Applied Mechanics, 2019, pp. 1-20.

[2] Halo, T. Biomechanika - Samonavíjecí infuzní hadičky (Biomechanics - Automatic Roll-up Infusion Hoses), unpublished bachelor thesis (written in Czech language), VSB – Technical University of Ostrava, 2019, head of thesis Karel FRYDRÝŠEK

[3] Frydrýšek, K. Biomechanika 1 (Biomechanics 1), monograph written in Czech language, VSB – Technical University of Ostrava, Faculty of Mechanical Engineering, ISBN 978-80-248-4263-92019, (in print).

[4] Frydrýšek, K., Dvořák, L. Biomechanics - Testing of Mechanical and Utility Properties of Bottles for Redon Drainage, Engineering Mechanics 2017, ISBN:978-80-214-5497-2, ISSN: 1805-8248, Svratka, Czech Republic, 2017, pp. 318-321.

[5] Liu, X.T., The Design of the Infusion Alarm, Proceedings of the 2017 7th International Conference on Mechatronics, Computer and Education Informationization (MCEI 2017), ACSR-Advances in Comptuer Science Research, ISBN:978-94-6252-430-9, ISSN: 2352-538X, Peoples Republic of China, 2017, pp. 491-495.

EAN 2019
57th conference on experimental stress analysis

Influence of loading speed on yield strength value of S235 and C45 steels

FUSEK Martin[1,a], PORUBA Zdeněk[1,b], HALAMA Radim[1,c] , FRANTIŠEK Ondřej[2,d], SCHNEIDER Markus[2,e] , BIAGINI Veronica[2,f] and SADÍLEK Marek[1,g]

[1]Department of Applied Mechanics; Faculty of Mechanical Engineering; VSB – Technical University of Ostrava; 17. listopadu 15, 708 00 Ostrava, Czech Republic

[2]ABB Corporate Research Center; Wallstadter Strasse 59, 685 26 Ladenburg, Germany

[a]martin.fusek@vsb.cz, [b]zdenek.poruba@vsb.cz, [c]radim.halama@vsb.cz, [d]ondrej.frantisek@de.abb.com, [e]markus.schneider@de.abb.com [f]veronica.biagini@de.abb.com, [g]marek.sadilek@vsb.cz

Keywords: Yield strength, Loading speed, Static loading, Dynamical loading, Digital image correlation

Abstract. The aim of the paper is to evaluate the dependency of the yield strength of steels S235 and C45 on loading speed. In order to quantify this phenomenon, static and dynamical tension tests of both steels were realized and evaluated. Static tension tests realized on flat specimen confirmed values of yield strength under static loading given my material sheets of both steels. Dynamical testing has shown different behavior of steels under increased loading speeds. While the steel S235 has shown a significant growth of yield strength with the increased loading speed in the interval from 0 m/s till 3 m/s, the steel C45 has shown this ability in the interval from 0 m/s till 1 m/s only and the subsequent growth of yield strength has been very slow. In order to quantify the strain distribution in the specimen during dynamical tests using digital image correlation method, experiments were recorded using high-speed camera as well. Unfortunately, none of the results could be used since the colour applied on the surface of the specimen was eroded from the surface due to its fast deformation.

Introduction

During different types of experiments, the dependency of yield strength value on loading speeds was observed [1]. In order to quantify this phenomenon, static and dynamical tension tests of S235 and C45 steels were performed and evaluated.

Values of the yield strength of S235 steel were measured and evaluated under four different loading speeds – static loading, 1 m/s, 2 m/s and 3 m/s.

Values of the yield strength of C45 steel were measured and evaluated under ten different loading speeds – static loading, 1 m/s, 1.5 m/s, 2 m/s, 2.5 m/s, 3 m/s, 4 m/s, 6 m/s, 10 m/s and 20 m/s.

Results measured can be subsequently used for design of devices working under higher speeds, such as electric contactors, switches, etc.

Laboratory Equipment

The measurement of yield strengths under static loading was realized using tension testing machine Testometric M500-50CT (50 kN), dynamical tests were realized on the high-speed tensile testing machine Instron VHS 65 (available testing speeds from 1 m/s till 20 m/s, maximal dynamical loading force 80 kN). The testing machine Instron VHS 65 and detail view on fixed specimen are depicted in Fig. 1 and Fig. 2 respectively.

In addition, the dynamical experiment was recorded using high-speed camera (max. 30 kHz) with the intention of its evaluation using digital image correlation method [2], see Fig. 3.

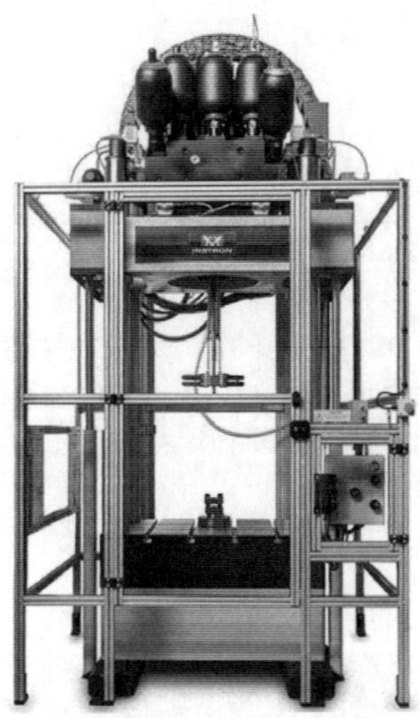

Fig. 1: High-speed tensile testing machine Instron VHS 65

Fig. 2: Jaws with installed specimen

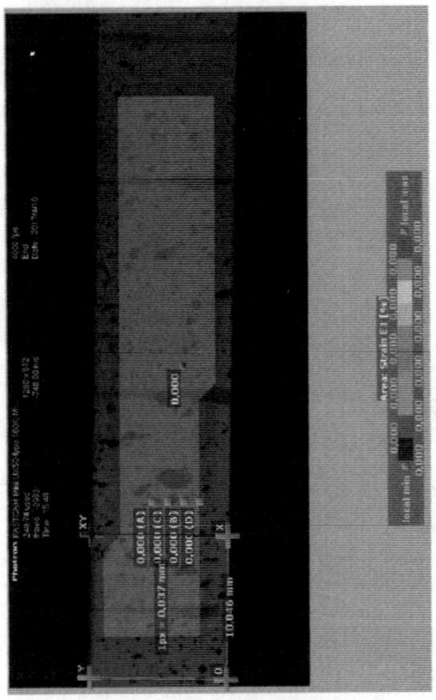

Fig. 3: Measurement of strain distribution using optical correlation method

Results Measured

Results measured for steel S235 and C45 are depicted in the form of stress vs. strain and yield strength vs. loading speed relations in Fig. 4 and Fig. 5 respectively. The course of yield strength vs. loading speed was extrapolated using linear Eq. (1) and exponential Eq. (2) for steel S235 and C45 respectively.

$$f_y = 84.172v + 349.7 \quad (1)$$

$$f_y = 508.56v^{0.0384} \quad (2)$$

where f_y [MPa] is the yield strength and v [m/s] is the loading speed.

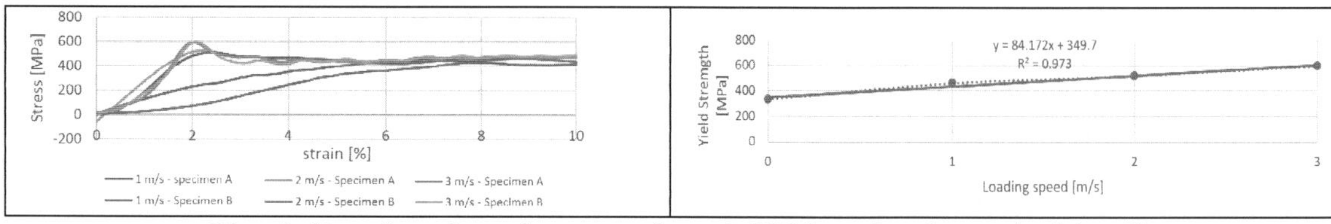

Fig. 4: Results of tension test of steel S235 under dynamical loading

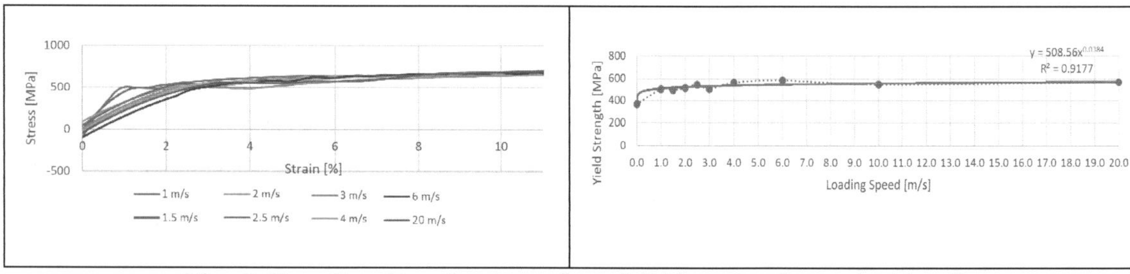

Fig. 5: Results of tension test of steel C45 under dynamical loading

As mentioned above already, the intention of authors was to measure the strain distribution on the surface of the specimen during the dynamical testing using digital image correlation method. To do so, the surface of the specimen was provided by spotted spraying and subsequently recorded by high-speed camera during the loading. Unfortunately, the fast loading caused the peeling-off the color sprayed on the specimen surface and made so the measurement impossible. The peeling-off process can be seen in series of images depicted in Fig. 6.

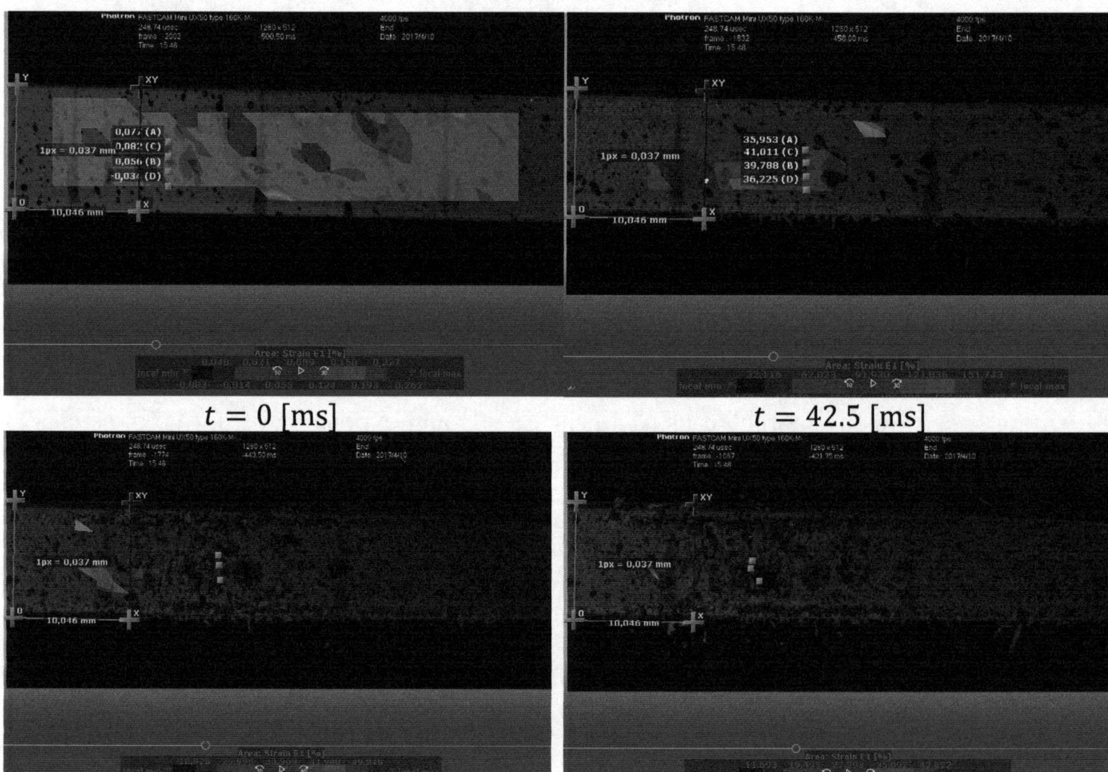

$t = 0$ [ms] $t = 42.5$ [ms]

$t = 57$ [ms] $t = 78.75$ [ms]

Fig. 6: Digital image correlation: peeling-off process of sprayed color

Conclusions

According to the results measured, following conclusions can be stated:

- Values of yield strengths of S235 and C45 steels under static loading correspond to the values defined in material sheets – min. 235 MPa and 340 MPa respectively,
- Steel S235 shows a significant growth of yield strength with the increasing loading speed in the interval from 0 m/s till 3 m/s,
- Steel C45 shows the growth of yield strength with the increasing loading speed in the interval from 0 m/s to 1 m/s, the growth of yield strength above 1 m/s is very slow,
- Results from high-speed camera could not be used since the colour applied on the surface of the specimen was eroded from the surface due to its fast deformation.

Acknowledgement

This work was supported by the Czech projects of Ministry of Education, Youth and Sports under No. SP2019/100.

Article has been done in connection with project Innovative and additive manufacturing technology - new technological solutions for 3D printing of metals and composite materials, reg. no. CZ.02.1.01/0.0/0.0/17_049/0008407 financed by Structural Founds of Europe Union.

This article has been elaborated under support of the project Research Centre of Advanced Mechatronic Systems, reg. no. CZ.02.1.01/0.0/0.0/16_019/0000867 in the frame of the Operational Program Research, Development and Education.

This article has been created under the state support of Technological Agency of the Czech Republic under the project TN1000071 National Competence Centre of Mechatronics and Smart Technologies for Mechanical Engineering of the program National Centers of Competence 1.

References

[1] MAINSTONE, R. J. Properties of materials at high rates of straining or loading. Matériaux et Constructions. 1975, 8(2), 102-116 ISSN 0025-5432.

[2] KUBIAK, M., DOMAŃSKI, T., DEKÝŠ, V. and SAPIETOVÁ, A. Measurement of Strain During Tension Test of Welded Joint Using Multi-camera 3D Correlation System. Procedia Engineering. 177(2017), 107-113. ISSN 18777058.

Set-up of the standard 2D-DIC system for quantification of residual stresses

HAGARA Martin[1,a], PÁSTOR Miroslav[1,b] and DELYOVÁ Ingrid[1,c]

[1]Department of Applied Mechanics and Mechanical Engineering, Faculty of Mechanical Engineering, Technical University of Košice, Letná 9; 042 00 Košice, Slovak Republic

[a]martin.hagara@tuke.sk, [b]miroslav.pastor@tuke.sk, [c]ingrid.delyova@tuke.sk

Keywords: Digital Image Correlation, residual stress, full-field analysis, hole-drilling method

Abstract. The paper deals with the set-up of the standard single camera digital image correlation (DIC) system for the quantification of residual stresses. To find proper parameters of the camera some measurements were performed including founding for a convenient distance of the camera from analyzed specimen, creating of speckle-pattern with adequate speckle size and density, analysis of the calibration parameters influence on the measurement results obtained by camera calibration performed by calibration targets of various sizes. Moreover, the investigation of optimal correlation parameters used for evaluation of strains is present in the paper. Finally, the paper includes information about the results obtained by simulating the change of positions between the hole-drilling device and the camera during the incremental drilling process.

Introduction

Recently, non-contact optical methods are increasingly used to stress-strain analysis. It is caused by a fact that the full-field investigation provides a rich source of data in a relatively short time without the need for using any additional devices. Digital image correlation method can be considered as one of the youngest optical methods used in experimental mechanics because its development relates with the rapid development of computer vision noticed since the beginning of the nineties of the 20th century.

DIC is a full-field experimental method, based on the correlation of digital images, captured during the loading process of the analysed specimen. For the reason that the images are not correlated as wholes, but along small picture elements called facets, it is necessary to create a random black and white speckle-pattern on the specimen surface to make all its parts unique. There are two types of analysis performed by DIC systems:

- 2D analysis realised by a single camera system,
- 3D analysis realised by a system with minimally two cameras.

DIC method has been used in many modern, interesting fields of mechanical engineering e.g. biomechanics [1-3] or modal analysis [4-7], however, it can also be used in vibration [8-11], thermal or any other analysis [12-14].

The development in a field of DIC application for quantification of residual stresses has been noticed since 2005 when McGinnis et al. [15] performed the measurements using two different optical methods – photogrammetry and 3D-DIC – combined with the core-drilling method and obtained just a small error in normal stress. Since then, a lot of analysis has been carried out in the field of using DIC combined with a hole-drilling method to quantify residual stresses [16-18], leading to design special devices combining DIC with hole-drilling method [19-21].

In 2019 Hagara et al. [22] came with the idea to use standard (commonly delivered) 3D-DIC system Q-400 Dantec Dynamics combined with the standardized hole-drilling method. The uniform, residual stresses simulated in flat steel specimen, were calculated from strains obtained at three small areas, which locations corresponded to the measuring grids of strain gage rosette HBM 1-RY21-3/120 0°/45°/90°, due to a methodology given in standard ASTM E837-13a. The same procedure was realised on the opposite part of the specimen using the strain gage rosette. The relative difference between the simulated normal stress and normal stress quantified by DIC was ca. 3%, which can be considered as a sufficient correspondence.

However, the use of a standard 3D-DIC system for quantification of residual stresses has some drawbacks. The first one is that the cameras are not directed straightaway to the specimen, which causes a distortion of the images leading in many cases to correlation errors. The second one, a higher distance of the cameras from the analysed specimen, can cause problems with the correlation of the images. Then the higher facets have to be used, which leads to the loss of spatial resolution in the surrounding of the drilled hole edge. For these reasons, the author decided to realise some preparatory measurements to qualify the possibility of using their 2D-DIC Q-400 Dantec Dynamics system for quantification of residual stresses.

Preparatory measurements relating to the use of 2D-DIC system for quantification of residual stresses

To obtain the parameters needed for the realisation of 2D deformation analysis performed in the surrounding of the small hole the following experiment was realised. A specimen with 3.05 mm thickness, which shape and dimension are depicted in Fig. 1, was made from photoelastic material PSM-1 (Young's modulus of elasticity $E = 2500\text{MPa}$ and Poisson's ratio $\mu = 0.38$)

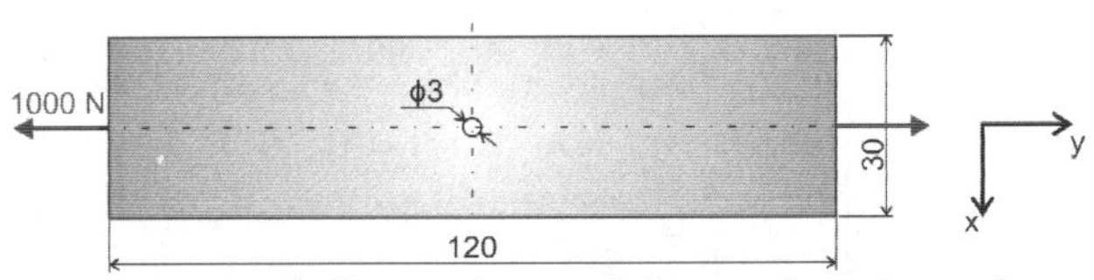

Fig. 1: Shape and dimensions of the analysed specimen

The reference image with a resolution of 2452x2056 px (the pixel density 111.45 px/mm) of the analysed part of the specimen can be seen in Fig. 2.

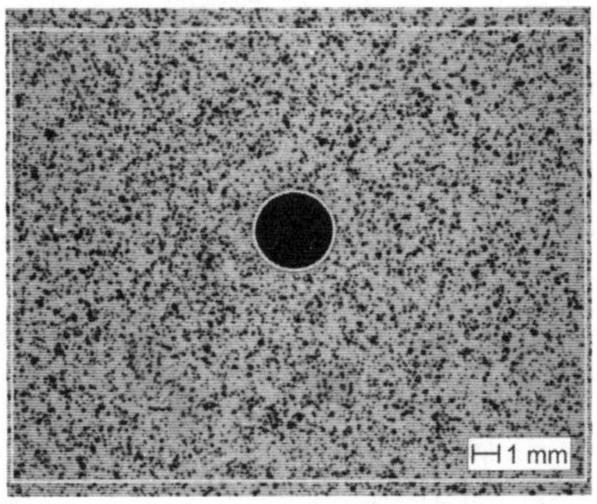

Fig. 3: Reference image of the specimen obtained by single camera DIC system and evaluation mask

In this case, a single-camera DIC system Q-400 Dantec Dynamics was arranged on a tripod in such a way, that the distance of its sensor to the specimen was approximately 130 mm (Fig. 3). Since the camera image plane is parallel to the object surface, distortion of the image does not appear. This phenomenon is hard to overcome when using a multi-camera system.

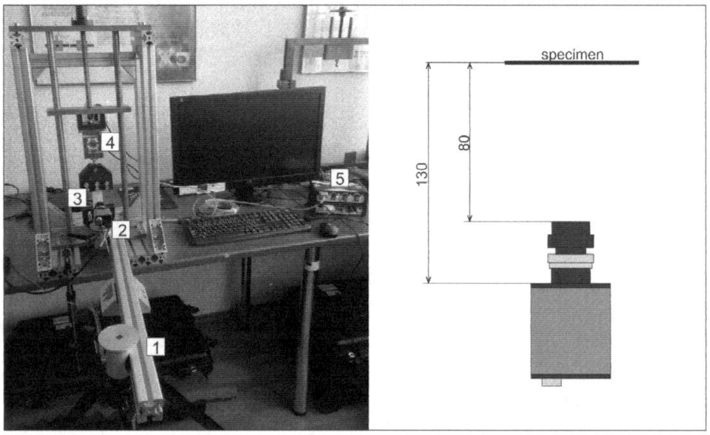

a) b)

Fig. 2: a) The measurement chain: 1) tripod, 2) single CCD camera, 3) specimen, 4) force sensor, 5) HBM QuantumX CX22-W with amplifier QuantumX MX410-B, b) Proposed distance of the camera from the specimen

To situate the camera closer to the specimen was not possible, because in this case, the image of the specimen surface was not sharp enough. On the other hand, it was not suitable to increase the distance between the camera and the specimen, because the higher distance, the smaller pixel density, which could lead to the loosing of important data.

The specimen was subsequently loaded by tension force acting in *y*-direction (see Fig. 1) until its magnitude reached 1000 N. Deformation of the specimen was measured by correlation system after each increase in the force of 100 N.

The calibration of the camera was realised using two calibration targets of different sizes – Gl-01-WMB_9x9 with and Gl-01_5-WMB_9x9 with 9x9 black and white fields of the particular fields' size 1x1 mm and 1.5x1.5 mm. Comparison of their sizes with respect to the analysed field of view can be seen in Fig. 4.

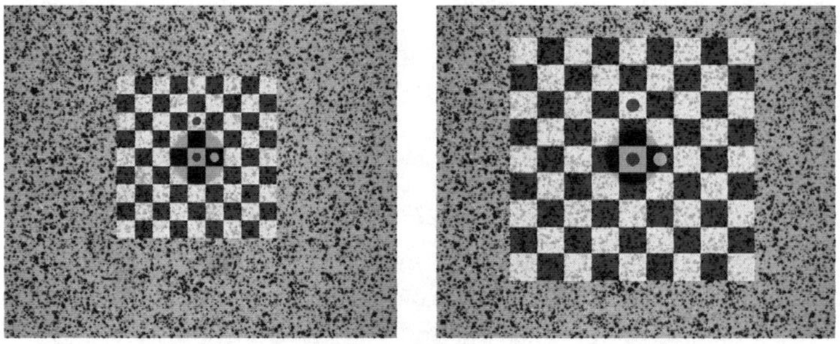

a) b)

Fig. 4: Analyzed field of view in comparison with the size of calibration target: a) Gl-01-WMB_9x9, b) Gl-01_5-WMB_9x9

The calibration of the DIC system used for measurement is performed in such way that calibration target is captured in different positions and rotations until the information about the intrinsic and extrinsic parameters of the camera are not assessed. The manufacturer of the correlation systems states that for sufficiently accurate calibration just eight different positions of the calibration target are necessary. However, it was proven that the higher the number of different positions, the higher the accuracy of the measurement [23]. For that reason, twenty different positions of the calibration targets were captured during both calibrations. The accuracy of the calibration is given by so-called calibration residuum. Approximately the same calibration residuum was achieved (0.276 px for the target with 1x1 mm fields and 0.283 px for the target with 1.5x1.5 mm) by both processes. Despite it, the obtained calibration parameters were not the same (see Fig. 5 and Fig. 6).

Intrinsic Parameters Camera Position 1 image:	
Focal length {x; y}:	**{1e+004 ± 3e+002; 1e+004 ± 3e+002}**
Principal point {x; y}:	**{750 ± 130; 1230 ± 100}**
Radial distortion {r^2; r^4}:	**{-0,173 ± 0,011; 0,8 ± 0,3}**
Tangential distortion {t_x, t_y}:	**{0,0025 ± 0,0009; -0,0039 ± 0,0011}**
Extrinsic Parameters Camera Position 1 image:	
Rotation vector {x; y; z}:	**{-3,105 ± 0,011; 0,004 ± 0,006; 0,06 ± 0,02}**
Translation vector {x; y; z}:	**{4,0 ± 1,1; -2,2 ± 0,9; 97 ± 2}**
Calibration residuum:	**0.276**

Fig. 5: Calibration parameters assessed by the DIC system using calibration target Gl-01-WMB_9x9

Intrinsic Parameters Camera Position 1 image:	
Focal length {x; y}:	**{1e+004 ± 2e+002; 1e+004 ± 2e+002}**
Principal point {x; y}:	**{1090 ± 40; 550 ± 70}**
Radial distortion {r^2; r^4}:	**{-0,204 ± 0,013; 4,2 ± 0,8}**
Tangential distortion {t_x, t_y}:	**{-0,0044 ± 0,0005; -0,0014 ± 0,0003}**
Extrinsic Parameters Camera Position 1 image:	
Rotation vector {x; y; z}:	**{3,116 ± 0,007; -0,057 ± 0,004; -0,015 ± 0,008}**
Translation vector {x; y; z}:	**{1,7 ± 0,3; 4,1 ± 0,6; 105,1 ± 1,4}**
Calibration residuum:	**0.283**

Fig. 6: Calibration parameters assessed by the DIC system using calibration target Gl-01_5-WMB_9x9

To investigate the influence of obtained calibration parameters on the results an analysis was realised. The aim of the analysis was to compare the results obtained experimentally with the results obtained numerically, whereby the data were compared just in three small areas, representing the measuring grids of the strain gage rosette HBM 1-RY21-3/120 0°/45°/90° used at the authors' workplace for quantification of residual stresses using the hole-drilling method. The numerical model was created in Ansys Workbench 17.0 in such way that mentioned areas, labeled in Fig. 7a as areas *a*, *b* and *c*, were uniformly divided and meshed by quadrilateral finite elements with the corresponding size. Subsequently, the numerical model was loaded by tension force 1000 N acting in *y*-direction (Fig. 7b).

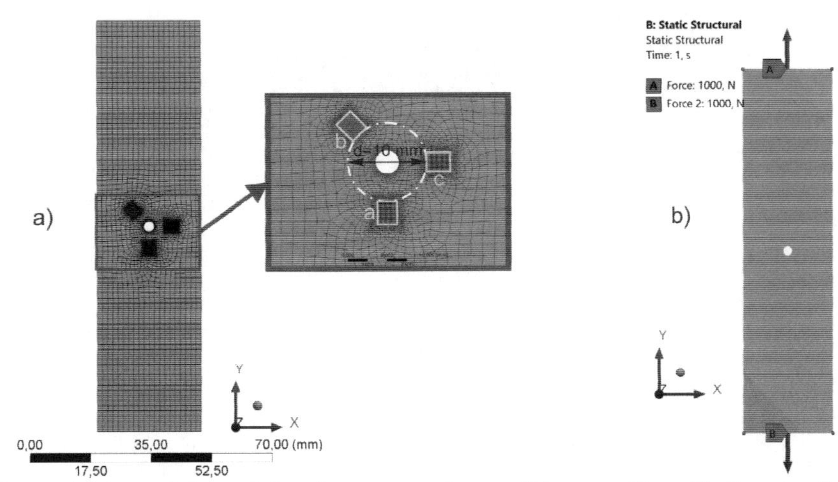

Fig. 7: Numerical model used for the comparison of the results with the results obtained experimentally

The mean values of equivalent von Mises stress calculated in areas a, b and c, were

$$\sigma_a^{Mises} = 9.73\text{MPa},$$
$$\sigma_b^{Mises} = 12\text{MPa},$$
$$\sigma_c^{Mises} = 11.27\text{MPa}.$$

In addition to the investigation of calibration parameters influence, the influence of the local regression smoothing level was analysed. The measurement was evaluated with different set facet sizes in the range from 15x15 px up to 30x30 px. Subsequently, the virtual gages in a form of polygons corresponding to the areas a, b and c were defined to calculate a parameter Δ_σ introduced by

$$\Delta_\sigma = \left(\frac{mean\left(\sigma_{i_{\exp}}^{Mises}\right)}{mean\left(\sigma_{i_{num}}^{Mises}\right)} - 1 \right) \cdot 100 \qquad (1)$$

where $mean\left(\sigma_{i_{\exp}}^{Mises}\right)$ are the mean values of equivalent von Mises stress obtained in areas a, b and c, by single camera DIC system and $mean\left(\sigma_{i_{num}}^{Mises}\right)$ are the mean values of equivalent von Mises stress obtained in areas a, b and c by Ansys Workbench 17.0. Each smallest Δ_σ was then associated with the optimal setting of the kernel size with respect to the corresponding facet size.

The bar graphs in Fig. 8 and Fig. 9 depicts the dependence between the facet size and the optimal kernel size used in local regression smoothing. As can be seen, the obtained dependences are the same for both evaluations. In the bottom parts of Fig. 8 and Fig. 9, the dependences between the facet size and corresponding Δ_σ on analyzed areas a, b and c can be seen. While the differences obtained on area b, where Δ_σ reached ca. 14-16%, are quasi-same for both evaluations, the biggest difference between the obtained results appeared by comparing equivalent stresses on the areas a and c. For a reason that much smaller Δ_σ was obtained on the area a by using calibration parameters assessed by calibration performed with smaller calibration target, it seems to be more convenient to use calibration target Gl-01-WMB_9x9 for the proposed arrangement of the 2D-DIC system.

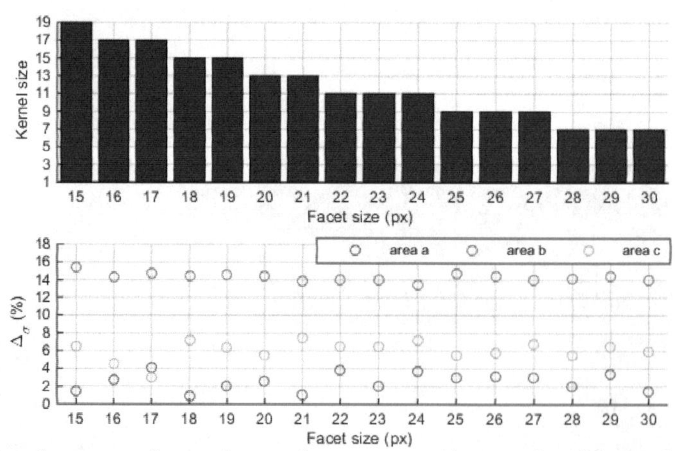

Fig. 8: Dependence between facet size and corresponding optimal kernel size used in local regression (upper graph), the dependence between the facet size and corresponding Δ_σ obtained on areas a, b and c (bottom graph) by evaluation using calibration parameters assessed by calibration with calibration target Gl-01-WMB_9x9

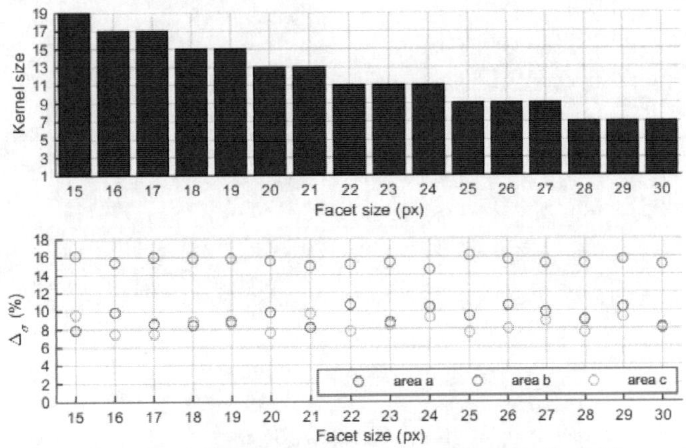

Fig. 9: Dependence between facet size and corresponding optimal kernel size used in local regression (upper graph), the dependence between the facet size and corresponding Δ_σ obtained on areas *a*, *b* and *c* (bottom graph) by evaluation using calibration parameters assessed by calibration with calibration target Gl-01_5-WMB_9x9

In the last part of work, the authors dealt with the analysis of the influence of the camera position change, which can occur in case of incremental hole-drilling, because it is necessary to change position between the camera and the hole-drilling device after each step of drilling. For the analysis authors decided to use a function of the virtual bearing tool, which is a part of Dantec Dynamics control software Istra4D. The mentioned tool allows knowing the reference as well as the actual position of the camera (see Fig. 10). To obtain the mentioned influence, after the last step of loading (1000 N) the tripod with the camera was 10-times rotated clockwise around its vertical axis and subsequently (approximately) back to its previous position.

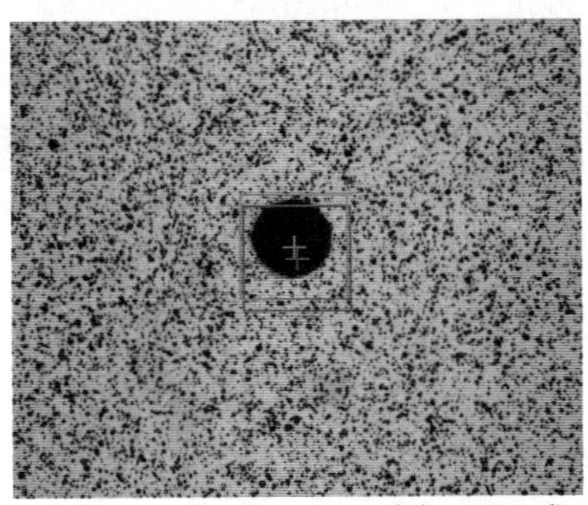

Fig. 10: Virtual bearing tool – reference position (red), actual position (green)

The evaluated strain fields corresponding to the last step of loading (1000 N) can be seen in Fig. 11. The facet size used for the correlation was set to 23x23 px, the level of local regression smoothing (Kernel size) was set to 11. The evaluation was performed using calibration parameters obtained with the smaller calibration target. As the data are obtained in the centres of facets, the diameter of the evaluated hole was approximately 3.5 mm, which still allows investigate the strains relatively near to the centre of the hole. By this phase of loading the strains on areas *a*, *b* and *c* reached the following mean values:

ε_a	ε_{b_x}	ε_{b_y}	$\varepsilon_{b_{xy}}$	ε_c
4.297e-3	-1.887e-3	5.617e-3	-0.128e-3	-1.512e-3

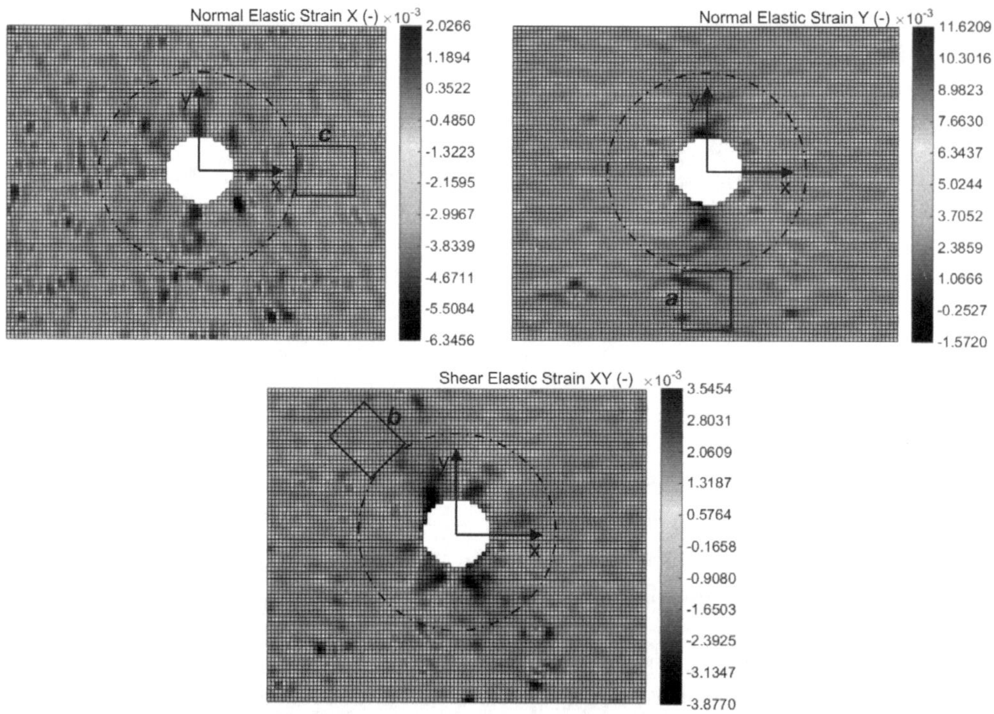

Fig. 11: Strain fields in the surrounding of the hole obtained by loading of 1000 N.

The relative differences in analysed quantities obtained after 10 rotations of the tripod with camera and its subsequent movement back to its previous position, by which the evaluated displacements do not exceed more than 0.06 mm, can be seen in Table 1.

Table 1: Relative difference in analysed quantities obtained after the camera's rotation and its subsequent movement back to its previous position

Reference values	Relative difference (%)									
	Rot. 1	Rot. 2	Rot. 3	Rot. 4	Rot. 5	Rot. 6	Rot. 7	Rot. 8	Rot. 9	Rot. 10
ε_a = 4.297e-3	-2.61	-7.29	2.26	7.12	5.27	4.39	-2.89	-4.64	6.27	6.16
ε_{bx} = -1.887e-3	-7.40	-18.39	-1.66	16.52	13.99	14.87	-9.12	-13.44	15.59	14.96
ε_{by} = 5.617e-3	-4.03	-3.73	1.41	5.71	7.61	3.15	-4.53	-4.11	4.89	6.12
ε_{bxy} = -0.128e-3	6.58	10.75	-19.17	-35.19	-18.59	-50.87	18.26	14.52	-31.01	-19.63
ε_c = -1.512e-3	-16.08	-22.94	5.99	30.63	17.68	31.92	-12.65	-20.25	25.78	18.52

Obtained differences could significantly influence the results of residual stress quantification. However, it has to be noted that the way of moving the camera back to its previous position was not very precise. For that reason, it will be a challenge for the authors to realise measurements and investigate, how the analysed data change if the displacement of the camera will be much smaller (e.g. in one order). For the realisation of such measurements, a unique automated hole-drilling device developed at the authors' workplace, allowing positioning in three mutually perpendicular directions with the accuracy of 1 μm, will be used. Recently, the device has been tested for the purposes of incremental hole-drilling combined with PhotoStress® method, however, it is assumed that it will be possible to use it also in combination with DIC system, attached on it.

Conclusions

The digital image correlation method allows a wide range of applications with high accuracy of the results. In the paper, some analyses lead to finding proper set-up of the 2D-DIC system combined with the hole-drilling method for the quantification of residual stresses, are described. Commonly used spray colours can be considered an effective way for creating random speckle-pattern with sufficiently small speckles and proper density. It seems that the results of stress/strain analysis depend markedly on the quality of camera calibration. For that reason, the authors propose to realise calibration of the camera more times to be sure that the calibration parameters are assessed correctly. The set of correlation parameters in a form of facet size and corresponding level of smoothing lead to the results, which do not differ markedly from the results obtained numerically. To ensure the accurate process of position changing between the camera and the hole-drilling device during incremental drilling can be considered as the biggest challenge. Moreover, in case if the released deformations are too small to be evaluated by DIC system with sufficient accuracy, the principle of the full-field measurement allows investigating displacements/strains with higher levels locating nearer to the centre of the hole. However, in this case, new calibration parameters used in the quantification process due to ASTM E837-13a will have to be found.

Acknowledgement

The paper is the result of projects implementation APVV 15-0435 and VEGA 1/0751/16.

References

[1] M. Lezaja et al., Bond strength of restorative materials to hydroxyapatite inserts and dimensional changes of insert-containing restorations during polymerization, Dental Materials 31 (2015) 171-181.

[2] V. Miletic et al., Local deformation fields and marginal integrity of sculptable bulk-fill, low-shrinkage and conventional composites, Dental Materials 32 (2016) 1441-1451.

[3] K. Čolić et al., Experimental and numerical research of mechanical behaviour of titanium alloy hip implant, Tehnički vjesnik 24(3) (2017) 709-713.

[4] R. Huňady, P. Pavelka, P. Lengvarský, Vibration and modal analysis of a rotating disc using high-speed 3D digital image correlation, Mechanical Systems and Signal Processing 121 (2019) 201-214.

[5] A.J. Molina-Viedma et al. 3D mode shapes characterisation using phase-based motion magnification in large structures using stereoscopic DIC, Mechanical Systems and Signal Processing 108 (2018) 140-155.

[6] A.J. Molina-Viedma et al. High frequency mode shapes characterisation using Digital Image Correlation and phase-based motion magnification, Mechanical Systems and Signal Processing 102 (2018) 245-261.

[7] C.M. Sebastian, E., López-Alba and E.A. Patterson, A comparison methodology for measured and predicted displacement fields in modal analysis, Journal of Sound and Vibration 400 (2017) 354-368.

[8] S. Barone et al., Low-frame-rate single camera system for 3D full-field high-frequency vibration measurements, Mechanical Systems and Signal Processing 123 (2019) 143–152.

[9] M. Svoboda et al., Effect of Impacts on Human Head, Manufacturing Technology 15 (2) (2015) 226-231.

[10] M. Svoboda et al. Force effect of strike and the possibility of causing a skull fracture of a human head, Springer Proceedings in Mathematics and Statistics : 13th International Conference on Dynamical Systems : Łódź, 7-10 December, 2016, 181 (2016) 353-360.

[11] F. Klimenda et al. Investigation of Vertical Vibration of a Vehicle Model Driving Through a Horizontal Curve, Manufacturing Technology 15(2) (2015) 143-148.

[12] M. Schrötter et al., Basic Thermal Field Analysis of a Small Turbo-Jet Engine, Proceedings of NTAD 2018 - 13th International Scientific Conference – New Trends in Aviation Development, (2018) 105-113.

[13] R. Halama et al., Comparison of ESPI and DIC Strain Contours Measurements under Three Loading Cases, Proceedings of Experimental Stress Analysis 2017 : 55th International Scientific Conference : Nový Smokovec, May 30-June 1, (2017) 359-363.

[14] Z. Paška et al. Strain Response Determination in Notched Specimens under Multiaxial Cyclic Loading by DICM, Proceedings of Experimental Stress Analysis 2017 : 55th International Scientific Conference : Nový Smokovec, May 30-June 1, (2017) 143-149.

[15] M. J. McGinnis, S. Pessiki, H. Turker, Application of three-dimensional digital image correlation to the core-drilling method, Experimental Mechanics 45 (4) (2005) 359-367.

[16] D. V. Nelson, A. Makino, T. Schmidt, Residual stress determination using hole drilling and 3D image correlation, Experimental Mechanics 46 (1) (2006) 31-38.

[17] M. J. McGinnis, S. Pessiki, H. Turker, Application of three-dimensional digital image correlation to the core-drilling method, Experimental Mechanics 45 (4) (2005) 359-367.

[18] J. Zhu, H. Xie, Z. Hu, P. Chen, Q. Zhang, Residual stress in thermal spray coatings measured by curvature based on 3D digital image correlation technique, Surface and Coatings Technology 206 (2011) 1396-1402.

[19] J. D. Lord, D. Penn, P. Whitehead, The application of digital image correlation for measuring residual stress by incremental hole drilling, Applied Mechanics and Materials 13-14 (2008) 65-73.

[20] A. Baldi, F. Bertolino, A low-cost residual stress measuring instrument, Residual Stress, Thermomechanics & Infrared Imaging, Hybrid Techniques and Inverse Problems 9 (2017) 113-119.

[21] T. Rief, J. Hausmann, N. Motsch, Development of a New Method for Residual Stress Analysis on Fiber Reinforced Plastics with Use of Digital Image Correlation, Key Engineering Materials 742 (2017).

[22] M. Hagara et al., Analysis of the aspects of residual stresses quantification performed by 3D DIC combined with standardized hole-drilling method, Measurement 137 (2019) 238-256.

[23] R. Huňady, M. Hagara, M. Schrötter, Impact assessment of calibration parameters on accuracy method of digital image correlation, Acta Mechanica Slovaca 16(2) (2012) 6-12.

EAN 2019
57th conference on experimental stress analysis

Optical analysis of low-floor tram drive motions

HAVLÍČEK Martin [1,a], DUB Martin [1,b] and LOPOT František [1,c]

[1] CTU in Prague, Faculty of Mechanical Engineering, Department of Designing and Machine Components, Technická 4, 166 07 Prague 6, Czech Republic

[a]martin.havlicek@fs.cvut.cz, [b]martin.dub@fs.cvut.cz, [c]frantisek.lopot@fs.cvut.cz

Keywords: Shift, Rotation, Gearbox, Traction electromotor, Optical analysis, Qualisys

Abstract: This paper is focused on special method for observation relative motions of bodies in space thanks to the optical analysis. In this case, relative motions are measured between the gearbox and the traction electromotor of low-floor tram. These components are marked by special markers. During the drive regime markers create trajectories, which are captured by cameras. All data is processed by system Qualisys. Each component gets own cartesian coordinate system and from them were found relative shifts and rotations between gearbox and traction electromotor.

Introduction

Trams are one of the most popular means of public transport, which are used worldwide. Due to this fact, specific needs are required. The main requirement is reliability. It means that the tram can work without any failures as long as possible. It is very complex to design tram traction electromotor because it challenges a lot of different types of loads. One of them are dynamic loads, which occur due to acceleration, deceleration, quality of the rail track and curves. These loads have a big impact on wearing of drive unit´s parts, e.g. gears, bearings etc.

The goal of this experimental measurement is to find out the relative motion between the traction electromotor and the gearbox. In case of a partly-sprung drive the traction electromotor is fully-sprung mass and gearbox is a partly-sprung mass due to its connection with the wheelset. These movements of a drive unit were measured by an optical cameras connected to system Qualisys.

Instrumentation

This experiment was realised by Qualisys. Qualisys is an optical system which can capture and analyse motions of bodies in space. This system was used to capture relative motion of the gearbox to the chassis and to the motor. To capture these data, at least three optical cameras are required. In this case, six cameras were used, divided into two groups. The first three of them were stationary and settled in the pit in the ground. The second group is composed of three optical cameras, which were bolted by special holders to the bottom of the tram. Both groups were calibrated and synchronized. To create the reference plane and cartesian coordinate system (see Fig.1) on objects which were monitored, three special spherical markers were put on the gearbox and three of them were put on asynchronous motor (see Fig.2). One spherical marker was attached to rotating shaft to determine speed of tram and synchronize time data.

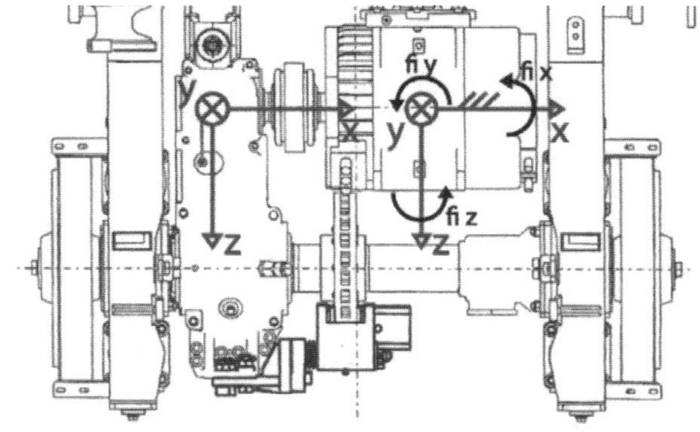

Fig. 1: Cartesian coordinate system of drive unit Fig. 2: Positions of markers on objects

Measurement

This measurement took place at testing loop in Prague. Three main drive regimes were measured: maximal acceleration, maximal deceleration and normal drive. A lot of data was obtained because of these three regimes. After these measurements, all the data were processed by Qualisys software and from this data can be found relative shift of asynchronous motor to the gearbox in all axis. Relative rotation of the gearbox to the asynchronous motor round all axis were obtained too. This paper focuses closely on maximal acceleration regime.

Maximal acceleration regime

One of the regime that was measured, was maximal acceleration. This means that the tram accelerated from zero speed as much as possible. In Fig.3 there is shown torque of the traction electromotor during acceleration. It can be seen that the tram started to accelerate from 2nd second with approximate torque 1000 Nm. This acceleration lasted till 10th second, at this time the tram started to break.

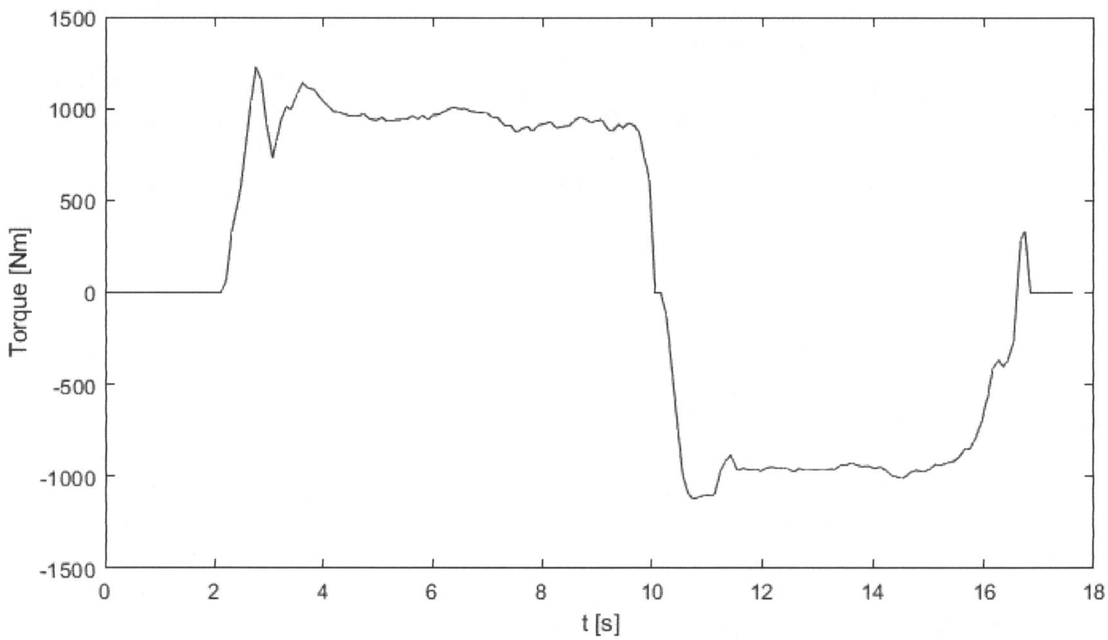

Fig. 1: Torque of traction electromotor during acceleration

In Fig. 4 there is shown relative shift in x-axis. From 2nd second it is seen that shift between the gearbox and the traction electromotor increased. From 10th second, the tram started to break and distance between bodies decreased. The values of relative shifts stayed in tolerance ±1 mm in comparison to zero speed.

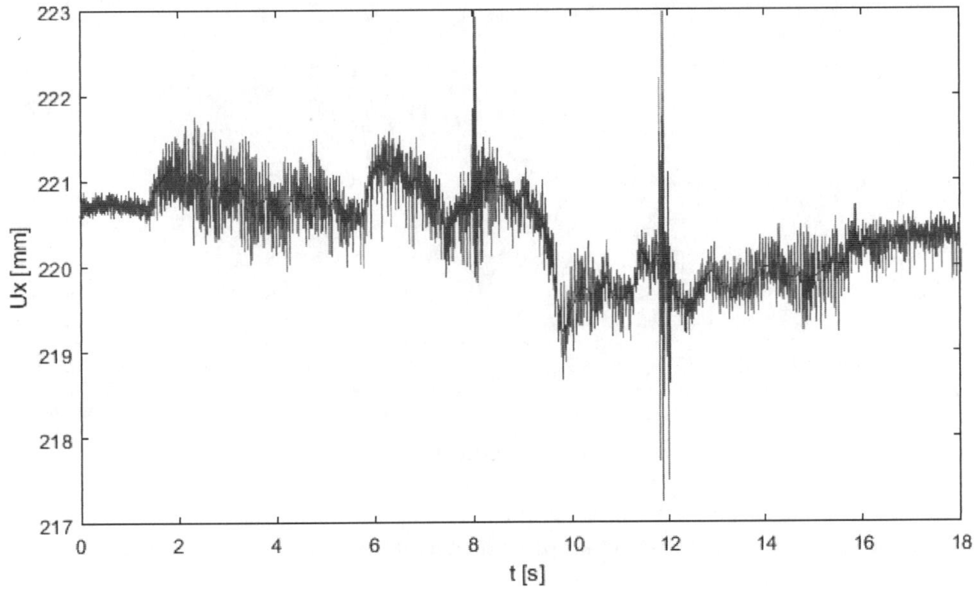

Fig. 2: Relative shift in x-axis

In Fig. 5 there is shown relative shift in y-axis. From 2nd second it is seen that shift between the gearbox and the traction electromotor increased. It is shown that this shift is affected by some oscillations. These oscillations increased during the acceleration and decreased during the breaking. These oscillations are probably caused by suspension of bodies. Traction electromotor is connected to bogie frame and gearbox is embedded to wheelset and its moment's support is catched by hinge connected to bogie frame. Gearbox is a partly-sprung mass and traction electromotor is fully-sprung mass. Total deflection of bodies was lower than ± 1,5 mm in comparison to zero speed.

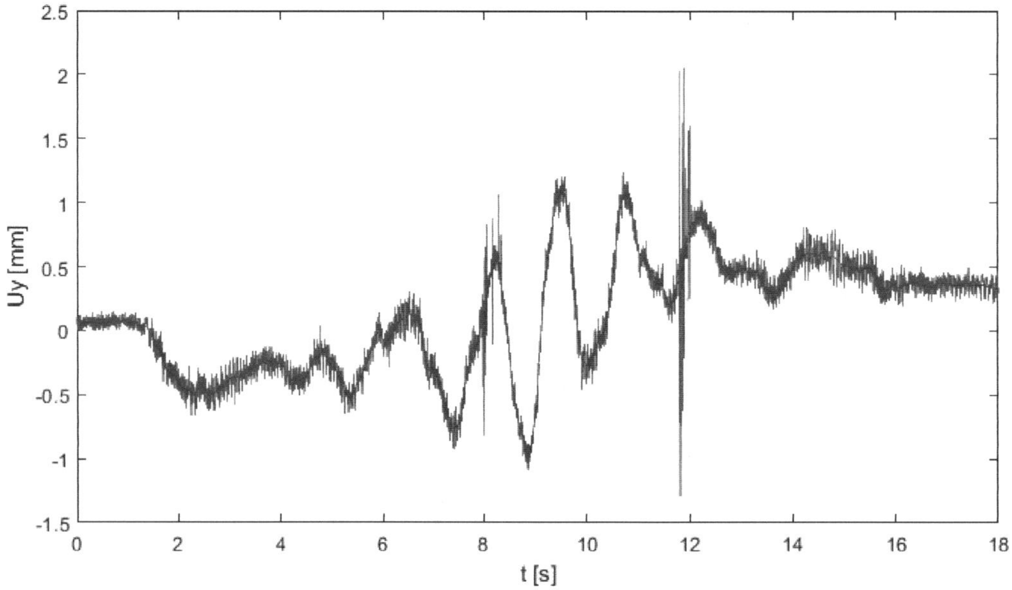

Fig. 3: Relative shift in y-axis

In Fig. 6 there is shown relative shift in z-axis. From 2nd second it is seen that shift between the gearbox and the traction electromotor increased. Deflection of the bodies was lower than ± 1 mm in comparison to zero speed.

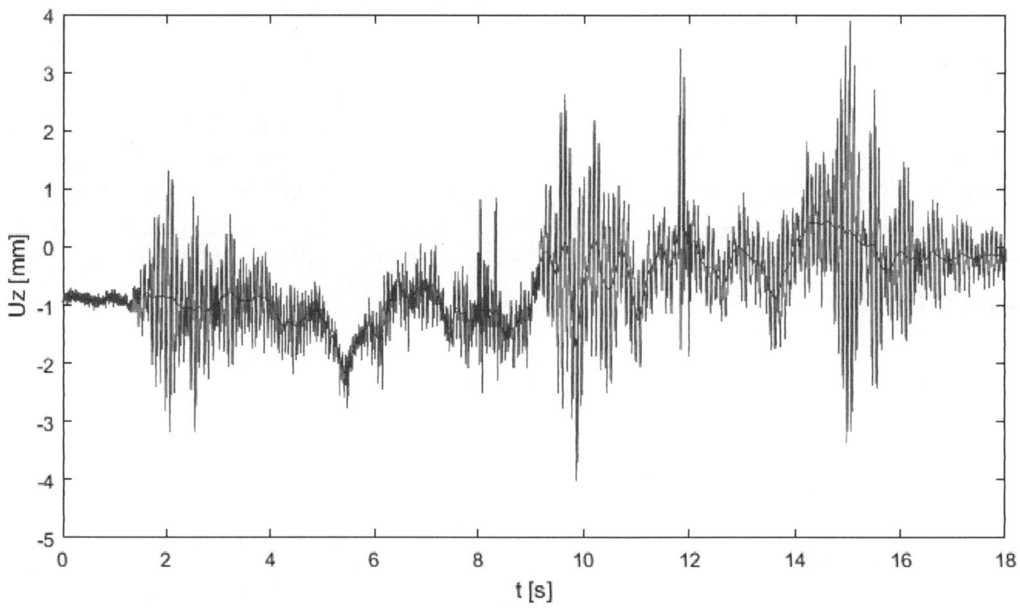

Fig. 4: Relative shift in z-axis

In Fig. 7, 8, 9 there are shown relative rotations round the x, y and z-axis. All these relative rotation between gearbox and traction electromotor have maximal deflection ±10° in comparison to zero speed.

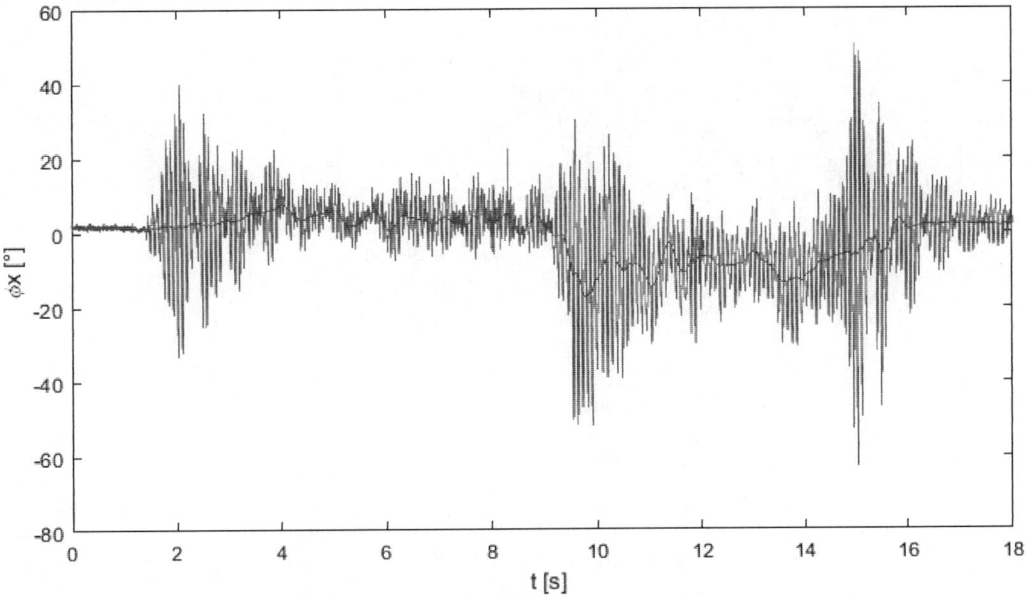

Fig. 7: Relative rotation round the x-axis

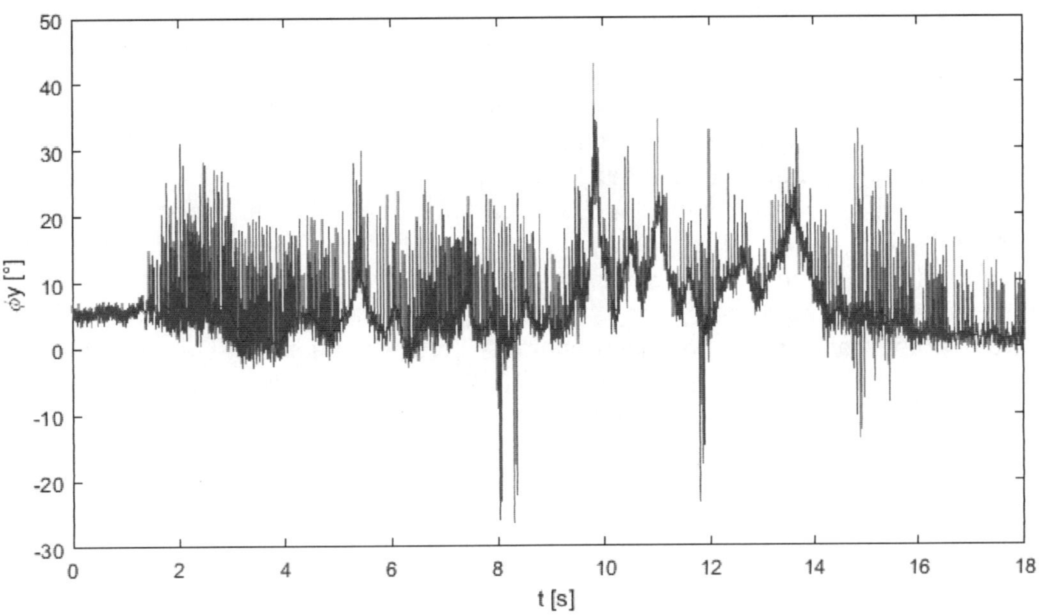

Fig. 8: Relative rotation round the y-axis

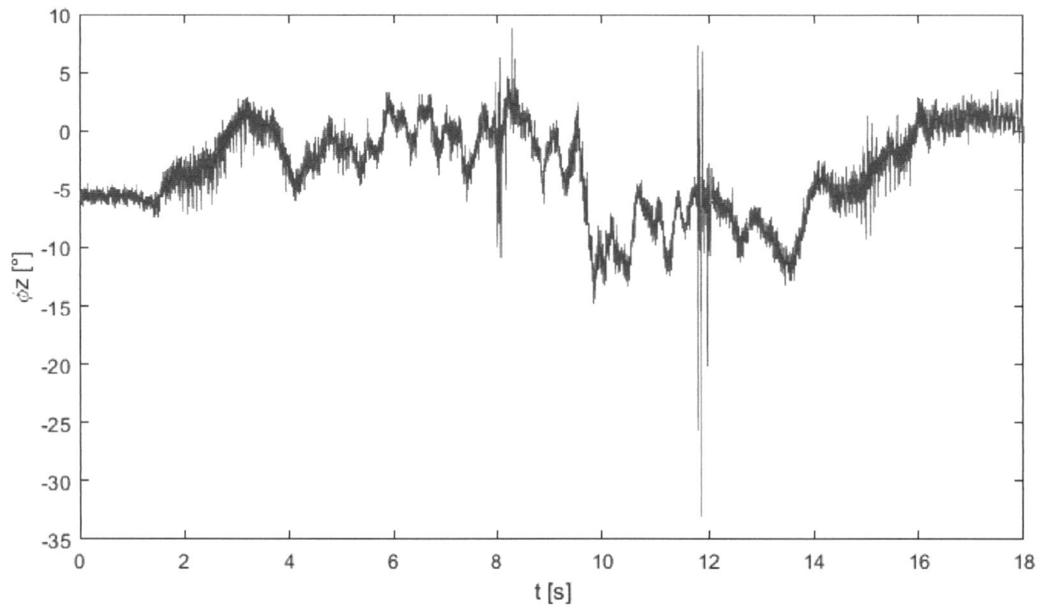

Fig. 9: Relative rotation round the z-axis

Conclusions

Thanks to this optical analysis there were analysed relative shifts of the traction electromotor to the gearbox in x-axis, y-axis, z-axis and relative rotations around x, y and z-axis. All results of the relative shifts had the maximal deflection ±1,5 mm in comparison to zero speed. Thanks to this analysis, there was found that the maximal deflections of the relative rotations are ±10° round all axis. This deflection is probably caused by different suspension of bodies. Traction electromotor is connected to the bogie frame and the gearbox is embedded to the wheelset and its moment's support is catched by hinge connected to bogie frame. It means that the gearbox is a partly-sprung mass and traction electromotor is a fully-sprung mass. The claw coupling that transmits torque from the traction electromotor to the gearbox is designed for relative rotation around 5° round y and z-axis. From these facts a hypothesis can be made. The relative rotation of the gearbox to the motor was higher than the allowed rotation of claw coupling and the coupling could not withstand this motion. In addition to this results were suggested two solutions which could help to remove this problem. Whereas the traction electromotor is fully connected to the bogie frame, the effort was concentrated to the connection of the gearbox. Gearbox is connected to the wheelset and moment's support is catched by hinge. This hinge is imposed in silentblock and the rigidity of this silentblock can be influenced by compression of hinge. If the rigidity of silent is higher, the motions of gearbox could decrease. Second solution is to use a different type of coupling.

References

[1] M. Dub, J. Kolář, F. Lopot, V. Dynybyl, O. Berka, Experimental Analysis of Stress State and Motion of Tram Gearbox Hinge, in: The 52nd International Conference on Experimental Stress Analysis, 2014.

[2] J. Kolář, F. Lopot, M. Dub, Vyhodnocení měření namáhání závěsky u nápravové převodovky tramvaje 14T, Technická zpráva ČVUT FS - U 12 120 - Z 14 - 01, 2014

EAN 2019
57th conference on experimental stress analysis

Determining coefficient of friction between oilseeds and rotor blade of hulling machine using equation of motion

HOIDEKR Jan[1,a], MACHAČ Martin[1,b] and TUREK Oskar[1,c]

[1]CTU in Prague, Faculty of Mechanical Engineering, Department of Designing and Machine Components, Technická 4, 166 07 Prague 6, Czech Republic

[a]jan.hoidekr@fs.cvut.cz, [b]martin.machac@fs.cvut.cz, [c]oskar.turek@fs.cvut.cz

Keywords: Oilseed hulling, Coefficient of Friction, Equation of Motion, Hulling Machine

Abstract. This work aims to describe a method of experimental determination of a friction coefficient between oilseeds, especially sunflower seeds, and a rotor blade placed in an impact hulling machine. Firstly, the movement of sunflower seeds in rotor during hulling process is described by equation of motion. Secondly, by analysing this description and simple experimental measuring the coefficient of friction is determined.

Introduction

The hulling of oilseeds, especially sunflower seeds, is a necessary part of the oil pressing technology. The sunflower seed consists of a kernel where the oil is contained and a hull mostly composed of cellulose. It is advisable to remove hulls before pressing, even though few hulls retained might be still necessary for the proper function of the pressing process. For hulling seeds, there are mostly used the impact hulling machines.

The impact hulling machine is based on cracking the hull by high-speed throwing of seeds against an obstacle. When the seed hits the obstacle, its hull cracks and separates from the kernel. The seeds are usually accelerated by a centrifugal force in a rotating rotor with blades and thrown against outer stator which can be smooth or specially shaped.

The displacement of the seed inside the hulling machine consists of two phases. In the first phase, the seed is accelerated in the rotor. This displacement is mathematically described by equation of motion. In the second phase, the seed leaves the rotor and is thrown against the stator. The direction of trajectory of thrown seed is known thanks to mathematical description of the first phase. Experimental measuring of this direction leads in the end to determining of friction coefficient between the oil seed and the rotor blade.

Determining an equation of motion

The equation of motion is stated for the first phase of hulling. There can be considered two coordinate systems in the rotor. The global coordinate system XY is fixed to the frame of the machine. There is an additional local Cartesian coordinate system xy which is fixed to the rotor blade. The kinematic diagram of moving seed in the rotor can be though as a point slidably set on a rotating rod with constant angular velocity ω. Absolute velocity of the seed v can be broken into components v_u and v_r. This is shown in Fig. 1 as well as the directions of acceleration vectors. Considered force diagram is shown in Fig. 2.

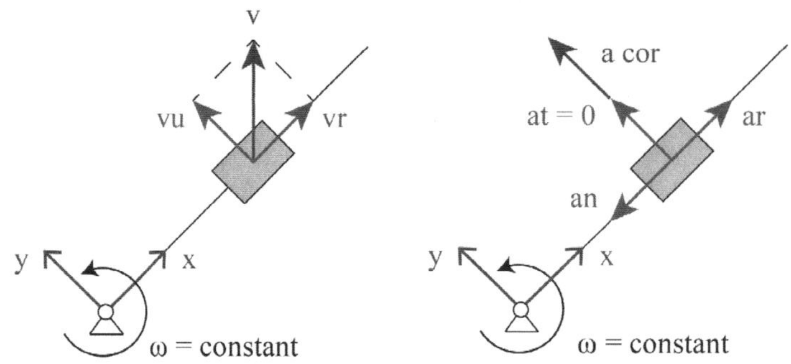

Fig. 1: Kinematic diagram of the sunflower seed displacement in the rotor

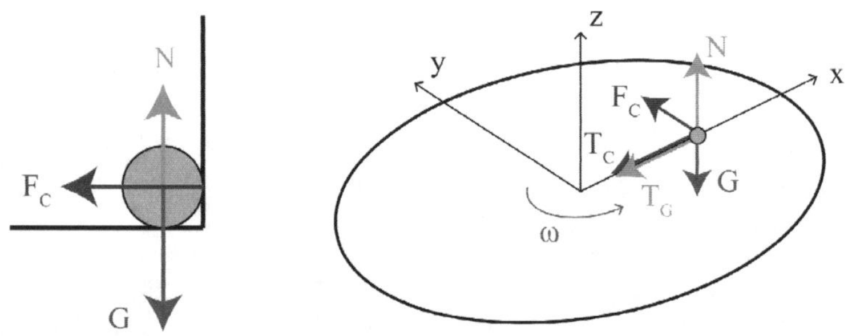

Fig. 2: Force diagram during displacement in the rotor

According to the Newton's second law of motion the equation Eq. (1) of motion is stated and expressed in local coordinate system xy [1].

$$\ddot{x} + (2\omega\mu) \cdot \dot{x} - (\omega^2) \cdot x = -g\mu \qquad x_{(t=0)} = x_0,\ \dot{x}_{(t=0)} = 0 \tag{1}$$

It is a second-order differential nonhomogeneous equation with constant coefficients. Solving this equation analytically out gives a seed trajectory description in the rotor as a function of a time

$$x_{(t)} = C_1 \cdot e^{\lambda_1 \cdot t} + C_2 \cdot e^{\lambda_2 \cdot t} + C_3 \tag{2}$$

where λ_1, λ_2, C_1, C_2 and C_3 are constants containing an initial conditions, machine dimensions, friction coefficient and angular speed of the rotor. The trajectory of the seed in the rotor in global coordinate system XY is shown in Fig. 3. There is a tangent of angle β defined as v_u divided by v_r which is generally dependent on time. For analyzing dependence of angle β_{out} on angular speed ω and friction coefficient μ there is a need to solve Eq. (3)

$$C_1 \cdot e^{\lambda_1 \cdot t} + C_2 \cdot e^{\lambda_2 \cdot t} + C_3 - x_{out} = 0 \tag{3}$$

where x_{out} equals to x coordinate of outer diameter of the rotor. Solving this equation numerically for certain friction coefficient and angular speed in MATLAB gives a t_{out} time of moment when the seed leaves the rotor blade [2]. Afterwards for the range of defined angular velocities and friction coefficients a Table 1 can be calculated. The region of interest for the speed of the rotor is between 1000 and 2000 rpm, for the friction coefficient it is between 0 and 1.

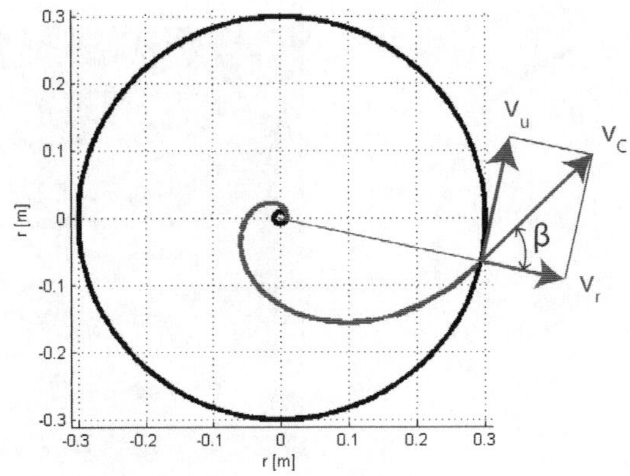

Fig. 3: Trajectory of the seed displacement in the rotor during the hulling process

Table 1: Dependence of angle β_{out} on the rotor speed and friction coefficient

coef. of friction [1]	angle β_{out} 1000 rpm [deg]	angle β_{out} 1100 rpm [deg]	angle β_{out} 1200 rpm [deg]	...	angle β_{out} 1900 rpm [deg]	angle β_{out} 2000 rpm [deg]
0.0	45.1793	45.1791	45.1790	...	45.1785	45.1784
0.1	47.9794	47.9794	47.9793	...	47.9792	47.9792
0.2	50.7328	50.7327	50.7327	...	50.7327	50.7327
0.3	53.3932	53.3932	59.3932	...	59.3933	59.3933
0.4	55.9225	55.9225	55.9224	...	55.9223	55.9223
0.5	58.2923	58.2922	58.2922	...	58.2921	58.2921
0.6	60.4859	60.4858	60.4858	...	60.4856	60.4856
0.7	62.4977	62.4976	62.4975	...	62.4973	62.4973
0.8	64.3307	64.3306	64.3305	...	64.3303	64.3303
0.9	65.9942	65.9941	65.9940	...	65.9938	65.9938
1.0	67.5006	67.5004	67.5003	...	67.5001	67.5001

Determining the coefficient of friction

According to the Table 1 it is obvious that for the region 1000 – 2000 rpm of speed of the rotor there are only small changes of angle β_{out}. Therefore it is acceptable to consider dependence of angle β_{out} on angular speed omitted for this region of velocities of the rotor. Considering this assumption the angle β_{out} is a function of only one variable, namely friction coefficient.

Experimental measurement of angle β_{out} then provides the friction coefficient between the sunflower seed and the rotor blade. The angle β_{out} is measured during the second phase of hulling process when the seed is thrown against the stator. In this phase the seeds continue in a straight line until they reach any obstacle, in this case until they hit the stator construction. In the case of stator construction according to Fig. 4 and considering that seeds are throwing equally all around the stator circumference, the bounced seeds will leave a footprint on the stator blade in affected green zone.

The angle β_{out} can be measured by optical methods from stator blades considering the real stator and rotor construction dimensions.

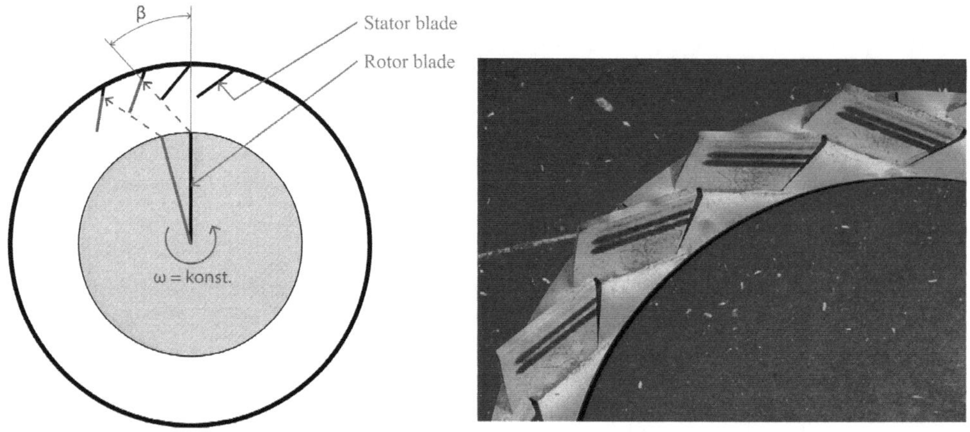

Fig. 4: Experimental determining of the friction coefficient by measuring the angle β_{out}

There was done one testing measurement of a specimen of sunflower seed with a moisture content of 5%. The material of rotor blade was hardened steel. Measured angle β_{out} was 59.5° what equals to friction coefficient of 0.55. This result is corresponding to range of expected values. The friction coefficient was measured by described method, however, there should be used another method in the next work to validate this result.

Conclusion

The description of displacement of the oilseed during the hulling process leads to stating a convenient method of determining friction coefficient between oilseed and rotor blade of hulling machine while. The advantage of this method is based on using directly industrial hulling machine for determining friction coefficient of certain specimen. In general, the coefficient of friction differs with a species of oil seed, but can also vary in many parameters, such a humidity or temperature of the specimen.

The knowledge of friction characteristics helps to construction designers or machine operators in local adjustments of hulling machine to reach maximal hulling rate since the parameters of seeds to hull are varying.

Acknowledgements

There is much appreciated sponsorship of grant TA04010992: Research and development of technologies for obtaining vegetable oils and cakes with an emphasis on quality of cakes as feed and the appropriateness of used of the construction materials. There is also much appreciated cooperation with Farmet a.s. where the experimental measurement was made.

References

[1] J. Hoidekr, J. Andrus, and V. Dynybyl, "The Mathematical Description of Sunflower Seed's Hulling Process Using Impact Huller", in The Latest Methods of Construction Design, 1st ed., Cham: Springer International Publishing, 2016, pp. 203-206.

[2] J. Hoidekr, "Analýza loupání slunečnicových semen", in Studentská tvůrčí činnost 2014, 2014, 1st ed.

Non-destructive analysis of concrete from Northwest Bohemia defence structure built in 1937

HOLČAPEK Ondřej[1,a], REITERMAN Pavel[1,b] and KOŤÁTKOVÁ Jaroslava[1,c]

[1]IExperimental Centre, Faculty of Civil Engineering, Czech Technical University in Prague; Thákurova 7, 169 00 Praha 6, Czech Republic,

[a]ondrej.holcapek@fsv.cvut.cz, [b]pavel.reiterman@fsv.cvut.cz, [c]jaroslava.kotatkova@fsv.cvut.cz

Keywords: Reinforced concrete, Compressive strength, Defence structure, Schmidt hammer, Historical structure.

Abstract. This contribution deals with the experimental investigation of compressive strength of over 80 years old reinforced concrete structure. In various regions of the Czech Republic we still have a large system of defence structures built in 1930th as the reaction for political development in Germany and the potential risk of war conflict. In the article the results of non-destructive analysis of concrete applied to a defence structure in northwest Bohemian region are presented. In 1937, the compressive strength measured destructively on test samples achieved 41.5 MPa. The non-destructive analysis was performed mainly in the interior part – roof and walls and also in one location in the exterior. The compressive strength measured by Schmidt hammer achieved 68.5 MPa in the exterior part, 53.2 MPa for roof in the interior part and the wall in the interior achieved between 49.4 MPa and 58.5 MPa. Performed non-destructive testing provides the approximate information about the quality of concrete without damaging the historical structure.

Introduction

In the conditions of political situation and development in Europe and especially in Germany during the 3rd decade of last century several defence military systems and structures were built in various regions across countries. The former Czechoslovakia government, in line with the French experts´ knowledge and experiences, decided to build an extensive system of defence reinforced concrete bunkers. The main burden of defence was based on the system of lightweight reinforced concrete bunkers called "ŘOP no.37" designed for 3 to 6 soldiers. The number 37 means year 1937 when these bunker were approved for widespread construction. The defence lines were planned and build especially in borders area with Germany. After the occupation of Austria this system was extended to the southern border. Over the more than nine thousands of bunkers were built and about over four thousands have been preserved to nowadays. Lastly, there is a growing interest in the preservation and further development of these by many persons, communities or companies.

Description of the analysed bunker. The defence structure analysed in this contribution is the type no. 37 with full designation C27/52a/A-140Z, see the photography on Fig. 1. "C27" means the line section of the defence system in Czechoslovakia, in this case Lišany, while "52a" is the number of bunker in line. The letter "A" means the type of bunker when this is the most widespread type of light defence structure. And the number "140" indicates the angle of shots in degrees (the used degree of shots varies between 120° and 220°) [2]. This type was originally designed for the side firing of two machine guns on right-hand and left-hand side. The crew consisted of seven soldiers. Investigated bunker is situated on the right bank of the river Ohře in cadastral area of Mradice. Fig. 2 shows the horizontal cross-section of the described bunker, while the Fig. 2 shows the photography of the bunker in-situ. The entrance stairs are covered by thin concrete wall around (marked with thin line). The photography on Fig. 2 also shows this technical solution. The reason for this precaution is to avoid the effect of floods from the river Ohře. The entrance is also protected by a machine gun. The fumes after firing were evacuated by hand operating fan.

The letter "Z" at the end of type designation means, that this specific bunker is "strengthened" to have higher ballistic resistance. The front wall has a thickness of 1200 mm, while the ordinary, non-strengthened bunkers have front walls with thickness of 800 mm.

Fig. 1: Photography of analysed bunker no 37 (C27/52a/A-140Z).

The front part, exposed to expected enemy´s attack, was protected by hand-placed rubble that also worked as camouflage. The rest of this wall can be seen on the left side of bunker on Fig. 1 and it is situated in a place of a growing tree.

It is necessary to mention that the bunkers of no. 37 were not intended for the crew´s long stay. The absence of any facilities confirms this statement. The soldiers used to stay inside only for the necessary time during the expected action of enemy. Within the fortification system built before the Second World War, we can also find larger better equipped and fortified bunkers or artillery fortress with higher ballistic resistance and designed for long stay of soldiers.

Experimental investigation. The quality of concrete from these historical bunkers means the main topic of this contribution. The non-destructive testing by Schmidt hammer [1] was used, with respect to the historical value of investigated structure. Fig. 1 shows the location of measurements performed on vertical structures – walls. There were five places in the interior part (marked as "I") and one location in exterior part (marked as "E").

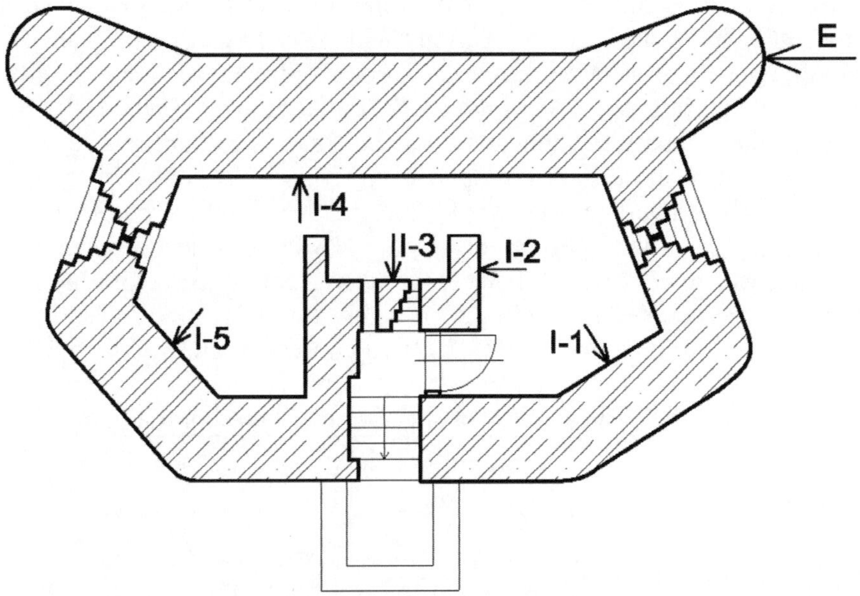

Fig. 2: A cross-section of bunker no 37 (C27/52a/A-140Z) with location of measurements.

Fig. 3: Different quality of concrete.

Fig. 4: Remains of the original plaster.

The properties of concrete achieve different characteristics while investigated in various locations – for example the denser concrete can be find in the lower part of walls (see Fig. 3). Therefore the measurement took place on three locations – in the top of the wall, in the centre and in the contact with the floor. Totally 45 measurements were performed on each interior locations. From horizontal structures the quality of roof from interior part was also investigated – totally over 50 measurements were performed. The exterior part has specific issue that limits using of Schmidt hammer. After the casting and unmoulding of formwork a slim layers of coloured plaster were additionally applied on the massive reinforced concrete structure. These plasters worked as a camouflage shield and in large number of bunkers have been preserved. In investigated bunker, remains of original plasters were identified in the contact of walls and earth (see Fig. 4). From this reason it is not possible to performed non-destructive testing by Schmidt´s hammer without removing original plasters. From this reason only one place from exterior side was analysed (marked as "E" on Fig. 1) where the original plasters has not preserved.

Results. From the visual point of view the concrete showed different qualities especially due to the used way of compaction. In 1930´s especially the manual compaction was used. The roof showed homogenous surface without caverns or similar imperfections. The surface layer of walls achieved various qualities. Most of the places showed homogenous surface of high quality. On the other hand there were locations with insufficiently compacted concrete, in extreme case with uncovered steel reinforcement. The significant differences in the quality of concrete were confirmed in study performed by Pazderka et. al. [3] where the bunker R-S74 was studied and the laboratory analysis showed that the bunker's concrete is very inhomogeneous and it has very different properties in terms of its permeability.

The table 1 summarizes average values of performed Schmidt´s hammer measurements including the standard deviation. During solving the NAKI II project an original report from strength test performed in 1937 by Klokner Institute was found, therefore the information of original strength was available. The strength in 1937 was calculated based on compressive test of cube specimens with dimensions 200×200×200 mm and achieved 41.5 MPa.

Table 1: Results of measurements performed with Schmidt hammer.

Location	Schmidt [MPa]	Standard deviation [MPa]
Exterior – E	68.5	± 2.7
Interior – I-1	57.8	± 5.2
Interior – I-2	52.9	± 5.4
Interior – I-3	58.4	± 5.3
Interior – I-4	58.5	± 4.8
Interior – I-5	49.4	± 4.7
Interior – roof	53.2	± 4.4
Strength calculated in 1937	41.5	N/A

Conclusions

The strength investigated by Schmidt hammer oscillates between 49.4 MPa and 68.5 MPa. The highest strength characterizes the exterior part, while the interior part reaches balanced values. The Schmidt´s hammer provided the tentative compressive strength of concrete, with carbonation consideration especially for the older concrete, which is the case under study. This process can cause the increase of strength in the surface layer. The concrete showed different quality especially caused by the used process of compaction.

Acknowledgement

This research work was financially supported by project of Ministry of Culture Czech Republic over project NAKI II – DG18P02OVV063.

References

[1] CSN EN 13 791: Assessment of in-situ compressive strength in structures and precast concrete components, 2007

[2] www.ropiky.net

[3] Pazderka J., Purkrtová M, Reiterman P., Moisture-related problems of historic concrete structures, Materials Science Forum Vol. 865, pp. 219-223, 2016, doi: 10.4028/www.scientific.net/MSF.865.219.

EAN 2019
57th conference on experimental stress analysis

Comparison of fatigue life of 2024 and 7475 friction welded joints with kissing bond defects

HOMOLA Petr[1,a], RŮŽEK Roman[1,b], KADLEC Martin[1,c] and HRON Robin[1,d]

[1] VZLU - Czech Aerospace Research Centre, Strength of Structure Department; Beranových 130, 199 05 Praha - Letňany, Czech Republic

[a]homola@vzlu.cz, [b]ruzek@vzlu.cz, [c]kadlec@vzlu.cz, [d]hron@vzlu.cz

Keywords: friction stir welding, fatigue, kissing bond, aluminium alloy

Abstract. The paper analyses and compares the static and fatigue properties of 7475 T7351 and 2024 T3 aluminum alloy plate materials with friction stir welded (FSW) joints in order to determine a critical size of the kissing bond (KB) flaw that will not significantly decrease the weld performance. While the static tensile strength of the FSW decreases with KB increase, the yield strain in not influenced by the KB presence. The flaw affects the material structure locally, which leads to a decrease in the fatigue properties. The crack initiation area (KB or out of KB initiation) provides information about significant decrease of fatigue properties of FSW joint. Comparison of both material sensitivity to KB defects lead to conclusion that 7475 material is more sensitive to KB defect in comparison to the 2024 plate material.

Introduction

Friction stir welding (FSW) is a high-performance joining technique used primarily for aluminium alloys. FSW technology contributes to increased joining speed, higher stress tolerance, better reproducibility and longer service life when compared to common riveting methods. FSW is joining method widely used mainly in aerospace and automotive, where fatigue performance is particularly crucial [1]. As the temperature gradients and heat affected zone of FSW welds are lower in comparison with traditional welding techniques, mechanical properties of FSW joints are higher compared with traditional welded joint. However, wider use of FSW among industrial applications is limited due the ability to detect flaws of a relatively small geometry. This problem limits the adoption of FSW for joining for critical components.

The FSW manufacturing process is influenced by many parameters. All parameter controlling in required tolerances is not so easy. For example, an inappropriate rate of heating can result in the formation defects such as lack of penetration, lack of fusion, tunnels, voids, surface grooves, excessive flash, surface galling, nugget collapse and kissing bonds [2]. The paper is focused on a typical weld defect occurring in joints made with FSW - the kissing bond (KB) defect. A kissing bond is a specific type of solid-state bonding defect where two previously separated regions of the material are in contact with little or no metallic bond present. KB flaws can occur in the root of the weld; in this situation, the materials are in close contact but are not chemically or mechanically bonded [3].

KBs can reduce the fatigue performance of joints and they are currently highly difficult to detect using existing non-destructive technique (NDT) methods. In recent years, several investigations have been conducted on the fatigue properties of friction stir welded joints at VZLU. Comparison of KB flaw geometry on reduction of the fatigue life of two typical aluminium materials used in aerospace – 2024 and 7475 is the main aim of this paper. Individual characterisation of 2024 and 7475 FSW joint is discussed in detail in [4,5].

Material and Test methods

Fatigue and static behaviour of two aluminium alloys (AA) – 2024 T3 and 7474 T7351 with or without KB were investigated. Plates with thickness of 6.35 mm were used. Several plates with different KB flaw sizes and a flaw free FSW joint were produced by TWI. Fatigue test and static tension specimens were machined from the same plates.

The test specimens were machined from the plates to the dimensions shown in Fig. 1a. The specimen design for static tension behaviour evaluation described in the AWS standard [6] was adopted. In the tests, values of yield strength, tensile strength, and elongation were evaluated. The specimen design described in the EN 6072 standard [7] was adopted. Test specimens were manufactured according to the drawing in Fig. 1b. Wöhler curve was obtained for each plate with varying KB sizes.

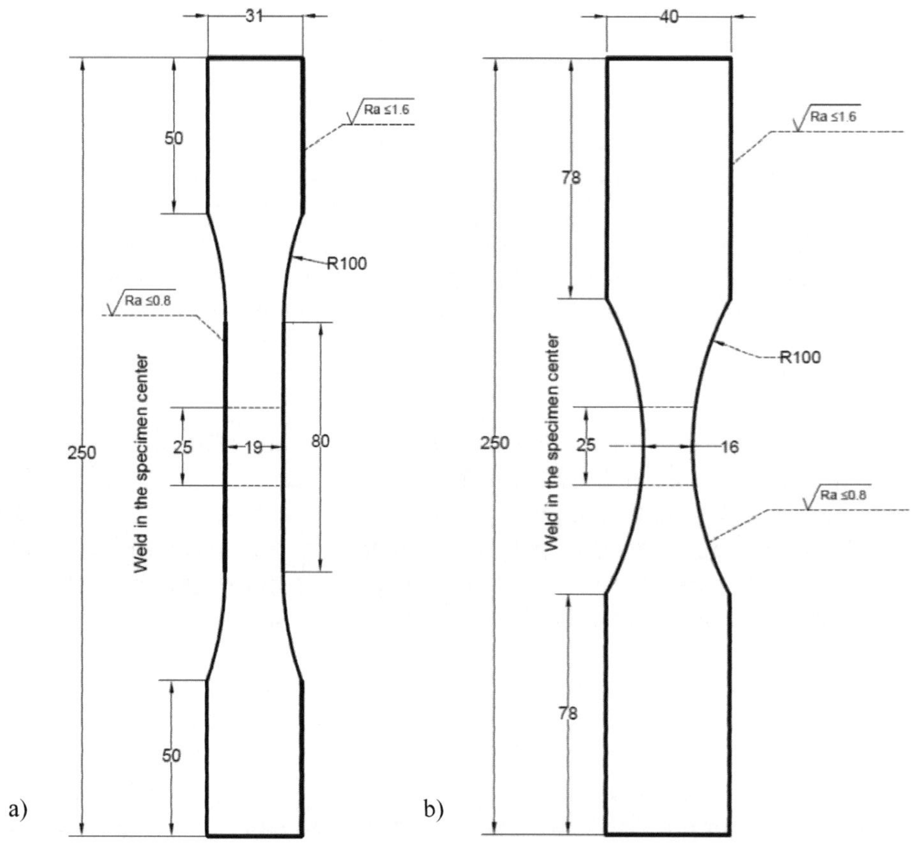

Fig. 1: Actual specimens' geometry based on the a) AWS standard [6] and b) EN standard [7]

Static tensile tests were performed on a mechanical test machine Instron 100 kN (Fig. 2 – left) and fatigue tests were performed on Hydraulic frame IST Hydropuls Sinus 100 kN (Fig. 2 – right). In the fatigue experiments using load control, data were generated using constant amplitude sinus loading with a stress ratio $R = \sigma_{min}/\sigma_{max}$ (max load/peak load) 0.1 and several different maximum stress levels σ_{max}. The test frequency range was between 5 Hz and 15 Hz. A fatigue curve was generated by performing a series of experiments at prescribed maximum stress levels in order to obtain the fatigue lives between 10^4 and 3.10^6 cycles.

Several scraps of each plate were cut in order to evaluate the weld microstructure, grains distribution and internal defects. The samples were prepared using the standard metallographic method. Microstructure of the samples was revealed by electrolytic Barker's etching method with subsequent optical microstructural contrasting in polarized light using Olympus GX51 light optical microscope.

Basic failure modes were identified by a visual examination which can provide a considerable amount of information about the failure process. The microfractographic analyses of the fracture surfaces were carried out using the TESCAN 3SBU Scanning Electron Microscope (SEM).

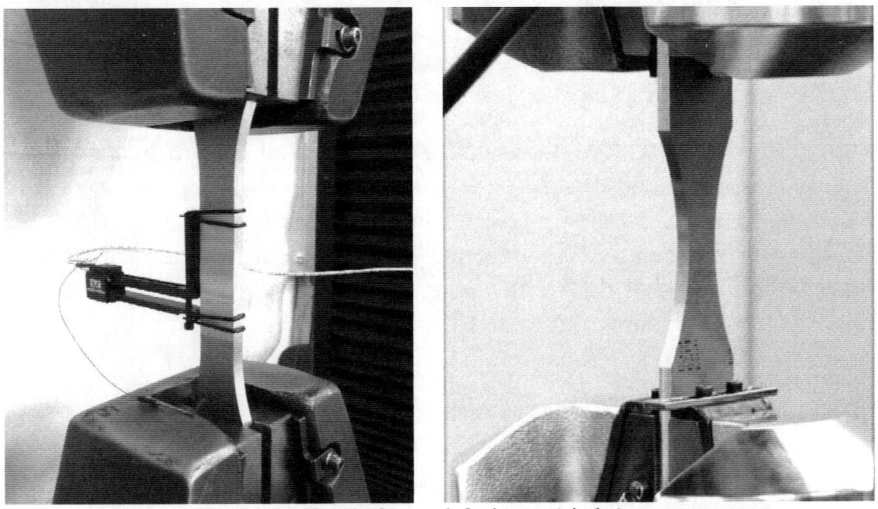

Fig. 2: Static tensile (left) and fatigue (right) test set-up

Static properties

The tensile strength (maximum stress), elongation (maximum achieved strain before fracture) and yield strength (offset stress point at 0.2% of the strain – $R_{p0.2}$) were evaluated for each specimen. Mean values of tensile strength and yield strength and respective sample standard deviations are compared for the both plate materials in the column graphs – Fig. 3 and Fig. 4.

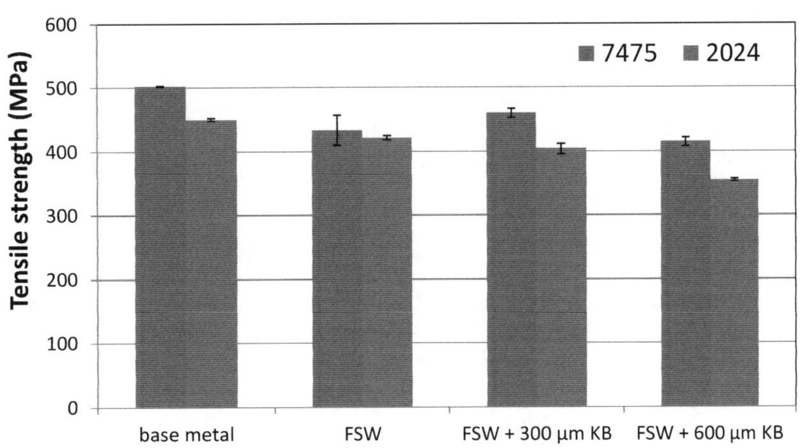

Fig. 3: Tensile static strength comparison of 2024 and 7475 materials

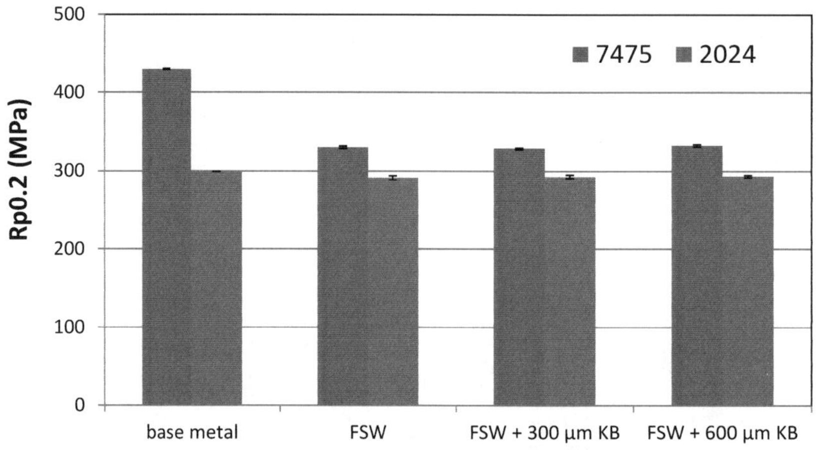

Fig. 4: Yield strength comparison of 2024 and 7475 materials

It can be stated that while the strength of the FSW decreases with KB depth increase, the Yield strain in not influenced by KB presence. Very slight differences of static strengths may be caused due to the fact that different batches of plates were used for the experiment. The batch-to-batch difference shows the sensitivity of properties on the weld performance conditions.

Fatigue properties

The values of fatigue life for individual specimens with selected load levels were obtained. The FSW specimens without and with KB with different sizes were compared. Comparison of evaluated data in a graphical form is shown in Fig.5. The maximal stress values σ_{MAX} presented in the graph are relative to the net cross section in the narrowest part of the specimen.

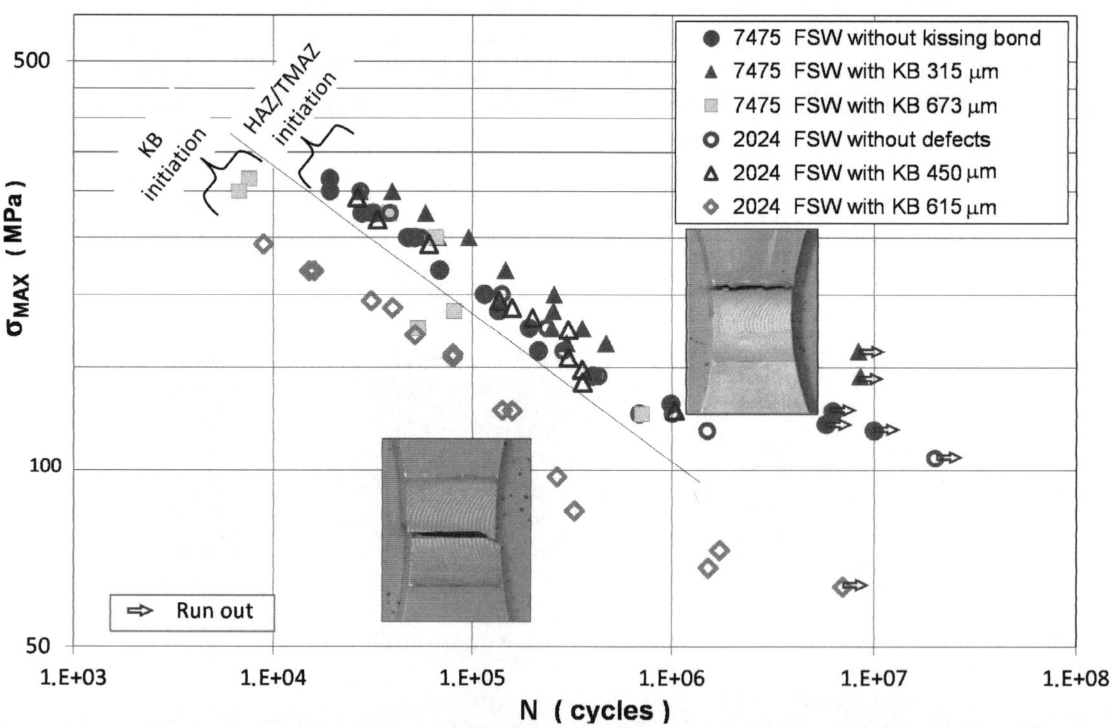

Fig. 5: Fatigue strength comparison of 2024 and 7475 materials

The full and empty circle points in Fig. 5 represent the reference results corresponding to the FSW of 2024 and 7475 material without any flaw, respectively. For all these specimens, the fracture initiation locations were detected in the transition area between the weld zone and parent material or out of weld initiation. Based on the statistical analyses it can be stated no difference between the fatigue behaviour of both materials was found.

The full and empty triangle points in Fig. 5 represent the FSW with KB flaw of average depth equal to 315 μm for 7475 and average depth equal to 450 μm for 2024 material. Similarly to flaw free welds all the specimens with "small" KB defects (<315 μm resp. <450 μm), no difference between the fatigue behaviour of both materials was found. No decreasing of fatigue properties were detected. By contrast, the fatigue lives of 7475 material with KB flaws of 315 μm were higher both for higher load levels and fatigue limit compared with flaw free FSW specimens. This fact can be caused by the fact that different batches of plates were used for the experiment. The possible reason for this behaviour could be different heat treatment response. All the fracture initiation locations were detected in the transition area between the weld zone and parent material. Based on the statistical analyses it can be stated the fatigue behaviour of specimens with the "small" KB flaws is not affected by presence of the flaw.

The third group of results generated the specimens with "large" size of KB flaws. It represents FSW with the 673 μm KB in case of 7475 material and FSW with the 615 μm KB in case of 2024 material. In case of 7475 material the results with a large data scatter in fatigue lives were obtained. Therefore, fatigue specimens manufactured from the third plate were categorised into the two groups based on the location of initiated fatigue cracks in the specimens. The results with crack initiation in KB are separated in Fig. 5 by the sloping thin black line. The 7475 specimens with cracks initiation in the KB had significantly lower fatigue lives decreased by one order of magnitude in comparison with the 7475 defect-free weld or with specimens with "small" KB flaws. The other group of 7475 "large" KB flaw specimens with initiation in in the transition zone had fatigue lives similar to pure FSW. By contrast the all fatigue cracks of 2024 specimens with "large" KB flaws of 673 μm initiated in KB defect and at the same time the fatigue lives were significantly (by one order of magnitude) shorter in comparison with the defect-free weld or with specimens with "small" KB flaws.

The results and discussion above indicate a higher sensitivity to KB defect of 2024 material as compared with 7475 material.

Metallographic and Fractography analyses

The crack initiation mechanism from the KB flaw notches is very similar for both materials. Fatigue cracks that initiated from a kissing bond are visible in Figure 6. The figure shows a crack initiation from the KB notch which was observed in most samples. The fatigue crack initiation locations were dependent on specimen widths and crack lengths. This was different in every specimen, which have resulted in the large scatter in fatigue lives graph for the large KB series. Fig. 7 shows an example of fatigue crack formation observed by scanning electron microscopy in 2024 FSW joint. A multiple crack initiation was usually observed with typical ratchet marks initiated at the kissing bond. The cracks interconnected into one major crack that propagated by the striation mechanism until a final fracture (typical ductile dimples observed).

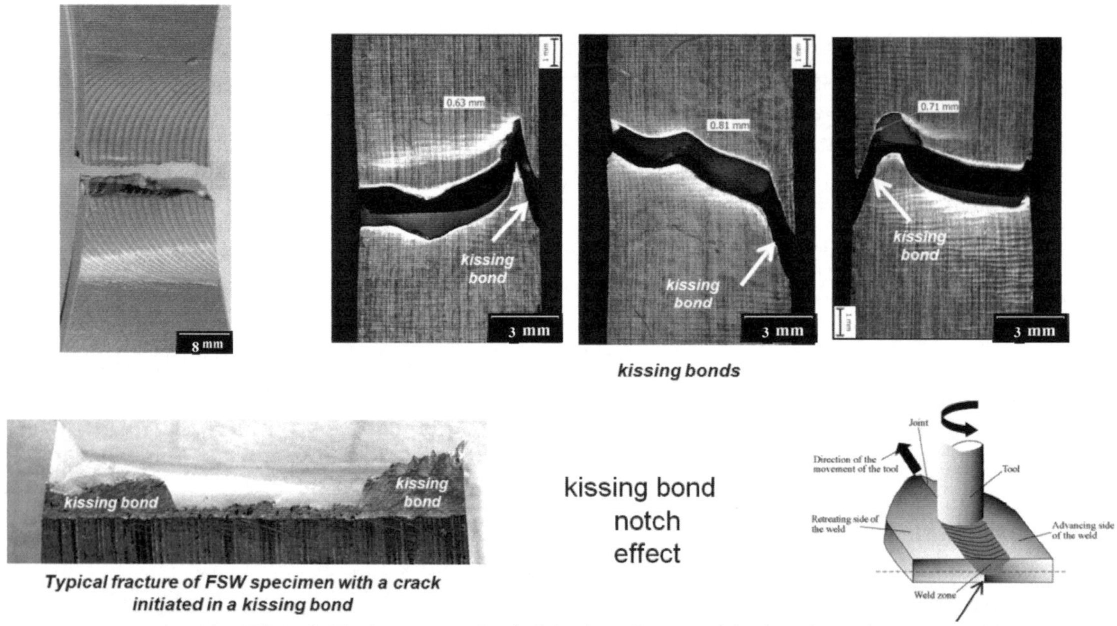

Fig. 6: Fatigue cracks initiation from a kissing bond

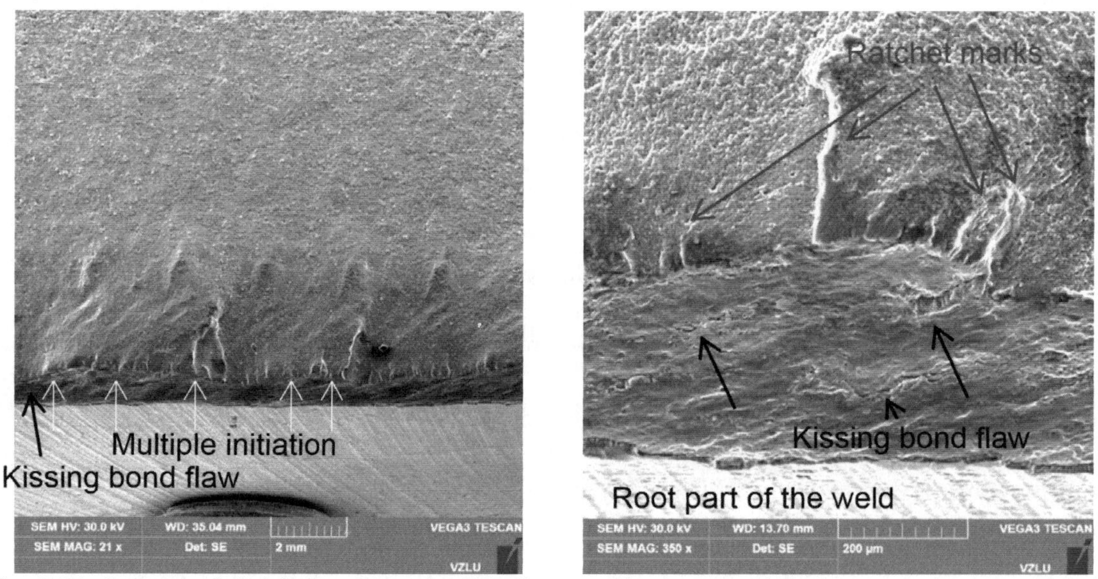

Fig. 7: Fatigue crack initiation from the KB flaw notch observed by electron microscopy: left) multiple crack initiation in KB; right) detail of KB surface.

Conclusions

While the static tensile strength of the FSW decreases with KB increase, the yield strain in not influenced by KB presence. Very slight differences of static strengths may be caused due to the fact that different batches of plates were used for the experiment. The batch-to-batch difference shows the sensitivity of properties on the weld performance conditions.

The experimental program and associated metallography and fractography analyses on a series of FS welded joints of AA 2024 T3 and AA 7475 T7351 materials that included various kissing bond sizes provided the following results:

- The flaw affects the material structure locally, which leads to a decrease in the fatigue properties.

- FSW of 2024 T3 material with 615 μm KB defect caused a significant decrease in the fatigue life and initiation from the KB was observed in all cases. Certain of the FSW specimens with 450 μm KB in depth failed out of the KB, in the HAZ/TMAZ area, and the fatigue behaviour was comparable with flaw-free FSW specimens. This finding means that the criterion and simultaneously the minimum detectable size for AA2024 T3 alloy must be below 400 μm.
- FSW of 7475 T7351 with 315 μm KB defect was demonstrated to be insignificant with respect to the fatigue life.
- Comparison of both material sensitivity to KB defects lead to conclusion that 2024 material is more sensitive to KB defect in contrast to the 7475 plate material.

Acknowledgement

This work was funded by the Ministry of Industry and Trade of the Czech Republic in the framework of Institutional support of Research organizations.

References

[1] E.D. Nicholas, W.M. Thomas. A review of friction processes for aerospace applications, Int J Mater Prod Technol 13(1) (1998) 45–55.

[2] J P. Podržaj, B. Jerman, D. Klobčar, Welding Defects at Friction Stir Welding, Metalurgija 54 (2015) 387-389.

[3] J O. Oosterkamp, L.D. Oosterkamp, A. Nordeide. Kissing Bond' Phenomena in Solid-State Welds of Aluminum Alloys, Welding Journal 83 (2004) 225-231.

[4] M. Kadlec, R. Růžek, L. Nováková. Mechanical Behaviour of AA 7475 Friction Stir Welds with the Kissing Bond Defect. International Journal of Fatigue, 74 (2015) 7-19. doi:10.1016/j.ijfatigue.2014.12.011.

[5] R. Růžek, M. Kadlec, L. Nováková. Influence of the Kissing Bond Defect on the Fatigue Life in Friction Stir Welds of 2024 Aluminium Alloy. Ciência & Tecnologia dos Materiais, 2015, 27 (2015) 92-99. http://dx.doi.org/10.1016/j.ctmat.2015.03.012

[6] AWS D17.3/D17.3M:2010. Specification for Friction Stir Welding of Aluminum Alloys for Aerospace Applications. Miami (USA), American Welding Society, 2009.

[7] EN 6072: Aerospace series: Metallic materials – Test methods, Constant amplitude fatigue testing. AECMA standard, 2010.

Peel resistance of geopolymer sandwich with foam and honeycomb cores

HRON Robin[1,a], MARTAUS František[1,b] and KADLEC Martin[1,c]

[1]VZLU – Czech Aerospace Research Centre, Beranových 130, Prague, Czech Republic

[a]hron@vzlu.cz, [b]martaus@vzlu.cz, kadlec[c]@vzlu.cz

Keywords: Composite, Geopolymer, Peel strength.

Abstract. Geopolymers are amorphous aluminosilicate materials which combine low temperature, polymer-like processing with high temperature stability and fire resistibility without toxic smoke generation. This work compares a conventional (glass/ phenolic resin) sandwich structure with a new sandwich structure (carbon/ geopolymer resin). In addition, three different manufacturing methods (hot bonding, cold bonding and curing/bonding in single operation) using four different adhesives were compared. Several types of organic and inorganic adhesives were used to bond the laminate skin to foam and honeycomb cores. The peel strength was two times higher than the reference glass/phenolic prepreg sandwich. In the group of honeycomb core specimens, the best results showed Resbond® 989 bonded specimens. The peel strength was three times higher than the reference sandwich. Fracture surfaces were evaluated considering the peel strength and load-displacement curves. This research can establish new fire resistible sandwich materials to be used in aerospace constructions.

Introduction

At present, there is an increasing need for materials eliminating fatal results of fire in case of aircraft accident [1]. A disadvantage of currently used composites is that they cannot be used in high temperatures, because their mechanical properties significantly degrade with increasing temperature [2]. Materials known as geopolymers have shown potential as a low-cost, environmentally friendly structural material with the ability to maintain strength at elevated temperatures [3]. Geopolymers withstands temperatures of over 1000 °C and can be utilized as matrix in fibre reinforced composites. The main advantage of geopolymers is the excellent temperature stability; fire resistance with no generation of toxic fumes and smokes, low thermal conductivity and good specific strength [4,5]. Generally the sandwich structures are capable of absorbing large amounts of energy under impact loads with results in high structural crashworthiness [6] Combination of these properties predetermines geopolymer sandwich materials for use in aircraft structures. For larger expansion of geopolymer composites into the aircrafts constructions, it is necessary to know comparison of these materials with commonly used materials [7] and environmental impact (hot-wet/ cold conditioning, influence of the operating fluids, etc.) on mechanical properties [8].

This work compares a conventional (glass/ phenolic resin) sandwich structure with a new sandwich structure (carbon/ geopolymer resin). In addition, three different manufacturing methods (hot bonding, cold bonding and curing/bonding in single operation) using four different adhesives were compared.

Materials

To compare commonly used (glass/phenolic resin) and new geopolymer (carbon/geopolymer) sandwich panels, the effect of various cores and adhesives to the average peel strength was examined.

As the skin, three plies of carbon fabric Kordcarbon Industry (200 g/m^2, 3K, plain) in 0°/90° warp/weft orientation were used. The cores had thickness of 10 mm. Specimens had dimensions of 76 x 465 mm. In the group of foam core specimens, the best results were obtained for single operation sandwich bonding using uncured GPL 30 geopolymer resin that was manually impregnated with the carbon fabric and placed on the uncured foam core. Standard vacuum bag assembly was applied.

For the sandwich core, aramid honeycomb and thermoplastic polymer foam were used. Glass/phenolic prepreg (Gurit PHG 600-68-37 T2, style 7781) and carbon fabric (200 g/m^2 3K) manually impregnated with GPL 30 geopolymer resin was used for the skin of the sandwich. Geopolymer resin GPL 30 was fully developed in Czech Aerospace research Centre (VZLU). The components of the GPL 30 are shown in Table 1. The table also shows the weight and weight ratio of the individual components. In the first step, pure geopolymer resin was prepared (by mixing components 1-5) and then short ceramic and carbon fibres were added.

Table 1: Geopolymer resin composition

	Component	m (g)	wt. %
1	SiO_2 (Thermal silica)	513.24	35.24
2	$Al(OH)_3$	453.32	28.74
3	H_2O	158.48	10.05
4	NaOH	111.16	7.05
5	KOH	163.80	10.39
6	Ceramic fiber LYTX-311-L	84	5.33
7	Carbon chopped fiber Tenax® – A HT C124 3 mm	93.24	5.91
Σ		1577	100

Table 2: Test matrix

Set	Core	Adhesive	Skin	Manufacturing
REF	aramid honeycomb	none	glass/ phenolic resin	bonded in single operation
L30				
K84		Promat® K84	carbon/ geopolymer GPL 30	cold bonded
RESB		Resbond® 989		cold bonded
LFX		Letoxit® LFX 062		hot bonded
PH600		Gurit PHG 600-44-50 T2		hot bonded
REF	foam	none	glass/ phenolic resin	bonded in single operation
L30				
K84		Promat® K84	carbon/ geopolymer GPL 30	cold bonded
RESB		Resbond® 989		cold bonded
LFX		Letoxit® LFX 062		hot bonded
PH600		Gurit PHG 600-44-50 T2		hot bonded

Sandwich panels were made by three types of procedures. First - the skin was bonded with the core in single operation, second – precured skin was bonded by the adhesive by the cold bonding procedure, the matrix is shown in Table 2. Four types of adhesives were used.

Methods

Tests were performed on mechanical test machine Instron 55R1185 in accordance with ASTM D1781 - Standard Test Method for Climbing Drum Peel for Adhesives [9], at room temperature conditions. The average peel strength was determined using the integral method along 127 mm of the specimen length.

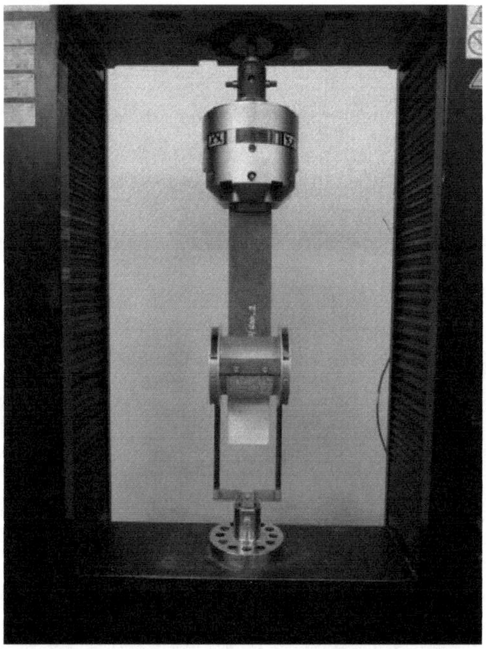

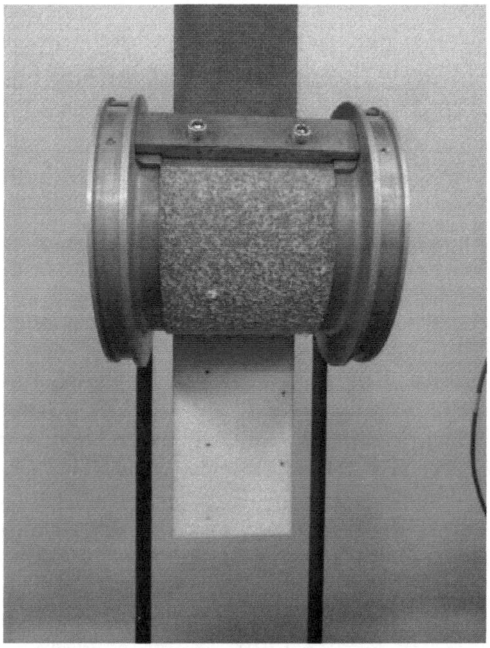

Figure 1: Test assembly of the drum peel test, overview and detail.

Results

Measured values of average peel strength and basic statistical evaluation (mean value, standard deviation – SD and coefficient of variation – CV) are shown in Table 3. All measured results were tested by Dixon's Q test on presence of outliers. Test was performed for significance level 0.05 and outlying values were not included into the test evaluation. A T-test was performed for determination whether the individual sets are significantly different from each other. Statistical significance 0.05 was chosen, Table 4.

Table 3: Measured average peel strengths in Nm/m, with statistical evaluation

	HONEYCOMB						FOAM					
	REF	*L30*	*K84*	*RESB*	*LFX*	*PH600*	*REF*	*L30*	*K84*	*RESB*	*LFX*	*PH600*
Mean	6.89	8.52	4.33	23.26	9.95	10.93	14.05	34.60	6.61	23.97	26.39	33.07
S.D.	1.60	1.14	0.21	2.07	2.44	1.18	0.19	7.96	0.32	1.24	2.31	1.04
C.V.	23.19	13.38	4.84	8.92	24.57	10.81	1.32	23.00	4.84	5.17	8.77	3.15

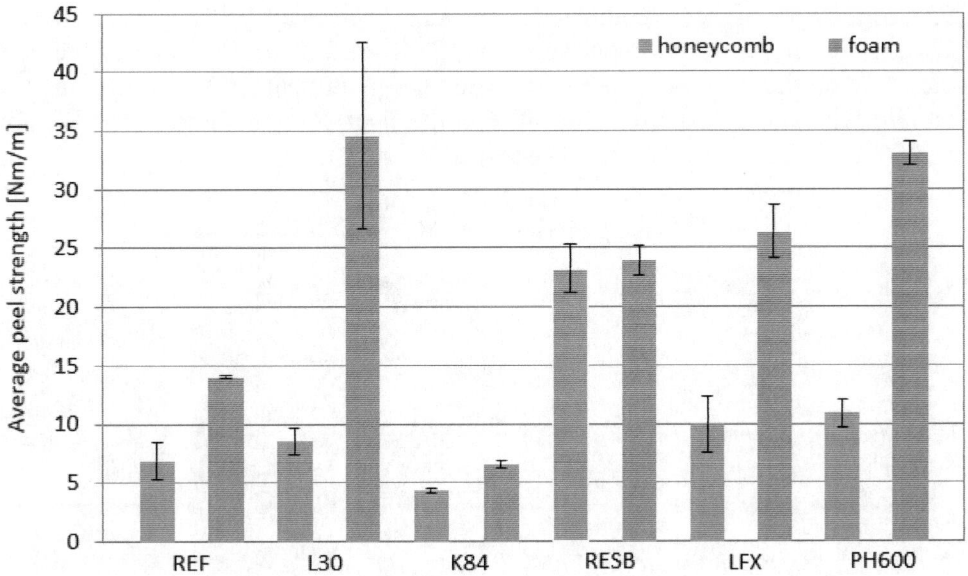

Figure 2: Measured average peel strengths in Nm/m.

In all cases, the average peel strength of foam sandwich samples was higher than the average peel strength measured on honeycomb samples. With exception of RESB series, the difference was evaluated as statistically significant. In the group of honeycomb core specimens, the best results showed set RESB (23.26 Nm/m) which was prepared by cold bonding. The peel strength was three times higher than the reference sandwich (6.89 Nm/m). In the group of foam core specimens, the best results showed set L30 (34.6 Nm/m), but on this set, the highest coefficient of variation (23 %) was also measured. Set PH 600 had the average peel strength less only of 4 % (33.07 Nm/m), but the CV was only 3.15%.

In comparison of REF and L30 sets (phenolic resin/ glass x geopolymer resin/ carbon), the measured values of average peel strengths showed that for honeycomb core was the difference statistically insignificant and for foam core was the difference statistically significant, see Table 4.

Table 4: Comparison of the results. ND: sets are not statistically significantly different, D: sets are statistically different.

	HONEYCOMB							FOAM					
	REF	*L30*	*K84*	*RESB*	*LFX*	*PH600*		*REF*	*L30*	*K84*	*RESB*	*LFX*	*PH600*
REF		ND	ND	D	ND	D	*REF*		D	D	D	D	D
L30	ND		D	D	ND	ND	*L30*	D		D	ND	ND	ND
K84	ND	D		D	D	D	*K84*	D	D		D	D	D
RESB	D	D	D		D	D	*RESB*	D	ND	D		ND	D
LFX	ND	ND	D	D		ND	*LFX*	D	ND	D	ND		D
PH600	D	ND	D	D	ND		*PH600*	D	ND	D	D	D	

Fractographical analysis was also performed. From each set was selected one referenced sample, macrophotography of the failure was made and compared with each other.

Figure 3 (comparing of honeycomb sets) shows a significant difference of the fracture surface between the set RESB and other sets. At RESB set, 75% of the adhesive remained on the core and 25 % on the skin. For other series the adhesive remained on mostly on skin. This is in agreement with the measured value of average peel strength, where the RESB series strength was 50 to 80% higher than for other series.

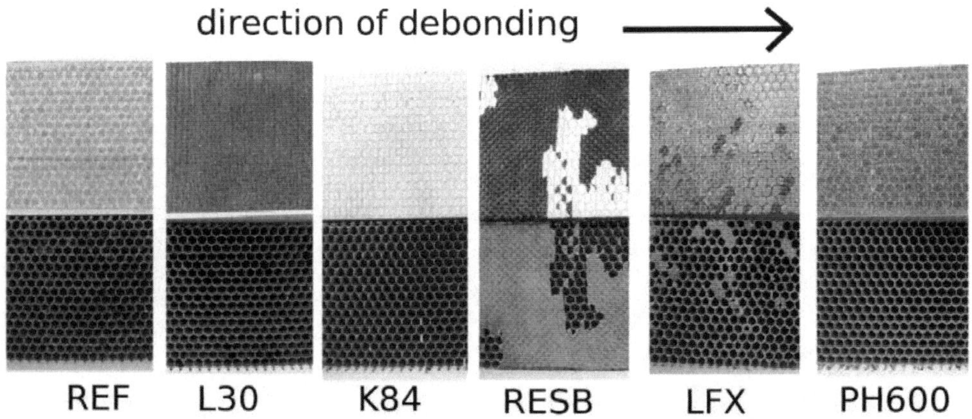

Figure 3: Fracture surfaces for honeycomb core specimens

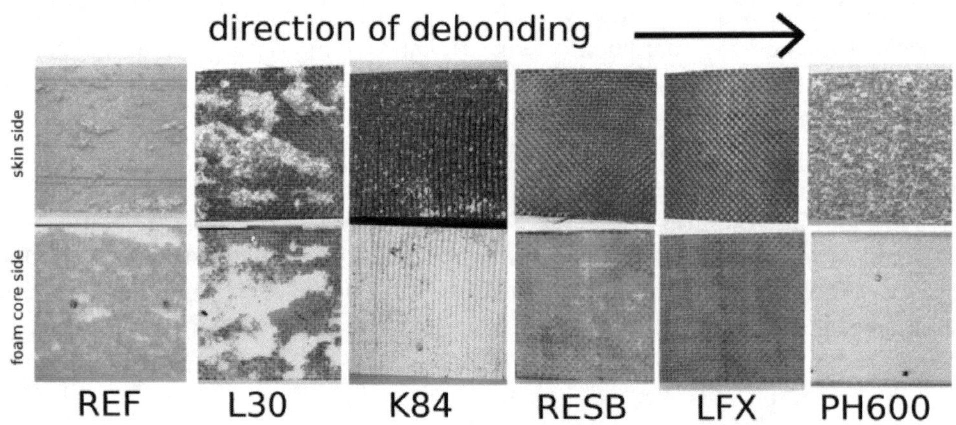

Figure 4: Fracture surfaces for foam core specimens

Figure 4 shows fracture surfaces for foam core specimens. When we compared the fracture surfaces with measured value of the peel strength, we can state that the major cohesive fracture in the core resulted in the highest peel strength (highest strength was measured on series RESB and PH600 where the cohesive failure dominated).

Conclusion

Highest strength values were measured on specimens with foam core and except the series where the Resbond 989 adhesive was used, the differences were statistically significant. From the honeycomb sets, the best results showed samples bonded by Resbond 989 - cold bonding. The peel strength was three times higher than the referenced set and one time higher than the second-best series PH600. In the group of foam core samples, the best results showed L30 series, but it is necessary to say that this set had big variation of the results (coefficient of variation was 23%). For second highest series was the average peel strength only 4% less and the variation of the results was approximately 3%.

Fractographic analysis of the fracture surface showed that for the honeycomb core has the strongest interface between the core and adhesive for the Resbond bonded samples and for foam core for the PH 600 series.

This paper showed that geocomposite sandwich panels can replace the commonly used sandwich panels.

Acknowledgement

This work was funded by the institutional support for the long-term conceptual development of a research organization provided by the Ministry of Industry and Trade of the Czech Republic.

References

[1] R.E. Lyon, Fire Safe Aircraft Cabin Materials, in: Fire and Polymers, ACS Symposium Series Number 599, G.L. Nelson, ed., American Chemical Society, Washington, D.C., p. 618, 1995.

[2] J. Krystek, V. Laš, V. Pompe, P. Hájková, Influence of Temperature on Selected Mechanical Properties of Geopolymer Composites, in: Proceedings of 55th Conference on Experimental Stress Analysis EAN2017, May 30th - June 1st, 2017, Nový Smokovec, Slovakia.

[3] I. Petríkova, B. Marvalová, A. Hrouda, L. Paur, Properties of Composite with Geopolymer matrix reinforced by Basalt Fabric, in: Proceedings of 55th Conference on Experimental Stress Analysis EAN2017, May 30th - June 1st, 2017, Nový Smokovec, Slovakia.

[4] P. Duxson, et al., Geopolymer technology: the current state of the art. Journal of Materials Science 42 (2007) 2917-2933.

[5] T.D. Hung, et al., New Generation of Geopolymer Composite for Fire-Resistance, Advanced in Composite Materials – Analysis of natural and Man-Made Materials, InTech, (2011)

[6] P. Bělský, M. Kadlec, Non-destructive Methods for Damage Assessment of Composite Sandwich Structures, MATEC Web Conf. Volume 188, 2018, 5th International Conference of Engineering Against Failure (ICEAF-V-2018).

[7] J. Davidovits, 30 Years of Successes and Failures in Geopolymer Applications - Market trends and Potential breakthroughs. Geopolymer 2002 Conference, Melbourne, Australia, Geopolymer Institute, pp. 1-16, 2002.

[8] R. Hron, M. Kadlec, F. Martaus, Mechanical Properties of Fibre Reinforced Geopolymer Composite Exposed to Operating Fluids, Solid State Phenomena, Volume. 278, 2018.

[9] ASTM D1781-98 (2012) - Standard Test Method for Climbing Drum Peel for Adhesives

EAN 2019
57th conference on experimental stress analysis

Undercut shape optimization with genetic algorithm and FreeFem++

HRUŠ Tomáš[1,a] and HERMANN Martin[1,b]

[1]Department of Mechanics; Faculty of Mechanical Engineering; Technical University of Liberec; Studentská 1402/2, 461 17 Liberec, Czech Republic

[a]tomas.hrus@tul.cz, [b]martin.hermann@tul.cz

Keywords: Genetic algorithm, Optimization, Shape of undercut, Finite Element Method, FreeFem++

Abstract. Our study examines an algorithm for searching ideal shapes of parts according to their functions and lifetime. Due to finite element method (hereinafter referred to as the „FEM") we obtain maximal equivalent stress at critical part of component. This is the value of fitness function for genetic algorithm, which is searching for ideal shape of part according to requirements.

Introduction

By the shape optimization is possible to obtain significant decrease of stress or decrease of strain without mass grow, eventually decrease of mass without increasing of stress or decreasing of strain.

General shape optimization is a complex problem with many parameters and huge difficulty. In this paper is presented optimization of a simple construction element – an undercut. Undercut works as a stress concentrator. Stress was calculated using FreeFem++ and the shape of the undercut was optimized. As optimization criterion was stress concentration used. Undercut with circular shape is used as reference.

Description of the undercut geometry

Undercut shape was described by two different ways – as a Bezier curve with three parameters and as a polyline with n parameters (see Fig. 1). Undercut width w and depth d are fixed.

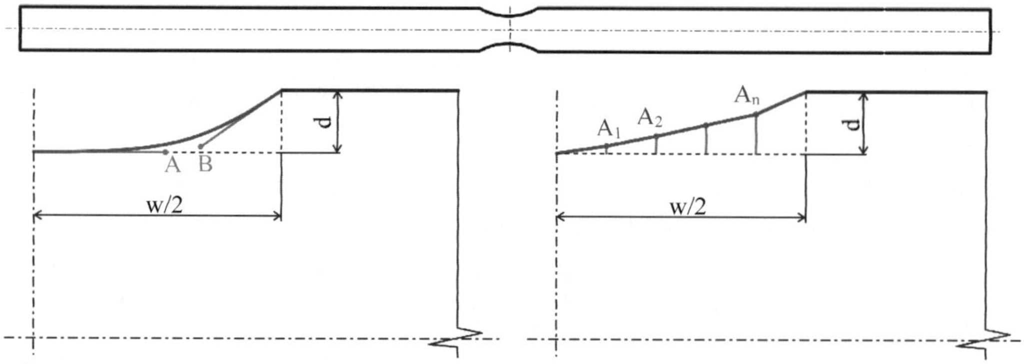

Fig. 1: *Top*: Bar with a undercut, *Bottom Left*: undercut modelled as a Bezier curve and described by *x*-coordinate of point A and *x* and *y*-coordinates of point B. *Bottom Right*: undercut modelled as polyline described by *y*-coordinates of points A_i.

The first method with only three parameters allows fast optimization, but results may be only inside the class of Bezier curves. The second method allows to describe any shape of undercut, but more parameters are necessary.

Optimization algorithm

Genetic algorithm was chosen as optimization method, because there is no possibility to explicit derivations formulation of the objective function. Objective function is stress concentration - maximal equivalent stress divided by nominal stress.

Implementation

FEM was implemented as a script in FreeFem++. Simple scripting language of FreeFem++ (similar to C++) allows fast and effective implementation of required functions.
Genetic algorithm was implemented in Matlab with FreeFem++ as external objective function and in FreeFem++ themselves. Matlab allows more operative modifications of genetic algorithm, FreeFem++ is faster.

Bar dimensions

A set of bar and undercut dimensions was used for testing – see Fig. 2 and Table 1. Undercut shape was optimized for each specimen using two undercut geometry descriptions (Bezier curve and polyline) and for two types of load – tension and bending.

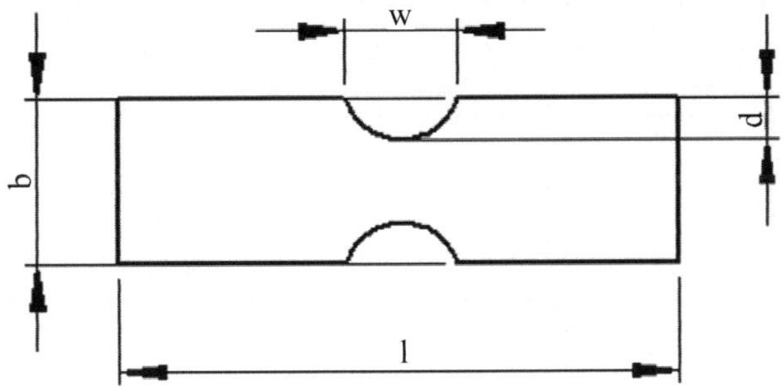

Fig. 2: Dimensions in general

Table 1: List of dimensions

Label of rod	l [mm]	b [mm]	w [mm]	d [mm]
w40d05	200	60	40	5
w40d10	200	60	40	10
w40d15	200	60	40	15
w50d05	200	60	50	5
w50d10	200	60	50	10
w50d15	200	60	50	15
w60d05	200	60	60	5
w60d10	200	60	60	10
w60d15	200	60	60	15

Results

Results for different undercut dimensions for bars loaded by tension are in Table 2. Results for bars loaded by bending are in Table 3.

A typical example is shown in Fig. 3. Bar is loaded by axial tension. Optimized undercuts has lower stress concentrations than reference circular undercut.

Table 2: Axially loaded bar: stress concentrations for circular and optimized undercuts

	d = 5 mm	d = 10 mm	d = 15 mm
		w = 40 mm	
Circular	1.3727	1.5310	1.4913
Bezier curve	1.2198	1.2972	1.2846
Polyline - 10 points	1.2926	1.3561	1.2887
		w = 50 mm	
Circular	1.2643	1.3869	1.3653
Bezier curve	1.1535	1.2035	1.1863
Polyline - 10 points	1.2365	1.3360	1.2115
		w = 60 mm	
Circular	1.1943	1.2837	1.2727
Bezier curve	1.1110	1.1441	1.1256
Polyline - 10 points	1.1656	1.2098	1.1297

Table 3: Bending loaded bar: stress concentrations for circular and optimized undercuts

	d = 5 mm	d = 10 mm	d = 15 mm
		w = 40 mm	
Circular	1.2168	1.2858	1.2615
Bezier curve	1.1130	1.1356	1.1041
Polyline - 10 points	1.1242	1.1443	1.0812
		w = 50 mm	
Circular	1.1463	1.2048	1.1984
Bezier curve	1.0718	1.0791	1.0607
Polyline - 10 points	1.0465	1.0799	1.0382
		w = 60 mm	
Circular	1.1053	1.1524	1.1536
Bezier curve	1.0441	1.0475	1.0368
Polyline - 10 points	10496	1.0526	1.0366

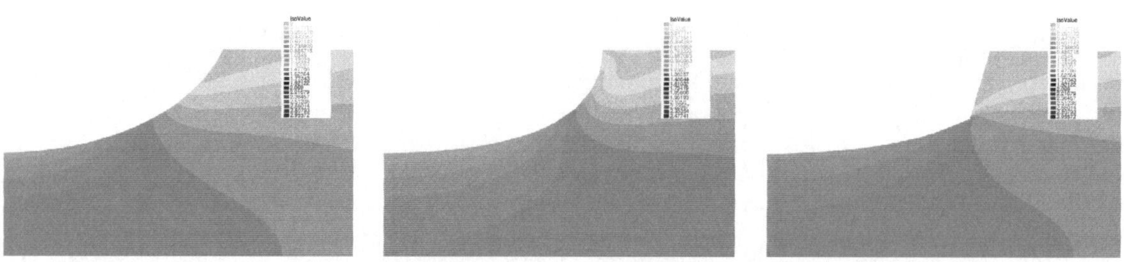

Fig. 3: Axially loaded bar. *Left*: Stress concentration in circular undercut (Max. = 1.478). *Center*: Stress concentration in undercut described by an optimized Bezier curve (Max. = 1.238). *Right*: Stress concentration in undercut described as an optimized polyline with 10 parameters (Max. = 1.270).

Conclusions

Genetic algorithm is useful and efficient for significant minimization of stress concentration by optimization undercut shape. It is efficient even by using Bezier curve, driven by only three parameters. This can be important for parts lifetime.

Acknowledgement

This publication was written at the Technical University of Liberec as part of the project "Research, development and application of advanced methods and technologies in technical mechanics, biomechanics and strength and elasticity" with the support of the Specific University Research Grant, as provided by the Ministry of Education, Youth and Sports of the Czech Republic in the year 2019.

References

[1] HECHT, Frédéric. New development in FreeFem++, Journal of numerical mathematics, 2012, vol. 20, no 3-4, p. 251-266.

[2] MOEZI, Seyed Alireza, ZAKERI, Ehsan, BAZARGAN-LARI, Yousef, ZARE, Amin. 2d Shape Optimization Via Genetic Algorithm. The Journal of Mathematics and Computer Science (JMCS), 2014. vol. 11, p. 209-217

[3] VALACKAITE, Laisvune, MACIUNAS, Darius, GLEBIENE, Elena. Shape Optimization of Two-dimensional Body Utilizing Genetic Algorithms. Conference of Informatics and Management Sciences, 2013, p. 449-452

[4] WOON, S.Y., QUERIN, O.M., STEVEN, G.P. Structural application of a shape optimization method based on a genetic algorithm. Structural and Multidisciplinary optimization. 2001, vol. 22, p. 57-64.

EAN 2019
57th conference on experimental stress analysis

Homogenization of material properties of a combined rubber/steel sheet using modal analysis

HUŇADY Róbert[1,a], PAVELKA Peter[1,b] and LENGVARSKÝ Pavol[1,c]

[1]Faculty of Mechanical Engineering, Technical University of Košice
Letná 9, 042 00 Kosice, Slovak Republic

[a]robert.hunady@tuke.sk, [b]peter.pavelka@tuke.sk, [c]pavol.lengvarsky@tuke.sk

Keywords: homogenization, equivalent material properties, experimental modal analysis

Abstract. The paper deals with determining of material constants of a homogenized material. The introduced homogenization technique is based on experimental modal analysis and subsequent parametric optimization using finite element method. The paper describes the basic principle of the proposed technique and presents its practical application, in which the equivalent material constants of a combined rubber/steel sheet are estimated.

Introduction

When analysing the mechanical behaviour of structures, it is necessary to determine their geometry, boundary conditions and material properties. Modern, unconventional materials, such as particulate and fibrous composites, laminates, honeycombs, metal foams, various porous materials, etc., are increasingly being used today. Additive technologies are also very popular. Subject to the build direction and machine resolution, material properties of an additively manufactured part often exhibit varying material distribution (heterogeneity) and direction dependent behaviour (anisotropy), i.e. they differ from the properties of the base material. Since, the elastic constants of the materials are fundamental to analysing the mechanical behaviour of structures, significant attention is paid to their determination in engineering practice. Replacing such materials with so-called equivalent or representative material with homogenized properties is one way to effectively describe their behaviour, especially for numerical calculations. Many homogenization techniques are known and others are still being developed [1], [2], [3], [4], [5], [6]. Homogenization techniques are very effective especially in combination with FEM for modelling of 2D-heterogeneous structures such as laminates or honeycomb plates [7]. However, their application is limited by knowing the material properties of the individual components. If these are unknown, the material has to be subjected to material tests. The paper introduces the alternative approach that is based on experimental modal analysis, the results of which are subsequently used in FEM analysis combined with parametric optimization. This approach has been successfully used to determine the Young modulus of homogeneous isotropic material and to estimate the constraint stiffness of the cantilever beam [8]. In this study, the method is applied to estimate material constants of the homogenized laminated sheet that consists of two steel plies and three rubber plies. The nominal material properties and ply layup of the sheet stated by the manufacturer is shown in Fig. 1a.

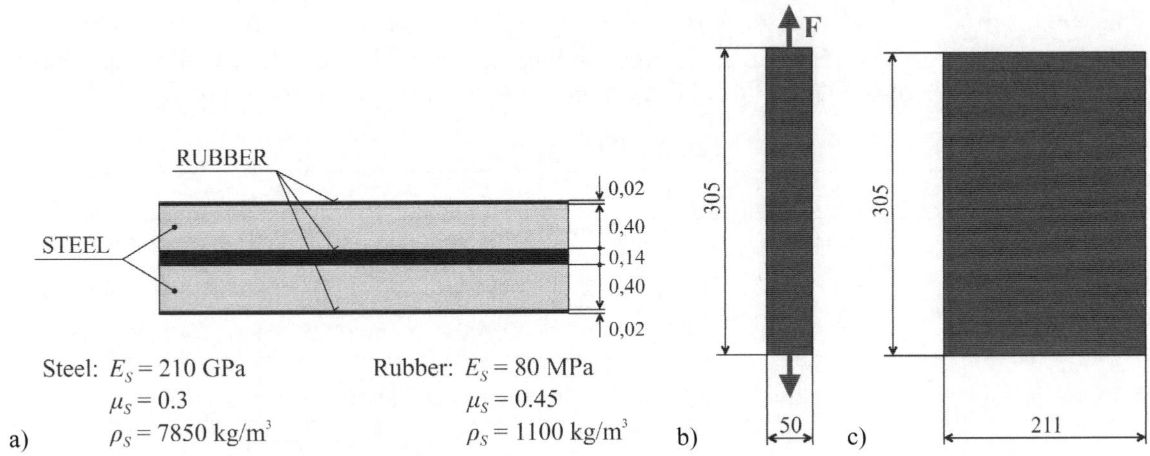

Fig. 1: a) Material properties and ply layup of the sheet, b) Dimensions of a tensile specimen, c) Dimensions of the sheet.

1. Principle of the proposed homogenization technique

The proposed homogenization technique is based on modal analysis of a material specimen. At the first phase, modal parameters, such as natural frequencies and mode shapes, are determined experimentally. The specimen should be planar and rectangular. It has to be freely supported to avoid the problem of constraint stiffness in the numerical model. The free suspension can be achieved by putting the body on a very soft pad (e.g. foam) or suspending it on soft springs. Such support is considered to be free if the highest natural frequency of rigid body modes is less than 10% of the value of the lowest deformation natural frequency [9]. At the second phase, the equivalent FE model of the same dimensions as the specimen is build. Its material properties are subsequently determined by the parametric optimization, the task of which is to tune the model so that its natural frequencies match the frequencies obtained by the measurement. Multicriterial optimization is ideal for this purpose. In this study, only one objective function is used. The optimization is based on frequency analysis, when the first natural frequency is used as an objective function. The frequency of any other mode is used as a constraint. The design variables are Young modulus and Poisson ratio that are estimated in the user defined range. However, material density must be known.

2. Verification of the principle

The principle of the homogenization technique is verified by numerical simulation of the process described in previous chapter. As the first, the homogenized properties of the equivalent material were determined from the results of FE analysis, in which the specimen shown in Fig. 1b was subjected to tensile loading. By calculation, we determine: E = 171 428 MPa and μ = 0.3276. These reference values will be used for verification of the proposed technique.

The natural frequencies and mode shapes of the laminated sheet (Fig. 1c) were determined by Abaqus. These frequencies can be considered as the results of an experimental modal analysis of the real sheet. The equivalent density of material was calculated by Abaqus based on nominal values. Nominal density was calculated to be 6348 kg/m^3. In practice, this can be determined based on volume and weight of a specimen.

The parametric optimization was performed in program NX Nastran, in which the FE model of homogenized sheet was built. The optimization process was based on the frequency analysis SOL103 Real-Eigenvalues. The frequency of the first mode was defined as objective function. The frequency of the second mode was used as the constraint of the optimization.

The Young modulus and Poisson's ratio were design variables. The values of the objective function and design variables in the corresponding design cycles are listed in Table 1. The natural frequencies of the laminated sheet and homogenized sheet are shown in Table. 2.

Table 1: Design process

Design cycle	0	1	2	3	4
f_1 [Hz]	48.49044	45.27167	47.98682	49.40555	49.55881
E [MPa]	155 000	135 818	155 818	170 576	171 637
μ [-]	0.25	0.25	0.28	0.327738	0.327738

Table 2: Natural frequencies of the sheet

Mode	1	2	3	4	5	6	7
Laminated sheet	49.561	56.630	113.21	120.98	143.46	164.02	206.96
Homogenized sheet	49.5588	55.1549	115.482	120.201	144.747	167.784	216.448

The results obtained by numerical optimization of material parameters are in good agreement with the references values. Differences in the natural frequency values are also acceptable.

3. Analysis of the real sheet

The practical use of the proposed technique is demonstrated on the real sheet. As the first, the natural frequencies of the sheet have been determined by experimental modal analysis. The sheet was putting on two very soft ribbons. Responses were measured by laser vibrometer Polytec PDV-100. The sheet was excited by impact hammer Bruel&Kjaer 8206 in 48 points. The experimental setup is shown in Fig. 2. The modal parameters are listed in Table 3.

Fig. 2: Experimental setup

Table 3: Natural frequencies of the sheet

Mode	1	2	3	4	5	6	7
Frequency	45.781	80.397	108.52	120.16	141.55	191.49	205.42
Damping (%)	6.15	2.85	4.14	2.63	3.48	2.53	3.23
Complexity	0.101	0.284	0.202	0.108	0.060	0.184	0.307

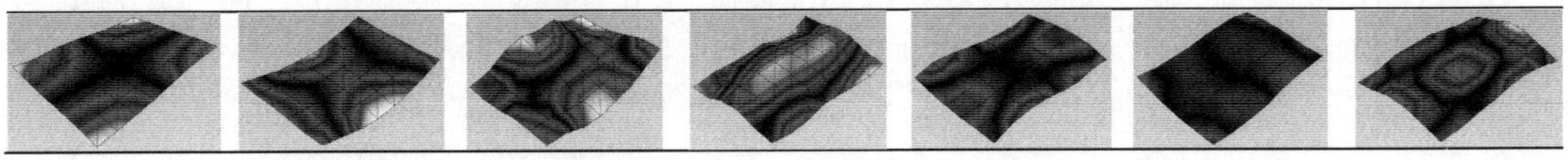

It is obvious that the natural frequencies of the sheet differ from the frequencies obtained by FE analysis of the reference model that is built on nominal values. The sheet shows the relatively higher damping.

The frequencies determined by EMA were used in the parametric optimization of the homogenized models. Density of the material was determined based on volume and weight of the sheet. Its value was calculated to be 6779 kg/m^3. The initial design values were $E_0 = 190\ 000$ MPa and $\mu_0 = 0.3$.

The parametric optimization was performed the way described in the previous chapter with one difference, the frequency of fourth mode was used as constraint. The values of the objective function and design variables in the individual design cycles are listed in Table 4.

Table 4: Design process

Design cycle	0	1	2	3	4	5
f_1 [Hz]	49.677	50.878	51.189	48.224	46.331	45.699
E [MPa]	190 000	180 000	190 000	170 419	169 905	169 989
μ [-]	0.3	0.3	0.326	0.378	0.378	0.378

Conclusion

The paper described the homogenization technique that is based on experimental modal analysis and numerical parametric optimization where the experimentally determined value of the first natural frequency is used as an objective function and the frequency of any other mode as constraint. This two values are needed to determine two elastic material constants (Young modulus, Poisson ratio). Effective alternative is multicriterial optimization. The proposed method allows to estimate material constant of homogenized material. Its functionality was verified by numerical simulation. Practical application was presented on the combined rubber/steel sheet.

Acknowledges

This research was supported by project VEGA 1/0355/18.

References

[1] X. Liu, V. Shapiro, Homogenization of material properties in additively manufactured structures, Computer-Aided Design, Volume 78, 2016, Pages 71-82, ISSN 0010-4485.

[2] Z. He, G. Wang, M.J. Pindera, Multiscale homogenization and localization of materials with hierarchical porous microstructures, Composite Structures, 2019, 110905, ISSN 0263-8223.

[3] M. Kamiński, Homogenization of particulate and fibrous composites with some non-Gaussian material uncertainties, Composite Structures, Volume 210, 2019, Pages 778-786, ISSN 0263-8223.

[4] G. Martínez-Ayuso, M.I. Friswell, S. Adhikari, H.H. Khodaparast, H. Berger, Homogenization of porous piezoelectric materials, International Journal of Solids and Structures, Volumes 113–114, 2017, Pages 218-229, ISSN 0020-7683.

[5] D.E.S. Rodrigues, J. Belinha, F.M.A. Pires, L.M.J.S. Dinis, R.M. Natal Jorge, Homogenization technique for heterogeneous composite materials using meshless methods, Engineering Analysis with Boundary Elements, Volume 92, 2018, Pages 73-89, ISSN 0955-7997.

[6] M. Acosta-Flores, E. Jiménez-López, M. Chávez-Castillo, A. Molina-Ocampo, J.J. Delfín-Vázquez, J. A. Rodríguez-Ramírez, Experimental method for obtaining the elastic properties of components of a laminated composite, Results in Physics, Volume 12, 2019, Pages 1500-1505, ISSN 2211-3797.

[7] A. L. Kalamkarov, I. V. Andrianov, and V. V. Danishevsky, “Asymptotic homogenization of composite materials and structures,” Applied Mechanics Reviews, vol. 62, 2009.

[8] A. Kaľavský, R. Huňady, P. Pavelka, Determining the Young’s modulus of a specimen using experimental modal analysis, In: Proceedings of Experimental Stress Analysis, EAN 2018, June 5th - 7th, 2018, Harrachov, Czech Republic.

[9] A. Bilošová, Modal testing, VŠB-TUO, Ostrava, 2011.

Advanced material model for numerical simulations of fine blanking

JÁNSKÁ Danuše[1,a], URBÁNEK Miroslav[1,b] and MAŠEK Martin[1,c]

[1]COMTES FHT, Metals Research Institute; Průmyslová 995, 334 41 Dobřany, Czech Republic

[a]danuse.janska@comtesfht.cz, [b]miroslav.urbanek@comtesfht.cz, [c]martin.masek@comtesfht.cz

Keywords: fine blanking, 1.4301 steel, FEM, material model

Abstract. Advanced manufacturing processes, including fine blanking, are widely used in the mass production of sheet metal parts. In the present article, numerical modelling and a real-world test of fine blanking are discussed with a focus on material characterization. The material was 1.4301 stainless steel and its model was constructed using measured mechanical properties. Tensile tests, plane strain tests and shear tests were carried out to determine steel characteristics under various stress states. All the tests were performed at room temperature and under quasi-static conditions. Local strains were determined using the ARAMIS digital image correlation (DIC) system from GOM company. After testing, metallographic analysis of the specimens was conducted for characterizing their fracture surfaces. By correlating the data from the DIC system, results of numerical modelling and metallographic examination, the instant of failure initiation in a specimen can be determined. When a ductile failure model is calibrated against tests under various stress states and used for modelling of blanking, it improves the description and the accuracy of the computational model of the process. The choice of the failure model has a substantial impact on the calculated magnitude of fine blanking force. To validate the material models chosen, an additional fine blanking test and metallographic examination were performed in order to assess the creation and shape of the sheared edge. A material model developed and validated by this procedure becomes useful in the design and optimization of real-world blanking processes.

Introduction

Fine blanking is a rapid and effective material parting method, which is employed in the series production of sheet parts. In fine blanking, a punch exerts a shearing action, whereby a blank is separated from a sheet or sheet strip by shear deformation. Given today's increasing demands on the quality and lifetime of engineering parts, it becomes necessary to focus on the blanking process and on the effects of its key aspects, such as the clearance between the die and the punch, their geometries, the punch speed and the hold-down force, on sheared edge quality. Burrs and distortions, which may occur on the sheared edge, require additional finishing operations, such as grinding, and therefore longer production times and higher costs.

Blanking is most often studied experimentally, which is an expensive undertaking, the results of which are generally impossible to apply to other parts with different shapes. For these reasons, finite element (FE) calculations are very useful because their predictions enable engineers to reduce the number of actual tests needed. Using computational methods, one can predict the shape and quality of the sheared edge, the strains, residual stresses and failure and distortion in the sheet stock and the product.

This paper explores the process of fine blanking from a 2 mm sheet of ordinary chromium-nickel austenitic stainless steel 1.4301.

Fine Blanking

Fine blanking (Fig. 1) is a shearing process which has been enhanced to produce blanks of high precision and surface quality which can be used directly for assembly of products. The clearance between the punch and the die is of major importance. Whereas the clearance per side in conventional shearing is normally 3–10% of the sheet thickness, the clearance in fine blanking is 0.5% of the thickness of the work material. In the experiment reported here, the clearance was 0.03 mm per side. The smaller is the clearance, the closer is the stress state to pure shear and the smaller are the undesirable tensile stress components under the bending load. The sheared edge radius is 0.2 mm. Experimental validation of the material models was performed with the use of a test tool which conformed to the above-detailed requirements for precision. The test tool had been developed for servo-electric test machine Zwick 250 (Fig. 2).

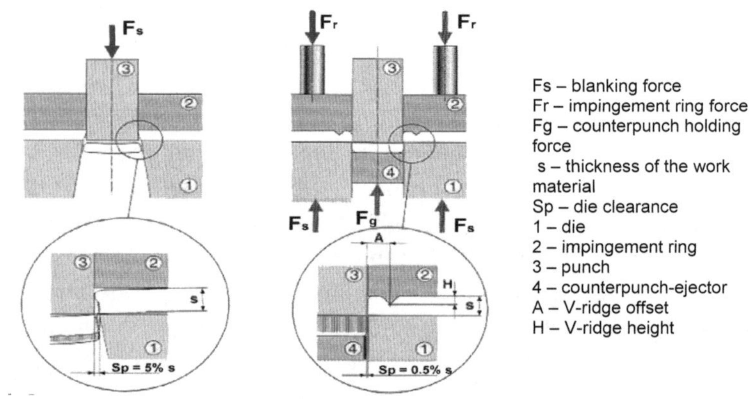

Fig. 1 Illustration of differences between conventional punching and fine blanking

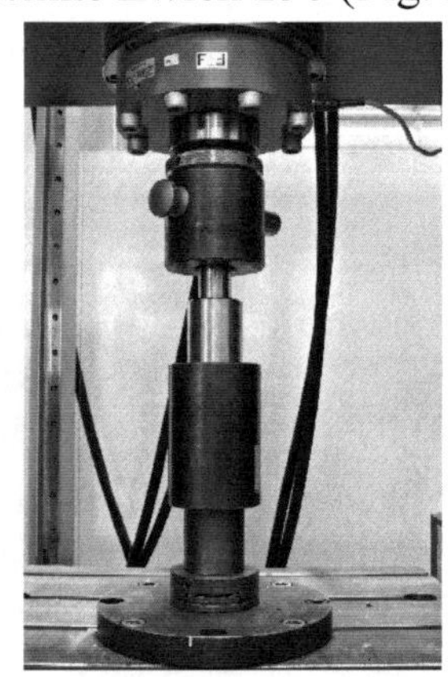

Fig. 2: Experimental test set-up in ZWICK machine

Material models

Considering the toughness of stainless steel, an appropriate material model should be chosen for numerical modelling of its behaviour. In this work, uncoupled models were used, in which plasticity and damage are accounted for independently.

Plasticity

When it comes to characterizing plasticity of material, the most-widely used formula is the Johnson-Cook (J-C) plasticity model (1). It is implemented in most commercial simulation codes. Many publications were devoted to its calibration, [8] . Its equation includes the following material constants: A – yield strength; B and n – material constants which characterize the real-world stress-strain curve. Together, these constants characterize the work-hardening curve under quasi-static conditions.

$$\sigma_y = \left(A + B\bar{\varepsilon}^{p^n}\right) \qquad (1)$$

Fracture

In this work, four different fracture models were considered. Depending on formulation, each of them requires a different number of mechanical tests for calibration. Ductile fracture is a process which leads to macroscopic separation in material. It is understood as a sequence of formation, growth and coalescence of voids which are present in the material. In continuum mechanics, it is described through stress state-related quantities. These quantities include triaxiality, a characteristic of three-dimensionality of the stress state, and the Lode parameter (2) which accounts for the effects of shear stress state.

$$\bar{\theta} = 1 - \left(\frac{6\theta}{\pi}\right) = 1 - \frac{2}{\pi}\arccos\xi \qquad \xi = \left(\frac{r}{q}\right)^3 = \cos(3\theta) \tag{2}$$

Below, the individual fracture models are described, including the minimum number of mechanical tests required for their calibration.

Johnson-Cook model

This model characterizes fracture behaviour of material under various stress states. Equation (3) includes material constants D_1, D_2 and D_3, which characterize the fracture curve for various stress triaxiality values under quasi-static conditions. Triaxiality values vary with the stress state. Generally, they are in the range from -1 to 1.

$$\varepsilon_{f\,JC} = \left(D_1 + D_2 e^{D_3\sigma^*}\right)[1 + D_4 * \ln \dot{\varepsilon}^*][1 + D_5 T^*] \tag{3}$$

Bao-Wierzbicki model

The second model was the Bao-Wierzbicky (B-W) fracture model. It is calibrated against tensile test data. Bao-Wierzbicky ductile fracture model comprises three regions which differ according to stress state. Region I ($1/3 < \eta$) involves high triaxiality values. Stress triaxiality induces formation, growth and coalescence of voids, a failure mechanism which occurs in tension. In region II ($0 < \eta < 1/3$) triaxiality values are positive but low, which is reflected in combined failure modes.

Region III ($-1/3 < \eta \leq 0$) comprises negative stress triaxiality values, at which shear failure occurs.

The failure initiation locus for the Bao-Wierzbicki model is defined in terms of equations (4).

Region I	**Region II**	**Region III**
$\bar{\varepsilon}_{f\,BWI} = a\frac{\eta_0}{\eta}$	$\bar{\varepsilon}_{f\,BWII} = b + (a-b)\left(\frac{\eta}{\eta_0}\right)^2$	$\bar{\varepsilon}_{f\,BWIII} = \frac{a}{1+3\eta}$

(4)

Constant a is the plastic strain at fracture in pure shear ($\eta = 0$). Constant b is the plastic strain at fracture in uniaxial tension ($\eta = 1/3$). Parameter b can be derived from tensile test data using formula (5), where A_0 denotes the cross-sectional area before the test and A_f is the fracture surface area. Constant a is calculated using equation (6).

$$b = \ln\left(\frac{A_0}{A_f}\right) \tag{5}$$

$$a = b\left(\frac{\sqrt{3}}{2}\right)^{\frac{1}{n}} \tag{6}$$

MAX shear model

The MAX shear model is calibrated against data from shear test specimens. The maximum shear stress criterion, known as the Tresca criterion, postulates that failure occurs when the largest shear stress reaches critical value $\tau_{max} = C_1$.

Hosford-Coulomb model

Hosford-Coulomb (H-C) model is an expansion of the classical Mohr-Coulomb criterion. The H-C model was proposed by Bai-Wierzbicky and presented by Mohr (Mohr and Marcadet, 2015). In this model, the Tresca stress has been replaced with Hosford equivalent stress (9) σ_{HF}. When this model is mapped to the triaxiality-Lode parameter space, a relationship is obtained, which was derived in [14] (10). The letters f denote Lode parameter-dependent functions. Values a, b and c must be determined by experiment.

$$\sigma_{HF} + C(\sigma_I + \sigma_{II}) = b \quad (9)$$

$$\bar{\varepsilon}_{f\,HC}[\eta, \bar{\theta}] = b(1+c)^{\frac{1}{n}} \left(\left\{ \frac{1}{2} (f_1 - f_2)^a + (f_2 - f_3)^a + (f_1 - f_3)^a \right\}^{\frac{1}{a}} + c(2\eta + f_1 + f_3) \right)^{-\frac{1}{n}} \quad (10)$$

In order to validate the above-described plasticity and fracture models, an experimental programme was developed and carried out. The experimental material was a cold-rolled sheet of 1.4301 stainless steel 2 mm in thickness.

Experimental programme

The purpose of the programme was to characterize stainless steel by determining its yield strength, ultimate strength, elongation, reduction of area and the force-displacement curve measured by means of an extensometer. The force-displacement data were input into the material models. Material constants were determined on the basis of and calibrated against mechanical test data for specimens shown in Fig. 3. The values obtained from this set of tests were used for characterizing the material in various stress states by means of the models of material behaviour under load: Johnson-Cook, Bao-Wierzbicky, Max shear and Hosford-Coulomb. The mechanical tests are described in detail below.

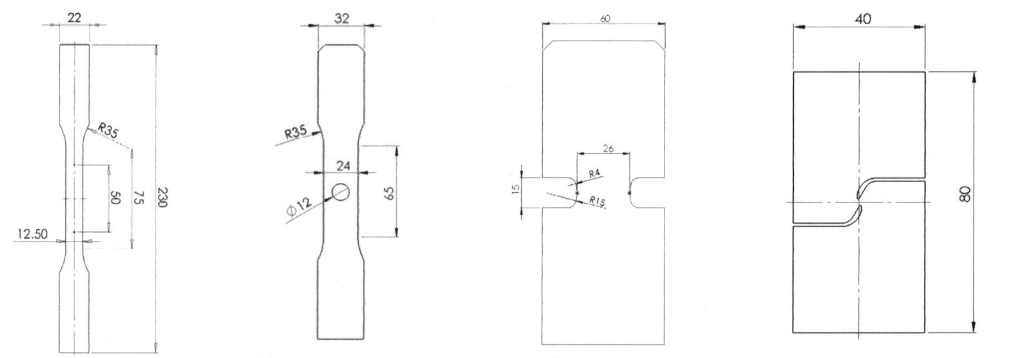

a) Specimen for tensile test | b) Specimen for tensile test with a hole | c) Plane strain test specimen | d) Shear test specimen

Fig. 3: Geometries of test specimens

Tensile test data were used for characterizing the plastic behaviour of the material. The data from the other tests provided description of fracture in the material. All the tests were conducted under quasi-static conditions and at room temperature. Three test specimens were used for each of them to ensure that the results are statistically relevant. Local strains were evaluated using an ARAMIS DIC system with 12-MPix cameras (Fig. 4, Fig. 5, Fig. 6, Fig. 7)

Fig. 4: Tensile test set-up

Fig. 5: Test set-up for the specimen with a hole

Fig. 6: Plane strain test specimen

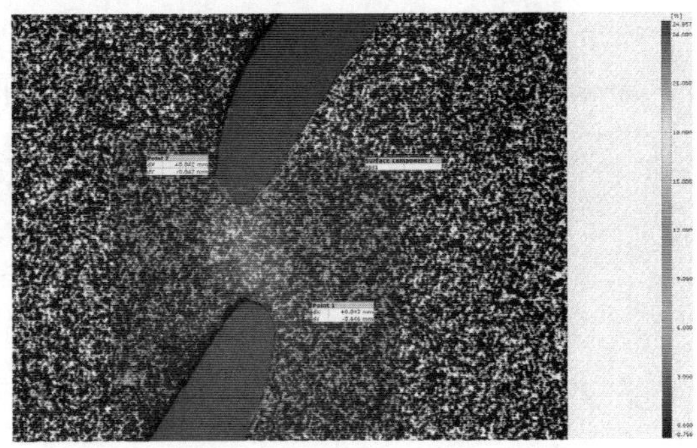

Fig. 7: Shear test specimen

Calibration of material models

The material model which comprised the Johnson-Cook plasticity model was developed with the use of an optimization script. Using the script, material constants A, B and n were determined from the data for unnotched solid test specimens. The values of the constants were refined with the aid of numerical modelling to ensure they lead to an agreement between the simulation model and the real-world test data (Fig. 8). The criterion was the match between the force-displacement plots. The simulation-based refinement of plasticity characteristics did not take into account damage.

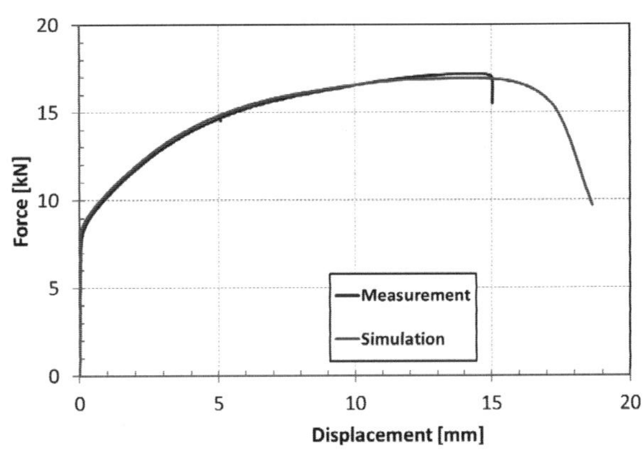

Fig. 8 Diagram of tensile tests and comparison with numerical simulation

Fracture behaviour was simulated by successively applying the Johnson-Cook fracture, Bao-Wierzbicky, MAX shear, and Hosford-Coulomb models. Based on triaxiality and fracture strain values found from mechanical tests and calculations, threshold values were determined and the curves for individual models were fitted to them. The material constants for the Johnson

Cook plasticity and Johnson Cook fracture models and the Hosford-Coulomb model were calibrated using an optimization script. They are given in Table 1.

Table 1: Calibrated simulation parameters

Johnson-Cook plasticity	A	B	n	
	190	1210	0.48	
Johnson-Cook fracture	D1	D2	D3	
	-0.45	1.8	-0.6	
Bao-Wierzbicky fracture	b	n		
	0.67	0.48		
MAX shear fracture	C	n		
	5.5	0.48		
Hosford-Coulomb fracture	a	b	c	n
	0.9908	1.1748	0.2	0.2743

For illustration, the graphs below show the comparison between mechanical test diagrams and curves from numerical modelling using Bao-Wierzbicky model. The diagram from tensile tests on specimens with a hole is shown in Fig. 9. The plane strain test diagram and evaluation are presented in Fig. 10. The shear test is evaluated in Fig. 12.

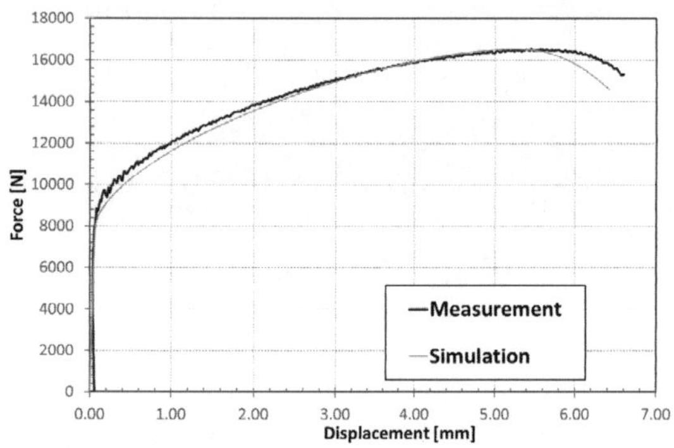

Fig. 9 Tensile test diagram for specimens with a hole and a numerical model curve

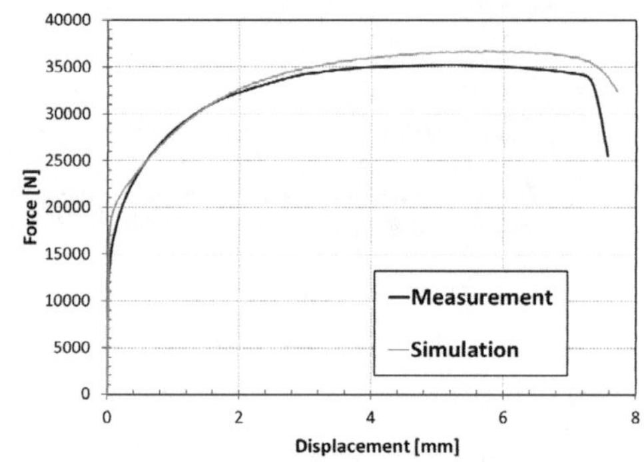

Fig. 10 Diagram of plane strain tests and comparison with numerical simulation

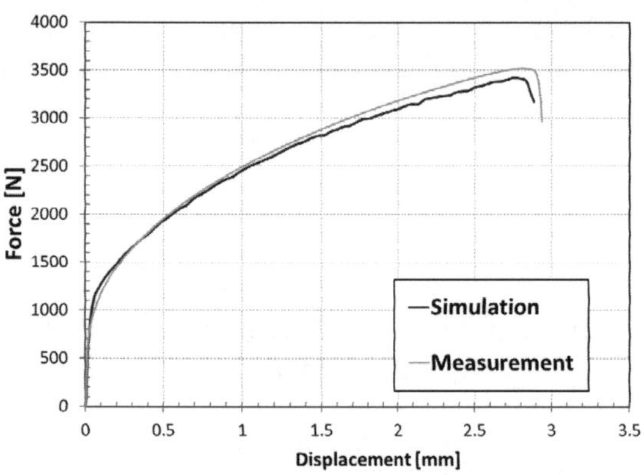

Fig. 11 Diagram of shear tests and comparison with numerical simulation

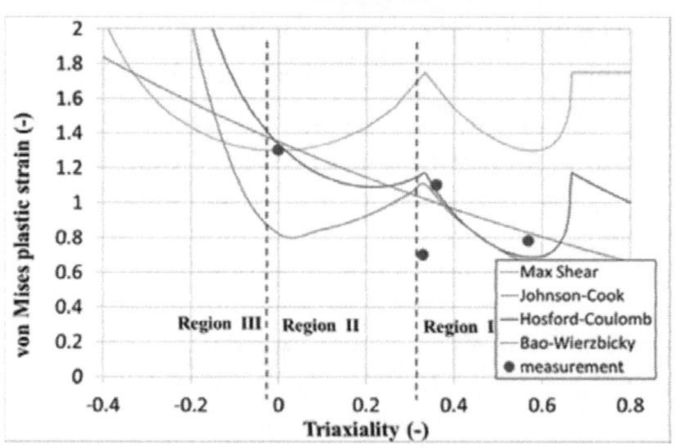

Fig. 12 Graph of fracture curves

Numerical modelling was performed successively with each of the four models. The fracture curves for various triaxiality values are plotted in Fig. 12

Fine blanking experiment

A real-world fine blanking experiment involving a steel washer was carried out in order to validate the material models. The tool used is shown in Fig. 3. The material for fine blanking was laser-cut round specimens of 1.4301 stainless steel of 2 mm thickness.

Four values of tool displacement were used. Three of them led to pre-defined incomplete fine blanking. The fourth specimen was punched out completely. In specimens T1, T2 and T3, the fine blanking operation stopped at displacements 2.1 mm, 3.1 mm and 3.65 mm, respectively; the values include the engagement of the impingement ring. Specimen T4 was blanked completely. Force [kN] vs. displacement [mm] curve was measured, as shown in Fig. 13. The complete fine blanking of the last specimen T4 was simulated because it was most relevant for modelling.

Abaqus/explicit solver was used for FE analysis. An axisymmetric 2D model with controlled displacement was developed. Mesh elements were CAX4R type with a size of 0.05 m. The tools were defined as rigid bodies. The hold-down force of the impingement ring and counterpunch were exerted by springs. The frame of the machine was simulated as a 1D element disposed between the point of application of force and the tool.

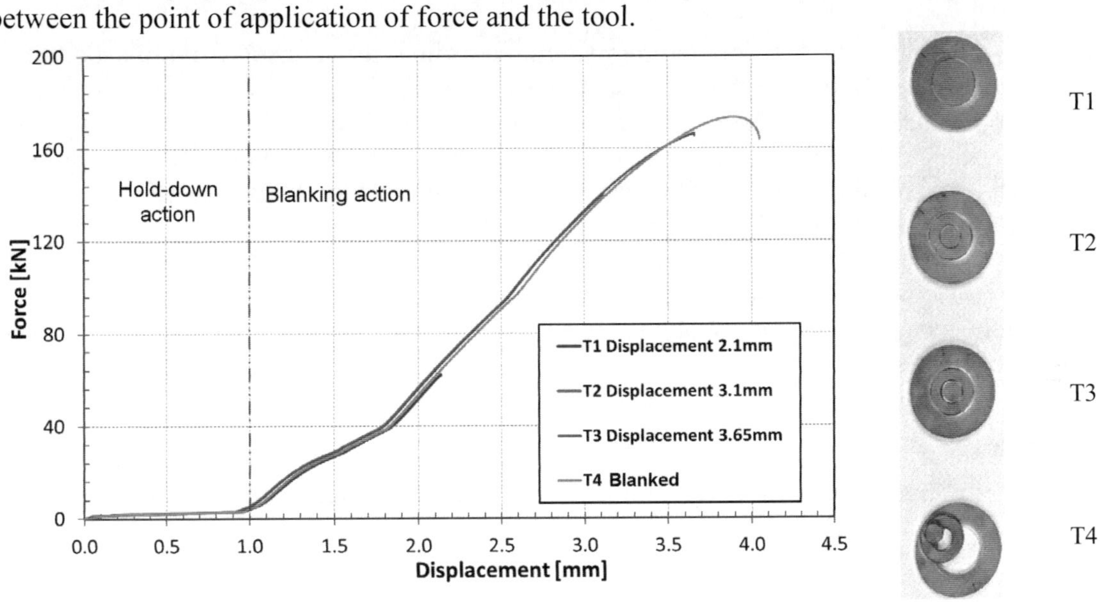

Fig. 13 Force-displacement curves for specimens T1 through T4

Fig. 15 compares measured data and the numerical model. The undefined compliance of the test tool and its parts, such as springs, was simplified in the numerical model, which led to differences between the simulation and the measurement. Although there were small displacement discrepancies, the real-world experiment and the simulation were in a very good overall agreement. The differences between the results of fracture calculations were given by specific model definitions. The dip in the J-C model data is due to damage in compression, where the J-C model reports lower fracture values, whereas the B-W model reports almost infinity.

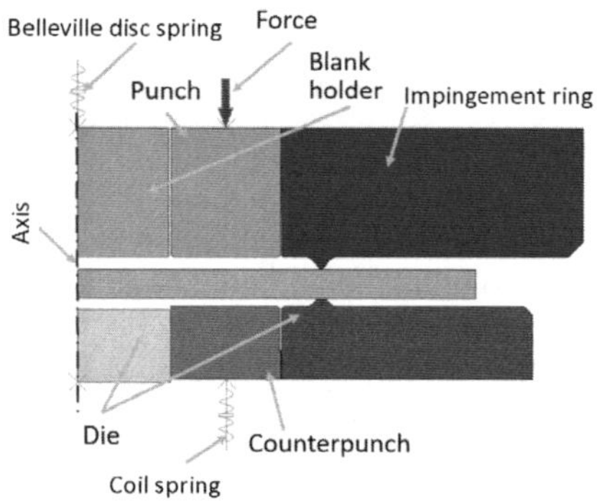

Fig. 14 Fine blanking

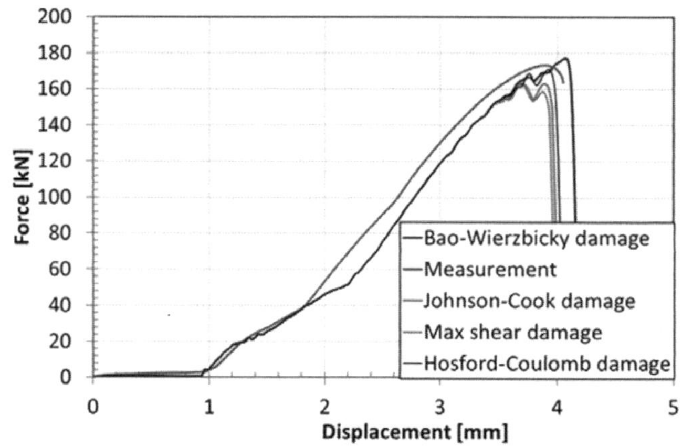

Fig. 15 Results of fracture modelling using individual models and comparison with real-world data

Conclusions

In the present paper, a method of finding material constants for Johnson-Cook plasticity and fracture models and Bao-Wierzbicky, Max shear and Hosford-Coulomb fracture models, as well as their validation by means of a technological test is described. Several types of test specimens suitable for calibrating these material models were presented. The models were validated by fine blanking of a precise stainless steel washer. Good agreement was achieved between the measurement and simulations. This means that the calibrated constants offer an accurate description of material behaviour. A material model developed and validated by this procedure is useful in the design and optimization of real-world shearing processes. It was found that the model which was the plainest in terms of calibration data was still fully sufficient for practical engineering use.

Acknowledgement

This study was created by project Development of West-Bohemian Centre of Materials and Metallurgy No.: LO1412, financed by the MEYS of the Czech Republic.

References

[1] Ł. Bohdal et al., Three Dimensional Finite Element Simulation of Sheet Metal Blanking Process. Applied Mechanics and Materials. 474, 430–435 (2014).

[2] T. Wesner∗ and P. Hora, Investigation of ductile fracture in fine blanking processes and virtual prediction of scar formation by means of 3d ale simulations, XIII International Conference on Computational Plasticity. Fundamentals and Applications (2015) 562 – 575.

[3] Rund M, Mašek M, Džugan J, Konopík P, Janovec J. Experiment and finite element analysis of U-profile subjected to dynamic loading. Buzaud E, Cosculluela A, Couque H, Cadoni E, eds. EPJ Web of Conferences 2018; 183:2056.

[4] Dalloz, A., Besson, J., Gourgues-Lorenzon, A.F., Sturel, T., and Pineau, A. (2009). Effect of shear cutting on ductility of a dual phase steel. Engineering Fracture Mechanics 76, 1411–1424

[5] Gutknecht, F., Steinbach, F., Hammer, T., Clausmeyer, T., Volk, W., and Tekkaya, A.E. (2016). Analysis of shear cutting of dual phase steel by application of an advanced damage model. Procedia Structural Integrity 2, 1700–1707

[6] Wang, K., and Wierzbicki, T. (2015). Experimental and numerical study on the plane-strain blanking process on an AHSS sheet. International Journal of Fracture 194, 19–36.

[7] Canales, C., Bussetta, P., and Ponthot, J.P. (2017). On the numerical simulation of sheet metal blanking process. International Journal of Material Forming 10, 55–71.

[8] URBÁNEK, Miroslav, TIKAL, Filip. Effective Preparation of Non-linear Material Models by Using Programmed Optimazation Script for Nurimerical Simulation of Sheet Metal Processing. In: Materiali in technologije, 49(2). Ljubljana, 2015. p. 5/291-295. ISSN 1580-2949.

[9] Peirs, J., Verleysen, P., Van Paepegem, W., and Degrieck, J. (2011). Determining the stress-strain behaviour at large strains from high strain rate tensile and shear experiments. International Journal of Impact Engineering 38, 406–415.

[10] Roth, C.C., and Mohr, D. (2018). Determining the strain to fracture for simple shear for a wide range of sheet metals. International Journal of Mechanical Sciences 149, 224–240.

[11] Hosford WF, A generalized isotropic yield criterion. J Appl Mech 1972; 39:607

[12] Wierzbicki, T., Bao, Y., Lee, Y.-W., and Bai, Y. (2005). Calibration and evaluation of seven fracture models. International Journal of Mechanical Sciences 47, 719–743.

[13] SCHMIDT, R.-A., P. HÖFEL, B. REH, F. BIRZER, M. HELLMANN, P. RADEMACHER a H. HOFFMANN. *Cold Forming and Fineblanking: a Handbook on cold processing, steel material properties and part, part design*. 1. Wien: Carl Hanser Verlag GmbH and Company, 2007. ISBN 9783446413504.

[14] Roth, C.C., and Mohr, D. (2016). Ductile fracture experiments with locally proportional loading histories. International Journal of Plasticity 79, 328–354.

EAN 2019
57th conference on experimental stress analysis

An alternative neo-Hookean material model with internal damping for finite-strain elastoplasticity

JERÁBEK Róbert[1,a] and ÉCSI Ladislav[1,b]

[1]Institute of Applied Mechanics and Mechatronics; Faculty of Mechanical Engineering; Slovak University of Technology in Bratislava; Námestie slobody 17, 812 31 Bratislava 1, Slovak Republic

[a]robert.jerabek@stuba.sk, [b]ladislav.ecsi@stuba.sk

Keywords: Nonlinear continuum theory for finite deformation of elastoplastic media, objective and thermodynamically consistent material model, Neo-Hookean material with internal damping

Abstract. At present, multiplicative plasticity theories are used to model material degradation of hyperelastic materials within the framework of finite-strain elastoplasticity. The theories assume that the intermediate configuration of the body is locally unstressed, for which no deformation exists that meets the conditions of compatibility. The assumption, however; is in contradiction to the theory of nonlinear continuum mechanics, as it violates proper stress transformations, resulting from the invariance of the internal mechanical power. As a result, the theories and their related material models are not continuum based. In this paper, an alternative neo-Hookean material model with internal damping for finite-strain elastoplasticity is presented. The model, based on the first nonlinear continuum theory of finite deformations of elastoplastic media, can imitate material degradation of the neo-Hookean material under excessive dynamic loading.

Introduction

Hyperelastic materials, such as rubber, vulcanized elastomers, various types of polymers, biomaterials, etc. are nonlinear elastic materials. The materials have a wide range of applications in industry. Hyperelastic materials can withstand large strains without undergoing permanent deformations or being fractured. The materials are considered to be isotropic under elastic loading and their constitutive equations are derived from strain energy density functions [1], defined in terms of the stored energy in the material per unit of the reference volume at each time instant [2]. In computational mechanics, there are available a number of different strain-energy functions to model hyperelastic materials. Without the need for completeness, let us mention a few concrete examples of hyperelastic materials and their applications. The Mooney-Rivlin material is used to model rubber-like materials and polymers [1], whose special case is the neo-Hookean material [1]. The Yeoh material model [3] was developed to model carbon-black filled rubber vulcanizates. The Ogden material has widespread use in modelling biomaterials and soft tissues [4, 5]. The Blatz-Ko [6] and the Arruda-Boyce [7] material models were proposed to model foamed or compressible elastomers. The Ogden-Storakers material has widespread use in modelling highly compressible foamed elastomers [8]. From among the newer materials, one should mention the Bower material [9], the Gent material [10] and the Dill material [11] for modelling rubber-like solids and polymers.

Contemporary material models for material degradation of hyperelastic materials use hyperelastic-plastic based multiplicative plasticity models to model the behaviour of the materials within the framework of finite-strain elastoplasticity. Though contemporary multiplicative plasticity models are considered to be continuum based, in reality they are not if one studies them strictly from the nonlinear continuum mechanics point of view. The theories assume that the intermediate configuration of the body is stress-free or locally unstressed, for which no deformation exists that meets the conditions of compatibility [12]. The assumption, however; is in contradiction to the theory of nonlinear continuum mechanics, as it violates proper stress transformations resulting from the invariance of the internal mechanical power.

The aim of this paper is to present a study of material degradation of a silicone specimen in uniaxial tension using a thermodynamically consistent neo-Hookean material model with internal damping. The model, which is based on the first nonlinear continuum theory of finite deformations of elastoplastic media [13], can imitate material degradation of the neo-Hookean material under excessive dynamic loading. In this paper a few selected analysis results are presented and briefly discussed.

Theory

The constitutive equation of the material. Hyperelastic materials are nonlinear elastic materials, whose constitutive equations are derived from strain energy density functions [1]. When in addition, they undergo material degradation during their mechanical loading, the materials are often modelled as elastoplastic, using the accumulated plastic strain as a measure of the material degradation. Considering the nonlinear continuum theory of finite deformations of elastoplastic media [13, 14], the elastic part of the deformation gradient can be expressed in the intermediate configuration of the modelled body as [14, 15]

$$\mathbf{F}^{el} = \mathbf{F} - \frac{\partial^{0}\mathbf{u}^{pl}}{\partial \mathbf{X}}. \tag{1}$$

The definition of the left isochoric Cauchy-Green tensor is based on the Lagrangian multiplicative split of the elastic deformation gradient $\mathbf{F}^{el}$ into a volumetric part $\mathbf{F}^{el}_{vol}$ and an isochoric part $\mathbf{F}^{el}_{iso}$. The left elastic isochoric Cauchy-Green tensor then takes the form [1, 12]

$$\mathbf{B}^{el}_{iso} = \mathbf{F}^{el}_{iso} \cdot \left(\mathbf{F}^{el}_{iso}\right)^{T} = \det\left(\mathbf{F}^{el}\right)^{-\frac{2}{3}} \cdot \mathbf{B}^{el}. \tag{2}$$

Since the material remains elastic in the intermediate configuration, there is no need to modify its strain energy density function. In this research neo-Hookean material model is used to model the plastic behaviour of hyperelastic-plastic material, whose strain energy density function takes the following well-known form [12, 13]

$$\overline{\psi}^{*}\left(I_1^{*}, J^{el}\right) = G \cdot \left(I_1^{*} - 3\right) + \frac{1}{d} \cdot \left(J^{el} - 1\right)^{2}, \tag{3}$$

where $I_1^{*} = I_1^{*}\left(\mathbf{B}^{el}_{iso}\right) = tr\left(\mathbf{B}^{el}_{iso}\right)$ is the first invariant of the characteristic equation of the corresponding eigenvalue problem using the left isochoric Cauchy-Green deformation tensor $\mathbf{B}^{el}_{iso}$ and $J^{el} = \det\left(\mathbf{F}^{el}\right)$ is the Jacobian of elastic deformation. The model uses two material parameters, the shear modulus G and a parameter d, which controls bulk compressibility. If d is zero, the material is incompressible. The relationship between the bulk modulus and the parameter d is then defined as [1]

$$K = \frac{2}{d}, \tag{4}$$

where the former can also be expressed in terms of the shear modulus G and the Poisson's ratio ν of the material as follows [16]

$$K = \frac{2 \cdot G \cdot (1+\nu)}{3 \cdot (1 - 2 \cdot \nu)}. \tag{5}$$

The stress constitutive function of the material in the intermediate configuration of the body then can be expressed as a Kirchhoff stress measure in the form

$$\boldsymbol{\tau}^{el} = \frac{\partial \overline{\Psi}^*}{\partial \mathbf{F}^{el}} \bullet \left(\mathbf{F}^{el}\right)^T = G \cdot dev\left[\mathbf{B}_{iso}^{el}\right] + \frac{2}{d} \cdot J^{el} \cdot \left(J^{el} - 1\right) \cdot \mathbf{I}, \tag{6}$$

where the deviator of the left elastic isochoric Cauchy-Green tensor is defined as

$$dev\left[\mathbf{B}_{iso}^{el}\right] = \mathbf{B}_{iso}^{el} - \frac{tr\left(\mathbf{B}_{iso}^{el}\right)}{3} \cdot \mathbf{I}. \tag{7}$$

The corresponding elastic 2nd Piola-Kirchhoff stress tensor then takes the form [14, 17]

$$\mathbf{S}^{el} = \left(\mathbf{F}^{el}\right)^{-1} \cdot \boldsymbol{\tau}^{el} \cdot \left(\mathbf{F}^{el}\right)^{-T}. \tag{8}$$

In order to take into account the internal/material damping, the material model is modified as follows

$$\mathbf{S} = \mathbf{S}^{el} + \mathbf{S}^{vis} = \left(\mathbf{F}^{el}\right)^{-1} \cdot \boldsymbol{\tau}^{el} \cdot \left(\mathbf{F}^{el}\right)^{-T} + {}^{mat}\mathbb{C}^{vis} : {}^{*}\dot{\mathbf{E}}^{el} \tag{9}$$

where ${}^{*}\dot{\mathbf{E}}^{el}$ is the time derivative of the elastic Green-Lagrangian strain tensor, derived from the elastic deformation gradient (Eqn. (1)) as follows [17]

$${}^{*}\dot{\mathbf{E}}^{el} = \frac{1}{2} \cdot \left[\left(\dot{\mathbf{F}}^{el}\right)^T \cdot \mathbf{F}^{el} + \left(\mathbf{F}^{el}\right)^T \cdot \dot{\mathbf{F}}^{el}\right], \tag{10}$$

and ${}^{mat}\mathbb{C}^{vis}$ is the fourth order material viscosity tensor, which can be calculated using the following formula

$${}^{mat}\mathbb{C}^{vis} = 2 \cdot G^{vis} \cdot \mathbb{I} + \lambda^{vis} \cdot \mathbf{I} \otimes \mathbf{I}, \tag{11}$$

$$G^{vis} = \frac{E^{vis}}{2 \cdot \left(1 + \upsilon^{vis}\right)}, \quad \lambda^{vis} = \frac{\upsilon^{vis} \cdot E^{vis}}{\left(1 + \upsilon^{vis}\right) \cdot \left(1 - 2 \cdot \upsilon^{vis}\right)}. \tag{12}$$

In Eqns. (11), (12) G^{vis} is the viscous shear modulus, λ^{vis} is the viscous Lamé constant, $\mathbb{I}$ denotes the symmetric fourth-order identity tensor, $\mathbf{I}$ is the second-order identity tensor, E^{vis} is the viscous Young's modulus and υ^{vis} is the viscous Poisson's ratio.

The stress constitutive function of the modified neo-Hookean material with internal damping then takes its final form in the current configuration as

$$\boldsymbol{\tau} = \mathbf{F} \cdot \mathbf{S} \cdot \mathbf{F}^T = \mathbf{F} \cdot \left[\left(\mathbf{F}^{el} \right)^{-1} \cdot \boldsymbol{\tau}^{el} \cdot \left(\mathbf{F}^{el} \right)^{-T} + {}^{mat}\mathbb{C}^{vis} : {}^{*}\dot{\mathbf{E}}^{el} \right] \cdot \mathbf{F}^T. \quad (13)$$

The reference definition of the yield surface. In this study J2 plasticity with isotropic hardening was considered. The yield surface in the material model was defined in the Kirchhoff stress space as follows [16, 17]

$$^{\tau}\psi = {}^{\tau}\sigma_{eq}\left(\boldsymbol{\tau}\right) - \tau_y \leq 0, \quad (14)$$

where the von-Mises equivalent Kirchhoff stress is defined as

$$^{\tau}\sigma_{eq}\left(\boldsymbol{\tau}\right) = \sqrt{\frac{3}{2} \cdot \sum \boldsymbol{\tau} : \sum \boldsymbol{\tau}}, \qquad \sum \boldsymbol{\tau} = \boldsymbol{\tau} - \frac{tr\left(\boldsymbol{\tau}\right)}{3} \cdot \mathbf{I}. \quad (15)$$

In Eqn. (14) τ_y denotes the Kirchhoff yield stress, which in the analysis was expressed as

$$\tau_y = \sqrt{r^2 - \left(a \cdot \frac{\partial^0 u^{pl}_{UT11}}{\partial X} \right)^2}, \quad r = {}^{\tau}\tau_y + {}^{\tau}Q, \quad center = \sqrt{r^2 - {}^{\tau}\tau_y^2}, \quad a = \frac{center + r}{b}, \quad (16)$$

where ${}^{\tau}\tau_y$ is the constant yield stress and ${}^{\tau}Q$ is the maximum hardening stress. Here, the parameter $\partial^0 u^{pl}_{UT11} / \partial X$ controls the material hardening/softening. It stands for the axial component of the material gradient of the Lagrangian plastic displacement field and is determined during uniaxial tensile testing of the material. In Eqn. (16) b denotes the maximum value of $\partial^0 u^{pl}_{UT11} / \partial X$, at which the material loses its integrity, i.e. when $\tau_y = 0$. In Eqn. (17) the spatial and the material gradients of the plastic velocity field are defined as follows, where $\mathbf{N}$ is a unit outward normal of the yield surface and $\dot{\lambda}$ is the plastic multiplier [14, 15]

$$\frac{\partial \dot{\mathbf{u}}^{pl}}{\partial \mathbf{x}} = \mathbf{d}^{pl} = \dot{\lambda} \cdot \frac{\partial^{\tau}\Psi}{\partial \boldsymbol{\tau}} = \dot{\lambda} \cdot \sqrt{\frac{3}{2}} \cdot \mathbf{N}, \quad \mathbf{N} = \frac{\Sigma\boldsymbol{\tau}}{\left\| \Sigma\boldsymbol{\tau} \right\|} = \frac{\Sigma\boldsymbol{\tau}}{\sqrt{\Sigma\boldsymbol{\tau} : \Sigma\boldsymbol{\tau}}}, \quad \frac{\partial^0 \dot{\mathbf{u}}^{pl}}{\partial \mathbf{X}} = \dot{\lambda} \cdot \sqrt{\frac{3}{2}} \cdot \mathbf{N} \cdot \mathbf{F}. \quad (17)$$

The corresponding formulas at a material point of the specimen during uniaxial tensile testing of the material then can be determined as

$$\frac{\partial \dot{u}^{pl}_{UT11}}{\partial x} = \dot{\lambda} \cdot \sqrt{\frac{3}{2}}, \quad \frac{\partial^0 \dot{u}^{pl}_{UT11}}{\partial X} = \frac{\partial \dot{u}^{pl}_{UT11}}{\partial x} \cdot F_{UT11} = \dot{\lambda} \cdot \sqrt{\frac{3}{2}} \cdot F_{UT11}, \quad \frac{\partial^0 u^{pl}_{UT11}}{\partial X} = \int_0^t \frac{\partial^0 \dot{u}^{pl}_{UT11}}{\partial X} \cdot dt. \quad (18)$$

In the analysis, numerical uniaxial tensile testing was used to assess the values of the elastic deformation gradient $\mathbf{F}_{UT}^{el}$, the axial component of the elastic deformation gradient F_{UT11}^{el}, the Jacobian of the elastic deformation gradient J_{UT}^{el} and the deformation gradient $\mathbf{F}_{UT}$ of the modelled material. Normally these would be determined experimentally, but since contemporary tensile testing does not provide the time-histories of the variables, we used the following formulas to determine them numerically [14, 15]

$$\mathbf{F}_{UT}^{el} = diag\left[F_{UT11}^{el}, \quad 1, \quad 1\right], F_{UT11}^{el} = F_{UT11} - \frac{\partial\, {}^0u_{UT11}^{pl}}{\partial X},$$
$$J_{UT}^{el} = \det\left(\mathbf{F}_{UT}^{el}\right), \mathbf{F}_{UT} = diag\left[F_{UT11}, \quad 1, \quad 1\right]. \tag{19}$$

Then during the analysis, the numerical uniaxial tensile testing too was realized by solving the following equation

$$\tau_{UT11} - \tau_y = 0. \tag{20}$$

Here τ_{UT11} is the axial component of the $\boldsymbol{\tau}_{UT} = \boldsymbol{\tau}$ stress constitutive function of the material defined by Eqn. (13), in which the deformation gradient $\mathbf{F}$ is replaced by $\mathbf{F}_{UT}$ and the elastic deformation gradient $\mathbf{F}^{el}$ by $\mathbf{F}_{UT}^{el}$. There was no change in the definition of the yield stress τ_y.

Numerical experiment

As a numerical experiment, material degradation of a silicone specimen in uniaxial tension was studied. The specimen dimensions were $0.05\text{m} \times 0.05\text{m} \times 0.6\text{m}$. The elastic material properties of the neo-Hookean material model were determined by the collective K.B. Putra et. al. [18]. One end of the specimen was fixed, while the second end underwent axial deformation by imposing tensile surface traction on it as a product of constant $p = 0.5\ MPa$ pressure and a Heaviside step function. In order to model the grip of the testing machine, the specimen was guided in transversal directions by imposing zero displacements in X and Y directions at its moving end as shown in Fig. 1. The figure also shows the spatially discretized body of the specimen where the arrows indicate the loading.

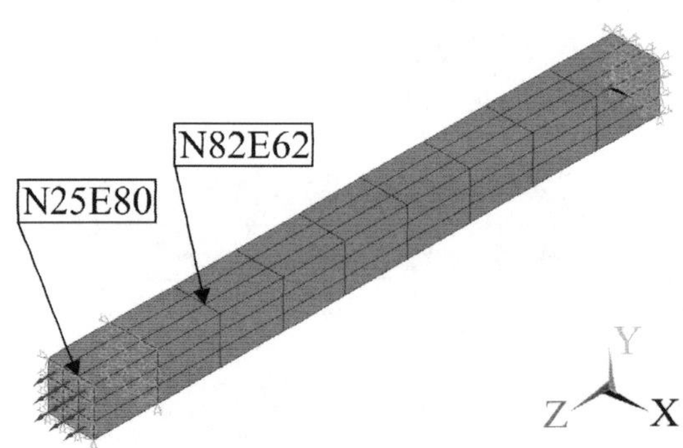

Fig. 1: Spatially discretized specimen in the analysis

The specimen was initially at rest. The analysis was run as transient dynamic until it failed to converge using implicit time integration and a $0.00001\ s$ time step size. Table 1 outlines the material properties of the specimen used in the finite element analysis.

Table 1: The material properties of the silicone specimen

$\rho_0 \left[kg \cdot m^{-3}\right]$	1520
$G\left[Pa\right]$	229 600
$d\left[-\right]$	0.00000624054
$E^{vis}\ \left[Pa \cdot s\right]$	7310
$\upsilon^{vis}\ \left[-\right]$	0.33
${}^{\tau}\tau_y \left[Pa\right]$	350 000
${}^{\tau}Q\left[Pa\right]$	50 000
$b\left[-\right]$	2.0

Numerical results

Fig. 2 shows a few selected results at the end of the numerical analysis. These are the axial displacement distribution, the accumulated plastic strain distribution, the 1st principal stress distribution as a Cauchy's stress measure, the von Mises stress distribution as a Cauchy's stress measure, the parameter $\partial\,{}^{0}u^{pl}_{UT11} / \partial X$ and the axial component of the deformation gradient F_{UT11} at the end of the analysis in the current configuration of the body. The values of the parameter $\partial\,{}^{0}u^{pl}_{UT11} / \partial X$ and of the axial component of the deformation gradient F_{UT11} are calculated from the numerical tensile test of the material.

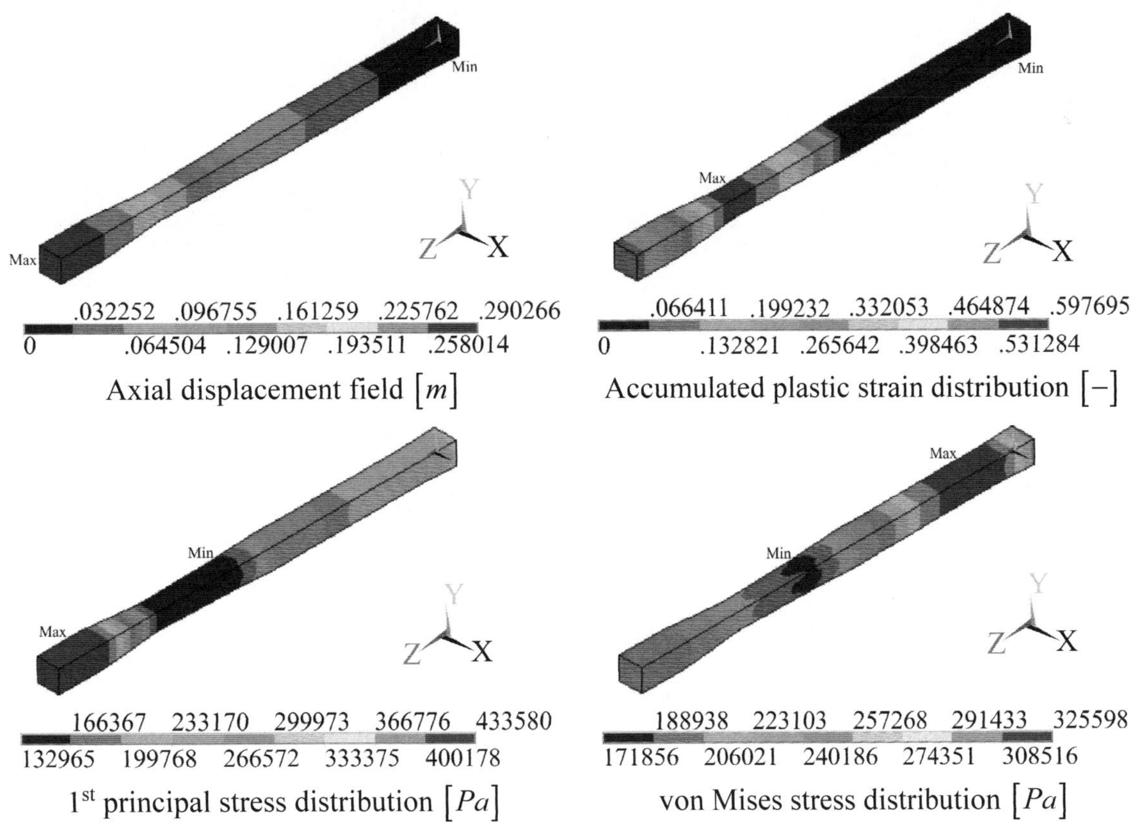

Axial displacement field $[m]$

Accumulated plastic strain distribution $[-]$

1st principal stress distribution $[Pa]$

von Mises stress distribution $[Pa]$

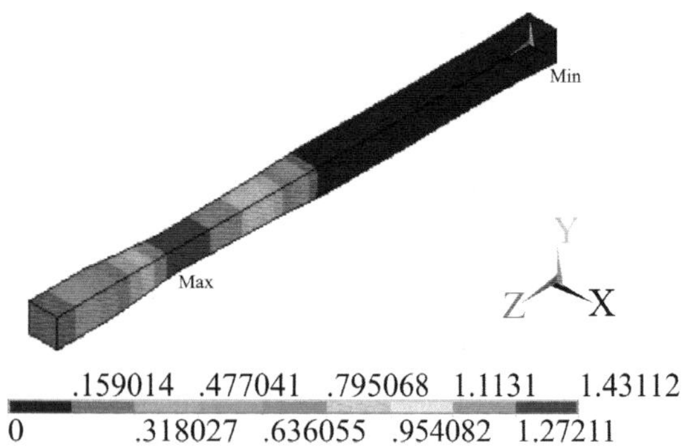

Parameter $\partial^0 u^{pl}_{UT11} / \partial X$ $[-]$

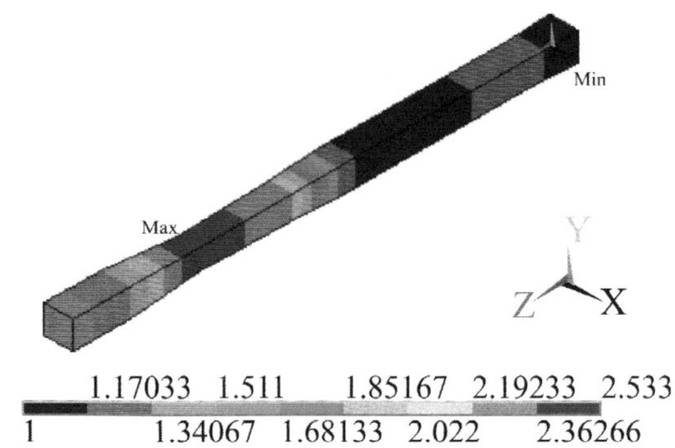

Axial component of the deform. gradient $[-]$

Fig. 2: A few selected results at the end of the analysis

As can be seen in the above figure, the maximum axial displacement of the moving end is $0.29\ m$ at the end of the analysis, i.e. when the analysis stopped to convergence. The maximum elongation is 48.3 %, which approximately corresponds to 60 % of the accumulated plastic strain and 143 % of the parameter $\partial^0 u^{pl}_{UT11} / \partial X$ that controls the isotropic hardening process. The axial component of the deformation gradient coming from the numerical uniaxial tensile test of the material is about 2.53 at the end of the analysis.

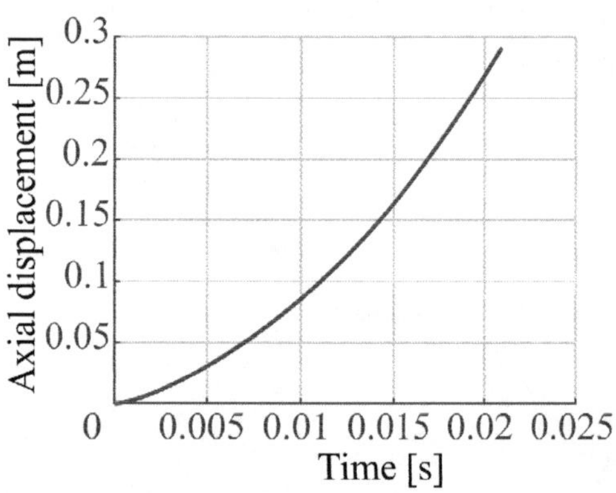

Axial displacement time-history curve at node *N25E80*

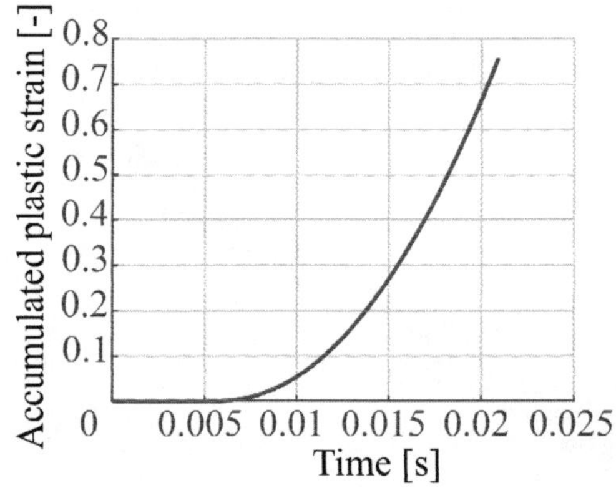

Accumulated plastic strain time-history curve at node *N82E62*

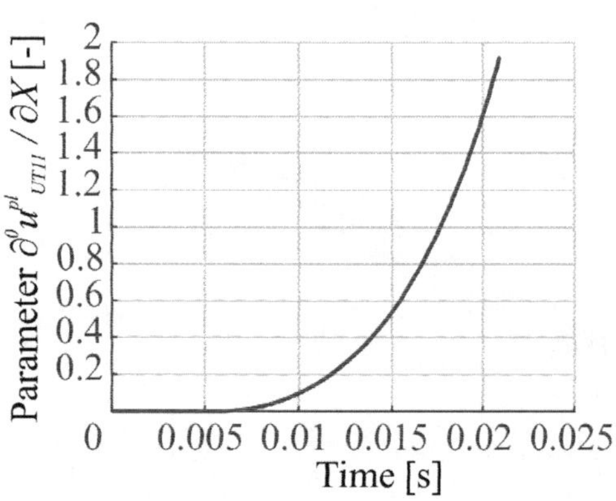

The time-history curve of the parameter $\partial^0 u^{pl}_{UT11} / \partial X$ at node *N82E62*

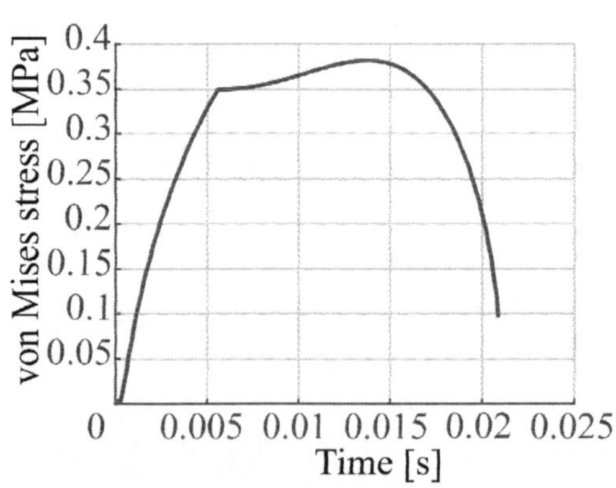

von Mises stress time-history curve at node *N82E62*

Fig. 3: A few selected time-history curves

Fig. 3 depicts a few selected time-history curves. These are the axial displacement time-history curve at node *N25* of element *E80* at the moving end of the specimen, the accumulated plastic strain time-history curve, the time-history curve of the parameter $\partial^0 u_{UT11}^{pl} / \partial X$ and the von Mises stress time-history curve at node *N82* of element *E62*. The locations of the nodes *N25E80* and *N82E62* are depicted in Fig. 1. As can be seen in the above figure, the material degradation starts approximately at 0.0056 s and continues until the end of the analysis, which can distinctly be seen in the time-history curves of the accumulated plastic strain and the parameter $\partial^0 u_{UT11}^{pl} / \partial X$ respectively. Similarly, the von Mises stress value rapidly increases in elastic loading and also during material hardening, but at a decreasing pace in the hardening stage. When the stress reaches its maximum at 0.014 s, i.e. when the maximum load-carrying capacity of the body has been reached, the von Mises stress starts to decrease until the analysis fails to converge. Such behaviour, shown on the von-Mises stress time-history curve of the Cauchy's stress of the neo-Hookean material, is characteristic of all materials under excessive loading. Material softening is caused by internal damaging processes in the material, which in phenomenological numerical analyses are taken into account by the uniaxial stress-strain curve of the material.

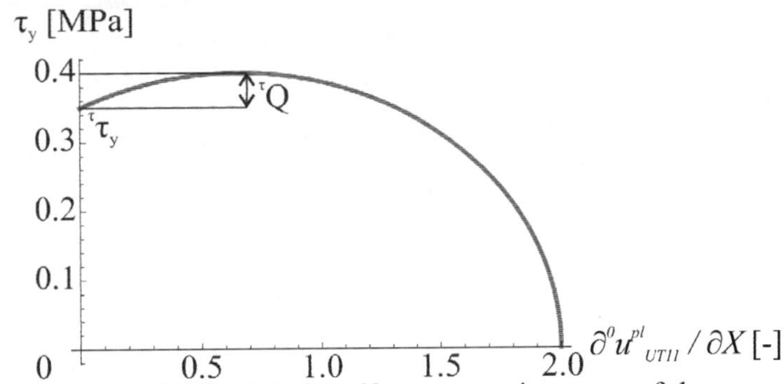

Fig. 4: The uniaxial Kirchhoff stress-strain curve of the material

Fig. 4 shows the uniaxial stress-strain curve of the neo-Hookean material used in the analysis. As it was mentioned in the above, the parameter $\partial^0 u_{UT11}^{pl} / \partial X$ controls the hardening/softening process. When its value exceeds 0.6, the material softens.

It also should be noted that the von Mises stress and the 1st principal stress values as Cauchy's stress measures are lower than their values as Kirchhoff stress measures. This is due to the fact that the Jacobian of deformation is $J > 1$ during the whole loading process, where considering that $\boldsymbol{\sigma} = \boldsymbol{\tau} / J$, caused the aforementioned decrease [15].

Conclusions

In this research uniaxial tension of a silicone specimen under dynamic loading was studied using a modified neo-Hookean material model with internal damping. The material model is based on the first continuum theory for finite deformations of elastoplastic media, which allows for the development of objective and thermodynamically consistent material models. The results of such material models are independent of the model description and the particularities of the model formulation. Although our analysis results seem to be reasonable, the proper verification of the model requires thorough material testing in order to determine all material parameters of the presented neo-Hookean material model. In addition to this, uniaxial tensile testing has to be extended by deformation gradient determination, which in general is not the subject of current material testing procedures.

References

[1] R.M. Hackett, Hyperelasticity primer, Springer, NY, 2016.

[2] A. Ali, M. Hosseini, B. Sahari, A review of constitutive models for rubber-like materials, American Journal of Engineering and Applied Sciences. 3 (2010) 1:232–239.

[3] O.H. Yeoh, Characterization of Elastic Properties of Carbon-black-filled rubber vulcanizates, Rubber Chem. Technol. 63 (1990) 792–805.

[4] R.W. Ogden, Non-Linear Elastic Deformations, Chichester, Ellis Horwood, 1984.

[5] R.W. Ogden, Large Deformation Isotropic Elasticity – On the Correlation of Theory and Experiment for Incompressible Rubberlike Solids, Proc. R. Soc. Lond. A, 326 (1972) 565–584.

[6] P.J. Blatz, W.L. Ko, Application of Finite Elasticity to the Deformation of Rubbery Materials, Trans. Soc. Rheol. 6 (1962) 223–251.

[7] E.M. Arruda, M.C. Boyce, Three-dimensional Constitutive Model for the Large Stretch Behavior of Rubber Elastic Materials, J. Mech. Phys. Solids. 41 (1993) 389–412.

[8] B. Storakers, On material representation and constitutive branching in finite compressible elasticity, J. Mech. Phys. Solids. 34 (1986) 125-145.

[9] A.F. Bower, Applied Mechanics of Solids, CRC, Boca Raton, 2010.

[10] A.N. Gent, A New Constitutive Relation for Rubber, Rubber Chem. Technol. 69 (1996) 59-61.

[11] E.H. Dill, Continuum Mechanics: Elasticity, Plasticity, Viscoelasticity, CRC, Boca Raton, 2007.

[12] E.A. de Souza Neto, D. Perić, D.R.J. Owen, Computational methods for plasticity, Theory and applications, John Wiley & Sons Ltd., Singapore, 2008.

[13] L. Écsi, P. Élesztős, An alternative material model using a generalized J2 finite-strain flow plasticity theory with isotropic hardening, Int. J. Applied Mechanics and Engineering, 23 (2018) 2:351-365

[14] L. Écsi, P. Élesztős, An Alternative Method for Modelling the Degradation of Hyperelastic Materials within the Framework of Finite-strain Elastoplasticity. In. Engineering Design Applications II: Structures, Materials and Processes. Eds. A. Öchsner and H. Altenbach, Springer, 2019 (In press).

[15] R. Jerábek, L. Écsi, Numerical Study of Material Degradation of a Silicone Cross-shaped Specimen Using a Thermodynamically Consistent Mooney-Rivlin Material Model, Novel Trends in Production Devices and Systems V, Materials Science Forum. 952 (2019) 258–266.

[16] M.A. Crisfield, Non-linear finite element analysis of solids and structures, Essentials, Vol. 1, John Wiley & Sons Ltd., Chichester, 2000.

[17] L. Écsi, P. Élesztős, R. Jerábek, R. Jančo, B. Hučko, An Alternative Framework for Developing Material Models for Finite-Strain Elastoplasticity, IntechOpen Ltd., London, DOI: 10.5772/intechopen.85112, 2019.

[18] K.B. Putra, J. Plott, A.J. Shih, Biaxial Mooney-Rivlin coefficient of silicone sheet by additive manufacturing, Procedia CIRP. 65 (2017) 189–195.

EAN 2019
57th conference on experimental stress analysis

Trabecular structures as efficient surface of dental implants

JÍRA Aleš[1,a], HÁJKOVÁ Petra[1,b] and ŘEHOUNEK Luboš[1,c]

[1]Department od Mechanics, Faculty of Civil Engineering, Czech Technical University in Prague, Thákurova 7, 166 29 Prague 6, Czech Republic

[a] jira@fsv.cvut.cz, [b] petra.hajkova@fsv.cvut.cz, [c] lubos.rehounek@fsv.cvut.cz

Keywords: 3D printing, Porous layer, Trabecular structures, Dental implant, Gyroid

Abstract. Fast development of additive manufacturing methods provides for production of advanced porous structures which form external layers of implants. Since implants usually replace parts of missing or damaged bone, its external layers are close to bone structure. Today, most of the porous structures which are used for implants are trabecular structures, composed of beams. Design of the structures depends on the type of implants, width of layers etc. Small implants, such as dental implants, have very thin external layers (~ 200 μm), which means that the structure is very fine. The quality of 3D printed structures therefore depends heavily on accuracy of current 3D printers (SLM method). Problems of 3D printing which can lead to implant failure are cracks inside of beams, their poor interconnection, etc.

The study is focused on the quality control of 3D printed trabecular structures, especially with respect to implant stability in bones and minimization of a risk of structural damage. The quality of the structures was evaluated by optical and microscopic analysis using SEM. Based on the results, we suggest substituting the trabecular structures by a gyroid structure composed of a system of walls. Gyroid structures are not prone to local discontinuities and have other advantages such as absence of sharp corners and edges, which contribute to better proliferation of bone cells.

Introduction

Development of additive manufacturing methods provides for production of complex trabecular structures. The purpose of trabecular structures, as an external layer of implants, is to improve osseointegration. Advantages of the structures lead to rapid scientific and technological development in the field of 3D printed metallic structures [1]. The main advantages are growth of contact area between implants and bone tissue, and a possibility to modify global mechanical properties, which helps to eliminate the stress shielding effect – the main problem of conventional implants. Furthermore, complex morphology of the structures allows the ingrowth of bone cells, which leads to better stability of dental implants in bones.

However, designed mechanical properties of 3D printed structures can be significantly affected by the quality of 3D printing. The current technological possibilities of the SLM method (Selective Laser Melting) are limited by fineness of the powder, which is used for the structures' production, and the machine accuracy. The accuracy of 3D printing is very important especially for dental implants because of its size [2]. Dental implants are ~3–5 mm in diameter, which means that the surface layer is very thin.

Authors [3, 4] who have dealt with the quality of 3D printed porous structures mentioned a mismatch between designed and as-produced width of beams in fine structures. It affects the porosity and therefore mechanical properties of the structures. Furthermore, they implied that there was a problem with production of the structures composed of horizontal and angled beams, for example dodecahedron, because of greater amount of sintered plaque, which is the result of melting. The melting of poor quality can also cause production of small pores (50–100 μm) which were not designed [5]. Bael et al. [6] pointed out that the structures with beams forming corners were often filled with excessive material.

The study is focused on the quality control of trabecular structures, especially with respect to implant stability in bones and minimization of a risk of the structure damage. For this purpose, we produced samples of six different trabecular structures. The beams thickness was ~200–300 μm. After the quality control, the samples were intended for use in mechanical compression and tension tests. Extensive defects were detected by a basic optic analysis. The defects were situated mainly near the interface of the trabecular and homogenous part of the samples. This finding led to a modification of a geometric shape of the samples for the tension tests, and a detailed microscopic analysis.

Materials and Methods

Design of Trabecular Structures. Trabecular structures usually take a form of spatial truss structures composed of beams. For the tests we designed 6 models of trabecular structures with different morphology, i. e. a size and a shape of pores, a width of beams, and orientation of beams in a 3D matrix. The structures were composed of basic units in the shape of Diamond, Dode Thick 30 %, and Rhombic dodecahedron 30 % (Fig. 1). Each of the basic units was used in two sizes for the design of the trabecular structures. Table 1 shows a summary of the structure parameters.

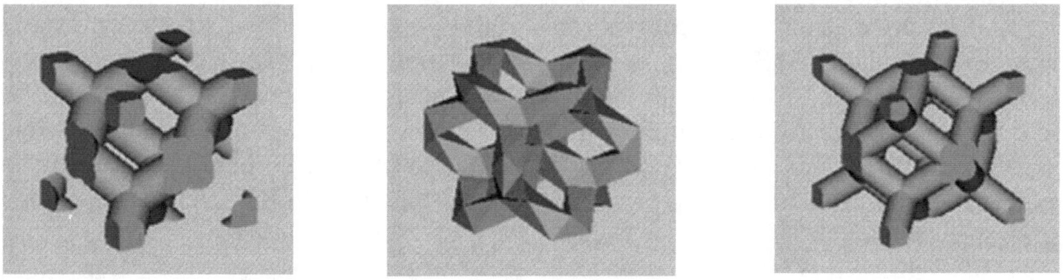

1) Diamond, relative density 30 %   2) Dode Thick [MSG]   3) Rhombic dodecahedron, relative density 30 %

Fig. 1: Basic units of the trabecular structures which are composed of beams.

The samples were composed of the trabecular and homogenous parts. The trabecular part of the samples was a cube (14x14x14 mm), which was connected to the homogenous parts from two opposite sides. Shapes of the homogenous parts were set up for the compressive and tension tests, which came after the microscopic analysis.

Tab. 1: The parameters of the trabecular structures which were designed for the microscopic analysis, compression and tension tests.

Trabecular structure	Shape of basic units	Size of basic units [mm]	Width of beams [μm]	Size of pores [μm]
1	Diamond	0,75	200	350
2	Diamond	1,00	260	450
3	Dode Thick	1,00	200	500
4	Dode Thick	1,25	250	630
5	Rhombic Dodecahedron	1,25	230	640
6	Rhombic Dodecahedron	1,50	290	800

Production of Samples. Basic material for the samples production was an alloy of titanium Ti6Al4V, also called Rematitan (Concept Laser, Germany), in the form of powder with maximal granularity of 63 μm. The samples were produced by an additive method, specifically by the method of SLM (Selective Laser Melting). The machine used for the samples production was M2 Cusing (Concept Laser, Germany). The 3D printing was performed in argon atmosphere, the argon purity of 4,6 (Linde, Germany). During the printing, the atmosphere in the welded chamber contained up to 0,5 % of oxygen.

After the samples were printed, it was necessary to perform a heat treatment. The aim of the heat treatment was elimination of inhomogeneities and internal tensions, which came from the metal powder sintering. The heat treatment was performed in a vacuum furnace. A temperature in the furnace increased gradually up to 840 °C in 4 hours. After that, the temperature was held constant for 2 hours. Finally, the samples were cooled down to minimum of 500 °C. During the cooling, the vacuum was substituted by an argon atmosphere [7].

Optic and Microscopic Analysis for Quality Control. The basic optic analysis provided an optical analysis of the surface morphology and quality of the beams. The surface morphology was analysed particularly from the point of view of defects, initiation of discontinuities, and possible local complications. For this purpose, photographs from the camera Canon 6D Mark II and the lens Canon MP-E 65mm f/2,8 Macro 5x were used.

Based on the results of the basic optic analysis, we decided to complete the results by a detailed microscopic analysis. The detailed analysis was performed on an electron microscopy SEM Phenom XL (Phenom World, Netherlands).

Results

The basic optic analysis showed significant uncertainties about a quality of connection between the individual parts of the samples. On the interface between the trabecular and homogenous part, many cracks in the beams and its poor interconnections were detected (Fig. 2). These defects can be a consequence of temperature changes during the 3D printing and the heat treatment. When the trabecular part of the samples is printed earlier than the homogenous part, the fine trabecular structure cools down faster. The difference in cooling down rate causes a difference in shrinkage of the individual parts. As the result of this process, the beams prone to shorten and break away from the warmer homogenous part. This observation is cause for concern that implants which are composed of homogenous internal body and trabecular external layer will contend with the same problem. During development of 3D printed implants with trabecular external layer, it is necessary to remove or eliminate the problem as much as possible.

Fig. 2: Defects of the samples on the interface between the trabecular and homogenous part.

Based on the results of the basic optic analysis, we performed a detailed microscopic analysis using SEM. Many discontinuities of the beams (Fig. 3a) and clusters of slag (Fig. 3b) were detected inside the structures. The capillary defects can cause a break of microparticles, a subsequent genesis of necrosis, and aseptic release of implants. The question is whether this phenomenon can be limited by greater thickness of the beams (>300 μm). Another problem is that the beams with greater thickness cannot be used for all types of implants. If thickness of the beams is greater and the mechanical properties stay the same, it is necessary to increase a size of the basic units, which means to increase a size of the whole implants. Therefore, it is possible to increase thickness of the beams of the structures for bigger implants, such as implant for hip and knee replacement, but not for small implants, such as dental implant.

a)

b)

Fig. 3: The detailed microscopic analysis showed: a) discontinuities of the beams, and b) clusters of slag inside the trabecular structures. The photo was conducted by an electron microscope (SEM).

Discussion

The basic optical and the detailed microscopic analysis showed the significant uncertainties and the problems about trabecular structures composed of beams. Apart from the connections between trabecular and homogenous parts of the samples being of poor quality, many discontinuities of the beams and clusters of slag were detected inside the structures. It is necessary to consider if trabecular structures are appropriate for external layer of implants, or it is better to substitute it by other type of porous structures.

A potential solution can be porous structures based on a wall system, which also meet fundamental requirements for bone ingrowth. One of the best examples is the structure composed of walls in the shape of gyroid. This structure has a system of interconnected pores (open porosity), which is necessary for bone ingrowth, and sufficient permeability for proliferation of bone cells [8]. Furthermore, gyroid structures do not comprise sharp edges and corners, which are spanned by cells, and thus eliminate areas prone to damage.

For the verification of the hypotheses that gyroid structures could have greater reliability with respect to crack initiation and other defects, a preliminary optic and microscopic analysis was performed. The preliminary analysis confirmed that gyroid structures were less prone to defects which had been detected at the trabecular structures. Any local discontinuities were visible from the microscopic photos (Fig. 4). It results in elimination of the risk of microparticles release, which can lead, in the worst case, to aseptic release of implants. In addition, porous and homogenous parts are not joined just in local points but linearly. Greater area of joint can prevent breaking away of the parts from each other. All in all, we also reached greater stability of the structure by the transition from a trabecular to wall system. A general analysis of gyroid structures, which includes a microscopic analysis, macro/micro mechanical testing, and a numeric analysis, will be aim of our future research.

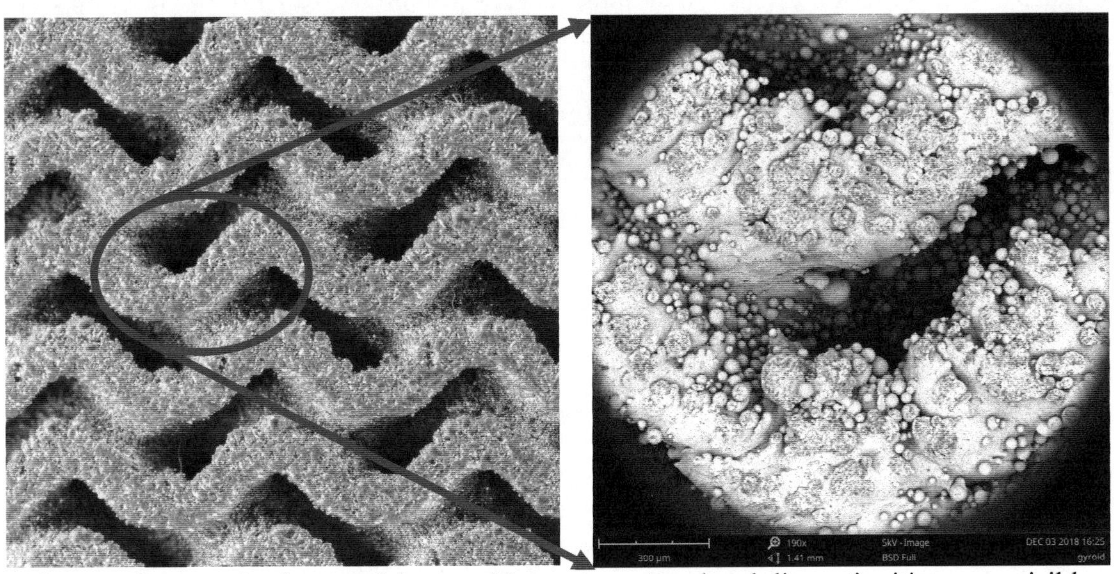

Fig. 4: Microscopic photos of the gyroid structure. Any local discontinuities were visible.

Conclusions

The uncertainties in the trabecular structures make us think over whether trabecular structures are an appropriate external layer for implants. We consider the cracks of the beams near the interface between porous and homogenous material, and inside the structures the biggest problem. The reason is that discontinuities of beams can cause failure of the whole structure, or release of microparticles into patient body. The resolution can be replacement of trabecular structures with porous structures based on a system of walls without sharp changes. The structure fulfilling the requirements is, for example, a gyroid structure. The preliminary optic and microscopic analysis proved that the gyroid structures were less prone to crack initiation and other defects, and therefore it is highly appropriate for inbone parts of implants (dental, acetabular, hip and knee replacements, etc.)

Acknowledgment

The financial support provided by the Technology Agency of Czech Republic, project n.TJ01000328 and by the Faculty of Civil Engineering, CTU, Prague, project n. SGS17/168/OHK1/3T/11 is gratefully acknowledged. The work was supported by Civil Engineering – Scanning Electron Microscope Laboratory, CTU Prague (CE-SEM Lab).

References

[1] M. M. Barak and M. A. Black, A Novel Use of 3D Printing Model Demonstrates the Effects of Deteriorated Trabecular Bone Structure on Bone Stiffness and Strength, J. of the Mech. Beh. of Biomed. Mat., 78 (2018) 455–464.

[2] Z. J. Wally et al., Selective Laser Melting Processed Ti6Al4V Lattices with Graded Porosities for Dental Applications, J. of the Mech. Beh. of Biomed. Mat., 90 (2019) 20–29.

[3] L. Mullen et al., Selective Laser Melting: A Regular Unit Cell Approach for the Manufacture of Porous, Titanium, Bone In-Growth Constructs, Suitable for Orthopedic Application, J. Biomed. Mater. Res. Part B Appl. Biomater., 89.2 (2009) 325–334.

[4] P. H. Warnke et al., Rapid Prototyping: Porous Titanium Alloy Produced by Selective lase Melting for Bone Tissue Engineering, Tissue Eng Part C Methods, 15.2 (2008) 115–124.

[5] N. Taniguchi el al., Effect of Pore Size on Bone Ingrowth into Porous Titanium Implants Fabricated by Additive Manufacturing: An In Vivo Experiment, Mater Sci Eng C Mater Biol Appl 59 (2016) 690–701.

[6] S. Van Bael et al., The Effect of Pore Geometry on the In Vitro Biological Behaviour of Human Periosteum-derived Cells Seeded on Selective Laser-melted Ti6Al4V Bone Scaffolds, Acta Biomater 8 (2012) 2824–2834.

[7] L. Řehounek et al., Geometrical Optimalization of Dental Implants with Regard to Osseointegration, Nano&Micro Mech. 2017, Acta Polytechnica CTU Proc. 13 (2017) 97-101.

[8] F. P. W. Melchels et al., Effects of the Architecture of Tissue Engineering Scaffolds on Cell Seeding and Culturing, Acta Biomater 6 (2010) 4208–4217.

EAN 2019
57th conference on experimental stress analysis

Measurement and simulation of dynamic systems parameters

JÍROVÁ Radka[1,a] and PEŠÍK Lubomír[1,b]

[1]Faculty of Mechanical Engineering, Technical University of Liberec, Czech Republic

[a]radka.jirova@tul.cz, [b]lubomir.pesik@tul.cz

Keywords: Dynamic system, Operating deflection shapes, Force simulation, Suspension

Abstract. Linear guiding systems are usually dynamically loaded. The similar situation emerges in the bearings of compactors drums. During the tests of the compactors are bearings dynamically loaded and reaction forces are transmitted through the test facility to the ground. For minimization of transmitting dynamic forces and emerging vibrations, is desirable to test the compactors on the suspended test facility. The aim of this paper was to identify initial conditions for the test facility suspension considering minimization of the emerging vibrations. We first identified operating conditions of the compactors - especially a loading uniformity and force amplitude transmitted to the ground. Force distribution we determined by ODS measurement. Force amplitude was calculated through mathematical model as a solution of differential equations. These results were further used for dynamic analysis of the suspension. The dynamic analysis performed showed efficiency of designed suspension. The force amplitude transmitted to the ground decreased significantly.

Introduction

Soil and asphalt compactors are usually designed with two tandem drums. During the operation those drums are excited by centrifugal force of the unbalanced mass, which is powered by hydraulic engine. Value of the centrifugal force and the oscillation amplitude are one of the most important machine parameters. For proper soil and asphalt compaction, it is necessary to inspect these parameters via experimental testing. The experimental tests are based on measuring the force and the amplitude on test facility that allows simulation of the compactor operating conditions. However, during tests the dynamic forces of the machine drums are transmitted through the test facility to the ground. It leads to high-intensity vibrations of the building construction. These vibrations affect negatively the building service life and labourers health. Minimization basic principles of the emerging vibrations are described in [1, 2].

Operating deflection shapes

Operating deflection shapes point to the asymmetric loading and mass distribution on the base desk. The machine tends to tilt especially around the transverse axis (Fig. 1).

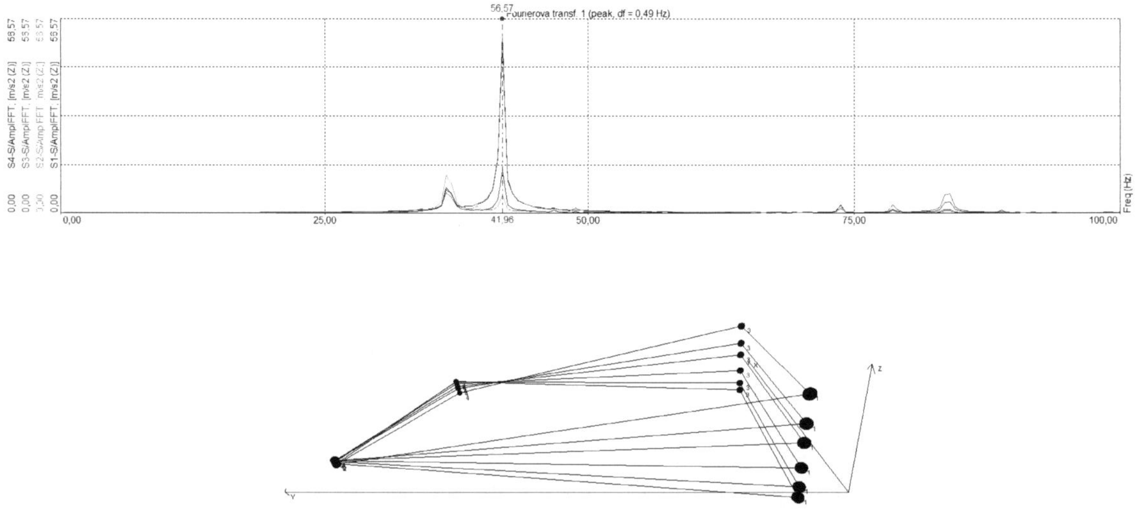

Fig. 1: Operating deflection shapes of base desk

Current state of the test facility with the compactor represents a dynamic system, wherein an elastic and damping links between the machine and the test facility and between the test facility and the ground are made of rubber elements. In the first case the links are rubber wheels on that the compactor drums are seated. Second elastic and damping link is rubber sheet between the test facility frame and the ground.

One of the most important criterion of a suspension system is the value of the force amplitude FZR transmitted to the ground. Fig. 2 shows amplitude frequency response of that force FZR and its detailed view.

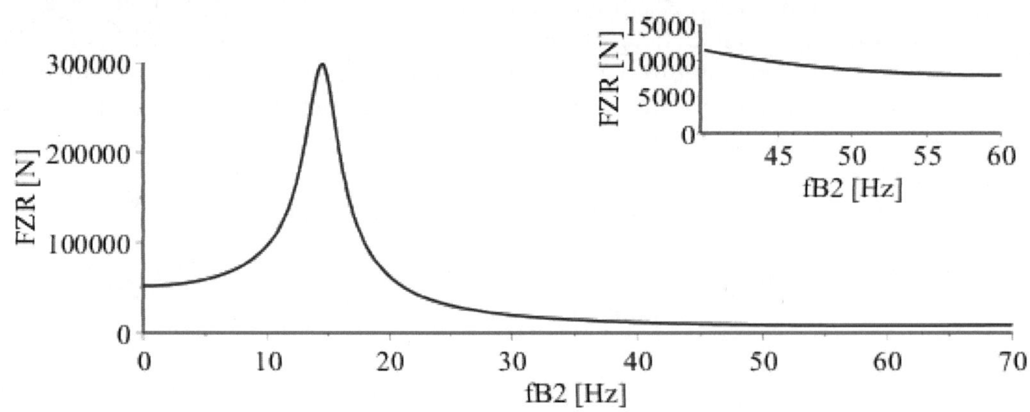

Fig. 2: Force amplitude transmitted to the ground

Results presented of the mathematical model solution shows relatively high values of the force amplitude transmitted to the ground.

Calculation of suspension

Suspension system calculation is based on solution of the mathematical model as a result of the motion differential equations. The analysed dynamic system may be in simplified way represented by dynamic model of two mass with two degrees of freedom.

Fig. 3 shows influence of the basic dynamic parameters on force amplitude transmitted to the ground. The basic parameters referred to Fig. 3 are stiffness of the machine and test facility connection kSR, the base desk weight mZ and operating frequency fB2.

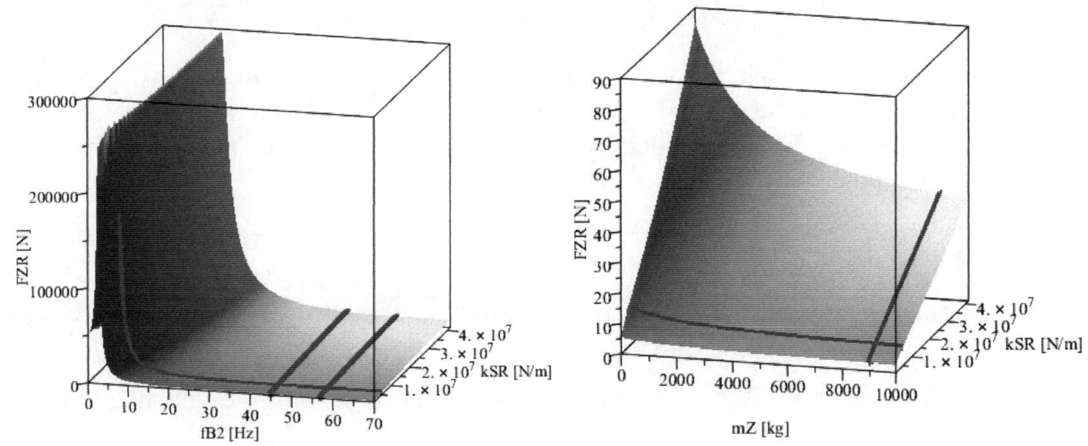

Fig. 3: Force amplitude transmitted to the ground

Higher value of the machine and test facility connection stiffness increases the force amplitude. However, the sufficient value of stiffness is needed regarding the simulation of the machine operating conditions.

Increase of the force amplitude may be partially compensated by the higher weight of the base desk.

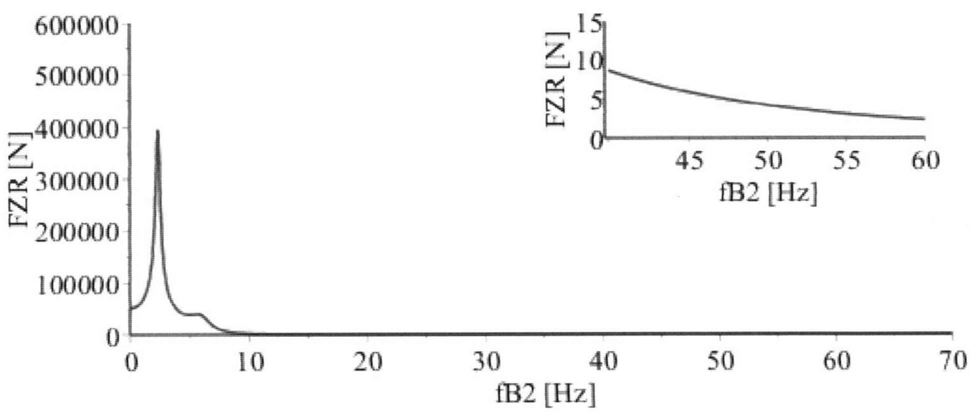

Fig. 4: Force amplitude transmitted to the ground

Fig. 4 shows amplitude frequency response of the force transmitted to the ground FZR and its detailed view with proposed optimal dynamic parameters.

Conclusions

The aim of this paper was to identify the basic dynamic parameters and describe dynamic conditions of the test facility. Dynamic analysis is necessary for efficient suspension design. Especially the stiffness of connection between the machine and the test facility frame and the base desk weight have the effect on minimization of dynamic reaction forces and vibrations. Results of the dynamic analysis showed efficiency of the suspension proposed. The force amplitude transmitted to the ground decreased significantly.

Acknowledgements

This publication was written at the Technical University of Liberec as part of the project "Vibration damage identification of linear guiding system" with the support of the Specific University Research Grant, as provided by the Ministry of Education, Youth and Sports of the Czech Republic in the year 2019.

References

[1] J. Homišin a kolektív, Súčasné trendy optimalizácie strojov a zariadení, C-PRESS. Košice, 2006. ISBN 80-8073-656-1

[2] H. Dresig, F. Holyweißig, Maschinendynamik, 8. auflage, Springer. Berlin, 2008.

EAN 2019
57th conference on experimental stress analysis

Damage Analysis of Dissimilar Metal Welds of Power Equipment

JUNEK Lubomír[1,a], SIEGL Jan[2,b], HAUŠILD Petr[2,c], Miroslava Ernestová[3,d] and BYSTRIANSKÝ Jaroslav[4,e]

[1]Institute Applied Mechanics Brno, Resslova 972/3, Veveří, 602 00 Brno, Czech Republic

[2]Department of Materials, Faculty of Nuclear Sciences and Physical Engineering, Czech Technical University in Prague, Břehova 7, 114 19 Prague

[3]ÚJV Řež, a. s., Hlavní 130, 250 68 Husinec – Řež, Czech Republic

[4]University of Chemistry and Technology Prague, Technická 5, 166 28 Prague 6 - Dejvice

[a]junekl@uam.cz, [b]jan.siegl@fjfi.cvut.cz, [c]petr.hausild@fjfi.cvut.cz, [d]emi@ujv.cz, [e]jaroslav.bystriansky@vscht.cz

Keywords: Dissimilar metal welds, damage of welds, stress concentration factors, measures

Abstract. Heterogeneous dissimilar welds (DMW) are always problematic and critical areas of pressure vessels and piping. A critical structural area is created by combining two materials with different chemical composition and with completely different mechanical and physical properties, especially different thermal expansion. This critical detail very often undergoes intensive degradation during operation and very often, the operational degradation of the junction of the two materials with different properties ends with the loss of integrity of the pressure boundary. There are three main factors leading to failures – higher operation temperature than design parameters, inappropriate design of a heterogeneous welded joint to areas with higher stress levels and poor erection quality, where defects are already in the welded joint from assembly. The cause of the failures is also other factors that are not known at the design stage of the pressure units or are not sufficiently evaluated. Susceptibility to stress corrosion cracking (SCC) of one of the used DMW materials during long time operation is one such unknown factor in the design stage. Stress corrosion cracking is the growth of crack formation in a corrosive environment. It can lead to unexpected sudden failure of normally ductile metals subjected to a tensile stress. The stresses can resulted from the crevice loads due to stress concentration, or can be caused by the type of assembly or residual stresses from fabrication.

Introduction

It is common practice in the power industry to produce robust pressure vessels from carbon steel and connecting piping systems from austenitic steel. A critical structural detail is created by combining two materials with different chemical composition and with completely different mechanical and physical properties, especially different thermal expansion. This critical detail very often undergoes intensive degradation during operation and very often, the operational degradation of the junction of the two materials with different properties ends with the loss of integrity of the pressure boundary. The cause of the damage is usually a number of factors that are not known at the design stage of the pressure units or are not sufficiently evaluated.

Description of Dissimilar Metal Welds (DMW) Damage

The two materials with different material properties connection is done in the power industry by welding. Welding of materials with different structural phases leads to degradation of the weld joint in heat-affected zone already within manufacturing. During welding and subsequent heat treatment at temperatures above 600 ° C, significant diffusion processes occur between the base material and the weld metal, mainly when a filler material is not stabilized. Carbon diffusion from a carbon steel (0,2C-1,0Mn-0,3Cr-0,3Ni) to austenitic filler material (0,08C-15Cr-25Ni-5Mo) results in the formation of decarburized and the carburized zone in the melting area and formation of carbides, which usually segregate at grain boundaries. This, together with the action of the media process and especially the high level of stress due to the very different thermal expansion of the carbon and austenitic steels during operation condition, leads to a significant reduction in the reliability, durability and safety of the weld joint as can be seen in Fig.1. Manufacturing process of the DMW joints so created the conditions for crack propagation along the fusion line of the first buttering layer.

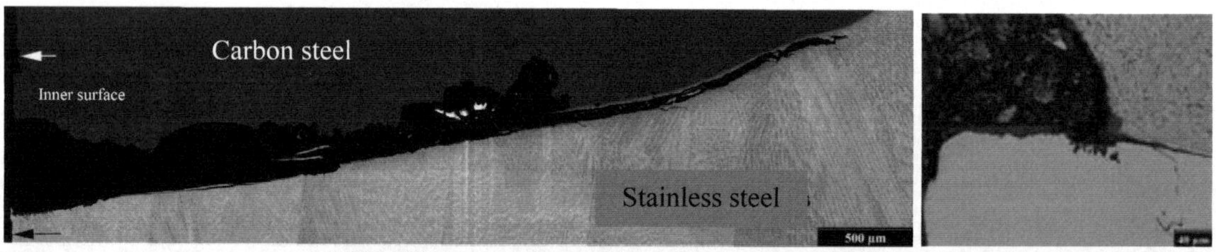

a) DMW damage on the pressure vessel [1]

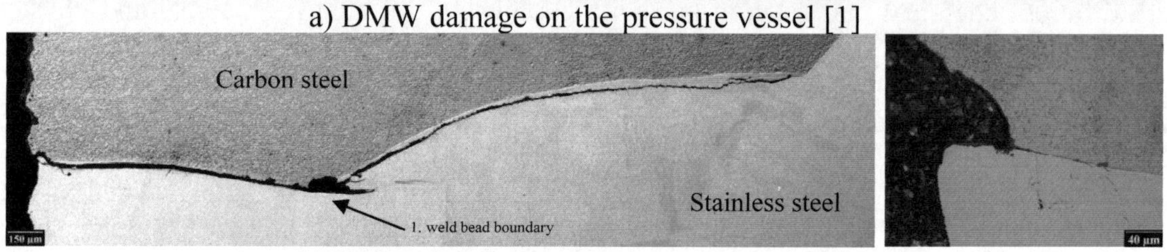

b) DMW damage on the piping system [2, 3]

Fig. 1: Damage of DMW along diffusion line

Initial Phase of DMW Degradation. It can be seen from presented figures 1 that DMW damage always start from corrosion defect on the inner surface, where the corrosion sensitive parts of weld joint has been (revealed) in a contact with the working medium. Corrosion of this part can produce pitting which can be significant as stress concentrator. The concentrator on the inner surface, additionally significantly increases the stress gradient from the different thermal expansion of the carbon and austenitic steels that is approximately 33%. Typically, a failure of DMWs usually occurs during heating startup operation condition and not during normal operation condition. The stress gradient during heating due to different thermal expansion of used materials increases extremely at a corrosion defect tip. Conditions are created for the second part of the failure. The stress level at a corrosion defect tip is dependent on the geometry of the formed stress concentrator.

The different thermal expansion of the used materials also damages the high temperature protective oxide layer during cooling. This high temperature oxide is formed on the inner surface during long-term operation of the pressure equipment at higher temperatures [5]. The metal is exposed by damaging the passivation layer and the exposed bare steel surface is immediately oxidized by increased velocity. The passivation layer damage due to the different expansion of the materials increases the corrosion rate of the inner surface damage. Accelerated corrosion damage creates a higher stress concentrator sooner than operation lifetime and it creates the conditions for the second phase of damage - the crack propagation from the corrosion defect. Damage time of DMW depends on the time of creation of the suitable stress concentrator.

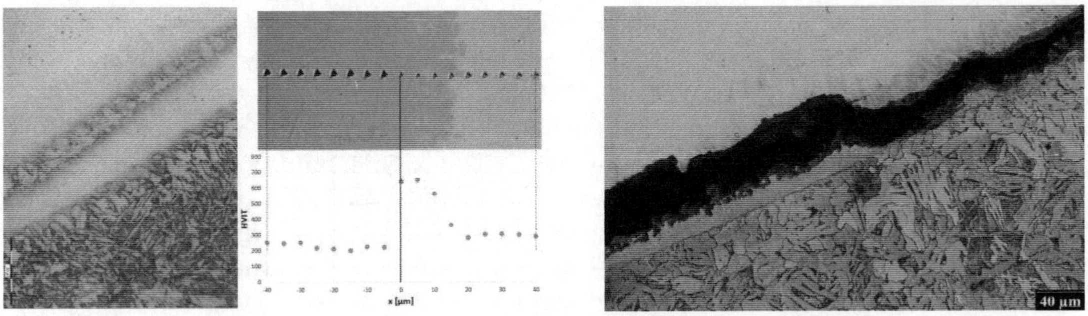

Fig. 2 Path for stress corrosion cracking degradation [4]

Phase of Crack Propagation by DMW. A very high hardness region was found by measuring the micro hardness across the weld joint (Fig.2).. During the manufacturing welding of used materials with different chemical composition and subsequent post weld heat treatment, a narrow path of material with a very high hardness is formed on the boundary of the diffusion layer at a depth of about 50 micrometers from a significant fusion line. Measured hardness up to 750 $HV_{0,01}$ (Fig. 2) indicates a very hard and brittle material without significant plasticity. There are three materials with completely different plastic behaviour in this DMW type - carbon decarburized material, narrow area with created carbides with very high hardness and austenitic plastic material. The damage always occurs in a region, which is most susceptible to cracking. When the corrosion damage on the inner surface connects with this narrow path at the diffusion boundary, conditions for initiation and subsequent crack propagation from this corrosion defect are created. Practice shows that DMWs are damaged only during transients. As soon as the tension due to the change of temperature and different thermal expansion of the used materials reaches the fracture stress of the narrow path, initiation and fast crack propagation occur through this path. Initially, the crack driving force is extremely high due to the stress concentrator, therefore the material is also damaged perpendicularly to the main crack where carbides on the grains boundaries are located. Crack growth is realized by stress corrosion cracking. Stress relaxation occurs during crack propagation and because stress relaxation is faster than the temperature change during the heating (or cooling) of pressure vessels to operation temperature, the crack arrests on the any material barrier, which is usually the subsequent weld bead. The narrow path with very high hardness is not continuous along the fusion line, which is another reason for the crack arresting. After crack is stopped and the second weld joint damage phase is terminated, the initial damage phase begins again as written above [6].

Corrective Measures

The main principle in design of welded structures in the case of DMW has to be kept. Location of a welded joint in places with lower stress areas should be designed. However, the different thermal expansion of the used materials reduces the importance of this essential requirement. Therefore, further corrective measures must be taken.

Post Weld Heat Treatment (PWHT). Post weld heat treatment on residual stress removal in the manufacture procedure of welded components is commonly designed and implemented. Especially when welding two pieces with a larger wall thickness over 20mm, where an austenitic buttering is welded onto the base carbon steel. In the case of DMW, the PWHT is not entirely optimal. Compressive residual stress is in heat-effected zone (HAZ) of DMW before

PWHT application and residual stress should be minimal after PWHT applying. But in case of DMW it is just on opposite. PWHT is usually conducted at 620-640°C. The structure is heated to this temperature and is kept several hours at this temperature. Residual stress is minimal or neglected at the end of dwell. But very high residual tensile stress is created in HAZ during cooling from PWHT temperature to room temperature due to different thermal expansion of austenite and carbon steel. Compression barrier against cracking is replaced by very high tensile residual stress. This effect of PWHT is very negative in case of structures with DMWs. It is recommended to not perform PWHT on DMW.

Humidity. Very high humidity increases metal corrosion rate mainly when the humidity is over 50% after drainage of pressure equipment. The high humidity inside the pressure equipment during long-term shutdowns or outages is very negative if the passivation protective layer of the created oxides is destroyed during cooling in DMW areas. High humidity increases the speed of the stress concentrators creations and thus accelerates the initial phase of the DMW degradation. It is recommended to keep the humidity inside pressure equipment under 50% during shutdowns or outages when drainage was done.

Oxygen. Very often, the pressure equipment is not drained during shutdown and the boiler water remains inside pressure vessels. The oxygen content of the boiler water during operation is very carefully monitored. However, during the shutdown, the oxygen content of the water is not controlled and it is completely ignored. The presence of oxygen in the water during two months shutdown significantly increases the corrosion attack of DMW materials, especially when the DMW passivation layer is broken after cooling. The oxygen saturation in water greatly accelerates the formation of stress concentrators in the DMW area and thus decreases the failure time by the crack propagation in filler material of DMW. It is recommended to control the oxygen content of the water during cooling and outages.

New Filler Material. Non-stabilized filler material with higher content of nickel (25%) and low content of chromium (15%) was used in our case. Used filler material also had a relatively high phosphorus and sulfur content, leading to increased weakening of the grain boundaries in HAZ after welding. In addition, during the welding, chromium reacted with carbon from the base material. Carbides together with the higher content of sulfur and phosphorus created areas on diffusion border with very hard quenched structures. All these led to the replacement of the filler material. Practice and experimental analysis showed that filler material with higher chromium content (25%) and lower nickel content (12%) is better suited for use as a transition wire. However, the replacement of the filler wire does not fully solve the root weld area susceptible to corrosion attack.

Protective Layers. As it was mentioned above, the replacement of currently used filler wire will not completely eliminate DMW problems. Even a change of DMW welding technology does not lead to the suppression of unwanted very hard structures. The only way to prevent stress concentrators (corrosion damage of the surfaces) in the DMW root area is to completely separate the material sensitive to stress corrosion cracking from the operation medium. It can be achieved in several ways: a) by welding anti-corrosion layers to the inner surfaces; b) galvanic nickel-plating of inner surfaces; (c) hot or cold sprays application on surfaces. Each of these methods has some advantages and disadvantages. From all mentioned methods, the method of galvanic depositing of the nickel layer on the inner surface was chosen and experimentally tested. Specimens of weld joints with nickel deposit were tested in air and in corrosion environment. The nickel protective layer of DMW appears stable even in cyclic temperature changes after solving the problem of adhesion of the nickel layer to the austenitic material. This method is currently applied on the sites.

Conclusion

The time of the power equipment operation has shown that the DMW with buttering layer is sensitive on intercrystalline corrosion and stress corrosion cracking. DMW with this buttering layer is one of the most critical areas of pressure equipment. DMW damage limits the lifetime of the power equipment, so regular attention must be given to these areas during the operation. One of the first tasks was to find all DMWs of this type and to include these DMWs in a regular operational inspection program. Volume NDT inspection had to be qualified along with their inclusion in the surveillance program since the appearance of weld joint heterogeneities makes detection of defects in the weld joint very difficult. An evaluation method of detected flaws in DMW had to be also developed. The newly developed evaluation method determines the acceptability of the defects and critical defect size in DMW. The critical size value determines the time until the DMW repair as well as the NDT inspection interval. New DMW repair technologies in dependence on the diameter and wall thickness of the repaired equipment had to be also developed due to the findings of defects in DMW. Fully automated welding equipment increasing DMW quality has been introduced. All of these activities have been aimed at maintaining the operability and reliability of power equipment.

References

[1] J. Siegl, P. Haušild, M. Karlík: Defect Analysis of Dissimilar Metal Weld of SG46 (in Czech), FJFI report No. V-KMAT-878/13, FJFI of ČVUT in Prague, FJFI Prague 2013.

[2] P. Haušild, J. Siegl: Defect Analysis of Dissimilar Metal Weld No. 305 (in Czech), FJFI report No. V-KMAT-990/16, FJFI of ČVUT in Prague, FJFI Prague 2016.

[3] M. Ernestová, D. Tonarová, P. Brabec, J. Burda: Evaluation of DMW taken from nozzels, (in Czech) report ÚJV Řež No. ÚJV-DITI 2302/437 Rev.1, Řež 2017.

[4] P. Haušild, J. Siegl: Damage Analysis of DMW from piping system (in Czech), report FJFI ČVUT No. V-KMAT-1034/18, FJFI of Czech Technical University in Prague, FJFI Prague 2018.

[5] Bystrianský J.: Destructive Evaluation of Piping Slice from Super Emergency Feed Water (in Czech), report VŠCHT No. z2_61-8064-e01, revision 2, Prague 02/2019.

[6] Junek L and. etc. : Degradation Mechanism of Dissimilar Metal Weld Joints on Steam Generator Collectors VVER 440MW, manuscript of Seventh International Workshop NEW METHODS OF DAMAGE AND FAILURE ANALYSIS OF STRUCTURAL PARTS November 1 - 4, 2016, Yokohama.

EAN 2019
57th conference on experimental stress analysis

Application of digital image correlation to vibration analysis of the composite air inlet during acoustic loading

KADLEC Martin[1,a], ZDĚBLO Jakub[2,b], ZDĚNEK Pavel[1,c] and BĚHAL Jiří[1,d]

[1]VZLU – Czech Aerospace Research Centre; Beranových 130, 199 05 Praha, Czech Republic

[2]Sobriety s.r.o.; Blanenská 1288/27, 664 34 Kuřim, Czech Republic

[a]kadlec@vzlu.cz, [b]jakub.zdeblo@sobriety.cz, [c]pavel.zdenek@vzlu.cz, [d]behal@vzlu.cz

Keywords: Polymer composite, Vibration, Acoustic loading, Digital image correlation, Aerospace

Abstract. The air inlet duct of the jet engine is loaded both by flight loads from acceleration and also by vibrations caused by high sound levels of the jet engine. The polymer composite is efficient due to good damping properties and resistance to fatigue cracks. In order to evaluate the properties of this structure, an experiment with acoustic loading by a set of speakers was performed. The vibrations were measured using digital image correlation (DIC) system with high speed cameras. Two frequencies were measured and displacement of the area of 200 x 90 mm was evaluated in three dimensions. The measured data will be used for the finite element model benchmark.

Introduction

Carbon fibre-reinforced polymer (CFRP) laminates are used in aerospace due to their effective strength to weight ratio. The airframe must fulfill requirements on aerodynamics, structural resistance and fatigue resistance [1]. The fatigue loading has two main sources, mechanical loading and acoustic loading [2]. The acoustically induced vibrations from the engine are important aspect that affects the durability of the part. The use of composite material is therefore a smart choice because of the better damping and fatigue resistance properties compared to metal [3]. Any design of airframe structures must be supported by strength analyses and composite material properties must be tested on different complexity levels [4]. The coupon level was already tested by the fracture toughness evaluation [5]. Flat panels were tested and evaluated numerically on the structural detail level [6, 7, 8]. Similarly, authors in paper [9] investigated numerically acoustic fatigue and dynamic response of composite panels under acoustic excitation.

This paper describes local vibration measurement during testing of the component level considering the acoustic loading. The vibration source of the jet engine was simulated using a set of speakers. The laminate vibrations can induce matrix cracking or delamination growth from potential manufacturing or in-service flaws caused for example by impact damage [10, 11]. The mapping of the displacement amplitudes is therefore needed to evaluate the resistance of the structure to the damage growth and to fulfil damage tolerance philosophy.

Materials and Methods

The air inlet (Fig. 1) to be used in a jet plane was manufactured from the prepreg Hexply 8552/AGP193-PW which consists of AS4 carbon fibre fabric and high performance tough epoxy matrix for use in primary aerospace structures. The average thickness in the measured critical region was 2.5 mm.

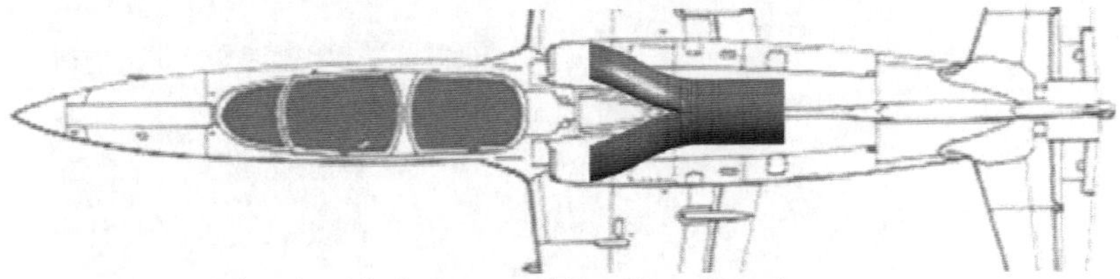

Fig. 1: Air inlet position in a jet plane

Experimental setup is visible in Fig. 2. It consists of the steel test fixture with clamped air inlet ducts and 4 speakers Tesla TVM ARA-389-00/4 with peak power of 300 W attached to the inlet by transition parts. Two amplifiers Dynacord S1200 were used, each with two channels.

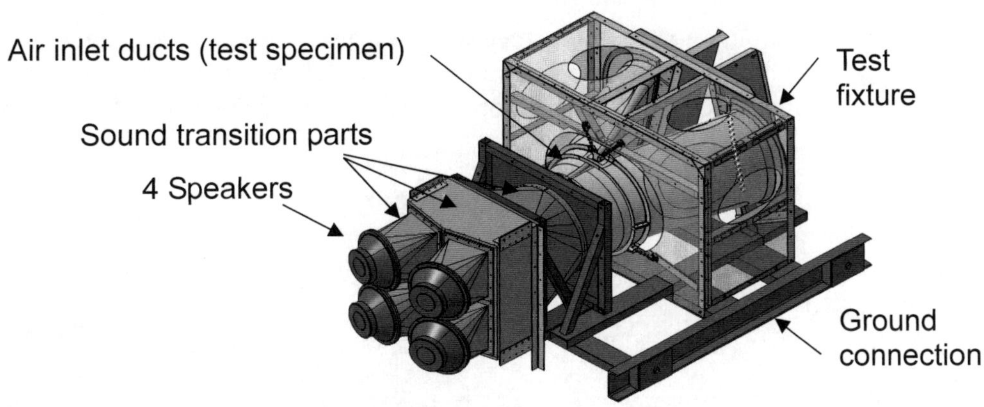

Fig. 2: Experimental setup of an air inlet in a test fixture

Two resonant frequencies were chosen based on a white noise measurement. First 84 Hz and the second was 118 Hz. These constant frequencies were used with sound levels needed to measure significant amplitudes of the laminate. An optical system used for the full-field displacement measurement of the measured part was based on digital image correlation (DIC). The Mercury software from Sobriety Company with two high-speed Phantom video cameras was used. The cameras had resolution of 1280 x 800 pixels. They were set to 2000 frames per second. The measured region with the dimensions of 200 x 90 mm is visible in Fig. 3. The Mercury software has several modules used for vibration evaluation such as Fast fourier transform (FFT), power spectral density, octave analysis and Campbell diagram.

Fig. 3: Air inlet in a fixture with a pair of high-speed cameras focused on the measured surface

Results

Displacement in the normal direction to the surface was the value needed to be evaluated. The displacement was evaluated on the whole measured region which provided the shape for the vibration for the chosen frequency (Fig. 4a, b). For selected points, displacement in time was evaluated (Fig. 4c, d).

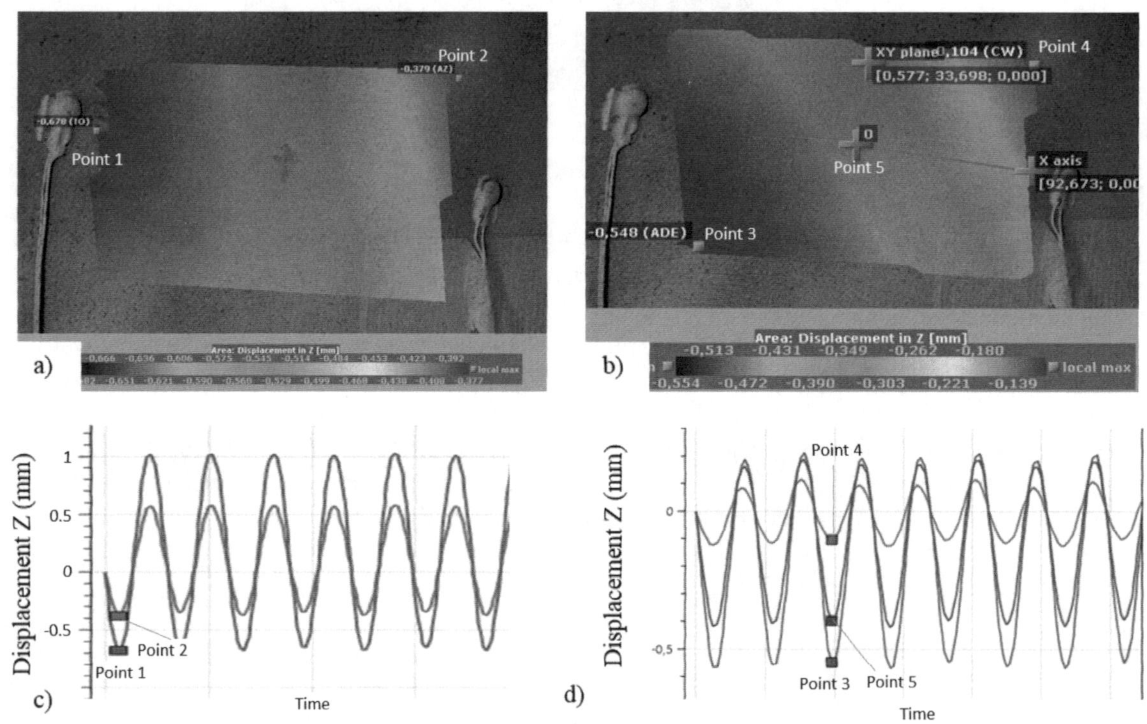

Fig. 4: a) Amplitude graph for a) 84 Hz and b) 118 Hz; c) example of selected point displacement in time for c) 84 Hz and d) 118 Hz.

The function of FFT was applied on the vibrations in the area which resulted in the amplitude graph (Fig. 5). A phase graph was also used to confirm that these frequencies hanged the vibration shape of the structure (Fig. 6). These variables will be used for whole inlet FE model calibration.

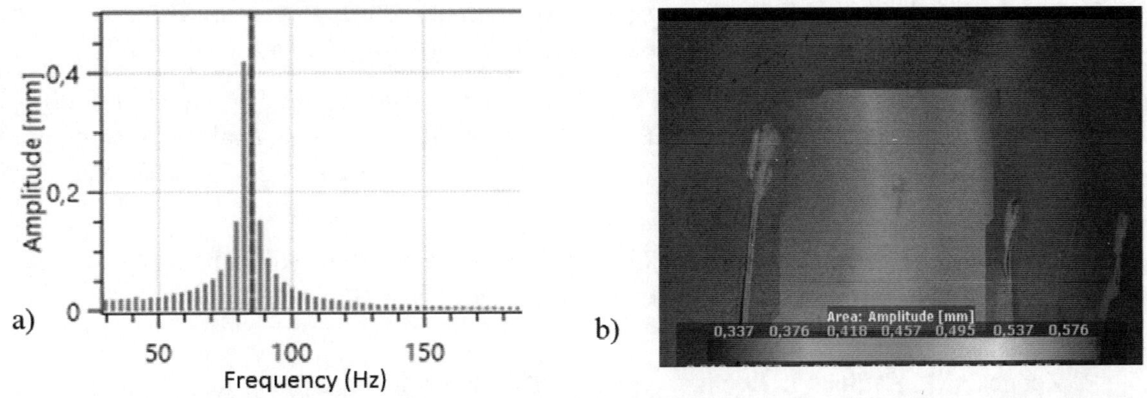

Fig. 5: a) Amplitude graph (displacement Z averaged over the are a) for 84 Hz and b) the full-field view

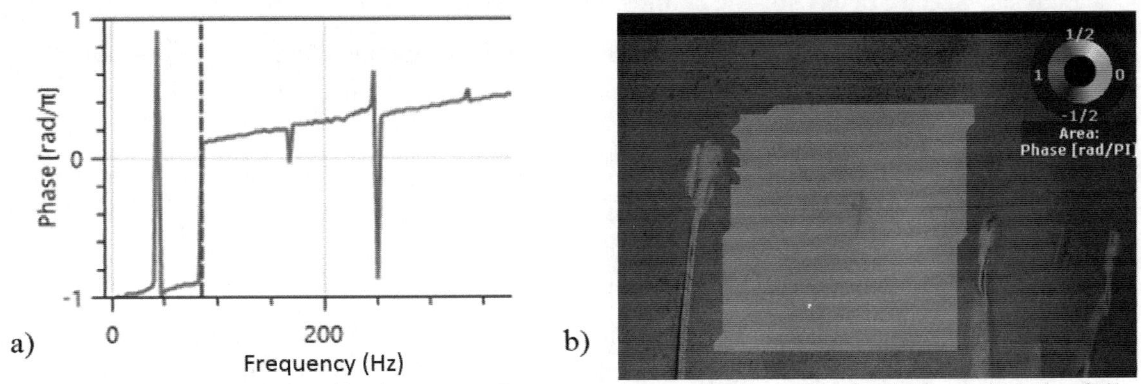

Fig. 6: a) Phase graph for displacement Z averaged over the area for a) 84 Hz and b) the full-field view

Conclusions

To conclude, the non-contact 3D vibrational measurement carried out using Mercury DIC software proved to be efficient in order to determine the vibration characteristics. Two resonance frequencies were evaluated and full-field local displacements were determined. The measured data will be used for the FE model benchmark.

Acknowledgement

The authors would like to thank the Technology Agency of the Czech Republic for supporting this research with project no. TE02000032.

References

[1] J.P. Arenas, R.N. Margasahayam, Noise and vibration of spacecraft structures. Ingeniare. Revista chilena de ingeniería, 14 (2006) 251-264.

[2] R.G. White, Developments in the acoustic fatigue design process for composite aircraft structures, Composite Structures, 16 (1990) 171-192.

[3] B. Benchekchou, R.G. White, Acoustic fatigue and damping technology in FRP composites, Composite Structures 37 (1997) 299-309.

[4] R. Růžek, J. Běhal: Certification programme of airframe primary structure composite part with environmental simulation, International Journal of Fatigue, 31 (2009) 1073-1080.

[5] M. Kadlec, J. Šedek, Delamination Growth under Mixed Mode I - Mode II Fatigue Loading in Polymer Composite, in: EAN 2018 Experimental stress analysis, Conference proceedings of the 56_{th} conference on experimental stress analysis, Czech Society for Mechanics, Prague (2018) 168-175.

[6] R. Doubrava, M. Oberthor, P. Bělský, J. Raška, Bird and hail stone impact resistance analysis on a jet engine composite air inlet, in: MATEC Web Conf. Volume 188 (2018), Art. No. 04006.

[7] P. Zděnek, M. Kadlec, J. Běhal, Application of laser shearography to vibration mode shape analysis of composite panel during acoustic load, in: EAN 2018 Experimental stress analysis Conference proceedings of the 56th conference on experimental stress analysis, Czech Society for Mechanics, Prague (2018) 457-463.

[8] J. Běhal, P. Zděnek, Effect of interlaminar flaw on composite panel behaviour under acoustic loading, MATEC Web of Conferences 188 (2018). https://doi.org/ 10.1051/matecconf/201818801005.

[9] C. Uz, T.T. Ata, Acoustic Fatigue and Dynamic Behavior of Composite Panels Under Acoustic Excitation, in: De Clerck J., Epp D. (eds) Rotating Machinery, Hybrid Test Methods, Vibro-Acoustics & Laser Vibrometry, Volume 8. Conference Proceedings of the Society for Experimental Mechanics Series. Springer, Cham (2016).

[10] R. Doubrava, M. Oberthor, P. Bělský, M. Dvořák, K. Doubrava, Experimental verification of jet engine composite inlet for bird and hail stone impact resistance, in: EAN 2018 Experimental stress analysis Conference proceedings of the 56th conference on experimental stress analysis, Czech Society for Mechanics, Prague (2018) 63-76.

[11] M. Kadlec, R. Růžek, A Comparison of Laser Shearography and C-Scan for Assessing a Glass/Epoxy Laminate Impact Damage, Applied Composite Materials 19 (2012) 393-407.

EAN 2019
57th conference on experimental stress analysis

Short overview of current methods for detecting fatigue limit

KAĽAVSKÝ Adam[1, a], HUŇADY Róbert[1, b] and KICKO Michal[1, c]

[1] Department of Applied Mechanics and Mechanical Engineering, Technical University of Košice, Faculty of Mechanical Engineering, Slovakia

[a]adam.kalavsky@tuke.sk, [b]robert.hunady@tuke.sk, [c]michal.kicko@tuke.sk

Keywords: fatigue limit, staircase method, modified staircase method, statistic

Abstract. The paper deals with determining the fatigue limit of a casual construction material. The paper described two different approaches to do it. The first approach is based on analytical calculation of fatigue limit evaluated by theory of elasticity and plasticity. However, these principles provide only a very rough estimate of the fatigue limit. In practice, the experimental principles described in the second part of the article are preferred and used more often.

Introduction

Machine components, structures, automobiles, planes, etc. are often subjected to a cyclic loading. Cyclic loading very often leads to the microscopic material damage. If the stress caused by this load is less than the strength of the material, the damage will accumulate until a crack develops or other damage leads to fracture of the component. Such a process of accumulation of failure that leads to fracture of a component due to cyclic loading is called fatigue [1] . Material fatigue has caused many catastrophic failures of planes, ships, bridges and other structures in history [2] , [3] . In the past fatigue curves (S-N curves) were discovered, for which the specific load amplitude (stress, plastic deformation, etc.) was called fatigue limit and it was assumed that there was no fatigue fracture below this value. The example of fatigue curve can be seen in Fig. 1.

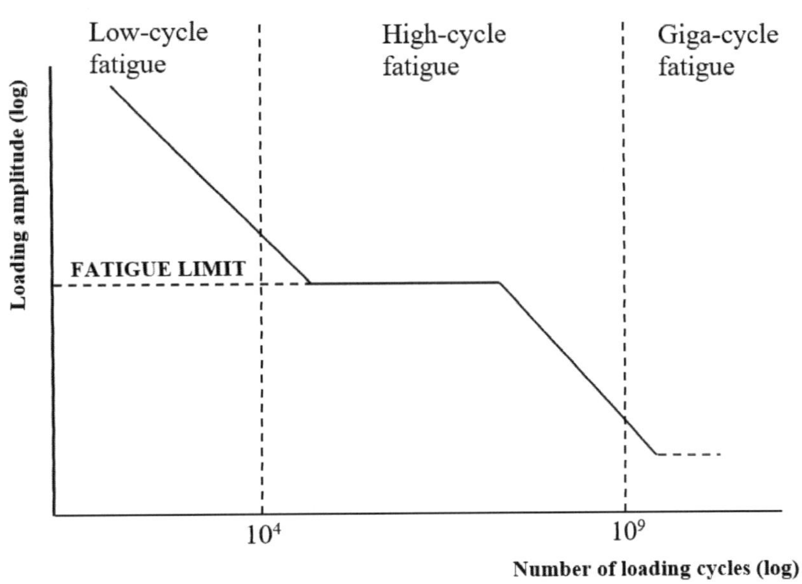

Fig. 1 – S-N curve

1. Analytic methods

This part shows how a design engineer could estimate the fatigue lifetime duration for a mechanical system (without knowing of a fatigue diagram). In Table 1 are listed basic formulas to estimation fatigue limit based on relation to main material characteristics like Yield Limit, Strength Limit, Hardness, True Fracture Strength.

Table 1 – Analytical evaluation of fatigue limit [4]

Evaluating value	Formula (R = -1, P = 50%) [MPa]	Condition of validity	Author
Strength Limit (R_m)	$\sigma = 0.46 \cdot R_m$ $\tau = 0.27 \cdot R_m$	Steel $R_m \geq 1200$ MPa	Žukov
Yield Strength (R_e, τ_k)	$\sigma = 0.45 \cdot R_e + 122$	Construction steel	Žukov
	$\tau = 0.448 \cdot \tau_k + 52$	Construction steel	Buch
True Fracture Strength (σ_f)	$\sigma = 0.35 \cdot \sigma_f - 10$	Construction steel	Žukov
Hardness (HB)	$\sigma = (0.128 \ldots 0.156) \cdot HB$	Carbon steel	Grebenik
$R_m + R_e$	$\sigma = 0.285 \cdot (R_e + R_m)$	Construction steel	Šapošnikov

Example:

Material properties of SAE-AISI 1040 Steel are known as $R_e = 460\,\text{MPa}$ and $R_m = 640\,\text{MPa}$. Problem formulation is to find the fatigue limit of investigated material by analytical calculation.

The SAE-AISI 1040 belongs to construction steel family. Therefore, formulas by Žukov and Šapošnikov can be used at problem solution (see Table 1).

$\sigma = 0.45 \cdot 460 + 122 = 329\,\text{MPa}.$ … Žukov

$\sigma = 0.285 \cdot (Re + Rm) = 0.285 \cdot (460 + 640) = 313.5\,\text{MPa}$ … Šapošnikov

Estimated fatigue limit should be close to the values 313.5 and 329MPa.

Experimental methods

In the engineering practice, the most used experimental methods based on statistics are staircase method and their modified version. Among other statistical methods for experimental evaluation of fatigue (strength) limit belong the methods listed in Table 2.

Table. 2 – Overview of fatigue strength evaluation methods.

Group	Method
Conventional	• S-N test method
Quantal response tests	• Probit method • Staircase method • Two-point method • Arcsin√P method
Accelerated stress tests	• Prot method • Step-loading method
More advanced statistical methods	• Random fatigue limit model • Bayesian method • Bootstrap method

Quantal response tests

Unlike the conventional S-N approach, quantal response tests do not consider the actual time to failure, but rather they are given by their nature a "pass/fail" approach. Thus, quantal response tests are designed to handle runout test data. In a quantal response test, specimens are tested at a certain load level and they either survive the specified number of cycles or they fail. Statistical analysis of test results allows an estimate of median and standard deviation of the fatigue strength at the specified number of cycles based on the proportion of failed specimens at each load level.

Staircase method

Fatigue limit σ_e is estimated by staircase method from Eq. 1 [5]

$$\sigma_e = \sigma_0 + d \cdot \left(\frac{A_n}{\sum n_i} - \frac{1}{2} \right), \tag{1}$$

where n_i is number of events (failures or survivals) for each stress level. Range between two stress levels is expressed by d. Lowest stress level is denoted as σ_0. Quantity A_n is obtained from

$$A_n = \sum i n_i \,. \tag{2}$$

In the staircase method, specimens with progressively increasing stress amplitude are tested until the failure occurs.

Example:

Problem formulation is to find the fatigue limit of investigated material at 10^7 cycles by staircase method. The results of fatigue tests of SAE-AISI 1040 Steel are shown in Table 3.

Table 3 – Dataset of SAE-AISI 1040 Steel used in staircase method.

Stress level	#																			
σ_i [MPa]	1	2	3	4	5	6	7	8	9	10	11	12	13	14	15	16	17	18	19	20
325							x												x	
320				x		o		x										o		o
315	x		o		o				x				x		x		o			
310		o								x		o		o		o				
305											o									
x fail																				
o pass																				

At the first step, it is necessary to perform data analysis and fill the values in Table 4.

Table 4 – Data extracted from Table 3.

Stress level	Level	Values	
σ_i [MPa]	i	n_i	$i \cdot n_i$
325	3	2	6
320	2	2	4
315	1	4	4
310	0	1	0
Σ	–	9	14

Subsequently, the average value of fatigue limit is calculated from Eq. 1

$$\sigma_e = 315 + 5 \cdot \left(\frac{14}{9} - \frac{1}{2} \right) = 314.72 \text{ MPa.}$$

Using the Matlab software, it is possible to obtain a curve of normal distribution of fatigue limit (Fig. 2). The normal distribution is calculated by standard formula as follows

$$f(\sigma_f) = \frac{1}{7.6349 \cdot \sqrt{2\pi}} \cdot e^{-\frac{(\sigma_i - 314.7222^2)}{2 \cdot 7.6349}},$$

where σ_i acquire values between lower fatigue limit and average fatigue limit in sum with estimated standard deviation.

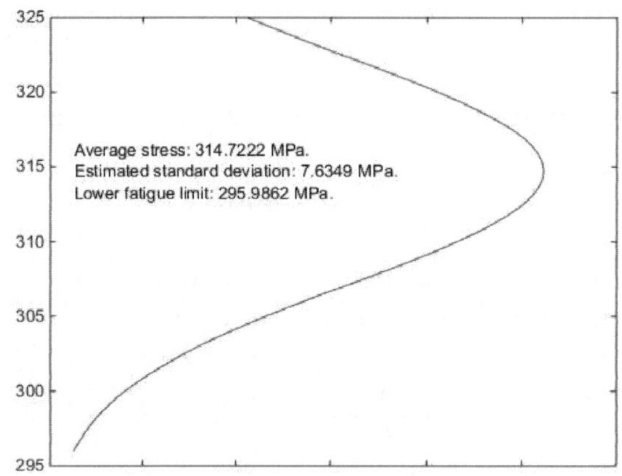

Fig. 2 – Normal distribution of fatigue limit values.

Average level of fatigue limit acquired from measured dataset achieved the value of 314.72 MPa. Estimated standard deviation was calculated at 7.63MPa and lower fatigue limit had approximated value ca. 296MPa (for 10% probability of failure at 95% confidence level).

Modified staircase method

Modified staircase method can be used only in the case if the standard deviation of fatigue strength is known (e.g. it is estimated from S-N curve). Minimum six specimens are necessary for testing in order to reliability of modified staircase method to be preserved. Optimum test is performed at 15 specimens [6] . Fatigue limit σ_e is estimated by staircase method from Eq. 3

$$\sigma_e = \frac{\sum_{i=2}^{n+1} \sigma_i}{n},. \quad (3)$$

where σ_i is stress level, and the number of experiments is denoted as n. In the process of measurement the value of σ_{i+1} is not obtained, but it is specify from previous stress level.

Example:
Problem formulation is to find the fatigue limit of investigated material at 10^7 cycles by modified staircase method. The results of fatigue testing of SAE-AISI 1040 Steel is shown in the Table 3.

Table 5 – Dataset of SAE-AISI 1040 Steel used in modified staircase method.

Level i	Measurement rank								
	1	2	3	4	5	6	7	8	9
σ_i [MPa]	315	310	315	320	315	320	325	320	315
Result	x	o	o	x	o	o	x	x	*

x	fail
o	pass
*	the result of test was specify from previous stress level

$$\sigma_e = \frac{(310+315+320+315+320+325+320+315)}{8} = 317.5\,\text{MPa}.$$

Conclusions

The paper presented a brief overview of current methods to determine fatigue limit using different analytical and experimental approaches. For analytical solution Žukov and Šapošnikov model of evaluation fatigue limit was used. Statistical model used in experimental method was Staircase method and its modified version. Deference of results between experimental and analytical model was under 4%. It may be noted that the analytical solution was quite accurate.

Acknowledges

This research was supported by project VEGA 1/0355/18.

References

[1] F. Trebuňa, M. Buršák. Medzné stavy - lomy. Košice : TUKE, 2002. p. 348.

[2] F. Trebuňa, P. Bigoš, et.al. Determination of reasons of cracking of outside shell of converter hood by means of experimental analysis methods of strength. Olomouc: Palacký University, EAN, 1997, p. 217-222.

[3] P. Bigoš, F. Trebuňa, E. Faltinová. Ussage of stress damage cumulation theories for assesment of heavz metallurgical bridge crane steel structure conditions. Košice: TUKE, 1997, P. 8-14, ISSN 1335-2334.

[4] M. Růžička, M. Hanke, M Rost. Dynamická pevnosť a životnosť. Praha : ČVUT, 1987. p. 213.

[5] S.R. Schmid, B.J. Hamrock, B.O. Jacobson. Fundamentals of Machine Elements. Third. Boca Raton : Taylor & Francis Group, 2014. p. 610.

[6] ČSN ISO 12107:2013. Kovové materiály – Zkoušky únavy – Statistické plánování a analýza údajů.

EAN 2019
57th conference on experimental stress analysis

Measurement of wheel-rail contact forces at the experimental roller rig

KALIVODA Jan[a] and BAUER Petr[b]

Czech Technical University in Prague, Faculty of Mechanical Engineering, Department of Automotive, Combustion Engine and Railway Engineering, Technická 4, 166 07, Prague 6, Czech republic

[a]Jan.Kalivoda@fs.cvut.cz, [b]Petr.Bauer@fs.cvut.cz

Keywords: Railway Bogie, Wheel-Rail Forces, Axle-Box Forces

Abstract. The article describes the newly developed system for the measurement of the forces acting in wheel rail contacts, which have been implemented on an experimental scaled railway bogie. The bogie is utilized for roller rig experiments focused on active controlled wheelset guidance. The measuring system consists of three parts: measurement of the lateral component of wheel-rail contact force, measurement of axle-box forces and measurement of overall tractive effort. The lateral component of wheel-rail contact force is determined by a measurement of the transverse deformation of roller disks. In order to determine axle-box forces, one component of the wheelset guidance was specifically designed as a three axis load cell. The overall tractive effort is determined by measurement of forces in the mechanism that connects the experimental bogie and the roller rig mainframe. The paper describes in detail the design and first test of the system.

Introduction

Force interaction in between rails and railway wheels is one of the most important issues in the development of the new rolling stock. Todays' effort to build an economic and environment friendly railroad brings a general and sustained demand to reduce wheel-rail contact forces below legislative limits as much as possible. Conventional methods of reduction wheel-rail contact forces are based on the optimization of suspension characteristics [1], or on mechanic or hydraulic linkages between the various components of the running gear. Because the possibilities of conventional methods are increasingly encountered at their limits, ideas of the utilization of active controlled elements in the wheelset guidance and railway vehicle suspension occur [2, 3].

The mathematical simulations of running behaviour of railway vehicles possessing active controlled running gears give a very promising results [4]. However, prior to the application of such systems in practice, they need to be thoroughly tested to meet the requirements of safety and operational reliability. Roller rigs can be advantageously used for these tests, because they allow testing of vehicle running behaviour in laboratory conditions, where it is possible to simulate extreme situations without a risk of railway accident [5, 6]. The principle of roller rig is in the replacement of a track by rotating rollers with a rail profile on their circumference. The tested vehicle is longitudinally fixed. Nevertheless, the creep conditions and forces in the wheel-roller contact points are analogical to the conditions in wheel-rail contacts of a vehicle running in a real track.

At the CTU in Prague a 1:3.5 scaled roller rig possessing an experimental bogie is utilized (Fig. 1). Each wheelset of the experimental bogie is equipped with an actuated steering mechanism for applying of a controlled torque to the wheelset in the yaw direction, or to steer wheelset to desired magnitude of yaw angle between the wheelset and the bogie frame [7]. The rig serves as an experimental facility for verification of results of mathematical simulations and demonstration of benefits of active controlled elements in railway bogies [8].

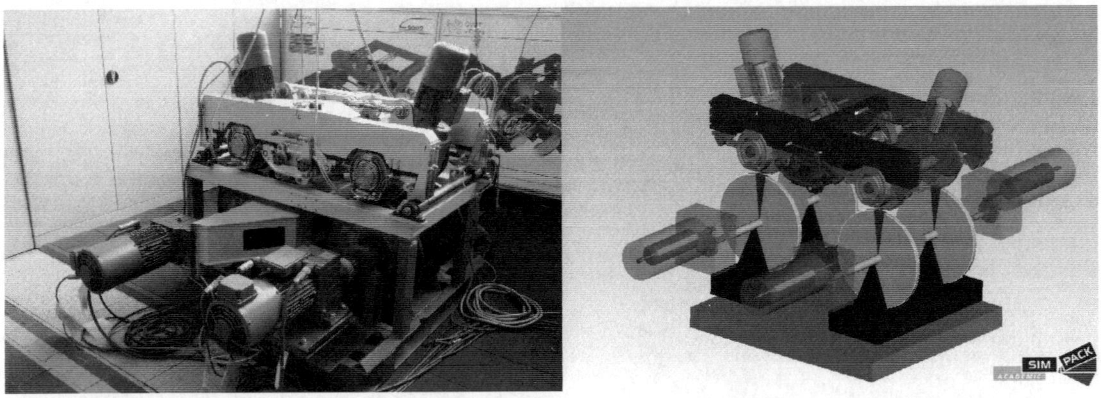

Fig. 1: The experimental railway bogie at the CTU roller rig (left) and its' MBS model (right)

An important operating parameter of rolling stock is the intensity the wear of wheels and rails. This wear is mainly affected by the magnitude of forces acting in wheel-rail contacts. Knowledge of these forces is therefore very important in assessing new types of railway running gears. However, a measurement of contact forces directly in the wheel-rail contact point is virtually impossible. In practice, indirect methods, usually based on the measurement of the strain on vehicle parts, are used [9, 10]. It is desirable to measure those strains as close as possible to the wheel-rail contact point.

Three systems of force measurement, which could be used for determination of wheel-rail contact forces are utilized at the CTU roller rig:

- measurement of the transverse deformation of roller disks,
- measurement of axle-box forces, i.e. forces transmitted between axle-boxes and the bogie-frame,
- measurement of bogie tractive effort, i.e. sum of longitudinal forces transmitted in the connection of the experimental bogie and the roller rig mainframe.

Measurement of the roller disc deformation

An important research tool in experimental testing of rail vehicles are strain gauge instrumented wheelsets, where the forces acting in the wheel-rail contacts are estimated by measurement of radial strains of wheel plates. In the case of the CTU roller rig a similar approach is used. Because the wheelsets of the experimental bogie are not applicable due to the small diameter and very high stiffness of the wheel disc, the system of measurement of disc strains was applied on the rollers (Fig. 2). This approach exhibits very good properties in terms of measuring the lateral component of wheel-rail contact forces. Its disadvantages are especially insufficiently low sensitivity for measurement of vertical and longitudinal components of wheel-rail contact forces and the necessity to transmit the output signal from rotating parts. At the CTU roller rig, several signal transmission systems were tested. Finally the own wireless signal transmission system has been developed in a cooperation with the CTU Faculty of Biomedical Engineering. The system provides synchronized wireless data transfer from all four roller in a sampling rate of 1 kHz and currently undergoes the final testing phase (fig 3).

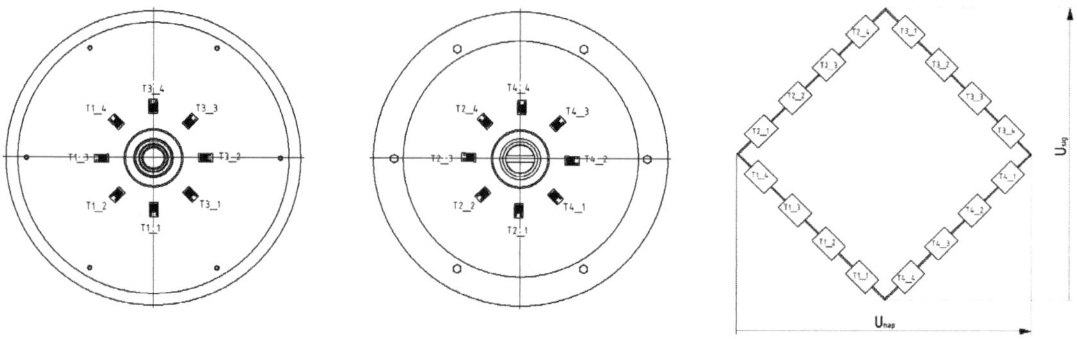

Fig. 2: Placement and connection of strain gauges

Fig. 3: Instrumented roller

Axle-box forces measurement

The possibility to measure wheel-rail contact forces was also considered in the design of the experimental bogie. In this case the contact forces are estimated by measurement of axle-box forces, i.e. forces transmitted between axle-boxes and the bogie frame. Considering several possible solutions, the development of specific component of the wheelset guidance acting as a 3-axis load cell was chosen. All forces between each axle-box and bogie frame are transmitted via a stirrup (Fig. 4), that is optimized for the strain gauge placement.

Fig. 4: Load cell for axle-box forces measurement

Each stirrup is instrumented with 36 strain gauges in three full bridges (Fig. 5). The wiring of strain gauges ensures the measurement of longitudinal F_x, lateral F_y and vertical F_z

component of axle-box force independently of each other. The parasitic resistive forces F_{R1}, F_{R2} and torques M_{R1}, M_{R2} in the connection of the stirrup and the bogie frame are eliminated from the measured signal [11].

Fig. 5: Instrumented stirrup

Information on the magnitude of axle-box forces are not regarded as an adequate substitute for information about the value of wheel-rail contact forces. They differ due to the inertia of wheelset. Moreover, the distribution of wheel-rail contact forces between the right and left wheel could not be determined by measurement of axle-box forces. The advantage of the measurement of axle-box forces is its easy realization without the necessity to transmit a signal from rotating parts.

Measurement of bogie tractive effort

The bogie is connected to the roller rig main frame using a Watt's mechanism (Fig. 6). The arm "1" is connected to the roller rig mainframe by a pair of ball-bearings and rotates around vertical pivot in its' centre. The bogie is connected to the arm "1" by two short linkages "2".

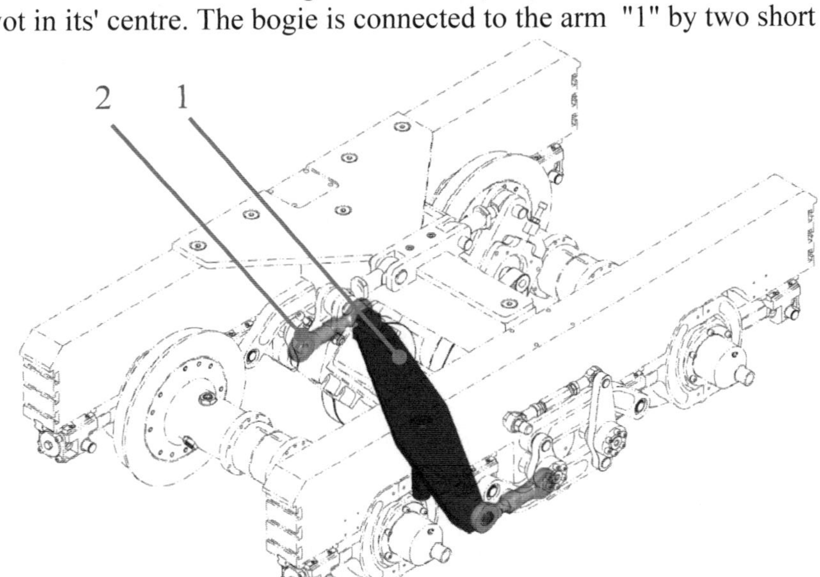

Fig. 6: The mechanism of connection between the bogie and the rig mainframe

Thus the bogie is longitudinally fixed with respect to the roller rig mainframe, whereas it can move in lateral and yaw direction without restrictions. The linkages "2" are instrumented with industrial single axis force transducer HBM U9B (Fig. 7) which measures tensile and compressive forces in the range of 1kN. The sum of the forces in the linkages "2" represents the total tractive effort of the bogie.

Fig. 7: Detailed photo of instrumented linkage

First tests

The first test of the measuring system was focused on the measurement of longitudinal forces. The system of the measurement of lateral roller disc deformation was not evaluated, as well as the vertical and lateral components of the axle-box forces. The setup of the experiment is depicted on the Fig. 8.

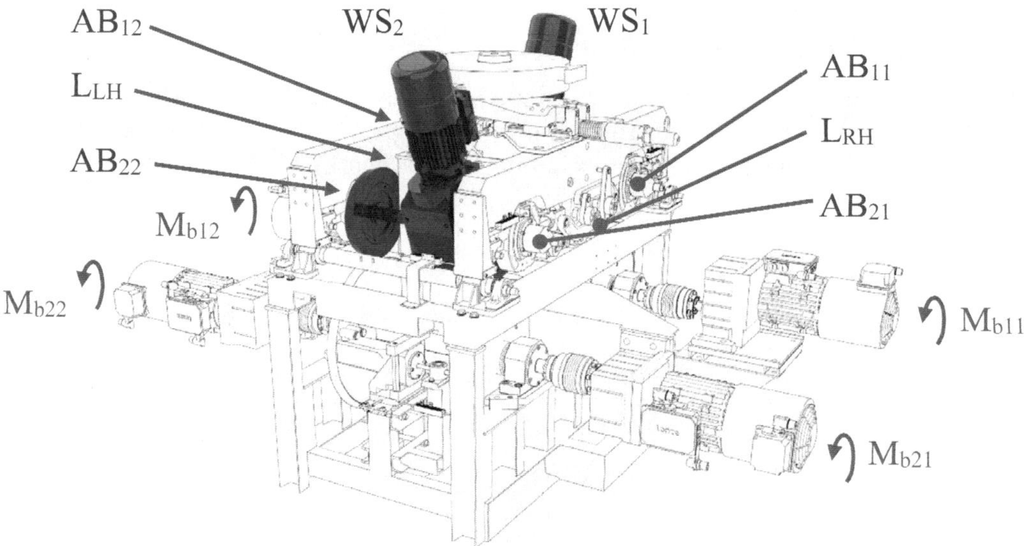

Fig. 8: Experiment setup

Wheelset traction motors WS_1, WS_2 were controlled to the constant speed of 1500 RPM, which corresponds to the vehicle running at a constant speed. The roller drives run in generator mode and simulated the vehicle's resistance forces. At the beginning of the experiment, the braking torques of the roller drives M_{b11}, M_{b12}, M_{b21}, M_{b22} were set to zero. The first index of the designation represents the wheelset number (1 … the first, 2 … the second wheelset of a bogie), the second index determines the side of the vehicle (1 … right side, 2 … left side). At a certain point in time, the braking torques of all four roller were abruptly changed to the constant non-zero value. The response of the axle-box force measurement system and tractive effort measurement system was monitored and compared to the simulation results obtained by a MBS model (Fig. 1, right). Measurements were made for different magnitudes of brake torque M_b. The results for $M_b = 50$ Nm are shown the Fig. 9 – 11.

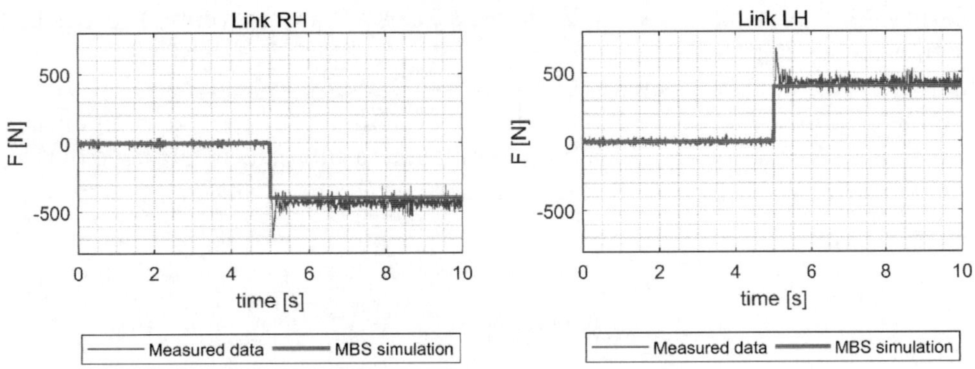

Fig. 9: Forces in linkages

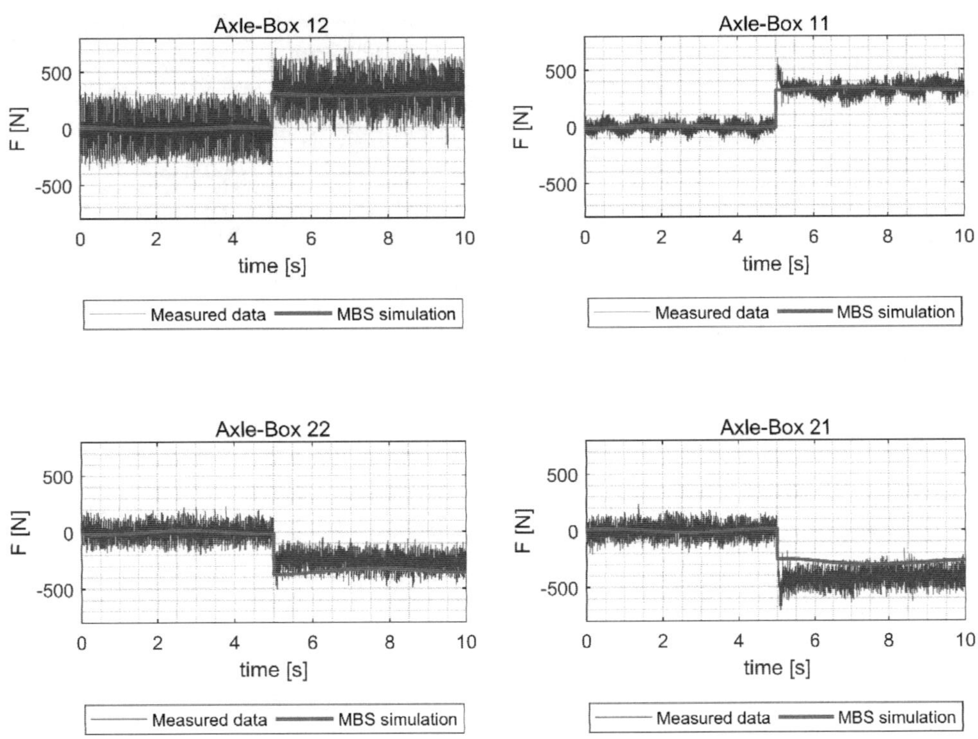

Fig. 10: Longitudinal component of axle-box forces

The measured forces in the linkages (Fig. 9) correspond very well with the simulation results. However, the simulation results give slightly lower values compared to the measured data. This is probably due to the fact that the simulation model do not incorporate the bearing and gear pairs resistances which increase the total brake torque of the roller drives. The mean value of the measured and calculated forces in linkages differs by a maximum of 7%.

The measured axle-box forces (Fig. 10) revealed some defects in the experimental bogie. On the axle-box 11 the force oscillations are apparent. Its' frequency corresponds to the wheelset RPM. This can be caused by a damage on the wheel running surface or by a damage of the axle-box bearing. Moreover, axle-box forces on the wheelset 2 are unevenly distributed. The axle-box 21 transmits grater longitudinal force than the axle-box 22. Based on this, the geometric deviation from the nominal position of the wheelset 2 setup can be deduced.

Nevertheless, the sum of longitudinal axle-box forces acting on individual wheelsets (Fig. 11) shows a very good match with the simulation results. The mean value of the measured and calculated sum of axle-box forces over all tests differs by a maximum of 7%.

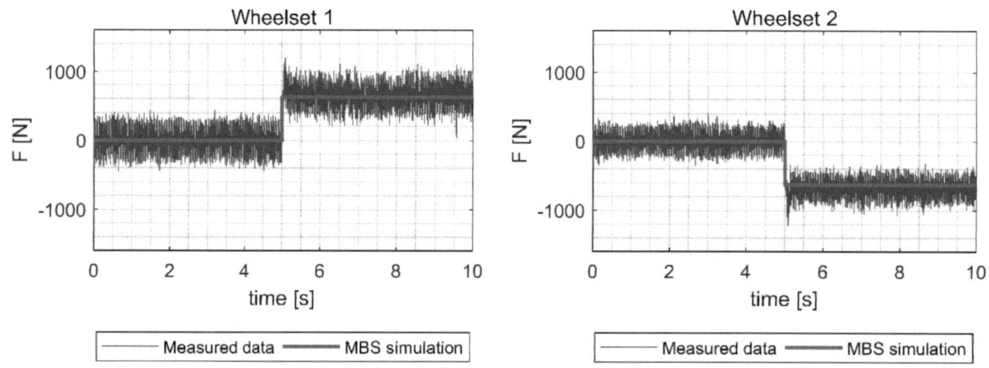

Fig. 11: Sum of longitudinal components of axle-box forces

Conclusions

The magnitude of the forces acting in the wheel-rail contact is an essential criterion for assessing the wear of wheels and rails. Because the wheel-rail contact forces are not measureable directly, the three different systems of force measurement, which could be used for determination of wheel-rail contact forces were implemented at the CTU roller rig. The first tests of these systems were focused on the longitudinal component of axle-box forces and overall bogie tractive effort measurement. The tests confirmed the functionality of the measuring system and showed a relatively good agreement of the simulation and the measured data. At the same time, these tests revealed some possible defects of the experimental bogie and pointed out the necessity to refine the simulation model.

Acknowledgement

This paper was created with the support of Technology Agency of the Czech Republic, project No TE01020038 „Competence Centre of Railway Vehicles“.

References

[1] I. Klimenko, J. Kalivoda, L. Neduzha, Parameter optimization of the locomotive running gear, in: Transport Means - Proceedings of the International Conference, 2018, pp. 1095-1098.

[2] R. M. Goodall, S. Bruni, T. X. Mei, Concepts and prospects for actively controlled railway running gear, in: Vehicle System Dynamics, 44 (suppl. 1), 2006, pp. 60-70, doi:10.1080/00423110600867374.

[3] N. Farhat, C. P. Ward, R. M Goodall, R. Dixon, The benefits of mechatronically-guided railway vehicles: A multi-body physics simulation study, in: Mechatronics, 51, 2018, pp. 115-126, doi:10.1016/j.mechatronics.2018.03.008.

[4] J. Kalivoda, P. Bauer, Roller rig tests with active stabilization of a two-axle bogie, in: Proceedings of the Third International Conference on Railway Technology: Research, Development and Maintenance. Edinburgh: Civil-Comp Press, 2016, paper 96, doi:10.4203/ccp.110.96.

[5] A. Jaschinski, H. Chollet, S. Iwnicki, A. Wickens, J. Von Würzen, The application of roller rigs to railway vehicle dynamics, in: Vehicle System Dynamics, 31(5-6), 1999, pp. 345-392. doi:10.1076/vesd.31.5.345.8360.

[6] S. Myamlin, J. Kalivoda, L. Neduzha, Testing of Railway Vehicles Using Roller Rigs, in: Procedia Engineering 2017(187), pp. 688-695, doi: 10.1016/j.proeng.2017.04.439.

[7] J. Kalivoda, P. Bauer, Mechatronic Bogie for Roller Rig Tests. In: Six, K., et al., eds. IAVSD 2015 - 24th International Symposium on Dynamics of Vehicles on Roads and Tracks. Graz, 17.08.2015 - 21.08.2015. Boca Raton: CRC Press. 2015.

[8] J. Kalivoda, P. Bauer, Scaled Roller Rig Experiments with a Mechatronic Bogie, in: Proceedings of the Second International Conference on Railway Technology: Research, Development and Maintenance. Edinburgh: Civil-Comp Press, 2014, paper 317, doi:10.4203/ccp.104.317.

[9] M. Dub, J. Kolář, F. Lopot, Axle gearbox hinge as a diagnostic member for rail vehicle drive, in: Experimental Stress Analysis - 56th International Scientific Conference, EAN 2018 - Conference Proceedings 2018, pp. 77-83.

[10] S. Myamlin, O. Lunys, L. Neduzha, O. Kyryl'chuk, Mathematical modeling of dynamic loading of cassette bearings for freight cars, in: Transport Means - Proceedings of the International Conference 2017, pp. 973-976.

[11] P. Bauer, J. Kalivoda, System of axle-box force measurement for experimental railway bogie, in: Experimental Stress Analysis - 56th International Scientific Conference, EAN 2018 - Conference Proceedings 2018, pp. 9-16.

EAN 2019
57th conference on experimental stress analysis

Influence of temperature on foam used in motorcycle protective equipment

KAŇÁKOVÁ Sandra[1,a], KOTTNER Radek[1,b] and BOŃKOWSKI Tomasz[2,c]

[1]Department of Mechanics, Faculty of Applied Sciences, University of West Bohemia, Technická 8, 301 00 Pilsen, Czech Republic

[2]New Technologies – Research Centre, University of West Bohemia, Univerzitní 8, 301 00 Pilsen, Czech Republic

[a]kanaksan@students.zcu.cz, [b]kottner@kme.zcu.cz, [c]tomasz@ntc.zcu.cz

Keywords: Drop test, Foam, Protective equipment, Temperature influence, Motorcycle

Abstract. This paper researches influence of the temperature on a foam, which is used in motorcycle protective equipment. The impact test was carried out on samples, which were cut from the center part of the protector that is suitable to protect shoulder, elbow and knee areas. Its main goal was to demonstrate an effect of three different temperatures, 22, 40, and 50 °C, on the foam. Six impacts, with impact velocities 1, 2, 3, 4, 5, and 6 $m \cdot s^{-1}$, were realized for each temperature. Surfaces of the impactor and the anvil were flat. The weight of the impactor was 5.25 kg. Obtained data were evaluated and subsequently can be used for a material model identification.

Introduction

Nowadays is a common matter to create simulations of the test of a protective equipment as shown in [1]. This trend is beneficial for both financial and time aspect of a protector designing process. However, it requires many experiments to acquire truly eligible material model, which can be used for these simulations. This is due to a lot of aspects affecting the real material behavior such as type of loading, loading velocity and temperature. It is the temperature that has potential to significantly affect mechanical behavior of an energy absorbing foam. Unfortunately, temperature changes are common matter during motorcycle sport.

This work follows previous research [2] and demonstrate the behavior of the energy absorbing foam in the course of both different impact velocity and different temperature. The tested foam SAS-TEC SCL-2 is commonly used in motorcyclist' protective equipment, for instance to protect shoulder, elbow and knee areas [3]. Used samples were cut from the central part of the protectors.

The aim was to determine temperature dependency of the force and the deformation responses during the impact test of the foam.

Obtained data can serve for comparison with different type of dumping material or for the material model identification. Thus obtained model will be coupled with the model of the leather [4] and together will complete model of the motorcyclist garment. Finally, this garment will be added to the model of a human body and used for motorcycle accident simulations, such as [5].

Experiment

The impact test was chosen as the best experiment to capture influence of the temperature on the foam. This experiment was selected mainly because its accuracy to the real accident situation.

The impact test was carried out in a drop tower designed and constructed by the authors. The weight of the impactor with a flat steel head was 5.25 kg. The reaction force was measured by the force cell KISTLER 9351B under a flat anvil (see Fig. 1). The impactor displacement was measured using two Micro-Epsilon optoNCDT 2300-50 lasers on the both sides of the impactor cross beam. The impactor deceleration was measured by the accelerometer KISTLER 8742A5. The sampling frequency was 26 kHz.

The experiment was carried out for six impact velocities: 1, 2, 3, 4, 5, and 6 $m \cdot s^{-1}$. The impactor releasing height was determined as

$$h_v = \frac{v^2}{2 \cdot g} + l_v \,, \tag{1}$$

where v is the impact velocity, g is the gravitational acceleration equal to 9. 81 $m \cdot s^{-2}$ and l_v is the sample height.

The test was conducted under three different temperatures. Namely 22, 40, and 50 °C. The samples were subjected to the particular temperature for one hour prior to the experiment. Each sample was used repeatedly at identical impact velocity (6 samples were used). Samples relaxed at least 24 hours between tests of different temperature (firstly, samples were tested at 22 °C, followed by 40 and 50 °C).

The rectangular samples were cut from the center part of the shoulder protector. The sample blocks had 80 mm per 80 mm base and their thickness was 11 mm. Their weight was 18 g.

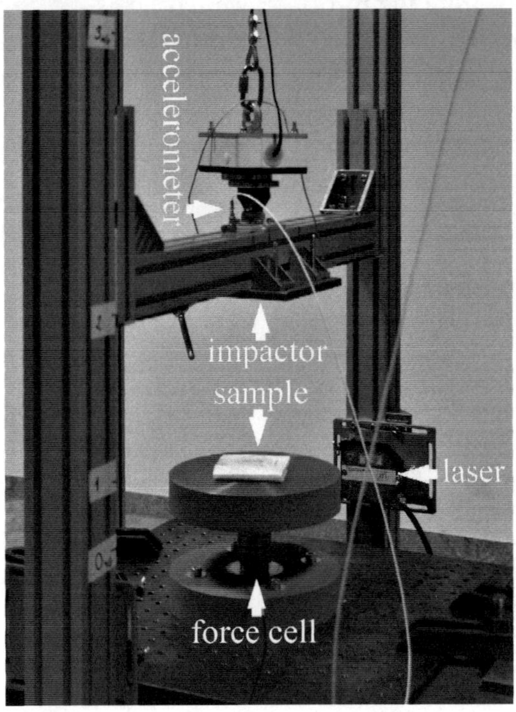

Fig. 1: Drop tower with the foam sample

Results

Obtained data are displayed in three graphs on Fig. 2 – 7. The first graph shows impactor displacement u ($u = 0$ represents contact of the impactor head and the anvil). Used lasers were able to register maximum 50 mm displacement. Moment, when the impactor is registered by the lasers, is $t = 0$. The second graph displays the reaction force F. The third graph displays impactor deceleration.

Impactor jumps are notable on Fig. 2 – 7. When the displacement u during the impactor jumps exceed 50 mm (the displacement graphs have staircase character), obtained data are evidently incorrect. In Fig. 7, the second landing in case of the temperature of 50 °C, the sample jump out of the center of the anvil and caused the impactor to collide directly with the anvil. Therefore, the measured data related to the second impactor landing were excluded from further evaluation.

Results summaries are displayed in Fig. 8 and Fig. 9. Fig. 8 shows maximum compressive deformations of the sample for six impact velocities in dependency on its temperature. It is obvious that the maximum deformation values increase with temperature increasing. The impactor and the anvil contact is notable in case of the impact velocity 6 $m{\cdot}s^{-1}$, were deformation equals 1. This collapse is given by the low stiffness of the foam caused by the temperature of 50 °C.

Fig. 9 shows maximum forces for six impact velocities in dependency on the sample temperature. The displayed values are determined in two ways, directly by the force cell (F in legend), indirectly as a multiple of impactor weight and maximum deceleration (A in legend). The forces show similar progressive character for increasing velocity for both ways of determination, however, there is also an increasing deviation between values of both forms of the determination. Forces from the force cell were used as the key results for future work.

As for result itself, it was already mentioned, higher temperature results in lower stiffness of the foam. Therefore, the maximum force decreased with increasing temperature in case the impact velocity was lower than 4 $m{\cdot}s^{-1}$. In the case of higher impact velocities than 4 $m{\cdot}s^{-1}$, the samples at 50 °C were compressed so much that maximum forces were higher than those tested at 22 °C.

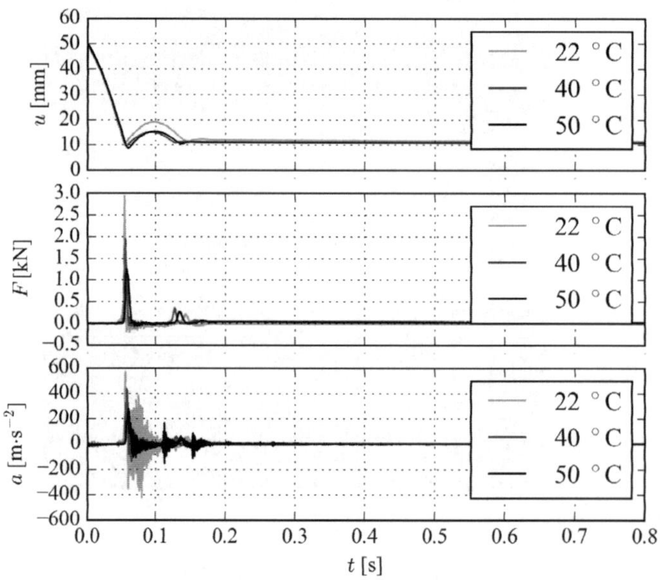

Fig. 2: Data from impact - 1 $m{\cdot}s^{-1}$

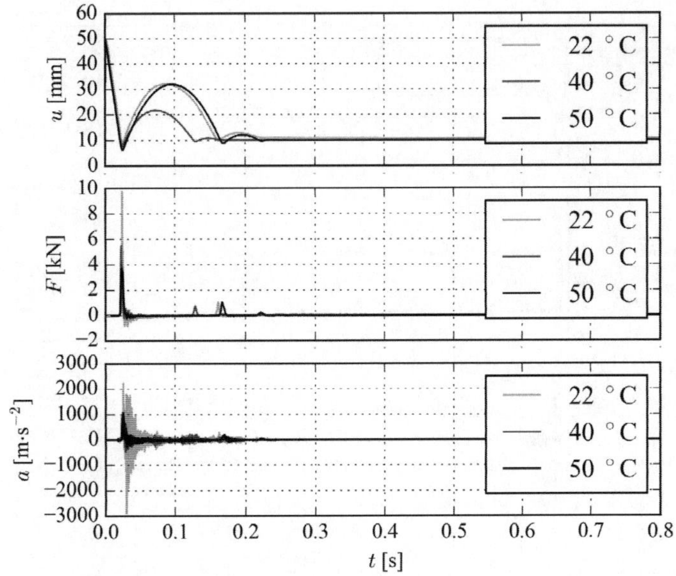

Fig. 3: Data from impact - 2 m·s^{-1}

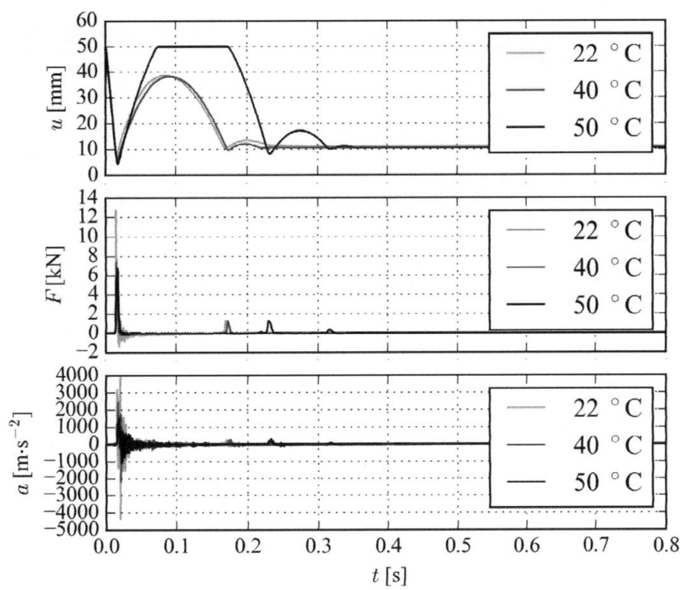

Fig. 4: Data from impact - 3 m·s^{-1}

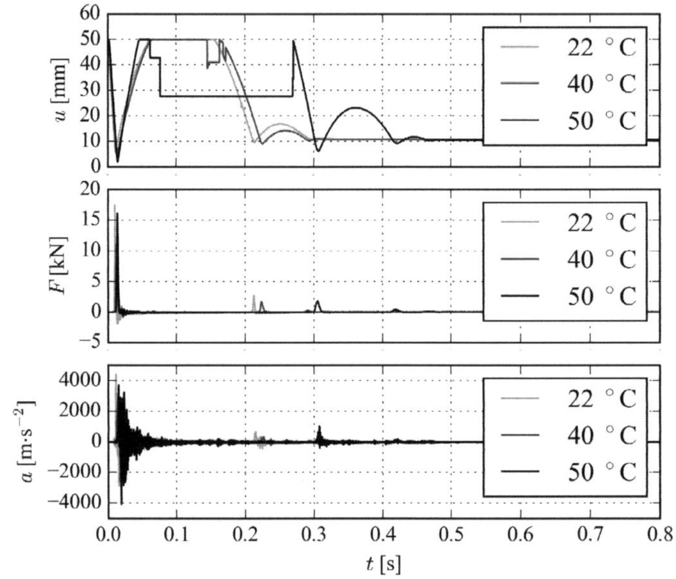

Fig. 5: Data from impact - 4 m·s^{-1}

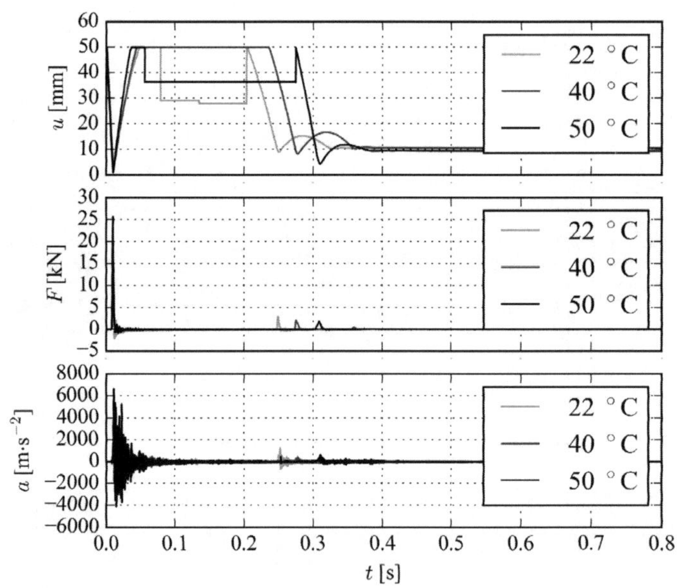

Fig. 6: Data from impact - 5 m·s^{-1}

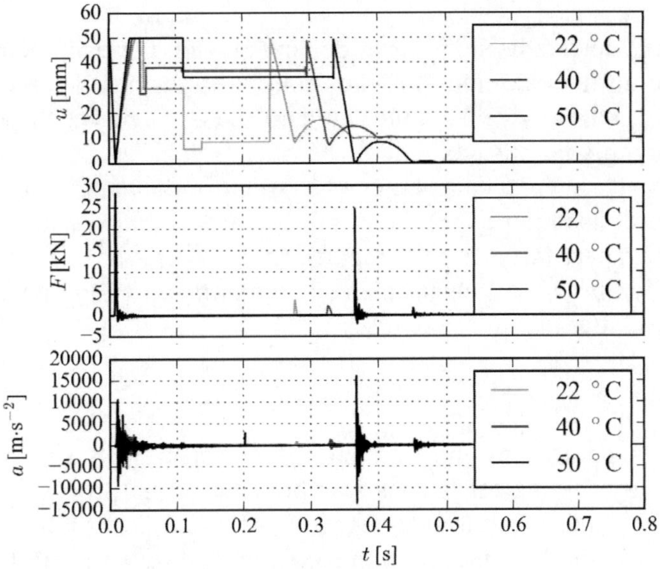

Fig. 7: Data from impact - 6 m·s^{-1}

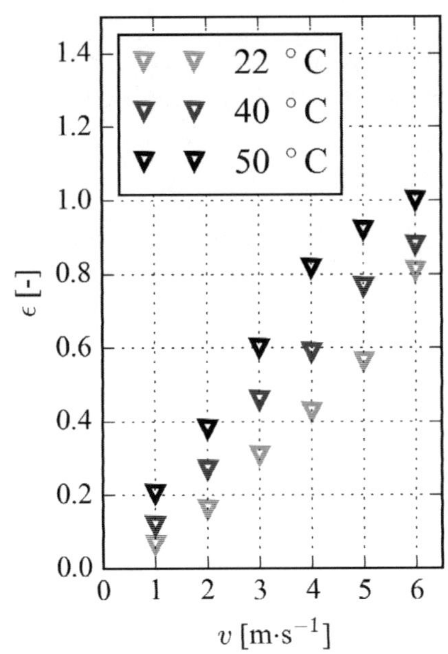

Fig. 8: Maximum compressive deformations

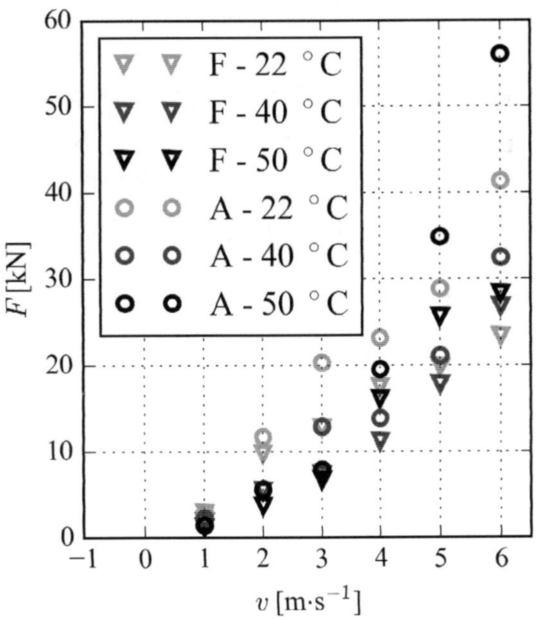

Fig. 9: Maximum forces

Conclusions

The foam significantly lowers its stiffness in dependency on the rising temperature, which has distinctive influence on its maximum compression and transmitted force. It is the maximum transmitted force that is the key value in the standard tests, and places the protective equipment in respective category, grading its quality.

The investigated experimental data serves as a target for the material model identification of the foam. In the future, a similar experiment will be carry out also with a spherical anvil. The concentration of the impact force and the significantly lower stiffness of the foam, caused by the temperature, could present serious problems in the protection of motorcyclists because both aspects are common matter in real accidents.

Acknowledgement

The work has been supported by the internal grant project SGS-2019-002 "Biomechanical models of human body, biological tissues and biomechanical processes with application in industry and medicine" and from European Regional Development Fund-Project „Application of Modern Technologies in Medicine and Industry" (No. CZ.02.1.01/0.0/0.0/17_048/0007280). The authors would like to thank PSí Hubík for providing foam samples.

References

[1] A. Fornells, N. Parera, J. A. Azpeitia, S. Ferris, J. J. Alba, Kid-shell: Safety System Protection for Child Passengers Travelling on Powered Two-wheeled Vehicles, Proceedings of the 6th Transport Research Arena, Warsaw, Elsevier, 2016, pp. 3886-3904.

[2] R. Kottner, S. Kaňáková, L. Šoltés, T. Bońkowski, J. Krystek, Mechanical Behaviour of Foam Used in Joint Protectors, 56th Conference on Experimental Stress Analysis (2018).

[3] Protectors. SAS-TEC Protektoren – Sicherheit aus Schaum [online]. SAS-TEC GmbH, ©2018. [Cit. 15. 4. 2019]. Accessible from: https://www.sas-tec.de/en/protectors/

[4] T. Bońkowski, L. Šoltés, L. Hynčík, R. Kottner, P. Kochová, Leather for motorcyclist garments: Multi-test based material model fitting in terms of Ogden parameters, Applied and Computational Mechanics 11 (2017) 129-136.

[5] L. Hynčík, T. Bońkowski, R. Kottner, Virtual assessment of motorcycle helmet contribution to decreasing injury risk in impact, 37th Fisita World Automotive Congress (2018).

EAN 2019
57th conference on experimental stress analysis

Experimental evaluation of thread locking methods for secure connections

KANAVAL Jan[1,a]

[1]Department of Designing and Machine Elements; Faculty of Mechanical Engineering; Czech Technical University in Prague; Technická 4, 166 07 Prague 6 - Dejvice, Czech Republic

[a]jan.kanaval@fs.cvut.cz

Keywords: Threaded fasteners; Prestressed bolted connection; Thread locking; Tightening torque; Untightening torque

Abstract. The term "Thread Locking" is used for methods of ensuring the threads stay in place without loosening. In some cases, the frictional forces from bolt axial preload are enough to hold the fastener together. In other cases, we need to use thread lockers. Thread locking can be separated into the following categories. Usage of mechanical parts to prevent loosening, further by increasing the coefficient of friction in the thread to prevent loosening, and finally by using adhesives in the thread to prevent loosening. This contribution describes analytical and experimental evaluation of untightening torque of bolted joints using Loctite thread lockers depending on surface degreasing quality.

Introduction

Threaded connections, such as prestressed bolted connections, are used in a wide variety of industrial applications. The term "Thread Locking" is typically used for methods of ensuring the threads stay in place without loosening. Bolts and nuts, this is probably the most common and low cost type of threaded connection. Screws and tapped holes, with this type of connection, we are installing a screw into a hole that has been drilled and tapped into a housing or mechanical component. The most commonly recognized difference between a screw and a bolt is that screws typically go into threaded holes while bolts go through 2 unthreaded components and fasten with a nut [1].

In some cases, the frictional forces from bolt axial preload are enough to hold the fastener together. But in other cases, we need to use thread lockers. Some of these cases are high vibration, expected stress relaxation (creep in plastics), and cyclic stresses [1]. Thread locking can be separated into the following three categories. Usage of mechanical parts to prevent loosening, further by increasing the coefficient of friction in the thread to prevent loosening, and finally by using adhesives in the thread to prevent loosening. This contribution describes analytical and experimental evaluation of selected thread locking methods to prevent loosening.

Standard Thread Locking Methods

1) By using of mechanical parts to prevent loosening of the bolt connection (see Fig. 1).
2) By increasing the coefficient of friction in the thread to prevent loosening of the bolt connection (see Fig. 1).
3) By using adhesives in the thread to prevent loosening of the bolt connection (see Fig. 1).

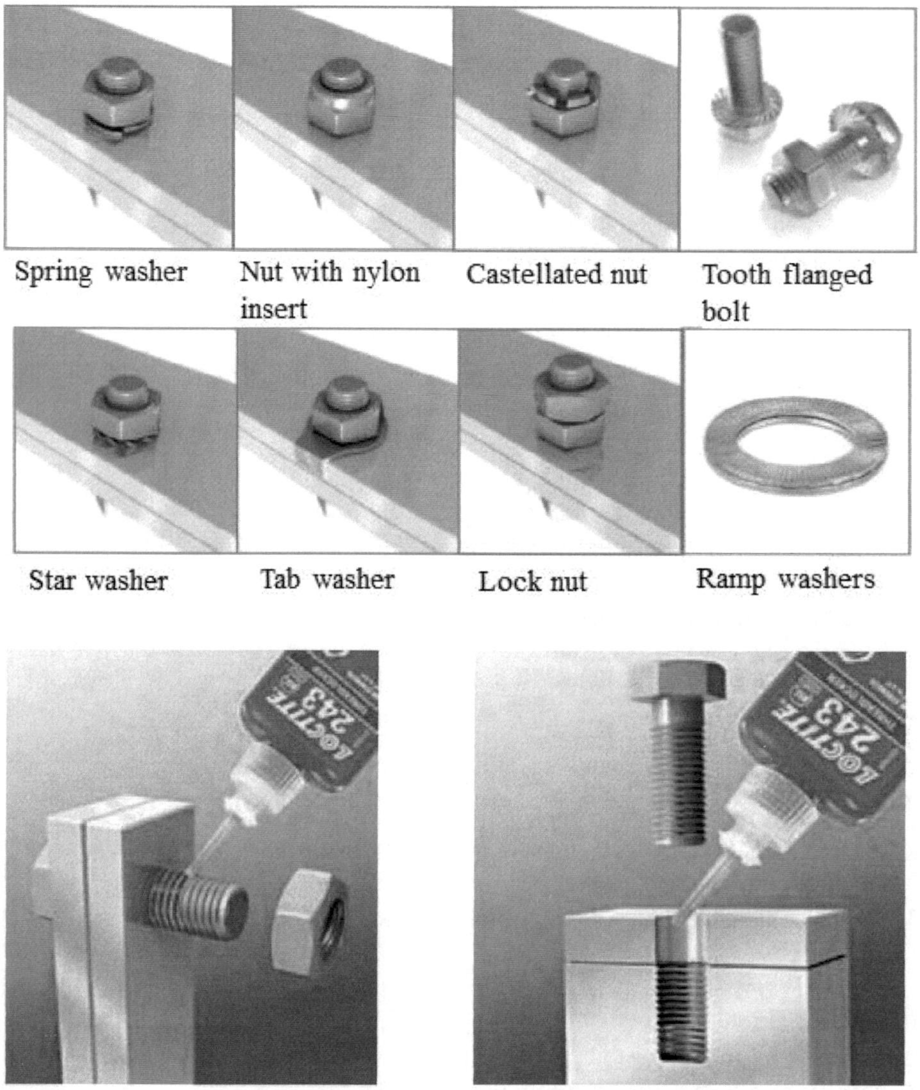

Fig. 1: Traditional threadlocking devices

Thread Locking Theory and Thread Locking Experimental Evaluation

To calculate the untightening torque moment of threaded connection Eq. (1) can be used:

$$M_{KK} = M_K + M_{TM} + M_{SL} + M_{LL} = Q_o \cdot \frac{d_2}{2} \cdot tg(\varphi' - \gamma) + Q_o \cdot \rho_{TM} \cdot f_{TM} + M_{SL} + M_{LL} \quad (1)$$

γ - thread pitch angle [°]; φ' - friction angle in thread (corresponding to the coefficient of adhesion) [°]; d_2 - flank diameter of the bolt thread [mm]; ρ_{TM} - friction radius under the nut [mm]; f' - coefficient of friction in thread (adhesion coefficient) [1]; f_{TM} - coefficient of friction under the nut [1]; M_{KK} - untightening torque [Nmm]; M_K - torque [Nmm]; M_{TM} - friction torque under the nut [Nmm]; Q_o - mounting - pretension force [N]; M_{SL} - self-locking torque of the nut [Nmm]; M_{LL} - Loctite locking torque of the nut [Nmm].

To verify the thread locking capability of selected thread locking methods to prevent loosening *Junkers Vibration Test* according to standard DIN 651151 can be used (see Fig. 2).

Another method to verify the thread locking capability may be measuring of the static untightening torque using a torque wrench. Measurements were performed on M8 screw connections (screw M8x50 - strength class 8.8; nut - strength class 8, galvanized - Zn; without the use of oil/grease; tightening torque 27.3 Nm) 24 hours after tightening. BMS PADDY, model TAW 200, (maximum capacity 200 Nm), torque wrench was used.

A simple testing fixture for measuring of the static untightening torque using a torque wrench is shown in Fig. 3.

The experimentally measured static untightening torque values for selected thread locking methods are shown in Table 1 and also graphically in Fig. 4.

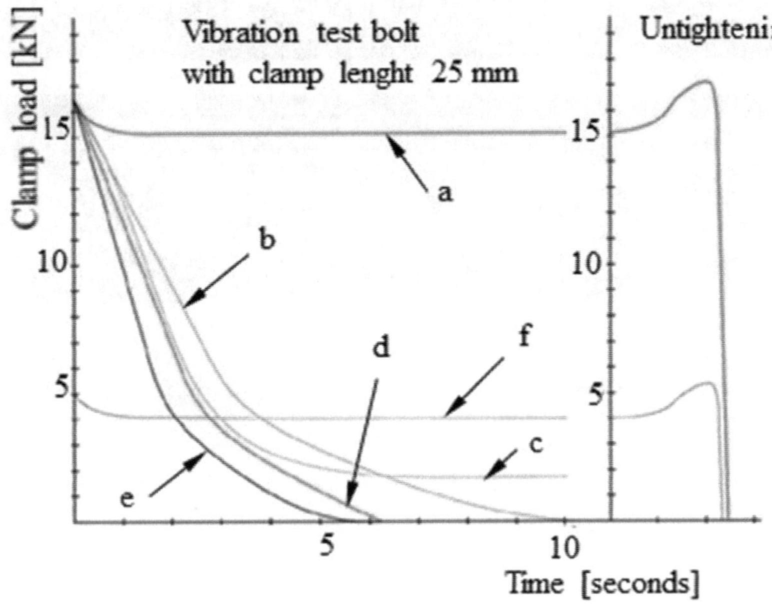

Fig. 2: Vibration test bolt size M8 with clamp length 25 mm (a - nut M8 on bolt with Nord-Lock at 70 % of yield point; b - nut M8 on bolt, clamp lenght 50 mm; c - nut M8 with nylon insert on bolt; d - nut M8 on bolt with split ring (spring washer); e - nut M8 on bolt; f - nut M8 on bolt with Nord-Lock at 25 % of yield point

Fig. 3: A simple testing fixture for measuring of the static untightening torque using a torque wrench BMS PADDY, model TAW 200, (maximum capacity 200 Nm)

Table 1: Untightening torque measurement results (using a torque wrench)

Thread locking method	Untightening static torque [Nm] (average of the measured values)
A) Screw + nut (without any washer)	23.6
B) Screw + nut with spring washer	20.2
C) Screw + nut with nylon insert	26.2
D) Screw + nut with Schnorr (VS) washer	25.2
E) Screw + RIPP LOCK nut	24.8
F) Screw + nut (DIN 6923)	28.5
G) Screw + nut with Heico-lock washers [2]	24.8
H) Screw + nut, Loctite 243 was applied [3]	25.8
I) Screw + nut, Loctite 270 was applied [4]	28.1

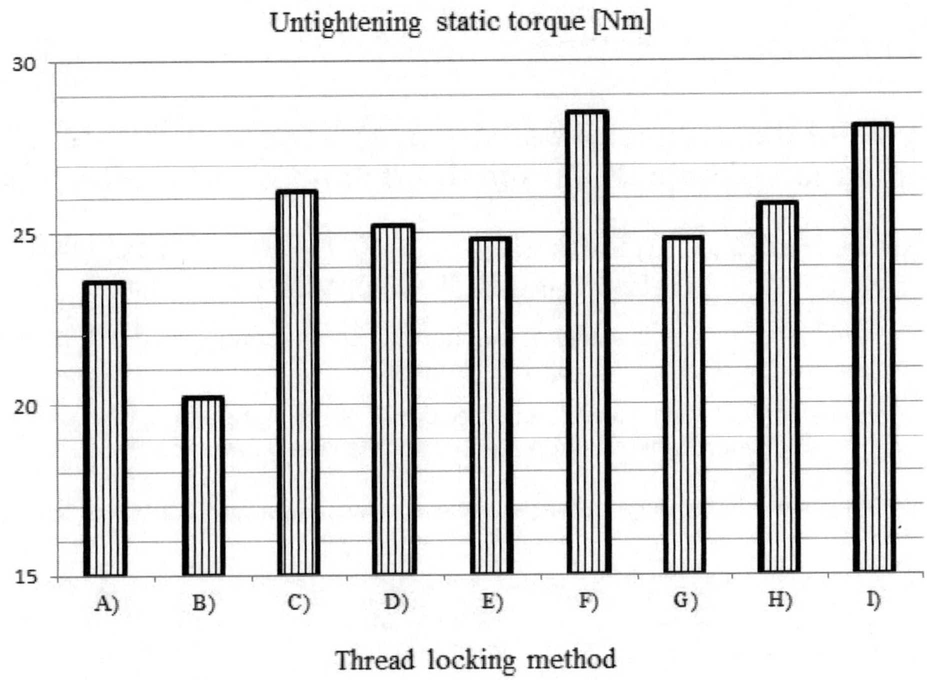

Fig. 4: Experimentally measured static untightening torque values for selected thread locking methods (A - Screw + nut (without any washer), B - Screw + nut with spring washer, C - Screw + nut with nylon insert, D - Screw + nut with Schnorr (VS) washer, E - Screw + RIPP LOCK nut, F - Screw + nut (DIN 6923), G - Screw + nut with Heico-lock washers, H - Screw + nut, Loctite 243 was applied, I - Screw + nut, Loctite 270 was applied)

Conclusions

Measurements of selected thread locking method's static untightening torques were realized. The measured static untightening torque values correspond in some cases to the results of the *Junkers Vibration Test* (see Fig. 2 - c, d, Fig. 4 and Table 1). The results of further experiments will be presented at the EAN 2019 conference.

Acknowledgment

Threaded fasteners and washers for experimental testing were provided by the company HEICO Czech s.r.o.

References

[1] Sean, T., Threaded Locking Methods for Secure Connections, Inf. on https://www.blog.fictiv.com/authors/sean-thomas, last accessed 2019-03-20.

[2] URL: https://www.heico-group.com/cs/heico-lock/, last accessed 2019-04-16.

[3] Prumex s.r.o., Technical sheet, Inf. on https://proprumysl.cz/soubory/20/243.pdf, Hrušovany nad Jevišovkou, last accessed 2019-03-20.

[4] Prumex s.r.o., Technical sheet, Inf. on https://proprumysl.cz/soubory/33/270.pdf, Hrušovany nad Jevišovkou, last accessed 2019-03-20.

Analysis of the causes of defects in structural elements with a combination of hole drilling methods and photoelasticimetry

KOSTKA Ján[1,a], PÁSTOR Miroslav[1,b], ČARÁK Peter[1,c], FRANKOVSKÝ Peter[1,d] and KULA Tomáš[1,e]

[1]Technical University of Kosice; Faculty of Mechanical Engineering; Letná 9, 042 00 Košice, Slovak Republic

[a]jan.kostka@tuke.sk, [b]miroslav.pastor@tuke.sk, [c]peter.carak@tuke.sk, [d]peter.frankovsky@tuke.sk, [e]tomas.kula@tuke.sk

Keywords: Optical methods, PhotoStress, Hole Drilling, Residual Stresses, Experimental Measurement

Abstract. The content of the paper is to acquaint with the problematics of structural failure, description of quantification of residual stresses by the drilling method and design of the methodology of using optical methods for their verification. The solution was based on lot of experience of the workplace (KAMaSI SjF TUKE) in the field of determination of residual stresses by the drilling method and the Ring-Core method, as well as the use of optical methods (photoelasticimetry and DIC) for full-field analysis of investigated objects. Although the optical methods have been used for stress and strain analysis for decades, with the continuous development of new and more accurate measuring instruments and devices, the possibilities of creating new application methodologies are opening up. To advisement the possibility of using the Optical PhotoStress method for quantifying residual stresses, has been designed an accurate positioning device to analyze the released deformations around the drilled hole in multiple steps as defined by ASTM E837-13a for drilling methods.

Introduction

Determination of causes and the prediction of the failure of the bearing members of structures is still a highly topical issue related to the assessment of the lifetime of structures, machines, equipment, etc. It should be pointed out that many failures of parts of structures, systems or machines are not only due to the stress induced by the load, but are also due to the occurrence of residual stresses. These also occur in unloaded structures or machine parts. They can be caused not only by the operating load, but also by the production technology (casting, rolling, pressing, etc.) as well as by welding. Determining the occurrence of residual stresses, their size and direction is problematic without carrying out experimental measurements. The risk of their occurrence is mainly related to the fact that they are superimposed on operating stresses, which can significantly affect the life of machines and equipment. For this reason, it is necessary to know and quantify these stresses to predict the occurrence of disorders. Currently, the drilling method is most commonly used to determine residual stresses in the material. This is a semi-destructive experimental method of quantifying residual stresses in a selected point. The authors use the SINT MTS 3000 drilling equipment or the RS 200 drilling machine to solve specific practical tasks. At the (Fig. 1) is an illustration of the determination of residual stresses in the deburring drum (Fig. 1a) in order to advisement their impact on cracking (Fig. 1b) [4-6].

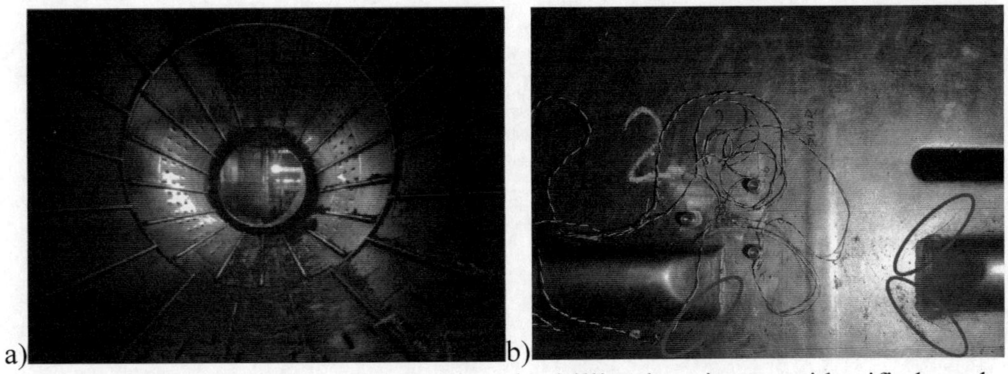

Fig. 1 Deburring drum: a) inside view, b) drilling location near identified cracks

Hole drilling method

The hole drilling method is the most common universal technique for measuring residual stresses in materials. They have standardized procedures, good reliability and accuracy. The principle of drilling methods is base on the chang of tension or deformation generated by drilling holes into materials or samples whitch include residual stresses. By knowing the size of the holes, directions and the size of the relaxed deformations and material properties, we can determine the residual stresses through the experimental analysis. We can rightly consider this method as a semi-destructive method, since the drilled hole can be negligible and can be repaired by welding or inserting a bolt or pin. It is often possible that this correction will alter the stress distribution and therefore blind holes are used instead of through holes [4].

Among the benefits of hole drilling method include:

1. application of the method is possible not only in laboratory conditions but also in practice,
2. usability for a wide range of materials - metal and non-metallic,
3. a fast and easy to apply process,
4. suitability to measure residual stresses even after surface treatment of component,
5. repeated measurements of residual stresses during various phases of component life,
6. nominal accuracy: steel - 30 MPa, aluminum - 10 MPa, titanium - 15 MPa [14].

The drawbacks of the drilling method include:

1. the resulting hole needs to be repair or welded again,
2. the maximum measuring depth is 2 mm,
3. measurement is not possible on very curved surfaces for attachment of the device - application on complicated components is limited,
4. strain gauges used for drilling are susceptible to noise and requires good surface preparation at the drilling site,
5. a strong deformation response rate from the material is required,
6. inaccuracies due to stress release after gradual drilling in depth [14].

At the Fig. 2a is a diagram of a uniformly distributed tension along the drilling depth. At the Fig. 2b, the structural view of the tension is unevenly distributed along the drilling depth. The individual drilling step profile corresponds to a "staircase" increment during progressive hole formation and residual stress measurement [1].

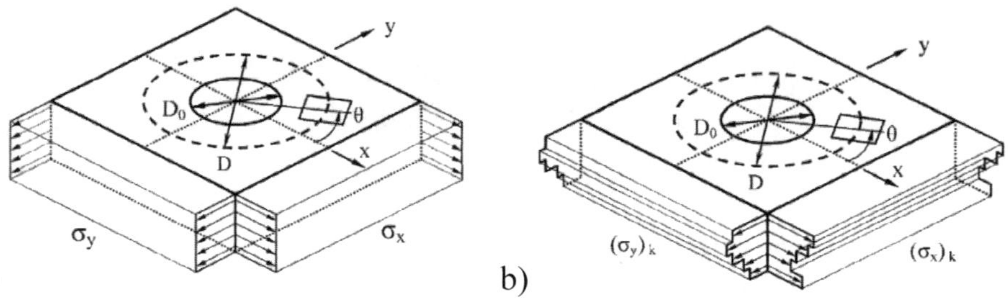

a) b)

Fig. 2 Uniformly (a) and unevenly (b) distributed stress

Blind hole analysis

In many cases, through the drilling process, through-holes cannot be realized due to the dimensions and different shapes of the components and therefore is used the shallow "blind" hole drilling method. But then there is the problem of local tension, for which there is no exact solution to date in the mathematical theory of flexibility. The previous equation can be applied only to the calculation if the appropriate coefficients $\bar{A}$ and $\bar{B}$ are used for the blind hole. These coefficients are obtained empirically (by experimental calibration or numerical modeling using FEM) because they cannot be obtained from theoretical considerations [4]. It is necessary to introduce another independent variable, namely the dimensionless hole depth Z/D. The generalized form of functions for coefficients can be expressed by:

$$\bar{A} = f_A(E, \mu, r, Z/D) \quad , \tag{1}$$

$$\bar{B} = f_B(E, \mu, r, Z/D) \quad . \tag{2}$$

In some casis of residual stress and the selected hole diameter, the released deformations grow when the hole depth increases. The specified process continues to a hole depth that is equal to its diameter or slightly larger. For the purpose of obtaining maximum deformation signals, the hole is drilled to a depth that corresponds to a ratio of at least $Z/D = 0{,}4$ [4,15].

It is also important to know the behavior of the local thin plate region when analyzing the blind hole. In ASTM E 837-13a, it is recommended to create holes in small depth increments [8]. This is done in order to obtain data to advisement whether the residual stress is uniform over the depth. If the residual stress varies, the calculated stress is always less than the actual maximum. At the Fig. 3 is a diagram for investigating relaxed deformations at any point in planar stress.

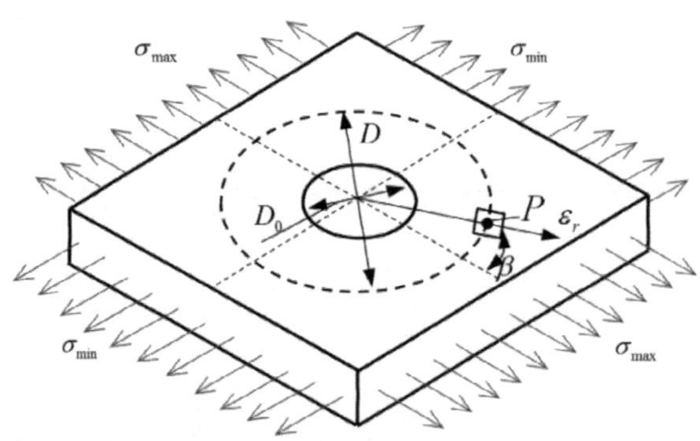

Fig. 3 Investigating loose deformations at any point at plane stress

Released surface radial deformations correspond to lightened main stresses as shown below:

$$\varepsilon_r = (\bar{A} + \bar{B}\cos 2\beta)\sigma_{max} + (\bar{A} - \bar{B}\cos 2\beta)\sigma_{min} \quad , \tag{3}$$

where:

ε_r	represents the released deformation in the radial direction by a strain gauge centered at point P,
$\bar{A}, \bar{B}$	are calibration constants,
β	is the angle measured against in the clockwise from the direction strain gauge to the σ_{max} direction,
σ_{max}	is the maximum main normal stress at the hole location before drilling,
σ_{min}	is the minimal main normal stress at the hole location before drilling,
D	is the mean diameter of the strain gauge,
D_0	is the diameter of the drilled hole.

The following relationships are used to quantify the constants $\bar{A}$ and $\bar{B}$ for a material with given elastic properties:

$$\bar{A} = \frac{-\bar{a}(1+\mu)}{2E} \quad , \tag{4}$$

$$\bar{B} = \frac{-\bar{b}}{2E} \quad , \tag{5}$$

where:

E	is Young's modulus of elasticity,
μ	is Poisson's ratio,
$\bar{a}, \bar{b}$	they are dimensionless and almost material-independent constants determine by the standard ASTM E837-13a.

Dimensionless and almost material-independent constants $\bar{a}$ and $\bar{b}$ vary with hole depth. $\bar{a}$ is independent of material, and $\bar{b}$ is dependent only to a small extent on Poisson's ratio. For blind holes in the Poisson's ratio in range from 0.28 to 0.33, the constants $\bar{a}$ and $\bar{b}$ change less than 1% [4].

Reflexive Photoelasticimetry

Utilization of reflexive photoelasticimetry (PhotoStress methods) is not only bound to the models of the objects under investigation, but also to the actual machine parts and assemblies. When the actual components have a high modulus of elasticity are produced their models with a lower modulus of elasticity, thereby achieving greater deformation at the same load. Neither the actual research object nor the model is made of optically sensitive material, but a thin film of optically sensitive material from the liquid components is applied thereto. Polariscop is use to investigate deformations and stresses on the surface of the test objects. After illumination with polarized light and subsequent loading can be observed through the analyzer isochromatic or isoclinic bands [13].

Design and verification of residual stress determination method using optical methods of mechanics

Despite the advances made in numerical methods, it is still not possible to unambiguously analyze the residual stresses using numerical methods. Therefore, even nowadays experimental methods in the area of detection and measurement of residual stresses have their irreplaceable place. It is necessary to know the individual stages of the experiment when stressing machine components or assemblies with drilling and optical methods and then determining stress and deformation sizes. The aim of this experimental part was to propose a methodology for determining residual stresses using optical methods, namely the PhotoStress method together with hole drilling method [11,12].

American norm ASTM E837-13a is currently the only generally accepted standard that deals with investigating residual stresses by experimental mechanics. However, there are a number of limitations in this standard. For direct drilling into the component and subsequent examination of the deformation values by optical methods, it was necessary to design and manufacture the device (Fig. 4 a,b), which would connect the two mentioned methods of detecting stresses and deformations in structural members. The idea was to design a device with an adjustable drilling head and an optical device for detecting deformation fields - the LF/Z-2 polariscope. The principle was to accurately drill into a defined depth of the component with a special cutting tool. Subsequently, by means of a stepper motor, is released by the horizontal displacement of the drilling head outside the opening the space for measuring the residual stresses by the LF/Z-2 polarisop with a compensator [8,15].

At the Fig. 4a is a view on one of the designs of a 3D model of a positionable drilling device, and at the Fig. 4b is the resulting real constructed device. The motors are used for horizontal and vertical movement of the drilling head and for rotating the drill or milling cutter.

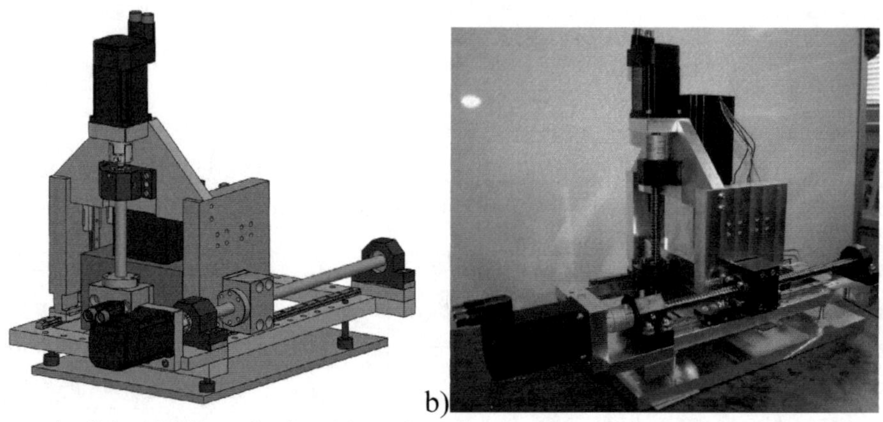

a) b)

Fig. 4 Proposal of the drilling device (a) and really configurable positioning drilling device (b)

The precise positioning drilling device was largely constructed from aluminum alloy due to its satisfactory strength and total weight of the device. Subsequently was tested the accuracy of the drilling device for horizontal and vertical movement of the drilling head and drilling through the unloaded samples with already applied photoelastic coating. Also tested was a hydraulic load device located in the newly built Prototyping and Innovation Center of the Faculty of Mechanical Engineering TUKE, which is capable of generating loads of up to 100 kN in two direction. The annealed samples were drilled under load and through the polarizing device and the photoelastic coating were examined the formed entities [16].

In the following section of experimental testing and investigatinf of stress fields, was designed an aluminum sample of dimensions and shape as shown in Fig. 5. The shape of the samples, resp. its 65 mm edges, were adapted for further testing by other methods for detecting stresses and deformation fields in the component.

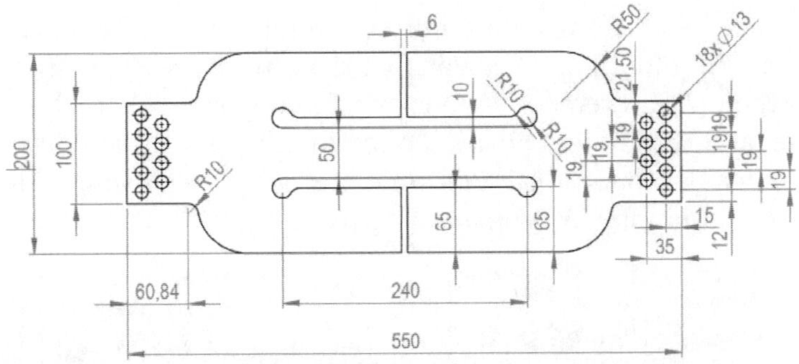

Fig. 5 Shape and dimensions of samples used for drilling detection and optical method PhotoStress

The sample was annealed to eliminate residual stresses. A photoelastic coating PS-1D was consist of a two-component adhesive (resin + hardener) was applied to the sample from aluminum as indicated by the manufacturer [15,16].

Furthermore, the simulated stresses were determined on a uniaxially loaded aluminum sample. It consisted in the gradual drilling of the blind hole and the subsequent observation of the emerging photoelastic entities in its vicinity. A strain gauge cross was applied to the test sample to check the stability of the induced load (Fig. 6). Each grid was connected to the quarter bridge. Time warp was recorded and evaluated by Catman Easy.

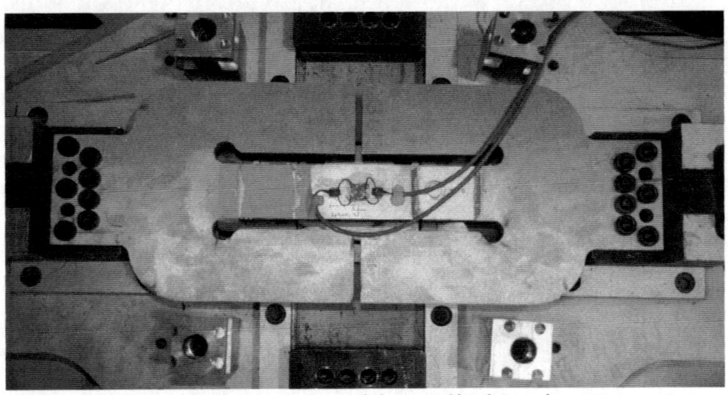

Fig. 6 Aluminium sample with applied strain gauge cross

At the Fig. 7 is a view on a drilling process on an aluminum sample loaded with a hydraulic loading apparatus and stress measuring by optical apparatus LF/Z-2 .

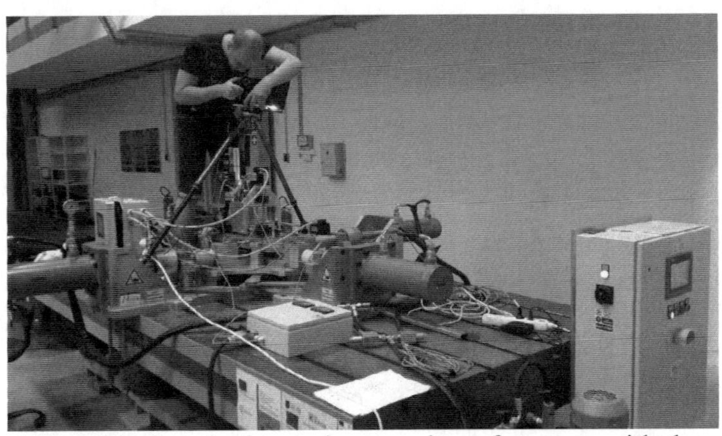

Fig. 7 Drilling with positioning device and measuring of stresses with the optical device LF/Z-2

The mill cutter speed with diameter 3.2 mm (or 18 mm) was 120 rpm. The released relative deformations around the blind hole were analyzed in twenty steps with an increment of 0.1 mm. During the whole measurement, the tensile force was 7.5 kN. By moving the drilling head in the horizontal direction to a position allowing the use of the polarisop, directions and differences of the main proportional deformations were observed. After each step were created images of isoclinic and isochromatic lines (Fig. 8).

Fig. 8 Process of registration of isoclinic and isochromatic lines with a polarisope LF/Z-2

The entire measurement phase without the application of strain gauge and photoelastic coating took approximately 2 hours. At the Fig. 9 is a more detailed time record of the vertical displacement of the drilling head into the annealed sample registered by the displacement sensor.

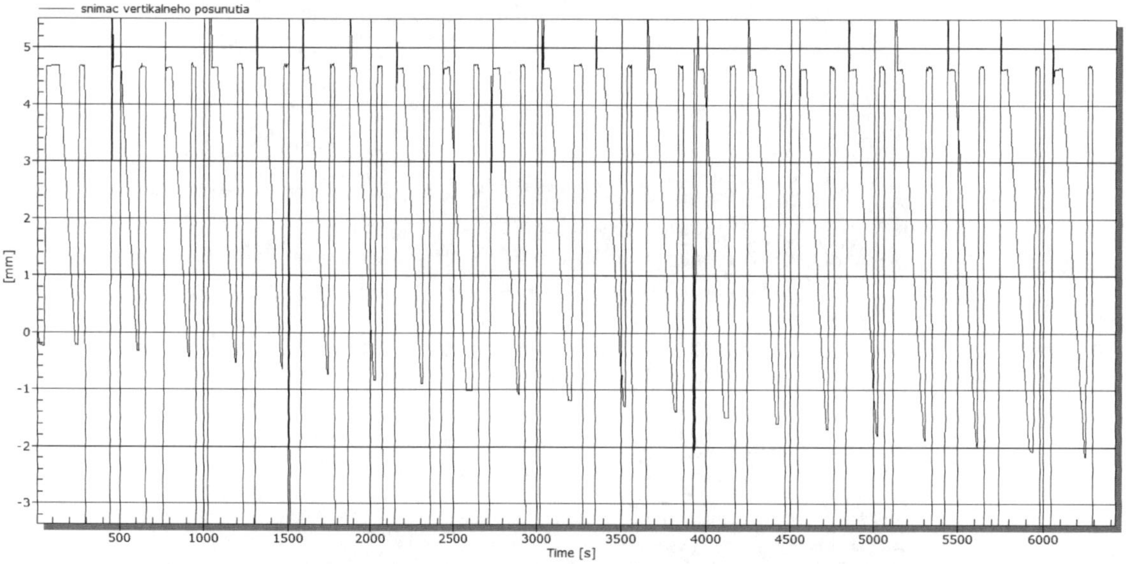

Fig. 9 Time recording from vertical movement of hole drilling

In the next section, isochromatic lines for a given depth of 0.5 mm (Fig. 10), 1.0 mm (Fig. 11), 1.5 mm (Fig. 12) and 2 mm (Fig. 13) are given for comparison of the released strain ratios. Blind hole drilling was not realized for larger depths (in teres of ASTM E 837-13a). From Fig. 10 - Fig. 13 it stands to reson that more pronounced bands (isochromatic lines) are more visible when drilling an annular groove, which corresponds to the results of numerical modeling. Despite this, it can be stated that these are relatively low stresses levels, their quantification is more demanding and the measurement inaccuracy is also higher in this case. On the other hand, low levels of simulated "residual" stresses do not pose a risk to the safe operation of machinery and equipment. Higher sensitivity of the simulated stress measurement is possible by using an optically sensitive coating with a greater thickness, which however limits the use of cutting tools (milling cutter or hollow cutter).

Within the proposed methodology of quantification of residual stresses using the PhotoStress method, it was due to the comparison of achieved results with the method of drilling, resp. Ring-Core method considered using identical cutting tools 3.2 mm diameter cutter used with RS-200 and hollow cutter used with SINT-MTS 3000 Ring-Core. The limiting factor is the shape of the cutting tools for drilling the blind hole, respectively. an annular groove, since the thickness of the adhesive and the coating used had to be added to the depth itself. Therefore was used the thinnest coating of PS-1D with a thickness of 0.5 mm.

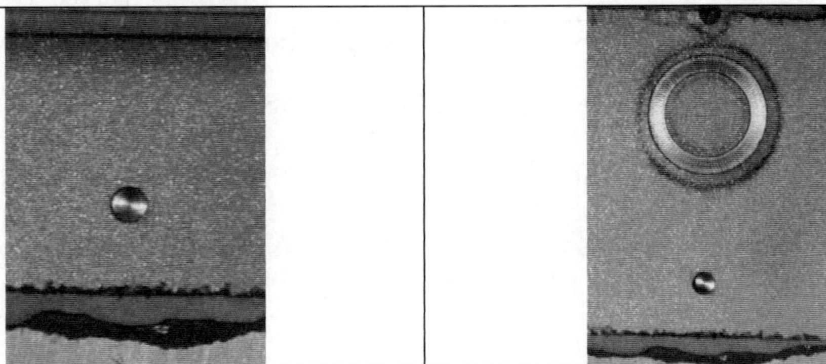

a)	b)

Fig. 10 Isochromatic lines for 0.5 mm depth a) blind hole, b) annular groove

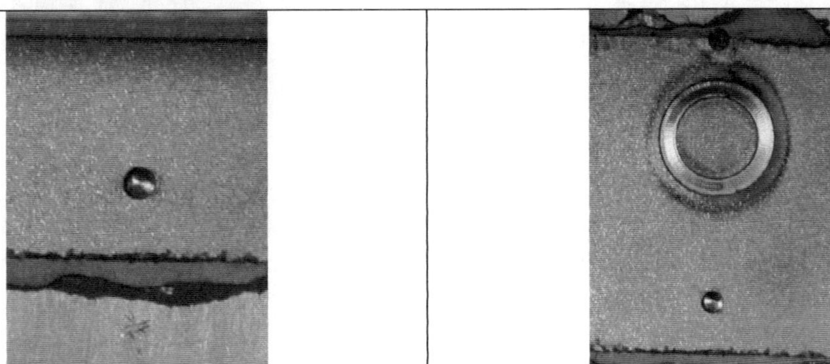

a)	b)

Fig. 11 Isochromatic lines for 1.0 mm depth a) blind hole, b) annular groove

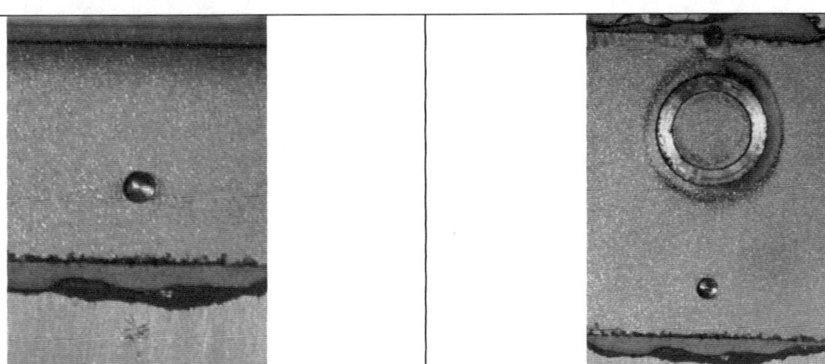

a)	b)

Fig. 12 Isochromatic lines for 1.5 mm depth a) blind hole, b) annular groove

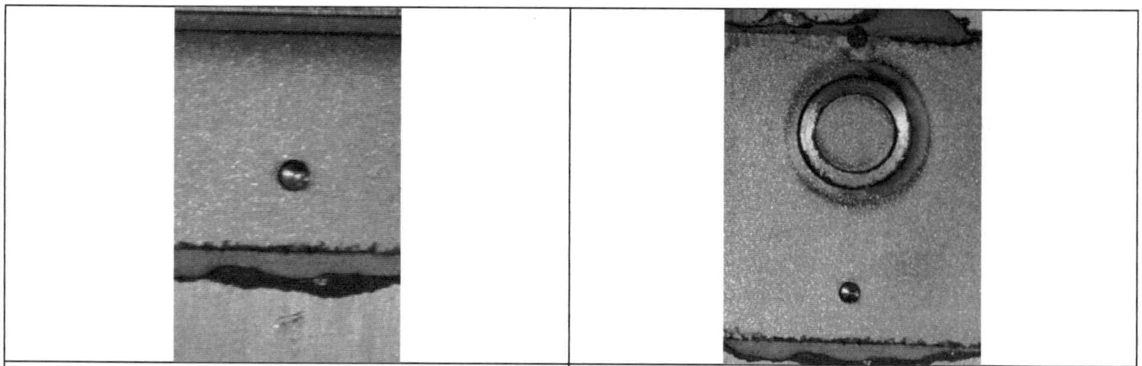

a) b)

Fig. 13 Isochromatic lines for 2.0 mm depth a) blind hole, b) annular groove

The time recording of the relative deformations in the test sample during the experimental measurements is shown in Fig. 14. It is clear from the figure that there were no significant variations during the measurement, the correct operation has been confirmed of the simulated load-causing hydraulic load device. The hydraulic loading device can be considered as one of the key elements of the measurement chain, since if the stress in the sample analyzed is changed in the experiment, the obtained results could not be considered as relevant.

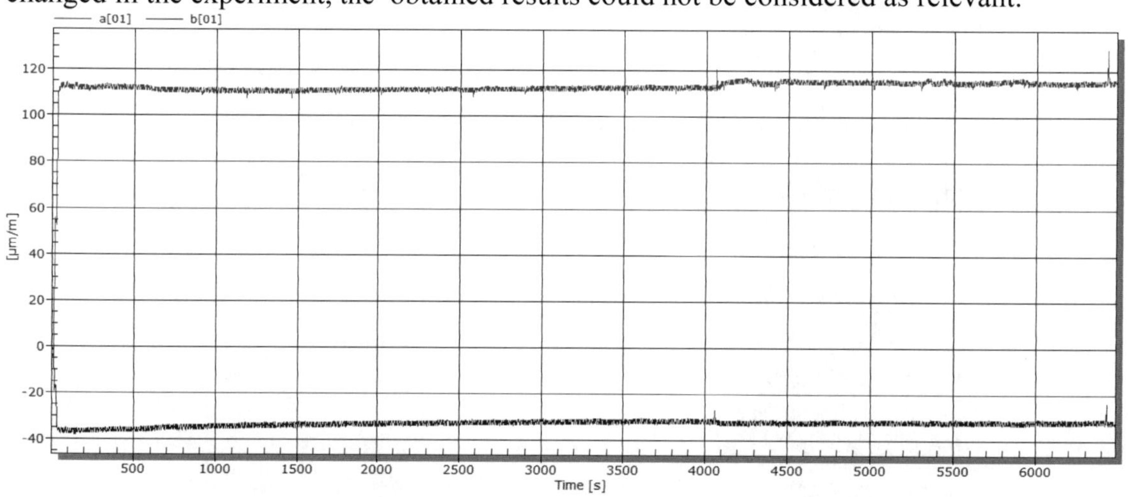

Fig. 14 Time recording of relative deformations on loaded test sample

Conclusions

In the submitted paper, the methodology of quantification of residual stresses was proposed and verified. On the basis of the results obtained, it can be stated that an accurate positioning device has been designed to evaluate the released relative deformations during the progressive drilling of the blind hole, respectively an annular groove using the PhotoStress optical method. These devices were used in experimental measurements on unloaded samples as well as annealed samples loaded with simulated loads. On the basis of the analysis of isochromatic bands around a 3.2 mm diameter blind hole and around a groove with an outer diameter of 18 mm and an internal diameter of 14 mm registered with a reflection polarisop, it can be concluded that the more visible deformations are around the annular groove. For experimental measurement, the PS-1D coating was chosen to achieve the lower adhesive and coating thickness. The disadvantage of said coating is less sensitivity to the deformation released. It is true that the thicker the coating, the more accurate the measurement. This fact will be dealt with in the next period, as the cutting tools used by RS-200 or SINT-MTS 3000 Ring-Core devices were used in the present work.

The advantage of applying the optical method for quantifying residual stresses is, among other things, the possibility of full-field deformation analysis in the investigated area on a real structure or device. Despite the good consensus of the results obtained by experimental measurement and numerical computation, which is part of the work [16], the results achieved cannot be considered final.

There is wide scope for:

- research into the possibility of using the proposed methodology for inhomogeneous, anisotropic, composite materials,
- optimizing the design of the pointing device not only to reduce its weight but also to position it in the third axis, allowing the device to be used for the drilling method, respectively Ring-Core,
- use of the proposed equipment not only for the PhotoStress method, but also in combination with the DIC (Digital Image Corelation) method, with which we also have extensive experience in the area of deformation and stress analysis in the vicinity of concentrators at the KAMASI workplace,
- automating the drilling and strip reading process by compensator, not only significantly reducing the length of the experimental measurement, but also eliminating the adverse effect of the human factor e.g. when positioning the device, etc,
- testing new cutting tools.

Acknowledgement

The authors would like to thank for financial support for this study the Slovak Research and Development Agency APVV-15-0435 and the Slovak Grant Agency VEGA under grants VEGA no. 1/0290/18 and VEGA no. 1/0751/16.

References

[1] KOBAYASHI, A. S.: Handbook on Experimental Mechanics. Society for Experimental Mechanics, Seattle, 1993, ISBN1-56081-640-6.

[2] PhotoStress® Coatings. Micro-Measurements. [online]. [cit. 2018-09-23]. Available on the Internet: <http://www.vishaypg.com/docs/11222/pscoat.pdf>

[3] PÁSTOR, M. - TREBUŇA, F.: Application of transmission photoelasticimetry for stress concentration analyses in construction supporting parts. In: Applied Mechanics and Materials. Vol. 611 (2014), p. 443-449. - ISSN 1660-9336

[4] TREBUŇA, F. - ŠIMČÁK, F.: Príručka experimentálnej mechaniky. TypoPress, Košice, 2007.

[5] TREBUŇA, F. - ŠIMČÁK, F. - BOCKO, J. - TREBUŇA, P.: Failure analysis of mechanical elements in steelworks equipment by methods of experimental mechanics. Engineering Failure Analysis 17 (2010), 787-801.

[6] TREBUŇA, P. - TREBUŇA, F. - ŠIMČÁK, F.: Analysis of possible causes of cracks initiation on barking drum. Engineering Failure Analysis 45 (2014), 106-117.

[7] Vishay Micro-Measurements. PhotoStress® Instruments. [online]. [cit. 2018-10-28]. Available on the Internet: http://www.vishaypg.com/docs/11212/11212_tn.pdf

[8] ASTM E 837-13a: Standard Test Method for Determining Residual Stresses by the Hole-Drilling Strain-Gage Method. [online]. [cit. 2019-02-05]. Available on the Internet: https://www.astm.org/Standards/E837.htm

[9] ERRAPART, A. – ANTON, J.: Photoelastic residual stress measurement in nonaxisymmetric glass containers. In: EPJ Web of Conferences 6, 32008 (2010).

[10] SCHAJER, G. S.: Practical residual stress measurement methods. John Wiley & Sons Ltd, The Atrium, Southern Gate, Chichester, West Sussex, PO19 8SQ, United Kingdom, 2013, ISBN 978-1-118-34237-4.

[11] NELSON, D.V.: Residual Stress Determination by Hole Drilling Combined with Optical Methods. Experimental Mechanics (2010) 50: 145–158; DOI 10.1007/s11340-009-9329-3.

[12] HARRINGTON, J.S. - SCHAJER, G.S.: Measurement of Structural Stresses by Hole-Drilling and DIC. Experimental Mechanics. (April 2017), Volume 57, Issue 4, pp 559–567.

[13] TREBUŇA, F. - PÁSTOR, M. - HUŇADY, R. - FRANKOVSKÝ, P. - HAGARA, M.: Optické metódy v mechanike. Košice : TU - 2017. - 550 s.. - ISBN 978-80-553-3168-3.

[14] VEQTER. [online]. [cit. 2019-3-12]. Available on the Internet: http://www.veqter.co.uk/residual-stress-measurement/centre-hole-drilling

[15] KOSTKA, J. - PÁSTOR, M. - ČARÁK, P. - FRANKOVSKÝ, P. - KULA, T.: Návrh metodiky zisťovania zvyškových a prevádzkových napätí odvrtávaním drážky pomocou dutej frézy. Novus Scientia 2017. - Košice: TU, 2017 S. 88-92. - ISBN 978-80-553-3080-8.

[16] KOSTKA, J.: Využitie optických metód experimentálnej mechaniky pri analýze príčin porúch nosných prvkov konštrukcií. Dissertation thesis. Košice, 2018

EAN 2019
57th conference on experimental stress analysis

The role of experiment in the proposal of new approach for the prediction of stiffness reduction of composite structures under fatigue loading

KRÁL Michal[1,a], CABRNOCH Bohuslav[1,b] and HOLÝ Stanislav[2,c]

[1]Czech Aerospace Research Centre, Department of Composite Technologies, Beranových 130,199 05 Praha - Letňany, Czech Republic

[2]Czech Technical University in Prague, Faculty of Mechanical Engineering, Department of Mechanics, Biomechanics and Mechatronics, Technická 4, 160 00 Praha 6, Czech Republic

[a]kral@vzlu.cz, [b]cabrnoch@vzlu.cz, [c]Stanislav.Holy@fs.cvut.cz

Keywords: Composite, Fatigue, Stiffness Reduction, Finite Element Method, Cyclic Loading

Abstract. The main goal of this paper is to present the results of the development of methodology for the prediction of stiffness reduction of composite structures under fatigue loading. The main idea of the proposed methodology is to estimate the stiffness reduction of composite structure with complex shape and stress field on the basis of data obtained during unidirectional tensile fatigue tests. To reach this idea it was necessary to find the solution of many problems such as methodology for the measurement of stiffness reduction during unidirectional tests, appropriate stiffness reduction model and the setup for the verification of proposed methodology.

Introduction

Despite a long-term research, the development of methods for the prediction of damage and degradation of mechanical properties of composite structures is still actual and rapidly developing topic. The motivation is based on the requirements of many industrial sectors including aerospace, automotive and others. During last decades many approaches and particular degradation models were developed to predict lifetime and residual mechanical properties of composite structures. Among existing models, it is rather complicated to find model, which can describe damage of structure with general shape and stress state. Almost all proposed models were developed on the basis of tensile experiments with prismatic test specimens. The application of these models in general engineering tasks is questionable.

Basic review of current state of the art

The research in this field lasts for more than 50 years. Developed degradation models are usually divided into three basic categories. The first category of models is usually called phenomenologically based models [1]. These models do not take into account a real damage mechanisms and describe damage on the basis of its outer manifestation, like reduction of stiffness and strength. These models can directly predict residual mechanical properties. A typical phenomenological model was proposed by Paepegem and Degrieck [2]. The second category of models is usually called fatigue life models. These models are based on generally known S-N curves and can be used for the prediction of fatigue life till the final fracture. A typical example is described in [3]. The classification of all existing approaches is beyond the scope of this review. More information can be found in [1].

All of the mentioned approaches are closely linked to experimental testing. Experimental measurement is necessary to obtain data for models identification and further for experimental verification of proposed methodologies. In terms of associated experimental testing, the literature is quite poor and in presented papers a significant improvisation is obvious. The reason of this fact stems from the high requirements to the size of statistical sample. During fatigue testing of composites, higher values of dispersion are reached because of inhomogeneous character of fibre composites and more complicated manufacturing process. Fatigue testing of polymer composites is usually performed with lower frequencies – generally up to 10 Hz. This is necessary because of the lower thermal conductivity and higher inner damping of polymer matrices. For more information see [4].

Description of proposed approach

The motivation of presented work was to propose methodology which can be used for general structures with complex stress state. This proposal consists of several key points including the proposal of methodology for experimental measurement of fatigue characteristics, the proposal of general degradation model, implementation of proposed model to FE code and verification of methodology using appropriately chosen experiments.

Stiffness of each lamina is defined by stiffness matrix C, see Eq. (1).

$$\sigma = C\,\varepsilon \tag{1}$$

Individual components of stiffness matrix C are expressed by five material characteristics including Young's moduli E_1, E_2 and shear modulus G_{12}. There are more material characteristics but only these are key for the description of lamina stiffness. It is expected that the reduction of composite stiffness can be described by the stiffness reduction of individual layers which can be described by the reduction of mentioned moduli E_1, E_2 and G_{12}. The reduction of these moduli can be measured during 1D tensile fatigue tests.

Experimental measurement of moduli E_1 and E_2 reduction

The methodology for measurement of the reduction of Young's moduli E_1 and E_2 is based on standard ASTM D3039/3039M-17 [5] for determination of Young's modulus during tensile static test. Proposed procedure is as follows. The cyclic loading of test specimen is stopped at defined numbers of cycles and static test according to above mentioned standard is performed. Fatigue characteristics are very often affected by the great dispersion. To minimize other inaccuracies which can arise during static tests, this static test is performed three times. Each fatigue test is repeated with four test specimens. Using this procedure, twelve values of residual Young's modulus are obtained for each number of cycles. In Table 1, a part of data for the stress level of 169 MPa is presented. In another step, the normality of data sets for each number of cycles was checked using Shapiro-Wilk's test of normality. The result of this test is statistics W, see Table 1. The normality assumption was met at the significance level of 0.05 in almost all cases. In some cases the normality assumption was not met because of outliers caused by measurement inaccuracy.

During testing of proposed methodology for the measurement of the decrease of Young's modulus, three stress levels were tested. To verify the significance obtained data another statistical testing was performed. In Fig. 1, the results of all three stress levels are presented as the dependence of mean values on the number of cycles.

Table 1: Example of measured residual Young's modulus data presented in normalized form

	n		0	200	10 000	50 000	100 000	200 000	300 000
	σ_{max} [MPa]	R [-]	E_I^n/E_I^0 [-]	E_I^n/E_I^0 [-]	E_I^n/E_I^0 [-]	E_I^n/E_I^0 [-]	E_I^n/E_I^0 [-]	E_I^n/E_I^0 [-]	E_I^n/E_I^0 [-]
Spec. 1	169	0.1	1.000	0.921	0.905	0.894	0.888	0.887	0.889
			1.000	0.915	0.902	0.886	0.885	0.885	0.889
			1.000	0.924	0.906	0.884	0.890	0.885	0.892
Spec. 2	169	0.1	1.000	0.917	0.911	0.899	0.897	0.888	0.876
			1.000	0.932	0.903	0.907	0.901	0.894	0.880
			1.000	0.918	0.894	0.882	0.889	0.882	0.883
Spec. 3	169	0.1	1.000	0.924	0.900	0.887	0.868	0.890	0.877
			1.000	0.932	0.908	0.894	0.886	0.896	0.889
			1.000	0.944	0.909	0.898	0.896	0.902	0.895
Spec. 4	169	0.1	1.000	0.938	0.909	0.896	0.910	0.906	0.902
			1.000	0.932	0.898	0.890	0.896	0.897	0.882
			1.000	0.938	0.911	0.901	0.901	0.908	0.895
Mean [-]			1.000	0.929	0.905	0.893	0.892	0.893	0.887
St. Deviation [-]			-	0.010	0.005	0.008	0.011	0.009	0.008
Var. Coefficient [%]			-	1.035	0.597	0.846	1.186	0.966	0.898
W statistics [-]			-	0.9404	0.9381	0.9711	0.9431	0.9340	0.9583

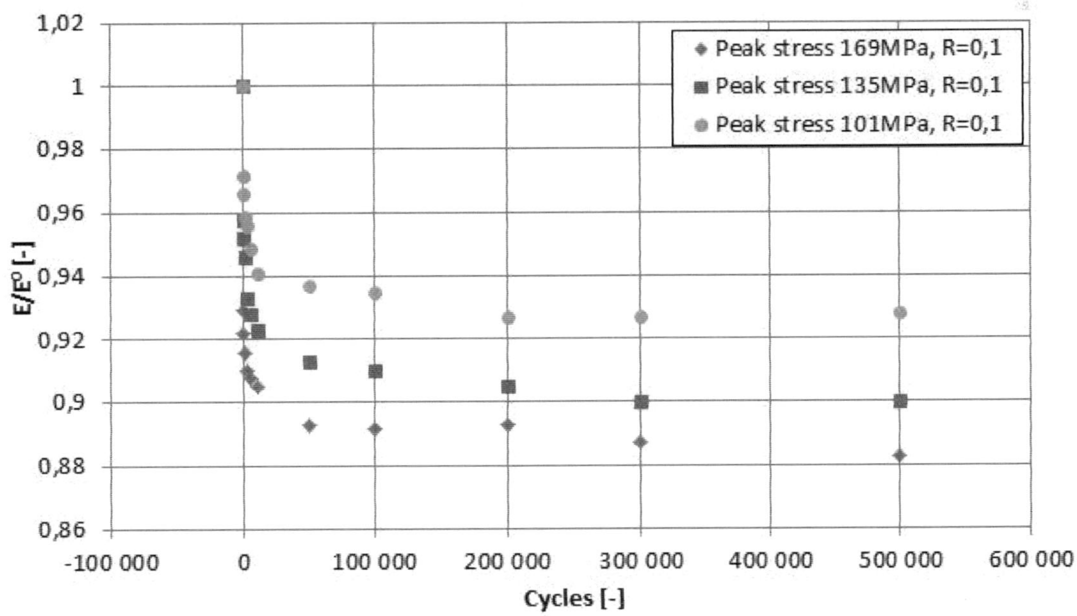

Fig. 1: Residual modulus E_I on three stress levels with peak stress equal to 101 MPa, 135 MPa and 169 MPa and cycle ratio coefficient equal to 0.1

To compare if there can be find a significant difference between these mean values, t-test was performed. On the significance level of 0.05 the assumption of mean values conformity was not met. Further, the hypothesis of the dispersion conformity was tested using F-test. On the significance level equal to 0.05 the assumption of the dispersion conformity was met in all cases.

Experimental measurement of the shear modulus G_{12} reduction

The methodology for the measurement of the reduction of shear modulus G_{12} is based on standard ASTM D3518/3518M-13 [6] for the determination of shear modulus during static tensile test, see Fig. 2. Methodology was modified and biaxial extensometer was used instead of strain gauges because of cyclic strain.

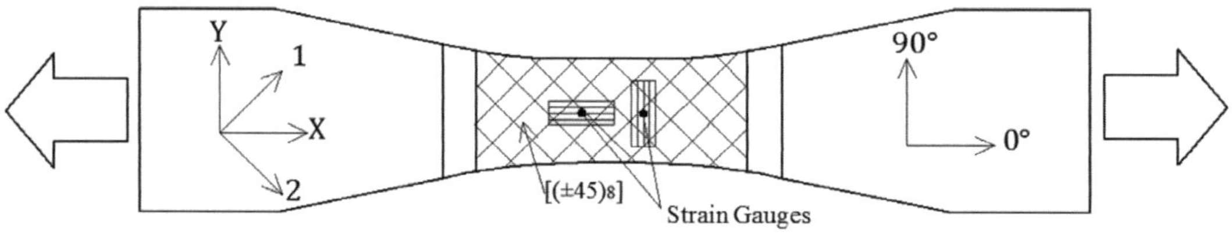

Fig. 2: Modified test specimen according to ASTM D3518 used for the determination of shear modulus in the plane of lamina

Shear modulus can be calculated using Eq. 2. Used coordinates systems are defined in Fig. 2.

$$G_{12} = \frac{\tau_{12}}{\gamma_{12}} = \frac{\sigma_x}{2(\varepsilon_x - \varepsilon_y)} \tag{2}$$

The methodology for the determination of reduction of shear modulus is similar to methodology used for Young's modulus. During research, it was shown, that modelling of the shear modulus G_{12} is not necessary on achievable load levels. The increase of shear modulus occurred instead of decrease, see Fig. 3. This is caused by the significant increase of matrix stiffness. More detailed study of this phenomenon is presented in [7].

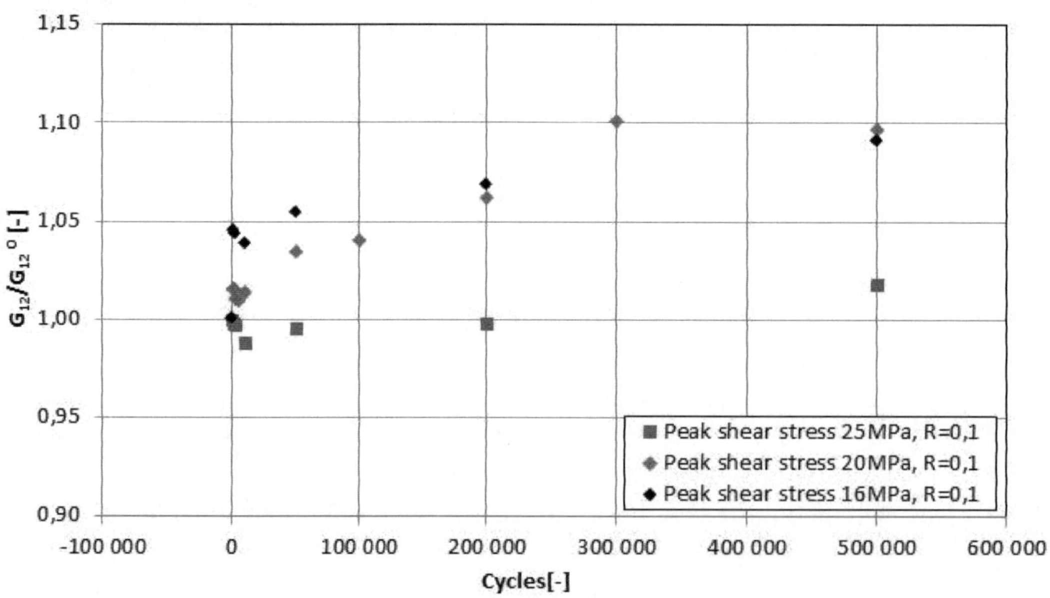

Fig. 3: Residual modulus G_{12} on three load levels with peak shear stress equal to 16 MPa, 20 MPa and 25 MPa

Degradation model

The main contribution of proposed model VZLU FDC Model stems from the possibility of damage prediction on arbitrary stress levels. Models proposed in the past did not have predictive capability and worked well only on stress levels closed to level used for the identification of the model. VZLU FDC Model can be expressed by the Eq. (3)

$$\frac{dD_i}{dn} = \frac{(\gamma_i \cdot \sigma_{imax} + \kappa_i)^{B_i} \cdot A_i \cdot (\sigma_{imax})^{C_i}}{B_i \cdot (D_i)^{B_i - 1}}, \quad D_i = 1 - \frac{E_i^n}{E_i^0}, \quad i = 1,2, \tag{3}$$

where D is damage parameter, σ_{max} is peak stress of the cycle, n is the number of cycles and the other variables are model coefficients. Using this model it is possible to predict stiffness reduction of elasticity moduli E_1 and E_2 of orthotropic layer. To identify model coefficients, it is necessary to measure the dependence of residual modulus on the number of cycles on minimally two stress levels.

Regression of experimental data using proposed model is shown in Fig. 4. Data on stress level 101 MPa and 169 MPa were used for the identification of model coefficients and stress level 135 MPa was used for the verification of predictive capability of the model. Predictive capability of the model can be rated as very good.

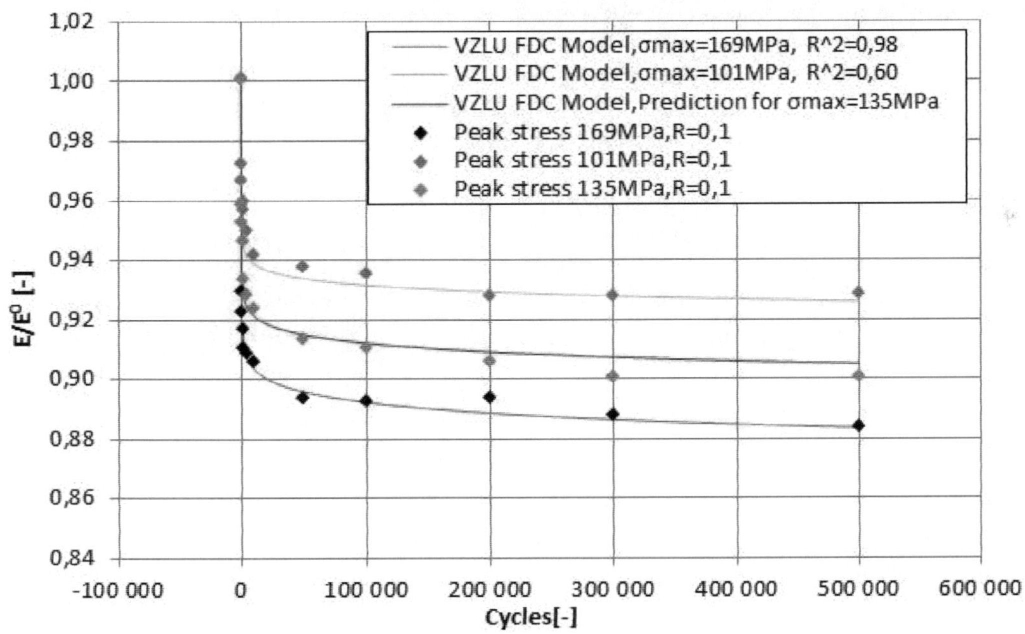

Fig. 4: Regression of experimental data using VZLU FDC Model on stress levels with peak stress equal to 101MPa and 169 MPa. Verification of predictive capability on stress level with peak stress equal to 135 MPa

Verification examples

Verification of the prediction accuracy was performed by the comparison of predicted and measured stiffness reduction. Proposed methodology allows the prediction of distribution of residual moduli in each layer. But it is impossible to verify these results because it is impossible to measure stiffness of individual layers of composite.

Verification was performed using model examples with increasing complexity of the test specimen shape and stress field. The first task was the tension of specimen with more complex stacking sequence [(0/90/ ±45/ 0/90/ ±45)s]. Testing was performed on two stress levels with peak stress of the cycle equal to 102 MPa and 127 MPa. Accuracy was determined as the fraction of predicted and experimental value. Obtained values of accuracy after 200 000 cycles are 0.95 for stress level 102 MPa and 0.82 for stress level 127 MPa. In Fig. 5 there are shown test specimen and the example of predicted residual modulus E_1 in layer 1 after 300 000 cycles.

In another step, fatigue tests in bending were performed. Experimental results were very affected by changes in stiffness of matrix and the comparison of experiment and prediction was impossible. More detailed discussion of this issue is performed in [8]. In Fig. 6 there is shown the setup of experimental measurement and numerical simulation.

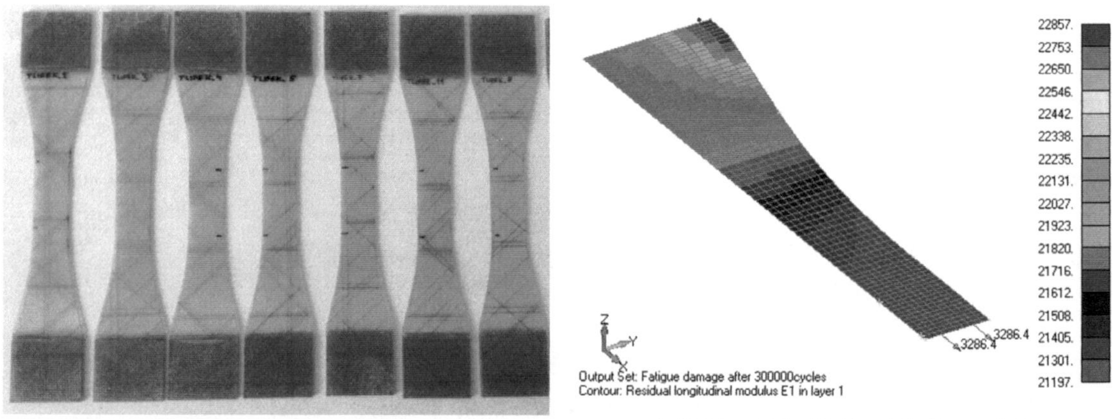

Fig. 5: Test specimens with stacking sequence [(0/90/ ±45/ 0/90/ ±45)s] used for experimental verification (left) and the distribution of residual modulus obtained by numerical simulation (right)

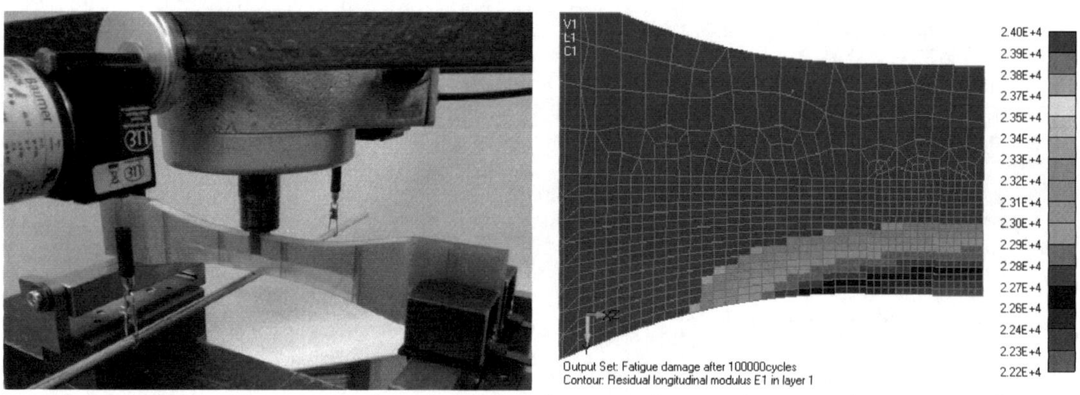

Fig. 6: Cyclic three - point bending test of composite specimen and corresponding numerical simulation using VZLU FDC Model

Conclusions

Proposed methodology shown to be an effective tool for the sizing of structures made of composite materials based on continuous glass fibres embedded in epoxy resin. During experimental measurement it was determined that cyclic loading can cause significant changes of mechanical properties of epoxy matrix affecting global response of composite structures. This newly discovered phenomenon will be the aim of following research. Obtained results show a great complexity of damage in composites modelling. This complexity is probably caused by a heterogeneous character of fiber composites and also by a great dispersion in quality and repeatability of manufacturing process.

Acknowledgements

This research was performed with financial institutional support from government budget through the Ministry of Industrial and Trade of the Czech Republic.

References

[1] J. Degrieck, W. van Paepegem, Fatigue damage modelling of fibre-reinforced composite materials: Review, Applied Mechanics Reviews. 2001, 54 (4) 279-300.

[2] W. van Paepegem, A new coupled approach of residual stiffness and strength for fatigue of fibre-reinforced composites, International Journal of Fatigue. 2002, 24 (7) 747-76.

[3] Z. Hashin, A. Rotem, A Fatigue Failure Criterion for Fiber Reinforced Materials, Journal of Composite Materials. 1973, 7 (4) 448-464.

[4] B. Harris, Fatigue in composites: science and technology of the fatigue response of fibre-reinforced plastics, Boca Raton: CRC Press, 2003. ISBN 18-557-3608-x.

[5] ASTM D3039 / D3039M-17, Standard Test Method for Tensile Properties of Polymer Matrix Composite Materials, ASTM International, West Conshohocken, PA, 2017.

[6] ASTM D3518 / D3518M-13, Standard Test Method for In-Plane Shear Response of Polymer Matrix Composite Materials by Tensile Test of a ±45° Laminate, ASTM International, West Conshohocken, PA, 2013.

[7] M. Král , B. Cabrnoch, S. Holý, Methodology for prediction of stiffness reduction of composite structures under fatigue loading, Strojírenská technologie, 2018, XXIII (2), ISSN 1211 – 4162.

[8] M. Král, The proposal of methodology for prediction of stiffness reduction of composite structures under fatigue loading, Dissertation, Faculty of Mechanical Engineering, CTU in Prague, 2018.

EAN 2019
57th conference on experimental stress analysis

Motion tracking in mixed reality

KRCMARIK David[1,a], CYRUS Jindrich[1,b], PETRU Michal[1,c] and KOCI Jan[1,d]

[1]Technical University of Liberec, Studentská 1402/2, 461 17, Czech Republic

[a]david.krcmarik@tul.cz, [b]jindrich.cyrus@tul.cz, [c]michal.petru@tul.cz, [d]jan.koci@tul.cz

Keywords: Tracking, Microsoft HoloLens, Communication unit, Industrial robots.

Abstract. We propose a viable approach for motion tracking of real objects within a holographic scene. Mixed reality applications are aspired to close cooperation of real and virtual worlds. For cooperation and interaction, we mainly need to know the precise position of real and virtual objects. These positions are likely to vary during the time. The movement can be tracked using traditional image recognition. However, this approach needs persistent visual contact with observed tracking image or a 3D object. Many mixed reality scenarios aim at industrial premises where tracked objects are equipped with a communication unit that can provide relevant information about movement and actual position. We propose a duplex communication interface based on TCP or UDP protocols that essentially helps with motion tracking of objects in the scene. The whole system can be refined by means of motion estimation based on the path already made and repetitive movements.

Introduction

The field of virtual and mixed reality devices is in recent years considerable widening. Most of the devices (HTC Vive, Oculus Rift) support virtual reality which does not require such a massive real-time image and sensor data processing as in the mixed reality. With the advent of Microsoft HoloLens [1] in 2016 new dynamically evolving environment started to gain attention. There are also other competitive platforms in the field of mixed reality (e.g. Magic Leap, Meta 2) which however does not provide such a rich experience as HoloLens. Most of the things with which human interacts are 3D objects. Commonly used 2D display screens can provide only a limited experience with such objects. Mixed reality provides true 3D visualization within real world hence deepens the perception of 3D objects and spatial interaction among them. Augmented reality by contrast is kind of mixed reality without interaction of virtual and real objects (e.g. Pokémon GO where real world is enhanced with new layer of perception).

Mixed reality combines real world objects with virtual objects into one scene. Head-mounted devices (HMD) are most commonly used since it is feasible in terms of technology and intuitive for users. The principle of combining real and virtual objects can be based either on video see-through (VST) [2] or optical see-through technology (OST). In video see-through approach the enviroment which is scanned with a RGBD (red-green-blue-depth) camera is enriched with virtual objects. The superimposed scene is then projected as in the case of virtual reality. OST devices are in our opinion more matured and natural with the advent of Microsoft HoloLens. User is able to see the surrounding environment immediatelly after placing the device on the head and the there is nearly no distortion of real world environment. The pros and cons of both concepts are described in many papers especially aimed to medical discipline [3].

When comparing mixed reality to virtual reality we think the future lies in mixed reality. The user is in direct contact with environment, there are significantly less reported sickness cases (characterized by eye strain, disorientation or even nausea) due to use of the device. OST-MDs are often self-contained and untethered, which means they does not need any additional computers, sensors or infrastructure.

Current challenges

In this paper we focus on OST-HMD Microsoft HoloLens because it is currently the most advanced device for mixed reality – Fig. 1. The device is used in many areas. Most of the applications are aimed to medicine and education.

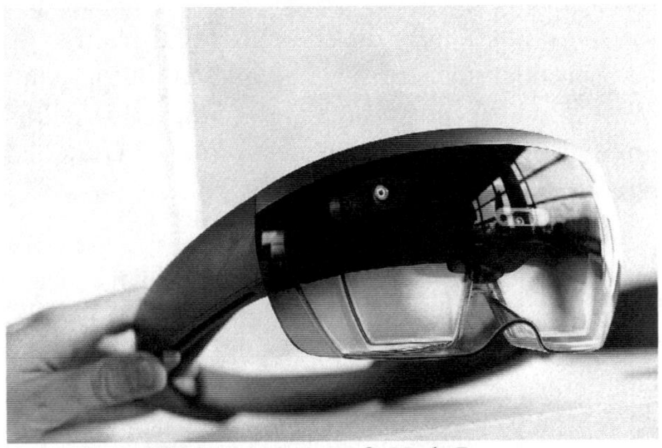

Fig. 1: Microsoft HoloLens

In [4] an autopsy was performed with the help of HoloLens. As reported useful information was shown during the gross examination – radiographs, annotations, 3D scanned pathology specimens, whole slide image viewer, laboratory information system, etc. During the anatomic pathology various ways of communication with device were used (gestures, voice commands). Tele-collaboration (telepathology) allowing remotely access a senior pathologist for instructions was found very beneficial and possibly usable for other scenarios. The usage of mixed reality seems to have the biggest impact on the medicine – cases for neurosurgery [5] or cardiology [6] to name few are described. As already mentioned HoloLens proves to be especially rewarding in education. Many concepts in e.g. physics are hardly described since they are highly abstract – thermal conductivity, entropy. A control group of 59 students was used to test how the visualization of abstract physical quantities helps with cognitive understanding [7]. Results show small positive effect and authors pointed out that particular augmentation could significantly help for insight.

Particularly huge attention is made to other so far unusual fields of mixed reality usage. The term Industry 4.0 and Internet of things are often coined as future directions to which we need readjust our thinking in order to be competitive in 21 century. The number of devices connected to the Internet rapidly increases. Many companies incorporate sensors within their products sending information to servers in order to monitor the state of the device and possible predictive maintenance using so called big data [8]. This can save significant amount of resources and time. At the same time devices with bidirectional communication units can not only provide information but they can be controlled remotely by the end user or configured and adjusted by the provider for better performance. Since the amount of the devices increases mixed reality can provide a suitable way for their easy-to-use configuration even by untrained personnel [9]. HoloLens can help with discovering and selecting desired devices. The personnel has free hands to manipulate and adjust some features and at the same time needed manuals or teleassistance is available. Mixed reality with its ability of interconnecting real and virtual objects, connection to the Internet and large databases with information collected by Internet of Things (IoT) and availibility of remote technical assistance is a key component in developing Industry 4.0. An interesting project called Shipyard 4.0 was deveeped which helps the engineers with interaction with product datasheets, maintenence procedures or quality control forms in real-world scenarios [10]. Both fog computing system and cloudlet approach were tested resulting in the cloudlet solution having better responsibility under high loads.

Important key to follow when designing immersive application for mixed reality are tele-immersion concepts – spatial immersion and presence [11]. In the paper we partly build on Minsky's tele-operated robot [12]. Mixed reality HoloLens provides two types of feedback to the user – visual and acoustical. We found that the most important aspect for visual feedback is location precision. It means that very disruptive for user is when the location of virtual object is not stable (in the terms of expected movement or stillness). On the other hand it can be beneficial when the user can immediately distinguish a real from a virtual via some cartoonish appearance or not perfectly rendered colors. 3D acoustical output significantly helps user with localization of virtual objects in the scene which are not directly visible and contribute to overall immersive feeling. Although there are motion controllers [13] available – Fig. 2, they do not provide haptic feedback to the user.

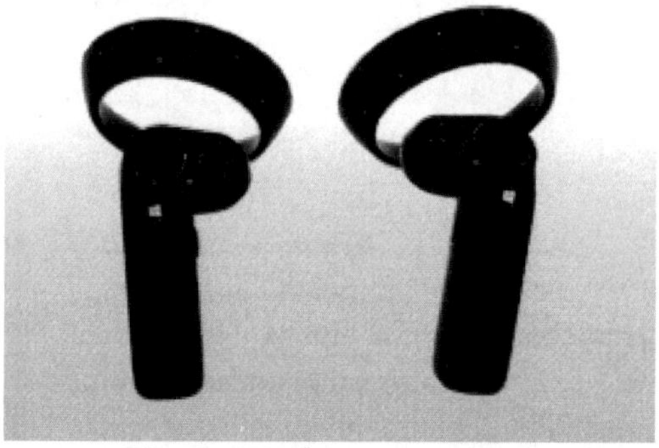

Fig. 2: Motion controllers

Haptic feedback [14] could be important for its intuitiveness and comfort. It allows the user to have also better performance when interaction with virtual content. In [15] and interesting concept enabling feeling of the material from which was made virtual object enabled the user to feel the weight. They have developed hRing which was put on the finger and they tracked the fingertips with the use of ARToolKit. A current state of the art in the field of HoloLens with haptic feedback is revised in [16]. It could be used for manipulation of virtual interfaces appended to real surfaces in the environment [17]. As already stated HoloLens are capable of recognition several gestures (build in gestures are 'tap', 'double tap' and 'drag and drop', system gesture is 'bloom') made by the user or even by other people in the scene. Since this feature is a software based more gestures could be possible added. In human-human interaction facial emotions are very important. Very interesting concept of detection user defined gestures is based on WiFi scanning There are even efforts [19] for gesture recognition of emotions based on databank of known correspondences of human feeling and their facial expression based on deep learning.

Security of usage HoloLens is often neglected. There is a common understanding that the data from device sensors have to be secured in order to avoid misuse from different applications possibly installed in the device or even from third subjects connected via WiFi or Bluetooth to HoloLens. It is important to realize that the users field of view can be also filled with much of information and objects coming from HoloLens and can possible be obtrusive or even dangerous since other information could be hidden (from real world or from other applications in HoloLens). This is called an virtual output security [20].

It was already mentioned that partly exceptional are HoloLens due to its ability precise itself localization and tracking of other real object in the scene. The fundamental concept is described in [21]. Unfortunately the whole thing is more complicated for OST devices [22] since one have to take into account the movement of eye. AR device have to be capable of 6-degree-of-freedom (6 DoF) movement tracking (x, y, z, roll, pith, yaw). Every pixel in the virtual scene (with index B) has to be transformed using position and rotation into a real scene (with index A) according to an equation

(1)

where R stand for rotation and t for translation. In the case of moving eye the coefficients for transformation (1) have to be heavily and periodically computed and updated. For that reason the field of view in HoloLens is limited in order to minimize eye movement – Fig. 3.

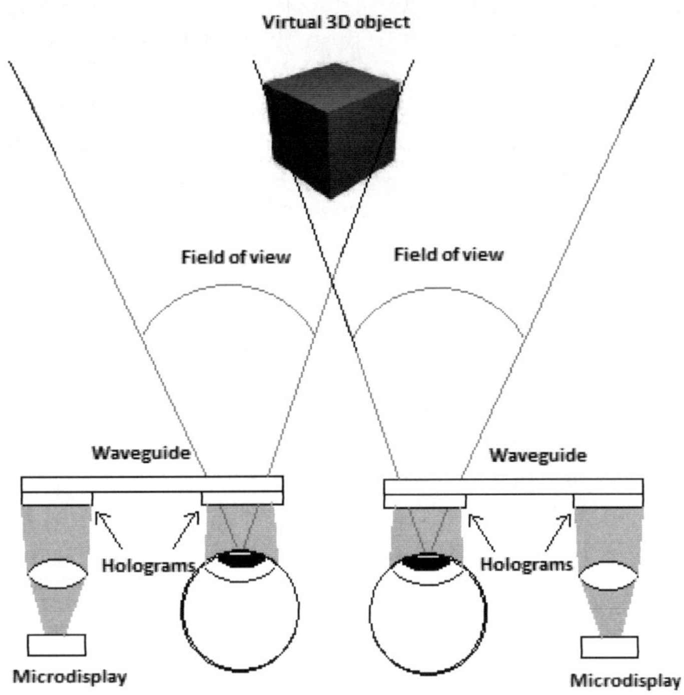

Fig. 3: Schematic view of moving eye problem for OST

If we want to used mixed reality platform even for applications where extremely accurate positioning and stability is demanded [23] we have to improve overall location error – for example using calibration [24]. Online temporal tracking [25] or well-know Kalman filter based solutions [26] seem to give promising results. Descriptor utilization is a concept where objects are compared from different views, classified and the system is in this way trained in such a way that only 16-dimensional descriptors can help with object recognition and their 3D pose estimation [27].

It is necessary to mention that there are also attempts to use low-cost RFID based devices which can to a certain extent track 3D objects in space [28]. A signal strength indicator (RSSI) was measured using commercial off-the-shelf device called RFTrack which uses discrete wavelet transform for detection target motion. Unfortunatelly precision of only 32 cm was reported. Tele-rehabilitation was tested using Microsoft Kinect (predecessor of HoloLens) based on optical system with the help of calibration [29]. The calibration is provided as complementary information concerning bone lengths, position of elbows etc. A Google Tango based project with additional visual markers [30] is another way of precise localization of moving devices in the scene which does not require any additional infrastructure.

As shown above there are many challenges in newly appearing platform of mixed reality. The thing of paramount importance remains precise localization which is not processor demanding and without needed ambient infrastructure.

Precise position and motion tracking

It is important to realize that in many situations where HoloLens are used there are usually a few devices, which can communicate with the rest of world. They can be intelligent devices robot-type or vehicle-type. HoloLens is a vital platform for simulation interaction between these devices and virtual world saving a lot of resources and time.

We can exploit such natural functionality to build a one-to-many bidirectional communication link possessing needed information about the scene. The whole setup assumes some kind of network for wireless connection – either WiFi or Bluetooth. Other communication standards can be easily implemented using separate converter like black-boxes. It is also necessary to count with the scenario that communication is temporally unavailable. In this case we use standard tracking camera for defined feature detection with the Vuforia backend for processing images. Previously the depth information was reported to be acquired using Kinect Fusion [31, 32]. In our setup we used only an industrial camera.

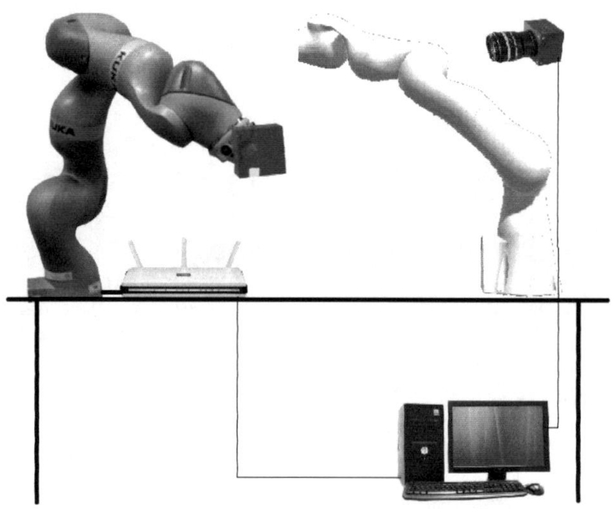

Fig. 4: Schematic view of setup

There was used high-throughput UDP communication in the HoloLens application to instantly report position of the end point of moving object. The goal of the whole project was to instruct the virtual robot using a gaze to pick particular cube (red or blue), handle the cube from a bowl to the real robot and the real robot took the virtual cube and placed it on belt. During the experiment, we have simulated network instability.

We have compared TCP and UDP base communication. Average time delay exhibited using TCP protocol was about 24 ms while for UDP communication it was only 7 ms.

Here is the example of the C# code used to establish the UDP communication:

```
private async void StartClient(bool sendWelcomeMessage = false) {
        try {
                //Create the DatagramSocket and establish a connection.
                clientDatagramSocket
                        = new Windows.Networking.Sockets.DatagramSocket();

                clientDatagramSocket.MessageReceived
                        += ClientDatagramSocket_MessageReceived;

                AddLog("client is about to bind...");
                await clientDatagramSocket
                                .BindServiceNameAsync(Cfg.UdpPort.ToString());
                AddLog(string.Format(
                                "client is bound to port number {0}", Cfg.UdpPort));

                if (sendWelcomeMessage) {
                        // Send a request to the echo server.
                        string request = "HoloLens are connected to you!";
                        // The server hostname that we connect to.
                        // only for confirmation message
                        var hostName = new Windows.Networking.HostName("SERVER");
                        using (var serverDatagramSocket
                                = new Windows.Networking.Sockets.DatagramSocket()) {
                                using (Stream outputStream
                                        = (await serverDatagramSocket.GetOutputStreamAsync
                                                (hostName,Cfg.UdpPort.ToString()))
                                                .AsStreamForWrite()) {
                                                using (var streamWriter
                                                        = new StreamWriter(outputStream)) {
                                                await     streamWriter.WriteLineAsync(request);
                                                                await streamWriter.FlushAsync();
                                                }
                                }
                        }
                        AddLog(string.Format(
                                "client sent the request: \"{0}\"", request));
                }
        }
}
```

The camera was an RGB Basler acA1600-20gc with Ethernet interface connected to a standard PC. We used Vuforia object tracking and this information helped during a loss of the communication between robot and HoloLens.

We have simulated also the loss of the communication on both lines – from camera and from robot to HoloLens. HoloLens immediately recognized this outage and switched to the previously learned trajectory. We assume that this scenario can apply to many cases with industrial robots, because the moving objects carry out periodical movements in the scene very often. The device is tracking the objects of interest in the scene and buffers the movements to the specific buffers (6 DoF described movement). If there is broken the communication line to HoloLens and the data about the actual robot position is missing, then the device tries to extrapolate the movement in order seamlessly provide visual content to the user.

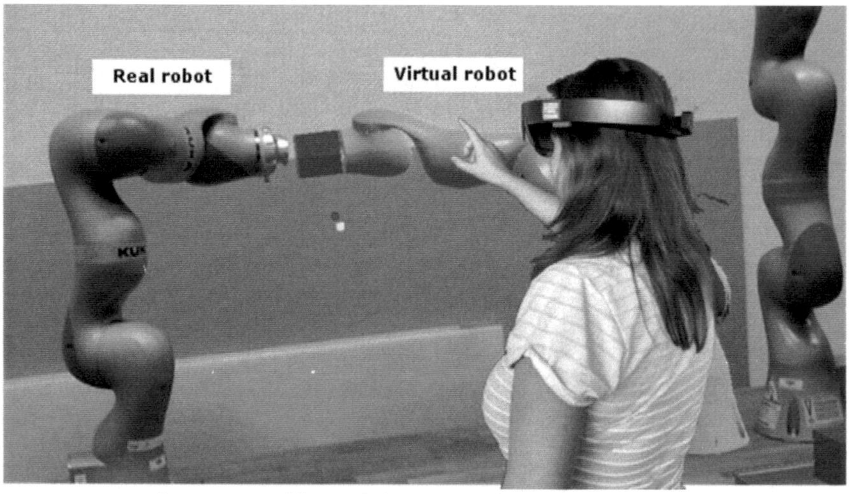

Fig. 5: Handling a virtual cube to the real robot

Summary

We presented a way of tracking moving object in the scene based on our already created infrastructure. Many applications with HoloLens are in environments, which are enriched with intelligent devices, or IoT enabled devices. Such environments can be readily exploited to provide necessary information of real object current location or movement directly to mixed reality devices. Hereby it is not necessary to perform demanding tracking using built-in sensors and process the data using processor-intensive algorithms directly within the mixed reality device. The strongest side of HoloLens is to visualize the virtual objects and to process spatial mapping in order to obtain the smooth and stable scene in the real time. Use of external precise data on mixed reality devices leads to the applications that consume fewer resources but are more efficient and produce breathtaking experience. Proposed method of very fast communication using UDP protocol can be a very handy tool, which is easily embedded and provides satisfying results. The time lag was as reported minimized so that user was unable to see any unnatural movements of virtual cube which was moved by the real robot.

Acknowledgements

The result was obtained through the financial support of the Ministry of Education, Youth and Sports of the Czech Republic and the European Union (European Structural and Investment Funds - Operational Programme Research, Development and Education) in the frames of the project "Modular platform for autonomous chassis of specialized electric vehicles for freight and equipment transportation", Reg. No. CZ.02.1.01/0.0/0.0/16_025/0007293.

References

[1] M.A. Hoffman, The future of three-dimensional thinking, Science 353 (2016) 876.

[2] J.C. Wang, et al., Video see-through augmented reality for oral and maxillofacial surgery, International Journal of Medical Robotics and Computer Assisted Surgery 13 (2017).

[3] J.M. Carmen, J. Calatrava, An augmented reality system for the treatment of phobia to small animals viewed via and optical see-through HMD: Comparison with a similar

system viewed via a video see-through HMD, International Journal of Human-Computer Interaction 27 (2011) 436-449.

[4] M.G. Hanna, et al., Augmented reality technology using Microsoft HoloLens in anatomic pathology, Archives of Pathology and Laboratory Medicine 142 (2018) 638-644.

[5] F. Incekara, et al., Clinical feasibility of a wearable mixed-reality device in neurosurgery, World Neurosurgery 118 (2018) E442-E427.

[6] V. Garcia-Vasquez, et al., Navigation and visualization with HoloLens in endovascular aortic repair, Innovative Surgical Sciences 3 (2018) 167.

[7] M.P. Strzys, et al., Physics holo.lab learning experience: Using smartglasses for augmented reality labwork to foster the concepts of the heat conduction, European Journal of Physics 39 (2018).

[8] J. Maktoubian, Proposing a streaming Big Data analytics (SBDA) platform for condition based maintenance (CBM) and monitoring transportation systems, EAI Endorsed Transactions on Scalable Information Systems 4 (2017).

[9] G. Alce, et al., AR as a user interface for the internet of things – comparing three interaction models, in: 16th IEEE International Symposium on Mixed and Augmented Reality (ISMAR), France, 2017.

[10] T.M. Fernandez-Carames, et al., A fog computing and cloudlet based augmented reality system for the industry 4.0 shipyard, Sensors 18 (2018).

[11] S. Ohl, Tele-immersion concepts, IEEE Transactions on Visualization and Computer Graphics 24 (2018) 2827-2842.

[12] M. Minsky, Telepresence, Omni 4 (1980) 45-51.

[13] K.R. Moser, J.E. Swan, Evaluation of user-centric optical see-through head-mounted display calibration using a leap motion controller, in: IEEE Symposium on 3D User Interfaces (3DUI), South Carolina, USA, 2016.

[14] C. Pacchierotti, et al., Wearable haptic systems for the fingertip and the hand: Taxonomy review and perspectives, IEEE Transactions on Haptics 10 (2017) 580-600.

[15] S. Scheggi, G. Salvietti, D. Prattichizzo, Shape and weight rendering for haptic augmented reality, in: 19th IEEE International Conference on Robot and Human Interactive Communication (RO-MAN), Italy, 2010.

[16] L. Meli, et al., Combining wearable finger haptics and augmented reality: User evaluation using and external camera and the Microsoft HoloLens, IEEE Robotics and Automation Letters 3 (2018) 4297-4304.

[17] R. Xiao, et al., MRTouch: Adding touch input to head-mounted mixed reality, in: 25th IEEE Conference on Virtual Reality and 3D User Interfaces (IEEE VR), Germany, 2018.

[18] H. Abdelnasser, M. Youssef, K.A. Harras, A ubiquitous WiFi-based gesture recognition system, in: 34th IEEE Conference on Computer Communications (INFOCOM), Hong Kong, China, 2015.

[19] D. Mehta, M.F.H. Siddiqui, A.Y. Javaid, Facial emotion recognition: A survey and real-world user experiences in mixed reality, Sensors 18 (2018).

[20] K. Lebeck, et al., Securing augmented reality output, in: 38th IEEE Symposium on Security and Privacy (SP), California, USA, 2017.

[21] W. Robinett, R. Holloway, The visual display transformation for virtual reality, Presence Teleoperators and Virtual Environments 4 (1995).

[22] J. Grubert, et al., A survey of calibration methods for optical see-through head-mounted displays, IEEE Transactions on Visualization and Computer Graphics 24 (2018) 2469-2662.

[23] R. Vassallo, et al., Hologram stability evaluation for Microsoft® Hololens™, in: Conference on Medical Imaging – Image Perception, Observer Performance, and Technology, Florida, USA, 2017.

[24] M. Garon, et al., Real-time high resolution 3D data on the HoloLens, in: 15th IEEE International Symposium on Mixed and Augmented Reality (ISMAR-Adjunct), Mexico, 2016.

[25] D.J. Tan, et al., A versatile learning-based 3D temporal tracker: Scalable, robust, online, in: IEEE International Conference on Computer Vision, Chile, 2015.

[26] Q. Yang, D.G. Taylor, G.D. Durgin, Kalman filter based localization and tracking estimation for HIMR RFID systems, in: IEEE International Conference on RFID, Florida, USA, 2018.

[27] P. Wohlard, V. Lepetit, Learning descriptors for object recognition and 3D pose estimation, in: IEEE Conference on Computer Vision and Pattern Recognition (CVPT), Massachusetts, USA, 2015.

[28] L. Liyao, et al., Accurate device-free tracking using inexpensive RFIDs, Sensors 18 (2018) 2816.

[29] K. Yu, et al., On the accuracy of low-cost motion capture systems for range of motion measurements, in: Conference on Medical Imaging – Imaging Informatics for Healthcare, Research and Applications, Texas, USA, 2018.

[30] B. Marques, et al., Evaluating and enhancing Google Tango localization in indoor environments using fiducial markers, in: 18th IEEE International Conference on Autonomous Robot Systems and Competitions (ICARSC), Portugal, 2018.

[31] R.A. Newcombe, et al., KinectFusion: Real-time dense surface mapping and tracking, in: 10th IEEE/ACM International Symposium on Mixed and Augmented Reality (ISMAR), Switzerland, 2011.

[32] S. Izadi, et al., KinectFusion: Real-time 3D reconstruction and interaction using a moving depth camera, in: ACM Symposium on User Interface Software an Technology (UIST), California, USA, 2011.

EAN 2019
57th conference on experimental stress analysis

Analysis of internal forces in chassis of bucket wheel excavator

KŘIČKA Jaroslav[1,a]

[1]Department of Designing and Machine Components; Faculty of Mechanical Engineering; Czech Technical University in Prague; Technická 4, 166 07 Praha, Czech Republic

[a]jaroslav.kricka@fs.cvut.cz

Keywords: Analysis, Chassis, Bucket Wheel Excavator

Abstract. Modern and high-power machines for surface mining of energy-producing raw materials require mechanisms of thoroughly perfect design and dimensioning. The theory of structure of mechanisms and methods of mechanics provide an indispensable theoretical foundation.

The presented research work deals with static analysis of bucket wheel excavator chassis. The knowledge of internal and external forces and their size is important for dimensioning of all parts of the chassis which have a direct impact on reliability and lifetime of the chassis.

Introduction

The crawler chassis are used in various types of machines, for example agricultural machinery, earth moving machines and machines for surface mining including large-sized excavators [1]. The chassis are manufactured in different variants of crawler design. Chassis serves several important functions:

- the transmission of weight and external forces to terrain,
- safeguarding the working stability while moving forward, backward and turning around,
- safeguarding that the allowed load of terrain in not exceeded,
- the transfer to a new locality of mining.

The connection of chassis to terrain is a complex issue. Thus, the research of chassis connection to its surroundings is planned for the next part of the project.

Headings. This type of chassis is of a new construction. According to the manner of excavator to its own chassis there are four-point supports. The movement of pair of crawlers is dependent by means of transverse beam. Each chain belt has its individual and reversible controled driving device which enables the machine to move in random bend or turn on the spot.

Fig. 1 shows very simplified 3D model of chassis we will analyse.

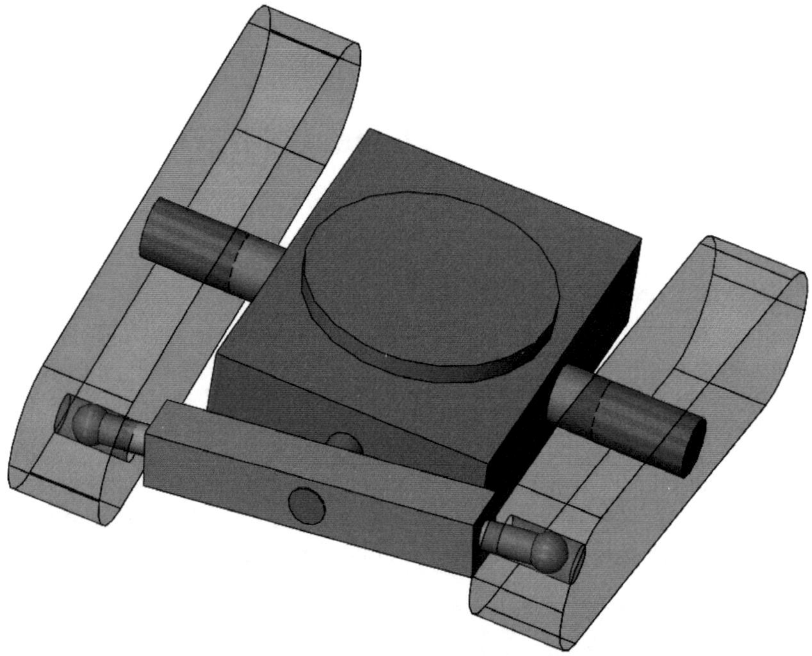

Fig. 1: Bucket wheel excavator chassis

Fig. 2 shows the kinematic diagram of simplified chassis and type of connections compiled in accordance with fig. 1. The illustrated spatial mechanism is assembled from five rigid bodies including frame. Each body has six degrees of freedom in space.

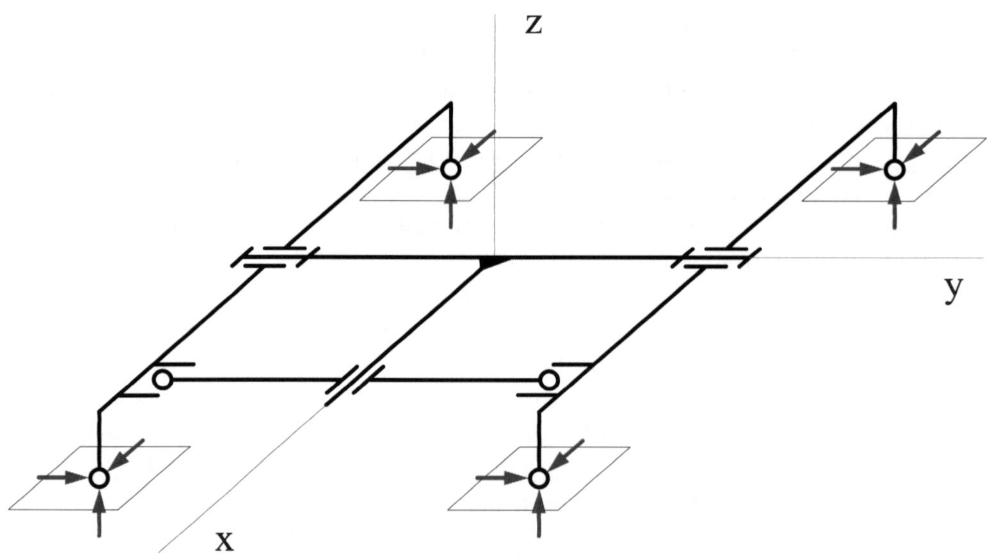

Fig. 2: Kinematic diagram of chassis

To find a solution it necessary to describe every body by the system of six linear algebraic equations [2]. The system of equations is driven by the extent of symplification of the real machine which might by performed on various levels [3].

The load of chassis is based on the assumption that the workload is situated in the level of ball trajectory by space system forces and moments, fig. 3.

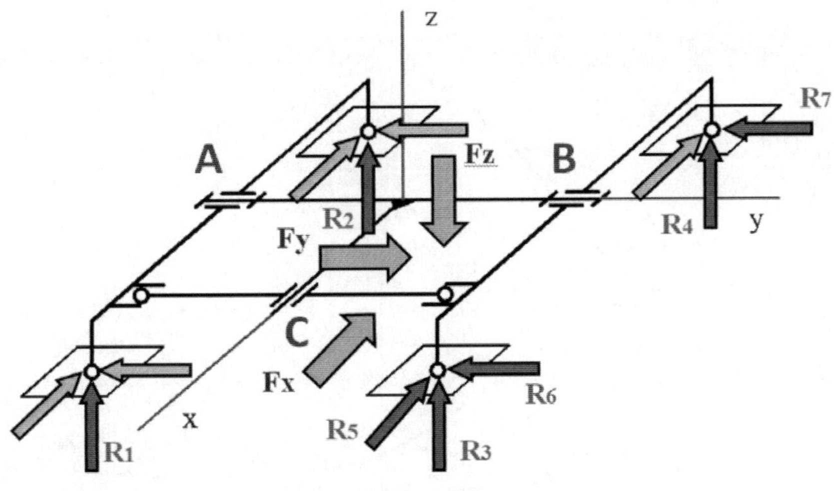

Fig. 3: Workload of chassis

The load of chassis is caused by its own weight, wind, breakout forces, forces of inertia and resistance to movement[1].

Numerical computation of unknown forces and moments has to executed for a great number of different operational situations. Acquired results will be used for designing and dimensioning of all chassis parts and for power determination of driving motors.

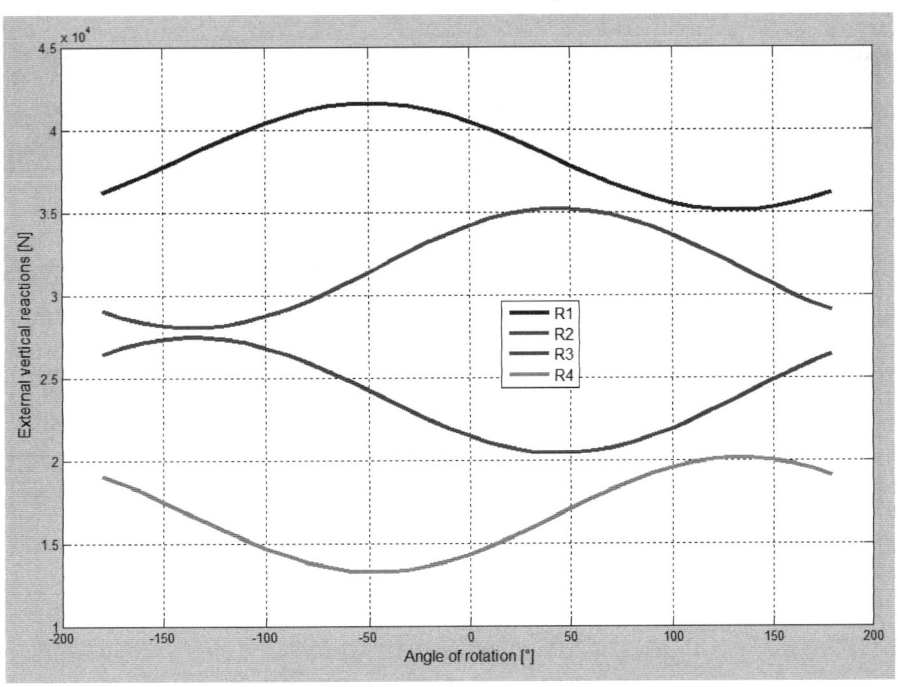

Fig. 4: Vertical forces in chassis contact points

Fig. 4 and fig. 5 show course of calculated force effects for one external load variation depending on bucket wheel position.

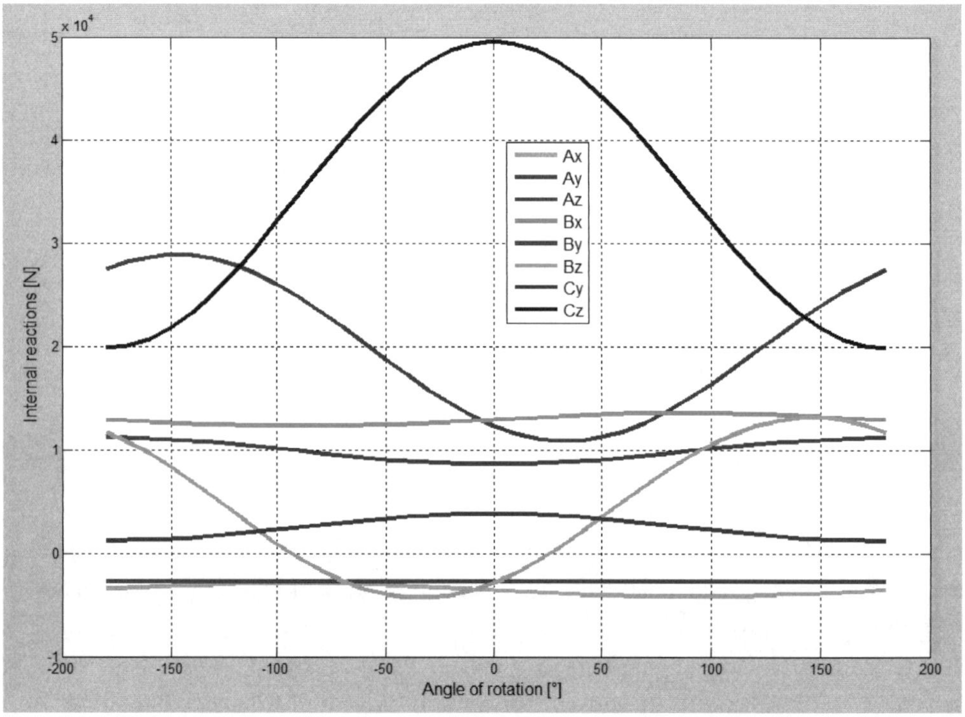

Fig. 5: Force effects in main journal of chassis

Conclusions

The contact of chassis with the surrounding muddy or rocky terrain presents an interesting issue. This will be the subject of further research.

References

[1] J. Hojdar, F. Helebrant, H. Gondek, Povrchové dobývací stroje, VŠB v Ostravě, 1991.

[2] V. Stejskal, J. Březina, J. Kněžů, Mechanika I, ČVUT v Praze, 1998.

[3] V. Brát, Maticové metody v analýze a syntéze prostorových vázaných mechanických systémů, Academia Praha, 1981.

EAN 2019
57th conference on experimental stress analysis

Development of material law for simulation of ultrasonic welding

KUKLÍK Martin[1,a], KLOUDA Filip[1], RULC Vojtěch[1] and KRIŽAN Jiří[1]

[1]Advanced Engineering s.r.o.; Na Ostrohu 16, 160 00 Praha, Czech Republic

[a]mkuklik@advanced-eng.cz

Keywords: Material, Ultrasonic Welding, Viscoelasticity, Manufacturing Process, Simulation

Abstract. It is important to involve influence of manufacturing process of parts and assemblies which are later analysed and developed by using CAE simulations. We are recently able to successfully simulate so many different kinds of manufacturing processes and bring their results as boundary condition for further calculation. Ultrasonic welding is one of commonly used physical technology, but possibilities to simulate this process are currently very limited. The project is about development of viscoelastic material model suitable for ultrasonic welding CAE simulation. It is related to usage of user subroutine in Radioss Solver and programming of user defined material model.

Introduction

New and new manufacturing technologies are implemented into practice. Computational simulations must follow this progress and should be prepared to react on it. Simulations of manufacturing processes can save a lot of expenses and accelerate development of products.

Our team decided to react to expanding use of ultrasonic welding in production of plastic and metal products as well as missing appropriate material law which would be able to simulate this process in explicit solver Altair RADIOSS.

The research's aim is to develop a material law for Altair RADIOSS, which enables to simulate a complete process of ultrasonic welding from contact of parts up to cool down.

CAE Solver

Altair RADIOSS is one of the high-performance world-wide used explicit solvers in the field of structural analysis of nonlinear dynamic load problems. It includes tools for multi-physical simulation and modelling of complex materials such as composites also. RADIOSS is used across industries to increase impact resistance, safety and manufacturability of structural designs. It includes almost 100 material laws and 20 failure material models which define physical behaviour of simulated materials [1]. Unfortunately, up to now no of them can be used for accurate simulation of ultrasonic welding.

Theory

Ultrasonic welding is manufacturing technology which uses ultrasonic waves for connecting parts of assembly. It transforms high-frequency ultrasonic acoustic vibrations into heat that increases temperature of contact layer between two parts up to melting point. Then the material of two parts is blended and the assembly cooled down.

Few different methods which differs in principle of heat production can be used for the welding. Our team focused on a method which uses visco-elasto-plastic properties of materials, which are characteristic for almost all real materials

Viscoelastic behaviour means that when the material is under stress, part of energy is transformed into heat and another part is stored for shape restoration when the stress is lost. This is due to hysteresis in stress strain dependency of the material. This hysteresis behaviour causes that the strain is delayed in comparison with the stress, i.e. stress–strain curve is different during loading and unloading process. A phase shift between stress and strain is significant. These effects are described in Eq. (1) [2].

$$\sigma = E_{RE} \cdot \varepsilon_0 \cdot \cos(\omega \cdot t) + E_{IM} \cdot \varepsilon_0 \cdot \cos\left(\omega \cdot t + \frac{\pi}{2}\right) \tag{1}$$

where σ is stress, ε_0 is strain, E_{RE} is real part of complex Young's modulus and E_{IM} is imaginary part – i.e. loss modulus which determines energy transformed to heat during material deformation.

Then loss factor *LF*, ratio of transformed and non-transformed energy, is a dimensionless quantity defined as Eq. (2) [3].

$$LF = {}^{E_{IM}}/_{E_{RE}} = \tan\varphi \tag{2}$$

where φ is a phase shift and it is a material characteristic that can be measured. This quantity can describe ability of material to generate the heat as well as damp mechanical vibrations. One can see the value of the phase shift is a crucial factor here.

Visco-elasto-plastic behaviour then means that when elastic limit, i.e. yield stress, is exceeded, the plastic changes are taken into account also. This effect is very important for correct simulation of welding of plastic materials, because part of the heat is also generated by plastic deformation.

Implementation of theory into the material law

It was decided that the implementation of theory into our material law will be sequential. We started with rather great disregards of physical effects which were taken in care only in next following phases of development. One of the simplifications of the solved problem was a linearization of elastic part of stress-strain curve which is shown at Figure 1. The amount of energy transformed into heat is proportional to area of hysteresis loop that characterizes dependency between stress and strain during one cycle of harmonic loading and unloading of the visco-elastic materials, see Figure 1 on the left. As dependency between stress and strain is nonlinear, the Hook's law does not hold for visco-elastic materials.

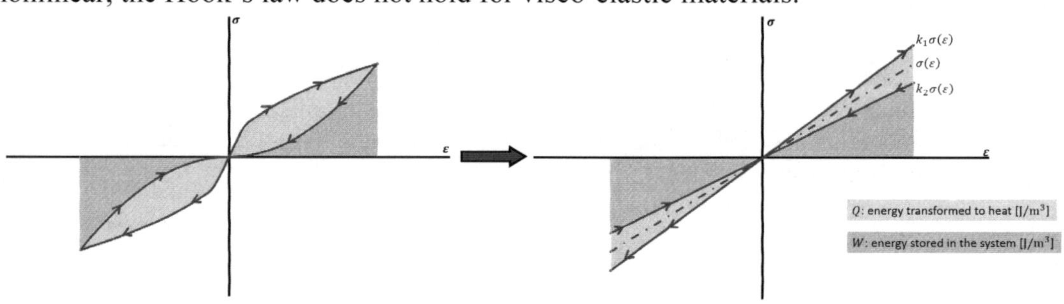

Fig. 1: Linearization of stress-strain curve

The basic relation for the energy is:

$$U = \frac{1}{2} V \sigma \varepsilon \tag{3}$$

where, U is internal energy (of element) and V is volume (of element). During unloading process part of the internal energy W is used for shape restoration and remaining energy Q is dissipated to the heat. The relation (3) holds for all components of stress and strain tensors and overall energy is sum of energy for all of these tensor components. The slopes k_1 and k_2 of ascending (loading) and descending (unloading) line are chosen so that after linearization:

$$\frac{Q}{W} = tan\,\varphi, \tag{4}$$

which corresponds to Eq. (2) and therefore the heat produced by linearized model is in accordance with real material behaviour.

When the elastic behaviour is linearized, then the plastic behaviour can be described by different material laws. As the best one for our purposes the Johnson-Cook model was chosen. It describes the material by Eq. (5).

$$\sigma = \left(a + b\varepsilon_p{}^n\right)\left(1 + c\ln\frac{\dot{\varepsilon}}{\dot{\varepsilon}_0}\right)\left(1 - (T^*)^m\right) \tag{5}$$

where, a, b, c and n are material parameters, ε_p is the plastic strain, $\dot{\varepsilon}$ is the strain rate, $T^* = (T - T_r)/(T_{melt} - T_r)$, T is the temperature, T_r is the ambient temperature, and T_{melt} is the melting temperature. For our current purposes the strain rate and temperature dependency of stress is neglected in this phase of development. Overall heat produced can be expressed as sum of heat due to the loss factor in the elastic region and heat due to the plastic deformation in the plastic region of the stress-strain curve. This is depicted in Figure 2.

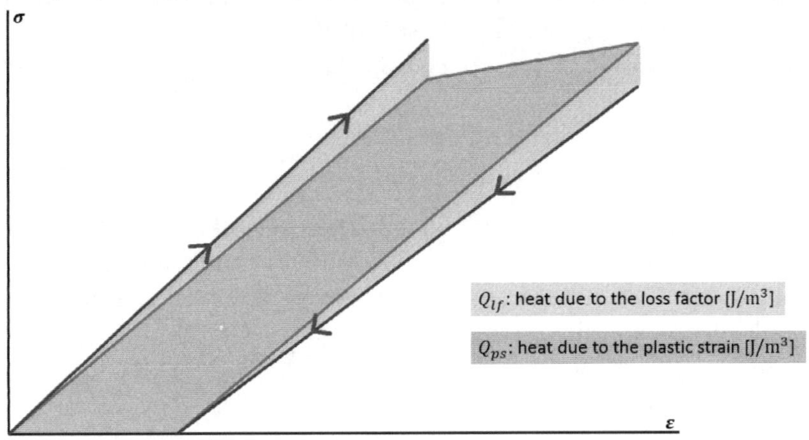

Fig. 2: Linearization of stress-strain curve with plastic behaviour

After few internal tests we decided that in this phase of development a heat transfer by conduction will be disregarded. We can presume that the heat transfer will be minimal, because the welding process is very fast (complete process lasts tenths or maximum seconds) as can be seen in Figure 3. Nevertheless, if there is a need of basic heat transfer, it can be modelled (in very simplified way) in postprocessing by averaging values on elements.

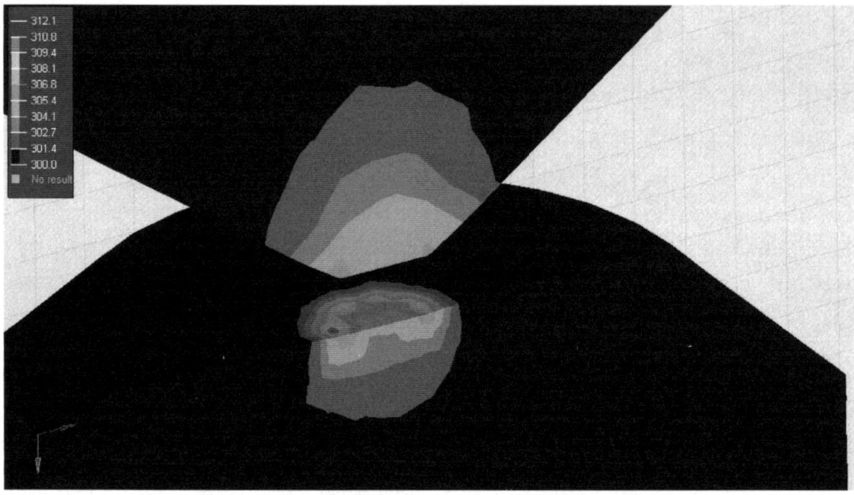

Fig. 3: Thermal distribution at time t = 7 ms

There was found out in benchmarks that mesh size has very great influence on result accuracy. On the other hand, size of the smallest element in mesh determines size of time-step of simulation (and computational time demands too) in RADIOSS solver, or any explicit solver. Due to it, the mesh size must be chosen very prudently and the whole mesh must be result of compromise between accuracy and speed even more than in common structural analysis.

Conclusions

A development of new material law is not ended. There is still a space for enhancing code and implementing other minor physical effects. We will continue in our effort. For example, because of simulation of post welding cooling, we would like to implement heat transfer model which is disregarded at this moment. Nevertheless, the current state looks promisingly enough to be sufficiently accurate to successfully simulate typical process of ultrasonic welding after finishing validation process.

References

[1] ALTAIR. *RADIOSS Reference Guide 2017.*

[2] VAŠINA, Martin. *Environmentální fyzika: Viskoelasticita materiálů* [online]. Zlín, 2010 [cit. 2019-03-14]. Available from: http://ufmi.ft.utb.cz/texty/env_fyzika/EF_06.pdf. Text materials for lectures. Ústav fyziky a materiálového inženýrství, Fakulta technologická, Univerzita Tomáše Bati ve Zlíně.

[3] Dynamická mechanická analýza. *Viskoelasticita materiálů* [online]. [cit. 2019-03-14]. Available from: https://www.viskoelasticita.cz/inpage/dma/

EAN 2019
57th conference on experimental stress analysis

Influence of defects in carbon nanotubes on their buckling behaviour

LENGVARSKÝ Pavol[1,a], BOCKO Jozef[1,b] and SHIMANOVSKY Alexandr[2,c]

[1]Department of Applied Mechanics and Mechanical Engineering; Faculty of Mechanical Engineering; Technical University of Košice; Letná 9, 04200 Košice, Slovakia

[2]Technical Physics and Engineering Mechanics, The Belarusian State University of Transport, Kirova str 34, 246653 Gomel, Belarus

[a]pavol.lengvarsky@tuke.sk, [b]jozef.bocko@tuke.sk, [c]tm.belsut@gmail.com

Keywords: Nanotube, Buckling, Defect, Beam

Abstract. The paper studies a buckling behaviour of simply-simply supported single-walled carbon nanotubes (SWCNTs) without and with defects. The buckling of carbon nanotubes without and with defects was investigated by the finite element method (FEM). The carbon nanotubes were modelled as beams. The defects were created by removing of randomly determined carbon atoms and beam elements connected to these nodes. The increasing number of defects leads to the decrease of the critical buckling force of SWCNTs. The decrease of the critical buckling force was almost the same for SWCNTs with the same diameter but a different chirality.

Introduction

A nanostructure is a tiny body of size between molecular and microscopic structures. Nanostructures can be divided into several categories of objects: nanobeams, nanowires, nanorings, nanoparticles, nanoribbons, nanoplates, and nanotubes. The discovery of carbon nanotubes (CNTs) in 1991 was a significant breakthrough in the field of nanostructures [1]. Extensive research of CNTs started simultaneously with the discovery of CNTs. It was found that CNTs have extraordinary mechanical, electrical, optical and thermal properties. These properties have been verified by experimental, analytical and numerical methods. The numerical methods include the molecular dynamics simulation (MD), which is a suitable tool for a description of behavior of carbon nanotubes. MD computation has a disadvantage because it puts high demands on computer resources (processors, memory) as well as the computational time. This is a reason why researchers started to use the FEM to describe the behavior of CNTs. The approach was based on the creation of the connection between the MD and the continuum mechanics. This treatment was established by Li and Chou [2] for simulation of graphene sheets and carbon nanotubes. They used a beam element for modelling of interatomic bonds between two adjacent carbon atoms. Then Li and Chou [3] used this approach for a study of elastic moduli of multi-walled carbon nanotubes (MWCNTs), where the van der Waals forces between tubes were represented by nonlinear truss model. The nonlinear behavior of the truss was obtained by the Lennard-Jones potential. Simultaneously with these papers [2, 3] started the investigation of Young's modulus and other material properties of graphene sheets [4] and CNTs by FEM. Hartmann et al. [5] investigated the material properties of graphene by the atomistic simulation approach. They determined

Young's modulus, Poisson's ratio and thickness of graphene from the finite element computations. The beam elements for modelling CNTs were used by Tserpes and Papanikos [6]. The mechanical properties of MWCNTs were also studied in [7], where authors used for modelling CNT beam elements and the van der Waals forces between tubes were modeled by linear spring elements. The stiffnesses of the spring elements were obtained from the equivalent force concept. The same approach for modelling and estimation material properties of CNTs was used in [8, 9]. The researchers used for modelling CNTs the spring elements instead of beam ones. The difference between beams and springs lies in absence of definition of the cross section area as the input parameter for computations. Moreover, different material parameters of Young's modulus, Poisson ration and wall thickness are used in literature so the comparison of inputs and outputs is not easy. Brcic et al. [10] and Sakharova et al. [11] investigated the influence of waviness or vacancy defects on mechanical properties of CNTs. A few years later started an investigation in the area of buckling behavior of CNTs. Researchers have used the experimental, analytical and finite element method with beam, spring, and shell elements. Zhang et al. [12] analyzed buckling of multi-walled carbon nanotubes (MWCNTs) under radial pressure. The molecular structural mechanics approach was used by Hu et al. [13]. The 3D beam elements for modelling interatomic bonds of C-C and rod elements for modelling van der Waals forces were used in this approach. They concluded that their model is adequate for prediction of the buckling load of single- and multi-walled carbon nanotubes with the bigger length. In the paper written by Rahmandoust et al. [14] buckling and natural frequencies of SWCNTs were studied. C-C bonds were represented by an Euler-Bernoulli beam, and they were modelled by the FEM. The results were in good agreement with the analytical solution. The perfect and the geometrically modified CNTs under combined loading were investigated in [15]. The authors concluded that geometric changes reduce the critical load. The buckling behavior of CNTs with junctions were studied in papers [16, 17], where the authors used the MD and FEM simulations. In paper [16] they investigated the influence of strain rate on the magnitude of load force. Xin et al. [18, 19] investigated buckling of defected SWCNTs and DWCNTs by the MD simulation. The CNTs were loaded by axial compression and results were compared with CNTs without defects. They studied the effects of the position of defects on the buckling behavior of CNTs. In [20] the results of investigation of buckling behavior for defected nanotubes can be found. Similar research was accomplished by Ghavamian and Öchsner [21], as well as by Poelma et al. [22]. Torabi et al. [23] investigated the buckling of DWCNTs under three basic load states (tension, bending and torsion) and for perfect and defective nanotubes with the armchair chirality. They used a rigid sphere for modelling carbon atoms, nonlinear springs for modelling C-C bonds and linear springs for the modelling van der Waals forces. The influence of defects in geometry of SWCNTs and MWCNTs on buckling was subject of research that is described in paper [24]. The additional mechanical properties, application areas and behavior of carbon nanotubes are described in publications [25, 26].

In the current paper buckling of simply-simply supported single-walled carbon nanotubes under axial load is investigated using FEM. SWCNTs are considered as tubes without and with defects. The commercial program ANSYS Mechanical APDL is used for modelling SWCNTs, where the BEAM4 element with two nodes and with six degrees of freedom at each node is used for modelling the beam. The computations were realized for the nanotubes with 1 %, 5 % and 10 % missing carbon atoms, respectively. The defects were created by removing carbon atoms as nodes and beam elements attached to these nodes. Finally, the critical buckling forces of defected SWCNTs were compared with perfect SWCNTs. The main goal of the paper is to investigate buckling behavior of single-walled carbon nanotubes without and with defects under axial loading. In comparison to the reviewed published papers we have investigated nanotubes with significant higher number of defects. Moreover, we have studied the influence of locations of defects on the magnitude of buckling force.

Theoretical background of carbon nanotubes modelling

The MD approach is usually used for modelling carbon nanotubes. A problem of the method is, as was mentioned above, that at this time also with using supercomputers we can simulate the behavior of carbon nanotubes only with small lengths (several micrometres). Moreover, finding of the solution is an extremely time-consuming process. For this reason, it is better to use the FEM for a description of behavior of carbon nanotubes. The modelling of carbon nanotubes can be realized by different types of finite elements, i.e. bar elements, beam elements, linear and nonlinear spring elements and shell elements. Of course, all models consisting of these different element types lead to different approaches for modelling as well as for defining material properties, boundary conditions and loadings.

The carbon nanotubes are modelled using beam elements, whose properties are obtained by making the connection between the molecular and continuum mechanics. Interatomic interactions in molecular mechanics as bond stretching, bond angle variation, dihedral angle torsion and out-of-plane torsion are replaced by beam elements that represent the load by pure tension, bending and torsion. The following relations between parameters in molecular mechanics k_r, k_θ, k_τ and parameters in continuum mechanics EA, EI, GJ were obtained using this connection [6]:

$$k_r = \frac{EA}{L}, \tag{1}$$

$$k_\theta = \frac{EI}{L}, \tag{2}$$

$$k_\tau = \frac{GJ}{L}, \tag{3}$$

where k_r, k_θ, k_τ are force constants (stiffness) of bond stretching, bond bending and torsional resistance, respectively, and E, G, L, A, I, J are Young's modulus, shear modulus, length of beam, cross-section area, quadratic moment of inertia and polar moment of inertia, respectively [2, 3, 6].

The Eqs. (1) – (3) represent the basic relations for the application of continuum mechanics for solving of carbon nanotubes. By assuming a circular beam section with diameter d and corresponding parameters $A = \pi d^2/4$, $I = \pi d^4/64$ and $J = \pi d^4/32$, then Eqs. (1) – (3) give [6]:

$$d = \sqrt{\frac{k_\theta}{k_r}}, \tag{4}$$

$$E = \frac{{k_r}^2 L}{4\pi k_\theta}, \tag{5}$$

$$G = \frac{{k_r}^2 k_\tau L}{8\pi k_\theta^2}. \tag{6}$$

Here, the parameters are given as [2, 3, 6]: $k_r = 6.52\times10^{-7}$ Nnm^{-1}, $k_\theta = 8.76\times10^{-10}$ Nnmrad^{-2}, $k_\tau = 2.78\times10^{-10}$ Nnmrad^{-2}, $L = a_{C\text{-}C} = 0.1421$ nm. The input data as diameter $d = 0.1466$ nm, Young's modulus $E = 5.488$ TPa and shear modulus $G = 0.871$ TPa are obtained for the beam element modelled using FEM. The simply-simply supported and axially loaded SWCNTs with these properties of the beam elements (BEAM4 – Euler-Bernoulli beam) are modelled in the ANSYS Mechanical APDL. The boundary conditions are applied to the boundary nodes on the beam elements of the SWCNTs. The top level boundary nodes are allowed to move in the axial direction of the nanotube. The force is distributed to the upper nodes and the total force magnitude is 1 nN. In order to know, at least approximately, number of nodes for modeled nanotubes we can give one representative example with chirality (6, 6) and length 3 nm. Such a model consists of 300 nodes and 438 nanobeams. At the beginning, the static analysis with active prestress effect is performed and then the buckling analysis is performed.

Buckling of carbon nanotubes

The simply-simply supported carbon nanotubes with defects are modelled only using the beam models of CNTs. The defects are modelled by removing of random carbon atoms (nodes) and removing corresponding beam elements attaching to these atoms (nodes). SWCNTs have 0 %, 1 %, 5 % and 10 % removed carbon atoms (nodes) of the total amount (Fig. 1). The defects are made using quasi-random pattern. At the beginning, the defective pattern with 1 %, 5 % and 10 % removed carbon atoms was created for every diameter of nanotube which was copied several times in order to have desired length of whole nanotube. Accordingly, the patterns have certain periodicity. In this section, an axial buckling of simply-simply supported single-walled carbon nanotubes is investigated.

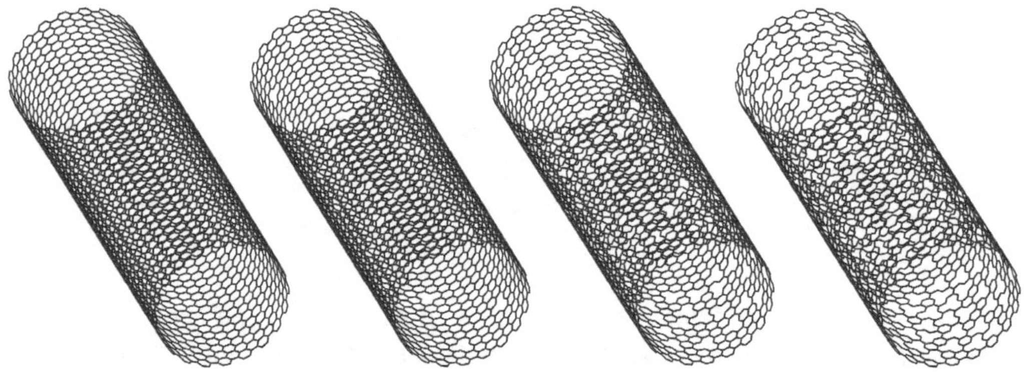

a) b) c) d)

Fig. 1: The armchair carbon nanotubes (20, 20) with defects a) 0 %, b) 1 %, c) 5 % and d) 10 %

Since the location of the defects is a very important factor, their influence to the magnitude of critical force was studied on models divided into 6 parts (Fig. 2(b)). Due to symmetry, the calculations were accomplished only for 4 models – the models had defects in the right bottom part (Fig. 2(c)), whole bottom part (Fig. 2(d)), right middle part (Fig. 2(e)), and whole middle part (Fig. 2(f)). If we want to compute the influence of defect positions we have to remove the same percentage of atoms as before, it means the number that is related to the whole nanotube. In Fig. 3(a) up is given distribution of defects for 1 % of removed atoms. In Fig. 3(a) down is given distribution of defects for 5 % of removed atoms. However, the picture shows that the structure consists of many big gaps so the computations do not have any sense. Accordingly, only computation for 1% defects inside considered part of structure were realized. The results of computations are given in Fig. 4, where the decrease of buckling force in comparison to the nanotube without defects is presented. The semi-logarithmic graph was chosen in order to stress differences between computed values for the whole extent of graph. As the graphs represent decreasing of critical force, the highest values are related to the more danger situations from the point of view of nanotube stability. Accordingly, for the simply-simply supported nanotubes it is more danger to have defects in the middle part of nanotube than on their ends. The concentration of defects on one side lead to smaller critical force as the concentration of defects on the whole part.

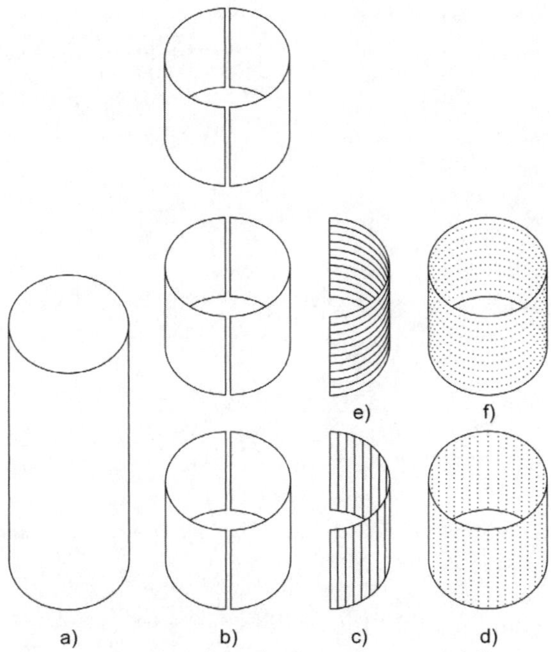

Fig. 2: The scheme of nanotube for description of areas with removed atoms a) whole nanotube b) partitioned nanotube c) bottom right part of nanotube d) whole bottom part of nanotube e) middle right part of nanotube f) whole middle part of nanotube

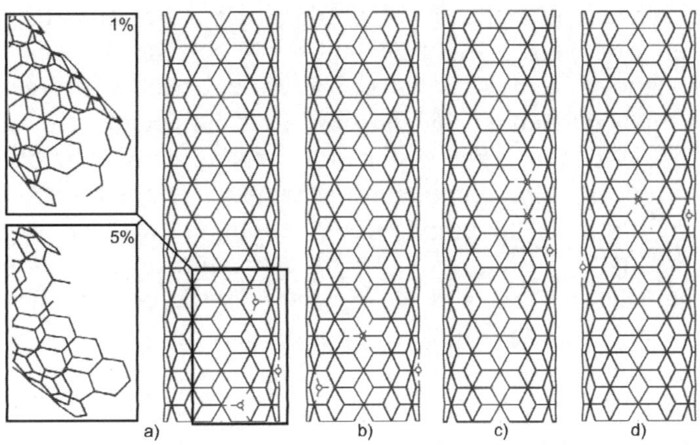

Fig. 3: The scheme of nanotube for description of areas with removed atoms a) whole nanotube b) partitioned nanotube c) bottom right part of nanotube d) whole bottom part of nanotube e) middle right part of nanotube f) whole middle part of nanotube

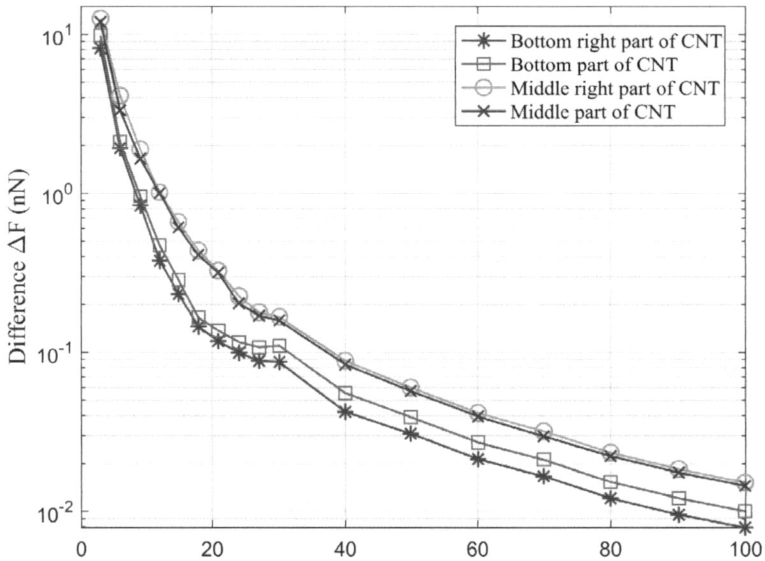

Fig. 4: Differences of critical buckling forces between perfect and defective carbon nanotubes (6, 6) in different parts of nanotube

The critical buckling forces F_{cr} of the single-walled carbon nanotubes (Table 1) under axial loading with vacancies 1 %, 5 % and 10 %, and length from the range $L = 3$ nm to 100 nm are shown in Fig. 5 – Fig. 6. The tendency of decreasing of the critical buckling force F_{cr} is almost the same in all cases, but a slight deviation can be seen for the curves that represent different levels of vacancies in the structures. The critical buckling force F_{cr} is the same for SWCNTs with the equal diameters, that is, the F_{cr} is independent of used chirality types (armchair, zigzag) of SWCNTs. The graphs show quantitative relations for magnitude of buckling force with respect to the number of vacancies in the nanotube of a given chirality and length. On the right side of the graphs (long nanotubes) we observe the Euler buckling. The shell buckling begins for nanotubes of length approximately 27 nm (at the beginning of plateau part) moving to the left side of graphs. However, on the basis of previous computations we can state that the shell buckling modes are not adequate to corresponding shell modes and accordingly the computed buckling forces are smaller. The graphs show obvious fact that increasing number of vacancies leads to the moving of the Euler buckling to the left side.

Table 1: The modelled carbon nanotubes with basic parameters

Nr.	Armchair SWCNT		Zigzag SWCNT	
	Chirality	Diameter (nm)	Chirality	Diameter (nm)
1.	(6, 6)	0.8142	(10, 0)	0.7834
2.	(10, 10)	1.357	(20, 0)	1.5669
3.	(15, 15)	2.0354	(26, 0)	2.0369
4.	(20, 20)	2.7139	(35, 0)	2.742

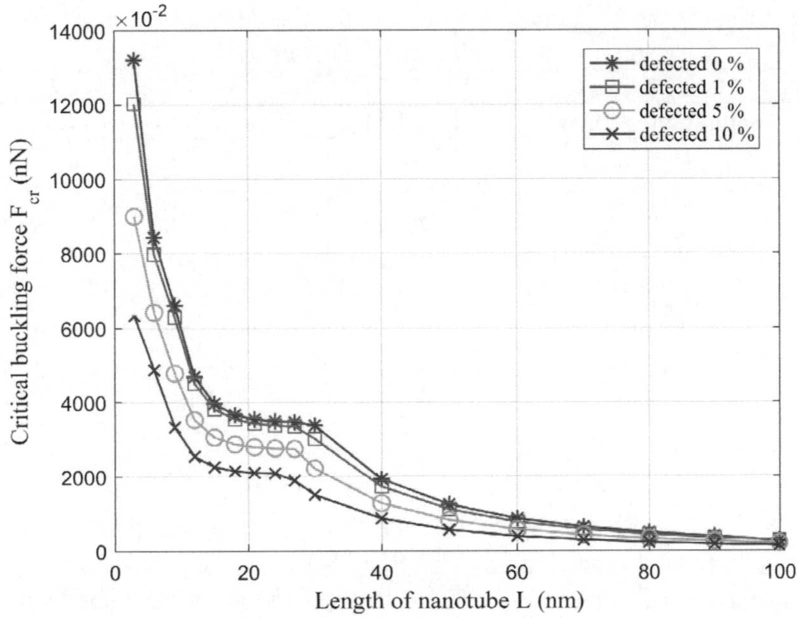

Fig. 5: The critical buckling force F_{cr} of the defected armchair carbon nanotubes (20, 20) with diameter D = 2.7139 nm

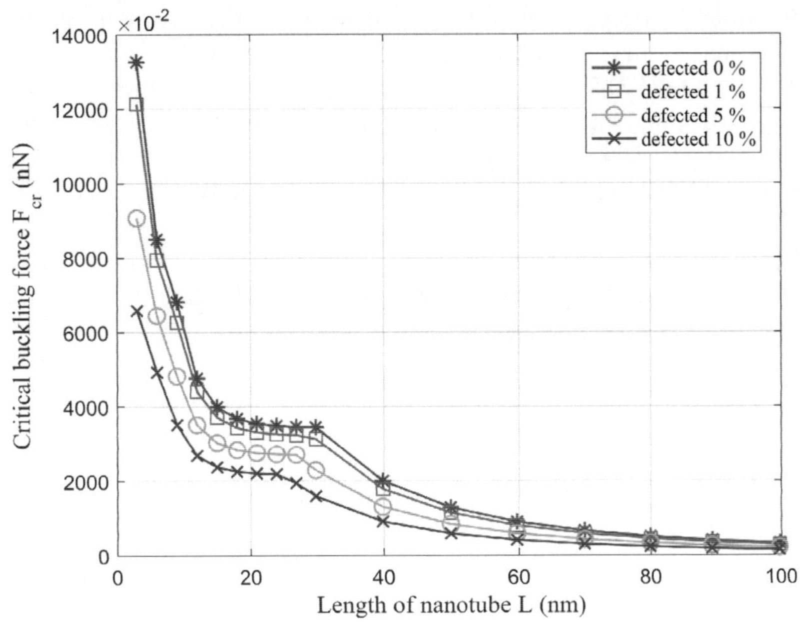

Fig. 6: The critical buckling force F_{cr} of the defected zigzag carbon nanotubes (35, 0) with diameter D = 2.742 nm

The average values of decrease of critical buckling force F_{cr} for the modelled simply-simply supported CNTs are given in Table 2.

Table 2: The decrease of critical buckling force F_{cr} for the defected carbon nanotubes with respect to perfect carbon nanotubes

Carbon nanotube		Decrease of critical buckling force F_{cr} (%)		
Chirality	Diameter (nm)	Defect 1 %	Defect 5 %	Defect 10 %
(6, 6)	0.8142	13.48	41.03	54.49
(10, 10)	1.357	7.57	33.09	56.58
(15, 15)	2.0354	8.72	33.79	55.89
(20, 20)	2.7139	6.96	31.55	53.35
(10, 0)	0.7834	12.25	34.62	74.27
(20, 0)	1.5669	11.77	40.53	65.55
(26, 0)	2.0369	11.66	34.43	53.08
(35, 0)	2.742	9.61	34.01	54.09

Conclusions

The buckling of axially loaded armchair and zigzag single-walled carbon nanotubes without and with defects was investigated. The SWCNTs were modelled by the beam finite elements. The properties of the beam finite element model were determined from the connection of the molecular and continuum mechanics. The critical buckling forces were calculated for four armchair and four zigzag carbon nanotubes with various lengths. The critical buckling forces of these defected SWCNTs were investigated. Defects were realized by removing of random carbon atoms (nodes) and beam finite elements attached to these atoms (nodes). It is apparent that the increasing number of defects leads to the decrease of the critical buckling force of SWCNTs. The decrease of the critical buckling force was almost the same for SWCNTs with the same diameter and different chirality. The small difference was caused by the random location of relatively homogeneously positioned vacancies.

The results show that the buckling of carbon nanotubes can be simulated by beam model. The all computations are not more time consuming. The next step can be the application of Eringen nonlocal theory for considering the small scale effect of CNTs and modeling of carbon nanotubes by springs elements because some authors state that spring models are more useful for carbon nanotubes [8, 9], especially in case of bending.

Acknowledgment

This research was supported by a grant from the Slovak Grant Agency VEGA No. 1/0731/16 - Development of Modern Numerical and Experimental Methods of Mechanical System Analysis.

References

[1] S. Iijima, Helical microtubules of graphitic carbon, Nature, 354 (1991) 56-58.
[2] C. Li and T. W. Chou, A structural mechanics approach for the analysis of carbon nanotubes, International Journal of Solids and Structures, 40 (10) (2003) 2487-2499.
[3] C. Li and T. W. Chou, Elastic moduli of multi-walled carbon nanotubes and the effect of van der Waals forces, Computational Science and Technology, 63 (11) (2003) 1517-1524.
[4] G. Cao, Atomistic studies of mechanical properties of graphene, Polymers, 6 (9) (2014) 2404-2432.
[5] M. A. Hartmann, M. Todt, F. G. Rammerstorfer, F. D. Fisher and O. Paris, Elastic properties of graphene obtained by computational mechanical tests, Europhysics Letters, 103 (6) (2013) 68004.
[6] K. I. Tserpes and P. Papanikos, Finite element modeling of single-walled carbon nanotubes, Composites: Part B, 36 (5) (2005) 468-477.

[7] C. W Fan, Y. Y. Liu and Ch. Hwu, Finite element simulation for estimating the mechanical properties of multi-walled carbon nanotubes, Applied Physics A, 95 (3) (2009) 819-831.
[8] G. I. Giannopoulos, P. A. Kakavas and N. K. Anifantis, Evaluation of the effective mechanical properties of single walled carbon nanotubes using a spring based finite element approach, Computational Material Science, 41 (4) (2008) 561-569.
[9] M. Rossi and M. Meo, On the estimation of mechanical properties of single-walled carbon nanotubes by using a molecular-mechanics based FE approach, Composites Science and Technology, 69 (9) (2009) 1394-1398.
[10] M. Brcic, M. Canadija and J. Brnic, Influence of waviness and vacancy defects on carbon nanotubes properties, Procedia Engineering, 100 (2015) 213-219.
[11] N. A. Sakharova, A. F. G. Pereira, J. M. Antunes and J. V. Fernandes, Numerical simulation study of the elastic properties of single-walled carbon nanotubes containing vacancy defects, Composites Part B: Engineering, 89 (2016) 155-168.
[12] Y. Q. Zhang, G. R. Liu and X. Han, Effect of small length scale on elastic buckling of multi-walled carbon nanotubes under radial pressure, Physics Letters A, 349 (5) (2006) 370-376.
[13] N. Hu, K. Nunoya, D. Pan, T. Okabe and H. Fukunaga, Prediction of buckling characteristics of carbon nanotubes, International Journal of Solids and Structures, 44, (20) (2007) 6535-6555.
[14] M. Rahmandoust and A. Öchsner, Buckling behaviour and natural frequency of zigzag and armchair single-walled carbon nanotubes, Journal of Nano Research, 16 (2011) 153-160.
[15] S. I. Yengejeh, S. A. Kazemi and A. Öchsner, Influence of combined loading on the structural stability of carbon nanotubes, Materials Chemistry and Physics, 158 (2015) 96-106.
[16] Z. Kang, M. Li and Q. Tang, Buckling behavior of carbon nanotube-based intramolecular junctions under compression: Molecular dynamics simulation and finite element analysis, Computational Materials Science, 50 (1) (2010) 253-259.
[17] S. I. Yengejeh, M. A. Zadeh and A. Öchsner, On the buckling behavior of connected carbon nanotubes with parallel longitudinal axes, Applied Physics A, 115 (4) (2014) 1335-1344.
[18] H. Xin, Q. Han and X. Yao, Buckling and axially compressive properties of perfect and defective single-walled carbon nanotubes, Carbon, 45 (13) (2007) 2486-2495.
[19] H. Xin, Q. Han and X. Yao, Buckling of defective single-walled and double-walled carbon nanotubes under axial compression by molecular dynamics simulation, Composites Science and Technology, 68 (7-8) (2008) 1809-1814.
[20] V. Parvaneh, M. Shariati and A. M. M. Sabeti, Investigation of vacancy defects effects on the buckling behavior of SWCNTs via a structural mechanics approach, European Journal of Mechanics – A/Solids, 28 (6) (2009) 1072-1078.
[21] A. Ghavamian and A. Öchsner, Numerical investigation on the influence of defects on the buckling behavior of single- and multi-walled carbon nanotubes, Physica E, 46 (2012) 241-249.
[22] R. H. Poelma, H. Sadeghian and G. Q. Zhang, Effects of single vacancy defect position on the stability of carbon nanotubes, Microelectronics Reliability, 52 (7) (2012), 1279-1284.
[23] H. Torabi, M. Shariati, E. Sedaghat and A. L. Zadeh, Buckling behaviour of perfect and defective DWCNTs under axial, bending and torsional loadings via structural mechanics approach, Meccanica, 48 (8) (2013) 1959-1974.
[24] M. Eftekhari, S. Mohammadi and A. R. Khoei, Effect of defects on the local shell buckling and post-buckling behavior of single and multi-walled carbon nanotubes, Computational Material Science, 79 (2013) 736-744.
[25] I. Elishakoff, Carbon Nanotubes and Nanosensors: Vibration, Buckling and Ballistic Impact, Wiley, New York, USA, (2012).
[26] R. Rafiee and R. M. Moghadam, On the modeling of carbon nanotubes: A critical review, Composites Part B: Engineering, 54 (2014) 435-449.
[27] F. I. Niordson, Shell Theory, North Holland, Amsterdam, Netherlands (1985).

EAN 2019
57th conference on experimental stress analysis

Improvement in accuracy of strain gauge measurements for using in high precision measurements of gearboxes efficiency

LOPOT Frantisek[1,a], DUB Martin[1] and KUBOVY Petr[2]

[1]Czech Technical University, Faculty of mechanical engineering, Technicka 4, 166 07, Praha, Czech Republic

[2]Charles University, Faculty of sport and physical education, Jose Martiho 31, 162 52, Praha, Czech Republic

[a]frantisek.lopot@fs.cvut.cz

Keywords: Notch, Strain gauge, Stress, Torque, FEM, Measurement, Gearbox, Efficiency

Abstract

The paper describes further development in our research from the previous year [1]. We are still focused on finding a method of measuring the torsion on shafts of gearboxes installed in a real machine to determine their actual efficiency. The biggest problem is the limited accuracy of our strain gauge apparatus, because of over-sized design of shafts and thus their very small deformations. We still believe that use of various technological and structural notches is an acceptable solution, whereby the level of recorded signal could be amplified by this way.

Introduction

As part of our work, we try to determine the actual efficiency of gearboxes installed in a real fabric in normal operating conditions. In addition to warming and vibration monitoring, the strain gauge measurements which can be used to detect torques on input and output shafts, seem to be an effective approach. The importance of our measurements increases with the total power transferred when in case of large industrial units a small loss of efficiency may reach tens of kW of lost power. To make our measurement really meaningful, we need to get below 1% of relative error. A reliable way to minimize the relative measurement error is to have a sufficiently strong signal in relation to the sensitivity of the apparatus. Because in real installations it is not possible to make adjustments on the shafts that would "boost" the signal, we need to use some alternative means. These are mostly structural and technological notches that act as natural mechanical stress concentrators. For our measurements, we assume that a higher value of mechanical stress means also a higher corresponding deformation and that we work in the zone of elastic response. In the literature, we can find quite extensive information about stress in the notches and their surroundings and based on this we believe that our assumption on its applicability in functional structures such as gearboxes is justified and true. On the basis of stress analysis, we can also define the requirements for the orientation and location of strain gauges so that the notch is best used from our point of view, i. e. to generate the strongest signal. Because analytical solution of this problem could be very complicated, we used the finite element method (FEM) here. The previous work [1] focused on verification that FEM is also useful for calibrating the strain gauge apparatus when a real calibration is not possible, and we have introduced an approach that allows to use this option in real condition. The current laboratory measurements are primarily used to determine the uncertainty of such a calibration.

Experiment - method

A new measuring stand has been designed for our work. The main change compared to the previous version is a statically certain test shaft mounting (Fig. 1).

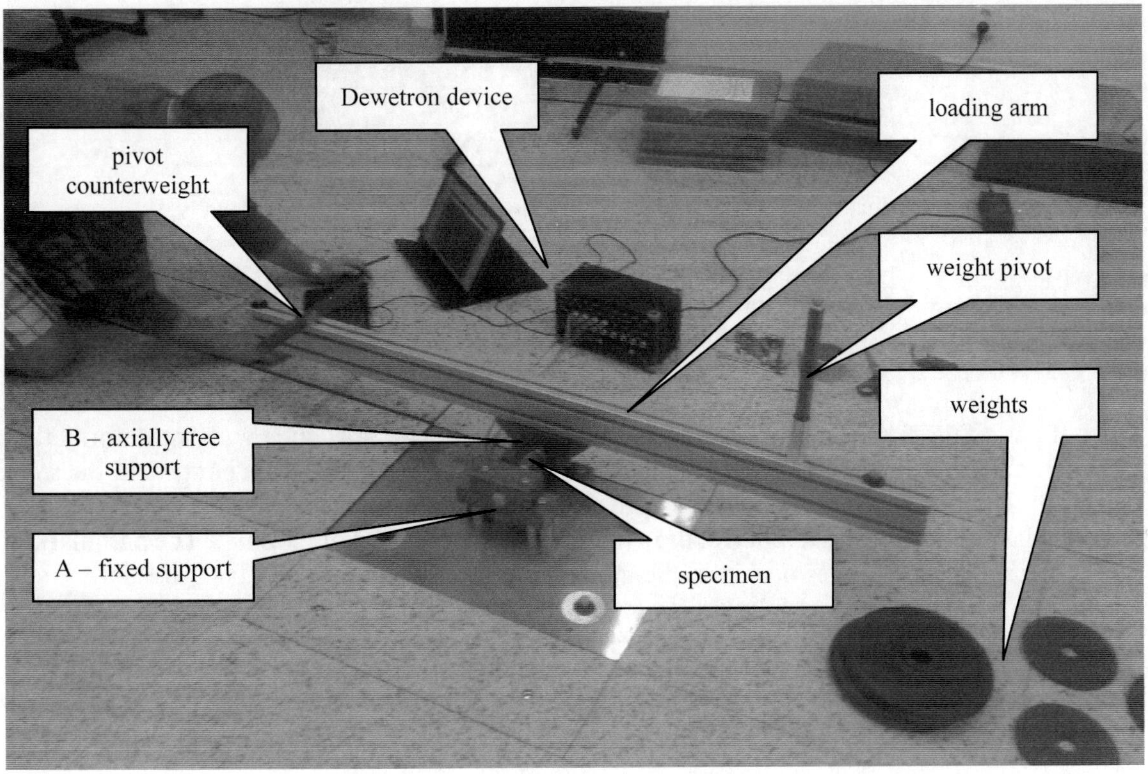

Fig. 1: Measuring place

One end (A) is mounted to a fixed part of the stand while the other end (B) is mounted on ball rollers with rolling contact. Thus, the measured shaft is loaded by torsion only. The loading of the specimens was induced by weights in steps of 0, 10, 15, 17.5 and 20 kg placed on the arm of defined length.

Three specimens of E355 steel [2] without chemical-theraml treatment were prepared for the measurement – see Fig. 2.

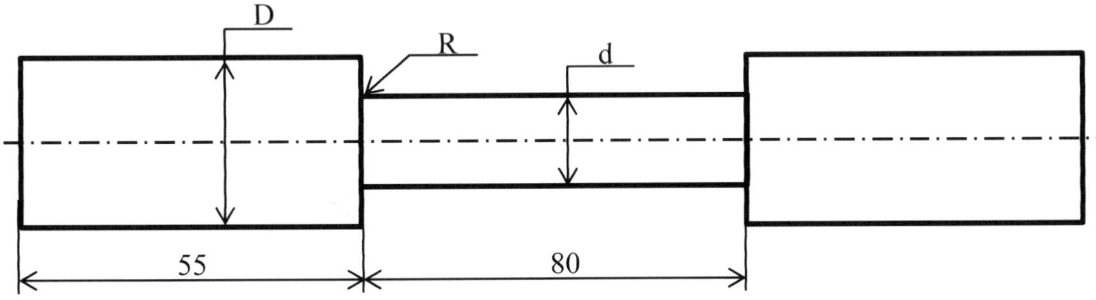

Specimen	D [mm]	d [mm]	R [mm]
1	25	15	2
2	25	15	4
3	25	20	2

Fig. 2: Specimen specification

Measurements were made using HBM LY11-0.6/120 strain gauges [3] installed by special application foil on the specimens acc. to Fig. 3.

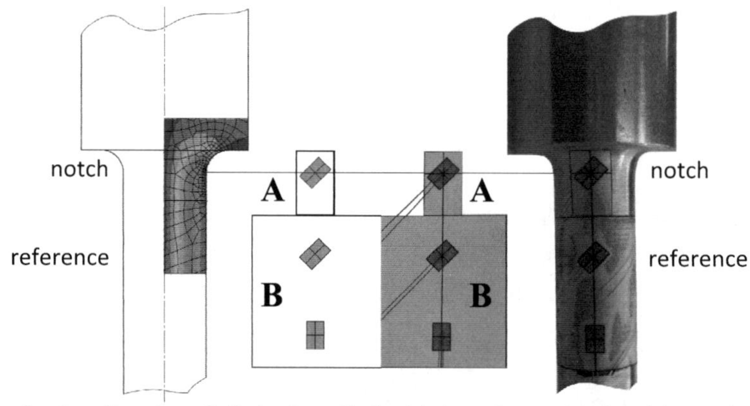

Fig. 3: Application foil design (left side) and use (right side – photo)

The left part of the figure shows a technical drawing of the shaft specimen with the deformation field determined by FEM and the drawing of the application foil with strain gauges. In the right part of the picture, there is real arrangement. The correct position of the foil is determined by the part marked B, which is applied to the available cylindrical part of the shaft. The part marked with A carries strain gauges placed in the area of the measured notch. In the B-part, the strain gauges for checking the data from the smooth cylindrical part are seen. From them, we get the magnitude of the torsional deformation. There are also other strain gauges for checking presence of pressure, tension and bending in the load applied on the tested shaft.

After several testing measurements, it became obvious that the halfbridge is an entirely satisfactory solution for the given purpose because of its installation simplicity and lower price. Also, strain gauges for checking the presence of tension or bending were found unnecessary and were not used in the final measurements.

The strain gauges were connected to the Dewetron Sirius station. The sampling frequency was set to 500 Hz with low pass filter on 10 Hz. The excitation voltage was 10 V and the amplifier sensitivity was 1 mV/V.

Experiment – data processing

The data processing procedure is presented on an example of one specimen. Fig. 4 presents an example of the recorded data (specimen No. 2, Fig. 2).

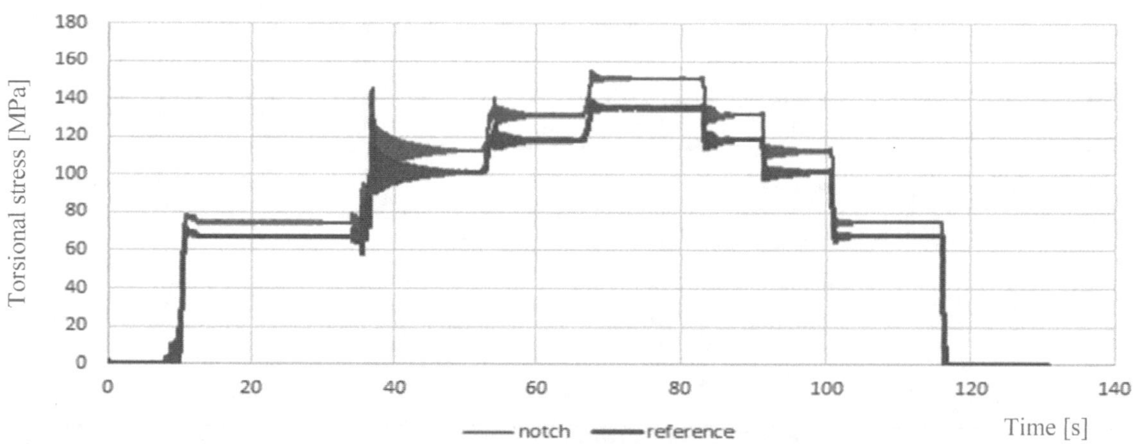

Fig. 4: Obtained data

From the measured quantity, which was the half-bridge voltage U [mV] related to the excitation voltage Un [V], it is possible to calculate the deformation (strain) of active part of the strain gauge using the definitional formula:

$$\text{Strain} = 2 \text{ x } (U/Un) \,/\, (\text{Gauge Factor from [3]} = 1.74) \qquad (1)$$

Then, using next definitional relation:

$$\text{Torsional stress} = G \text{ x Strain}, \qquad (2)$$

where G is appropriate modulus of elasticity for shear loading of measured shaft (81 GPa, [2]), the presented torsional stress is determined. For its calculation, it is, of course, possible to use the relation between torque and stress in the smooth middle section of the measured shafts as well:

$$\text{Torsional stress} = \text{Torque} \,/\, Wk, \qquad (3)$$

where Wk is modulus of cros-ssection for torsion.

Based on comparison the results of experimental data processing (2) with the results of the analytic calculations based on the known applied torque (3) and the outputs of the FEM analysis, the correctness and accuracy of the measurement is checked.

In the graph in Fig. 4, the distance of the stress levels measured at the notch and at the smooth cylindrical part of the specimen can be seen very well. This difference is described as the ratio of the values in the notch to the values of the smooth part. In terms of applicability of our approach for measuring different levels of gear shafts loading by torque, it is very important that this ratio remains constant as the graph in Fig. 5 shows.

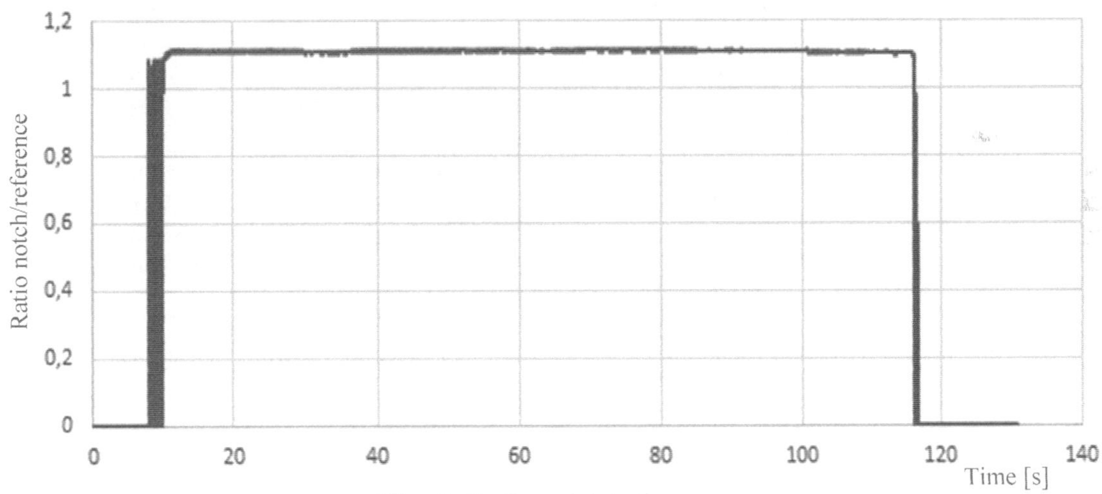

Fig. 5: Ratio notch/reference

Since we are very far from the yield point of the shaft material in our measurements, this graph can be taken for granted. It is, in fact this shape that is important for practical measurements. Any deviation from that pattern means that the data we would be obtaining becomes unreliable and their processing would be difficult if not impossible. As for possible causes, deviation from its individual but constant values would suggest that the notch becomes plasticized and further measurement and data processing would be problematic.

Results

Tab. 1 presents an overview of data processing outputs for measured specimens. The data in the table shows the average values from stable parts for each loading level. The right column gives the "Stress concentration factor" Kt values determined from the Peterson's charts [4] at the eFatique.com portal [5].

Tab. 1: Results - overview

Specimen	Loading [Nm]	Experiment Smooth part [MPa]	Experiment Notch [MPa]	FEM analysis Smooth part [MPa]	FEM analysis Notch [MPa]	Analytics Smooth part [MPa]	Kt
1	0	0	0	0	0	0	1.30
	47.6766	70.5525	89.7071	70.5255	89.6975	70.5252	
	71.5149	105.7734	134.5606	105.7883	134.5463	105.7879	
	83.4341	123.4007	157.0221	123.4197	156.9707	123.4192	
	95.3532	141.0601	179.4306	141.051	179.395	141.0505	
	83.4341	123.4108	157.0208	123.4197	156.9707	123.4192	
	71.5149	105.7879	134.5700	105.7883	134.5463	105.7879	
	47.6766	70.5606	89.7105	70.5255	89.6975	70.5252	
	0	0	0	0	0	0	
2	0	0	0	0	0	0	1.18
	47.6766	67.7899	75.3401	67.796	75.3348	67.7955	
	71.5149	101.7003	113.0111	101.694	113.0022	101.6933	
	83.4341	118.6393	131.8325	118.6431	131.836	118.6422	
	95.3532	135.6026	150.6803	135.592	150.6696	135.5911	
	83.4341	118.6400	131.8376	118.6431	131.836	118.6422	
	71.5149	101.6999	113.0099	101.694	113.0022	101.6933	
	47.6766	67.7912	75.3399	67.796	75.3348	67.7955	
	0	0	0	0	0	0	
3	0	0	0	0	0	0	1.35
	47.6766	29.0308	36.6728	29.026	36.6947	29.0260	
	71.5149	43.5309	55.0092	43.539	55.04205	43.5390	
	83.4341	50.8008	64.1774	50.7955	64.21576	50.7955	
	95.3532	58.0399	73.3456	58.052	73.3894	58.0520	
	83.4341	50.7997	64.1774	50.7955	64.21576	50.7955	
	71.5149	43.5401	55.0092	43.539	55.04205	43.5390	
	47.6766	29.0311	36.6728	29.026	36.6947	29.0260	
	0	0	0	0	0	0	

Presented results show very good agreement between measured data, results of the FEM analysis and from analytical calculation in a smooth cylindrical part of samples. The FEM analysis is perfectly matched also to the measured data in the notch - the worst ratio of these values does not exceed 0.06 % of the FEM values (Tab. 2).

Tab. 2: Results – comparison, average values

Spc.	Experiment/FEM analysis Smooth part [MPa] average	Experiment/FEM analysis Smooth part [MPa] worst	Experiment/FEM analysis Notch [MPa] average	Experiment/FEM analysis Notch [MPa] worst	Experiment/Analytics Smooth part [MPa] average	Experiment/Analytics Smooth part [MPa] worst
1	1,000082189	1,000497692	1,000197136	1,000327328	0,999914053	0.999498304
2	0,999997100	0,999910024	1,000048839	0,999973611	0,999995785	0,999915193
3	1,000022527	0,999791566	0,999403012	0,999402585	0,999977324	1,000208477
total	1.000033939	1.000497692	0.999882996	0.999402585	0.999962387	0.999498304

None of the measurements achieved the values resulting from the above mentioned Stress Concentration Factor. This, if fact, further supports relevance and reliability of data obtained by the described method.

Conclusions

Based on the above results, it appears that the aim less than 1% of the uncertainty of the strain gauge output is achievable, thus our method can actually be used to monitor the efficiency of big industrial gearboxes.

The results confirm that the use of structural and technological notches may bring the necessary "amplification" of the strain gauge signal.

With regard to small deviations found, we can also confirm applicability and feasibility of the FEM analysis as a method of virtual calibration of strain gauge set-outs.

It is also interesting to note that for stepped shafts with simple radiuses in steps loaded by torsion, the maximum mechanical stress is located at the transition of the smaller cylindrical part and the notch radius.

References

[1] F. Lopot, M. Dub, M. Janda, O. Berka, Use of notches in the shafts loaded by torque for improvement in accuracy of strain gauge measurements, EAN 2018 proceedings, Liberec, 2018.

[2] http://www.metalmax.cz/znacky.php (29. 4. 2019).

[3] HBM Strain gauges catalogue, available from (29. 4. 2019): https://www.hbm.cz/wp-content/uploads/S01265.pdf.

[4] W.D. Pilkey, Peterson's Stress Concentration Factors, 2nd edition. John Wiley & Sons, Inc., New York, 1997.

[5] https://www.efatigue.com/constantamplitude/stressconcentration (29. 4. 2019).

Crash test of a tow hitch for car trailers

LUFINKA Aleš[1,a]

[1]Department of the Design of Machine Elements and Mechanism; Faculty of Mechanical Engineering; Technical University of Liberec; Studentská 2, 46117 Liberec, Czech Republic

[a]ales.lufinka@tul.cz

Keywords: Tow hitch, Crash test, Test device

Abstract. Each new type of the tow hitch must be tested for its homologation. The standard test simulates only its loading by a classic car trailer but some manufacturers do other additional tests [1]. Simulating the crash of a tow bar to a fixed barrier (such as a garage wall) when reversing is one of them. The test device for these tests was built in the University's laboratory in collaboration with the tow bar manufacturer. The testing methodology is described in this paper.

Introduction

The tow hitch crash test simulates a situation where the driver crashes to a wall when reversing (see Fig. 1).

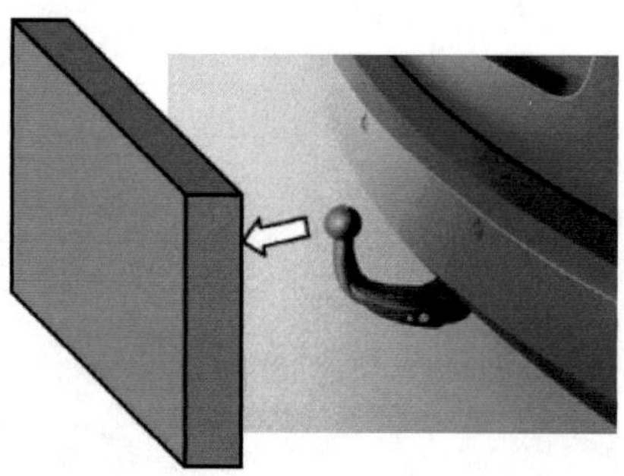

Fig. 1: The tow hitch crash test principle

The test is not carried out with a real car, the tow hitch is mounted on a special trolley that has the weight of the car. The trolley is accelerated to the prescribed speed and then in inertia movement it crashes to the barrier. It is clear that the reversing speed is not very high, the test parameters are based on this. Prescribed crash test parameters are shown in Table 1.

Table 1: Prescribed tow hitch crash test parameters

Parameter	Prescribed value	Tolerance
crash speed	4 km/h	+ 0.2 km/h
test trolley weight	identical to the car	±5 %
recorded quantities	real crash speed	-
	tow hitch beam deformation	-
	high-speed video in the horizontal direction	-
	high-speed video in the vertical direction	-

The new special testing device for this crash test was built in collaboration between the university and the towing device manufacturer. The device is located in the university lab and all crash tests are done there.

The test device design

The principle of the test device is shown in the Fig. 2.

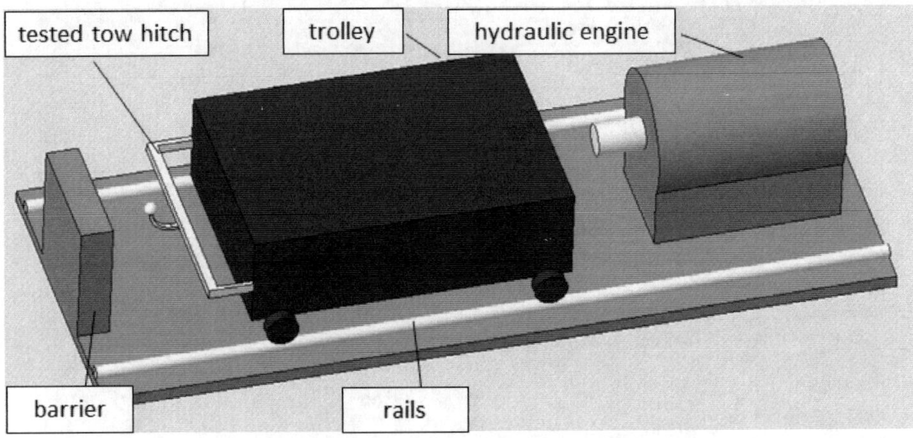

Fig. 2: The testing device principle

There are two rails on the base plate and the trolley with the attached tow hitch moves on them. The trolley is accelerated by a linear hydraulic motor. The barrier is attached to the base plate at the end of the rails. Two basic assumptions had to be fulfilled when designing the test device. Firstly, the weight of the trolley must be easily adjustable, as testing of the towing devices for vehicles of different weights is assumed. Secondly, the trolley must be able to accelerate to the prescribed speed with a hydraulic engine available in the laboratory. The required trolley weight is up to 2000 kg, the hydraulic motor has a piston rod stroke of 400 mm and its maximum force is 35 kN. The feasibility of the design was verified using the following simple equations for uniformly accelerated motion. The necessary acceleration to reach 4 km/h (i.e., 1.11 ms^{-1}) on the 400 mm track is:

$$a = \frac{1}{2} * \frac{v^2}{d} = \frac{1.11^2}{2*0.4} = 1.54\ [ms^{-2}] \tag{1}$$

The 35 kN hydraulic engine is capable of accelerating the weight 2000 kg to acceleration:

$$a = \frac{F}{m} = \frac{35000}{2000} = 17.5\ [ms^{-2}] \tag{2}$$

This simple calculation has shown that the required parameters are met with a large margin. So, it will not be a problem to keep them in a real test, where the conditions will be worse (trolley friction, engine start delay, etc.).

The real test device instrumentation

Two cylindrical rails are mounted on the base plate. A trolley moves on the tracks. Movement along a defined path and minimization of friction during movement is determined by the use of rails. The tested tow hitch is attached to the trolley in the same way as in the vehicle. The weight of the trolley is set using a steel sheet weight set. The crash barrier is attached to the base plate in front of the trolley, the hydraulic engine to the trolley acceleration is anchored to the base plate behind it (see Fig. 3). The hydraulic engine is controlled by a computer system so that the piston rod movement can be defined very precisely.

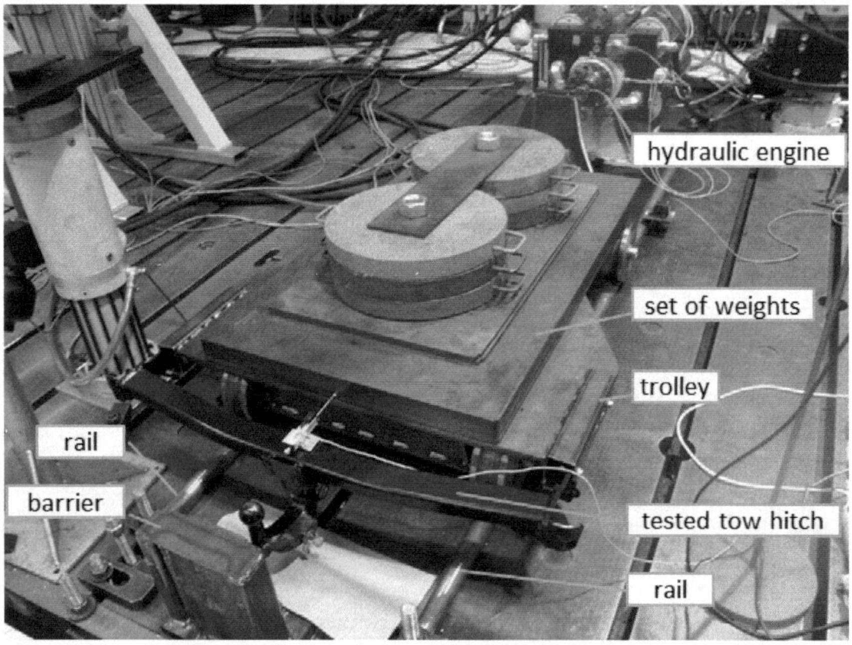

Fig. 3: The real testing device

The movement and speed of the truck should be monitored during the test. Both quantities are measured by the special cable-extension incremental transducer MT2E. This type allows measurements at very high accelerations and it is specifically designed for crash tests. The tow hitch deflection during crash is measured by the linear position sensor with restoring spring TR050. This sensor is designed for high acceleration and speed too (see Fig.4). Dewetron components [2] are used for measuring and recording signals and video. The measurement device DEWE5000 is the basis for the measuring system. Both sensors and the hydraulic engine piston rod position signal from the engine control system are connected to this measurement system. The sampling frequency is set to 2 kHz. Two DS CAM GiGE cameras capture the crash site at speed of 500 frames per second. The first is placed horizontally and the second one is vertical. The Dewe 5000 allows synchronous recording of camera images with measured signals. Unfortunately, it is too old and its hardware is not able to process the data streams from both cameras simultaneously. Therefore, only a horizontal camera is attached to it. The vertical camera is recorded to the separate measuring unit Dewesoft Sirius which is synchronized with the DEWE5000 and both records are linked together to one file after the test. The test device measurement and control system block scheme are shown in the Fig. 5.

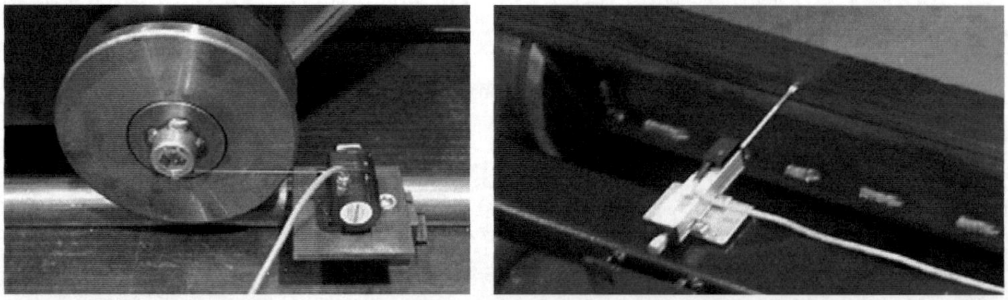

Fig. 4: The trolley position and speed sensor (left) and the beam deflection sensor (right)

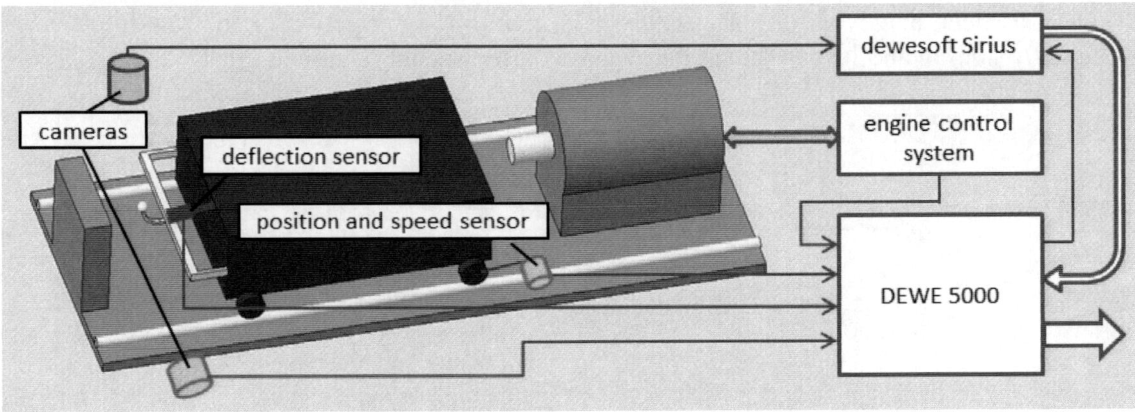

Fig. 5: The measurement and control system block scheme

The tow hitch crash test

The crash test is performed according to the following scheme. The trolley is accelerated by the hydraulic engine at the start of the movement. The cosine ramp movement of the piston is used to prevent the starting shock. When the prescribed speed is reached, the engine piston rod starts to brake and the trolley continues by its inertia to the obstacle. Of course, the starting speed must be slightly higher than the desired impact speed, because the trolley slows down due to the friction during inertia movement. The necessary speed increasing was determined by several attempts. Than the trolley crashes to the obstacle and bounces back. Meanwhile, the engine piston rod changes the direction of its movement and then it gradually brakes the returning trolley to prevent its further bouncing from the engine piston and its further impact to the obstacle. The engine control scheme is shown in the Fig. 6.

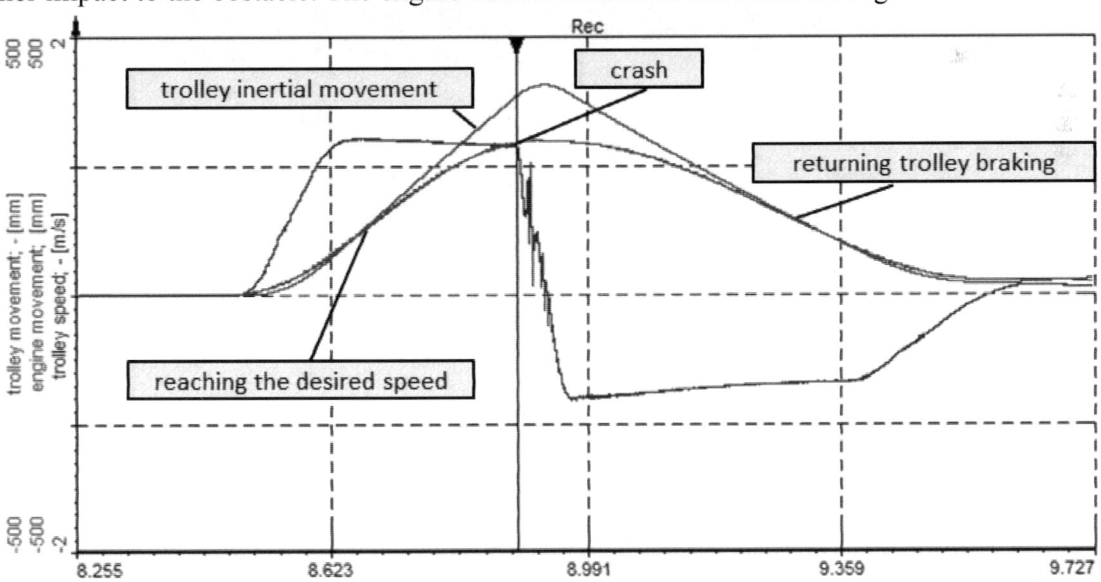

Fig. 6: The hydraulic engine control scheme

Crash test result example

All crash tests are done with prototypes of new tow hitches. It is therefore clear that the test results are the property of their manufacturer and cannot be published here. Therefore only the test result of tow hitch used to tune the motion of the hydraulic engine is used as an example of the test result in this article (see Fig.7). The graph shows that the prescribed crash speed of

4 km/h has been achieved very accurately. The maximum deflection of the tow hitch beam was nearly 20 mm and its permanent deformation after impact remained greater than 10 mm.

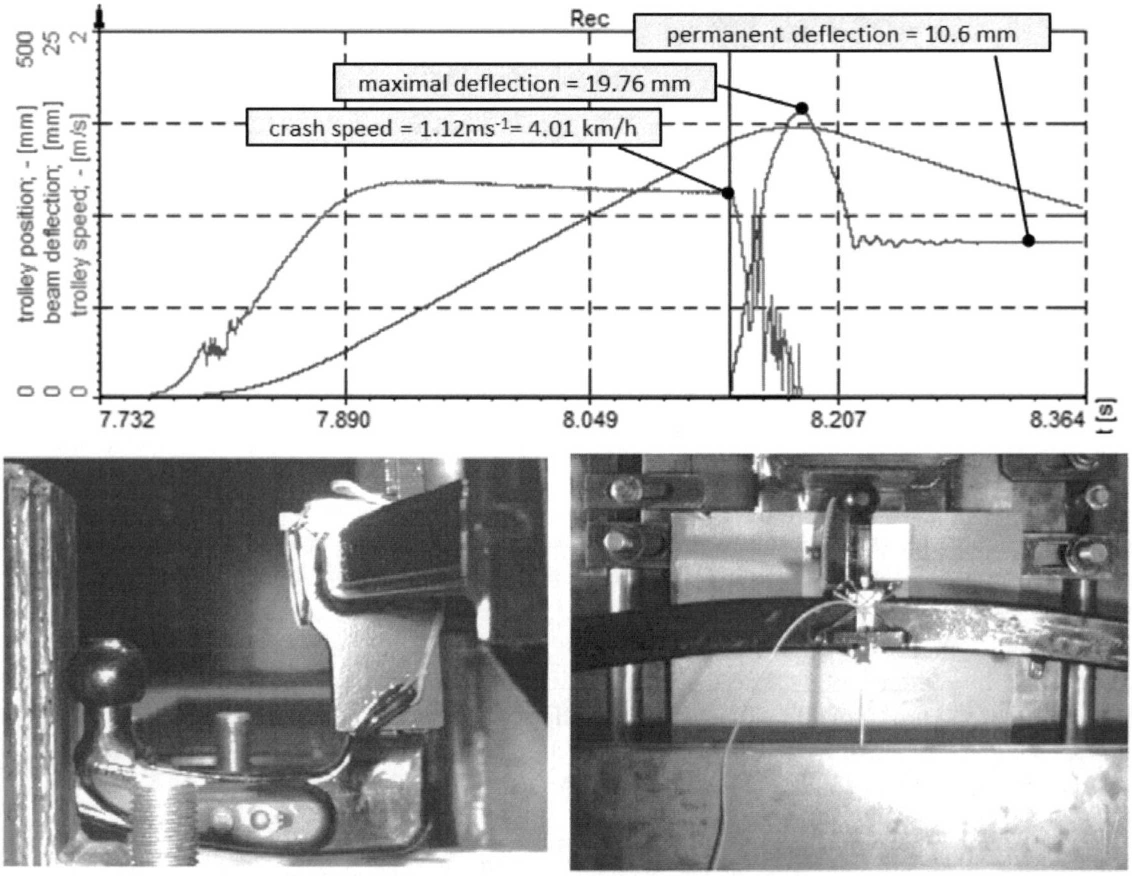

Fig. 7: The crash test result example

Conclusions

The test device built at the Technical University of Liberec, in collaboration with the tow hitch manufacturer, allows the testing of tow hitch prototypes. The results of crash tests are used to develop new types of tow hitches and the manufacturer can effectively compare their proposed parameters with the real properties. Plus the tow hitches safety during their use has been increased.

Acknowledgment

This publication was written at the Technical University of Liberec as part of the project "Innovation of the products, equipment and processes in engineering practice" with the support of the Specific University Research Grant, as provided by the Ministry of Education, Youth and Sports of the Czech Republic in the year 2019.

References

[1] A. Lufinka, Testing methodology of a tow hitch for car trailers, Book of Proceedings of the ICMD2016, (2016) 443-446, ISBN 978-80-261-0609-8.

[2] Dewetron measurement devices, Available at: www.dewetron.com/products.

EAN 2019
57th conference on experimental stress analysis

FEA of NSW structure to predict deflection during a large assembly

MALY Pavel[1,a], SINGH Shuvay[2,b] and VUILLERMET Raphael[1,c]

[1]ATLAS Experiment, CERN, CH-1211 Geneva 23, Switzerland

[2]University of Cape Town, Rondebosch 7701, South Africa

[a]pavel.maly@cern.ch, [b]shuvay.singh@cern.ch, [c]raphael.vuillermet@cern.ch

Keywords: FE Analysis, ATLAS detector, New Small Wheel, Support structure

Abstract. The article describes the application of FE analysis as a prediction of deflection, and a survey of measurement methods to control the real deflections during the assembly steps of New Small Wheel (NSW) support structure. The FEA method of additive mass and stiffness was used to simulate assembly steps and was therefore the important tool for planning the whole assembly procedure. To achieve the required detector performance, the geometry is controlled, the deflections are measured using survey techniques and after that compensated in specific assembly steps.

Introduction

The project of the NSWs consists of the replacement of the forward regions of the ATLAS detector and is part of the muon system upgrade due to the increase of the instantaneous luminosity in the next upgrades, [1], [2]. The ATLAS detector is one of the four major experiments of the LHC (Large Hadron Collider) at CERN (European Organization for Nuclear Research), and with the dimensions at 46 m long, 25 m in diameter and 7000 tons in weight is the largest particle detector, as seen in Figure 1.

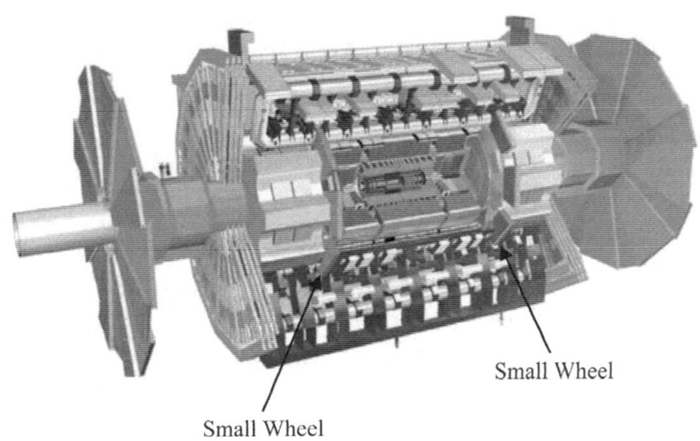

Fig. 1: ATLAS detector

The design of the NSWs and the choice of material was done to achieve optimal weight and strength of the NSW support structure to meet physics requirements. Because of the size, shape and weight of the structure, great attention is given to the assembly procedure of the NSW. In general, the analysis of structures using different methods is crucial in CERN, example can be found in [3].

Method

The mechanical study was used for dimensioning and calculating of the final deflection shape of the structure. The rigidity of the structure is permanently changing during each assembly step. Therefore, it was of interest to predict the influence of each assembly step on the final deformation so that it could be stepwise compensated. The compensation is done based on the analysis of the results from FE and measurement by lifting up the extremity of the hub with the crane while the feet spokes are used for the shimming.

Investigated structure. The analysis and measurement was carried out on two support structures of New Small Wheel. The NSW is a disk-shape configuration of two detector technologies, i.e. sectors, placed on the support structure that consist of the shielding disc (NJD), hub, spokes and alignment system. The whole NSW is approximately 11 m in diameter and 100 tons in weight.

NSW structure assembly procedure. The first step of the NSW assembly procedure is the assembly of NJD, i.e. shielding disc, that consists of internal and external segments joined together by means of pined and bolted connections. The second step consists of the assembly of the shielding hub which is the main supporting structure of the spokes. Then the spokes are assembled on the hub. Each detection chamber is attached to the spokes by means of adjustable kinematics mounts. The last steps of the assembly are the installation of the A-plate and hub extension. The main assembly steps are shown in Figure 2.

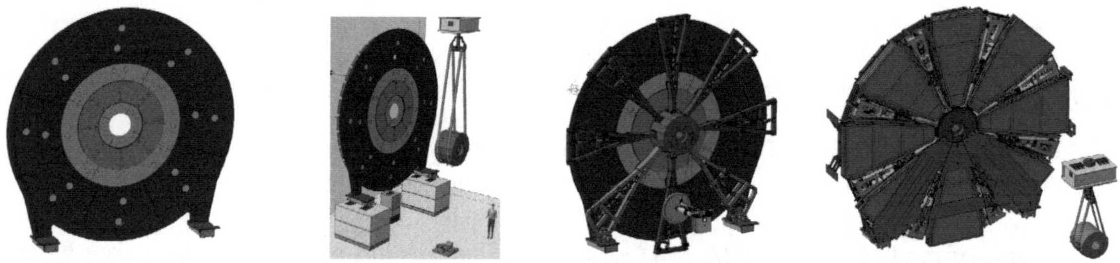

JD dics | HUB | Spokes | Sectors, A-frame, Hub extension

Fig. 2: NSW assembly steps

FE analysis

Specific FEA method was used to achieve the best possible information about the behaviour of the structure. Different loading states corresponding to the assembly steps were used for the simulations and to extract individual deflection contribution of each assembly step. Therefore, only the newly added components maintained the real density while the other components only retained the stiffness properties with a zero density. This method is the only one that can give the deformation and strength information of a structure during the different steps of the construction as its matrix of rigidity is changing at each step. From this method one can extract individual deflection contribution.

Simulation model. Based on the method one simulation model was used for each assembly step. The FE model of NSW was created as a deformable 3D solid model consisting of the components with material properties of steel or aluminium alloys. The example of the simulation model with the meshing and boundary conditions is in the figure 3. The used elements for meshing are 10-node tetrahedrons or 8-node hexahedrons with element size of approximately 50 mm. The boundary conditions are applied in poles that are fixed at the extremities and at the level of JD feet where one is remote displacement with free rotations and fixed translations and the other is longitudinal (vertical) with defined stiffness.

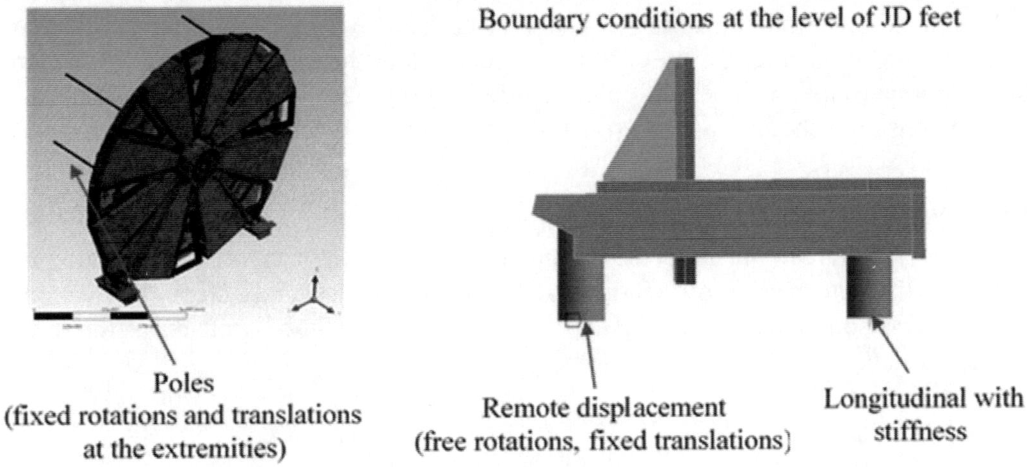

Fig. 3: Simulation model with boundary conditions

FE analysis results. Vertical deformation measured on the specific place on the hub was used for the subsequent analysis of simulation results. This corresponds to the place where the deflections are measured using survey techniques.

The loading states corresponding to the assembly steps are numbered from 1 to 9 are summarized together with the results in table 1. Hub deflection show the individual contribution to the deformation of each assembly step. Cumulative hub deflection gives information of the total increasing deformation during the assembly procedure.

Table 1: Loading states

Loading state	Description of the structure	Hub deflection [mm]	Cumulative hub deflection [mm]
1	JD + Poles	-1.06	-1.06
2	JD + Poles + Hub	-1.27	-2.33
3	JD + Poles + Hub + Small Sector Spokes (SSS)	-0.05	-2.38
4	JD + Poles + Hub + SSS + Small Sectors (SS)	-0.05	-2.44
5	JD + Poles + Hub + SSS + SS + Large Sector Spokes (LSS)	-0.08	-2.52
6	JD + Poles + Hub + SSS + SS + LSS + Large Sectors (LS)	-0.13	-2.65
7	JD + Poles + Hub + SSS + SS + LSS + LS + A-Plate	-0.02	-2.67
8	JD + Poles + Hub + SSS + SS + LSS + LS + A-Plate + Hub Extension	-0.11	-2.78
9	JD + Hub + SSS + SS + LSS + LS + A-Plate + Hub Extension	-0.42	

The results analysis assumes the compensation of both NSWs between the assembly steps 2 and 3. Using the specific FE method of additive mass and stiffness, the contribution of the deformation by the removing the poles (step 8 → step 9) is -0.32 mm and the total cumulative hub deflection is -3.09 mm. In order to achieve the needed value for the compensation of deflections by the hub lifting, the premise is considered that the deformation of the assembled JD disk was compensated with help of poles, i.e. the loading state 1 is used as a reference. Then the total compensation should be approximately 2 mm of hub lifting.

Compensation of the deflections

The schematic procedure of the deflection compensating is shown in figure 4. The main purpose is to eliminate the deformation of the final NSW during the assembly steps in advance. The compensation was conducted by pulling up the extremity of the hub by means of the overhead crane of the building.

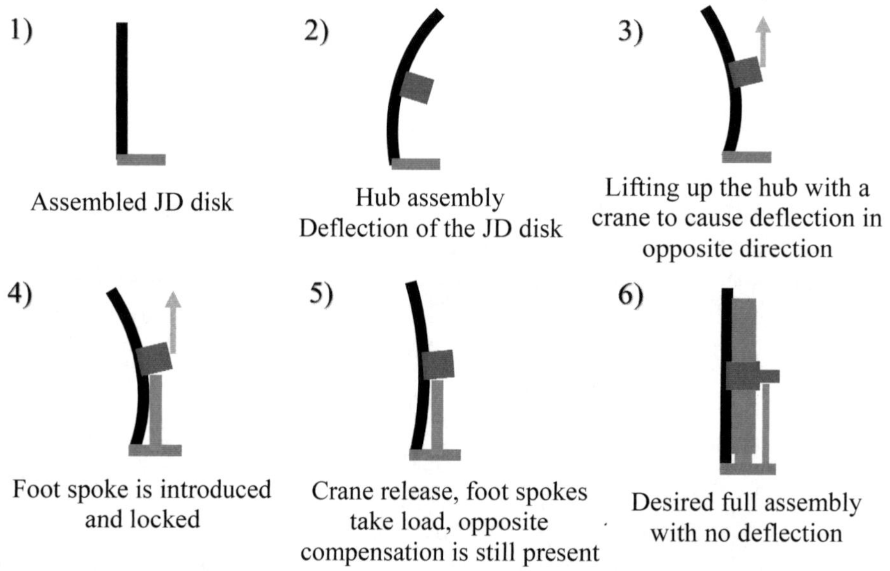

Fig. 4: Compensation of the deflections

Conclusions

The actual results and compensations of the deformation obtained from the ongoing NSW assembly process correspond to the requirements. It concludes that appropriate use of FEA methods together with precise control with survey techniques can be used to predict and deformations compensation during the assembly of large objects.

References

[1] B. Stelzer, The New Small Wheel Upgrade Project of the ATLAS Experiment, http://dx.doi.org/10.1016/j.nuclphysbps.2015.09.182.

[2] ATLAS Collaboration, New Small Wheel Technical Design Report, CERN-LHCC-2013-006.

[3] M. Rachac, M. Dub, M. Janda, P. Maly, J. Kamenicky, O. Berka, V. Dynybyl, P. Kubový, F. Lopot, Experimental analysis of welding tooling for ITk ATLAS, 55th Conference on Experimental Stress Analysis, pp. 620-624.

EAN 2019
57th conference on experimental stress analysis

Testing of honeycomb sandwich panel with potted insert

MANDYS Tomáš[1,a], LAŠ Vladislav[1], ZEMČÍK Robert[1] and KRYSTEK Jan[1]

[1] University of West Bohemia in Pilsen, NTIS – New Technologies for Information Society, Technická 8, 301 00 Pilsen, Czech Republic

[a]tmandys@kme.zcu.cz

Keywords: sandwich panel, honeycomb, insert, pull-out

Abstract

This work is focused on the experimental and analytical determination of the load-carrying capability of the inserts potted in the sandwich structure with honeycomb core. The pull-out test of the sandwich structure specimens with two types of inserts were performed and load-carrying capability of inserts determined and compared with the results obtained via analytical solutions. The parametrical study of influence of the parameters describing the potting geometry on load-carrying capability was performed in case of analytical solution.

Introduction

Sandwich panels with honeycomb core are widely used in different structural applications in lightweight transportation systems such as marine, aeronautic and astronautic industries because of their high specific strengths and stiffness.

A sandwich structure is made of two face sheets (skins) that are adhesively bonded to a lightweight core material. The combination of different bonded materials utilizes the properties of each separate component to the structural advantage of the whole assembly. The face sheets are usually fibre-reinforced composite laminates. Core material can be classified as being cellular, corrugated or honeycomb. Honeycomb cores are usually with hexagonal shaped cells made from aluminium or aramid. Honeycomb panel face skins are very thin which discourage direct mechanical attachment to other components. Therefore, the installation of special types of fasteners is required. Threaded inserts are universally employed to accept the attachment screws of various other components (brackets, holders, etc.).

The knowledge of the load capability of particular insert is important for the design of individual sandwich panels and for usage of particular type of inserts. The load capability is in case of insert not only dependent on the type and geometry of the insert, but also on the geometrical conditions associated with its installation (bonding) into the sandwich structure. The effect of the insert shape on the mechanical characteristics of composite sandwich panel and the load transfer characteristics are investigated experimentally in [1]. Besides the experimental investigation, the numerical simulation and the analytical calculation are other possibilities for determination of the load capability for particular insert in sandwich structure [2, 3, 4]. The theoretical background of the analytical solution of the load-carrying capability of the inserts is summarized in [2]. The comparison of the results of the numerical simulations and the performed experiments in case of pull-out loading is shown in [3, 4]. The parametric study investigating the effect of design variables such as core height and honeycomb core density and the face sheet thickness of the sandwich structure on the failure loads of the sandwich inserts is published in [5].

Analytical solution of insert load capability

Failure of through-the-thickness type inserts of sandwich plates usually occurs in the honeycomb core next to the interface between honeycomb and the potting. In this place the core shear stress τ_C in the honeycomb reaches the maximum value $\tau_{C\max} = \tau_C(r = b_P)$. The failure occurs when $\tau_{C\max}$ reaches the core shear strength $\tau_{C\text{crit}}$. Static load-carrying capability of through-the-thickness insert P_{crit} in a sandwich plate can be estimated by the formula [2]

$$P_{\text{crit}} = 2\pi \cdot b_P \cdot d \cdot \tau_{C\text{crit}}, \tag{1}$$

where d is distance between the face sheet middle surfaces and b_P effective potting radius. Distance d is defined

$$d = \frac{t_{top}}{2} + c + \frac{t_{bot}}{2}, \tag{2}$$

where c is the core height and t_{top}, t_{bot} are thicknesses of top and bottom face sheets.

Effective potting radius b_P is an analytical dimension describing the radial influence zone of the potting and it is defined as the average distance of the nearest single cell wall surrounding the potting from the center of the insert. The value of the effective potting radius b_P depends on the combination of the insert radius b_i and size of core cell (cell size) S_C and on the certain position of the insert center within the hexagonal cell. The value of the effective potting radius b_P can be determined by

$$b_P = \frac{1}{n}\sum b_n, \tag{3}$$

where b_n is the radius of the corresponding effective potting geometry based on the position of double cell walls and failing single cell walls between hexagonal core and potted area. The effective potting geometry is red marked on Fig. 1, where is evident its meaning. The minimum $b_{P\min}$ and the average effective potting radius $b_{P\text{avg}}$ can be given by the empirical formulae [2]

$$b_{P\min} = 0.93192 \cdot b_i + 0.874 \cdot S_C - 0.66151, \tag{4}$$

$$b_{P\text{avg}} = 1.002064 \cdot b_i + 0.940375 \cdot S_C - 0.7113. \tag{5}$$

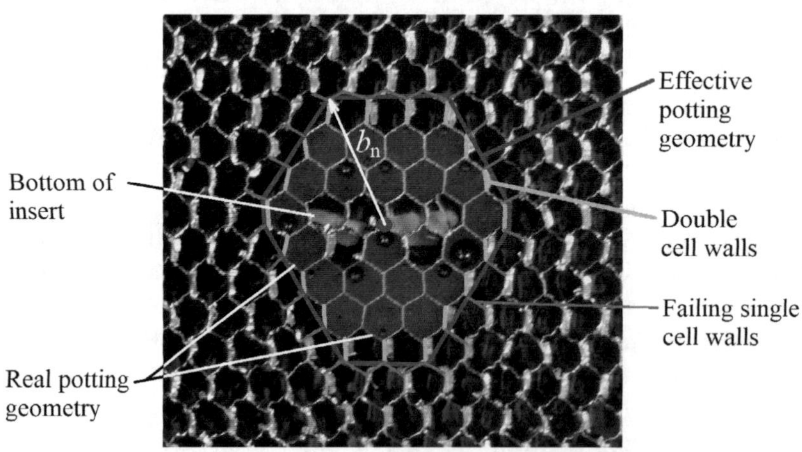

Fig. 1: Potting geometry of specimen with M8 insert

In case of potted type inserts (blind inserts) of sandwich plates the out of plane static load of potted insert P_P can be divided into three parts

$$P_P = P_f + P_S + P_n, \tag{6}$$

where P_f is load part carried by the upper face sheet, P_S load part carried by shear stresses in the core around potting and P_n load part carried by normal stresses in the honeycomb core underneath the potting material. Load part P_S is limited by the core shear strength $\tau_{C\,\mathrm{crit}}$ and P_n is limited by core strength $\sigma_{C\,\mathrm{crit}}$, that can be represented by tensile $\sigma^T_{C\,\mathrm{crit}}$ or compressive strength $\sigma^C_{C\,\mathrm{crit}}$, depending on the type of loading of partially potted insert. The shear rupture and tensile or compressive core rupture of honeycomb core in case of plane loading of insert occur usually simultaneously. The critical load $P_{P\,\mathrm{crit}}$, determining the load-carrying capability of potted insert in a sandwich plate, can be expressed in form [2]

$$P_{P\,\mathrm{crit}} = \frac{P_{\mathrm{crit}} - 2\pi \cdot b_P \cdot c \cdot \tau_{C\,\mathrm{crit}}}{2} + 2\pi \cdot b_P \cdot h_P \cdot \tau_{C\,\mathrm{crit}} + \pi \cdot b_P^2 \cdot \sigma_{C\,\mathrm{crit}} . \tag{7}$$

In Eq. (7) h_P is the potting height that represents average depth to which the core cells are filled with potting resin. The average value of potting height $h_{P\,\mathrm{avg}}$ can be determined empirically in the form [2]

$$h_{P\,\mathrm{avg}} = h_{P\,\mathrm{min}} + A \tanh\left(\frac{c - h_{P\,\mathrm{min}}}{h_{P\,\mathrm{min}}}\right) , \tag{8}$$

where $h_{P\,\mathrm{min}}$ denoted the minimal allowable potting height in the core that depends on the insert height h_i and the depth of the milled hole into sandwich structure (obviously 3-4 mm deeper than insert height). The meaning of the mentioned parameters is shown in Fig. 2. In Eq. (8) A is the geometrical parameter dependent on cell size S_C and that matches the size of one sixth of hexagonal cell circumference.

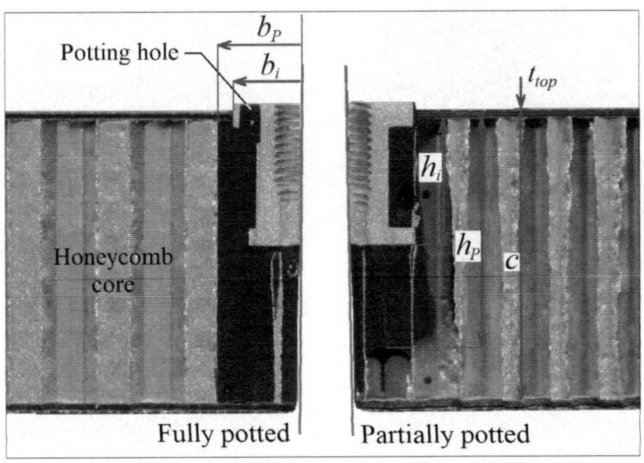

Fig. 2: Cross-section areas of specimens with fully and partially potted M5 inserts

Experimental determination of insert load capability

The blind inserts with inner threads M5 and M8 installed (bonded) into sandwich structure were subjected to out-of-plane tensile pull-out tests. The sandwich specimens of the total dimension of 110 mm × 110 mm and the total thickness of 30 mm consisted of the face sheets made from 8 layers of unidirectional carbon prepregs with orientation [0/45/90/–45]s and total thickness of 0.74 mm and honeycomb hexagonal-type core made from 5056 aluminium alloy with cell size 3/16" and single wall thickness of 0.001". The inserts were centrally located in selected face sheets of the specimens. The universal test machine Zwick-Roell Z050 was used. When the insert was loaded, the upper face sheet was pressed against the plate which has a central inner hole with diameter of 80 mm to ensure a sufficient free area around the insert. Six specimens for each insert type were tested. The loading speed was v = 1.0 mm/min. The testing setup is displayed in Fig. 3. The loading of specimens was carried out until visible damage. Force-displacement curves were the output from testing (see Fig. 4). Typical failure of the top and bottom face sheets of sandwich arising during the pull-out test is displayed on Fig. 5. Load capabilities were determined from aforementioned curves. In case of testing of specimens with M5 inserts, two failure modes of potted area (listed below) were observed depending on whether the inserts were only partially or fully potted. This fact is evident in the results of force-displacement curves, when the partially potted inserts reach the lower forces.

The occurrence of failure in case of partially potted M5 insert is underneath the potting material in the honeycomb core. This is the case of core strength $\sigma^T_{C\,\mathrm{crit}}$. The failure in case of fully potted M5 insert occurs at the interface of adhesive (potting) and bottom face sheet – corresponding strength $\sigma^T_{F\,\mathrm{crit}}$. The strength $\sigma^T_{F\,\mathrm{crit}}$ was additionally obtained from flatwise tension test of sandwich structure performed according to standard [6].

Fig. 3: Tensile pull-out fixture setup

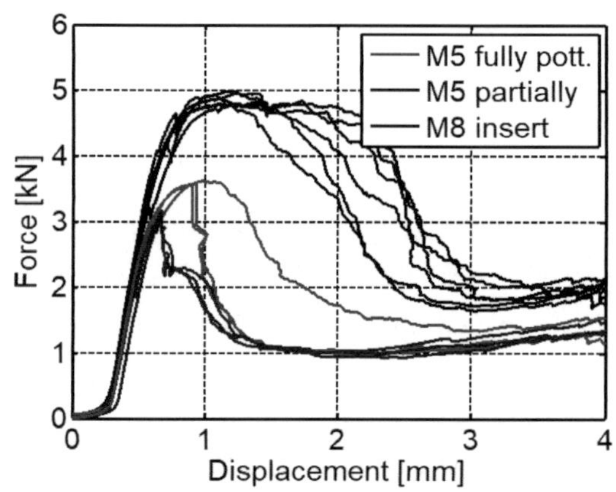

Fig. 4: Force-displacement dependencies

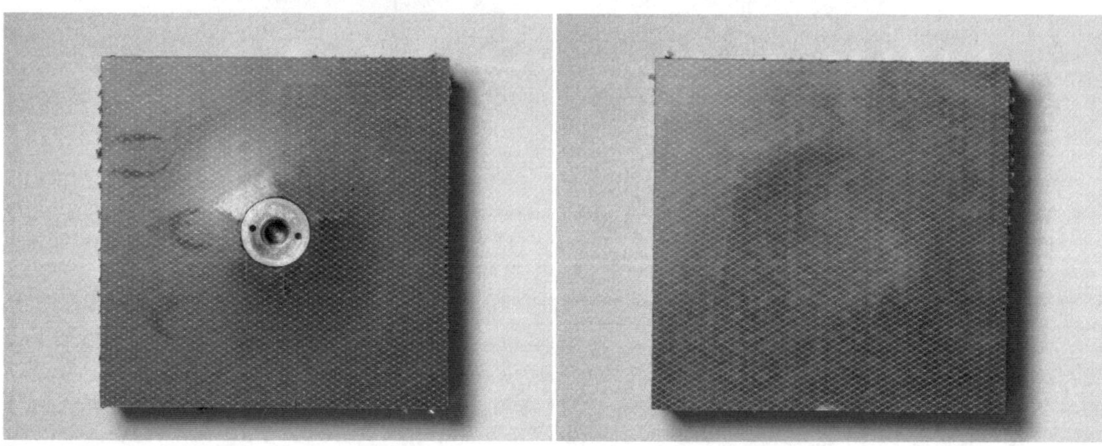
Fig. 5: Typical failure of tested specimens on top skin (left) and bottom skin (right), presented specimen with blind insert M8

Comparison of experimental and analytical results

The minimum and the average load capability of appropriated tested inserts in sandwich structure were analytically calculated according to the above equations. Analytical solution was used for both cases of potted area of M5 inserts detected during the experiment. The material and geometrical parameters needed for the calculation were obtained by the literacy [2, 7], by experiment and by the measuring of the specimens and are summarized in Table 1. In case of core strength were used two different values in dependence of type of failure caused due to occurrence of two different potting types (partially and fully potting) in case of M5 insert. The comparison of experimentally and analytically obtained load capabilities of inserts is performed in Table 2.

Table 1: Geometrical and material parameters

Sandwich structure
t_{top}, t_{bot} = 0.738 mm, c = 28.52 mm, S_C = 4.763 mm, A = 5.0 mm
$\tau_{C\,\mathrm{crit}}$ = 1.07 MPa, $\sigma^T_{C\,\mathrm{crit}}$ = 5.85 MPa, $\sigma^T_{F\,\mathrm{crit}}$ = 2.52 MPa
Insert
M5: b_i = 7.15 mm, h_i = 12.60 mm
M8: b_i = 10.70 mm, h_i = 19.60 mm

Table 2: The comparison of experimentally and analytically obtained load capabilities P_P

Type of insert and installation	Experiment [N]	Analytical solution min/avg. value [N]
Blind M5 partially potted	3099	3098 / 3913
Blind M5 fully potted	3589	3501 / 3831
Blind M8 fully potted	4889	4993 / 5486

The parametric study of influence of parameters describing the potting geometry, namely potting height h_P and potting radius b_P, on load-carrying capability of insert was performed. The potting height h_P was limited by the insert height h_i and by core height c. The upper value of insert radius b_i for calculation of potting radius b_P was considered to be additional 50% of its dimension. In this case, height h_i and radius b_i are the dimensions of the milled hole for the insert installation that may not be the same as the inserts dimensions. The graphical representation of influence of resultant potting height h_P and potting radius b_P on the on load-carrying capability of M5 and M8 inserts is shown in Fig. 6. The influence of the widening of the insert radius b_i on the load-carrying capability is summarized for case of fully potted inserts in Table 3.

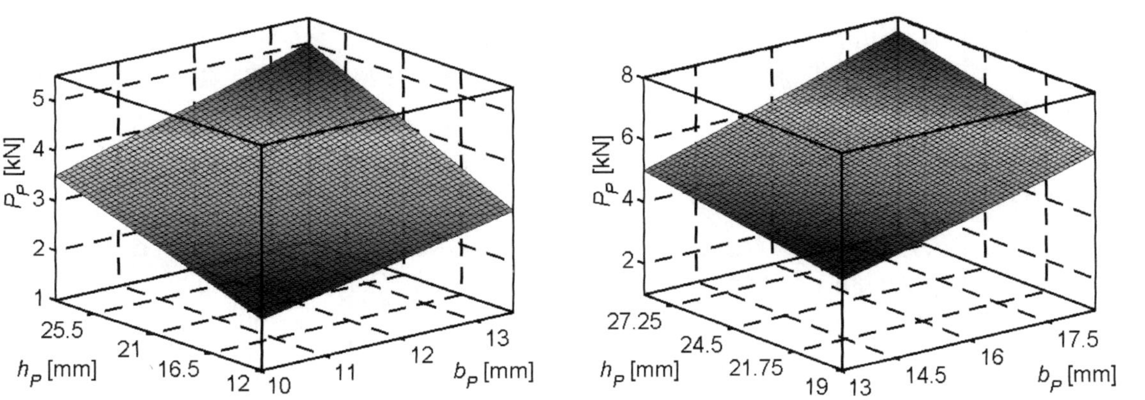

Fig. 6: Influence of the potting geometry on the load carrying capability of the blind insert with M5 thread (left) and M8 thread (right)

Table 3: The comparison of the dependency of insert radius on the load capability of the insert

Type of insert and installation	h_i [mm]	b_i [mm]	h_P [mm]	b_P [mm]	P_P [N]
Blind M5 fully potted	28.52	7.15	28.52	10.16	3501
		10.72		13.50	5004
Blind M8 fully potted	28.52	10.70	28.52	13.47	4993
		16.05		18.46	7569

Conclusions

The load capability of insert in sandwich structure was experimentally and analytically determined and the results were compared. The experimental data show a reasonable agreement with minimum values of load capabilities obtained from analytical solution. Based on this observation, a parametric study of influence of potting geometry on the load capability of insert was performed using analytical solution. This geometry depends on dimensions of the milled hole in form of radius and height for the insert installation in the sandwich core that may be different from the insert geometry. The load carrying capability of the inserts can be increased by the widening of the potting radius in the honeycomb core. The other possibility of increasing load carrying capability depends on the potting height and from this derived the partial or full potting, when the fully potted inserts have the higher load carrying capability than the partially.

Acknowledgement

This publication was supported by the project LO1506 of the Czech Ministry of Education, Youth and Sports under the program NPU I and by project TRIO FV10400 "Development of processes applicable to development and production of components for the space industry" of Czech Ministry of Industry and Trade.

References

[1] B.J. Kin, D.G. Lee, Characteristics of joining inserts for composite sandwich panels, Composite Structures, 86, pp. 55-60, 2008.

[2] Space Engineering – Insert design handbook, ECSS-E-HB-32-22A, ESA Requirements and Standards Division, Netherlands, 2011.

[3] R. Roy, K.H. Nguyen, Y.B. Park, J.H. Kweon, J.H. Choi, Testing and modelling of Nomex^{TM} honeycomb sandwich panels with bolt insert, Composites: Part B, 56, pp. 762-799, 2014.

[4] S. Heimbs, M. Pein, Failure bahaviour of honeycomb sandwich corner joints and inserts, Composite Structures, 89, pp. 575-588, 2009.

[5] K.I. Song, J.Y. Choi, J.H. Kweon, J.H. Choi, K.S. Kim, An experimental study of the insert joint strength of composite sandwich structure, Composite Structures, 86, pp. 107-113, 2008.

[6] Standard ASTM C297, Test Method for Flatwise Tensile Strength of Sandwich Constructions.

[7] E.W. Kuenzi, Mechanical properties of aluminium honeycomb cores, Forest Products Laboratory, U.S. Department of Agriculture, Forest Services, 1962.

EAN 2019
57th conference on experimental stress analysis

Using technical seismicity to identify some modal characteristics of a bridge

MELCER Jozef[1,a]

[1]Department of Structural Mechanics and Applied Mathematics, Faculty of Civil Engineering; University of Žilina; Univerzitná 8215/1, 010 26 Žilina, Slovak Republic

[a]jozef.melcer@fstav.uniza.sk

Keywords: Modal characteristics, Technical seismicity, Experimental test, Numerical modelling, Bridge structure

Abstract. There are characteristics which define the dynamical individuality of every structure from dynamical point of view. The knowledge of these characteristics is important especially in case of bridge structures. The presented paper is dedicated to the numerical simulation and experimental validation of some modal characteristics of a bridge structure. The random bridge response on the excitation due to technical seismicity from rail transport is analysed. The results of analyses of vibration records offer values suitable for verifying numerical procedures.

Introduction

Natural frequencies and natural modes represents the basic characteristics defining the dynamical individuality of all dynamical systems [1, 2]. In bridge construction, it is good to know the values of at least the basic natural frequencies [3]. These characteristics can be obtained by numerical or by experimental way. Currently, it is effective to use the FEM computational model and the required characteristics to obtain numerically [4]. However, the experiment is the only way to verify numerically obtained results. Therefore, the role of experiment in mechanics is indispensable. The present article describes one of the possibilities of experimental verification of some natural frequencies of a bridge.

Numerical solution

The subject of the analysis is the bridge construction with three fields. Each field acts statically as a single supported beam with a span of 29 m. Bridge width is 11.50 m. The bridge is made of prefabricated elements I73, Fig. 1.

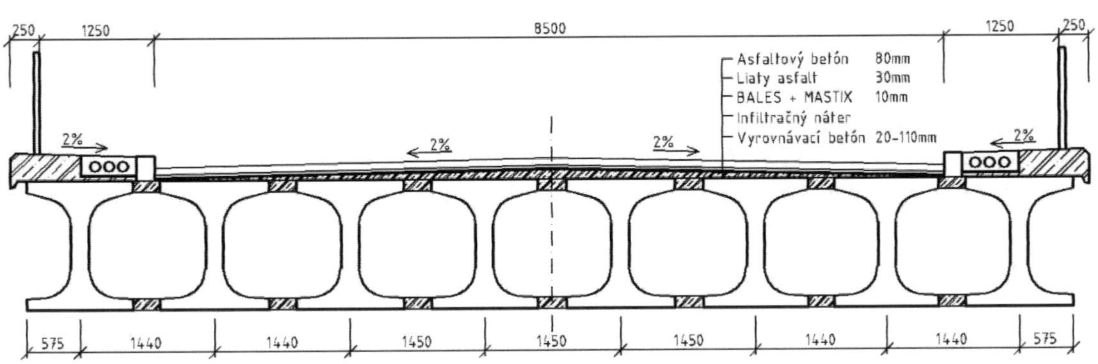

Fig. 1: Cross section of the analysed bridge

The FEM computational model using the deck-wall elements was created for the bridge. A quadrangular, eight nodal, Mindlinian element is used [4]. The element has nodes in the quadrilateral corners and at the centre of the sides. The centre nodes have like unknowns the components of displacement vector $\{\mathbf{u}\} = [u, v, w]^T$, the corner nodes have like unknowns the components of displacement vector $\{\mathbf{u}\} = [u, v, w]^T$ and the components of rotation vector $\{\boldsymbol{\varphi}\} = [\varphi_x, \varphi_y, \varphi_z]^T$. The model had 812 finite elements. On this computational model, natural frequencies and natural modes were calculated. As an example, Fig. 2 shows the first four natural modes and the corresponding natural frequencies.

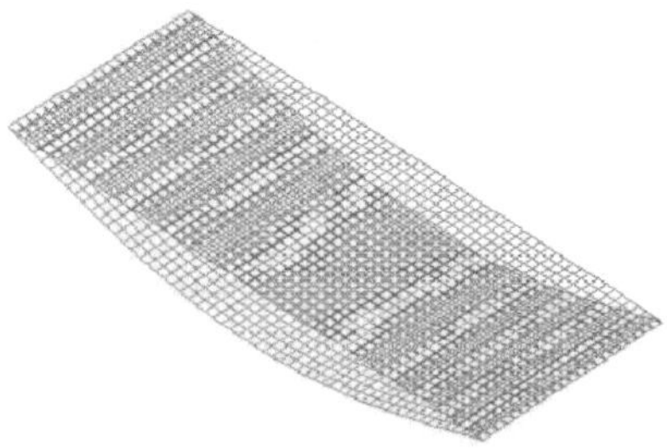

$j = 1$, $f_{(1)} = 4{,}04$ Hz

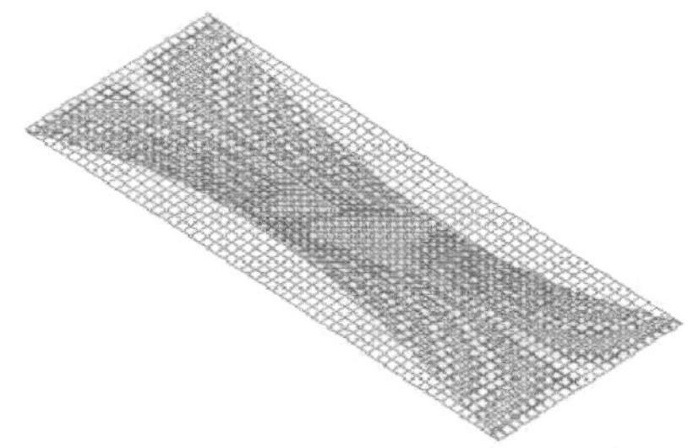

$j = 2$, $f_{(2)} = 10{,}59$ Hz

$j = 3$, $f_{(3)} = 14{,}89$ Hz

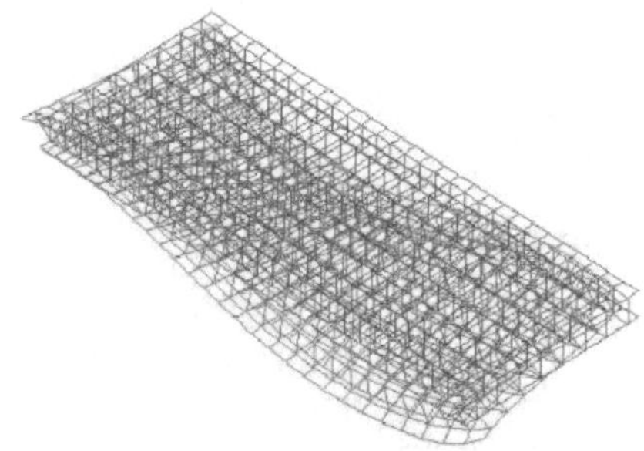

$j = 4$, $f_{(4)} = 19{,}44$ Hz

Fig. 2: The first four natural modes and the corresponding natural frequencies

Experimental test

To verify the results of numerical calculation the experimental test on the bridge structure was carried out. As a source of kinematic excitation, subsoil vibration caused by the passage of trains under the first bridge field was used. It is the analysis of the bridge's response to technical seismicity induced by rail transport. Acceleration sensorsVB12VD 150040 from MMF (Metra Mess Frerquenztechnik) were used to measure bridge vibration. The sensors operate linearly in the frequency range 0.15 - 260 Hz. The signal is amplified by the charge amplifier M68 (also from MMF), Fig. 3. The amplifier serves as an integrator and low-pass filter. The voltage signal is digitized in an analogue - digital converter and it inputs to the computer in digital form. The NI 9215 converter from the firm National Instruments was used, Fig. 4. The signal is stored and further processed by the computer, Fig. 5. DYSIS system was used to analyse the measured signals. The second bridge field was measured. The sensors were placed on the sidewalk in half and a quarter of the span, Fig. 5. The view of the tested bridge and the measuring centre located under the bridge is shown in Fig. 6 Comparison of numerically and experimentally obtain natural frequencies can be seen in Table 1.

Fig. 3: Sensor VB12VD 150040 and charge amplifier M68

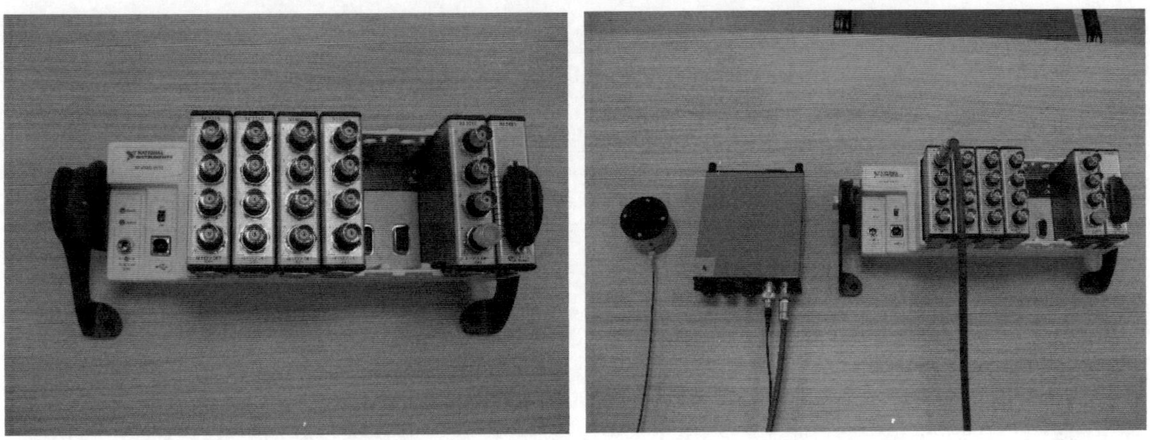

Fig. 4: A/D converter and complete assembly

Fig. 5: Final stage – computer and placing sensors on the bridge

Fig. 6: View of the tested bridge and the measuring centre

Table 1: Comparison of numerically and experimentally obtain natural frequencies

Natural frequencies $f_{(j)}$ in [Hz]			
j	FEM model	Experiment	Difference Δf in [%]
1st bending	4.04	4.12	- 1.94
1st torsional	10.59	9.15	+ 15.74
2nd bending	14.89	14.29	+ 4.20
2nd torsional	19.44	19.38	+ 0.31

Conclusions

In the case of bridge structures, experimental verification of numerical results is needed. Some modal characteristics of the bridge can be verified, for example, by analysing the random vibration induced by the technical seismicity. As the results of experimental measurements show the harmony between calculation and experiment is good. The only difference is in the 1st torsional mode of the vibration. The reasons may be more. Their analysis goes beyond the possibilities of this article.

References

[1] D. Kuchárová, J. Melcer, Dynamic of structures (in Slovak), first ed., EDIS, University of Žilina, 2000.

[2] Jimin He, Zhi-Fang Fu, Modal analysis, first ed., Butterworth-Heinemann, Oxford, 2001.

[3] J. Melcer, et. all, Dynamic of transport structures (in Slovak), first ed., EDIS, University of Žilina, 2016.

[4] I. Němec et al., Finite element analysis of structures, principles and praxis, Shaker Verlag, Aachen, 2010.

EAN 2019
57th conference on experimental stress analysis

Different types of damages of glued laminated timber beams

MELZEROVÁ Lenka[1,a] and ŠMÍDOVÁ Eliška[1,b]

[1]CTU in Prague, Faculty of Civil Engineering; Department of Mechanics; Thákurova 7, 16629, Prague 6, Czech Republic

[a]melzerov@fsv.cvut.cz, [b]eliska.smidova@fsv.cvut.cz

Keywords: Glued Laminated Timber, Two Types of Damages, Four-Point Bending Tests, Finger Joints

Abstract. Twenty-six glued laminated timber beams were tested in four-point bending until failure. Two types of damages have been identified. The first type of failure is a longitudinal crack. This crack usually begins at the point of knot or other wood defects always in the tensile area. The second, more frequent type of failure is the destruction of the finger joint, which is located at the lower edge of the beam in its middle third.

Introduction

Twenty-six glued laminated timber beams were tested in the laboratory of the Faculty of Civil Engineering. Beams have been tested in several series since 2005. Two types of damages have been identified. This work complements research conducted at the other universities with beams of other dimensions. One of the papers on this subject was published by Erik Serrano [1,2]. The investigated beams had cross-sectional dimensions of 0.1x0.32 m and were composed of eight 4 cm lamellas. The beam length was 4.5 m and the support spacing was 4.2 m. The beams were subjected to a destructive four-point bending tests with loads in thirds of the span.

Types of damages

Two detected types of glued laminated timber beams failure are as follows. The first type of failure is a longitudinal crack. This crack usually begins at the point of knot or other wood defects always in the tensile area (Fig. 1). Cracks have always propagated through individual laminations and their transition from one ply to the other was smooth regardless of the horizontal glued surfaces.

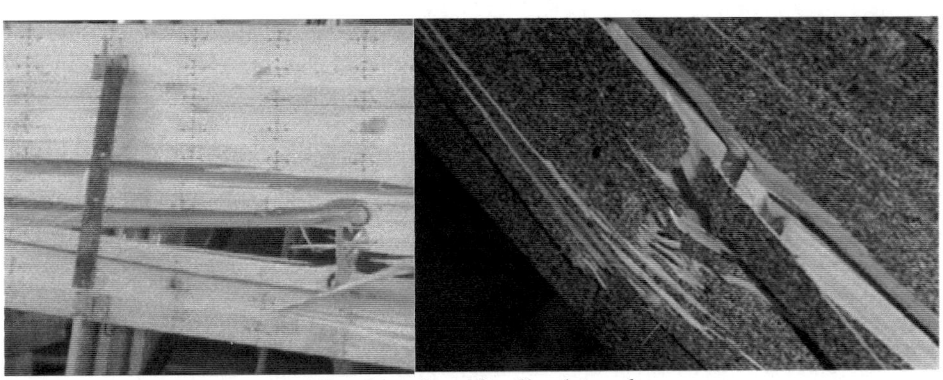

Fig. 1: A longitudinal crack

This type of beam failure is exceptional. Large knots are removed during the production of glued laminated timber. The load needed for its creation is extreme. The second, more frequent type of failure is the destruction of the finger joint, which is located at the lower edge of the beam in its middle third (see Fig. 2).

Fig. 2: Destruction of the finger joint

Both types of damages were obtained by the evaluation of the camera recording made during the bending tests for six beams. For the previous twenty beams, a plot of cracks after collapse of the structure is available. In these drawings, cracks are seen through the finger joints. This is the case with thirteen beams. The collapse of these beams occurred at a lower load that corresponds to this failure mechanism [3,4]. The following table (tab. 1) shows the distribution a beam numbers from the presented research program according to their failure, together with the load at the moment of collapse. The load here is the sum of the two loads, in thirds of the span at the four-point bend (see Fig. 3).

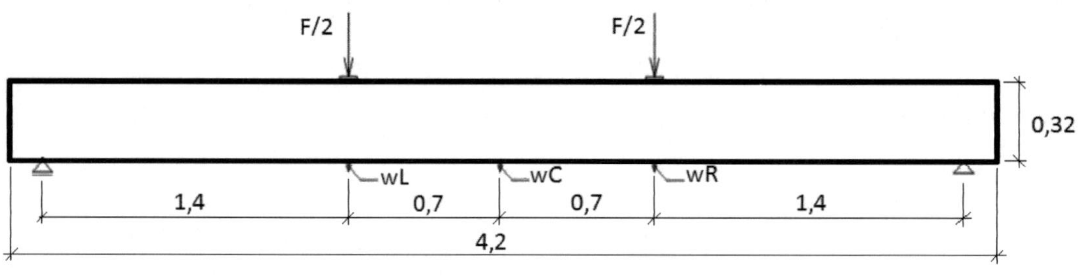

Fig. 3: Example of tested beam

Due to the existence finger joints in the lower lamella, together with a relatively low load level during the collapse of the beam, it is possible to assume damage to the other seven beams through the finger joints.

Table 1: Numbers of beams both damages types

Type of damage	Number of beams	Average loading [kN]
Out of joint	2	138
Through joint	17	72
Unknown	7	85

Further information, which are necessary for the most accurate assessment of beams damage are the results of non-destructive tests. In principle, these tests measure the density of wood. The measured density of wood can also be used to calculate the distribution of modulus of elasticity and wood strength. Non-destructive tests were performed on all 26 beams.

Owing to a considerable heterogeneity of laminated timber structures a large number of local measurements is needed. At present, only one such experimental method, which builds upon driving an indenter with the help of Pilodyn 6J device in into the wood, is available. In particular, a spike 2.5 mm in diameter is shot into the wood with the enforced energy of 6 J. The local elastic modulus in the fiber direction is then evaluated empirically based on the depth of indentation as,

$$E = -564{,}1 \times t_p + 19367, \quad (1)$$

where E is the searched Young's modulus in MPa and t_p is the measured indentation depth in mm.

Research has shown that material with damage, localized at critical finger joints, could be used to model the damage of glued laminated timber beams. The degree of this damage could be determined from the beams presented.

The damage caused by the oblique cracks is always initiated by the knot or other damage to the wood. The knot area ratio (KAR) is significant for this type of damage, which is part of existing research following the well-known Karlsruhe model.

Conclusions

The presented article is the basis for the creation for the detailed models of the failure of glued laminated timber beams. The main goal of the paper is to evaluate the adequacy of available experimentally determined data. It is also necessary to put research into the context of previously published works.

Acknowledgement

The support by the SGS project No. SGS17/168/OHK1/3T/11 is gratefully acknowledged.

References

[1] E. Serrano, Rational Modelling and Design in Timber Engineering Applications Using Fracture Mechanics, in: WCTE 2016 CD-ROM Proceedings, Vienna University of Technology, 2016.

[2] E. Serrano, P.J. Gustafsson, H.J. Larsen: Modeling of Finger-Joint Failure in Glued-Laminated Timber Beams, Journal of Structural Engineering, 127, (2001).

[3] P. Kuklík, L. Melzerová, Kompozitní materiály na bázi dřeva. ČVUT, Praha, 2011.

[4] L. Melzerová, Studie přetváření a porušování nosníků z lepeného lamelového dřeva. ČVUT, Praha, 2018.

[5] L. Melzerová, M. Šejnoha, Results of Penetration Tests Performed on Timber GLT Beams, in: World Conference on Timber Engineering, Proceedings, 49-50, (2014).

[6] L. Melzerová, M. Šejnoha, Influence of Distribution of Finger Joints and Timber Flaws on the Damage Evolution of Laminated Glued Timber Beams during Four Point Bending, in: WCTE 2016 CD-ROM Proceedings, Vienna University of Technology, 2016.

On a possible use of neutron three axis diffractometer for studies of elastic and plastic deformation of polycrystalline materials

MIKULA Pavol[1,a] RYUKHTIN Vasil[1,b] and ROGANTE Massimo[2,c]

[1]Nuclear Physics Institute ASCR, v.v.i., 25068 Řež, Czech Republic

[2]Rogante Engineering Office, I-62012 Civitanova Marche, Italy

[a]mikula@ujf.cas.cz, [b]ryukhtin@ujf.cas.cz, cmain@roganteengineering.it

Keywords: Neutron diffraction, instrumentation, strain measurements

Abstract

Feasibility of using a high-resolution three axis neutron diffractometer performance for elastic and plastic deformation studies of metallic polycrystalline samples is presented. The method consists of unconventional set up employing bent perfect crystal (BPC) monochromator and analyzer with a polycrystalline sample in between (see Fig. 1). After the realization of focusing conditions in real and momentum space at the neutron wavelength of 0.162 nm, a high angular resolution up to $FWHM(\Delta d/d)$=2x10^{-3} was achieved on the standard α-Fe(110) sample (2 mm diameter) which then opened the possibility for the measurements of small lattice parameter changes of samples. The feasibility of the instrument for macro- and micro-strain as well as grain size studies is demonstrated on the polycrystalline samples of low carbon shear deformed steel wires and the NiTi plates subjected to heat treatment.

Introduction

At present, the investigations of residual strains/stresses are usually carried out at the dedicated double axis diffractometers (strain scanners) with a BPC focusing monochromator situated on the first axis, a sample situated on the second axis and with a position sensitive detector (PSD) [1-5]. However, the $\Delta d/d$ resolution of these dedicated scanners derived from the *FWHM* of the diffraction lines is sufficiently high for small sample gauge volumes (when the width of the irradiated part of the sample is about 2 mm or less) but rarely better than 8x10^{-3} for bulk samples and therefore they are used for the measurements of elastic strain

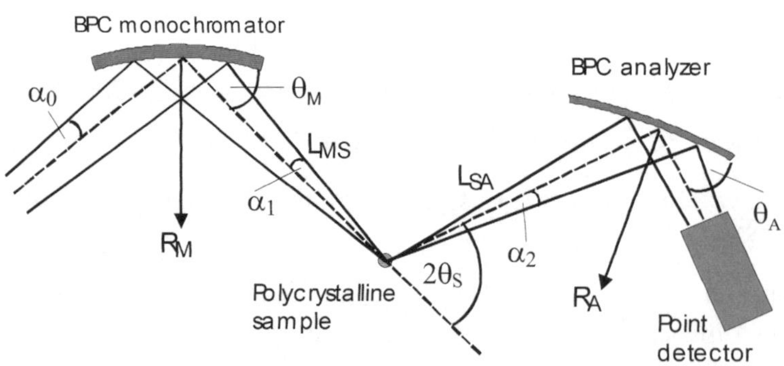

Fig. 1: Three axis diffractometer setting employing BPC monochromator and analyzer as used in the experiment (R_M, R_A - radii of curvature, θ_M, θ_A - Bragg angles).

effects resulting in the changes of lattice constants and angular shifts of the diffraction lines. For the investigation of micro-strain effects resulting in a change of the *FWHM* and shape of diffraction profiles a substantially higher resolution is required which can be achieved just by the proposed 3-axis diffraction set-up.

Experimental details

The set-up shown in Fig. 1 was realized on the 3-axis neutron diffractometer installed at the Řež research reactor LVR-15. Si(111)-monochromator and Si(400)-analyzer (or Si(311)-analyzer) single crystals had the dimensions of 200x40x4 mm^3 and 20x40x1.3 mm^3 (length x width x thickness), respectively. The monochromator Si(111) providing the neutron wavelength of 0.162 nm had a fixed curvature with a radius R_M of about 12 m. The curvature of the analyzer was R_A= 9 m, where the best resolution was found. For a practical demonstration of the feasibility of using the three axis set-up we used two types of polycrystalline samples: First, non-deformed as well as deformed α-Fe(110) wires (low-alloyed steel) of the diameter of about 5 mm with accumulated shear deformation as a result of rolling with the shear of the metal ingot and conventional wire drawing (see Tab. 1 and 2). The fourth column in Tab. 1 shows the reduction degree in drawing deformation. Then, we used 3 Ni(50%)Ti(50%) alloy samples (untreated, treated at 900 °C for 1.5 hour and treated at 900 °C for 3 hours). They were in the form of discs of the diameter of 25 mm and the thickness of 5 mm.

Table 1: Description of deformation of the samples.

Sample number	ϕ [mm]	Shear def. [%]	Drawing def. [%]
1	5.10	0	0
2	4.28	8	23.2
3	5.35	0	0
4	4.28	16.6	23.2
5	5.57	0	0
6	4.28	23	23.2

Table 2: Chemical composition of low-alloyed structural steel grade 08G2S GOST 1050 element

Element	C	Mn	Si	S	P	Cr	Ni	Cu	N2
wt %	0,08	1,87	0,82	0,020	0,022	0,02	0,02	0,02	0,007

Low-carbon shear deformed steel wires

In this case, a Si(400) bent perfect slab was used as the analyzer. The description of the non-deformed as well as accumulated shear deformed samples is shown in Table 1. Table 2

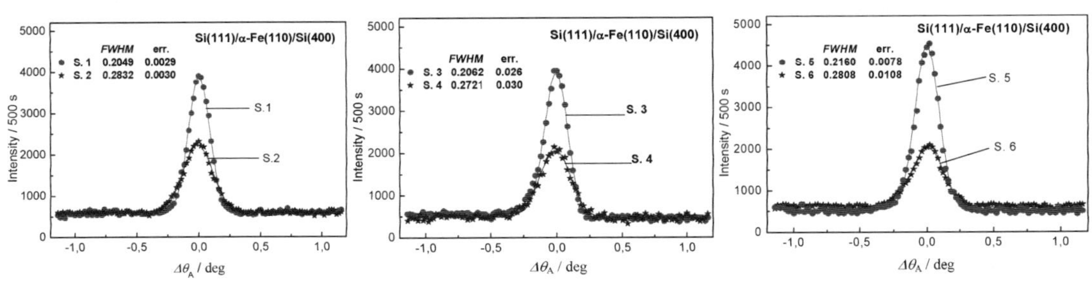

Fig. 2: An example of diffraction profiles related to the samples S.1-S.6 situated in the vertical position as analyzed by BPC Si(400) analyzer (strain measurements of the radial components).

describes the chemical composition of the steel. For the measurements of the radial component of strains the steel samples were situated on the second axis of the diffractometer in the vertical position. The width of the incident beam was 8 mm and therefore, the whole volume of the sample was irradiated (within the whole diameter). The height of the incident beam was 20 mm. The obtained results of the analyzer rocking curves are shown in Fig. 2. From the introduced individual profiles (see Fig. 2), we can detect the following features: the peak intensities and *FWHM*s related to the deformed samples differ very little. This is brought about by the fact that the diameter of the deformed samples were equal and that the shear deformation had a negligible effect. Very close values of *FWHM* point out the fact that the lattice deformation in the radial direction was basically brought about by drawing deformation. The integrated intensity under the peak profiles related to the non-deformed samples S.1, S.3 and S.5 primarily corresponds to the irradiated volume of the sample, which is, of course, maximum for the sample S. 5.
In the next step the steel samples were situated on the second axis of the diffractometer in the horizontal position. The obtained results are shown in Fig. 3. In comparison with the previous

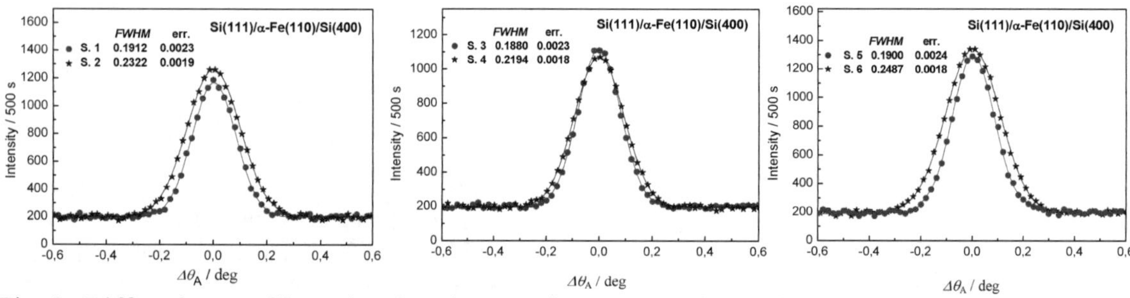

Fig. 3. Diffraction profiles related to the samples S.1-S.6 situated in the horizontal position for measurement of the axial component as analyzed by the bent perfect Si(400) analyzer.

case, for the samples in the horizontal position their irradiated volume is much smaller and correspondingly neutron signal was smaller. In this case, it was found out that the resolution was dependent on the width of the incident beam impinging the sample and therefore, we used the slit width of 5 mm. It can be seen from Fig. 3 that the resolution represented by *FWHM* was practically the same for all non-deformed samples S.1, S.3 and S.5. Small differences in *FWHM* can be seen for plastically deformed samples. It points out the fact that the lattice deformation was basically brought about by drawing deformation and influencing the radial strain component and much less the axial strain component.

NiTi shape memory alloy samples

In this case we tested the feasibility of study of structure changes as a result of a thermomechanical external load of NiTi alloy samples. It is well known that the outstanding

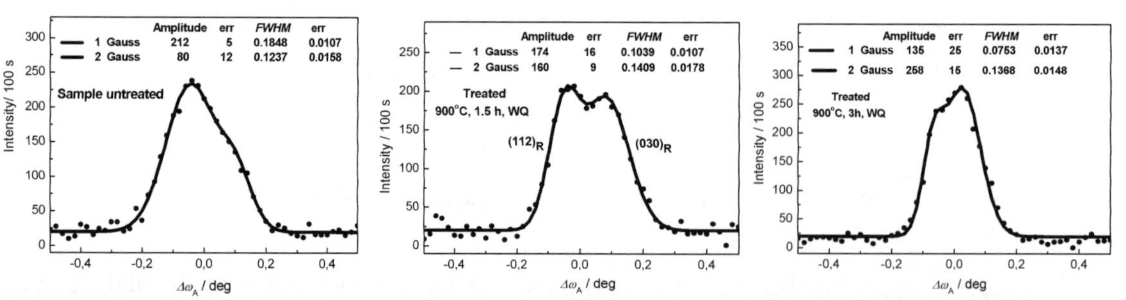

Fig. 4: An example of diffraction profiles related to the NiTi samples situated in the vertical position as analyzed by BPC Si(311) analyzer. The curves are fitted by two Gaussians.

properties of NiTi super elastic shape memory alloys which are accompanied with high mechanical properties make them suitable for a variety of applications. NiTi alloys can also contain such other chemical elements as e.g. Co, Fe, V, etc. NiTi alloys have in principle two distinct phases: austenite phase having high symmetric B2 structure that exists at high temperatures and martensite phase having low symmetric orthorhombic (or monoclinic) $B19^0$ structure that exists at low temperatures. When cooling the austenite phase then right before the transformation to martensite phase in the vicinity of 20 °C under certain circumstances an intermediate R-phase with rhombohedral crystal structure can be formed [6,7] Fig. 4 shows an example of the analyzer rocking curves obtained on the mentioned samples. The measurements were carried out at the ambient temperature of 18.5 °C (the temperature in the reactor hall). It can be seen from Fig. 4 that the temperature treatment had an influence on the quality of the NiTi alloy and that two close reflections of the R-phase could be clearly identified, namely, for the treatment at 900 °C for 1.5 hour. Then, by warming this sample to 45 °C, R-phase was transformed to the austenite B2 one as shown in Fig. 5.

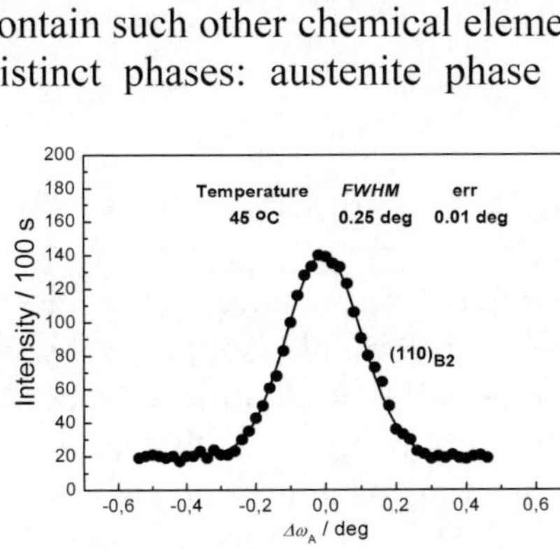

Fig. 5: Austenite phase diffraction profile obtained at 45 °C.

Summary

After similar preliminary instrumentation studies [8] another 3-axis neutron diffractometer setting employing BPC monochromator and analyzer with the studied polycrystalline samples between them was tested for practical application. It has been proved that the diffractometer setting provides a sufficiently high $\Delta d/d$ resolution (d is the lattice plane distance) permitting macro- and micro-strain studies as well as the grain size distribution after applying shape analysis of neutron diffraction peaks [9,10]. The neutron diffraction results obtained on the samples of low-alloyed quality structural steel and NiTi shape memory alloys document the feasibility of this unconventional 3 axis set-up for studies of structure properties of polycrystalline samples in the plastic deformation region. It can be seen from Tab. 3 that

Table 3: Summary of the *FWHM*s as calculated in ($\Delta d/d$)-scale

Sample number	S.1	S.2	S.3	S.4	S.5	S.6
FWHM($\Delta d/d$) - Radial [10^{-3}]	4.09	5.66	4.12	5.44	4.32	5.61
FWHM($\Delta d/d$) - Axial [10^{-3}]	3.82	4.64	3.76	4.38	3.80	4.97

though the *FWHM*-effects resulting from the applied deformation on the samples are very small and cannot be studied by the conventional strain scanner, thanks to the high resolution of the present experimental set-up, they are clearly measurable. In particular, such results can be used as an additional support to complement the information achieved by using the other characterization methodologies. As to NiTi samples it should be pointed out that major characteristic peaks of different phases (martensite, austenite, R-phase) are within only a few $\Delta\theta_A$ degrees in the vicinity of $2\theta_A$ scattering angle [7], where the best resolution of the neutron diffraction set-up was adjusted.

Finally, it should be pointed out that contrary to the conventional double axis strain scanner, this three axis setting provides a high resolution for bulk samples e.g. of the diameter of 5-10 mm whilst for a conventional scanner such diameters themselves introduce, to the resolution, an uncertainty in *FWHM* of the diffraction profile at least of 1×10^{-2} rad. Thanks to the high resolution property of the 3 axis set-up, it can be, of course, used also for the measurements of residual elastic deformation macro-stresses, however, less efficiently in comparison with the conventional two axis neutron diffraction measurement [1-5] which has not so high resolution requirements.

Acknowledgement

Measurements were carried out at the CANAM infrastructure of the NPI CAS Řež supported through MŠMT project No. LM2015056. The presented results were also supported in the frame of LM2015074 infrastructural MŠMT project "Experimental nuclear reactors LVR-15 and LR-0". Bragg diffraction optics investigations are in the Czech Republic supported by the ESS project LM2010011: "Contribution to Partnership in Large Research Infrastructure of Pan European Importance" and in Italy by the COST Action CA16122. We thank Ms. B. Michalcová from NPI ASCR for a significant help in the measurements and basic elaboration of the data.

References

[1] I.C. Noyan, J.B. Cohen, Residual stress: Measurement by diffraction and interpretation, 1st edn; 1987, New York, Springer-Verlag.

[2] M.T. Hutchings, and A.D. Krawitz, (Eds.), 1992, Measurement of residual and applied stress using neutron diffraction, NATO ASI Series, Applied Sciences, Vol. 26, Kluwer Academic Publisher, 1992.

[3] P. Mikula, M. Vrána, P. Lukáš, J. Šaroun and V. Wagner, High-resolution neutron powder diffractometry on samples of small dimensions, Materials Sc. Forum, 228-231 (1996) 269-274.

[4] P. Mikula, M. Vrána, P. Lukáš, J. Šaroun, P. Strunz, H.J. Ullrich, V. Wagner, Neutron diffractometer exploiting Bragg diffraction optics - A high resolution strain scanner. In Proc. of ICRS-5, June 16-18, 1997, Linköping, T. Ericsson, M. Odén and A. Andersson (Eds.), 2 (1997) 721-725.

[5] B.S. Seong, V. Em, P. Mikula, J. Šaroun, M.H. Kang, Unconventional performance of a highly luminous strain/stress scanner for high resolution studies, Materials Sc. Forum, 681 (2011) 426-430.

[6] T. Yoneyama, S. Miyazaky, Shape memory alloys for biomedical applications, Woodhead Publ., Cambridge 2009.

[7] Z. Lekston, M. Zubko, X-ray diffraction studies of the reversible phase transformation in NiTi shape memory alloy, Acta Physica Polonica A, 130 (2016) 1059-1062.

[8] P. Mikula, M. Vrána, J. Šaroun, J. Stammers, V. Em, Dispersive double bent crystal monochromators Si(111)+Si(311) and Si(111)+Si(400) with a strongly asymmetric diffraction geometry of the analyzer for powder diffractometry, Powder diffraction, (2019), in print.

[9] R. Delhez, T.H. de Keijser, E.J. Mittemeijer, 1982, Determination of crystallite size and lattice distortions through X-ray diffraction line profile analysis, Fresenius' Zeitschrift für analytische Chemie, 312, 1 (1982) 1-16.

[10] V. Davydov, P. Lukáš, P. Strunz, R. Kužel, Single-line diffraction profile analysis method used for evaluation of microstructural parameters in the plain ferritic steel upon tensile straining, Materials Sc. Forum, 571-572 (2008) 181-188.

EAN 2019
57th conference on experimental stress analysis

Compression resilience and impact resistant properties of fibre reinforced sandwich composites

MISHRA Rajesh[1,a], MILITKY Jiri[1,b] and PETRU Michal[2,c]

[1]Department of Material Engineering, Faculty of Textile Engineering, Technical University of Liberec, Studentska 2, 46117, Liberec, Czech Republic

[2] Department of Machinery Construction, Institute for Nanomaterials, Advanced Technology and Innovation, Technical University of Liberec, Studentska 2, 46117, Liberec, Czech Republic

[a]rajesh.mishra@tul.cz, [b]jiri.militky@tul.cz, [c]michal.petru@tul.cz

Keywords: Sandwich structures, Composites, Impact, Knife penetration, Compression

Abstract. This paper presents an experimental investigation on the compression behaviour of fiber reinforced sandwich composites. In this study, 5 different types of sandwich composites were prepared with spacer as middle layer. Also, 3 different types of woven kevlar fabric structures were used as outer layers. Sandwich composites were fabricated using epoxy resin by wet lay-up method under vacuum bagging technique. Compression behaviour, ball burst and knife penetration was tested and effect of weave type of these composites were studied using the compression stress-strain curves. We know, spacers have excellent compression elasticity and cushioning.

Introduction

Many studies have been carried out to find the contribution of aramid fibers in a hybrid carbon–aramid composite. Effect of type of weave also plays an important role. Twill weave was widely used with epoxy resin for impregnation. To evaluate the influence of the aramid fibers, a comparative study between carbon and carbon–aramid woven–reinforced composites was undertaken. A sandwich layer of 100% polyester nonwoven fabric was used. This was made into a composite by using two types of resins, polyester and epoxy resin. Maximum impact energy absorption was obtained by the blended fabric (polyester/glass) woven with satin structure and treated with polyester resin. In a few cases, carbon and carbon-aramid hybrid woven fabrics were studied. Woven fabric of different kinds such as, plain, twill 2/2, quasi-UD were used [1,2]. Although different types of sandwich composites are being developed with aramid fibres as the outer layer, there is no sandwich with knitted spacer as the middle layer. This study will give the effect of spacer as middle layer [3-6].

Experimental

Different types of sandwich composites were made for this study. They differed with the outer and middle layers. The skin or outer layers are made of Kevlar and basalt woven fabric of different weave structures, such as plain, 2x2 twill and satin. The woven kevlar fabrics were obtained from an industry.

The core or middle layer is of warp knit spacers made up of polyester fibers. Warp knit spacers are produced on double needle bar warp knitting machine. These spacers consist of two ground surfaces, which are bound through a pile of yarns, which are monofilaments with stitches. Both the warp and weft knit spacers were obtained from an industry. The composition of sandwich composites is given in table 1.

Table 1: Composition of sandwich composites

Sample No.	Composition of skin	Weave of skin	Thickness of skin (mm)	Areal Density of Skin (GSM)	Composition of Core	Thickness of Core (mm)	Areal Density of Core (GSM)
1	Kevlar	Plain	0.55	410	Polyester	11.1	517
2	Kevlar	Twill	0.51	438	Polyester	11.1	517
3	Kevlar	Satin	0.6	335	Polyester	11.1	517
4	Kevlar	Plain	0.2	115	Polyester	11.1	517
5	Basalt	Plain	0.45	377	Polyester	11.1	517

Sandwich composite manufacturing. The composites were prepared using epoxy resin L180 from the industry Havels composites. The resin to hardener ratio was 40:16 per 20 cm X 20 cm sample. Vacuum bagging technique was used to improve the quality of sandwich composite produced by wet lay-up method. Initially, the resin and hardener are mixed with the help of a magnetic stirrer. The mixture is then applied onto the surface of the woven kevlar fabric by help of a roller [7,8]. In-laboratory set up of vacuum bagging technique is shown in figure 1. A perforated layer of nonwoven on the glass makes the first layer. The sample is placed on the release film. The whole set up is covered with a flexible vacuum bag. Edge of the bag is sealed with vacuum sealing compounds or sealant. Vacuum pressure of 600bar is applied. When vacuum is applied, the trapped air and excess resin escapes out which further consolidates the laminate and eliminates voids. The vacuum pressure was set for 24 hours. After which the sample is put into oven at 60°C for 15 hours.

Fig. 1: Vacuum bagging for composite preparation

Measurement of compression. The measurement of compression for the prepared sandwich composites was done as shown in figure 2. The sandwich composite samples were tested according to the Standard ASTM D 575 on a TIRA Tester device set up with two compression circular plates. The size of all the specimens was 10cmX10cm. The compression tests were conducted at a speed of 10 mm/min up to a deformation 40% of the initial thickness for each composite sample in an environment of 20 0C and 65% relative humidity [9].

Fig. 2: Compression testing set-up for sandwich composites

Knife penetration test. Knife penetration test was carried out on Testometric as show in in figure 3. The sandwich composite was placed between the jaws. The traverse rate of the knife was set to 10mm/min. the size of the specimen was 10x10 cm. The environment was maintained at 20 0C and 65% relative humidity [10,11].

Fig. 3: Knife penetration test

Ball burst test. The ball burst test was carried out in the same testometric tester with the knife replaced with a ball head. ASTMD 3787 standard was followed for testing of the sandwich composite. The traverse rate of the ball head was set to 100mm/min. The diameter of the ball is 44.45mm [12-15]. The set up is shown in the figure 4 below.

Fig. 4: Ball burst test

Results and Discussion

The first and foremost parameter taken into consideration is the effect of spacer as core material used in the study. Generally a spacer fabric has good cushioning effect. The typical stress-strain curve of a spacer can be divided into four regions [16-18]. The initial stage, elastic stage, plateau or energy absorbing stage and the densification stage as shown in figure 5.

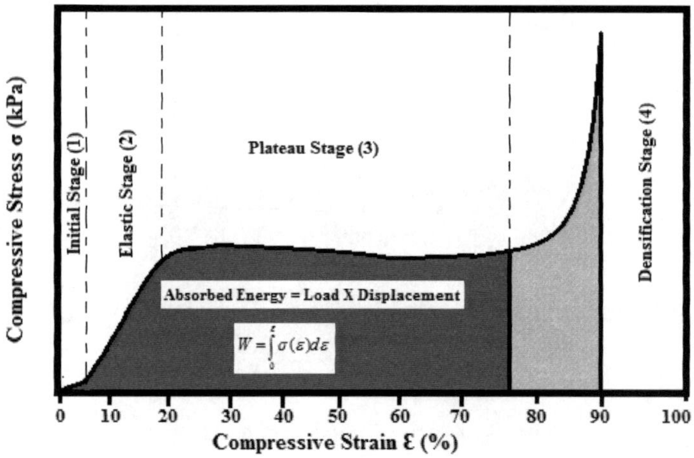

Fig. 5: Stages of compression for 3D spacer fabrics

A similar trend of stages in observed in the sandwich composites in figure 6.

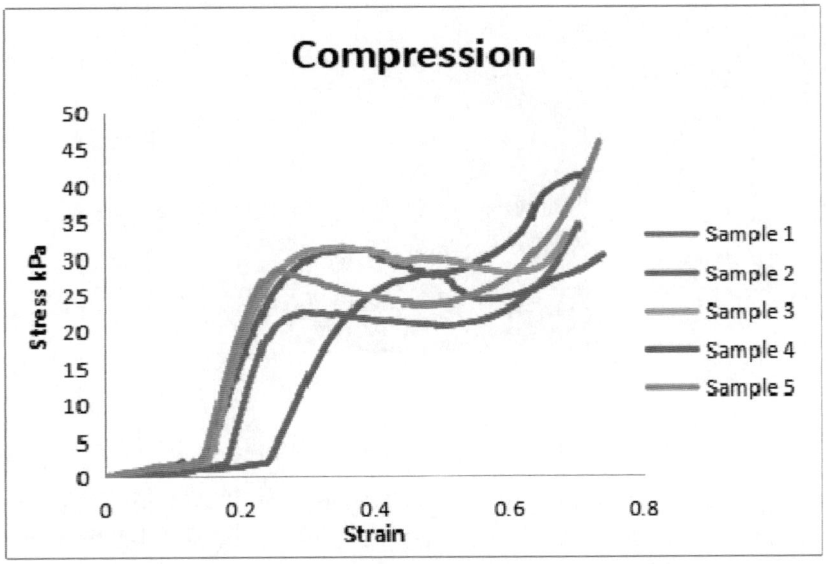

Fig. 6: Compression result for sandwich composite

Compression depends on mostly the core material. There is a minor difference due to structure of outer layer. Better adhesion of surface layer leads to more uniform compression cycle. Compression resilience is very high in all samples.

Maximum compression energy with minimum thickness of the surface fabric. Thus the overall compression is mostly dominated by the spacer fabric layer. Compression resilience maximum with thinnest surface layer. With similar middle layer, there is similar compression and recovery behavior.

The results of knife penetration resistance are shown in figures 7 and 8. Mostly dependent on cover of surface layer. Twill weave based sandwich composite shows maximum resistance due to better adhesion of yarn floats, impregnation of resin and adhesion with core layer.

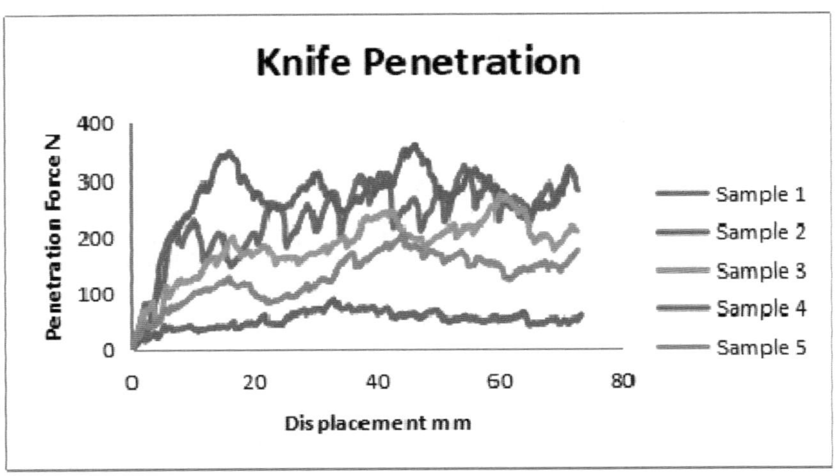

Fig. 7: Knife penetration curve

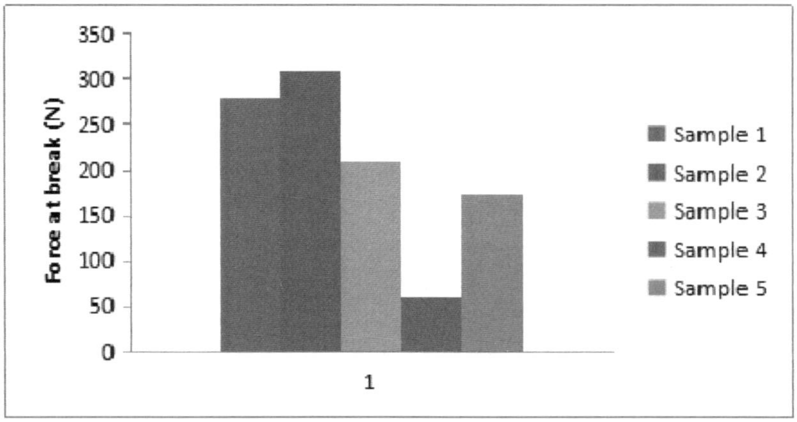

Fig. 8: Knife penetration force

In the knife penetration test, effect of weave structure and density of skin layer plays a significant role. From the graph it is clear that sample 2, with Kevlar twill woven fabric as skin has highest resistance to penetration compared to other weave structures. Sample 1 and 4 have the same plain woven Kevlar as skin but with different densities. Sample 1 with higher density plain weave shows better performance compared to sample 4 with lower density plain weave.

Results of ball bursting test are shown in figure 9.

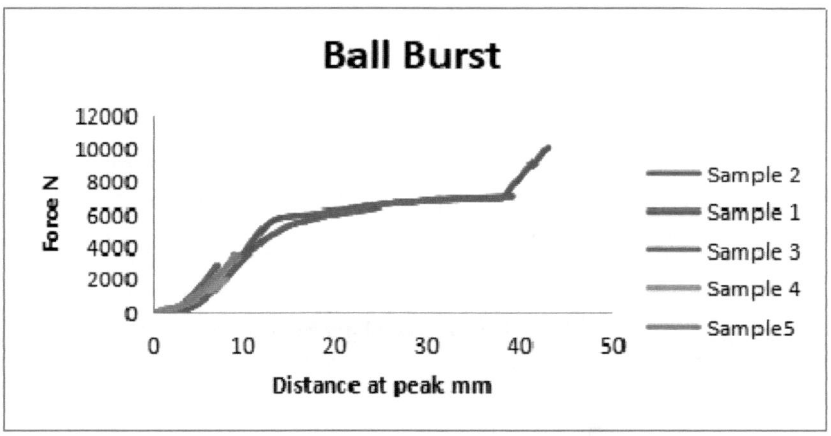

Fig. 9: Result of ball burst test

During the ball burst strength, sample 1 and 2 showed different trend compared to other samples. They did not break or crack like the other 3 samples.

From figure 10, it can be seen that sample 1 and 2 bulged out. While sample 3-5 cracked.

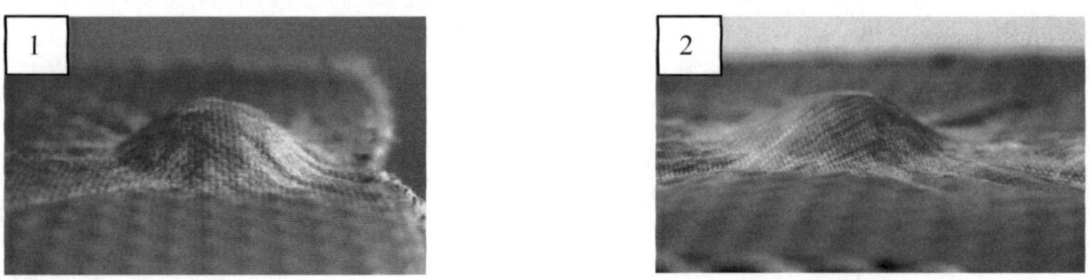

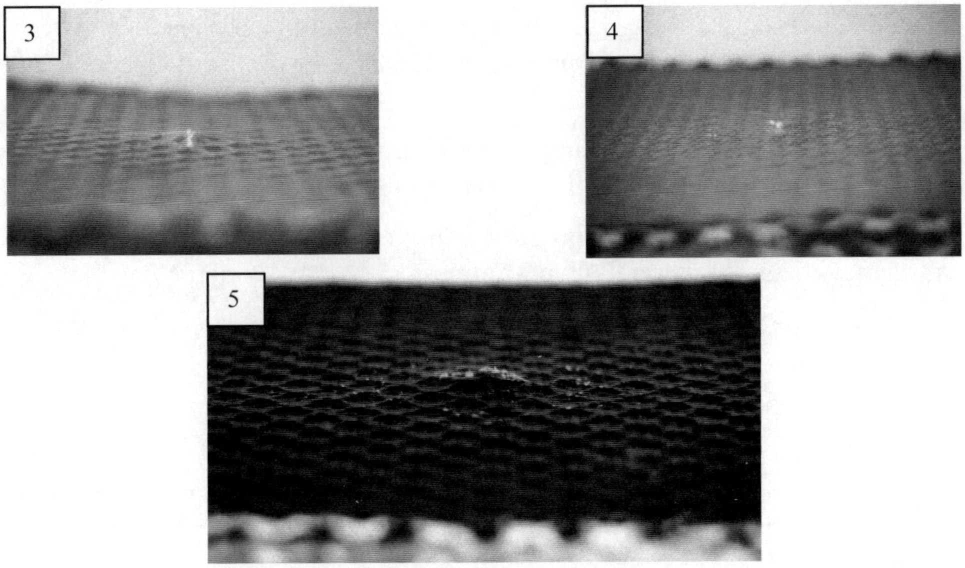

Fig. 10: Samples 1-5 after the ball burst test

The bulging deformations are recoverable after 30 minutes of relaxation. Maximum impact energy with plain woven surface layer Twill and Satin show similar energy level. As it is a high velocity impact test, the integrity of the woven structure or the inter yarn friction is most dominant in deciding the level of resistance

Conclusions

The sandwich composites were developed and evaluated for various mechanical properties. Compression resilience is excellent in sandwich composites with 3D knitted spacer as the core. As the resilience is mainly function of the bulk and middle layer, there is no significant influence of the outer layers of woven fabrics in the sandwich composite system.

Maximum knife penetration resistance is obtained with twill weave on surface due to maximum yarn cohesion and resin impregnation. Higher amount of cohesive friction results in higher resistance against penetration of sharp objects like the knife edge. Plain and twill fabrics offer sufficient resistance again ball burst. The yarn deformation allows formation of dome shape. Maximum impact resistance in ball burst is obtained for plain weave due to inter-yarn binding (friction).

Acknowledgement

This work was supported by the Ministry of Education, Youth and Sports of the Czech Republic and the European Union – European Structural and Investment Funds in the frames of Operational Programme Research, Development and Education – project Hybrid Materials for Hierarchical Structures (HyHi, Reg. No. CZ.02.1.01/0.0/0.0/16_019/0000843), project "Modular platform for autonomous chassis of specialized electric vehicles for freight and equipment transportation", Reg. No. CZ.02.1.01/0.0/0.0/16_025/0007293, SGS grant No. 21159 of Technical University of Liberec and project "Sophisticated hybrid tapes for fabrication of composites by precise winding" [Project no: TJ01000292, 14014/136].

References

[1] M. Karahan, Effect of Weaving Structure and Hybridization on the Low-Velocity Impact Behavior of Woven Carbon-Epoxy Composites, Fiber. Text. East. Eur. 22 (2014) 109-115.

[2] C.C. Lin, Elucidating the Design and Impact Properties of Composite Nonwoven Fabrics with Various Filaments in Bulletproof Vest Cushion Layer, Text. Res. 79(2009) 268–274.

[3] R. Mishra, B. K. Behera, J. Militky, 3D woven green composites from textile waste: mechanical performance, J. Text. I. 105 (2014) 460-466.

[4] R. Mishra, B. K. Behera, J. Militky, Recycling of textile waste into green composites: performance characterization, Polym. Compos. 35 (2014) 1960-1967.

[5] R. Mishra. Drape behavior of 3D woven glass-epoxy composites. Polym. Compos. 37 (2016) 472-480.

[6] H. Jamshaid, R. Mishra, J. Militky, S. Hussain, Basalt hybrid woven textile materials for advanced thermal applications, Ind. J. Fibre Text. Res. 44 (2019) 56-64.

[7] R. Mishra, H. Jamshaid, J. Militky, Basalt Nanoparticle Reinforced Hybrid Woven Composites: Mechanical and Thermo-mechanical Performance, Fiber. Polym. 18 (2017) 2433-2442.

[8] R. Mishra, H. Jamshaid, J. Militky, Mechanical, thermal and interfacial properties of green composites from basalt and hybrid woven fabrics, Fiber Polym. 17 (2016) 1675-1686.

[9] R. Mishra, V. Arumugam, J. Militky, M. Tunak, In-plane shear behavior of 3D knitted spacer fabrics, J. Ind. Text. 46 (2016) 868-886.

[10] R. Mishra, H. Jamshaid, Thermo-Mechanical Characteristics of Basalt Hybrid & Non-hybrid Woven Fabric Reinforced Epoxy Composites, Polym. Compos. 37 (2016) 2982-2994.

[11] R. Mishra, B. K. Behera, J. Militky, Impact simulation of 3D woven kevlar-epoxy composites, J. Ind. Text. 45 (2016) 978–994.

[12] R. Mishra, J. Militky, N. Gupta, R. Pachauri, B.K. Behera, Modeling and simulation of earthquake resistant 3D woven textile structural concrete composites, Comp. B: Eng. 81 (2015) 91-97.

[13] R. Mishra, V. Baheti, B. K. Behera, J. Militky, Novelties of 3D woven composites and nanocomposites, J. Text. I. 105 (2014) 84-92.

[14] R. Mishra, Specific functional properties of 3D woven glass nanocomposites, J. Comp. Mater. 48 (2014) 1745–1754.

[15] B.P. Dash, B. K. Behera, R. Mishra, J. Militky, Modeling of internal geometry of 3D woven fabrics by computation method for structural composites, J. Text. I. 104 (2013) 312-321.

[16] R. Mishra, Impact tolerence of 3D woven nanocomposites: A simulation approach, J. Text. I. 104 (2013) 562-570.

[17] R. Mishra, J. Militky, B.K. Behera, Modeling and simulation of 3D orthogonal fabrics for composite applications, J. Text. I. 103 (2012) 1255-1261.

[18] R. Mishra, R. Tiwari, M. Marsalkova, J. Militky, B.K. Behera, Effect of TiO_2 nanoparticles on Basalt/Polysiloxane composites: mechanical and thermal characterization, J. Text. I. 103 (2012) 1361-1368.

Hybrid fabrics for composites reinforcement

MISHRA Rajesh[1,a], MILITKY Jiri[1,b] and PETRU Michal[2,c]

[1]Department of Material Engineering, Faculty of Textile Engineering, Technical University of Liberec, Studentska 2, 46117, Liberec, Czech Republic

[2] Department of Machinery Construction, Institute for Nanomaterials, Advanced Technology and Innovation, Technical University of Liberec, Studentska 2, 46117, Liberec, Czech Republic

[a]rajesh.mishra@tul.cz, [b]jiri.militky@tul.cz, [c]michal.petru@tul.cz

Keywords: Hybrid woven composites, Impact energy, Flexural modulus, DMA, TGA

Abstract. In this study, experimental investigations are carried out to check the thermal and mechanical behavior of woven Basalt/PET and Basalt/PP fiber hybrid woven fabric epoxy composite laminates. Three types of weaves are used for all hybrid structures. Fabricated composite samples are subjected to mechanical and thermal characterization. Results are discussed in terms of fiber effect, weave geometry and the resin effect. The results reveal that the hybridization of basalt in different weaves leads to significant improvement in the static and dynamic mechanical properties of composites. Structure of weave, fiber type and resin properties have strong influence on mechanical properties of composites of basalt hybrid fabrics. The modulus is significantly increased in composites due to interfacial bonds between fibers and resin. Thermal behavior of fiber and composite was observed by Thermal Gravimetric Analysis and Differential Scanning Calorimetry. Thermal properties are also affected by hybridization. Thermal conductivity is strongly affected by resin properties. Fractography studies of the damaged surface are also demonstrated.

Introduction

The textiles are core ingredient of human life since ancient times. Use of fabrics ranges from clothing to industrial application based on the raw material for specific use. The development of different set of materials has spread its uses beyond expectations. Hi-tech composite reinforcing materials have become an integral part of many applications, e.g. IT, Electronics, Automotive, Sports, Health care, Energy generation, Energy storage etc. Fibers are used in different polymer matrices for enhancement of features like strength and rigidity. In aerospace and defense industries, there is a demand for materials with higher strength and lower weight. Thus fiber reinforced composites became an obvious choice. Recently, the composite products are increasingly used for sun-shading devices, plane wings, fuselages or any other beams and shells in modern applications [1,2]. As the demand for fiber reinforced composites increased for specific applications, a number of new fiber/resin systems were on offer for the designers. These composites exhibit brittle fracture and bear high cost but their extra ordinary properties like low density, high rigidity and high strength raise their demand for specialized solutions. Two or more types of fibers embedded in a matrix of reinforcing phase constitute a hybrid composite. Hybrid composites give the privilege to create a material bearing desirable properties among the combination of fibers, that is more cost effective and we can mitigate the non-desirable properties from the combination. It can help to tailor the requirement for specific materials. In comparison with conventional composites, hybrid composites exhibit balanced strength and stiffness, balanced thermal distortion stability, fire resistance behavior, reduced weight and/or cost, improved fatigue resistance, reduced notch sensitivity, improved fracture toughness and impact resistance [3-6].

The reinforcement fibers themselves are unmanageable and are therefore usually arranged into sheets which can be handled, oriented on a mould, shaped and cut. The methods of binding the individual fibers together are varied and have significant impact on both the manufacture of the component and structural performance of the composite. The fibers are bound in a variety of ways including weaving, stitching and knitting. Woven fabric composites particularly offer better dimensional stability when they are exposed to a large range of temperatures.

Woven structure is defined by orthogonal interlacement of at least two sets of yarns called warp and weft. Woven structures formulate an important part of technical textiles and their applications. Weave structure helps in providing better and balanced properties in plane of fabric area where as interlacement of yarns provides out of plane strength which results in take up of secondary load due to load path eccentricities, load buckling, tolerance and better impact resistance in comparison with unidirectional laminated composites. Considering that the woven fabric reinforced composite materials are not entirely homogeneous, large resin rich areas are formed by the interlacing of undulating warp and fill yarns. In the high performance fiber-polymer matrix system, the difference of damping is much larger than that of the stiffness. Large resin rich areas act as the built-in damper elements. Their distribution, depending on the architectures of the weave, determines the damping of the composite structure.

Glass fiber-reinforced composites were widely used to fabricate various applications in recent times but the aim to produce more environment friendly composites is leading to reduction of glass fiber use [7,8]. Basalt fiber has evolved as a replacement of glass fiber due to its environment friendly natural origin. Basalt rock is known for its thermal properties, strength and durability since long time. Basalt is known to possess better tensile strength in comparison with E-glass fibers, they have greater failure strain than carbon fibers, they have very good resistance to chemical attacks and they produce less poisonous fumes [9]. The recycling of basalt fiber can be done in much better environment friendly way than glass fibers (Behera and Mishra, 2008). When a resin with basalt fibers is recycled, same material of basalt powder is obtained so it is classified under sustainable material category.

Conventional reinforced composites are always problematic at the end of their service life during recycling. Due to environmental and legal issues, recycling process has become an increasingly difficult issue. The solution to the issues related to recycling was answered by use of green matrix based renewable resource or green composite which are generally formulated by embedding the fibers into bed of bio-polymers that can integrate more readily into natural biodegradation cycles, for example by CO_2-neutral incineration, which includes recovery of energy, or by composting.

We have a twofold aim of our research which is, to enhance use of green composites by adding natural basalt fibers in biodegradable polymeric resins and, to develop hybrid composites with most commonly used fiber i.e. polyester and polypropylene. Basalt woven hybrid composites are developed. An investigative study is conducted about role of different reinforcing yarns in composites with their properties and capabilities. PET and PP yarns have been used in the hybrid composites besides basalt yarn, and the mechanical characteristics of the composites have been investigated as a function of weave and fiber composition. Further aim of the work is to study the thermal properties of these hybrid composites. In this study, geometry of hybrid textile fabric was the prime focus and effect of weave structure on the mechanical behavior in textile composites was investigated.

Experimental

Basalt is a fiber originated from rocks and having excellent thermal resistance. PET and PP are thermoplastic polymeric materials. The idea was to use these different polymeric fibers in the weft along with basalt warp so as to create hybrid woven structures and to further investigate the compatibility with epoxy resin in a composite manufacturing. The prepared composites were evaluated for mechanical and thermal characteristics. The polyester (PET) and polypropylene (PP) yarns used in this study were available commercially. The basalt yarn was received from company Kamenny Vek (KV). The details of yarns are given in Table 1. Green epoxy resin CHS-Epoxy G520 and hardener TELALIT 0600 were supplied by Spolek, Czech Republic. It is a low molecular weight basic liquid epoxy resin containing no modifiers, certified by International Environmental Product Declaration Consortium (IEC).

Table 1: Properties of fibers and yarn

Properties	Basalt	Polyester	Polypropylene
Diameter of fibers (micron)	12	22	34
No. of filaments	890	900	300
Linear density of yarn (Tex)	295	250	292
TPM (Twists/m)	20	24	30
Tensile strength (N)	92.75	88.91	57.44
Tensile elongation %	1.29	12.55	12.27
Tenacity (N/tex)	0.315	0.305	0.23
Modulus MPa	9,378	1069	721

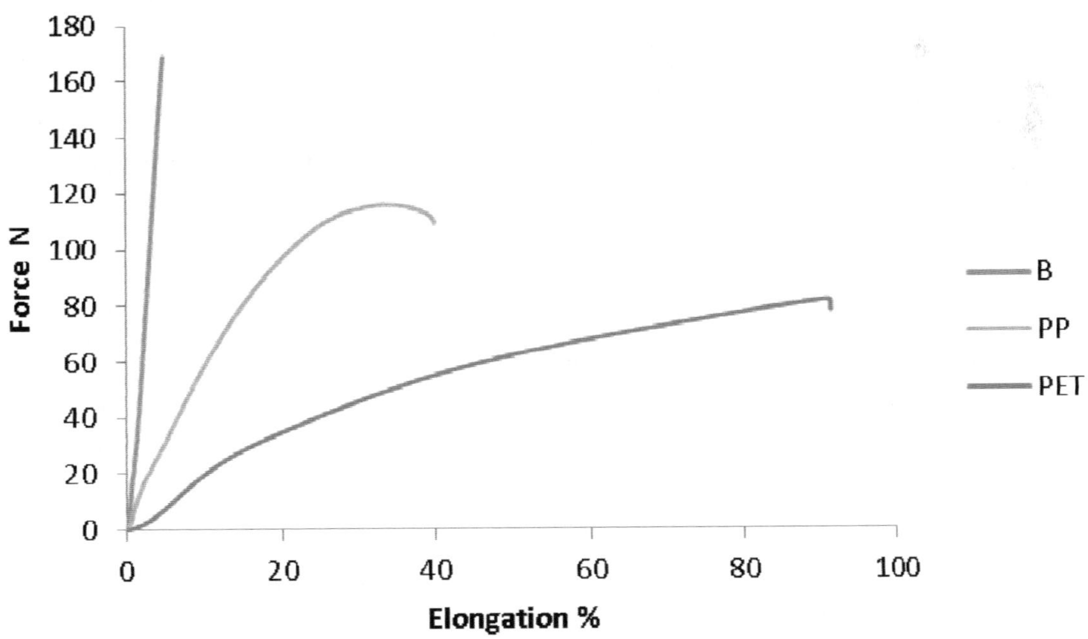

Fig. 1: Force & elongation curve for yarns

Methods. Hybrid fabric samples were developed with Plain, Matt, and Twill weaves from Basalt/PET and Basalt/PP yarns. Non-hybrid Basalt–Basalt fabrics were also developed. All fabrics were made on the CCI sample loom with the same thread density for all fabrics, 12 threads/cm in warp and 8 threads/cm in weft. All the fabric variants were measured, according to standard procedures. Yarn tensile properties were measured and shown in Fig. 1.

Preparation of composites. The Basalt hybrid (B/J) and non-hybrid (B/B) fabrics were reinforced in epoxy resin to prepare the composites with single ply of the reinforcing fabric. The Bio epoxy resin and corresponding hardener at weight ratio of 100:32 (by weight) according to manufacturer recommendations were mixed. The mix was stirred manually to disperse the resin and the hardener in the matrix. The mixing was done thoroughly before the fabrics were reinforced in the matrix body. The composite slabs were made by conventional hand-lay-up technique followed by light compression molding technique. A stainless steel mould having dimensions of 250 x 250 x 40 mm^3 was used as shown in Fig. 2. The prepared resin mixture was poured on fabric layers and spread out by a hand roller. The gentle rolling action of hand roller confirmed the wetting of fabrics and the excess resin was squeezed out of the panel layup by the roller. The composite layup along with Teflon sheets were sandwiched between a pair of steel plates and cured at 120°C for 1.0 h in mechanical convection oven with predetermined weight. The fiber volume fraction (Vf) of all composites was around 0.4.

Testing methods. The static and dynamic mechanical Properties of composites are characterized along with thermal stability characterization.

Impact test was performed according to standard EN ISO 527-5. The specimen size for tensile testing was 15 cm × 2 cm as per standard. Using aluminum tapes of 1 mm thickness to prevent gripping damage. Composites elastic modulus was quantified through measurements of samples for the tensile properties in warp and weft direction on a TIRATEST universal tensile tester at a crosshead speed of 2 mm/min at room temperature. For each type of sample, 5 tests were repeated.

Three points bending test is used to find the flexural strength and flexural modulus according to EN-ISO 14125 on TIRATEST universal testing machine. All tests were carried out with a span to thickness ratio of 32:1 and a cross-head speed of 2 mm/min. Load is applied at the midpoint of a beam until the specimen bends. The average values of 5 specimens for each sample in both warp and weft direction have been reported.

Impact tests are designed to measure the resistance to failure of a material to a suddenly applied force. The test measures the impact energy, or the energy absorbed prior to fracture. All impacted specimens were un-notched. Charpy impact test was conducted as per EN-ISO 14125 rectangular using CEAST RESIL 5.5 with a force of 22 J at a velocity of 2.9 m/s. The width and thickness of the specimen were measured and recorded. The work of fracture/impact strength values were calculated by dividing the energy in J recorded on the tester by the cross-sectional area of the specimen. The average values of 5 specimens for each sample in each direction have been reported.

Dynamic mechanical properties. Dynamic mechanical analysis was performed on a DMA 40XT RMI equipment. The samples were tested using three-point bending mode at a frequency of 1Hz in temperature scan mode. The DMA test was executed in the temperature range of 27 to 100 °C at a heating rate of 3 °C/min. This test was performed according to EN ISO 6721-1. For each sample, five measurements were done. Dynamic mechanical analysis (DMA) yields information about the mechanical properties of a specimen placed, usually sinusoidal, oscillation as a function of time and temperature by subjecting it to a small, usually sinusoidal, oscillating force. It allows the measurement of two different moduli of the materials, a storage modulus (E') which is related to the ability of the material to return or store mechanical energy and a loss modulus (E″) which is related to the ability of the material to dissipate energy as a function of temperature. Ratio of loss modulus to storage modulus is given by tan (delta).

(a)Tensile (b)Flexural (c)Impact

Fig. 2: Experimental set up for mechanical characterization of composites

Thermal properties. Thermal properties of materials indicate the physical response to the application of heat and resultant change in temperature in response to the applied heat. Physical impact of temperature on materials is reversible in case of short term application where as it is non reversible in long term application as it associates with the chemical changes. The long term effects at elevated temperature are aging/degradation resulting in affecting the mechanical, physical and chemical properties. There is a variety of standard testing methods for thermal properties. Various techniques include differential scanning calorimetry (DSC), thermo mechanical analysis (TMA), thermo gravimetric analysis (TGA) etc.

The Mettler Toledo TGA/SDTA851^{e} instrument was used to study the thermal gravimetric behavior (thermal stability and degradation) of the composites. Differential thermal and thermogravimetric analysis (DTG)/(TGA) was done. Thermo gravimetric analysis was performed under dynamic nitrogen atmosphere. The samples were heated from 25°C to 700°C at a heating rate of 10°C/min to yield the decomposition temperature, mass loss and maximum decomposition peak.

The Perkin Elmer Differential Scanning Calorimeter DSC6 was used. Samples weighing approximately 7±3 mg were placed in aluminum crucibles and sealed. The samples were heated at a constant rate of 15°C/min between temperature ranges of 25°C to 400°C and then cooled in nitrogen atmosphere with a flow rate of 20 ml/min.
Measurement of thermal conductivity by C-Therm TCi Thermal conductivity analyzer was performed. It allows determining accurate values for thermal conductivity of material without extensive sample preparation or damage to the sample. Measurement of the thermal conductivity of the composite samples was done with TCi Thermal conductivity analyzer by following EN 61326-2-4:2006 standard. To minimize the measurement error to 5%, each sample was repeatedly measured 10 times.

Morphology analysis was done by SEM and video macro scope (Navitar Macroscope) with CCD camera, software N/S-Elements. The SEM photographs of composite samples were taken using a scanning electron microscope TS5130-Tescan SEM at 20 kV accelerated voltage. The surfaces of the samples were coated with gold by means of a plasma sputtering apparatus prior to SEM investigation and were investigated at 2,000× magnification to observe the fiber matrix adhesion. Fractography studies of tensile and impact-damaged sample were carried out by scanning electron microscope. Prior to analysis, the fractured samples were also coated with gold and samples were examined with different magnification level for obtaining high resolution images. Broken laminates were also investigated by Macro scope.

Results and discussion

Mechanical properties describe the resistance of material against external forces. When evaluating the mechanical properties of hybrids, a general rule of mixtures approach may be utilized to quantify a material property with respect to the volume concentration of its constituents. Many researchers (Chou and Kelly, 1980) have however noted the existence of a hybrid effect in which the material property as predicted by the rule of mixtures differs to that observed in reality.

Measurement of tensile stress-strain properties is the most common mechanical measurement. It is used to determine the behavior of a sample while under an axial stretching load. The tensile properties of the composite material is dependent on tensile properties of fiber, matrix and interfacial bonding of resin and reinforcement material. Normally in fiber reinforced composites, the modulus of a composite material is dependent on the reinforcing fiber properties, whereas the tensile strength is a function of the matrix properties. During failure, initially the crack starts at interface (junction between fiber and resin) and then propagates to matrix. Cracking is followed by progressive failure of the fibers (Munikenche, Naidu, and Chhaya, 1999).

Tensile properties. It can be observed from Fig. 3, that non-hybrid basalt woven fabrics have highest tensile properties as basalt yarn has highest strength value, followed by hybrid structures with PET yarn. Tensile modulus indicates the stiffness of materials. Following 100% basalt based composites, Basalt/polyester based composites have high modulus followed by Basalt/PP composites. As polyester tows have higher number of filaments per crosssection, lower liner density and low twist, thus the warps come closer after relaxation and a higher level of crimp in warp direction is achieved which makes them more stiffer.

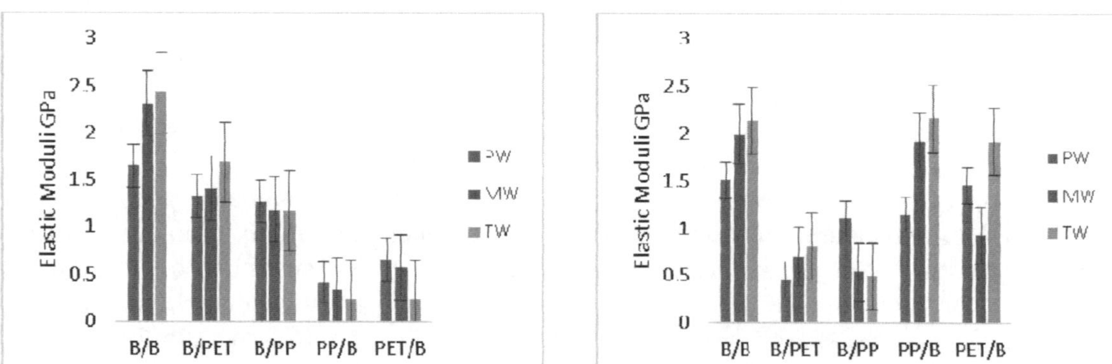

(a) Warp (b) Weft

Fig. 3: Elastic moduli of composites

The amount of crimp in a plain woven fabric is more than any other existing weave due to the presence of huge no of intersections. For such high amount of crimp, the yarn orientation is very poor resulting in to weaker in-plane tensile strength. Tensile strength of plain weave is less than twill, probably due to lesser orientation of yarns in the direction of the applied force in non- hybrid basalt as well as B/PET woven structures.

In case of B/PP this condition is reversed i.e plain woven composites have maximum strength. It may be due to weaker interfacial bonding between PP and bio epoxy resin. As PP yarn has compact structure due to higher amount of twist and slightly higher linear density. This leads to less porosity of yarn, decrease interfacial bonding between fiber and resin, so load is transferred to fabric, which follows the same trend of fabric although, effect is not very significant.

In structures where basalt is used in weft, it follows the same trend as fabric due to weaker interfacial bonding with the thermoplastic filament yarns having lower porosity which prevents efficient impregnation of resin into core of the tows. During weaving process of fabric, level of stresses on warp yarn is more which leads to decrimping, elongation and compactness of yarn causes contraction. This causes the floating adjacent yarns which are free from interlacement to join together to form jammed structure. The jammed structure has great influence on the properties of the fabric such as the porosity and it adversely affects resin penetration between fibers. Load is transferred to fabric directly from the resin. Also it can be seen that by using basalt in weft direction, the increase of moduli is not significant. In the fabric composite formation, there is a mechanical interlocking between the fibers and the resin due to frictional forces. Friction force is highly dependent on the area of contact between the resin and the fibers. In case of twill, contact area between resin and fiber is higher, bonding between fiber and matrix is high, thus higher modulus is observed. The plain weave (under almost the same condition of fabric weight/area) is less stiff than twill (due to higher amount of crimp). This is due to the fact that the resin transmits and distributes the applied stress to the yarns resulting in higher modulus. When load is applied to textile reinforced composites, resin transfers the stress to the reinforced fibers. Effective transfer of stress and load distribution throughout the interface is possible when strong adhesion exists at the interfaces. The reinforcing woven fabrics contain more number of yarns in warp direction than in weft direction. Due to higher thread density along warp direction, composites show contrast in mechanical behavior compared to weft direction. When basalt is used in weft direction moduli improvement in weft direction is prominent because of dominance of basalt yarn properties. The mechanical properties of basalt yarn are superior as compared to basalt in warp as the later gets abraded and strained to a higher extent during flexing action of shedding and during beat up action on the loom.

Flexural properties. The flexural modulus is a measure of the resistance to deformation of the composite under bending loads. Under the flexural loading, the surfaces of the specimen are subjected to greater strains than the specimen centre. Composite samples face compressive and tensile fracture during flexural test. Surface layer of composites face tensile fracture while the compressive mode works on the bottom layer [10-12].

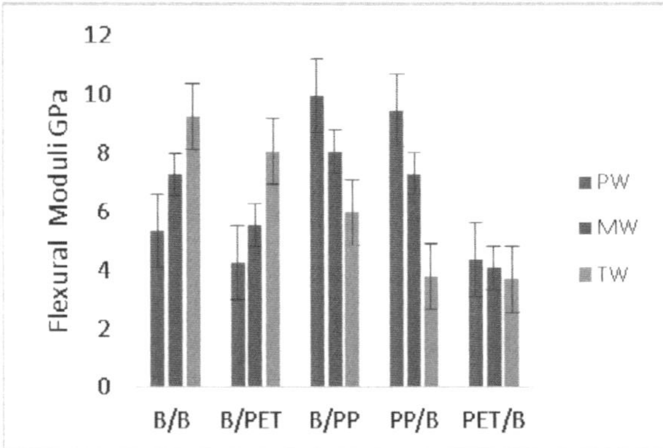

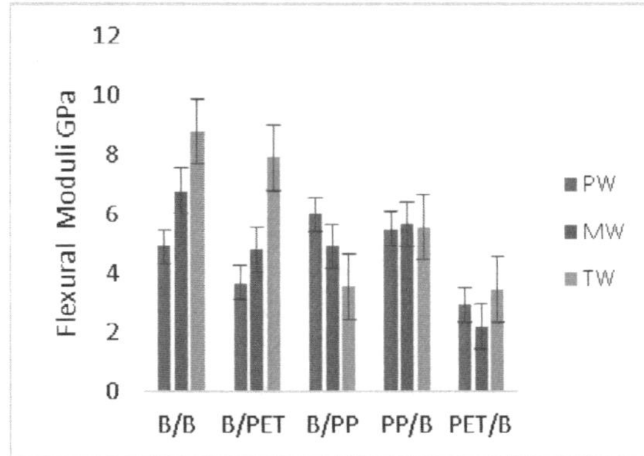

(a) Warp (b) Weft

Fig. 4: Flexural moduli of composites

It can be seen from Fig. 4 that, B/B composites have higher bending properties as it also dependent on fiber properties and it follows the same trend as tensile properties. The twill woven fabrics have higher interfacial bonding thus higher bending strength and modulus except B/PP, PP/B and PET/B, where interfacial bonding is not so strong. The resistance to bending is best for composites reinforced with twill fabric as they are stiffer than plain.

As like tensile properties, flexural moduli of composites are found to be higher in the warp direction than weft direction. Less number of yarns in weft direction of composites limits their tensile stress dispersion. Therefore, as the tensile stress tries to propagate upwards, delamination failure occurrs and thus reducing its flexural strength.

Moreover, this research work found that fiber orientation has maximum impact on flexural properties compared to the tensile properties. It is also evident that the hybrid material follows theory and experiment that has been reported by others. A main reason for the higher flexural strength of laminates is that the volume of material subjected to the maximum stress is smaller for three point bending test than the tensile test. So, the presence of the critical defects is much lower than in the tensile test. It may attribute the observed higher flexural strength to laminates [13, 14].

Impact strength. Impact Strength of textile reinforced composites is a measure of the ability of the composites to resist the fracture failure under stress applied at high speed and is directly related to the toughness of the composites. This is one of the ways to measure the relative toughness value of reinforced composites, but it will not reflect the overall toughness value of the composite material. Impact strength is a measure of the impact energy required to fracture a specimen. It is generally accepted that the toughness of a fiber reinforced composite is mainly dependent on the stress–strain behaviour of fiber. Strong fibers with high failure strain impart high work of fracture on the composites. Fibers play an important role in the impact resistance of composites as they interact with the crack formation and act as stress-transferring medium. Next to fiber, another important factor that influences the impact energy is the fiber-matrix interfacial shear strength. It can vary depending on the adhesion between fiber and matrix [15-17].

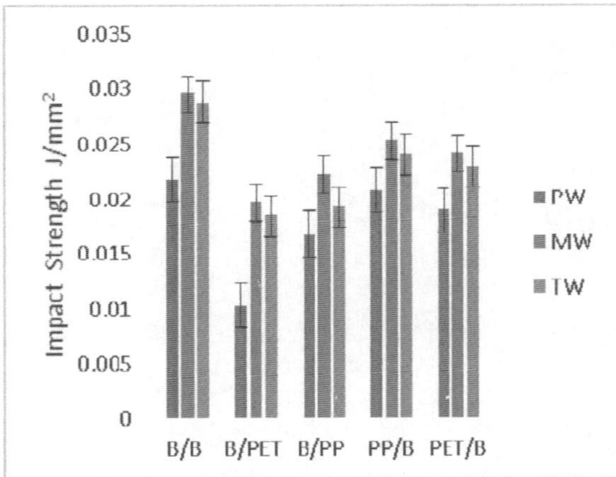

(a) warp

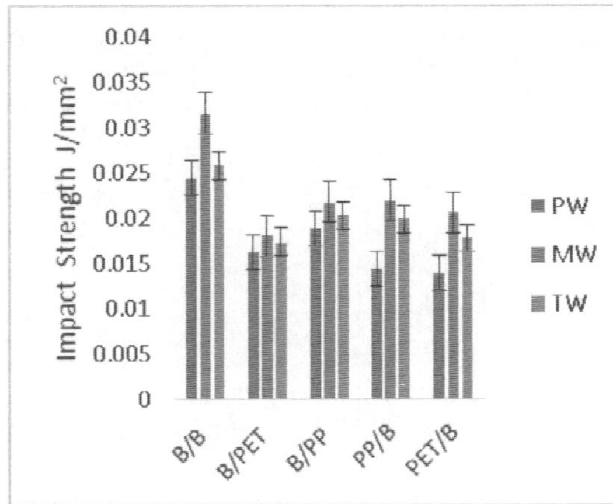

(b) weft

Fig. 5: Impact energy of composites

Unnotched specimens were not broken during Charpy tests. It can be observed from fig. 5, that non-hybrid basalt structures have higher impact strength followed by PP/B structures. The increase in the impact strength with the basalt fibers content can be attributed to the higher energy dissipation at the fiber/matrix interface required to detach the fibers from the matrix. It can be observed that PP yarn decreases the stiffness and rigidity of the composite, thus causing an increase of impact resistance.

Matt weave based structures have highest impact strength in all composites. During an impact situation, radial distribution of forces in all axes is desired, which is more probable in balanced weaves like the matt weave where groups of floats exist in both warp and weft. Containing higher amount of aligned fibers in warp direction, the composite produces higher resistance to impact stress.

Dynamic mechanical properties. The dynamic mechanical properties of the composite materials have been the objective of many investigators. In order to assess the performance of structural applications, the dynamic mechanical properties help in material evaluation with repect to temperature, time and frequency. In designing of the composite materials with desired dynamic properties, storage modulus and damping are the important features. The developed composite samples from Basalt hybrid and non-hybrid fabrics were subjected to a range of temperatures in order to predict and control composite processing behavior. The results are shown in Fig.s 6-7. Fig. 6 shows the *E'* of each sample in the warp direction. It is found that the values of *E'(storage modulus)* have decreased in varying degrees with the increasing temperature, indicating that the visco-elasticity of composite varies significantly from the fabric structures [18].

Storage modulus (*E'*) is related to elastic response of material and is the reflection of stiffness. It is related to its load bearing capacity. Since the epoxy resin exhibits temperature dependencies of visco-elastic properties, increased temperature results in the decrease of *E'*. It can be seen that there is a gradual fall in the storage modulus of all composites when the temperature is increased.

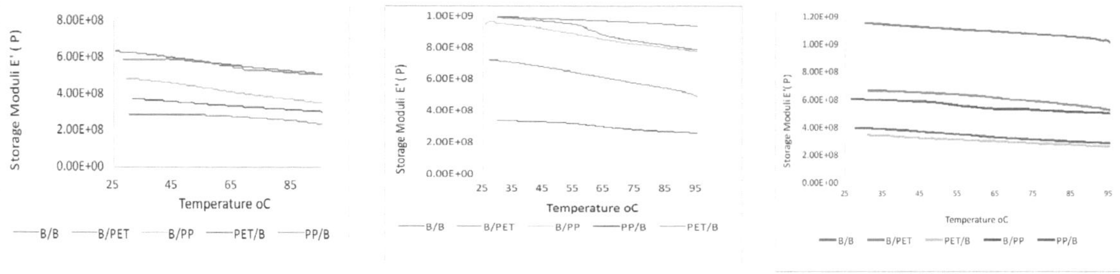

(a) Plain weave (b) Matt weave (c) Twill weave

Fig. 6: Storage moduli for composites of all weaves

Storage modulus of basalt fiber epoxy composite is attributed to the strength of fibers and also the fiber morphology, which aided the fiber to matrix adhesion thereby reducing the mobility of the polymer chains. It is clear that Twill structures have better stiffness than other structures at 1 Hz. Higher storage modulus means stiffer material. Storage modulus of the matrix affects the mechanical properties of the composite material. Hence, improvement in storage modulus also gives improvement in mechanical properties of the composite system, specifically the tensile properties. The results for storage modulus explain the highest gain in tensile modulus of the Twill weave based composites. Trends of graphs obtained for storage modulus and tensile modulus are similar.

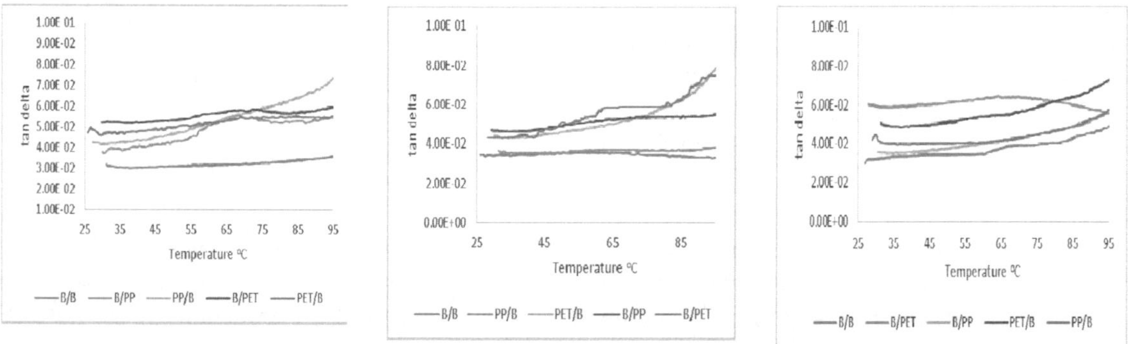

(a) Plain weave (b) Matt weave (c) Twill weave

Fig. 7: Loss factor of all woven composites

Damping factor is the ratio of the loss modulus (E″) to the storage modulus (E′). It is the ratio of the energy dissipated to the energy stored during a dynamic loading cycle. It is also known as loss factor (tan delta). Material with higher tan delta suggests that more heat was produced and more deformation could not recover when extgernal force was removed i.e., the elastic behavior decreases. In nature, tan delta is the response of internal friction forces. The mechanical damping factor can be utilized to show the impact resistance of the material. When a composite material consisting of fibers (essentially elastic), polymer matrix (viscoelastic) and fiber/matrix interfaces is subjected to deformation, the deformation energy is dissipated mainly in the matrix and at the interface (Mishra, 2014).

If one material has lower tan delta than other, it means better impact resistance and the lower tanδ value is shown in B/B exhibiting a better adhesion between basalt fibers and green epoxy matrix possible as compared to hybrid composites. This can be justified by the restriction of the motion of polymer chains resulting from the incorporation of rigid fibres. B/B have lower value of tan delta, and it indicates that more deformation for B/B composites could recover when external force is removed. The incorporation of basalt fibers for hybrid fabrics, which act as barriers to the mobility of polymer chain, led to a lower degree of molecular motion and consequently lower damping characteristics.

Thermogravimetric analysis. Thermogravimetric/Differential thermogravimetric (TG/ DTG) analysis and differential heat flow (DHF) were performed to study the thermal properties of hybrid and non-hybrid composites.

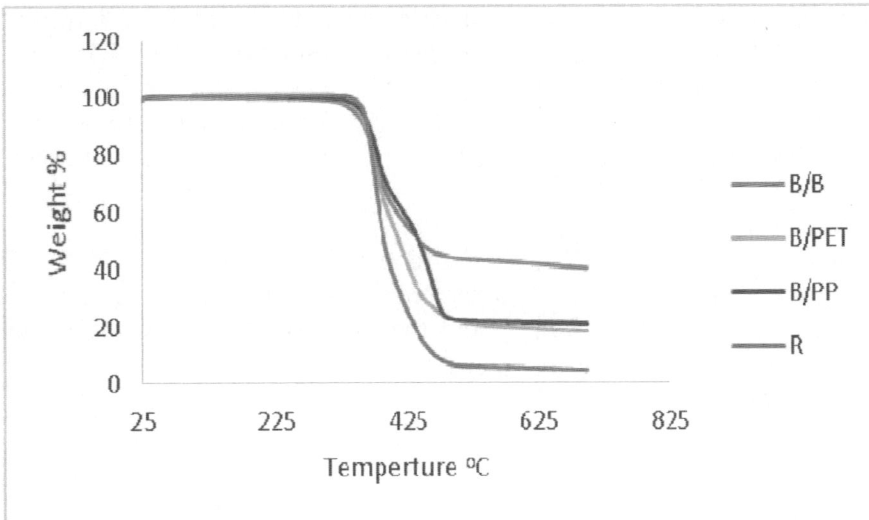

Fig. 8: TGA of composites

TGA is a useful technique for studying thermal behavior of composite materials. Thermal stability of the polymer and reinforcing materials is an important parameter because manufacturing of composites in most cases requires curing; therefore the degradation behavior of the reinforcing fibers helps in selecting the processing temperature and also the working temperature of the developed composite materials. The derivative results of TGA can be plotted to view clear picture of those events which occur at very close interval with each other. The point of greatest rate of change on mass loss in TGA curves i.e. the inflection point is indicated by peak on DTG curve that helps in getting identification of events easier and extraction of data becomes easier as well.

It can be seen from Fig. 8 that no degradation takes place until 325°C. The Thermogram of the neat green epoxy resin shows a gradual weight loss with increasing temperature which started around 320°C. From DTG thermogram it is observed that a broad single peak starts around 330°C with a maximum degradation at 375°C. The curves presented in Fig. 8 indicate the existence of only one main mass-loss region, always located between 330°C and 375°C. This region can be attributed to the thermal decomposition of the polymer matrix. From the thermograms TGA, it can be observed that, in fiber reinforced composites, two stage degradations occur: one stage is responsible for epoxy and other stage is for fiber. It can be viewed that B/B and B/PET composites degradation starts around 363°C and for B/PP at 330°C. The slightly lower onset of degradation is attributed to the decomposition of the fibers in the fabric microstructure.

Fig. 9 shows different thermograms from DTG where it is observed that, B/PP decomposition is a two-stage process characterized by a first step in the temperature range 330°C–380°C which may be attributed to degradation of resin, followed by second weight loss at 478°C and almost completely depleted. PP has branch points on alternative carbon atoms. The availability of reactive tertiary H atoms initiates its thermal degradation. Polypropylene is liable to chain degradation from exposure to heat. Oxidation usually occurs at the tertiary carbon atom present in every repeat unit. A free radical is formed here, and then reacts further with oxygen, followed by chain scission to yield aldehydes and carboxylic acids. Melting temperature of PP is around 168°C, which is observed by the peak of DHF.

It can be noticed from Fig. 8 that the polyester degradation follows a two-step reaction scheme characterized by a first step in the temperature range of 340°C–370°C followed by a second decomposition step located in the range 410°C–600°C. This behavior is determined by random scission of the polyester backbone (ester linkage) and to the oxidation and the breakage of the secondary bonds. Melting temperature of PET is around 258°C, which is observed by DHF in DSC.

The thermal analysis has illustrated that hybrid composites are stable until 325 °C. From the thermogram TGA, it can be viewed that the maximum degradation temperature T_{max} has significantly improved for B/PET fabrics and the thermal stability of the composites have been improved, which justifies the development of strong fiber-matrix interface in basalt hybrid woven composites.

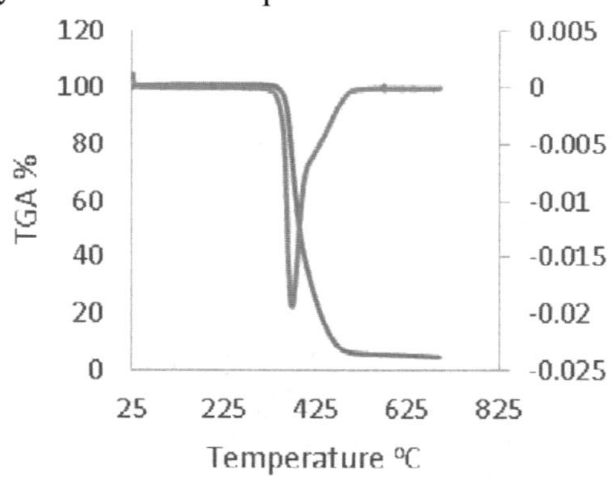

(a) Resin TGA & DTG

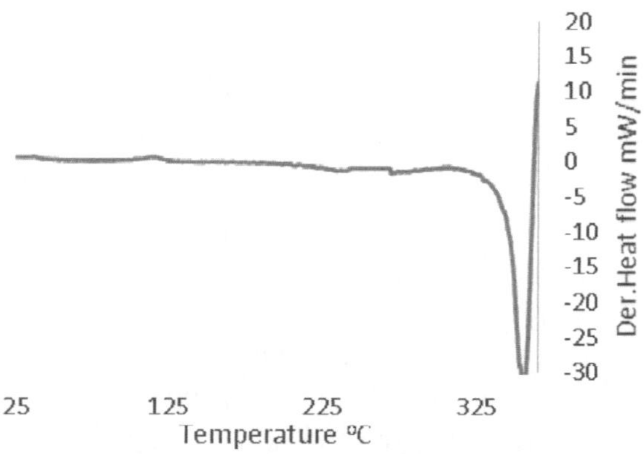

(b) Resin DHF

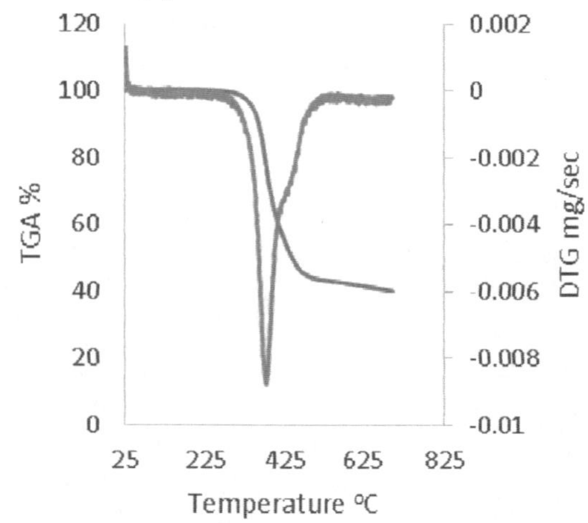

(c) B/B TGA & DTG

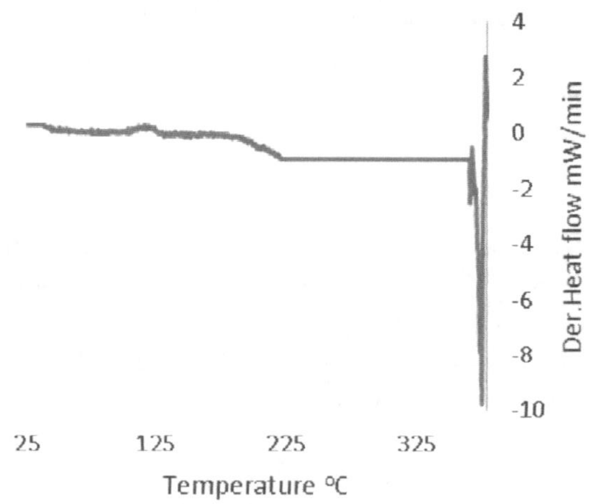

(d) B/B DHF

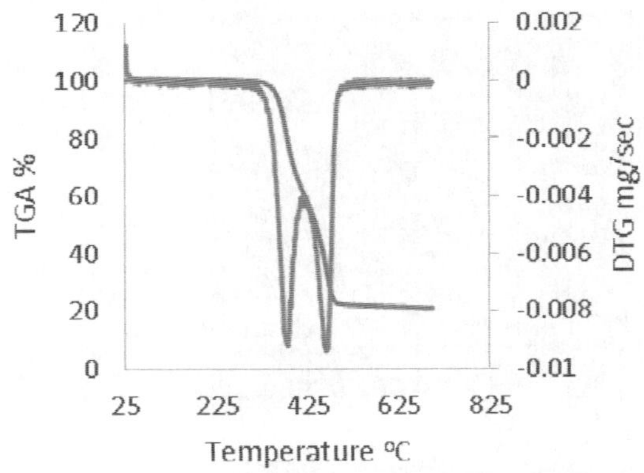

(e) B/PP TGA & DTG

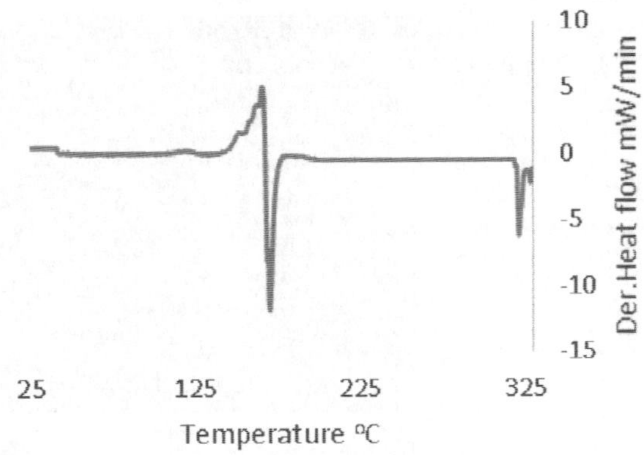

(f) B/PP DHF

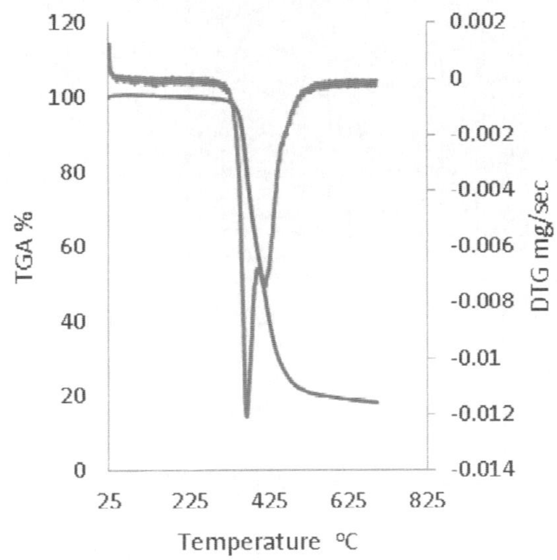

(g) B/PET TGA & DTG

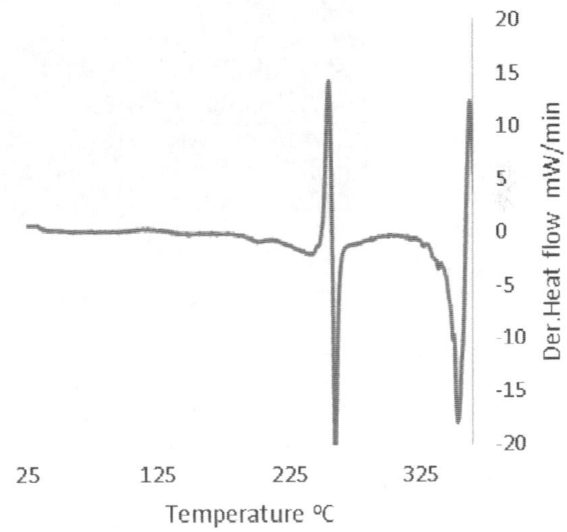

(h) B/PET DHF

Fig. 9: TGA, DTG and DHF of specimens

Morphology/interfacial properties of composites. To find out the fiber matrix adhesion inside the composites, SEM studies were carried out. SEM images of the tensile fractured surfaces of the composites are presented in Fig. 10. It is clearly indicated for PP and PET based composites that fiber-matrix interface is not so strong and the bonding between fiber and matrix is not good. Small gaps are evident in the matrix near to the fibers However, for Basalt based composites, the interface suggested better fiber matrix adhesion which is supported by the SEM images. The data presented above for the mechanical properties of composites, also supported the SEM observation.

It shows excellent interfacial strength, which means that fracture occurs rather in the body of the matrix and not at the interface. Even though photography image shows less adhesion between fiber and matrix, microscopy image could not find any fiber slippage in damaged samples. It confirms that the interface bonding of fiber and matrix are good for basalt pure and hybrid fabric composites.

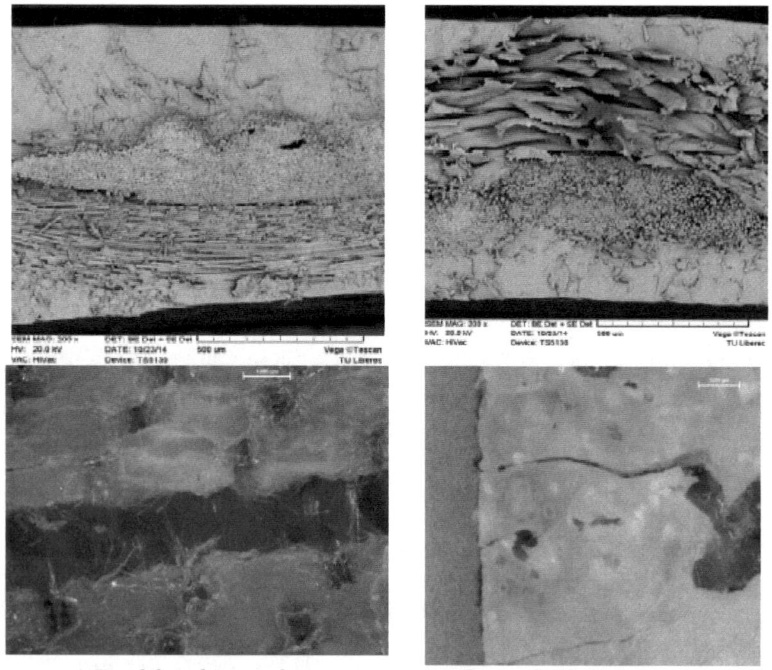

Basalt based composite Thermoplastic (PP) yarn composite

Fig. 10: SEM and macroscopic images

Conclusions

In the present work, the effect of weave and hybridization of basalt fiber on the mechanical and thermal behavior of woven fabric/green epoxy composites was investigated. Tensile, flexural and impact properties of composites were enhanced. It can be seen that 1/3 twill fabric composites have higher tensile and flexural moduli. Use of PET in weft direction increases tensile strength for all weaves. Use of PP in weft direction increases impact properties. Structure of weave, fiber type and resin properties have strong influence on mechanical properties of composites of basalt hybrid fabrics. It is concluded that, fiber type has strong influence on thermal properties of composites. The thermal conductivity of the composite depends on thermal conductivity of fiber and epoxy resin which is used for the matrix. The concept of hybrid polymer composites from basalt and synthetic fibers thus seems to be having promising application areas.

Acknowledgement

This work was supported by the Ministry of Education, Youth and Sports of the Czech Republic and the European Union – European Structural and Investment Funds in the frames of Operational Programme Research, Development and Education – project Hybrid Materials for Hierarchical Structures (HyHi, Reg. No. CZ.02.1.01/0.0/0.0/16_019/0000843), project "Modular platform for autonomous chassis of specialized electric vehicles for freight and equipment transportation", Reg. No. CZ.02.1.01/0.0/0.0/16_025/0007293, SGS grant No. 21159 of Technical University of Liberec and project "Sophisticated hybrid tapes for fabrication of composites by precise winding" [Project no: TJ01000292, 14014/136].

References

[1] H. Jamshaid, R. Mishra, and J. Militky, A green material from Rock: Basalt fiber, J. Text. I 107(2016) 923-937.

[2] H. Jamshaid, R. Mishra, and J. Militky, Thermal and mechanical characterization of novel basalt woven hybrid structures, J. Text. I 107 (2015) 462-471.

[3] R. Mishra, B. K. Behera, J. Militky, 3D woven green composites from textile waste: mechanical performance, J. Text. I. 105 (2014) 460-466.

[4] R. Mishra, B. K. Behera, J. Militky, Recycling of textile waste into green composites: performance characterization, Polym. Compos. 35 (2014) 1960-1967.

[5] R. Mishra. Drape behavior of 3D woven glass-epoxy composites. Polym. Compos. 37 (2016) 472-480.

[6] H. Jamshaid, R. Mishra, J. Militky, S. Hussain, Basalt hybrid woven textile materials for advanced thermal applications, Ind. J. Fibre Text. Res. 44 (2019) 56-64.

[7] R. Mishra, H. Jamshaid, J. Militky, Basalt Nanoparticle Reinforced Hybrid Woven Composites: Mechanical and Thermo-mechanical Performance, Fiber. Polym. 18 (2017) 2433-2442.

[8] R. Mishra, H. Jamshaid, J. Militky, Mechanical, thermal and interfacial properties of green composites from basalt and hybrid woven fabrics, Fiber Polym. 17 (2016) 1675-1686.

[9] R. Mishra, V. Arumugam, J. Militky, M. Tunak, In-plane shear behavior of 3D knitted spacer fabrics, J. Ind. Text. 46 (2016) 868-886.

[10] R. Mishra, H. Jamshaid, Thermo-Mechanical Characteristics of Basalt Hybrid & Non-hybrid Woven Fabric Reinforced Epoxy Composites, Polym. Compos. 37 (2016) 2982-2994.

[11] R. Mishra, B. K. Behera, J. Militky, Impact simulation of 3D woven kevlar-epoxy composites, J. Ind. Text. 45 (2016) 978–994.

[12] R. Mishra, J. Militky, N. Gupta, R. Pachauri, B.K. Behera, Modeling and simulation of earthquake resistant 3D woven textile structural concrete composites, Comp. B: Eng. 81 (2015) 91-97.

[13] R. Mishra, V. Baheti, B. K. Behera, J. Militky, Novelties of 3D woven composites and nanocomposites, J. Text. I. 105 (2014) 84-92.

[14] R. Mishra, Specific functional properties of 3D woven glass nanocomposites, J. Comp. Mater. 48 (2014) 1745–1754.

[15] B.P. Dash, B. K. Behera, R. Mishra, J. Militky, Modeling of internal geometry of 3D woven fabrics by computation method for structural composites, J. Text. I. 104 (2013) 312-321.

[16] R. Mishra, Impact tolerence of 3D woven nanocomposites: A simulation approach, J. Text. I. 104 (2013) 562-570.

[17] R. Mishra, J. Militky, B.K. Behera, Modeling and simulation of 3D orthogonal fabrics for composite applications, J. Text. I. 103 (2012) 1255-1261.

[18] R. Mishra, R. Tiwari, M. Marsalkova, J. Militky, B.K. Behera, Effect of TiO_2 nanoparticles on Basalt/Polysiloxane composites: mechanical and thermal characterization, J. Text. I. 103 (2012) 1361-1368.

EAN 2019
57th conference on experimental stress analysis

Experimental technique to enhance SLAM modelling of autonomous vehicle with Microsoft© Hololens

MOEZZI Reza[1,2,a], HLAVA Jaroslav[2,b], KRČMAŘÍK David[1,c] and KOČÍ Jan[1,d]

[1] Institute for Nanomaterials, Advanced Technologies and Innovation, Technical University of Liberec, Studentská 2, 461 17, Liberec, Czech Republic

[2] Faculty of Mechatronics, Informatics and Interdisciplinary Studies, Technical University of Liberec, Studentská 2, 461 17, Liberec, Czech Republic

[a]reza.moezzi@tul.cz, [b]jaroslav.hlava@tul.cz, [c]david.krcmarik@tul.cz, [d]jan.koci@tul.cz

Keywords: Localization, Mapping, Autonomous Vehicle, Extended Kalman Filter, HoloLens

Abstract. This presented paper, intends to introduce new approach to Simultaneous Localization and Mapping (SLAM) problem in autonomous vehicle by scanning the environment based on commercially available Microsoft HoloLens technology. Such a device is the first functional augmented reality gadget offered to a wide audiences. The whole mean of project is to propose an innovative hybrid architecture on SLAM. For certain landmarks extended Kalman filter (EKF) is applied. Therefore the computational complexity is minimized and remaining information for absolute reliable and accurate locations are achieved.

Introduction

In order to navigate an autonomous robot in indefinite environment, localization is necessary to know where the robot is. Localization is the estimation of robot position based on a definite map of environment. Contrarywise, mapping is the structure of environment's map, knowing the right pathway of the robot. [1] in an indefinite environment for autonomous robot, SLAM (Simultaneous Localization and Mapping) model is essential for navigation [2].

The Extended Kalman Filter (EKF) was generally implemented in the SLAM modelling for decades [3]. In mapping structure, map involves the categorizations of landmarks, which each one, characterizes the position of a barrier or a part of the barrier. New different approaches to SLAM have been developed during recent years [4]. FastSLAM is introduced by Montemerlo & Thruns [5] for stochastic SLAM. In this algorithm, the state variables are divided into two different sets, first is evaluated using particle filtering and second valued with EKF. FastSLAM significantly diminishes the problem sophistication [6]. Another approach is called L-SLAM [7] focused on orientation of each particle rather than fastSLAM, which models the orientation along the robot's position. In all methods, landmarks are fully covariant with each other and cause a quadric computational complexity according to their number in the map [8].

On the other hand, in AR (augmented reality) field, Spatial mapping feature in Micrrosft© holoLens delivers a detailed image of real-world surfaces in the environment around it (Fig 1). So, in this experimental project, we will preserve the map with the comparative locations of the landmarks and the corresponding path provided by HoloLens spatial mapping, and only a minor number of the landmarks are active in each EKF step. Therefore, the computational time and cost decreas and the absolute locations of the landmarks would be derived more reliable and accurate.

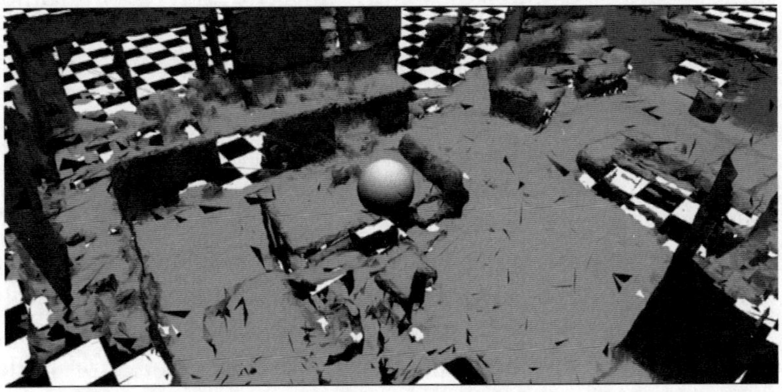

Fig. 1: 3D mapping by Microsoft© HoloLens

SLAM

The simultaneous localization and mapping problem is generally abbreviated as SLAM, define when the robot does not know its orientation and location, and also the map of environment is not accessible. In its place, control actions $u_{1:t}$ and measurements $z_{1:t}$ is given. The term "simultaneous localization and mapping" describes the follow-on problem: In SLAM, the vehicle obtains a map of its environment whereas simultaneously localize itself according to the map [9]. The complete SLAM problem would be defined in form of a probability distribution function. If the initial state of the vehicle x_0^R (state at t=0) and landmark observations Z_t and control inputs U_t up to and including time t are available, the joint posterior density of the landmark's locations and the state of the autonomous robot can be explained by the following probability distribution function:

$$P(x_t^R, m | Z_t, U_t, X_0^R) \tag{1}$$

If the state of the motion of the robot x_t^R , and the right locations of the landmarks **m** are given, the observation model describes the probability of seeing of all landmarks z_t, Consequently, we can state the sensing of the vehicle in probabilistic form [10]. For easiness, the motion of the robot and map of landmarks at time step t are noted as x_t in form of following augmented set [11] and it will be denoted as state of the whole system and the probabilistic observation model:

$$x_k = \{x_k^R, m\} \text{ (State of entire system)} \tag{2}$$

$$P(z_t | x_t) \text{ (observation model)} \tag{3}$$

An autonomous vehicle accomplishes actions in the environment that varies the position of it. This action is due to the order of control actions. If u_t is defined as control action, then $U_t = \{u_1, u_2, \ldots, u_t\}$ would be the set of actions performed by the robot at time t, transition density below expresses the way the location of the robot changes probabilistically:

$$P(x_t | x_{t-1}, u_t) \tag{4}$$

Above transition probability density acknowledge us that if at time step t-1 the vehicle was located at x_{t-1}, and performed a control action u_t, then it will stand at location x_t, at time step t. Mentioned transaction density is then motion model which defines how the control action u_t alters the location of the vehicle. SLAM can be described as a Bayesian estimation problem [10]. Therefore, location of the vehicle and landmark can be estimated with the consideration of disturbances. Regarding to probabilistic aspect, it can be said that the vehicle has a **belief** of where its pose and landmark position is. then the vehicle is not seeing one specific location but several positions for landmarks and itself at time t.

The **belief** of vehicle could be explained by a probability density total locations $x_t \in S$ where S is the set of all positions of the landmarks and the vehicle. The following formula states the belief of vehicle:

$$Bel(x_t) = P(x_t|X_{t-1}, U_t, x_0) \tag{5}$$

That is, with the reflection that the map of landmarks is m, the robot believes that it is at state x_t^R at time t, specified the initial state x_0 and all locations of the robot x_0^R, and the set of control actions U_t up to time t. This probability distribution function shows the maximum probability at which the whole system could be. The final goal of SLAM is to deliver this belief as adjacent as possible to the real system's distribution. There is only one maximum point out of this function at the true locations and elsewhere is zero. This description defines the unimodal system. Using total probability and the Markov localization [12] gives us an efficient formula for the next step as following.

$$\begin{gathered} Bel(x_t) = P(x_t|z_1, u_0, z_2, u_1, \dots, z_{t-1}, u_t, x_0) = \\ \int_S P(x_t|x_{t-1}, Z_{t-1}, U_t, x_0) \times P(x_{t-1}|Z_{t-1}, U_t, x_0)\, dx_{t-1} \end{gathered} \tag{6}$$

Markov assumption says that the prior state of the system is independent of the present state based on information of the present state. this assumption can be used, and It makes the problem simpler [9].

EKF SLAM

An explanation to the SLAM using EKF, with various motivating theoretical explanation, is comprehensively described in [9] . This algorithm is performed on autonomous vehicle, regardless of the recently reported variation of its estimation because it is an empirical for the nonlinear filtering problem. Gaussian noise assumption describes the EKF which expressively weakens the EKF SLAM's capability to deal with uncertainty. With a great uncertainty in the posterior, the linearization in the EKF sometime miscarries. Therefore, an extra hybrid model combined with Bayes filter EKF based, is demanded. the procedure has two steps, first prediction and second update to solve SLAM using the measured Lidar sensor data of the mobile vehicle. Figure 3 illustrates the result of simulation of EKF SLAM algorithm in Matlab based on webmap data collected with Lidar scanner.

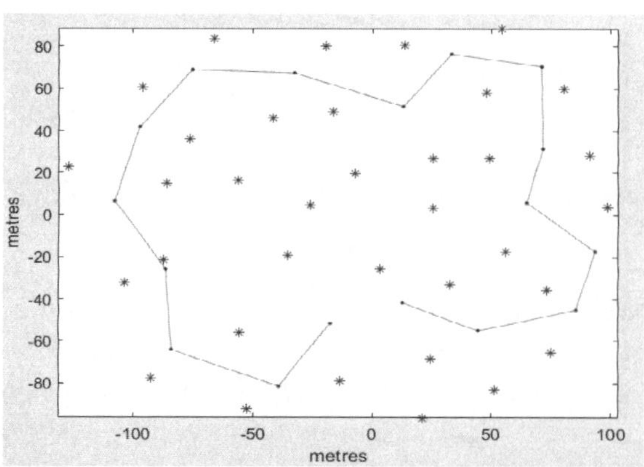

Fig. 2: EKF SLAM algorithm simulation result in Matlab

Mapping with HoloLens

HoloLens [13] with spatial mapping feature, is capable to map scene of its surrounding environment. The device performs mapping by scanning in a 70-degree cone type region between 0.8 and 3.3 m from it and sketch a mesh on what it observes. Spatial mapping is able to detect environmental objects like walls, ceiling, chairs, floor, tables and the collected information can be used to put objects in correct places (figure 3). The main purposes of this feature are holograms navigation, physics of object, occlusion, placement and visualization. Continuous mapping possibility is best suited for dynamic environments where user needs to react to the environment. Only disadvantage of continuous scanning is that it has higher CPU impact and consequently, it affects performance and heats up the device.

Fig. 3: Spatial Mapping with HoloLens in indoor environment

In mapping of environment, HoloLens applies three steps by programmed scripts, first Spatial Processing Test class allows application to scan the environment for a specified amount of time and then process the Spatial Mapping Mesh (find planes, remove vertices) after defined time has expired, this step prepare the device for mapping in a certain time.

```
public class SpatialProcessingTest : Singleton<SpatialProcessingTest>
{
    [Tooltip("How much time (in seconds) that the SurfaceObserver will run after being started; used when 'Limit Scanning By Time' is checked."
    public float scanTime = 30.0f;

    [Tooltip("Material to use when rendering Spatial Mapping meshes while the observer is running.")]
    public Material defaultMaterial;

    [Tooltip("Optional Material to use when rendering Spatial Mapping meshes after the observer has been stopped.")]
    public Material secondaryMaterial;

    [Tooltip("Minimum number of floor planes required in order to exit scanning/processing mode.")]
    public uint minimumFloors = 1;
```

Fig. 4: First step in HoloLens mapping, Spatial processing test class

The second step is Remove Surface Vertices which will remove any vertices from the Spatial Mapping Mesh that fall within the bounding volume. This can be implemented to create holes in the environment, or to help reduce triangle count after finding planes.

```
public class RemoveSurfaceVertices : Singleton<RemoveSurfaceVertices>
{
    [Tooltip("The amount, if any, to expand each bounding volume by.")]
    public float BoundsExpansion = 0.0f;


    /// <param name="source"></param>
    /// <param name="args"></param>
    public delegate void EventHandler(object source, EventArgs args);

    public event EventHandler RemoveVerticesComplete;

    private bool removingVerts = false;

    private Queue<Bounds> boundingObjectsQueue;

#if UNITY_EDITOR || UNITY_STANDALONE

    private static readonly float FrameTime = .016f;
```

Fig. 5: Second step in HoloLens mapping, Remove surface vertices

And the final step is Surface Meshes to Planes class that will find and create planes based on the meshes returned by the Spatial Mapping Manager's Observer. This class calls a sub class named Plane Types Enum Editor which enables multi-selection of the 'Draw Planes' and 'Destroy Planes' options in the Inspector.

```
public class PlaneTypesEnumEditor : Editor
{
    public SerializedProperty drawPlanesMask;
    public SerializedProperty destroyPlanesMask;

    void OnEnable()
    {
        drawPlanesMask = serializedObject.FindProperty("drawPlanesMask");
        destroyPlanesMask = serializedObject.FindProperty("destroyPlanesMask");
    }

    public override void OnInspectorGUI()
    {
        base.OnInspectorGUI();
        serializedObject.Update();

        drawPlanesMask.intValue = (int)((PlaneTypes)EditorGUILayout.EnumMaskField
                ("Draw Planes", (PlaneTypes)drawPlanesMask.intValue));

        destroyPlanesMask.intValue = (int)((PlaneTypes)EditorGUILayout.EnumMaskField
                ("Destroy Planes", (PlaneTypes)destroyPlanesMask.intValue));

        serializedObject.ApplyModifiedProperties();
    }
}
```

Fig. 6: Third step in HoloLens mapping uses plane type enum editor class

Enhanced SLAM Algorithm

An enhanced hybrid SLAM with EKF has been considered here, improved by HoloLens mapping data which acting as a spectator to acquire the on-line system uncertainty. An adaptive state estimation technique using EKF and HoloLens has been developed. In this study, the mobile vehicle with encoder values (w_{rk} ,w_{lk}) is learned using mapping Hololens data in the update-step. The mean μ_t and covariance matrix $\sum_t$, which are derived from environmental information values (x_t, y_t, θ_t) using the spatial mapping algorithm, is entered to the prediction-step, as illustrated in Figure 7.

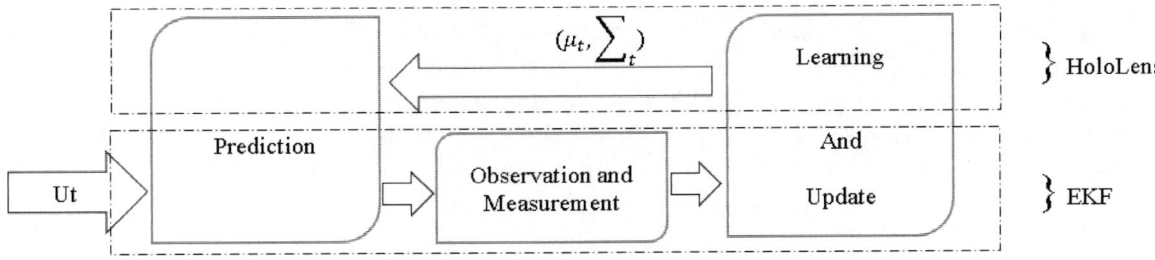

Fig. 7: The diagram of EKF and HoloLens Hybrid SLAM

Spatial mapping data particularly is very useful on measurement and observation step because it decreases calculation time. If the vehicle remains on a prediction step, it comprises all input information which may include unnecessary data. However, the combined algorithm helps to learn about only the necessary data through observation step. Through this technique, the calculation time get decreased based on less complicated problem definition and fewer iterations.

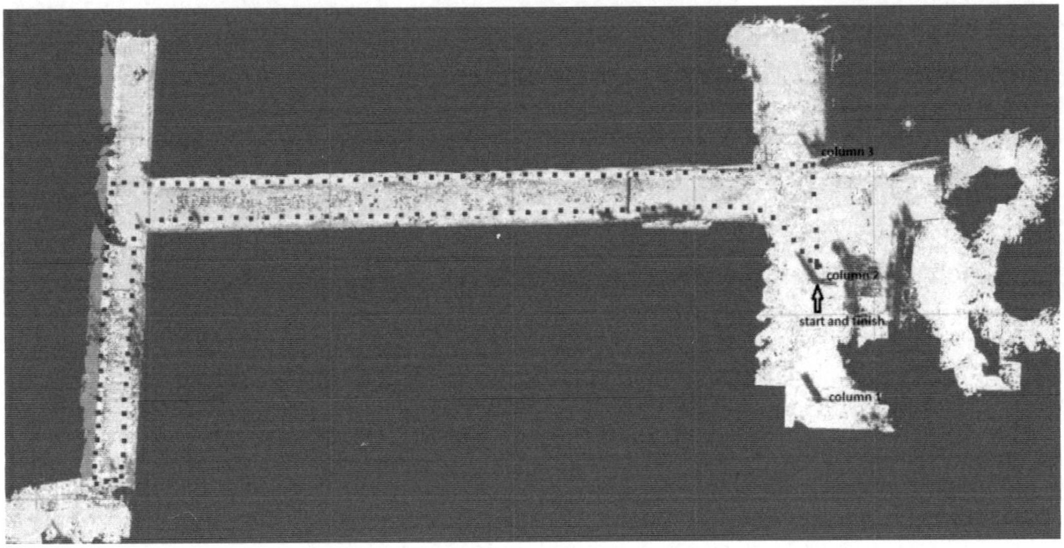

Fig. 8: Map aquired by autonomous robot in indefinite environment

Future work

According to this tremendous approach, we have reached interesting results which can be improved. To control vehicle in lateral path, fuzzy logic can be implemented for routing [14] of the Autonomous vehicle. Also, we can merge our approach with PID controller [15] based on the localization method to increase the accuracy. The camera of HoloLens can storage images of mapping to classify the most security routing using image processing tools described in [16]. The theoretical and experimental studies [17] in the field of fibre networks are rapidly increasing and there is room to carry out in autonomous vehicle. Our future work might be focus on the numerical model [18] of the Autonomous vehicle [19].

Conclusion

The SLAM is very interesting and essential problem in the field of autonomous mobile vehicle while it navigates desired path and simultaneously creating a map of the physical environment and the vehicle estimates position of itself on corresponding map. This paper illustrated an enhanced hybrid SLAM method which combines augmented reality device on a mobile robot, to minimize the EKF SLAM error which mostly caused by its linearization procedure and disturbance assumption. This algorithm, as writers have discovered, is the first innovative approach in SLAM problem which applies augmented reality capabilities. The simulation results of EKF SLAM is shown, the HoloLens classes for mapping are studied and the experimental result of hybrid mapping architecture is obtained.

Acknowledgment

The result is obtained through the financial support of the Ministry of Education, Youth and Sports of the Czech Republic and the European Union in the frames of the project “Modular platform for autonomous chassis of specialized electric vehicles for freight and equipment transportation”, Reg. No. CZ.02.1.01/0.0/0.0/16_025/0007293

References

[1] R. Siegwart, I. Nourbakhsh and D. Scaramuzza, Introduction to Autonomous Mobile Robots, London, England: The MIT Press, 2011.

[2] J. Sasiadek, A. Monjazeb and D. Necsulescu, "Navigation of an autonomous mobile robot using ekf-slam and fastslam," 16th Mediterranean conference on Control and automation, p. 517–522, 2008.

[3] J. Sol`a, Simulataneous localization and mapping, Barcelona: Institut de Robòtica i Informàtica Industrial, CSIC-UPC, 2014.

[4] Röfer, F. Udo and W. T. René, "A SLAM Overview from a User’s Perspective," Künstl Intell, vol. 24, p. 191–198, 2010.

[5] M. Montemerlo and S. Thrun, "Simultaneous localization and mapping with unknown data association using fastSLAM," Proceedings of IEEE International Conference on Robotics and Automation, ICRA, vol. 2, no. 3, p. 1985–1991, 2003.

[6] J. Sasiadek, A. Monjazeb and D. Necsulescu, "Navigation of an autonomous mobile robot using ekf-slam and fastslam," 16th Mediterranean conference on Control and automation, p. 517–522, 2008.

[7] V. Petridis and N. Zikos, "L-slam: reduced dimensionality fastslam algorithms," in WCCI, Barcelona, 2010.

[8] L. Wei, W. Tao and Z. Yachong, "A Relative Map Approach for Efficient EKF-SLAM," in Proceedings of 2014 IEEE Chinese Guidance, Navigation and Control Conference, Yantai, China, August 8-10, 2014.

[9] S. Thrun and W. Burgard, Probabilistic Robotics, London: The MIT Press, 2006.

[10] S. Thrun, "Probabilistic algorithms in robotics," AI Magazine, vol. 21, no. 4, pp. 93-109, 2000.

[11] J. Guivant, E. Nebot and S. Baiker, "Localization and map building using laser range sensors in outdoor applications," Journal of Robotic Systems, vol. 17, no. 10, pp. 565-583, 2000.

[12] S. Thrun, D. Fox and W. Burgard, "A real-time algorithm for mobile robot mapping with applications to multi-robot and 3D mapping," in IEEE International Conference on Robotics and Automation, San Francisco, 2000.

[13] M. Garon and P.-O. Boulet, "Real-time High-Resolution 3D Data on the HoloLens," in IEEE International Symposium on Mixed and Augmented Reality Adjunct Proceedings, USA, 2016.

[14] Moezzi Reza, Vu Trieu Minh, "Fuzzy logic control for non-linear model of the ball and beam system," 11th International DAAAM Baltic Conference “INDUSTRIAL ENGINEERING” 20-22nd April, Tallinn, Estonia, 2016.

[15] Hlava, Jaroslav, Nikita Zemtsov, and Galina Frantsuzova. "Application of PID controller based on the localization method for ancillary service provision." 2016 International Siberian Conference on Control and Communications (SIBCON). IEEE, 2016.

[16] Haythem, Bahri, et al. "Accelerating Fourier descriptor for image recognition using GPU." Applied Mathematics & Information Sciences 10.1: 297, 2016.

[17] Krčmařík, David, et al. "Theoretical and experimental analysis of long-period fiber gratings made directly into Er-doped active fibers." Journal of Lightwave Technology 27.13: 2335-2342, 2009

[18] Petru, Michal, et al. "Experimental and numerical analysis of crack propagation in light composite materials under dynamic fracturing." Communications-Scientific letters of the University of Zilina 16.3A : 82-89, 2014.

[19] Petru, Michal, Tomas Martinec, and Jaroslav Mlynek. "Numerical model description of fibres winding process for new technology of winding fibres on the frames." Manufacturing Technology 16.4 (2016): 778-785, 2016.

Theory vs. experiment

OKROUHLÍK Miloslav

Institute of Thermomechanics, Dolejškova 5, 182 00, Prague 8, Czech Republic

ok@it.cas.cz

Keywords: Stress waves, Singularities, Threshold, Model validity.

Abstract. As generally accepted today, the model is a purposefully simplified concept of a studied phenomenon invented with the intention to predict – what would happen if … Accepted assumptions (simplifications) consequently specify the validity limits of the model and in this respect, the model is neither true nor false. The model, regardless of being simple or complicated, is good, if it is approved by an appropriate experiment. When we are modelling particular phenomena of Mother Nature, the question of truth becomes thus irrelevant since the models we are designing, checking and using, either work or do not work to our satisfaction. So it is obvious that we rather strive for robust models with precisely specified limits of validity and not for philosophically defined categories of truth and falsehood. From it follows that it is the validity of models, theories and laws that is of primary importance.

Introduction

The author ponders about things that necessarily come into engineering mind when the results obtained by theoretical, numerical and experimental approaches in solid continuum mechanics are correlated and compared with a pious wish to ascertain which of them are 'truer' or closer to 'reality'. This invokes many questions. How is the truth related to consistency and validity of theoretical, numerical and experimental models we are inventing and employing? What is the role of thresholds in physics, engineering, computation and in an experiment? The doubts stemming from uneasy answers to above pertinent questions are complemented by discussing examples from theoretical, numerical and experimental results obtained by solving dynamical problems in solid continuum mechanics.

Model vs. experiment

To get rid of doubts we often claim that it is the experiment, which ultimately confirms the model in question. But experiments, as well as the subsequent numerical treatment of models describing the nature, have their observational thresholds. And sometimes, the computational threshold of computational analysis is narrower than those of an experiment. From this point of view, a particular experiment is a model of nature as well.

Another mental hindrance we might have, in our incessant quest for truth, is the lack of precise definitions of certain mechanical quantities. Definitions of conceptually defined quantities as force, stress, energy, etc. are rather intuitive and often circular.

Examples and hints

Practical examples from the field of stress wave propagation tasks will be presented with the intention to show remedies required to avoid 'well-conceived' ideas, errors and blunders.

Example_1. Wavefront singularities and threshold uncertainty

It is a good practice to have a solid theoretical background at hand. When the stress wave propagation tasks, induced by shocks and impact, are treated, one should be aware of the fact that there are two fundamental stress wave types – namely the longitudinal and transversal ones. This fact was explained by Christiaan Huygens, (1629 – 1695). The wavefronts of longitudinal (also called P, primary, irrotational, dilatational, extensional) and transversal (S, shear, rotational, equivolumetrical) are schematically depicted in Fig. 1 for an elastic halfspace loaded by a point force whose time distribution is prescribed by the Heaviside function. Also, the appearance of the wavefront of so the called Schmidt's wave, introduced by the surface points, which were already hit by the longitudinal wave, become sources of the slower – i.e. transversal – waves, is depicted.

In this respect, it should be emphasized that a frequently used term 'the sound velocity in solids' is wrong and unjust. Actually, we have distinct speeds for 1D, plane stress, plane strain, and 3D cases.

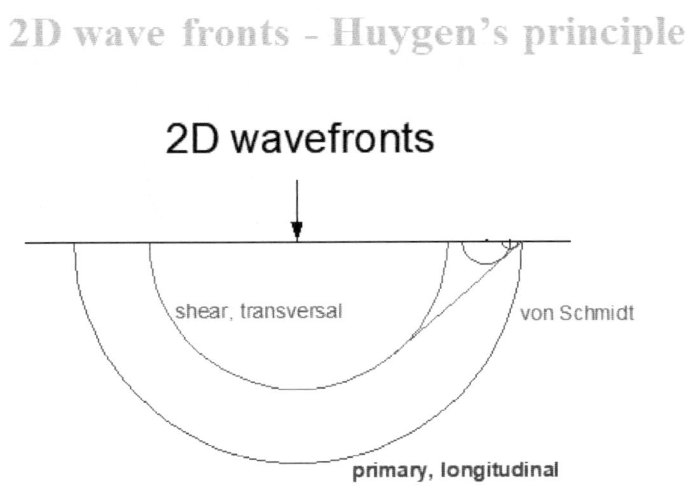

Fig 1. Wave fronts in elastic half space due to the sudden application of a point force

These phenomena are easily 'proved' by finite element computations as shown in Fig. 2. The upper part of the figure shows the velocity distribution of in an elastic solid due to a point loading, whose time distribution is given by the rectangular pulse. The depicted vector directions justify the accepted terminology – longitudinal and transversal waves. Also, the appearance of so-called Rayleigh waves, characterized by elliptic patterns of particles in the vicinity of the body's surface, is shown. See [1].

The lower figure, also a result of finite element computations, clearly shows the appearance of Schmidt's waves and the effect of reflections on the opposite boundary.

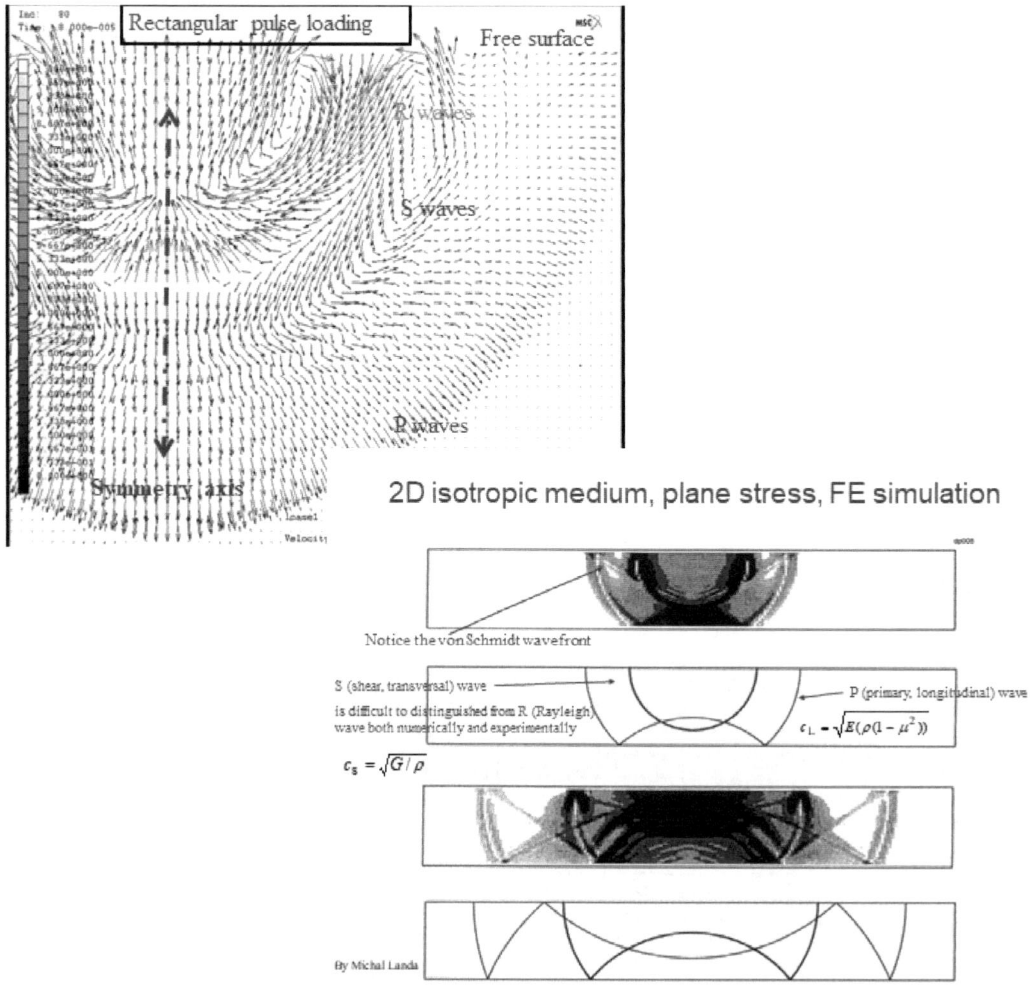

Fig. 2. Computed wavefronts

Lamb's problem

The task is schematically illustrated in Fig. 3. The original Lamb's analysis employs Fourier superposition of harmonic waves for the transient normal loading on a half-space. Later Cagniard presented a numerical method for the inversion of the Fourier's transform, and Brepta [2] and others were able to carry out the inversion analytically even for cases with more complicated boundaries. The analytical solution the Lamb's problem on a free boundary is relatively simple and is available in a closed form giving space-time distributions of radial and axial displacements. It is attributed to Pekeris, whose analytical formulae, presented in [3], are easy to evaluate. Computed results, depicted in space-time coordinates in Fig. 4, show two significant singularities as well as arrivals of longitudinal, shear and Rayleigh waves.

The existence of singularities in our models (and the solid continuum is a model) always indicates certain unjustified assumptions and/or simplifications accepted at the beginning of the modelling process. In this case, the singularities occur due to the application of the point force, which, strictly speaking, is a forbidden phenomenon in continuum mechanics.

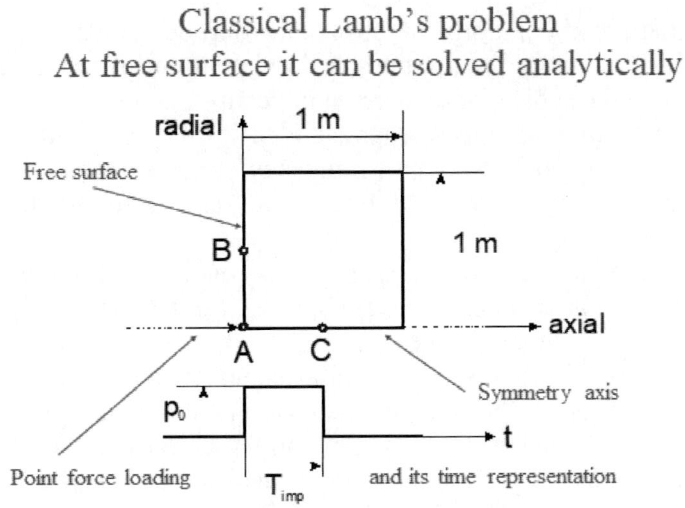

Fig. 3. Lamb's problem

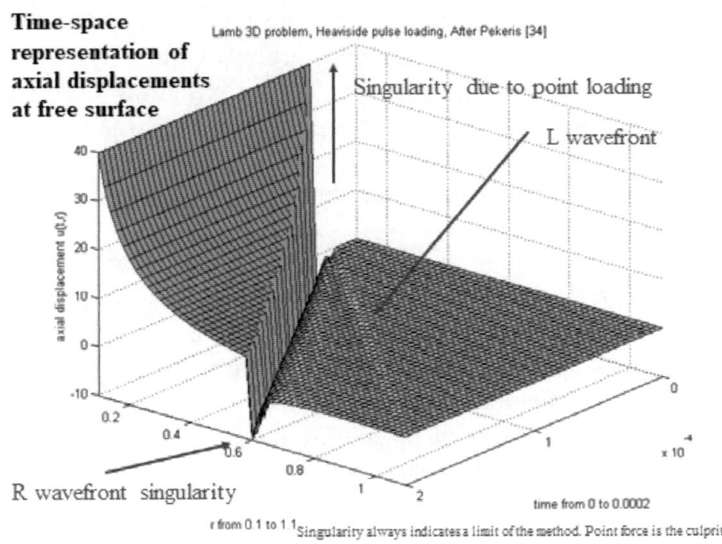

Fig. 4. Lamb's problem – analytical solution

Seemingly unproblematic model of elastic continuum has embedded singularities in it. For example a point force, a frequently used tool in engineering analysis, is a forbidden entity in continuum mechanics since it leads to a singularity response – this is manifested by the fact that the displacement under the application of a point force tends towards infinity. To a certain extent this property is retained when the continuum is treated by means of a FE model. Actually, it is smeared out by the existence of shape functions but with diminishing meshsize it is well observable by the increase of displacement under the application of a point (nodal) force. The FE mesh made of 'null-sized' elements would provide the infinite displacement under the application of a nodal force as the continuum model. So making a finer and finer mesh we are representing better and better those continuum properties that are mathematically correct but physically unattainable. This is a sort of paradigm we are used to live with. Singularity in displacement response to a point loading, Rayleigh waves and crack analysis are well-known examples both in continuum and its FE representation. See [4].

Example 2. – Dispersion

In finite element computations of stress wave propagation tasks, the dispersion phenomena significantly influence the accuracy and reliability of obtained results. That's why we focus our attention on the subject of dispersion in detail.

Dispersion is a medium property characterized by the fact that the velocity of the propagating harmonic wave depends on its frequency. Consequently, the wave packages, being composed of the whole spectrum of frequencies, are distorted during their passage through the medium.

A seemingly academic theoretical subject of spatial and temporal dispersion in discretized continuum solid mechanics still has significant practical consequences for current users of commercial finite element packages.

Studying the subject of dispersion allows understanding and minimizing the unpleasant side effects typical for the finite element computations of non-stationary dynamics tasks of solid computational mechanics. Also, the 'correct' determination of mesh size and the time step can be safely determined. It is shown that the Fourier analysis is an excellent tool for checking the obtained results and for the safe estimation of their validity.

In solid continuum mechanics, we distinguish *material*, *geometrical*, *spatial* and *temporal* types of dispersion. The *material dispersion* is induced by the damping properties of the medium through which the wave propagates. The *geometrical dispersion* occurs due to changing the geometrical shapes of waveguides through which the waves propagate, The *spatial dispersion* is typical for the numerical approaches based on discretization in space, in which originally continuum structure of material is replaced by small finite parts – elements, and finally, the *temporal dispersion* is typical for the numerical solution of transient tasks – the output quantities, instead of being described by continuous functions of time, are evaluated in discrete time intervals only.

In solid continuum mechanics, there is a group of the wave propagation tasks which are purely dispersionless. As examples, the propagation of stress waves in unbound 1D, 2D, and 3D spatial geometries could be mentioned. See [1], [5]. The theoretical cases described and solved there can be efficiently exploited as useful etalons for judging the seriousness and amount of dispersion errors in discrete media.

On the other hand, the FE model of continuum is generally of dispersive nature, see [7], [8]. To show and explain dispersion phenomena in detail, let's study the time response of a thin rod modelled by 1D constant strain elements, while the behaviour of idealized dispersionless solid continuum, represented by the wave equation, can serve as a suitable etalon.

Studies concerning the finite element dispersion analysis of 2D and 3D elements are numerous – see [7], [8], [9]. The conclusions, shown for the 1D wave equation and for its 1D finite element model, are quantitatively valid for 2D and 3D elements as well.

It should be emphasized that in 1D continuum, there is only a single speed of wave propagation, i.e. $c_0 = \sqrt{E/\rho}$.

However, due to the existence of normal and shear stresses in 2D and 3D continua, there are two kinds of stress waves that could propagate in an unbound medium. See [1], [5]. There are the *primary* (P, *longitudinal*) and *transversal* (S, *shear*) waves – their speeds are different. The speed formulas and values (in m/s), for a typical steel material ($E = 2.1\times10^{11}$ Pa, $\rho = 7600$ kg/m^3, $\mu = 0.3$), are as follows.

$c_0 = \sqrt{E/\rho} = 5189$ ... 1D wave, slender bar,

$c_1 = \sqrt{(2G+\lambda)/\rho} = 6020,$ ... P wave for plane strain and for 3D,

where $\lambda = \dfrac{\mu E}{(1+\mu)(1-2\mu)}$, $G = \dfrac{E}{2(1+\mu)}$.

$c_2 = \sqrt{G/\rho} = 3218$ ... S wave, shear, both for 2D and 3D,

$c_3 = \sqrt{E/(\rho(1-\mu^2))} = 5439$ ... P wave for plane stress.

In Fig. 5 the speeds of wave propagation for 1D constant strain elements with consistent and diagonal mass matrices (L1C and L1D) are compared with those of 2D constant strain elements for an equilateral triangle with consistent and diagonal mass matrices. The right-hand part of the figure depicts the propagation speeds the longitudinal (upper group of curves) and transversal (lower group of curves) waves as functions of the dimensionless wavenumber. As we have seen before, the elements with consistent mass matrix (plotted by dashed lines) overestimate the 'correct' speed of propagation (horizontal lines) while the speed of propagation for elements with the diagonal mass matrix is systematically underestimated. In the discretized 2D continuum the speed of propagation is furthermore influenced by the direction of the wave propagation, indicated by the δ parameter in the figure. This means that a discretized 2D medium exhibits the anisotropy behaviour, not existing in an ideal 2D model of solid continuum. See [10].

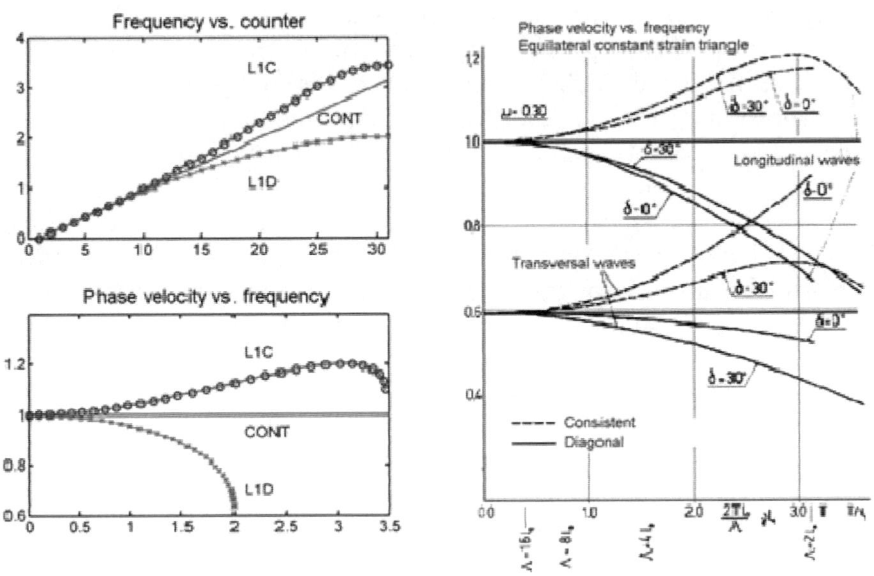

Fig. 5. Dispersion properties for constant strain elements in 1D and 2D

On the x-axis there are indicated values of another parameter, namely $\Lambda = k \times l_0$, where k is the number of elements that could be 'squeezed' into the considered wavelength. One sees that if there are 16 elements into a wavelength, then the dispersion error is negligible – for only 2 elements, the dispersion error is considerable. On should emphasize that the dispersion errors could be minimized, but not completely eliminated.

Generally, the results of modelling should be independent of the method being used. This is not always so as shown in Fig. 6, where the same task – propagation of a halfsine pulse – is treated using consistent and mass matrix formulations respectively, resulting in different positions of the dispersion side effects, i.e. the 'false' vibrations appearing – either in front of or behind the pulse.

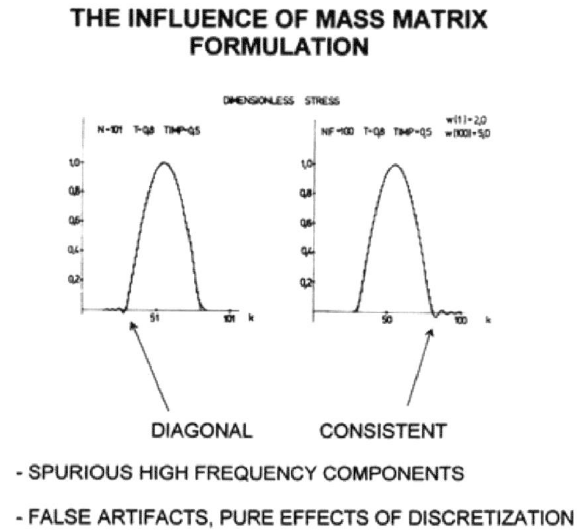

Fig. 6. Consequences of mass matrix formulation

The dispersion analysis of higher order 1D elements (i.e. quadratic and cubic), see [7] revealed that using these elements in 1D, we get the a chance to increase the cut-off limit of the highest available frequency but at the expanse of higher computing costs and of the appearance of embedded band pass filters. The dimensionless phase velocities as functions as dimensionless frequencies for 40 linear (L1), 20 quadratic (L2) and 13 cubic (L3) finite elements are depicted in Fig. 7. Approximately, the same number of degrees of freedom was used. One can observe that for the higher order elements, the first part of dispersion curves are 'improved', the dispersion curves, however, consist of distinct branches that are separated by certain frequency regions (called the cut-off filters) through which the particular frequencies cannot propagate. For the L2 element, the second part of the dispersion function is called the *acoustic branch*. For the L3 element, there is also the third part which is traditionally called the *optical branch*. The attributes C and D appearing by the L1, L2 and L3 identifiers signify the consistent and diagonal mass matrices formulations. Generally, using the higher order elements we get a longer spectrum with 'acceptable' dispersive properties, not differing much from that of the ideal continuum. But at the cost.

The effects of dispersion induced anisotropy in 2D modelling of stress wave propagation are shown in Fig. 8 for a body assembled of square elements with bilinear shape functions. You might notice it is direction dependent. Different meshes, characterized by the ratio of the mesh size (q) to the wave length (Λ), are considered.

The polar dispersion diagrams for bilinear and serendipity biquadratic finite elements with the consistent mass matrix and for four different dimensionless wavelengths of propagation wave are shown in Fig. 9.

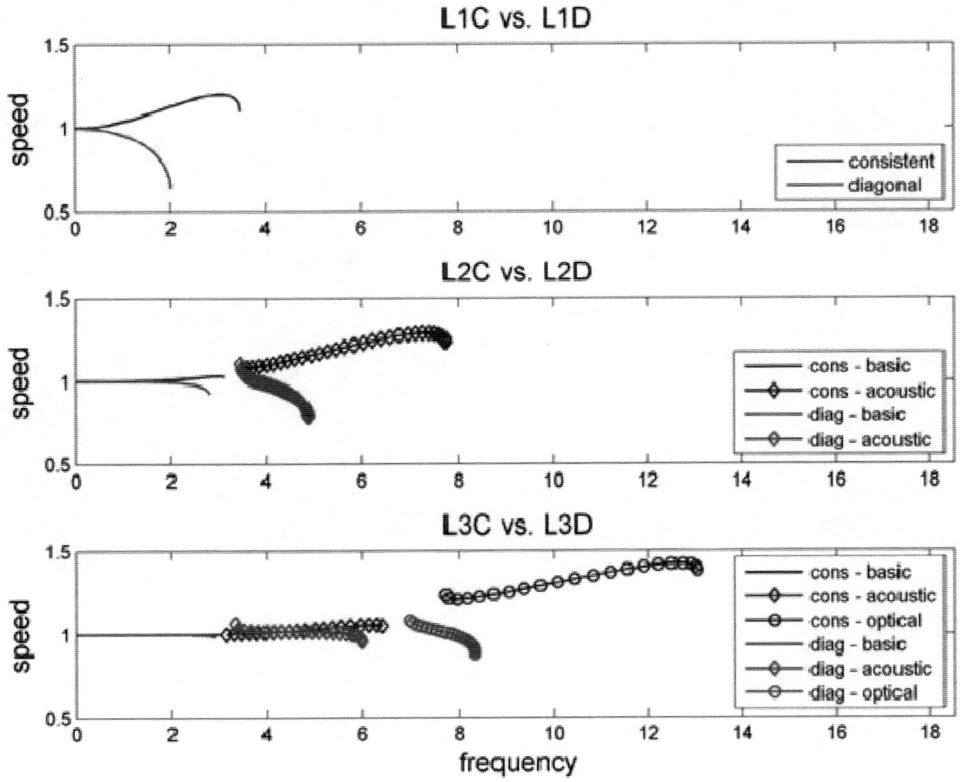

Fig. 7. The frequency spectra of higher order 1D elements

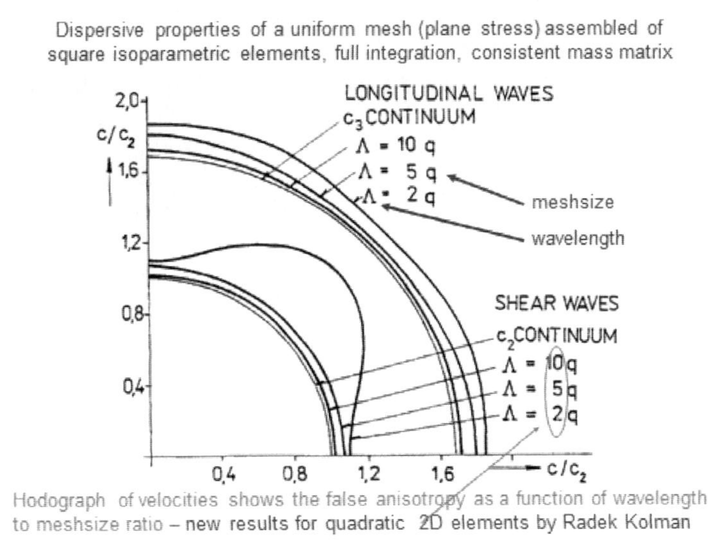

Fig. 8. Bilinear square element

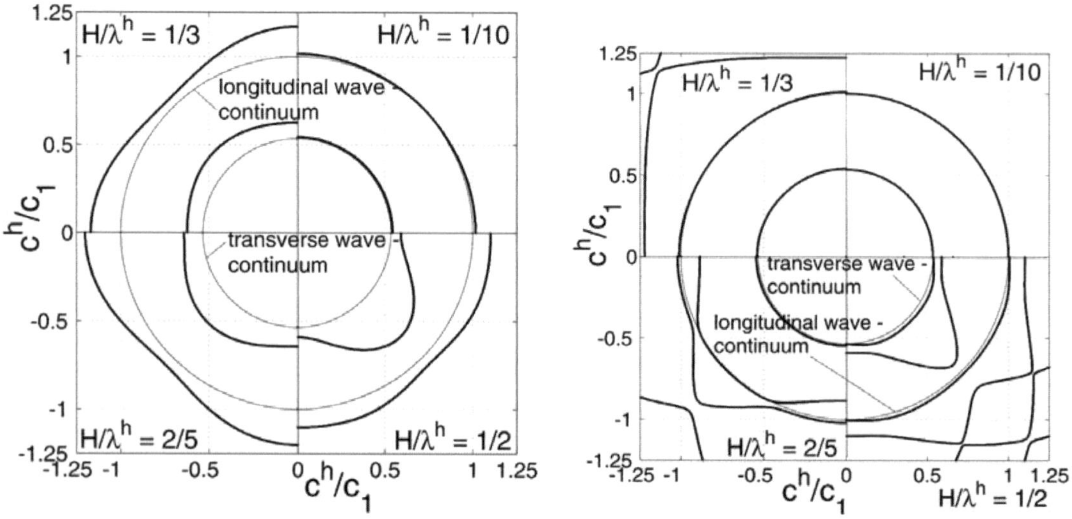

Fig. 9. Quadratic square elements

Polar dispersion diagrams for a plane square 4-node bilinear (on the left) and 8-node serendipity biquadratic (on the right). Finite elements with the consistent mass matrix for four different dimensionless wavelengths are presented. See [11].

Example 3. – Computational and experimental thresholds

Limits of all kind of analyses (analytical, numerical and/or experimental) depend on the observational resolution – threshold, which is typically a minimum value of a signal that can be detected by the system.

Validity limits are checked by comparison with analytical methods, comparison with experiment, self-assessment, and statistical tools and by FFT analysis.

In this paragraph, the investigation of the stress wave propagation in an impact loaded shell by Double Pulse Holographic Interferometry is compared with a corresponding finite element analysis. The task is schematically depicted in Fig. 10. The shell was loaded by an exploding wire.

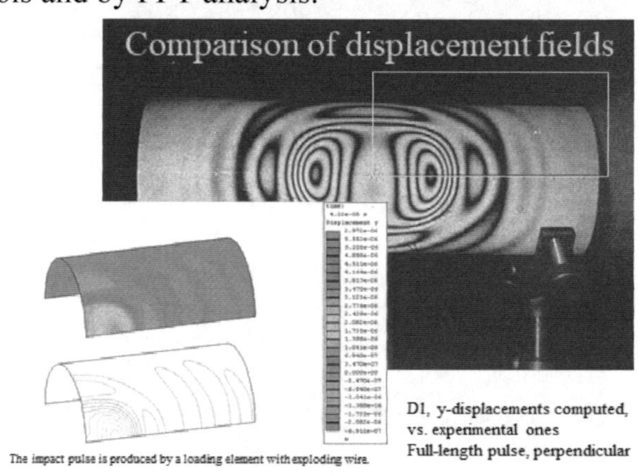

Fig. 10. Impact loaded shell

Velocities, as functions of time, were experimentally recorded and compared to those obtained by finite element computations with two different mesh densities. A rather unsatisfactory agreement or both approaches is presented in Fig. 11.

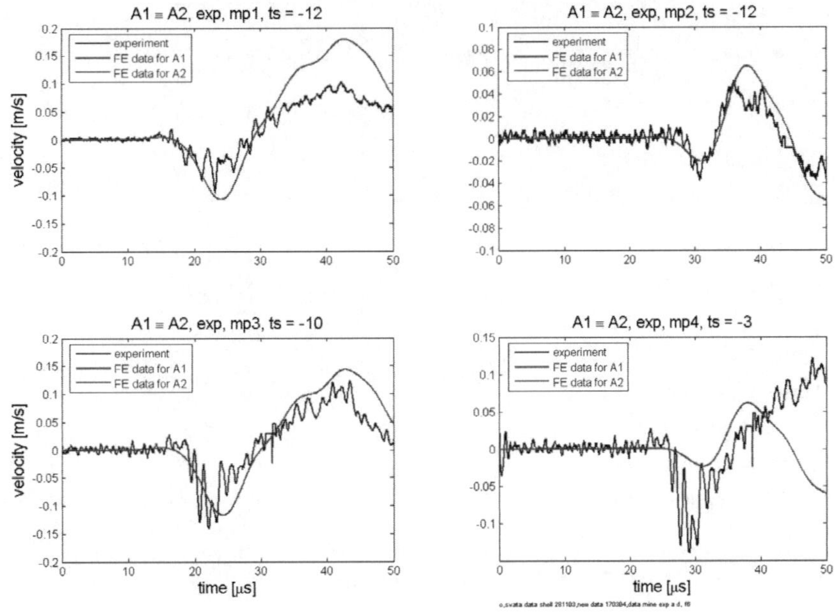

Fig. 11. Experiment vs. two FE meshes at four locations

Finite element limits

The FE shell results for the coarse (A1) and fine (A2) meshes are not distinguishable in the scale of the figure. Since there is no visible difference of results with mesh refinement one can conclude that time step and mesh size are set correctly. From this fact, however, we cannot deduce that the results are 'physically correct'. They are 'correct' within the validity range applicable to this particular FE model – the shell element a priori assumes what is happening within the shell thickness and does not take into account the actual wave processes occurring there.

Experimental limits – the first part

Applying the *Matlab improfile function* to experimental data allows determining both the absolute and relative thresholds of the measurement used. The procedure is outlined in Fig. 12.

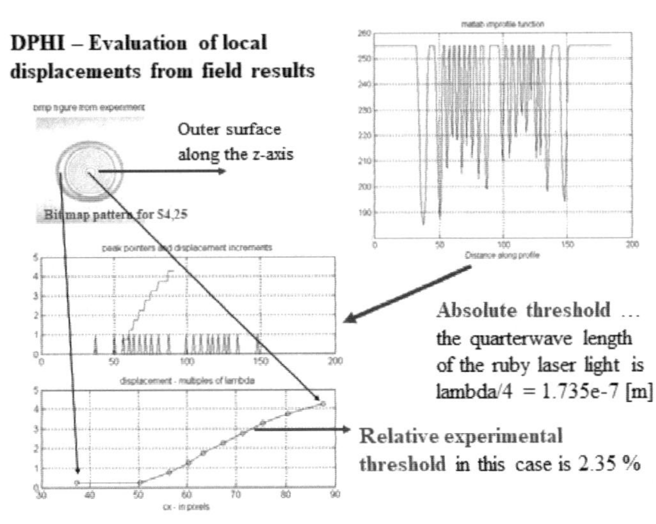

Fig. 12. Experimental threshold analysis

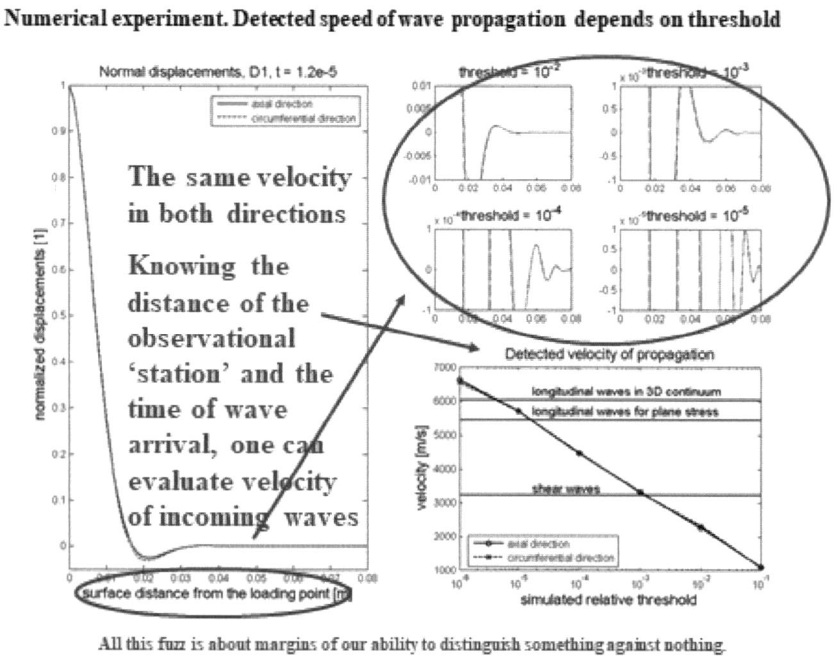

Fig. 13. Numerically simulated threshold. The registered speed of propagation depends on the threshold.

The threshold issue is important when the speed of propagation is to be determined by experimental means. Observing the first measurable response at a certain time in a given distance from the loading point, one can estimate the propagation velocity. This way the estimated value depends on the threshold value of an apparatus being used for the measurement of a particular physical quantity and is usually constant at the experiment site.

The observational threshold is usually constant for the considered experimental setup being used for the measurement of a particular physical quantity.

The influence of the threshold value on the measured results could be achieved by a numerical thought experiment.

What would be a common sense approach? Sitting at a certain observational node, whose distance from the loaded node is known, one would estimate the speed by measuring the time needed for the arrival of the 'measurable' or 'detectable' signal. And the measurable signal is such a quantity that is – in absolute value – greater than a 'reasonable' observational threshold. And what is a proper value of it is a good question.

Imagine a standard finite element double-precision computation giving at a certain time the spatial distribution of displacements at a node on the surface of a body. Assume that the distance of our observational node from the loading node is known. Now, let's set a 'reasonable' value of the threshold and apply a sort of numerical filter on obtained displacements, which erases all the data whose absolute values are less than the mentioned value of the threshold. This way, for a given threshold value, we get a certain arrival time and from the known distance we obtain the propagation speed. Working with displacements normalized to their maximum values allows us to consider the threshold values as the relative ones.

Varying the simulated threshold value in the range from 10^{-6} to 10^{-1} we will get a set of different velocities of propagation. As a function of the simulated threshold they are plotted in Fig. 13. Material constants for the standard steel were used. The horizontal lines represent the theoretical speeds for longitudinal waves in 3D continuum, for longitudinal plane stress waves in 2D continuum as well as for the shear waves. Obviously, the shear wave speeds are identical both for 3D and 2D cases.

It is known that the longitudinal waves carry substantially less amount of energy than these of the shear and Rayleigh waves and that the surface response, measured in displacements or strains, is of substantially less magnitude for the former case.

One can conclude that for a correct capturing of the longitudinal velocity value, the relative precision of at least of the order of 10^{-6} is required. This is a tough request. The relative threshold of the order of 10^{-3} is more common in experimental practice. However, in an experiment with the relative precision of the order of 10^{-3}, one would not detect the arrival of longitudinal waves and might wrongly conclude that the first arriving waves are of the shear nature and would wrongly estimate the velocity of propagation of the order of 3000 m/s.

All this fuzz is about margins of our ability to distinguish something against nothing. This is, however, crucial for any meaningful human activity.

Physical and technical aptness and reliability of the model depends not only on a threshold value but – and mainly – on application of the model within the limits for which it was constructed, i.e. proper satisfaction of initial and boundary conditions, material assumptions, mesh properties, etc.

Experimental limits – the second part

Trying to explain the differences in velocity distributions, presented in Fig. 11, we focused our attention to the data related to the experimental loading process. As mentioned before, the experimental loading was secured by exploding wires. Actually, four experiments were carried out to get velocity distributions at the above-mentioned four locations. The time distribution of four randomly taken pulses is depicted, together with an 'average pulse' that was used as an input for the numerical computation, in the first row of Fig. 14.

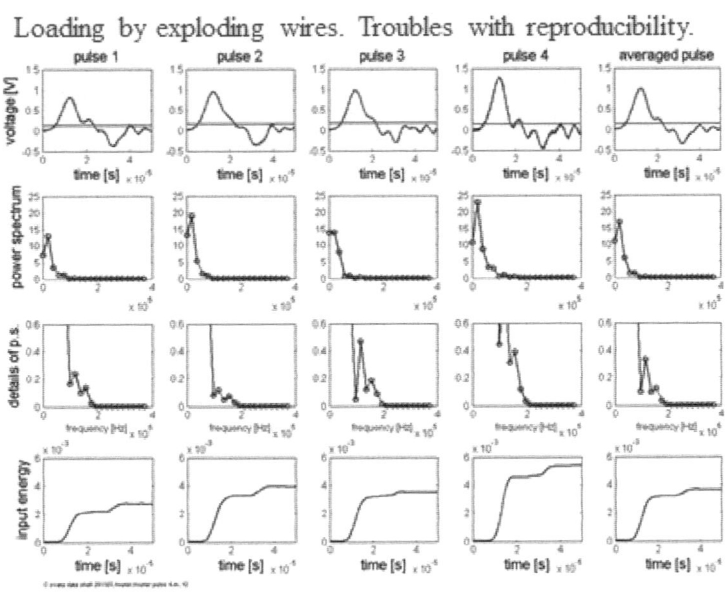

Fig. 14. Troubles with reproducibility

Shown discrepancies in time distributions of kinematic quantities, computed for individual pulses, indicate that differences in the loading distributions play a more important role than originally believed.

In two subsequent rows of Fig. 14 one can observe corresponding Fourier's power spectra of loading pulses in full and detailed views. The spectra were computed a posteriori by FFT. First twenty discrete FFT frequencies are plotted only. Having, however, a proper hardware or using a real time simulation package, the FFT analysis is readily available at the experimental site as well, and could be obtained immediately after the experiment (meaning the explosion of the wire) has been carried out.

The fourth line of subfigures presents the sum of potential and kinetic energies supplied by the particular pulse to the loaded body. This quantity – a result of FE computations – is not readily available to the experimentalist.

There are, however, other quantities – available to the experimentalist – that might help. Realizing that the first term of the Fourier's transform analysis is $c_0 = \int_{-\infty}^{\infty} P(t) e^{i0t}\, dt = \int_{t_{\text{initial}}}^{t_{\text{final}}} P(t)\, dt$, reveals that its physical meaning is the mechanical pulse. Denoting the discretized time evolution of reference pulse $\{\mathcal{P}\} = \{\mathcal{P}_1\ \mathcal{P}_2 \ldots \mathcal{P}_n\}^{\mathrm{T}}$ we can express its mean value $\mathcal{P}_{\text{mean}} = \frac{1}{n}\sum_{i=1}^{n}\mathcal{P}_i$ and the approximate the area under the pulse $\int_{t_{\text{initial}}}^{t_{\text{final}}} P(t)\, dt$ by $\mathcal{P}_{\text{mean}}\, \Delta T$. Which means the same.

For given pulses, this correlation is shown in the first column of Fig. 15. Without going to details one can show that there is another useful correlation, namely, between the Euclidian norm of the pulse quantities and the square root of input energy. For details see [4].

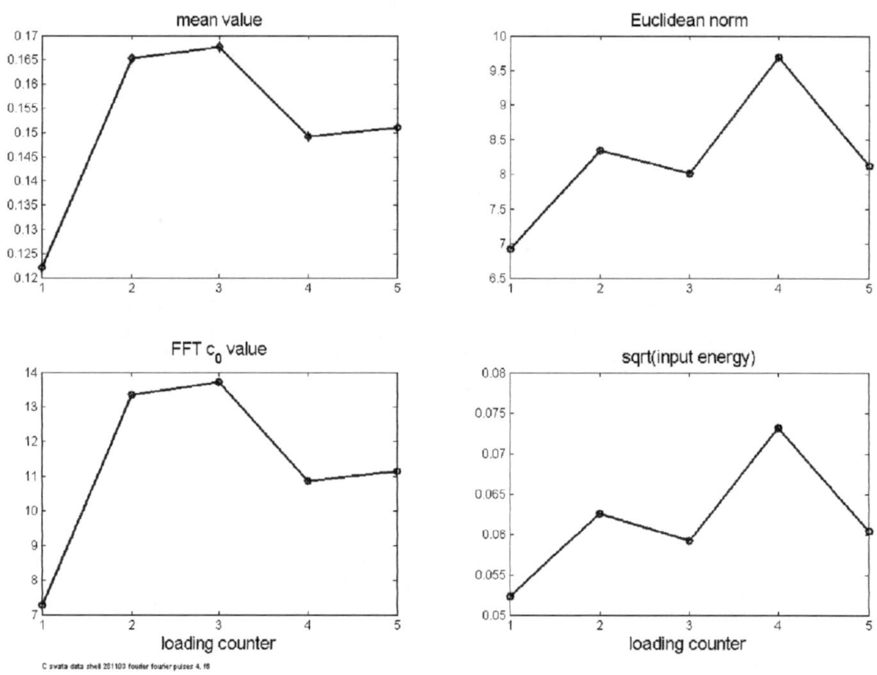

Fig. 15. A hint for analysis of experimental data – statistics and FFT

Example – 4. In FE analysis, the speed of stress wave propagation depends on the integration operator

In order to assess the reliability and precision of time integrating methods, a stress wave propagation task was computed using Newmark (denoted NM in the text) and central difference methods (CD). Comparison of axial strains at a certain location of the considered body, obtained by both methods, is presented in Fig. 23. The same time integration step (1e-7 [s]) was used in both cases. For the NM method the consistent mass matrix was employed, while the diagonal mass matrix was used for the CD method.

The left-hand subplot presents the strains in the whole computed time range showing the excellent agreement of results due to both approaches. The agreement documents that the accepted space and time discretizations are suitable for this geometry and loading – the differences are minimized. Two couples of horizontal lines indicate the areas that are shown in detail the in right-hand side subplots.

The difference between NM and CD solutions can be numerically assessed by means of a cantered correlation coefficient which can be geometrically interpreted as an angle between NM and CD time vectors by means of the cosine of an angle between two vectors.

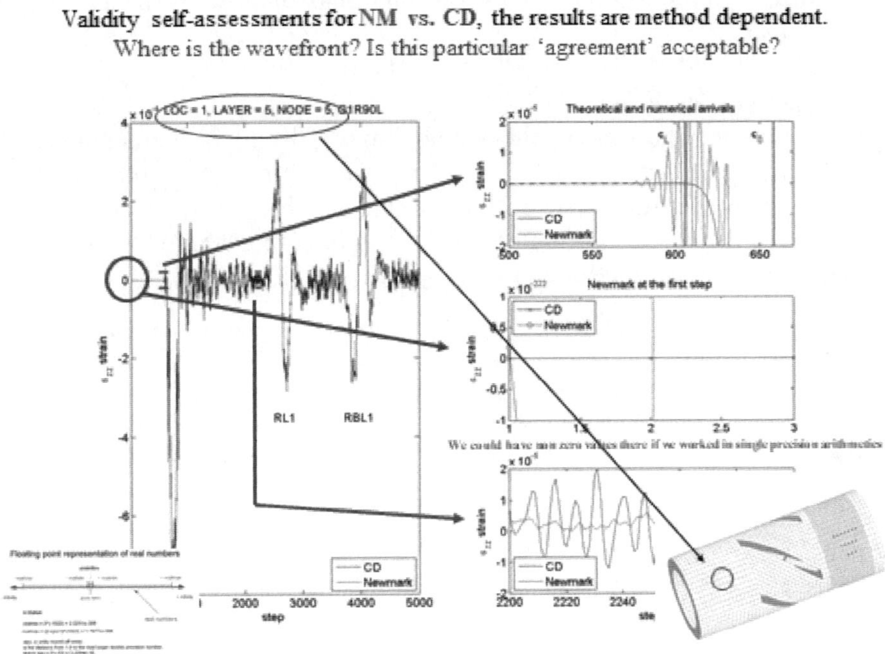

Fig. 16. Different speed of propagation for NM and CD operators

In the upper right-hand subplot of Fig. 23 the positions of theoretical arrivals of 3D longitudinal and 1D waves are indicated by red vertical lines. The actual axial strain distributions computed by the NM and CD methods are shown as well. From the analysis of dispersion properties of finite elements and that of time integration methods it is known that the computed speed of wave propagation for the CD approach with diagonal mass matrix underestimates the actual speed, while the NM approach with consistent mass matrix overestimate the actual speed. The presented results nicely show this.

What is less known is the fact, that the speed of propagating waves with NM-consistent modelling is actually 'infinitely' large.

To find the response of a body in time due to the impact loading requires solving the system of differential equations in which the inversion of stiffness or mass matrices (depending on the mass matrix formulation and on the time integration method being employed) is needed. [13]. This way, the sparse structure of relevant matrices might vanish, and a sort of unlimited range of influence (both in space and time) could prevail. If we worked with infinite number of significant digits we would get a nonzero response over the whole domain for any time greater than zero. Working, however, with standard precision arithmetic we will get no 'measurable' response in domains where the computed values are, in absolute value, less than the computational threshold – the smallest positive floating point number. Besides, the limited range of influence is secured by the fact that a substantial part of the energy is always carried by low frequency components. See [12].

Results of modelling should be independent of the method of solution, but in practice the agreement is both the method and the threshold dependent.

The shown infinite speed of propagation in FE analysis is an artefact of a particular method of treating differential equations in time by Newmark method. Even if this fact is fundamentally wrong, it does not devaluate the solution itself because its manifestation is well below the 'reasonable' threshold of observation.

From the engineering point of view the presented agreement is acceptable. Two methods are giving similar results, which a necessary but not sufficient condition for our belief that we are on the right track. From the point of view of numerical analysis theory it is important to be aware of these artefacts

Limits of finite element analysis

For fast transient problems as shock and impact the high frequency components of solutions are of utmost importance. In continuum, there is no upper limit of the frequency range of the response. In this respect continuum is able to deal with infinitely high frequencies. This is, however, a sort of singularity deeply embedded in the continuum model.

As soon as we apply any of discrete methods for the approximate treatment of transient tasks in continuum mechanics, the value of upper cut-off frequency is to be known in order to 'safely' describe the frequencies of interest.

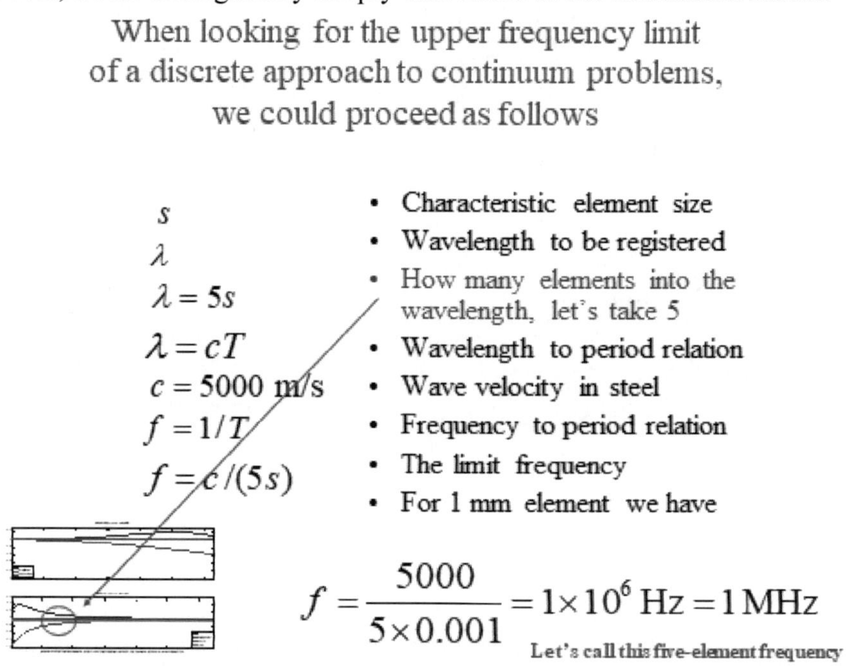

Fig. 17. Element size vs. frequency

A procedure leading to approximate correlation between the element size and the maximum frequency – the model assembled of elements of that size could safely transmit – is outlined in Fig. 17.

A rule of thump, which is easy to remember, is that a 1mm element is 'good' for 1 Mhz frequency.

Example – 5. Impact induced stress wave energy flux. Validation of FE and experimental approaches.

The temporal and spatial distribution of the stress wave energy flux in an axially impacted cylindrical tube, whose middle part contains four spiral slots, is studied experimentally and numerically. The high-speed recording of transient surface strains was used in the experiment, based on 1D theory, while the 3D finite element treatment was employed for the numerical analysis. The aim of the paper is to ascertain how reliable is the energy assessment based on transient recordings of surface strains and on subsequent 1D wave theory reasoning. The presented study is to determine how much of the impact energy, which is predominantly of axial (longitudinal) nature, is transferred into torsional (or shear) energy mode as well as into other energy modes not seen by the experiment. The task is sketched in Fig. 18.

Considering the linear elastic continuum and the validity of Hook's law in the form $\sigma_{ij} = C_{ijkl}\, \varepsilon_{kl}$, the internal stress power and the time rate of kinetic energy are as follows $W_{\text{int}} = \dot{E}_{\text{s}} = \dfrac{\mathrm{d}}{\mathrm{d}t}\int_V \tfrac{1}{2}\, C_{ijkl}\, \varepsilon_{kl}\, \varepsilon_{ij}\, \mathrm{d}V$ and $W_{\text{kin}} = \dot{E}_{\text{k}} = \dfrac{\mathrm{d}}{\mathrm{d}t}\int_V \tfrac{1}{2}\, \rho\, \dot{u}_i\, \dot{u}_i\, \mathrm{d}V$. The details, describing how these genetic formulae were evaluated in the FE analysis and in the experiment, are in [12].

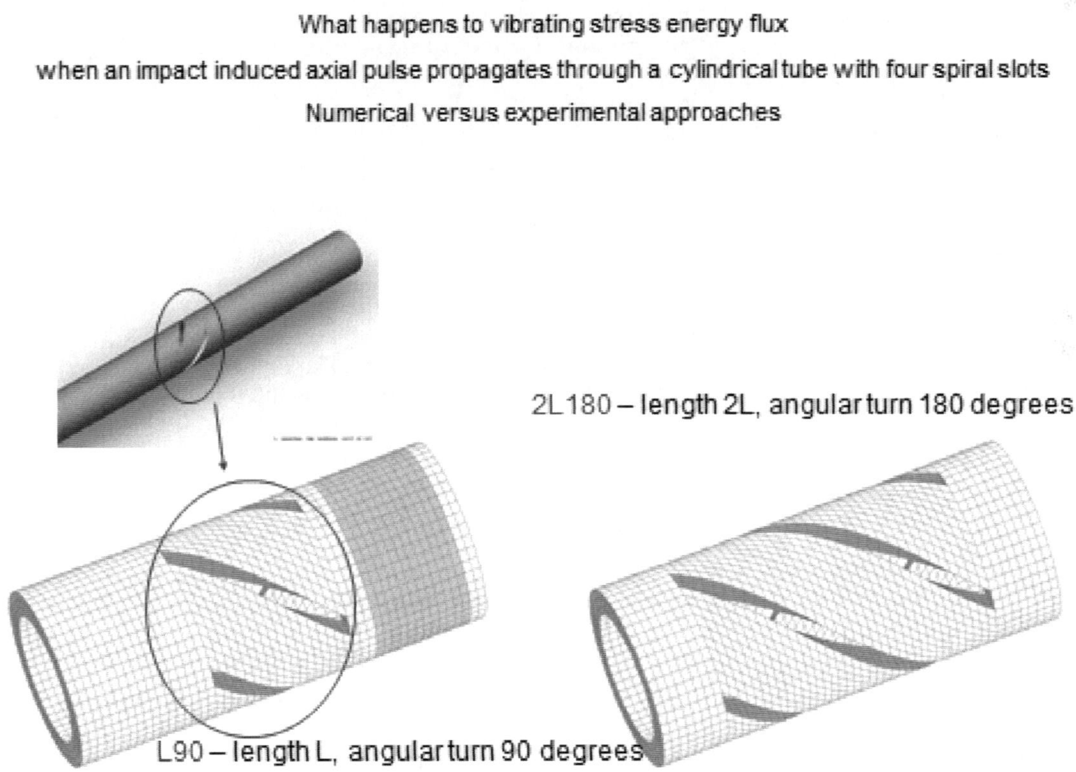

Fig. 18. FE approach

In FE analysis the energy flux computation can be more naturally assessed by analyzing the time history of energy in the whole body and in its parts.

In experimental analysis it is convenient to evaluate the amount of energy in the body as the cumulative energy flux through a cross sectional area, observed within a specified time interval.

The strain gauge rosette orientation and the gauge locations on the surface of the tube are indicated in Fig. 19. To avoid parasite bending signals the same strain gauges were positioned at the opposite parts of the surface. The 'experimental' tube is longer then the 'finite element' one, but the signals, reflected from the supported end of the tube, comes to the measurement locations long after the signal recording has been finished. A standard Wheatstone bridge set-up with and a fast transient recorder were employed.

The details of FE computations with emphasis on the proper choice of element sizes with respect to frequency contents to be transferred and of determination of integration methods and used time steps and an of experimental setup are in [12].

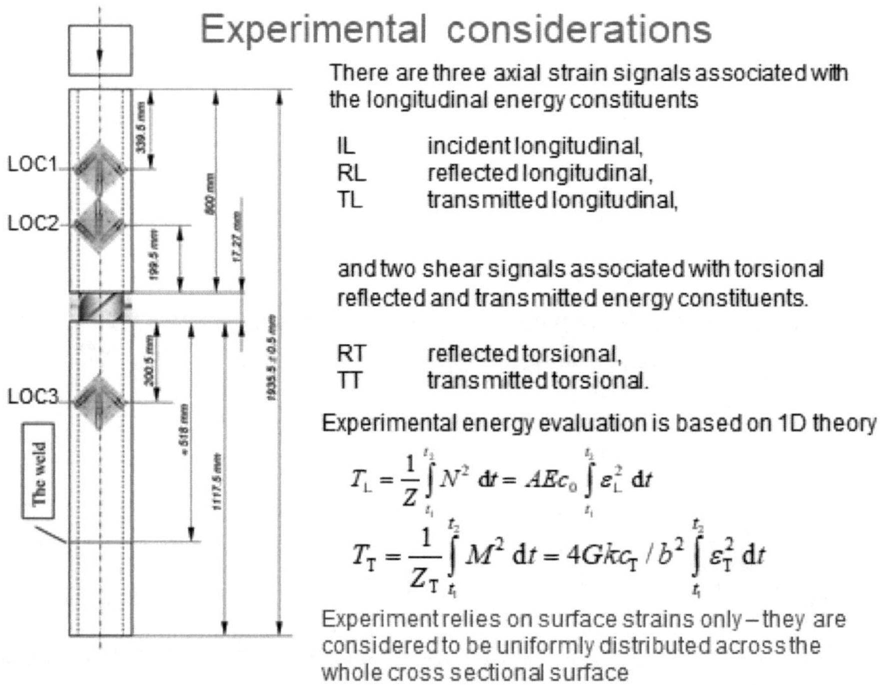

Fig. 19. Experimental approach

Comparison of axial strains obtained by an experiment and by the FE element analysis is in Fig. 20.

Is it a bad or a good agreement? Is experiment or FEA closer to reality? Where is the truth? Experimental people often say to FE analysts: All your high frequency components are just a numerical noise.

The cut-off frequency of experimental setup had the value of 100 kHz. The Nyquist frequency for the FE analysis, based on the integration timestep – which is a sort of experimental sampling interval – is 0.5 MHz.

There are many questions. Among them: What are the experimental limits in this particular case? Can all high-frequency components be attributed to numerical noise?

Possible answers: The 3D finite element analysis is compared with the experiment based on the 1D theory. Only axial and shear strains are measured – radial ones are not. The values of measured surface quantities (displacements, strains, velocities) are attributed to the whole cross-sectional area. 1D wave theory is used. Smaller frequency sampling rate is employed.

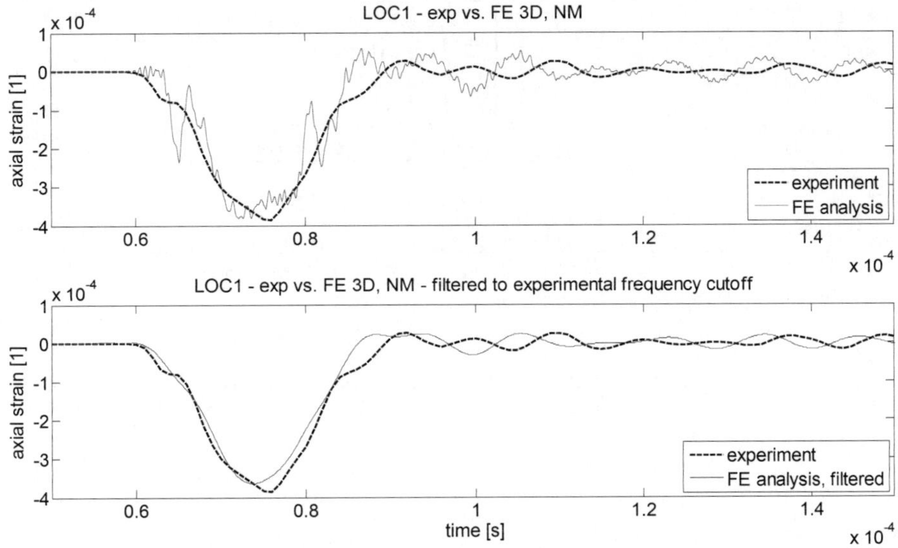

Fig. 20. Comparison of axial strains. Raw and filtered data.

The experiment, as conceived in this case, could not 'catch' the 'actual' frequency components higher than its upper frequency limit.

In this case the upper frequency limit of FE analysis is substantially higher, so the FE spectrum is longer.

A question arises what is the range of computed frequencies which are 'correct', especially in view of the fact that the FE transfer spectrum is method dependent. Notice the differences for NM and CD treatment.

Since an experiment with a finer time and frequency resolution is not available, the FE analysis should help itself to answer the question. Self-assessment by mesh- and timestep refinement could help. Remember the Richardson method, known from quadrature analysis, where the subsequent halving the integration increment is used for the quadrature error estimation.

Having small differences between two alternative approaches does not automatically imply that the results are correct. It only means that for a given loading and the employed time and space discretizations, there is almost no 'measurable' difference between results obtained by two types of approaches. One has to realize that the existence of close solutions, stemming from alternative approaches, is only a necessary, but not a sufficient, condition of 'correctness'. And what is 'correct', in the sense of correct modeling the Mother Nature, is difficult to define.

Limits of continuum, FE analysis and experiment

These limits are schematically depicted in Fig. 22, which is a sort of extension of Fig. 17, where element sizes vs. frequency are depicted. This time, the limit of continuum is limited from above by the austenite steel grain size, and far far above by the atom size. Maximum frequency that we are able to capture by the first class transient recorders is of the order of 1GHz. A range of a practical frequency range of FE analysis is depicted as well. This information might be complemented by a piece of wisdom attributed to S.C. Hunter who claims "To neglect corpuscular structure of matter the specimen should be at least 10^4 times larger than the interatomic distance.

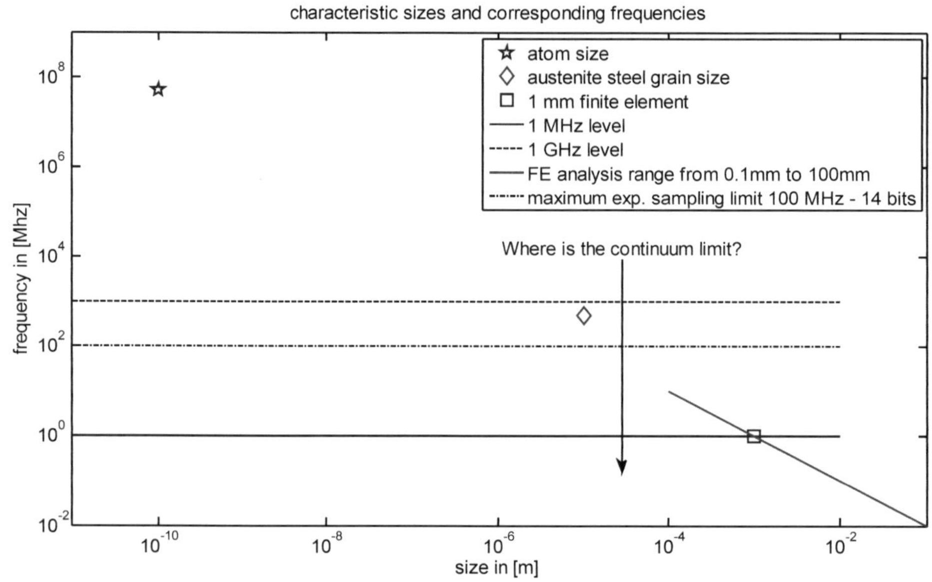

Fig. 22. Limits

Conclusions

Mechanical theories, principles, laws and models, used in engineering practice, cannot be proclaimed true or false. They are either right (working to our satisfaction) or wrong. Regardless of being simple or complicated, they are 'right', if approved by an appropriate experiment (i.e. conceived in agreement with accepted assumptions of the theory). History reveals that wrong theories might appear, but not being confirmed by experiments, are quickly discarded as ether or flogiston. Theories are right only within the limits of their applicability. We cannot claim that a theory being proved by an experiment is right. The only thing we can safely state is that such a theory is not proved wrong.

Generally, a singularity appearing in a model always means a serious warning concerning the range of validity of that model. Usually, a more general model – having a wider scope of validity – is invented removing that singularity. Very often there is no need to discard the older and simpler model, since it might perfectly work in the validity range for which it was conceived. The role of doubts on our way to understanding the nature is far from negative.

References recommended for further studies

Encyclopedia of Physics, Vol. III/1, Editted by Fluege, Springer, Berlin, 1960.

Eddington, Arthur S.: New Pathways in Science, Cambridge University Press,

Love, A. E. H.: A Treatise on the Mathematical Theory of Elasticity. (First edition 1892). New York, Dover Publications, 1944.

Rucker, R. von Bitter: *Infinity and the Mind: The Science and Philosophy of the Infinite*. Princeton, N.J.: Princeton University Press, 1982.

Belytschko, T., Liu, W.K. and Moran, B.: Nonlinear Finite Elements for Continua and Structures, John Wiley, Chichester, 2000.

Okrouhlík, M.: Impact induced stress wave energy flux. Validation of numerical and experimental approaches. International Conference on Vibration Problems 2011. September 5-8, 2011, Prague. Springer Proceedings in Physics 139, Springer New York, 2011, ISBN 978-94-007-2068-8, pp. 25 – 34.

Okrouhlík, M. (editor): Mechanics of Contact Impact. Applied Mechanics Review, 1994, vol. 47, No. 2.
Belytschko, T., Liu, W.K. and Moran, B.: Nonlinear Finite Elements for Continua and Structures, John Wiley, Chichester, 2000.
Okrouhlík, M., Pták, S., Lundberg, B., Valdek, U.: FE Assessment of an Experiment Employed for Analysis of the Transient Stress Energy Flux Through Spiral Slots in Axially Impacted Cylindrical Tube. Journal of Mechanical Engineering, 60, 2009, No. 4, pp. 181 – 209.
Love, A. E. H.: A Treatise on the Mathematical Theory of Elasticity. (First edition 1892). New York, Dover Publications 1944.
Bathe, K.J., Noh, G.: Insight into an implicit time integration scheme for structural dynamics, Computers and Structures 98–99, 1–6, 2012.
Bathe, K.J.: Finite element procedures in engineering analysis. Prentice-Hall, Englewood Cliffs, N.J., 1982.
Belytchko, T., Hughes, T.R.J: Computational method for transient analysis. Amsterdam: North Holland, 1986.
[Belytchko, 2000]: Belytschko, T., Liu, W.K. and Moran, B.: Nonlinear Finite Elements for Continua and Structures, John Wiley, Chichester, 2000.
Belytschko, T., Mullen, R.: On dispersive properties of finite element solutions, In Modern Problems in Elastic Wave Propagation, John Wiley, pp. 67 – 82, 1978.
Brillouin, L.: Wave propagation in periodic structures, McGraw-Hill, New York, 1946.
Brillouin, L.: Wave propagation in periodic structures: Electric filters, and crystal lattices, Dover Publications, New York, 1953.
Červ, J., - Adámek, V., - Valeš, F., Gabriel, D., Plešek, J.: Wave motion in a thick cylindrical rod undergoing longitudinal impact. Wave Motion. 66, November 2016, pp. 88-105. ISSN 0165-2125.
Dokainish, M.A., Subbaraj, K.: A survey of direct time-integration methods in computational structural dynamics - I. Explicit methods. Computers & Structures 32(6):1371–1386, 1989.
Dyna manuals, www.dynasupport.com/support/tutorial/users.guide.
Goudreau, G.L., Taylor, R.L.: Evaluation of numerical integration methods in elastodynamics. Computer Methods in Applied Mechanics and Engineering 2, 67–97, 1972.
Hilber, H. M. et al.: Improved Numerical Dissipation for Time Integration Algorithms in Structural Dynamics, Earthquake Engineering and Structural Dynamics, No. 5, pp. 283-292, 1977.
Houbolt J.C.: A recurrence matrix solution for the dynamic response of elastic aircraft, J. Aeronant. Sci. 17, 540–550, 1950.
Hughes, T.J.R., Liu, W.K.: Implicit-explicit finite elements in transient analysis: Stability Theory. Journal of Applied Mechanics 45, 371–374, 1978.
Hughes, T.J.R.: The Finite element method: Linear and dynamic finite element analysis, New York:Prentice-Hall, Englewood Cliffs, 1983.
Hulbert, G. M., Hughes, T. R. J.: Space-time finite element methods for second order hyperbolic equations, Computer Methods in Applied Mechanics and Engineering 84, 327–347, 1990.
Cho, S.S., Park, K.C., Huh, H.: A method for computation of wave propagation: Extension to two and three dimensional problems. In Book of Abstracts. EUROMECH Colloquium 540, Prague, October 1-3, 2012.
Chung, J., Hulbert,G.M.: A Time integration algorithm for structural dynamics with improved numerical dissipation: The generalized-method, Journal of Applied Mechanics, 60, 371–375,1993.
Lew, A., Marsden, J. E., Ortiz, M., West, M.: Variational time integrators, International Journal of Numerical Methods and Engineering 60,153–212, 2004.

Love, A.E.H.: A treatise on the mathematical theory of elasticity, New York, Dover Publications, 1944.
Marc manuals, MSC Software Corporation, 2 MacArthur Place, Santa Ana, CA 92707 USA, www.mscsoftware.com.
Newmark, N.M.: A method of computation for structural dynamic, Journal of the Engineering Mechanics Division 85, 67–94, 1959.
Okrouhlík, M., Pták, S.: Assessment of experiment by finite element analysis: Comparison, self-check and remedy, Strojnícky časopis, Vol. 56, 2005, No. 1.
Okrouhlík, M. (editor): Mechanics of Contact Impact. Applied Mechanics Review, Vol. 47, No. 2, February 1994.
Okrouhlík, M. and Pták, S.: FE and experimental study of axially induced shear stress waves in a tube with spiral slots. Part 1. Research Report of the Institute of Thermomechanics, Z 1417/2008, Prague, 2008.
Okrouhlík, M., Hoeschl, C.: A contribution to the study of dispersive properties of one-dimensional Lagrangian and Hermitian elements. Computers and Structures, Vol. 49, No. 5, pp. 770-795, 1994.
Okrouhlík, M., Pták, S.: Assessment of experiment by finite element analysis: Comparison, self-check and remedy. Strojnicky casopis, Vol. 56, No. 1, 2005.
Okrouhlík, M.: Dispersion and Time Integration Schemes in Finite-Element Analysis – A Practical Pictorial Manual. International Journal of Mechanical Engineering Education, Vol. 41, 1: pp. 44-71, 2013.
Okrouhlík. M. (editor): Numerical methods in computational mechanics, Institute of Thermomechanics, Prague 2008, http://www.it.cas.cz/cs/ok
Park, K.C., Lim, S.J., Huh, H.: A method for computation of discontinuous wave propagation in heterogeneous solids. Basic algorithm description and application to one-dimensional problems. International Journal of Numerical Methods and Engineering 91, 622–643, 2012.
Park, K.C.: Practical aspect of numerical time integration. Computers & Structures, 7, 343–353, 1977.
Stade, E.: Fourier Analysis, John Wiley and Sons, Hoboken, New Jersey, 2005.
Stein, E., de Borst, R., Hughes, T.J.R.: Encyclopedia of Computational Mechanics, ISBN: 978-1-119-00379-3, 2017, 4024 p.
Subbaraj, K. Dokainish, M.A.: A survey of direct time-integration methods in computational structural dynamics - II. Implicit methods. Computers & Structures 32(6): 1387–1401, 1989.
Zienkiewicz, O.C.: The finite element method in engineering science, McGraw-Hill, New York, 1971.
Zukas, J. A.: High Velocity Impact Dynamics. John Wiley and Sons, Inc., New York, 1990.

References

[1] Kolsky, H.: *Stress Waves in Solids*, Clarendon Press, Oxford, 1953.
[2] Brepta, R., Okrouhlík, M.: Disperzní vlastnosti obdélníkových a čtvercových prvků pro dvojrozměrnou oblast. Report Z998/86, ÚT ČSAV, Praha 1986.
[3] Graff, K.: (1975) Wave Motion in Elastic Solids, Clarendon Press, Cambridge.
[4] Okrouhlík, M., Pták, S.: Assessment of experiment by finite element analysis: Comparison, self-check and remedy, Strojnícky časopis, 2005, vol. 56, No. 1, pp. 18 – 39.
[5] Achenbach, J.D.: Wave propagation in elastic solids. North-Holland Series in Applied Mathematics and Mechanics, North-Holland, ISBN 0-7204-0325-1, 1973.
[6] Okrouhlík, M., Pták, S.: Pollution-Free Energy Production by a Proper Misuse of Finite Element Analysis, Engineering Computations, 2003, vol. 20, No. 5/6, pp. 601 – 610.

[7] Okrouhlík, M., Hoeschl, C.: A contribution to the study of dispersive properties of one-dimensional Lagrangian and Hermitian elements. Computers and Structures, Vol. 49, No. 5, pp. 770-795, 1994.

[8] Belytchko, T., Hughes, T.R.J: Computational method for transient analysis. Amsterdam: North Holland, 1986.

[9] Aboud, N.N., Pinsky, P.M.: Finite element dispersion analysis for the three-dimensional second-order scalar wave equation. International Journal of Numeric Methods in Engineering, Vol. 35, pp. 1183 – 1218, 1992.

[10] Brillouin, L.: Wave propagation in periodic structures: Electric filters, and crystal lattices, Dover Publications, New York, 1953.

[11] Kolman, R., Plešek, J., Okrouhlík, M., Gabriel, D.: Grid dispersion analysis of plane square biquadratic serendipity finite elements in transient elastodynamics. Int. J. Numer. Meth. Engng 2013; 96:1–28.

[12] Okrouhlík, M., Pták, S., Lundberg B., and Valdek, U.: FE ASSESSMENT OF AN EXPERIMENT EMPLOYED FOR ANALYSIS OF THE TRANSIENT STRESS ENERGY FLUX THROUGH SPIRAL SLOTS IN AXIALLY IMPACTED CYLINDRICAL TUBE, Engineering MECHANICS, Vol. 16, 2009, No. 2, p. 1–19

[13] Bathe, K.-J.: Finite Element Procedures, Prentice-Hall, Englewood Cliffs, New Jersey, 1996.

EAN 2019
57th conference on experimental stress analysis

Experimental method of the service life estimation of a general kinematic pair of a cam mechanism

ONDRÁŠEK Jiří[1,a] and HEJNOVÁ Monika[1,b]

[1]VÚTS, a.s., Svárovská 619, 460 01 Liberec, Czech Republic

[a]jiri.ondrasek@vuts.cz, [b]monika.hejnova@vuts.cz

Keywords: Cam mechanism, Service life, Test rig

Abstract. The paper deals with the experimental method of the service life estimation of the contact region of a general kinematic pair. In the case of a general cam mechanism, the general kinematic pair is formed by the contact of a cam and a follower working surfaces. We can determine the service life of the cam mechanism both experimentally and theoretically. The theoretical determination has an advantage that it is very quick, but can be inaccurate. Therefore we pay attention to the experimental determination of the service life. The test rig simulates the cam mechanism and allows the testing of the huge number of parameters. We can obtain information this way and use it for refining the mathematical model commonly used for the service life prediction.

Introduction

Part of the cam mechanism design is the determination of suitable material for production. It is possible to choose the theoretical procedures without deeper consideration of the properties of the selected material. This paper describes a test rig for testing materials suitable for cam production.

Rolling contact fatigue of the cam mechanisms

At the contact areas of the cam surface or below these points, contact stress becomes a periodical magnitude related to the angular cam displacement ψ. These transitory stresses are characterised by pulses with a periodicity of 2π (see Fig. 1). The stress limit σ_{hc} is in perfect agreement with the disturbance caused by the transitory stress. [1]

The reduced stresses σ_{red} are limited by the actual strength condition, written in the form (1).

$$\max \sigma_{red} < \sigma_{hc} \tag{1}$$

According to [1] it is a usual relation for steel between the transitory stress limit and the strength limit described by equation (2).

$$\sigma_{hc} = 0{,}66 R_m \tag{2}$$

Also applies (3), where R_e is the sliding limit of the steel.

$$R_e = (0{,}55 \div 0{,}8) R_m \tag{3}$$

Considering equation (2) a (3), we can write the condition (4) stating that no destructive action of plastic deformation is produced in the general pair under operation [1].

$$\max \sigma_{red} < R_e \tag{4}$$

This condition is the starting point for determining the material suitable for the cam mechanism production. Values of the strength limit, sliding limit, and tension fatigue limit for every steel can we found in the material sheets, eventually in the literature, e.g. [2].

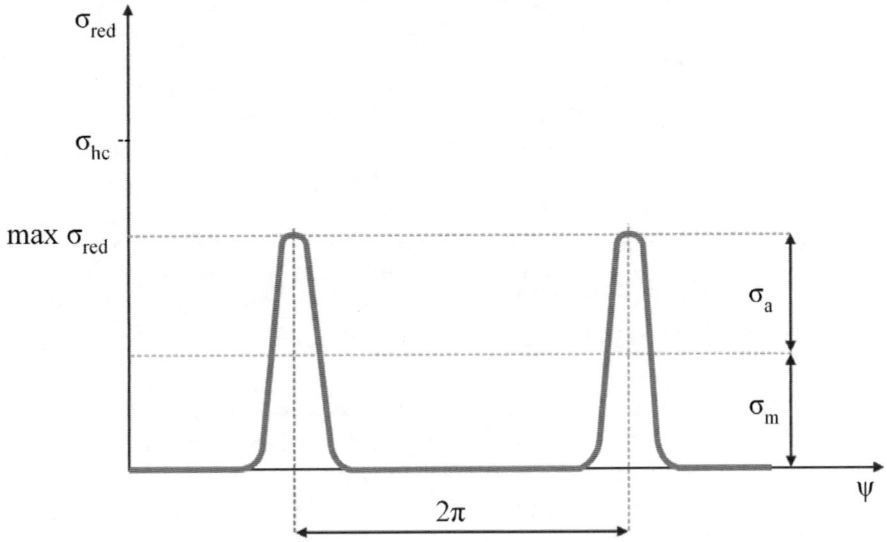

Fig. 1: Load course on the contact areas of the cam surface [1]

The rolling contact fatigue (RCF) occurs in functional surfaces that are exposed to repeated loading (high local pressure) during movement of the functional surfaces of the machine components. We can describe RCF by 3 stages of the material damage [3]:

1. Stage of the material mechanical properties changes
2. Stage of the cracks nucleation
3. Stage of the cracks propagation to fracture

At the loading of the contact surfaces, there are initial changes in properties, cyclic hardening, or softening [4]. As a general rule, the solid materials are reinforced, for example by cold forming, during the fatigue process cyclically softened and, on the contrary, soft materials cyclically hardened. Experiments have shown that cyclic softening occurs for materials with a ratio of strength limit to sliding limit $Rm/Re < 1.2$ and for cyclic hardening if $Rm/Re > 1.4$. [5]

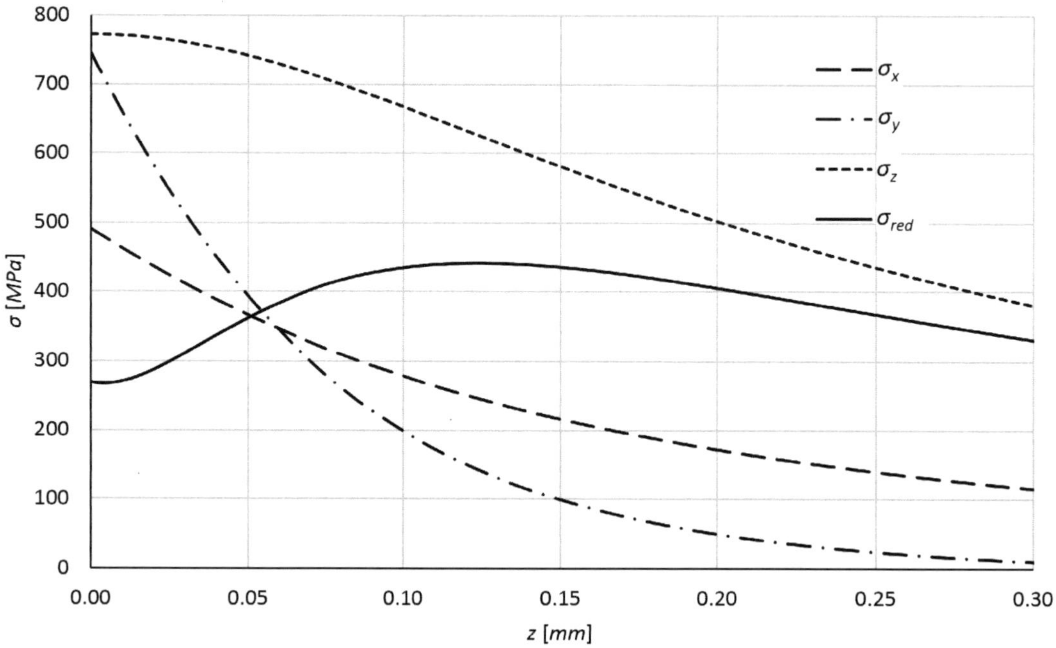

Fig. 2: The state of stress between the crowned roller and the cam in contact [6]

In the next stage, there are such changes in the structure of the material that the first cracks are formed. The crack formation below the surface is connected to the dependence shown in Fig. 2 [6]. There is a maximum of the reduced stress in this area.

If the stress below the surface is high, small plastic deformations can occur in the material and repeated loading may lead to crack formation. These cracks can join together with cracks on the surface and can lead to loosening of the small parts of the surface (see Fig. 3 and Fig. 4). This results in irreversible damage of the contact surfaces and the component must be changed.

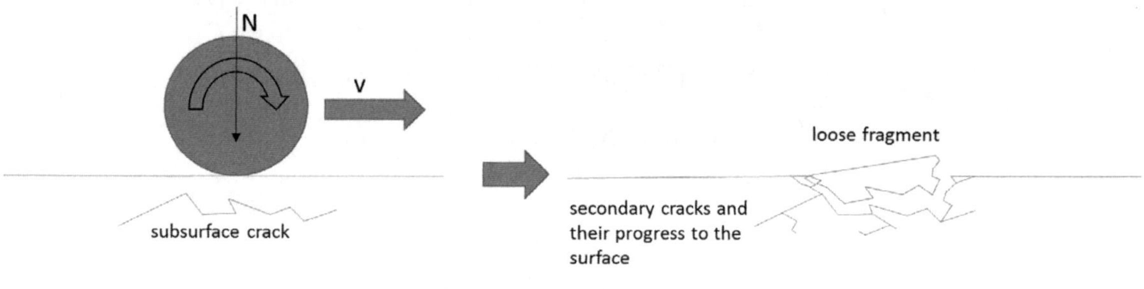

Fig. 3: Pitting formation

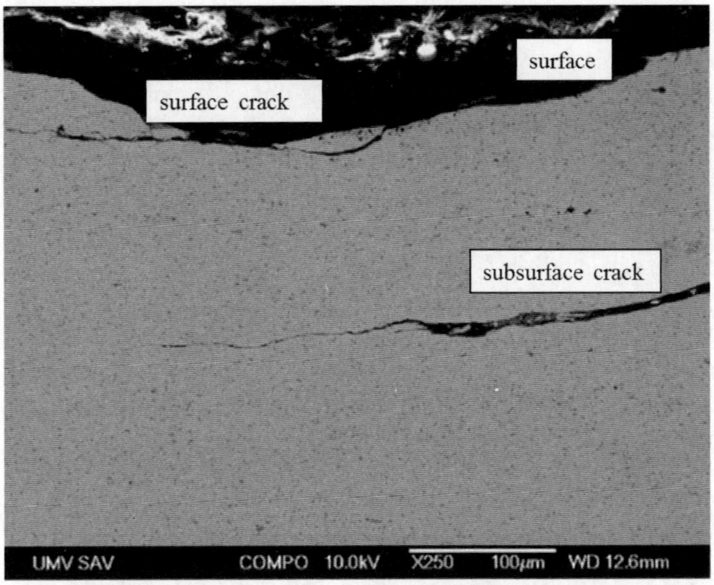

Fig. 4: Detail of the damaged part (acquired in own research)

RCF is very much depend on the mechanical properties of the surface, number and size of impurities and roughness of the surface. The effect of inclusions is showed schematically, acting as a stress concentrator, giving rise to a higher probability of cracking in Fig. 5.

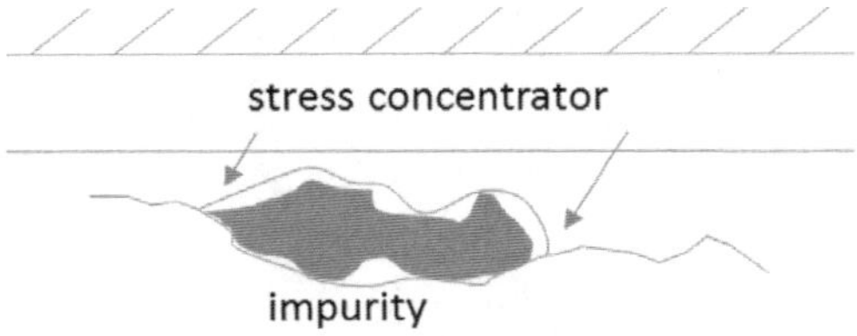

Fig. 5: Material imperfections

Sliding occurs in the operation of the cam mechanism, not just the rolling of the roll on the cam. According to literature [7], a suitable method to protect against fatigue wear is to reduce the coefficient of friction between the two contact surfaces to such a value that the tensile forces do not suffice for delamination during shear or contact fatigue during the shear. Another very important aspect in controlling of the contact fatigue during both sliding and rolling is the purity of the material. Materials with as few inclusions and imperfections as possible should be selected for rolling and sliding. Increasing of the material hardness may also help in some cases, but this method is limited by the embrittlement of the hardened material. Materials for mutual sliding should be selected carefully.

Factors affecting the service life

In general, contact fatigue arises as a result of material imperfection (a material has various constraints and imperfect structure), or working conditions (especially lubrication and load). At the rolling of the body its material must be of high quality, since any imperfection or inclusions manifests as a crack initiation site. The surface of the material should also be of high quality (i.e. as smooth as possible, without bumps) to prevent crack propagation from the surface. Lubrication also has a significant impact on contact fatigue. However, we must not forget other working conditions, such as changes in load (stress) and slippage during sliding, which are related to changes in the relative velocities of the bodies in contact. [7]

When the cam and roller are in contact, they do not only roll together, but also slide. The rolling causes repeated very high stress concentrated on a small area close to the contact site. Pressure stress or frictional stress generation in contact significantly contributes to contact fatigue of both surfaces, with both rolling and sliding. It is generally well known that contact fatigue is very sensitive to sliding. Even a small amount of slips accompanying the rolling process will reduce the service life. [7]

Many factors influence material fatigue. It is therefore necessary to consider the whole set of information not only about the material but also on other influences. There is an overview of this information in the Fig. 6.

Some factors can be described precisely by means of relationships, such as the effect of sample size, the stress gradient, the effect of surface quality, etc. Specific relationships can be found in the literature, e.g. [8].

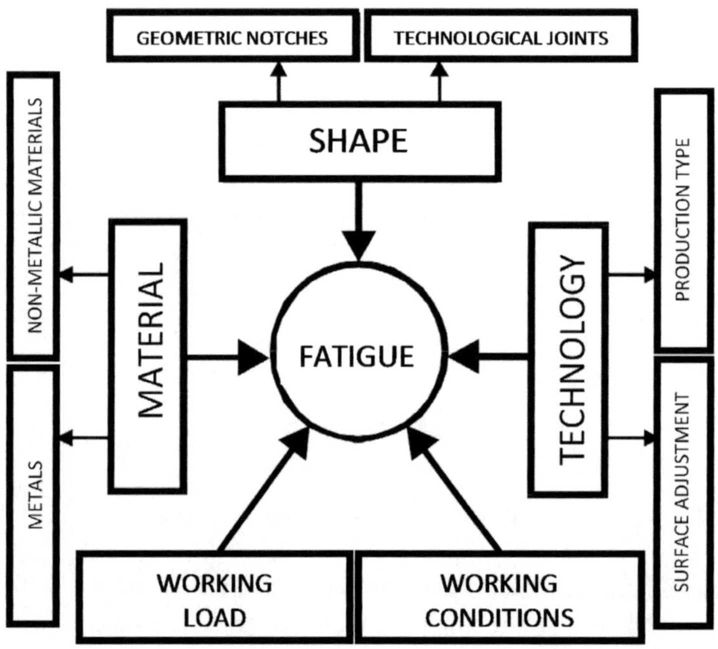

Fig. 6: Main factors affecting the fatigue process [8]

To complete the fatigue curves information, a description of the test conditions is required for the experimental determination. In addition to the type of loading, amplitude and asymmetry of the cycle, information about the sample shape, surface quality, material, and heat treatment is useful. Last but not least, the working environment, sample temperature, type of machine, course of testing, breaks and so on also affect the results of fatigue tests. [8]

Experimental determination of the service life

In addition to the theoretical method of determining the life of the functional surface of the cam [9], it is possible to use the experimental method. This is based on the loading of the test sample surface by the periodic force depending on the angle of rotation of the sample. In this case, the stress on the contact surfaces of the samples has a pulse character. Thus, it is a type of load transient. The loading body represents a cylindrical roller and the loaded body, in analogy with the cam mechanism with radial cam and roller follower corresponds to the cam. The loaded body is stressed by the periodic force N by means of the loading body, the amplitude takes on a size depending on the material of the loaded body and on the method of processing its contact surfaces. A kinematic scheme of the conceptual design of a test rig for testing samples is shown in Fig. 7.

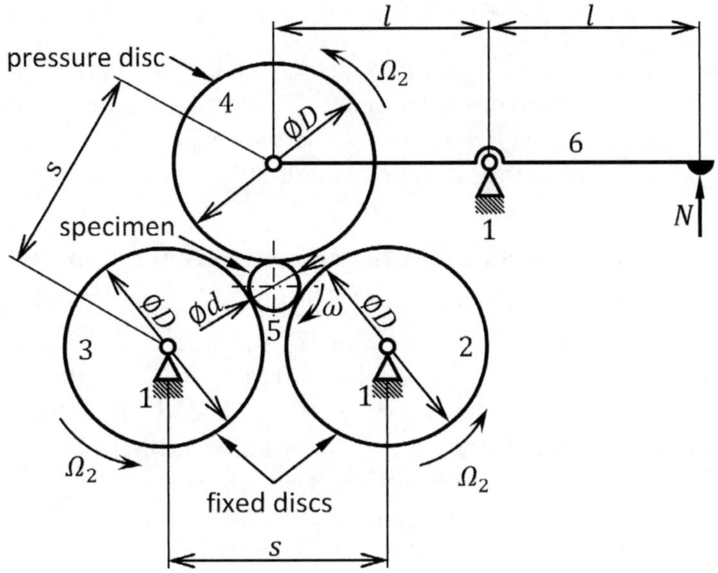

Fig. 7: Kinematic scheme of a test rig

The specimen 5 of the cylindrical shape with an exactly defined width is mounted on the shaft with three rotating discs around it. One of which is pressure (4) and two fixed (2 and 3). The discs are arranged in a common plane at the apices of the equilateral triangle with the sides s, with their rotation axes parallel. Pressure disc 4 is rotatably supported on one of the arms of a pivotally mounted two-arm lever 6 with a length of arms l. The two-arm lever is coupled with its second arm to the loading force N of a defined size. The entire system must be dimensioned due to the maximum load.

In terms of the geometrical arrangement of the load disks and the loaded sample, the sample is loaded with three equally sized N responses from each disk once.

Due to rotation of the test body 5 among three rotating discs rotated with the same angular velocity Ω_2, the loading force N has the character of pulses with a period of $2\pi/3$. Thus, there is a transient type of loading. The three-point contact reduces the testing time to $1/3$, it makes testing much easier as each sample is tested 10^7 to 10^8 cycles. The aim of the experiments is to achieve fatigue damage of the contact area of the specimen after a certain number of cycles.

Assuming the contact of two cylindrical bodies with parallel axes touching the length of l and the length of which is the uniformly distributed loading force N, the condition (5) can be derived for its size.

$$N \leq \frac{l\pi R_{eq}}{E^*}\left(\frac{K_R \cdot R_m}{0.6}\right)^2, \qquad K_R \in \langle 0.55; 0.8\rangle \tag{5}$$

In condition (5) the equivalent radius of curvature of a cylindrical roller of radius R_{eq} in contact with a cylindrical cam of radius r can be expressed as (6),

$$R_{eq} = \frac{Dd}{2(D+d)} \tag{6}$$

where d is the diameter of the sample and D is the diameter of the disks. The constant E^* defines the material elasticity of the bodies in the contact being defined by the formula (7),

$$\frac{1}{E^*} = \sum_i \frac{1 - \nu_i^2}{E_i}, \qquad i = 1,2 \tag{7}$$

where constants μ_i are Poisson´s ratio and E_i Young's modulus of elasticity of the given body i. In our case it is the specimen and discs. Parameter R_m represents the ultimate tensile strength of the material from which the specimen is made and the constant K_R expresses the ratio of the yield tensile strength $R_{p0.2}$ just to the ultimate strength. [9]

The influence of the shape of the discs profile on the stress of the specimens

The geometrical shape of the disc profile itself has a significant effect on the stress distribution, due to the load in the contact areas of the specimen. The contact of the disc and specimen forms the general kinematic pair. In this case, the contact stress on the surface of the contact surfaces of the bodies in contact with and under the respective surface is periodic, which results from the method of loading. The applied stresses are transient, with pulses having a period of $2\pi/3$ in the case of specimens and 2π in the case of discs.

It is possible to use both the conclusions of the contact mechanics [10] for the respective case of the contact of two elastic bodies as well as the possibility of the finite element method for the calculation of contact stresses. Assuming that the shapes of the specimens and discs would be cylindrical, it can be used to determine the distribution of the contact stress of the results of Hertzian contact theory for contacting cylindrical bodies with parallel axes [10], but with some limitations, as will be explained below. The finite element method is particularly useful when it comes to contact of the sample and the disc with a profile shape other than cylindrical or convex.

In the case of conventional cylindrical discs, there are discontinuities in the intersections of the cylindrical profile with the specimen profile, i.e. if one contact part is axially shorter than the other, as well as in the chamfer of the disc edges, see Fig. 8. In the vicinity of those profile discontinuities, the contact between the disc and the specimen cannot be considered to be straightforward to which Hertzian theory of contact can be applied, but to be a more complex three-dimensional type of contact. These discontinuities cause a very rapid increase in the pressure distribution in the respective contact area of the bodies. In fact, these local increases in pressure distribution can exceed the strength limit of the given material, thereby causing plastic deformations, residual stresses in the material or steel hardening. Furthermore, the area of question will be more susceptible to fatigue damage to the contact surfaces, i.e. to pitting or spalling of the material.

In order to reduce the excess edge stress in the case of the cylindrical discs, it is possible to achieve a crown axial cross section with such a shape that includes a straight line and a single circular arc or a combination of several circular arcs, see Fig. 8. However, such shape of a disc crown whose one segment is cylindrical and the subsequent segment thereof is convex, leads to a certain concentration of stress in the transition from the cylindrical section to the convex one.

According to [11], the logarithmic profile of the disc crown, it can achieve uniform stress distributions for different load levels of the general kinematic pair, Fig. 8. A characteristic feature of this profile is that it descends monotonously from its centre to the edge according to the logarithmic function. It should be noted here that the production of the general profile of the disc crown is a technologically demanding issue in the required precision and quality.

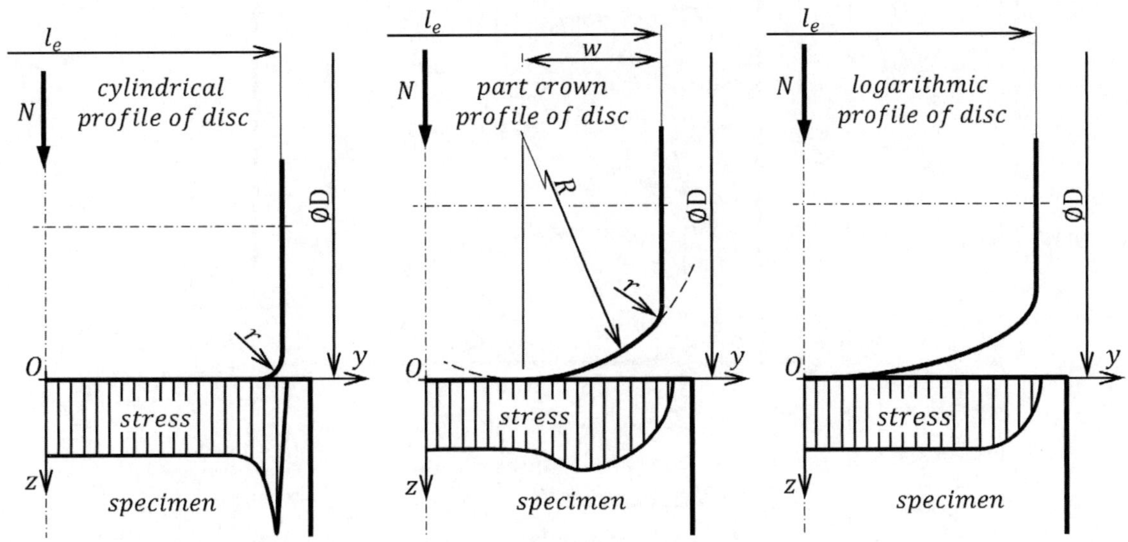

Fig. 8: Schematic drawing of the contact of the various discs profiles with the specimen

From the nature of the course and the results of the tests, it is necessary that the distribution of contact stresses in the contact area and its vicinity of the sample is as even as possible. This requirement is also placed on the stress in the vicinity of the shape discontinuities, which are apparent in the cylindrical profile of the disc, see Fig. 9. For this reason, attention was paid to the influence of the shape of the loading discs profile on the stress in the test specimen. The effect of the discs was analyzed by FEM. In the foregoing we have mentioned that the logarithmic profile of the disc crown is advantageous both in terms of uniform distribution of contact stress in the contact area but also in defining of the disc and specimen misalignment. However, this type of profile is extremely difficult to manufacture. A certain compromise solution seems to be the design of the disc with convex crown segments, with the middle crown section being cylindrical, see Fig. 9 - $Profile\ A$. The convex segments are characterized by their radius of curvature ρ_y in the plane of Oyz, whose size is among other things the object of optimizing the shape of the discs. The characteristic dimensions of such a disc are its diameter D, the radius of the convex part of the crown $\rho_y \equiv R$, the width of the convex part w and the radius of the rounding edge r. Another optimization parameter of such a crown profile is, besides the radius of the convex section $\rho_y \equiv R$, its width w. The optimization of the crown shape itself is carried out in order to distribute the contact stress as evenly as possible with the greatest possible load exerted by the effect of the force N.

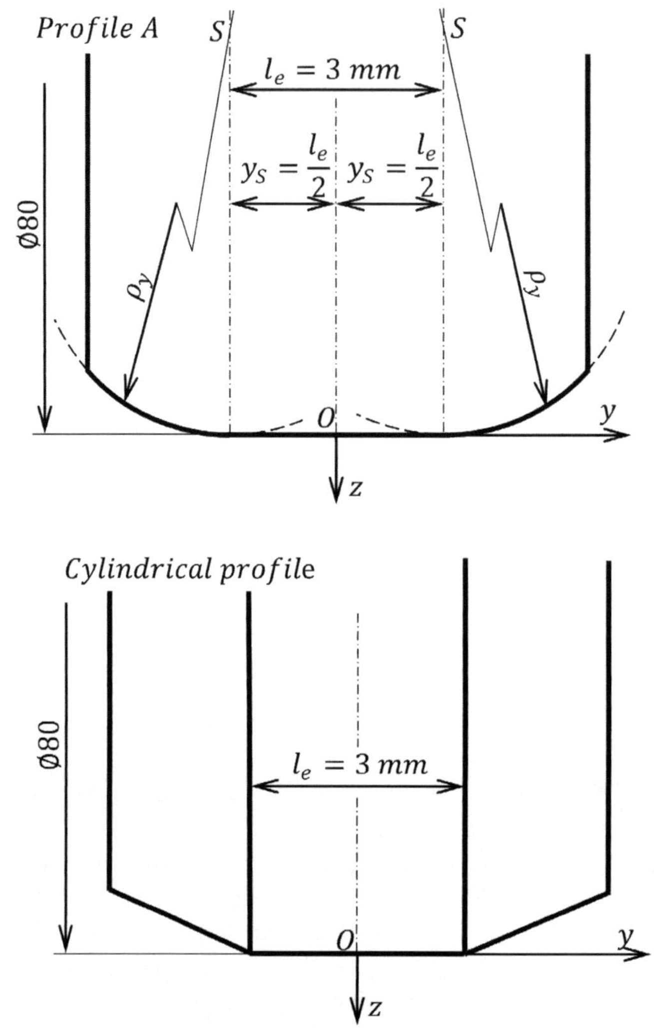

Fig. 9: Schematic drawing of the various discs profiles

The types of disc profiles according to the schematic representation shown in Fig. 9 were analyzed in detail. Each of the said disc types comprises of a cylindrical part, which is connected to either convex segments or conical segments. In summary, the typical disc dimensions are included in Table 1. The material parameters of the load discs and the test specimen are included in Table 2.

Table 1: Characteristic dimensions of the discs

		Profile A	Cylindrical profile
Diameter	$D\ [mm]$	82.4	82.4
Length of cylindrical part	$l_e\ [mm]$	3.0	3.0
Fillet radius	$\rho_y\ [mm]$	200	–

Table 2: Characteristic material parameters

		Disc: 90MnCrV8	Specimen: 16MnCr2
Density	$\rho\ [kgm^{-3}]$	7850	7850
Young's modulus of elasticity	$E\ [GPa]$	210	206
Shear modulus	$G\ [GPa]$	80	79
Poisson's ratio	$\nu\ [-]$	0.3125	0.3038

In the model, the respective surface of the disk was subjected to a surface load induced by the force effects of (8).

$$N = 4000\,N \tag{8}$$

Based on the analyses, the distribution of the contact stress was determined in the contact area and its vicinity induced effect of the load force N.

Fig. 10 shows the course of the maximum reduced stress in a depth of z_e depending on the half width of the specimen for all types of discs in contact with the cam. Depth z_e expresses the depth under the surface of the cam where just the maximum value of the reduced stress depending on the distance from the contact surface is reached.

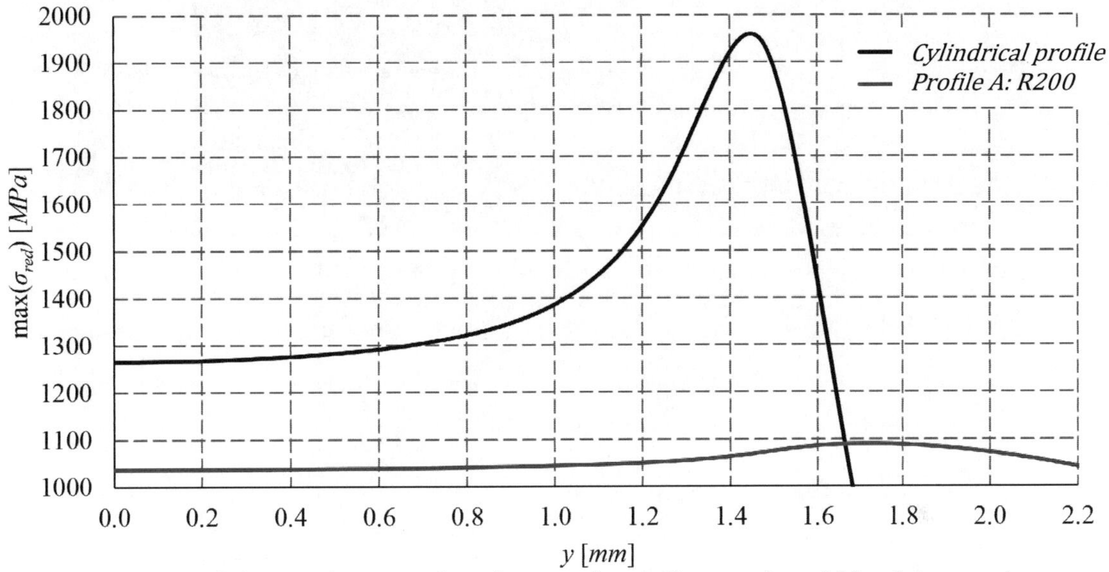

Fig. 10: Course of the maximum reduced stress depending on the width of the specimen

Fig. 11 shows the distribution of the reduced stress induced by the contact of the specimen and the disc with a radius of curvature $\rho_y = 200\,mm$ in the case of $profile\ A$ and the cylindrical disc. It can be seen from the above figures that in the vicinity of the profile discontinuities, the magnitude of reduced stresses increases. This is particularly evident in the case of a cylindrical disc, where the contact between the disc and the specimen cannot be considered simply straight, but rather a more complex three-dimensional contact type. It is also clear that a significantly more uniform stress distribution is achieved by the shape of the disc according to Fig. 9 - $Profile\ A$. We have also chosen this type in terms of acceptable technological demands of its production in the required quality and precision.

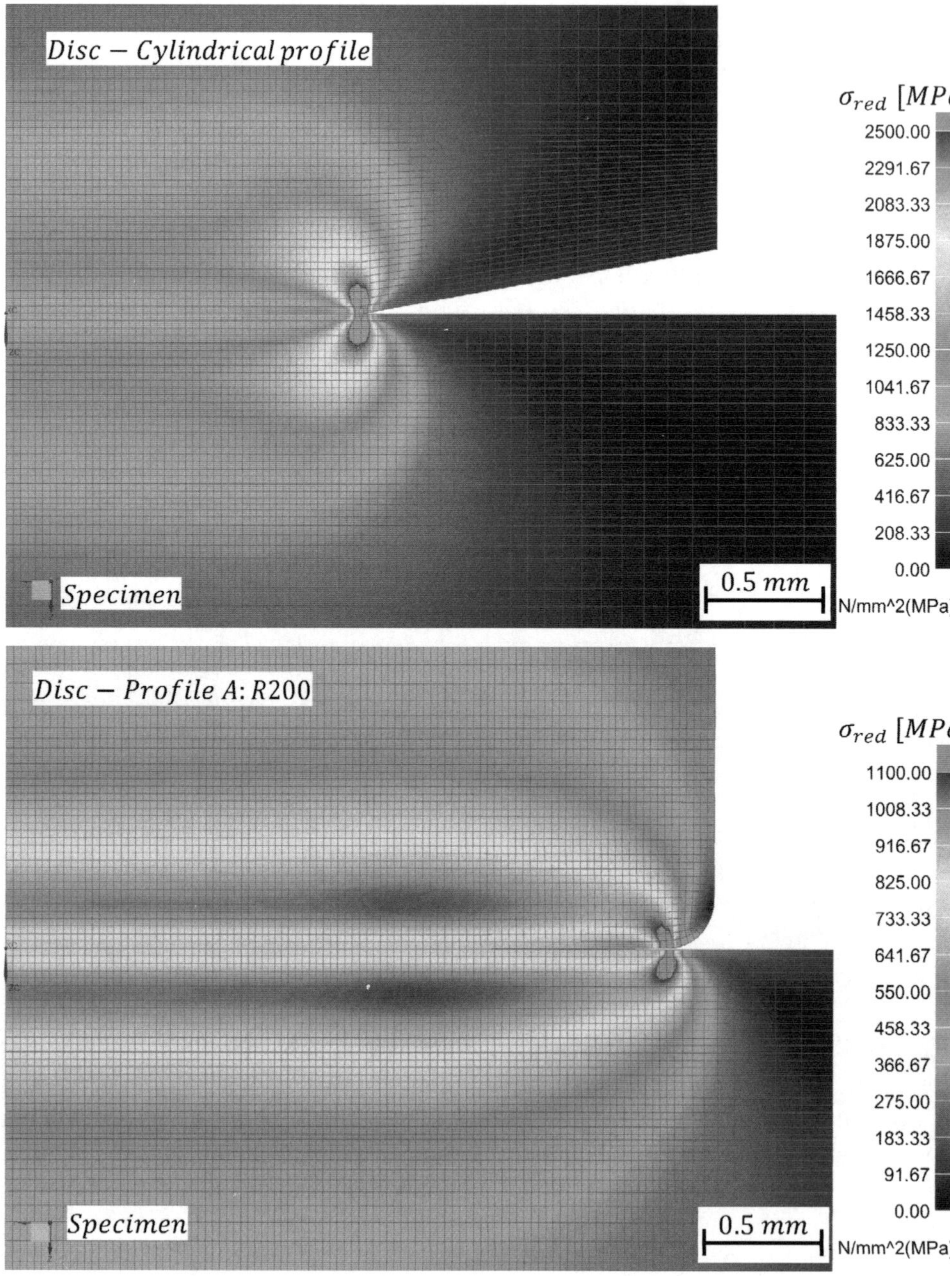

Fig. 11: Distribution of the reduced stress in the contact area of disc and specimen

Conclusions

This paper discusses a design of a test rig for testing materials intended for the manufacture of cam mechanisms. It allows us to respect and test a whole range of parameters. It simulates cam and roller contact. However the test runs three times faster than the real cam mechanism.

Suggested test rig allows testing under the following conditions:

- working conditions
  - Slide Roll Ratio (0% (pure rolling) to +/- 200% (pure sliding))
  - Lubrication (no lubrication or influence of various types of lubricants)
  - Temperature (at normal operating temperature or reduced / increased)

- steels (types, production technology, with / without (chemically-)heat treated layer)
- other materials (non-metallic materials, composites, etc.)
- coatings, finishes

Acknowledgments

This paper was created within the work on the project FV20235 – Project supported by the Ministry of Industry and Trade of the Czech Republic.

References

[1] Koloc Z, Václavík M. Cam Mechanisms, first ed., Amsterdam: Elsevier; 1993. 424s. ISBN 0-444-98664-2.

[2] Namáhání na tah, tlak [online], [cit. 2019-04-04]. URL: <https://docplayer.cz/7596316-Namahani-na-tah-tlak.html>

[3] Pekař, V.: Některé druhy poruch - Cyklické zatěžování konstrukce až k únavovému lomu [online], [cit. 2017-06-28]. URL:<http://www.reliability.estranky.cz/clanky/nektere-druhy-poruch/nektere-druhy-poruch---cyklicke-zatezovani-konstrukce-az-k-unavovemu-lomu.html>

[4] Hrivňák, I.: Fraktografia [online], Bratislava, 2009. Slovenská technická univerzita, Materiálovotechnologická fakulta so sídlom v Trnave. [cit. 2017-06-28]. URL:<http://web.tuke.sk/hf/data/knihy/fraktografia.pdf>

[5] Halama, R. a kol.: Vlastnosti a zkoušení materiálů [online], Ostrava, 2013. Vysoká škola báňská – Technická univerzita Ostrava, Fakulta strojní. [cit. 2019-03-28]. URL:<http://projekty.fs.vsb.cz/463/edubase/VY_01_014/Vlastnosti%20a%20zkou%C5%A1en%C3%AD%20materi%C3%A1l%C5%AF/02%20Text%20pro%20e-learning/Vlastnosti%20a%20zkou%C5%A1en%C3%AD%20materi%C3%A1l%C5%AF%2008.pdf>

[6] Ondrášek, J.: The Stress Distribution in the Contact Region of a Cam Mechanism General Kinematic Pair, Mechanism and Machine Science 52: Mechanisms, Transmissions and Applications, Proceedings of the 4th. MeTrApp Conference 2017, pp. 99-108, Springer International Publishing AG 2018, ISBN 978-3-319-60701-6

[7] Tomek, O.: Výpočtové modely kontaktní únavy strojních součástí [online], Brno, 2007. 47 s. Bakalářská práce. Fakulta strojního inženýrství. Vysoké učení technické v Brně. Vedoucí bakalářské práce doc. Ing. Martin Vrbka, Ph.D. [cit. 2019-03-27]. URL:<http://www.ustavkonstruovani.cz/FileDownload/getFile/130/Bakalarska_prace_Tomek_Ondrej.pdf/>

[8] Růžička, M: Únavové křivky a faktory, které je ovlivňují [online], [cit. 2018-04-12]. URL: <http://www.kmp.tul.cz/system/files/duz_2017_2_ru.pdf>

[9] Ondrášek, Jiří. Obecné kinematické dvojice vačkových mechanismů, first ed., Liberec: VÚTS, a.s., 2018. 90 s. ISBN 978-80-87184-77-6.

[10] Johnson, K., L.: Contact Mechanics, Cambridge University Press, Cambridge, 1985, ISBN 0-521-34796-3.

[11] Reusner, H.: The logarithmic roller profile the key to superior performance of cylindrical and taper roller bearings, Ball Bearing Journal, 230, 1987, pp. 2-10.

Comparison of shrinkage of the rammed earth prepared from illitic and illitic-kaolinitic clay

PADEVĚT Pavel[1,a], BITTNAR Petr[2,a] and MUŽÍKOVÁ Barbora[3,a]

[1]Department of Mechanics; Faculty of Civil Engineering; Czech Technical University in Prague; Thákurova 7, 166 29 Prague 6, Czech Republic

[a]pavel.padevet@fsv.cvut.cz, [b]petr.bittnar@fsv.cvut.cz, [c]barbora.muzikova@fsv.cvut.cz

Keywords: Rammed earth, Shrinkage, Creep. Illitic Clay, Illitic – Kaolinitic Clay, Mixture Water.

Abstract. The paper senses the comparison of the impact of different types of clay on the rheological properties. Illitic and Illitic kaolinitic clay have been used in this study. The creep and shrinkage of rammed material was observed. The influence of material composition on rheological behaviors was observed. Finally, comparison of both materials was realized.

Introduction

Material properties are needed for structural design [1, 2]. Supporting structures made of unburned clay can have great variability of properties. The paper attempts to contribute by comparing different blend designs to the type of clay and the amount of water in the blend design. Load-bearing capacity is a very important piece of information [3]. The creep under load and shrinkage from drying to such properties includes.

Materials The rammed earth was prepared from two types of clay. The first one has label 1 – KR5. It was prepared using the Illitic – kaolinitic clay [4]. The second one mixture marked 2 – AGL9 was prepared using the Illitic clay. Both clays differ in their ability to bind water to their structure.

The specimens were prepared from a mixture of water sand and appropriate clay. Table 1 reports the composition of mixtures and the amount of water in the mixture. The difference between the two mixtures was in the clay type and the amount of mixing water. Its amount is based on the weight of the clay used. The sand used was 0 - 2 mm with continuous grain size curve.

Creep and shrinkage of stewed clay is experimentally measured on specimens made of rammed earth. The specimen length is 70 mm. The bodies are made in the form of beams with an octagonal base with a distance of opposite sides of approximately 20 mm. The shape of the specimens is shown in Figure 1. The specimens are subjected to loads for the required time, which corresponds to the load in the elastic area of the material.

Table 1: Composition of both materials

Description	Clay	Sand	Clay / water
1 – KR 5	20	80	0.37
2 – AGL 9	20	80	0.45

Deformation increases at the start of the test in these cases. Drying of the material causes to stop shrinking and only a slight increase in creep rammed earth. Creep rammed clay prepared from Ilitic - kaolinitic clay exhibits sensitivity to temperature, which is seen on the right hand graph in Figure 2. Ambient temperature during the measurement is changed by ± 1 ° C.

Measurement A comparison between the two different mixtures is shown in Figure 2. The creep values of both graphs are in dimensionless quantities. The first mixture (left) achieved a very significant creep value immediately after the specimen load. Both remaining specimens shoved the same trend of creep growth. The mixture prepared from Illitic - kaolinitic clay behaved differently. The deformation increased suddenly after loading. Consequently, the deformation was still noticeably increasing for one day. Then the increase in deformation stopped.

Fig. 1: Specimens for creep and shrinkage tests prepared from Illitic clay.

Mixture No. 2 contained a higher proportion of water in the mixture as shown in Table 1. More plastic blend No.2 showed a sudden increase in strain under load. In contrast, mixture No.1 reacted gradually to the load. This is shown in Figure 2 (right).

The second phenomenon, creep size is also evident from the comparison of the both graphs of Figure 2. The drier mixture of No.1 reached several times lower creep values than the mixture No.2. This phenomenon is also related to the amount of mixing water. Although both mixtures contained the same amount of clay, the amount of water significantly affected the rheological properties of the rammed earth. The difference in the amount of water both mixtures was 0.08 weight relative to the amount of clay.

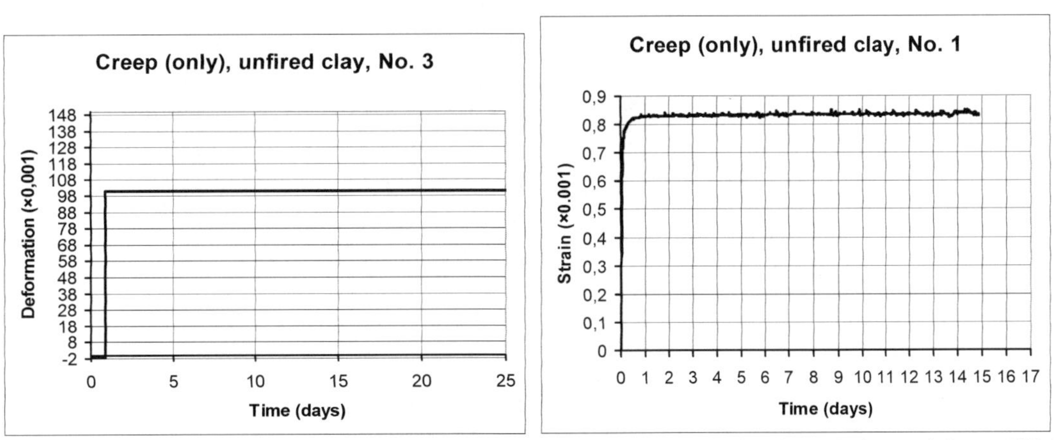

Fig. 2: Comparison of creep of rammed earth prepared from Illitic clay (left) and from Illitic-kaolinitic clay (right).

The result of the shape and size of the creep curve shows that a small change in the amount of mixing properties can significantly affect rheological properties. The creep of both mixtures did not increase after the initial increase in deformation. The phenomenon is related to the increase in the strength of rammed earth. Drying of the mixture causes an increase in strength.

Similar behaviour is observed on the shrinkage curves of both materials, see Figure 3. The left graph shows the shrinkage curve of the second Illitic clay mixture [5]. The rammed earth with a lower content of mixing water showed less shrinkage. Higher shrinkage was measured on rammed earth from Illitic clay with higher water content.

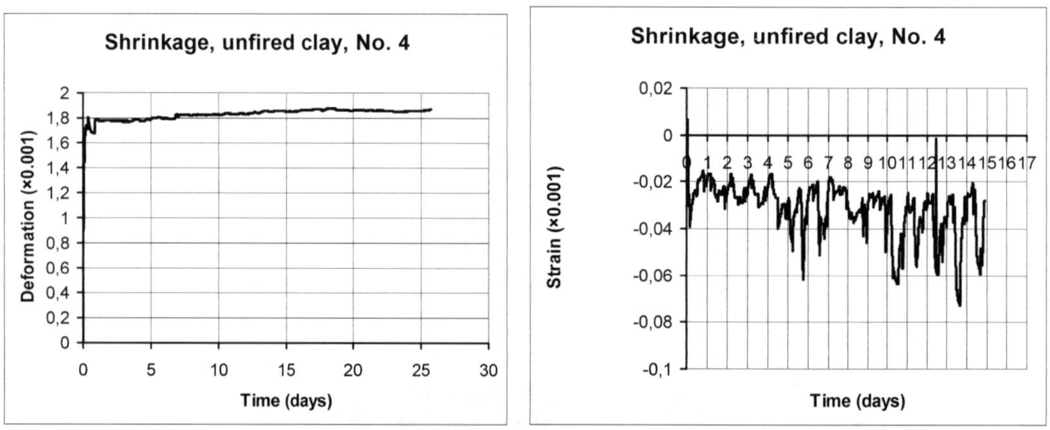

Fig. 3: Comparison of shrinkage of rammed earth prepared from Illitic clay (left) and from Illitic-kaolinitic clay (right).

Illitic - kaolinitic clay tends to expand instead of shrinking. This is shown in the graph in Figure 3 on the right. The shrinkage of this rammed earth is in the tens of microstrain. On the other hand, rammed earth No. 2 had shrinkage in milistrains. This mixture showed shrinkage. There was no expansion, as in the first case, see the graph on the left in Figure 3.

The mixture prepared from Illitic clay reacted more sensitively to the temperature changes shown in Figure 5 on the left. Decreasing the temperature by 1.5 ° C caused an increase in deformation around 13 days from the start of the test. The deformation increase was approximately 0.1 milistrain. The clay prepared from Illitic - kaolinitic clay was not affected by temperature [6].

Fig. 4: Specimen for material test prepared from Illitic – kaolinitic clay.

The temperature was steady for the duration of the test, as shown in Figure 5 on the right. The initial increase in clay deformation 2 is caused by instrumentation of the measurement. In both cases, the measurement principle was the same. The higher water content was more sensitive to the shrinkage measurement method as shown in the graph to the left in Figure 3.

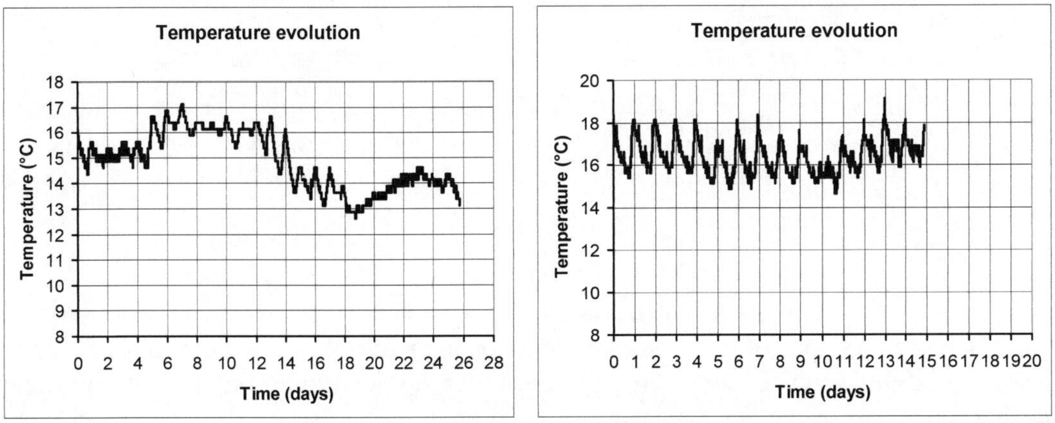

Fig. 5: Temperature evolution of ambient in laboratory during testing of materials; Illitic clay (left) and from Illitic-kaolinitic clay (right).

Figures 1 and 4 illustrate the difference in structure between the two materials. Illitic - kaolinitic clay with lower water content creates a surface structure with a higher proportion of sand grains, see Fig.4. On the other hand, a mixture with a higher water content better on the surface envelopes the sand grains, and the structure looks more homogeneous. The influence of the water content plays an important role for higher plasticity of the material. The mixture better fills the mould when ramming, see Figure 1.

The colour difference of bodies is caused by the use of two different types of clays. Illitic clay stains the material in light brick colour.

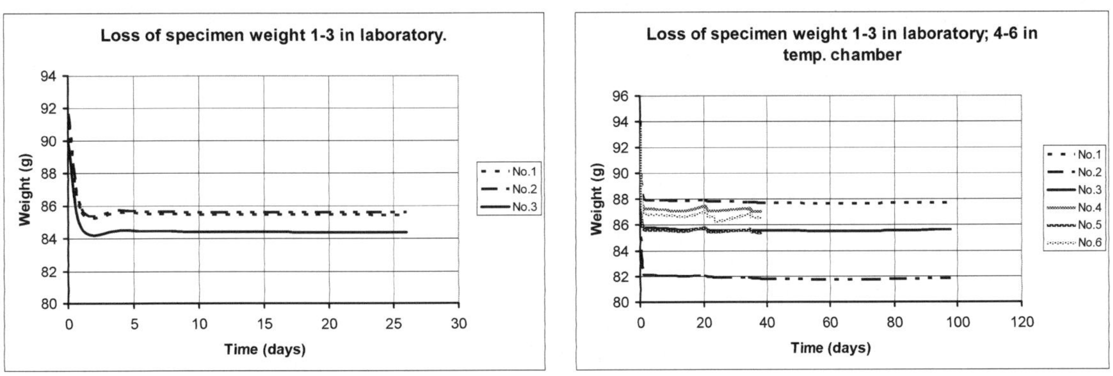

Fig. 6: Evolution of specimen weight during the test of materials; Illitic clay (left) and from Illitic-kaolinitic clay (right).

Figure 6 accompanies the idea of material drying rate. The weight of the specimens at the creep measurement point is measured throughout the tests. The main change in specimen weight of 20 × 20 × 100 mm occurred in the first two days of both measurements. Measurements of the weight change of illitic-kaolinic clay were accompanied by weight measurements under steady-state conditions in the air conditioning chamber. A temperature of 20 ° C and a relative humidity of 50% was maintained throughout the measurement. The laboratory measurement curves and measurements in the air-conditioning chamber are the same, see the graph to the right in Figure 6. The bulk density in both cases at the end of the test was 2150 kg / m^3.

Table 2: Size of creep and shrinkage.

Name of composition / specimen	1 (KR 5)	2 (AGL9)
1	0.831	38.106
2	1.363	-1.463
3	0.988	100.788
4	-0.028	1.858
5	0.009	1.453
6	-0.001	0.223

Conclusions Behaviour of rammed clay depends significantly on the amount of water. As can be seen in Table 2, the creep of the mixture with a higher water content was significantly higher than the creep of the No. 1 mixture. The creep and shrinkage values in the table are in microstrain. The shrinkage of the mixtures is also different.

Rheological properties of rammed clay are also significantly affected by drying the material and reaching the final strength.

Acknowledgement

This work was supported by Grant Agency of Czech Republic GACR No. 18 – 10884S "Unfired and rammed clay for construction".

References

[1] I. Žabičková, Clay buildings [online]. Brno, Era 21, 2002 [vid. August 18, 2015], ISBN 80-86517-21-7.

[2] G. Minke, Building With Earth [online]. 2006, [published August 18, 2015]. ISBN 978-3-0346-0822-0. Available from: http://archive.org/details/Gernot Minke-Building With Earth.

[3] H. Schroeder, Moisture transfer and change in strength during the construction of eathern buildings, Informes de la Constructión, Vol. 63, No. 523, 2011, pp. 107 – 113.

[4] P.Padevět, T. Otcovská, B. Mužíková, Material Properties, of Rammed Earth Prepared from Illitic - Kaolinitic Clay, WSEAS Transaction on Applied And Theoretical Mechanics, Vol. 13, 2018, pp. 181 – 186.

[5] P. Padevět, P. Bittnar, T. Otcovská, Shrinkage of Rammed Earth Based on Illitic Clay, 56th International Scientific Conference on Experimental Stress Analysis EAN 2018, 2018, Harrachov, Czech Republic, pp. 321 – 325.

[6] M. Jankula, T. Hulan, I. Śtubňa, J. Ondruška, R. Podoba, P. Šín, P. Bačik, A. Trnik, The influence of heat on elastic properties of illitic clay Radobica,/ Journal of the Ceramic Society of Japan, Vo. 123, Issues 1441, 2015, pp. 874 – 879.

EAN 2019
57th conference on experimental stress analysis

Testing of C/PPS structural element for aircraft

PADOVEC Zdeněk[1,a], SEDLÁČEK Radek[1,b] and ZÁMEČNÍKOVÁ Tereza[1,c]

[1]Department of Mechanics, Biomechanics and Mechatronics; Faculty of Mechanical Engineering; Czech Technical University in Prague; Technická 4, 16000 Prague, Czech Republic

[a]zdenek.padovec@fs.cvut.cz, [b]radek.sedlacek@fs.cvut.cz, [c]tereza.zamecnikova@fs.cvut.cz

Keywords: Composite, Fabrics, Thermoplastics, Aircraft structural element

Abstract. Presented work deals with testing and FE evaluation of composite structural element manufactured from C/PPS 5H satin fabric [1,2] used for junction of rib with keel beam lower panel in aircraft construction. Tested composite version should replace titanium one which is used nowadays. Loading of composite special element (bracket) was simulated in FE software Abaqus and it was proved that it should withstand the load without failure. After the manufacturing of three specimens an experiment was done to confirm the FE results.

Introduction

Many crucial connecting components in aircraft are manufactured from titanium because of the fact, that electrochemical corrosion can occur between steel and composite. Typical titanium part is analysed bracket used for junction of rib with keel beam lower panel. Its shape requires a variety of machining operations which are expensive, and they leave a lot of waste material. Therefore, composite version of the bracket was designed and manufactured from C/PPS 5H satin fabric by thermoforming technology and this version was simulated and tested.

Finite Element Computation. In Fig. 1 you may see the results for failure index according to maximal stress theory for given load of the bracket which is described in Table 1. Maximal failure index is equal to 0,3934 which means that the safety factor is around about 2,54. This means that composite bracket can replace titanium one without failure and with saving 58 % on the weight - Ti version weights 0,308 kg, composite one 0,129 kg.

Table 1: Load of the bracket

Location	F_x [N]	F_y [N]	F_z [N]
1	-276,1	-411,9	3 496,9
2	-345,6	-618,9	3 007,8
3	-168	-415,7	2 846,1

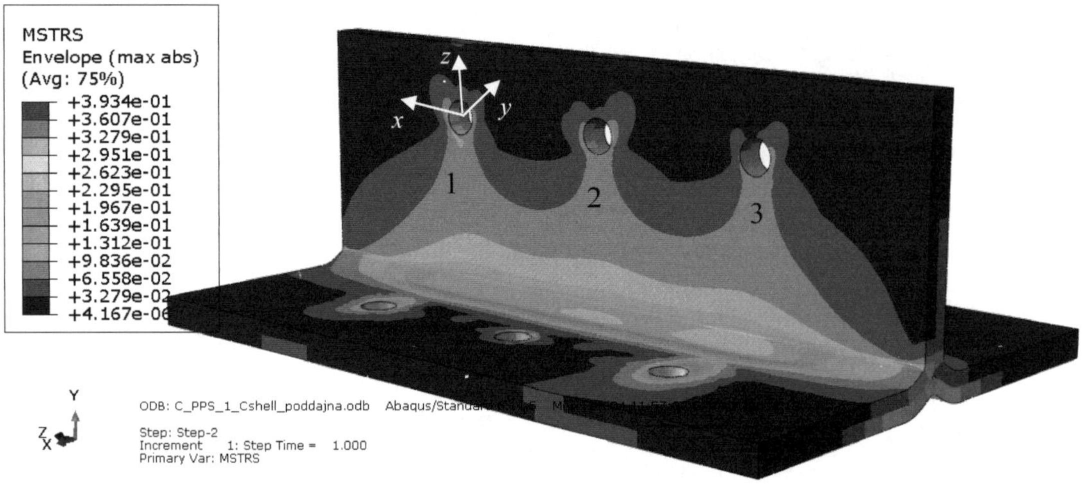

Fig. 1: Failure index according to the maximal stress theory for given load

Experiment. Experiment was done on TIRA 2300 universal testing machine with loading speed of 1 mm/min (respectively 2 mm/min). Tensile load was realized through the screws in the web of the element jointed with the jaws of the machine, just in the z direction (direction of the profile´s web) – when we look in Table 1, we may see that the load in this direction is much more higher than the load in. x and y direction. Specimen in the jaws can be seen in Fig. 2. Relationship between loading force and displacement for three tested specimens can be seen in Fig. 3. Comparison of failed specimen with FE prediction of failure index (according to maximal stress theory [3]) can be seen in Fig. 4.

Fig. 2: Specimen loaded in the testing machine

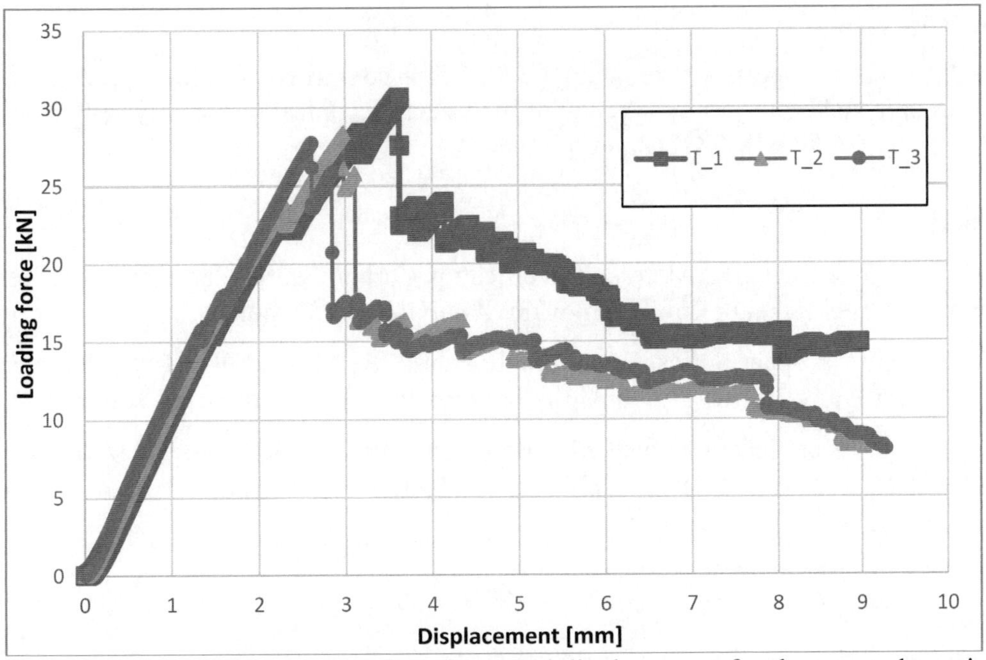

Fig. 3: Relationship between loading force and displacement for three tested specimens

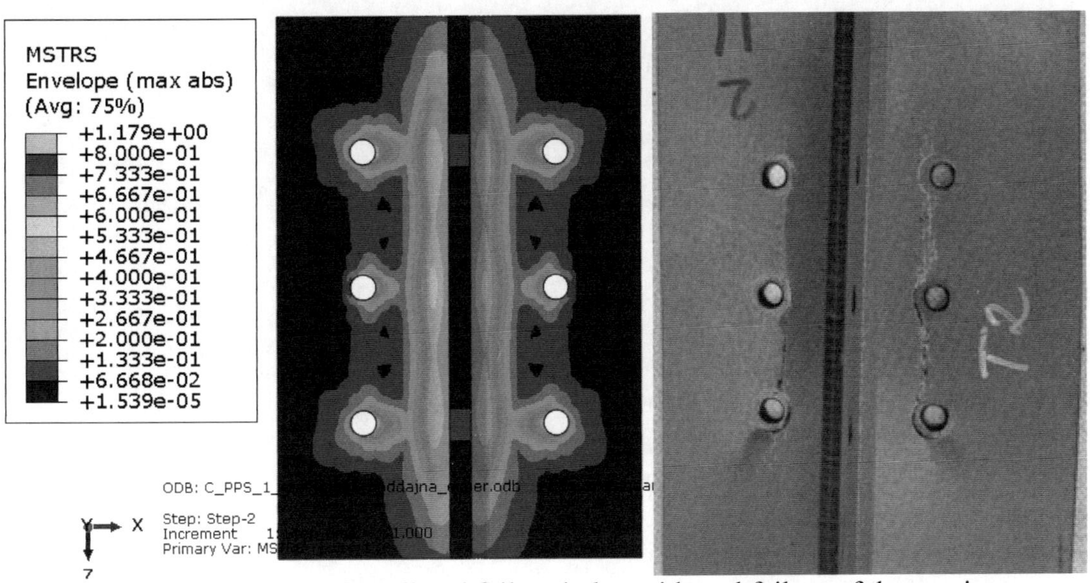

Fig. 4: Comparison of predicted failure index with real failure of the specimen

Conclusions

Three specimens of composite bracket were tested with achieving average maximal force 28,94 ± 1,25 kN.

Comparison of FE and experimental results shows that FE model predicts well areas of failure which are around the bolt holes and on transition radius between lower flange and web of the profile.

Future work will focus on re-design of composite bracket geometry, its testing and FE simulation and comparison with this geometry and original titanium solution.

Acknowledgements

This work has been supported by project FV30033 of the Ministry of Industry and Trade of the Czech Republic and by the Grant Agency of the Czech Technical University in Prague, under grant No. SGS18/175/OHK2/3T/12.

References

[1] AIMS05-09-002, Airbus Material Specification Carbon Fabric, 285 g/m2 fibre area mass with 43% PPS Resin or Equivalent resin Material Specification.

[2] ABS5045, Aerospace Series, Carbon Fibre Fabrics Reinforced PPS, Laminates w/ or w/o Lightning Protection, w/ or w/o Anti-Corrosion Protection, Structural Material.

[3] https://classes.engineering.wustl.edu/2009/spring/mase5513/abaqus/docs/v6.6/books/usb/default.htm?startat=pt05ch17s02abm04.html.

EAN 2019
57th conference on experimental stress analysis

Numerical and experimental analysis of high precision positioning system

PÁSTOR Miroslav[1,a], BOCKO Jozef[1,b], LENGVARSKÝ Pavol[1,c] and PRADA Erik[2,d]

[1]Department of Applied Mechanics and Mechanical Engineering; Faculty of Mechanical Engineering; Technical University of Košice; Letná 9, 04200 Košice, Slovakia

[2]Division of Machine and Devices, ZTS VVÚ KOŠICE, Južná trieda 95, 041 24 Košice, Slovakia

[a]miroslav.pastor@tuke.sk, [b]jozef.bocko@tuke.sk, [c]pavol.lengvarsky@tuke.sk, [d]erik.prada@ztsvvu.eu

Keywords: Positioning system, Precision, Stand, Actuator

Abstract. In this paper numerical and experimental analysis of high precision positioning system is investigated. The positioning system is designed with or without mechanical reductor. The functionality of mechanism and its stiffness and strength characteristics were checked by the finite element method. Magnitudes of displacement of reductor was checked by numerical and experimental method. It was found out that it is sufficient to use an actuator without mechanical reductor to achieve the required accuracy of the positioning system.

Introduction

Precise positioning is demand in almost every industry branch, manufacturing process automation, material division, product palletizing, etc. In most cases, positioning accuracy in tenth of millimetres is sufficient. In microelectronics, laboratory instruments, etc., precision requirements are increasing, but production costs are also increasing. However, in such cases, the positioning of small loads with masses of the order of few kilograms is concerned. In the precision positioning of larger loads is necessary to consider the deformation of the elastic members which are part of the positioning systems. The numerical modelling using the finite element method is the most commonly used to verify the required strength and stiffness. The outputs of numerical modelling are applied not only in the optimization of structural parts, but also in the kinematic analysis of movements of the whole supporting system. The basic aim of the kinematic analysis is to verify the range of reference point movements depending on the actuator movements, taking into account real bonds. The precise positioning system depends on the accuracy of the manufacturing of the individual components and at the same time it is necessary to perform a control measurement of it on real stands [1-6].

The paper presents the procedure of experimental verification of achieving the required accuracy of the precise positioning system without and with mechanical reductor. The key elements of the designed system consist from elastic members whose shapes and dimensions have been checked by numerical modelling [3]. The aim was to achieve positioning accuracy of load with mass 600 kg and repeatability of positioning. Experimental measurements aimed to verifying the accuracy of the positioning system were carried out on the designed stand without and with load.

Design of methodology for testing the positioning system

The aim of the methodology is to design the measurement system so that the results obtained by numerical modelling can be verified by experimental measurement on a real model. The stand should allow a vertical positioning of load with mass 600 kg. Two variants of precise positioning of load are analysed. The first variant consists of vertical actuator in combination with an elastic member. The second variant contains in addition to the horizontal actuator and elastic members mechanical reductor commonly used in precise positioning systems [1, 2]. Optimization of shape and dimensions of chosen parts of positioning system is realised by the finite element method [7]. The object of numerical modelling is to obtain the dependence between the horizontal displacement by the actuator in the range ±5 mm and vertical displacement of the selected point of the load. In the Fig. 1 is shown a model of the analysed positioning system with elastic members and mechanical reductor.

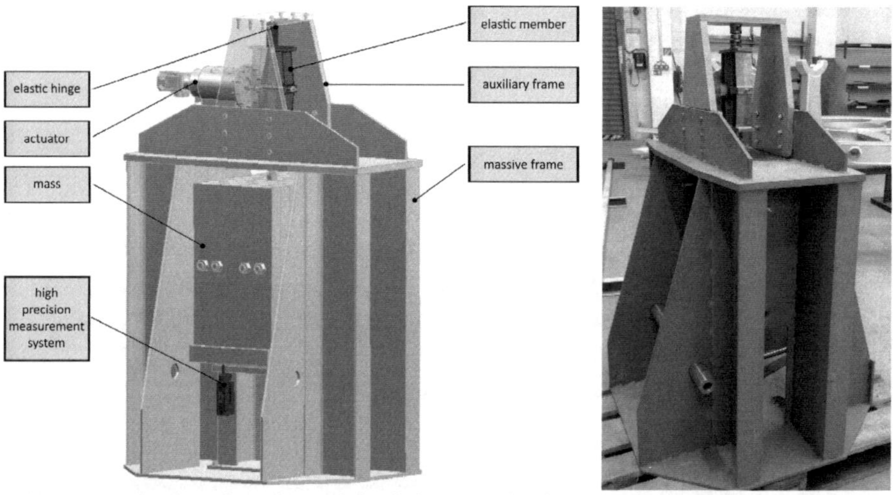

Fig. 1: Numerical and real model of the stand with elastic member and reductor

The displacement field in positioning system is shown in two directions – in the x-axis direction (Fig. 2 left) and in the y-axis (Fig. 2 right). If the system is loaded by the mass of 600 kg, then the vertical displacement of the bottom part is equal 0.2355 mm. This value is taken as reference value. In the Tab. 1 are given the values of displacements of bottom part for the values of displacement of vertical actuator. In addition to the analysis of displacement fields, a stress analysis is performed. In this analysis the critical points of the positioning system are analysed. The maximum values of von Mises stresses do not exceed the value 560 MPa, which is below the yield strength for the material 34CrNiMo6 (WNr. 1.6582) [3].

Table 1: Displacement of bottom surface of mass

Prescribed displacement in Y direction [mm]	Displacement of bottom surface of weight in X direction [mm]
0	0.0000
0.5	-0.1002
1	-0.2000
2	-0.3999
3	-0.5998
4	-0.7996
5	-0.9994

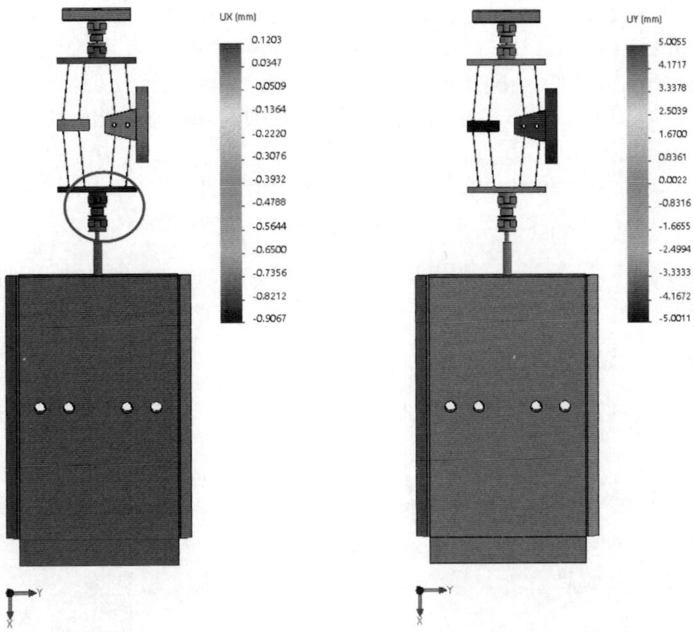

Fig. 2 Displacement fields of points in X and Y direction, respectively

The numerical simulations have not been performed for the positioning with the vertical actuator, since the mechanical movement of the actuator is identical as the movement of the mass.

Differences between the analysed variants can occur on real objects due to inaccuracy of the production of individual components, the backlash of the mechanical parts of the actuator and the like. For this reason, verification experimental measurements on real models are realized.

Experimental verification of accuracy of positioning system

To compare the above mentioned variants, a measurement chain is designed, which includes a stand allowing experimental verification of positioning accuracy with and without mechanical reductor under operating conditions for defined load states. In the Fig. 3 is shown 3D model of the stand with and without mechanical reductor.

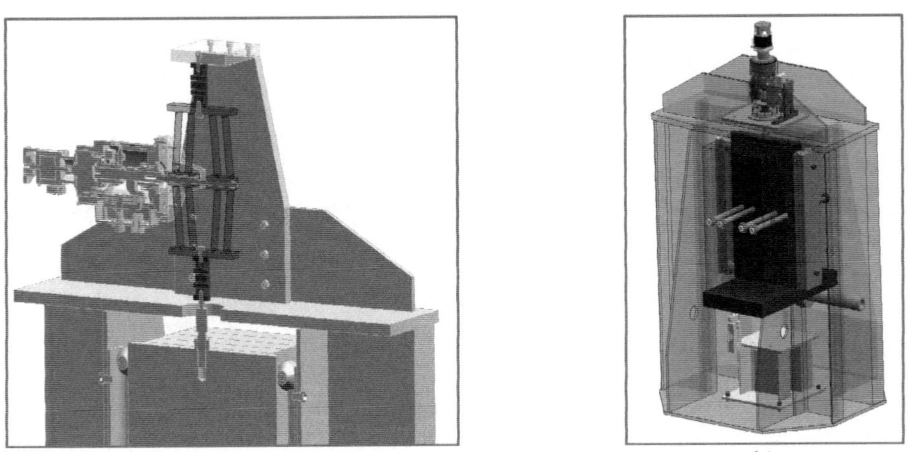

a) b)

Fig. 3 Model of stand a) with mechanical reductor, b) without reductor

In the Fig. 4 is shown the stand with and without mechanical reductor, respectively. For the stand without reductor is used the same massive frame, the difference is only in the design for mounting the actuator. In the case of positioning with the mechanical reductor, a radial actuator RLA with maximal loading 100 kg is used, in the case of mechanical reductor a vertical actuator VLA with maximal loading 500 kg is used. The basic parameter for selecting the actuator was the maximum load required for trouble-free positioning of the load with the prescribed weight. The load was caused by the weight of the steel plates themselves.

a)

b)

Fig. 4 Stand a) with mechanical reductor, b) without mechanical reductor

Fig. 5 Preview of used actuator

Mechanical sorting and calibration of linear actuator

The first phase after the mechanical production of components and the mechanical assembly of the linear actuator is its electro-mechanical recovery and alignment to the specified reference point. The reference point is the basis of every accurate measurement system and ensures repeatability of accuracy of the device even after its repairing or service interventions. A very important task of mechanical sorting and calibration of actuator is to eliminate partially or completely defective parts that would render the device inoperable over time. In any case, if a component's continuous test results appear outside the tolerances, the component must be removed and replaced with a new component to avoid significant future damage that occurs sooner or later as a result of this defective part. The result of the above process is a assembling protocol that contains all of the parameters that are essential to the proper operation of the actuator.

Complex test without load

The first test is performed without load. The test is run for approximately 45 minutes and ended automatically. At the beginning of the test, the actuator output piston center position is set. The test consisted of positioning the actuator within ±3 mm, with one step corresponding to 0.2667 μm (i.e. 3 mm = 11 250 steps). Three main parameters were registered during the test: repeatability, linearity and mechanical backlash of transmission system. In the Fig. 6 is shown the protocol from the test without load, where the maximal deviations for the respective actuator step are given.

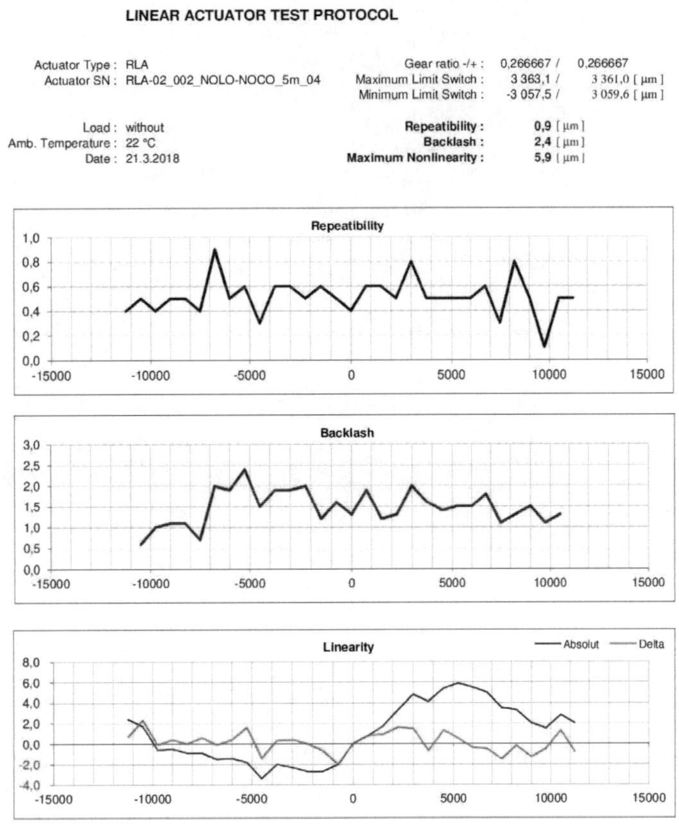

Fig. 6 Linear actuator test protocol without load

Complex test with load

After performing the test of the actuator without load, the tests with the load are performed (the ambient temperature is about 22-23°C). Vertical actuator VLA (Fig. 3b) is loaded by a mass of 600 kg and the horizontal actuator RLA (Fig. 3a) by mass of 120 kg. The masses represent a 20% overrun of the manufacturer's maximum values. The test of the actuator under full load verifies its performance parameters.

A precision displacement sensor (Fig. 7) is used to verify the real displacement of the lower load. In addition, the precision displacement sensors are used to identify the response of the mechanical parts of the actuator – backlash.

Fig. 7 The displacement sensor located at the bottom part of the stand

The results of the test of precisions actuators have confirmed that the parameters measured during the activation, mechanical sorting and calibration of individual actuators do not change due to the load of the actuator. In the Fig. 8 are shown the protocols from functionality and reliability testing of actuators.

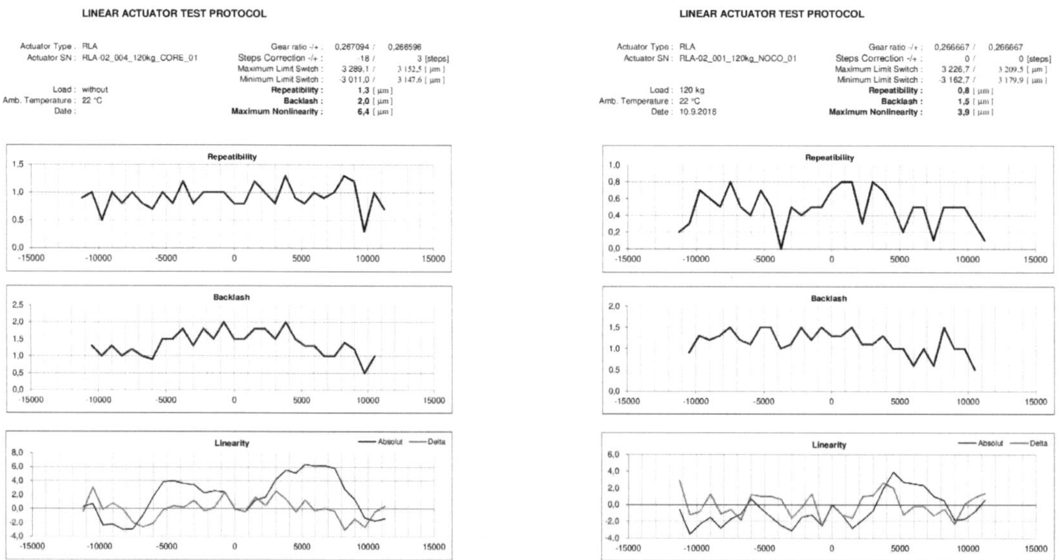

a) b)

Fig. 8 Test protocol of actuator RLA a) without load, b) with load

In the Tab. 2 are given the most unfavourable variations regarding of repeatability, mechanical backlash of the actuator gear system and linearity detected during testing.

Table 2: Variations during testing

	Actuator VLA without mechanical reductor		Actuator VLA with mechanical reductor	
	without load	with load	without load	with load
Repeatability [µm]	2.1	0.9	1.3	0.8
Backlash [µm]	2.6	1.6	2.0	1.5
Nonlinearity [µm]	6.5	4.9	6.4	3.9

From the test results, it is clear that the solution with the mechanical reductor gives a slightly better values than is the case without the mechanical reductor.

In Tab. 3 is given the displacement of the lower surface of mass. The results are obtained by numerical modelling and experimental measurements. From the results can be stated that the values are little different, which is probably caused by the production of a mechanical stand. Despite this, it can be stated that a very good agreement of results is achieved. In the production of a real device with the mentioned type of positioning mechanism, it is necessary to pay increased attention to the production of individual parts as well as the assembly itself. Since the welding process is used in the manufacture of the positioning device, it is also necessary to pay extra attention to residual stresses, which may adversely affect the desired positioning accuracy. Non-destructive [8] or semi-destructive residual stress [9-11] methods are most commonly used to quantification of residual stresses.

Table 3: Displacement of the lower surface of mass

Horizontal displacement (actuator RLA) [mm]	Vertical displacement	
	FEM [mm]	Experiment [mm]
0	0.0000	0
0.5	-0.1002	-0.1025
1.0	-0.2000	-0.2047
1.5	-0.2999	-0.3073
2.0	-0.3999	-0.4062
2.5	-0.4998	-0.5054
3.0	-0.5998	-0.6038

Discussion

The rapid development of hardware and software contributes significantly to experimentation. Computer support is objectively conditioned mainly by the level of electronization, objective social specifics and subjective factors of users, and thus also by the experiment. The computer application in the technology is affected the experiment from various perspectives:

- the computer becomes a direct part of the experiment, while changing their structure and level,
- there are new possibilities to prepare an experiment consisting of planning measurements and computer-aided preparation of multi-factorial and extremalization experiments,
- the quality, quantity and way of realization of computational works performed within individual stages of the experiment are changed by using experimental software,
- the performance parameters of the experiment have increased considerably, namely the speed and operability of the measurement itself, the evaluation of the measurement results, the processing of the extensive information gathered,
- it is possible to realize new forms of information processing and storage for the creation of databases of experimental quantities, direct transfer of experimental results as inputs for the following calculations, real-time experiment control, control of various metallurgical, technological, chemical and other processes, activity management (acquisition of material characteristics for various loads, etc.).

Conclusions

In the paper was described numerical and experimental analysis of high precision positioning equipment. The computations were accomplished for different types of loadings including self-weight of structure and prescribed displacement of parts due to movement of actuator. From the computations can be stated fact that the most stressed parts of structure withstand the applied loading. The aim of such analysis was to verify range of movement of reference points on positioning axes of the object. The problem of this analysis was that this system is statically undetermined and accordingly the stiffness of joining elastic members have to be considered for the computations. The main interest of authors was focused on verification of results obtained by numerical modeling and by experimental measurements on a real stand.

Moreover, the experimental analysis of movement of reference points of the stand with and without elastic member was realized. The measurements showed that the stand with elastic member improved the system behavior. In addition, the temperature tests of the actuator were also made, where its temperature sensitivity coefficient was determined. Its value ranged from 0.9 to 1.1 μm/°C.

Based on the experimental measurement, it can be concluded that the required positioning mechanism accuracy has been achieved. It should be noted that several mechanical details such as different concentrators and the like have to be solved when using a mechanical reductor. Due to the finding of relatively small differences in deviations when using a mechanical reductor and without it, it is sufficient to use an actuator without a mechanical reductor to achieve the required accuracy of the positioning system.

Acknowledgments

The work has been accomplished under the research project Stimuly No. Req-00169-003 financed by the Slovak Ministry of Education as well as grant VEGA No. 1/0731/16 and grant VEGA No. 1/0751/16.

References

[1] J. Torres, H. H. Asada, High-gain, high transmissibility PZT displacement amplification using a rolling-contact buckling mechanism and preload compensation springs, IEEE T Robot, 30 (2014), 781-791.

[2] M. Kelemen, M. Pástor, I. Virgala, A. Gmiterko, Precision positioning with mechanically amplified actuator, Ame. J. Mech. Eng. 5 (6) (2017) 316-320.

[3] F. Trebuňa, J. Bocko, M. Pástor, P. Lengvarský, Deformation and strength properties of elastic members of high precision positioning equipment, Ame. J. Mech. Eng. 5 (6) (2017) 263-268.

[4] P. Šlesar, R. Jančo, Press-fit evaluation and study of displacement in temperature changes, Stroj. časopis – J. Mech. Eng. 68 (1) (2018) 103-108.

[5] P. Élesztős, R. Jančo, V. Voštiar, Optimization of welding process using a genetic algorithm, Stroj. časopis – J. Mech. Eng. 68 (2) (2018) 17-24.

[6] M. Žmindák, Dynamic and sensitivity analysis general non-conservative asymmetric mechanical systems, Stroj. časopis – J. Mech. Eng. 68 (2018) 105-124.

[7] F. Trebuňa, J. Bocko, M. Pástor, P. Lengvarský, E. Prada, The finite element analysis of high precision positioning system J. Mech. Eng. – Stroj. časopis 68 (4) (2018) 41-48.

[8] M. Hagara, F. Trebuňa, M. Pástor, R. Huňady, P. Lengvarský, Analysis of the aspects of residual stresses quantification performed by 3D DIC combined with standardized hole-drilling method, Measurement, 137 (2019) 238-256.

[9] M. Beghini, L. Bertini, P. Raffaelli, Numerical analysis of plasticity effects in the hole-drilling residual stress measurement, J. Test. Eval. 22 (6) (1994) 522-529.

[10] C. Barile, C. Casavola, G. Pappalettera, C. Pappalettere, Residual stress measurement by electronic speckle pattern interferometry: a study of the influence of analysis parameters, Struct. Integr. Life 12 (3) (2012) 159-163.

[11] F. Trebuňa, F, Šimčák, J. Bocko, P, Šarga, P. Trebuňa, M. Pástor, J. Mihok, Quantification of residual stresses in the weld by the hole-drilling method, Metalurgija. 47 (2) (2008) 133-137.

EAN 2019
57th conference on experimental stress analysis

The application of motion magnification method in 3D digital image correlation

PAVELKA Peter[1,a], HUŇADY Róbert[1,b], LENGVARSKÝ Pavol[1,c] and KICKO Michal[1,d]

[1]Department of applied mechanics and mechanical engineering; Faculty of Mechanical Engineering; Technical University of Košice; Letná 9, 042 00 Košice, Slovak Republic

[a]peter.pavelka@tuke.sk, [b]robert.hunady@tuke.sk, [c]pavol.lengvarsky@tuke.sk, [d]michal.kicko@tuke.sk

Keywords: Phase based motion magnification, Experimental modal analysis, Digital image correlation

Abstract. The paper deals with an application of phase based motion magnification and its using in high speed digital image correlation. The aim is to use the magnification method to magnify vibration motion of an investigated structure at specific frequency when the excitation level or response level is low. Three measurements were performed. As the first one, experimental modal analysis was performed to find natural frequencies and mode shapes of the investigated object. Subsequently, the measurements of vibration responses of the object were performed by digital correlation system under the low and high-level excitation, when the object was excited by harmonic signal with frequency corresponded to natural frequency of a chosen mode. The video sequences captured under the weak excitation were subsequently magnified and the results were compared with the measurement done under the strong excitation.

Introduction

Optical methods are used more often than before in field of engineering. The optical methods have many advantages. The current trend and highly discussed topic is computer vision in interdisciplinary scientific field. Many scientists deal with this subject. In this paper, the phase-based motion magnification technique is used to magnify mode shape when the excitation is weak. The phase-based motion magnification method was researched to make subtle motion to be visible from video and it is an extension of techniques motion magnification and Eulerian motion magnification [1], [2], [3]. The base of this techniques is an analysis of motion or colour changes in complex-valued image pyramids. Authors present their approach as microscope for changes in videos. The application of phase-based motion magnification method in mechanical engineering can be found in publication [4], where authors determine operational deflection shapes and natural frequency of cantilever beam and pipe. The response is measured by accelerometer, laser vibrometer and displacement is derived from motion caught by high speed camera. The time series from accelerometer and laser vibrometer were integrated to obtain displacement. The all results are compared with each other. They rated the noise floor in the signal, too. A.J. Molina-Viedma et al. [5] determined mode shapes of cantilever beam excited by dynamic shaker by harmonic signal and they measured responses by digital image correlation system. In this work, the application of motion magnification in frequency domain is being studied.

Proposed procedure

As the first, the experimental modal analysis was performed order to find suitable mode for magnification. Then the structure was excited via vibration shaker by harmonic signal with frequency correspond to natural frequency of the chosen mode with two different excitation levels.

The process of measurement with DIC – Q450 Dantec Dynamics

The process of measurement with DIC consists of follow steps. A stochastic pattern has to be applied on the surface of investigated structure to ensure the correlation. It is necessary to set the cameras to the investigated object and to calibrate the device using a calibration target. Istra 4D is control software for the correlation device, where calibration, acquisition and correlation are done. Frames captured by two cameras are stored in cameras' memory and then transferred to the computer via LAN interface. The series of images with specific frame rate is a video. The region of interest is selected. The following image matching process is called correlation. The displacement and deformation fields are the results of the correlation process. In the case of experimental modal analysis, the measurement was started automatically after the hit of impact hammer. Manual acquisition was used for measurements with shaker.

Experimental modal analysis

The object of interest was steel rectangular plate with dimension 250 x 180 mm. The plate was attached to a vertical stand in the center of the plate through a screw connection. The responses were measured by DIC system Q450 Dantec Dynamics with two high speed cameras. Frequency rate of cameras was set up to 3600 Hz. The frequency range of measurement was 0-1800 Hz. A stochastic pattern was sprayed onto the surface of the plate to ensure image correlation process. The excitation was done by the impact hammer Bruel & Kjaer Type 8206 hitting on the rear side of the object. Measurement chain is shown in Fig. 1. The evaluations were performed in DICMAN 3D [6]. The complex mode indicator function (CMIF) plot obtained by experimental modal analysis is shown in Fig. 2. There are lot of peaks that represent the presence of natural modes. The mode with the frequency 880 Hz was chosen for further analyses. Its mode shape is also shown in Fig. 2.

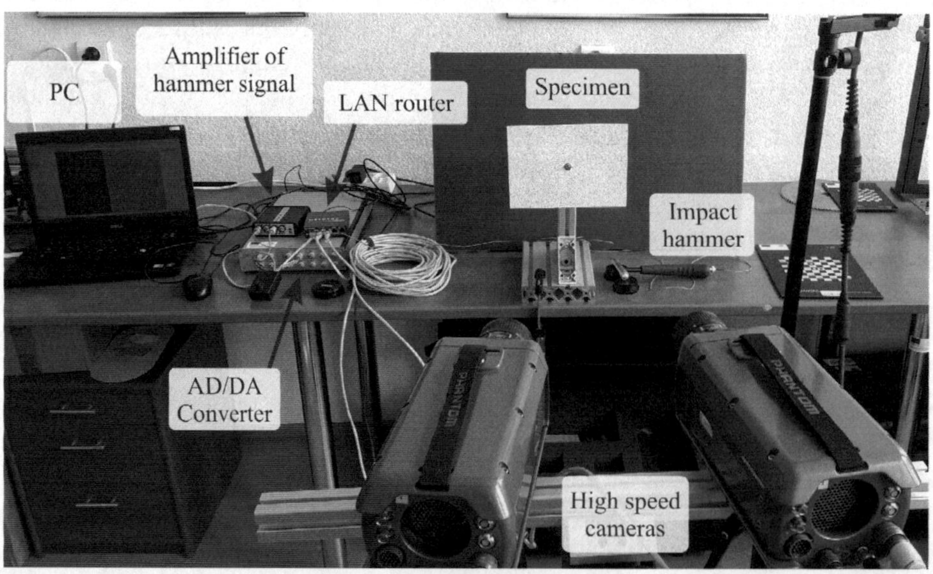

Fig. 1: Measurement chain

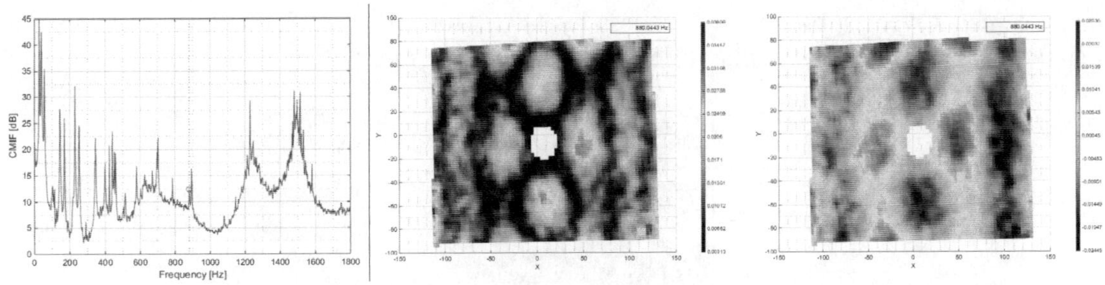

Fig. 2: Modal analysis results: CMIF function (left), absolute shape of the chosen mode (middle), real shape of the chosen mode (right)

Harmonic excitation of the object

The specimen was attached to dynamic shaker Veb Robotron type 11076 and excited by harmonic signal with the frequency 880 Hz. Two measurements were performed, when the structure was excited with high-level and low-level of excitation, separately. Both measurements were performed under the same conditions, only the amplitude of excitation was changed. The evaluation process was done in DICMAN 3D. The CMIF function from high-level excitation measurement is in Fig. 3. There is a significant peak on the frequency corresponding to the excitation frequency. Corresponding mode shape is shown in Fig. 3. This mode is excited very well, which is also evident from the clearly recognizable shape of vibration.

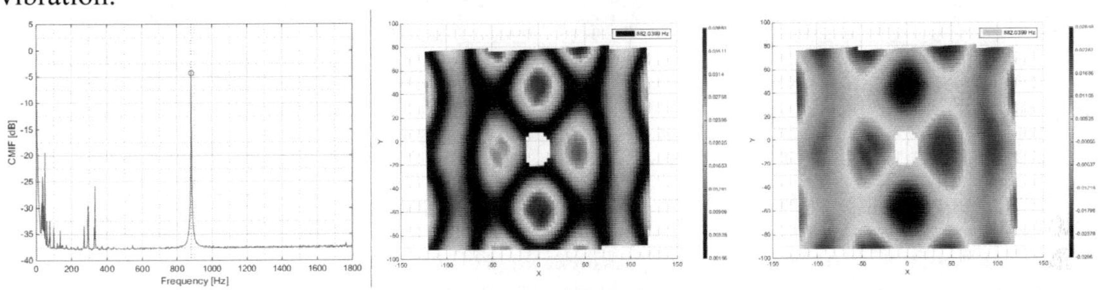

Fig. 3: The results for high-level excitation: CMIF (left), absolute mode shape (middle), real mode shape (right)

In Fig. 4, the results of the measurement under low-level of excitation are shown. As can be seen in CMIF plot, the magnitude of the peak at excitation frequency is much smaller than in the previous measurement. It is due to smaller level of the excitation. Since the mode is not sufficiently excited, its shape is noisier. Video sequence from this measurement was subjected to phase-based motion magnification.

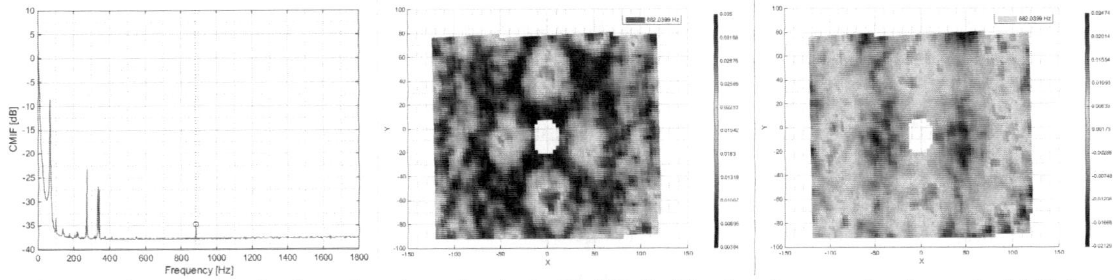

Fig. 4: The results for low-level excitation: CMIF (left), absolute mode shape (middle), real mode shape (right)

Motion magnification was applied on videos recorder by both cameras. After applying magnification, the videos have been subjected to correlation process in ISTRA 4D and fields of displacements were subsequently process by DICMAN 3D. CMIF function, absolute and real shape of vibration after the magnification are shown in Fig. 5.

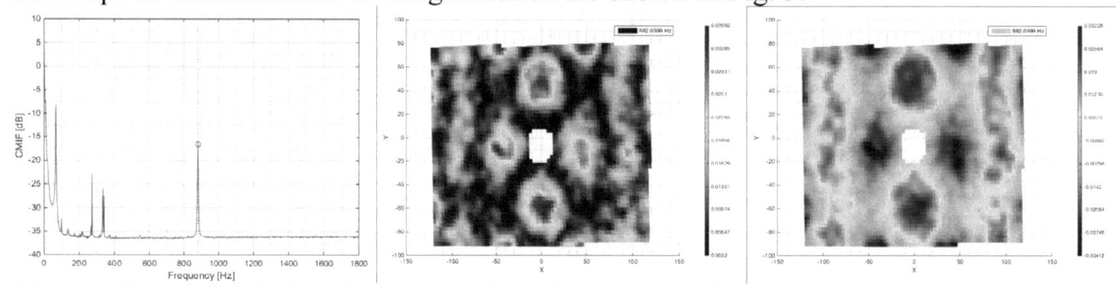

Fig. 5: The results for low-level excitation after the magnification: CMIF (left), CMIF (left), absolute mode shape (middle), real mode shape (right)

Modal phase collinearity (MPC) describes the spatial consistency of the identified mode shapes and it is an indicator which quantifies the collinearity of the phase angle in the identified mode shapes and determines their accuracy. The MPC values are ranged from 0 to 1, where 0 indicates a complex mode, and 1 represents a perfect in-phase (normal) mode. The MPC factors of the analysed mode for all three measurements are listed in Table 1.

Table 1: Comparison of MPC factors of the analysed mode from all measurements.

Mode: 880 Hz	Modal analysis	High-level excitation	Low-level excitation	After the magnification
MPC	0.53474	0.9818	0.6485	0.74445

Discussion and conclusions

The application of motion magnification in digital image correlation seems to be very effective technique that allows to improve the responses of the structure. It can be used in the cases, when the excitation is nonsufficient, or response is too weak or loaded with higher noise. In the paper, the phase-based magnification method was used to magnify the chosen mode of the analysed structure that was excited by low-level harmonic signal with the frequency of the mode. The magnified video sequences were imported back to correlation software and subsequently used to calculate the 3D displacement fields of the magnified vibration response. After the magnification, the shape and MPC factor of the given mode were improved. The resultant quality depends on the settings of parameters such as frequency width, gain, filter type, bandwidth, etc., by which the magnification is controlled. Their influence will be detaily investigated in the further work.

Acknowledgement

This research was supported by project VEGA 1/0355/18, VEGA 1/0731/16 and 1/0290/18.

References

[1] N. Wadhwa, M. Rubinstein, F. Durand, W.R. Freeman, Phase-Based Video Motion Processing, ACM Trans. Graph. vol.32 issue 4 (2013).

[2] C. Liu, A. Torralba, W. T. Freeman, F. Durand, E. H. Adelson, Motion magnification, ACM Trans. Graph. vol. 24 issue 3 (2005) 519–526.

[3] Eulerian video magnification for revealing subtle changes in the world, ACM Trans. Graph. vol. 31 issue 4 (2012).

[4] J. G. Chen, N. Wadhwa, Y.J. Cha, F. Durand, W. T. Freeman, O. Buyukozturk, Modal identification of simple structures with high-speed video using motion magnification, Journal of Sound and Vibration, vol. 345 (2015) 58-71.

[5] A.J. Molina-Viedma, L.Felipe-Sesé, E. López-Alba, F. Díaz, High frequency mode shapes characterisation using Digital Image Correlation and phase-based motion magnification, Mechanical Systems and Signal Processing vol. 102 (2018) 245-261.

[6] R. Huňady, M. Hagara, A new procedure of modal parameter estimation for high-speed digital image correlation, Mechanical Systems and Signal Processing vol. 93, (2017), 66-79

Using modal-parameter-based analysis of the behaviour of alkali-activated material with surface degraded by ultraviolet radiation

PAZDERA Lubos[1,a], TOPOLAR Libor[1,b], DVORAK Richard[1,c] HODULAKOVA Michaela[1,d], PLSKOVA Iveta[1,e], BILEK Vlastimil[2,f] and MIKULASEK Karel[3,g]

[1]Institute of Physics, Faculty of Civil Engineering; Brno University of Technology; Veveri 331/95, 602 69 Brno, Czech Republic

[2]Materials Research Centre, Faculty of Chemistry; Brno University of Technology; Purkynova 464/118, 61200 Brno, Czech Republic

[3]Institute of Mathematics, Faculty of Civil Engineering; Brno University of Technology; Veveri 331/95, 602 69 Brno, Czech Republic

[a]pazdera.l@fce.vutbr.cz, [b]libor.topolar@vutbr.cz, [c]dvorak.r1@fce.vutbr.cz,[d]hodulakova.m@fce.vutbr.cz,[e]plskova.i@fce.vutbr.cz,[f]bilek@fch.vut.cz,[g]mikulasek.k@fce.vutbr.cz

Keywords: Alkali-activated material, Ultraviolet radiation, Degradation, Analysis, Wave propagation

Abstract. Monitoring the behaviour of constructions, building structures, and materials caused by degradation processes is the main purpose of non-destructive testing methods. Most building materials are exposed to the action of environment. As some parts of building structures come into contact with external effects such as sunshine, they may be altered by ultraviolet radiation.

For building materials, monitoring their structure degradation is a much more difficult task than for metal structures mostly due the non-homogeneity. Since, for homogeneous structures, impulse (acoustic) methods can be used to advantage particularly because the monitoring they enable is non-destructive, they were used in this case, too. To monitor the degradation caused by ultraviolet radiation, the impact echo method was employed [1-4].

Introduction

To accelerate the development of new materials and get a better understanding of their behaviour in their application environment, it is necessary to carry out various experiments. One of the parameters influencing the internal material structure is, for example, ultraviolet radiation. In addition to the traditional, using the modal parameters may be another method suitable for monitoring changes in materials during impulse excitation.

Modal analysis can be defined as the process of characterizing the dynamic response of a system in terms of its modes of vibration. Because any (periodic) function can be represented as a series of sinusoidal functions, vibration of a real structure can be represented as a series of modal contributions. [5] Each mode is defined by its natural frequency, damping, and mode shape. The basic idea is to compute the Frequency Response Function (FRF), which requires a response of the structure to the applied force. Note that the force applied and the response are measured simultaneously. [6] Using the Fourier transform, the signals obtained are then transferred to the frequency region. [7] Thus, if measuring acceleration, the outcome is

Accelerance, which is proportional to the quotient of acceleration to force. The impulse force may be induced by a hammer with a piezo element. [8]

Ultraviolet radiation (UV) may be defined as electromagnetic radiation with a wavelength of 100 to 400 nm, which is shorter than the visible light but longer than X-rays. For humans, it is invisible. It can be divided into three radiation types UVA radiation with a wavelength of 315 to 400 nm, which forms about 90% of sunshine and is not captured by the atmosphere, glass or clothes. UVB radiation has a wavelength of 280 to 315 nm, forms about 10% of sunshine, is largely blocked by the atmosphere, captured by glass, clothes, and sunglasses. UVC radiation is the hardest UV radiation having a wavelength of less than 280 nm.

Experimental set-up

The size of the specimens monitored is 323 mm × 83 mm × 19 mm as shown in Fig. 1. Their density is 2100 kg/m^3. Composition of the mixture is in Tab. 1. For the measurement purposes, a regular 2-by-5 grid was created of points on both sides of each specimen. Two sensors (S1 a S2) were placed on a specimen at extreme points opposite to each other each on the opposite side of the specimen (Fig. 1 right side).

Table 1: Mixture

Materials	Ratio [%]
Slag	22.2
Crushed sand 0-2 mm	66.3
Water	7.9
50% NaOH solution	3.4
plasticizer (lignosulfonate-based)	0.2

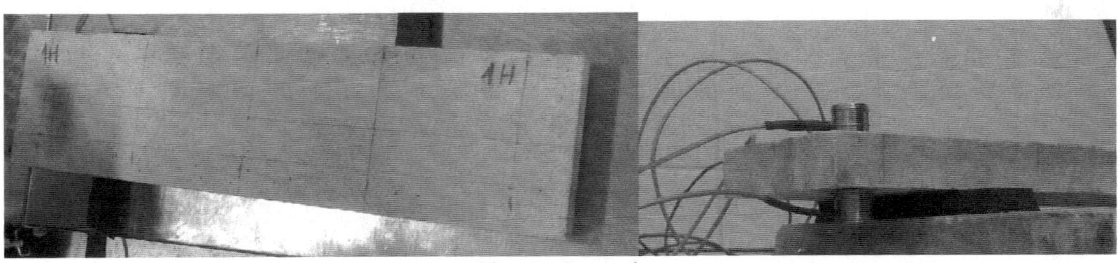

Fig. 1: Sample

The actual measurement was carried out with a sensor placed at the bottom corner on a rectangular grid (see Fig. 1) with individual mechanical impulses (strikes) transferred to other nodes of the grid. The evaluation is based on frequency analysis using modal parameters, that is, eigen-frequencies, attenuations, and transfer functions. The modal frequencies were calculated by [9] in Matlab. To estimate the natural frequencies and damping ratios, the least-squares complex exponential (LSCE) algorithm was used [10,11]. From the theoretical computation, we get the basic frequencies at approximate values of 755 Hz, 1.73 kHz, 2.04 kHz, 2.52 kHz, 3.55 kHz, etc. For the evaluation, however, we were interested in the frequency range above 20 kHz where increased sensitivity is expected to changes in the internal material structures [14].

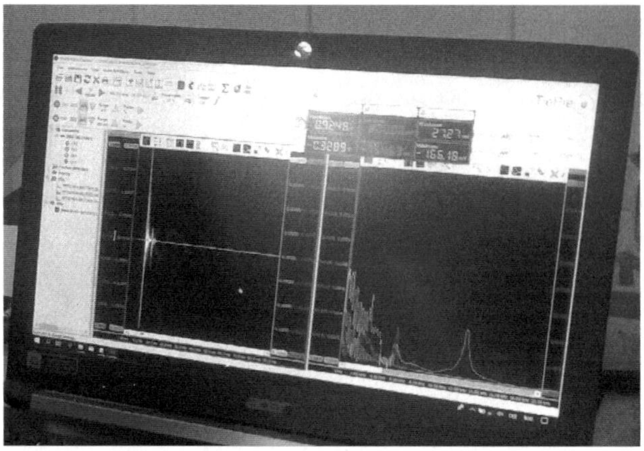

Figure 2: Measurement process

Since, because of waste material processing, alkali-activated materials are considered more environment-friendly than those based on the Portland cement, the specimens were made of such material. If this material is exposed to a degradation process, UV radiation in this particular case, its properties will change. Alkali-activated materials are based on suitable aluminosilicate precursors with sufficiently high content of amorphous phase, such as granulated blast furnace slag, fly ash and metakaolin. Compared to those based on the Portland cement, alkali activated materials are usually more durable in being equally or better resistant to aggressive media such as acid solutions, sulphates, chlorides, and others [12,13].

Each specimen was measured in a reference state (submerges in water at a laboratory temperature of 21° C) and in a state with one of its sides having been UV irradiated for 28 days and sprinkled with water. The irradiation intensity was chosen to correspond to three-month exposition to sunshine. Both specimen types were tested at the same age.

Results

The measurement was carried out as follows. A sensor (S1) was placed on the side that was free of the mould, the one with rough surface denoted by (h). Another sensor (S2) was placed on the side that was in the mould (d). A modal hammer was used to strike at the nine points chosen both on the upper side (h) and at the nine opposite points on the lower side (d).

The diagrams in Figures 3 to 6 show the dependency of the calculated modal attenuation (dr) on the modal frequencies (fn). The first four modal frequencies calculated were chosen for the evaluation in the frequency region above 20 kHz.

As this is one of the basic construction elements, concrete, its structure is non-homogeneous from the viewpoint of mechanical wave motion. Thus, the application of modal frequencies, which are expected to behave linearly, is not so simple and unequivocal as it is with homogeneous materials. From Figures 3 and 4, it can be concluded that, if the upper side is struck, the way the waves propagate with respect to a sensor placed on the same side (Figure 3 diagram hS1 and Figure 4 diagram h28S1) is different from the way the waves propagate with respect to the one placed on opposite side (Figure 3 diagram hS1 and Figure 4 diagram h28S1).

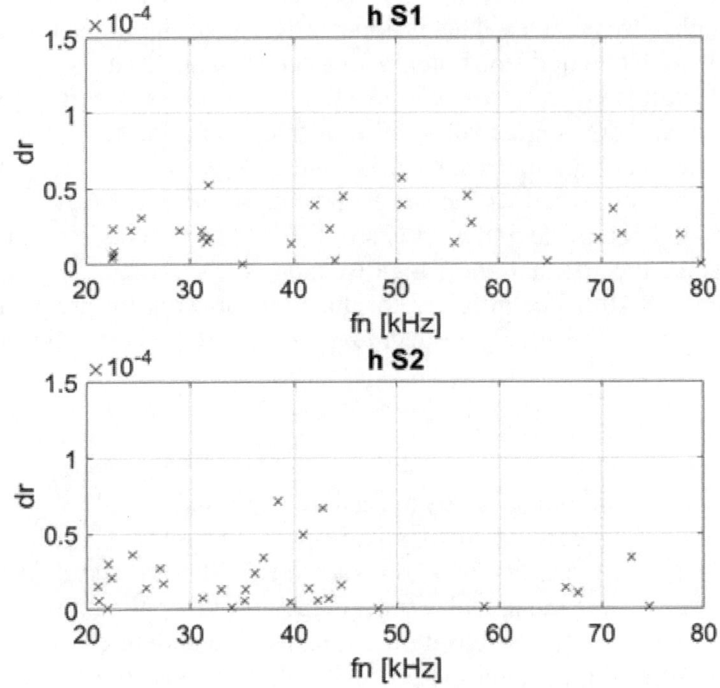

Figure 3 The dependency of the modal attenuation (dr) on the modal frequency (fn) after strikes on upper side of the original specimen

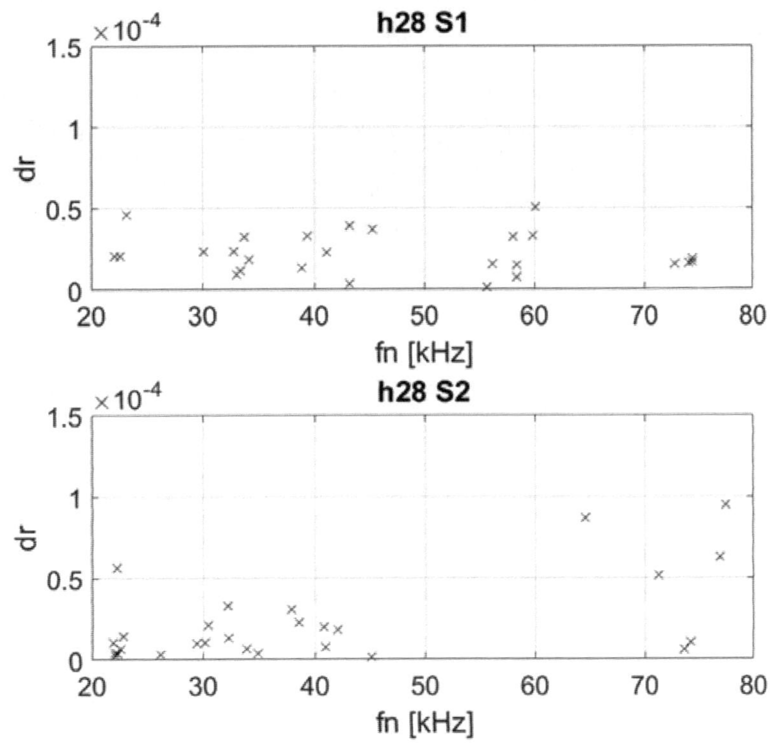

Figure 4 The dependency of the modal attenuation (dr) on the modal frequency (fn) after strikes on upper side of the specimen degraded by UV irradiation

The influence of the modal parameters, i.e., frequency and attenuation through irradiation is stronger for the higher frequencies. This is apparently due to the impact of the ageing of the specimen surface by the simulated sunshine. With a sensor placed on the side that was struck, the frequencies are cumulated in small regions (Figures 3 and 4 diagrams S1). Also in the non-degraded specimen, there is attenuation of high frequencies above 60 kHz up to a size of $5 \cdot 10^{-5}$. A similar character can be observed in a non-degraded specimen with the strikes carried out on the side on which a sensor is placed (Figure 4 diagram S1). It may be concluded that the influence of degradation is stronger for the waves moving through a non-degraded surface. Thus, the attenuation of high frequencies is stronger (Figure 4 diagram S2) achieving values above $5 \cdot 10^{-5}$. The influence of the attenuation in the frequency region about 40 kHz is opposite. Due to the specimen degradation, the frequency region about 50 kHz is freed.

Conclusions

The use of "high-frequency" modal analysis makes it possible to determine the impact of ultraviolet radiation on the degradation of alkali-activated material. A disadvantage of the impact-echo method is the necessity to know the initial, that is, the starting status of the material to be monitored. Its indisputable advantage is the application of non-destructive monitoring of the changes of the material structure, with an identical testing sample being monitored in the course of the degradation process making it possible to see the development of degradation by ultraviolet radiation. The main application area of this analysis is research and development because, for industrial applications, the measurement and evaluation are excessively demanding. This type of analysis may perhaps be used to judge the degree of degradation by sunshine of alkali activated concretes. It should be noted that, in this particular case, the term "modal" is not entirely compatible with its usual use.

Acknowledgment

This paper has been worked out under the project Czech Science Foundation **GACR No. GA 19-04703S**"The use of non-destructive methods for testing of the condition of degraded alkali activated concrete" supported by Czech Science Foundation and the project No. LO1408"AdMaS UP - Advanced Materials, Structures and Technologies", supported by Ministry of Education, Youth and Sports under the „National Sustainability Programme I".

References

[1] L. Pazdera, R. Dvorak, M. Hodulakova, L. Topolar, K. Mikulasek, J. Smutny, Application of acoustic emission method and impact echo method to structural rehabilitation, 19th International Conference on Rehabilitation and Reconstruction of Buildings, CRRB 2017; Prague; Czech Republic, Key Engineering Materials, 776 (2018) 81-85,

[2] L. Topolar, D. Stefkova, M. Hodulakova, The Assessment of Cement Mortars after Thermal Degradation by Acoustic Non-destructive Methods, 3rd International Conference on Innovative Materials, Structures and Technologies, IMST 2017; Riga; Latvia; IOP Conference Series: Materials Science and Engineering, 251, 1 (2017)

[3] R. Dvorak, M. Hodulaková, L. Topolar, Determination of condition of concrete samples of different mixture degraded by high temperatures by acoustic non-destructive method Impact-echo, EAN 2017, 55th Conference on Experimental Stress Analysis 2017, Novy Smokovec; Slovakia, (2017) 426-434

[4] M. Matysik, I. Plskova, Z. Chobola, Assessment of the impact-echo method for monitoring the long-standing frost resistance of ceramic tiles, Materiali in Tehnologije, 49, 4 (2015) 639-643

[5] T Marinone, P. Avitabile, J. Foley, J. Wolfson, Efficient computational nonlinear dynamic analysis using modal modification response technique, Mechanical Systems and Signal Processing, 31 (2012)67-93

[6] P. Avitabile, Modal analysis/dynamic systems, Experimental Techniques, 27(6) (2003), 32-33

[7] F. Chen, Y. Chen, HX Hua, Vibration analysis of a submarine elastic propeller-shaft-hull system using FRF-based substructuring method, Journal of Sound and Vibration, 443 (SI) (2019) 460-482

[8] R. Zenzen, I. Belaidi, S. Khatir, MA Wahab, A damage identification technique for beam-like and truss structures based on FRF and Bat Algorithm, Comptes Rendus Mecanique, 346 (12) (2018) 1253-1266

[9] H. Vold, J. Crowley, Gt. Rocklin, New Ways of Estimating Frequency-Response Functions, Sound And Vibration, 18(11) (1984) 34-38

[10] WL Zhou, D. Chelidze, Generalized eigenvalue decomposition in time domain modal parameter identification, Journal of Vibration and Acoustics-Transactions of the ASME, 130(1) (2008)

[11] P. Pavelka, R. Hunady, Determining modal parameters of mechanical system by using enhanced frequency response function, 55th International Scientific Conference on Experimental Stress Analysis 2017, EAN 2017; (2017) 81-86

[12] Z. Tang, WG Li, Y. Hu, JL Zhou, VWY Tam, Review on designs and properties of multifunctional alkali-activated materials (AAMs), Construction and Building Materials, 200 (2019) 474-489

[13] P. Rovnanik, H. Simonova, L. Topolar, P. Bayer, P. Schmid, Z. Kersner, Carbon nanotube reinforced alkali-activated slag mortars, Construction and Building Materials, 119 (2016) 223-229

[14] YY. Kim, An evaluation technique for high-frequency dynamic behavior of a sandwich microcantilever beam, Journal of Sandwich Structures & Materials, 21(3) (2019) 1133-1149

EAN 2019
57th conference on experimental stress analysis

Hybrid method suitable for measurement of residual stresses by hole-drilling technique

PEČENKA Luboš[1,a], HALAMA Radim[1,b], FOJTÍK František[1,c], PAŠKA Zbyněk[1,d] and ROJÍČEK Jaroslav[1,e]

[1]Department of Applied Mechanics; Faculty of Mechanical Engineering; VŠB – Technical University of Ostrava; 17. listopadu 2172/15, 708 00 Ostrava-Poruba, Czech Republic

[a]lubos.pecenka@vsb.cz, [b]radim.halama@vsb.cz, [c]frantisek.fojtik@vsb.cz, [d]zbynek.paska@vsb.cz, [e]jaroslav.rojicek@vsb.cz

Keywords: Hole-drilling Method; Hybrid Method; Residual Stresses; Inverse algorithm

Abstract. This contribution is focused on determination of residual stresses by new hybrid methods of mechanics. There is described the evaluation of residual stresses by hole drilling method based on newly proposed approach. Numerical simulation of gradually removing material and application of suitable load on inner area of the hole are the main principle of this technique. From experimental data can be derived values and directions of principal residual stresses by iterative computing and inverse algorithm. The results are compared with results from hole drilling method evaluated by standard ASTM E837.

Introduction

Residual stresses are present in all technical materials and their effects can be harmful or helpful. Compressive stresses are considered as helpful. Compressive stresses can be created by surface peening or another technique, which are used for parts subjected to cyclic loading. Influence of tensile stresses are always negative. Tensile stresses can cause growth of cracks and decrease of fatigue limit. This is the reason, why investigation of residual stresses take place in engineering practice.

Although there is progress in numerical simulation of residual stresses, the experimental measurements are still important for validation of results from computer modelling. Numerical methods can be successfully used by evaluation of measured results. There are arising new hybrid methods of mechanics.

The contribution explains basics of the newly proposed approach for evaluation of measurement by hole-drilling method, which is based on the inverse approach considering appropriate boundary conditions in the FE simulation of hole-drilling. The results of the hybrid approach are compared with results given evaluation of measurement by standard ASTM E837 [1].

Experiment – measurement of residual stresses

Investigation of residual stresses was carried out on the thermal affected specimen by hole drilling method and measured by the strain gauge (Fig. 1). Three tensometric apparatus P-3500 from Vishay company and apparatus RS-200 were used for measurement. There were used tensometric rosettes of type A with denotations EA-06-062RE-120. Circular opening with diameter 1.8 mm was gradually drilled with 0.05 mm increment in each step [2]. Released deformations were measured by tensometric rosette up to 0.3 mm depth. Measured values of strains on the strain gauges of tensometric rosette along the depth 0.3 mm are depicted on the table 1. Strain gauges of the tensometric rosette are denoted by first three letters of alphabet and corresponds to strain indices. Strain $\boldsymbol{\varepsilon_a}$ corresponds to the L direction according to the Fig. 1. ESPI optical measurement with apparatus Q-100 from Dantec Dynamics company was used on the same specimen with hole drilling methodology. Mounting of the laser-optical sensor to the measuring jig is shown on Fig. 2 [2]. Results of this measurement were influenced by the instability of the jig and by optical noise. For analysis in this study were used only results obtained from tensometric measurement with implementation of drilling hole method.

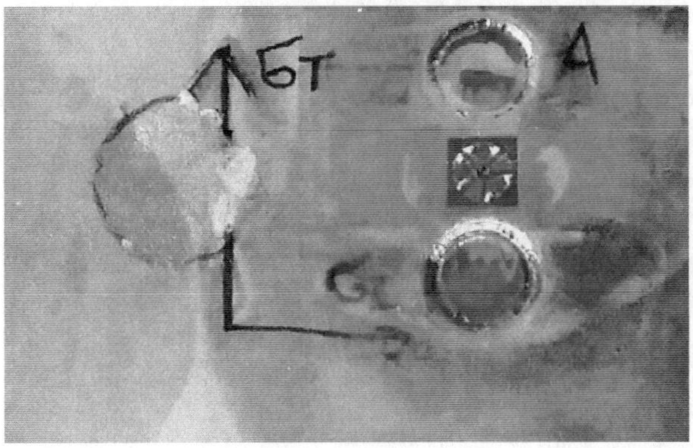

Fig. 1: Measured spot on the specimen [2].

Fig. 2: Measuring jig with laser-optical sensor Q100 [2].

Table 1: Measured values of deformation on the tensometers of tensometric rosette [2].

Z [mm]	ε_a [μS]	ε_b [μS]	ε_c [μS]
0.1	144	103	53
0.2	362	241	121
0.3	560	364	173

Evaluation of residual stresses by finite element method

New methodology of evaluation of residual stresses is based on the numerical simulation, when the depth of drilled hole increases gradually in many steps and there are simultaneously applied appropriate tangential and normal loads on the area of the hole. The most suitable parameters of the both functions are found out from the inverse algorithm, so the values converge (in particular location) closely to the measured values.

A parametric model was created in order to perform numerical solution. The modification of the geometry, material parameters and elements size is done easily. The model was created in APDL (Ansys Parametric Design Language) script in ANSYS program.

The model has shape of cubic with edge length 8 mm. In the middle of the surface is created a hole with diameter 1.96 mm. The depth of the hole can vary, and it is possible to modify it in six steps. For numerical solution, these elastic constants were used: Young modulus $E = 2.06 \cdot 10^5$ MPa and Poisson's ratio $\nu = 0.3$.

The discretization of this model to finite elements was performed using the SOLID95 six-part spatial element.

On Fig. 3, there is depicted detail of FEM model near the hole - near the site of interest. Because of the inverse algorithm, the size of elements was adjusted in order to reduce the computation time with regards to precision (max error 5 %).
The boundary conditions are fixed support of the bottom side of the cubic and application of appropriate loading to the inner cylindrical surface of the hole (see Fig. 3). The cylindrical surface was created by deactivation of the elements. For this purpose, it is necessary to create special surface elements (Surface Effect Elements) in software ANSYS.

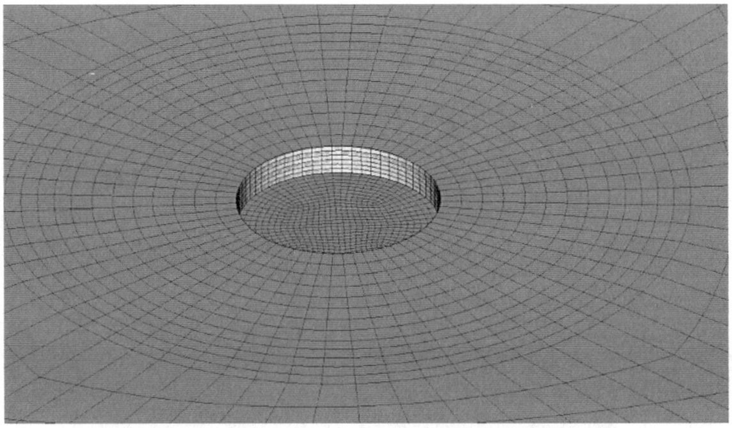

Fig. 3: Detail of FEM model near the hole.

From the strain gauge measurement performed with type of A rosette [1], deformation values were obtained at three specific points, which are located at a certain distance along the entire circumference of the hole. This is the reason, why symmetry cannot be used for this model and FEM analysis. The surface load distribution is defined by the relations that are based on the theory of elasticity for plane stress, namely the parametric equations of the circle in Mohr plane:

$$\sigma_\rho = P_1 + P_2 \cos(2\beta) + P_3, \tag{1}$$

$$\tau_\rho = P_2 \sin(2\beta) - P_3 \cos(2\beta). \tag{2}$$

Applied functions for examined stresses σ_ρ, τ_ρ in the plane ρ depend on the three unknow parameters P_1, P_2, P_3. Angle β is considered as a variable in a cylindrical coordinate system that is in the centre of the hole.
The aim is to find out the values of parameters P_1, P_2, P_3, so that the strain around the hole corresponds to the measured strains on the stain gauges of the tensometric rosette. Then the main stresses $\sigma_{1,2}$ and their rotation angle α can be determined from known relations:

$$\sigma_{1,2} = P_1 \pm \sqrt{{P_2}^2 + {P_3}^2}, \tag{3}$$

$$\alpha = \frac{1}{2} \arctan\left(\frac{P_3}{P_2}\right). \tag{4}$$

Inverse algorithm

The inverse approach serves, for example, to identify material parameters using a probabilistic algorithm, a genetic algorithm, a differential evolution method, or a neural network. However, the inverse algorithm can generally be applied to tune the computational model with the experiment.

First step is to find out the initial parameters (P_1, P_2 a P_3) and with these parameters is carried out the entrance calculation by FEM in the first layer in the depth of 0.1 mm. The calculated values of nodal deformations are compared with the measured values on strain gauges in the rosette. If the required minimum overall average deviation is not reached, new parameters are set, and the process is repeated. By repeated calculations of the inverse algorithm it is possible to achieve a good agreement between the experiment and the numerical calculation.

The listed load functions with the currently found parameter values remain in effect in the remaining steps. Furthermore, the second layer of elements is deactivated. New functions are introduced, and their parameters are found out in the same way. The process is repeated. All data obtained by this procedure, even with calculated and measured values, are tabulated, see Tab. 2, Tab. 3 and Tab. 4.

Table 2: Parameter values found.

Z [mm]	δ [-]	P_1 [MPa]	P_2 [MPa]	P_3 [MPa]
0.1	0.076	-1 239.9	430.7	12.4
0.2	0.104	-1 171.1	515.7	11.6
0.3	0.154	-1 088.3	482.7	10.2

Table 3: Evaluation according to the described new methodology.

Z [mm]	ε_3 [µS]	ε_2 [µS]	ε_1 [µS]	σ_1 [MPa]	σ_2 [MPa]	α [°]
0.1	98	146	45	-809.0	-1 670.8	90.8
0.2	344	220	142	-655.3	-1 686.9	90.6
0.3	559	352	247	-605.5	-1 571.1	90.6

Table 4: ASTM standard evaluation.

Z [mm]	ε_3 [µS]	ε_2 [µS]	ε_1 [µS]	σ_1 [MPa]	σ_2 [MPa]	α [°]
0.1	103	144	53	-716.4	-1 438.8	92.8
0.2	362	241	121	-742.2	-1 559.2	89.9
0.3	560	364	173	-677.8	-1 472.2	89.6

Results obtained from the numerical solution by FEM analysis and from experimental measurement are graphically depicted on Fig. 4.

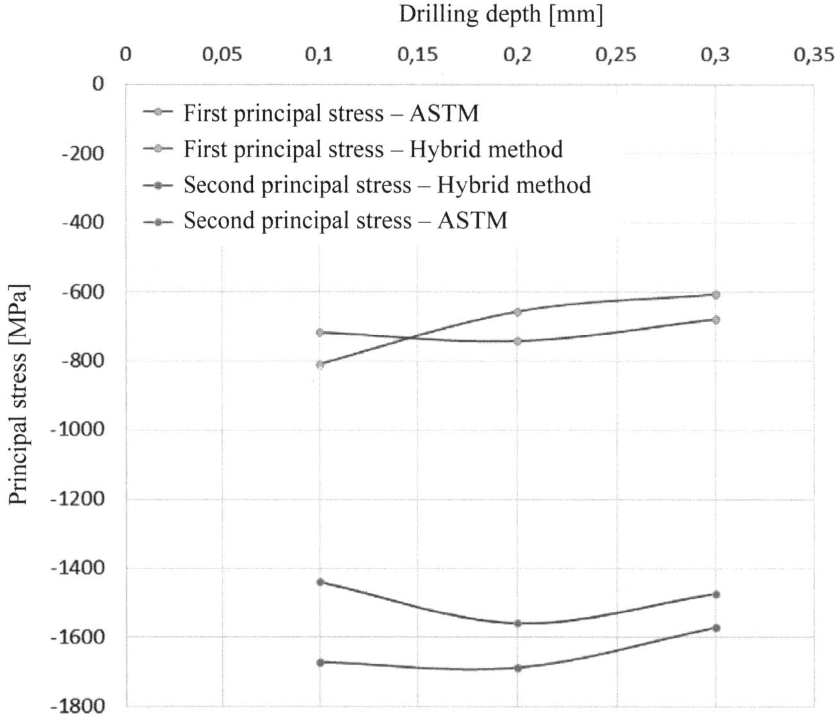

Fig. 4: Comparison of evaluated data.

To show what strains cause the applied functions around the hole, the results of the analysis are presented in the last step of the solution. Specifically, these are the contours of the first and second principal strains (Fig. 5 and Fig. 6).

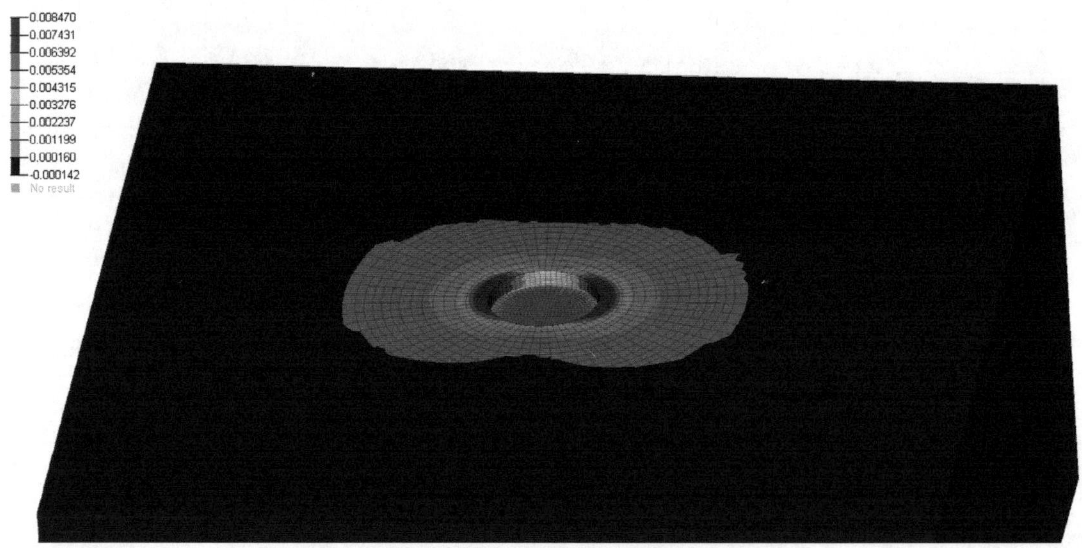

Fig. 5: The first principal strain contours.

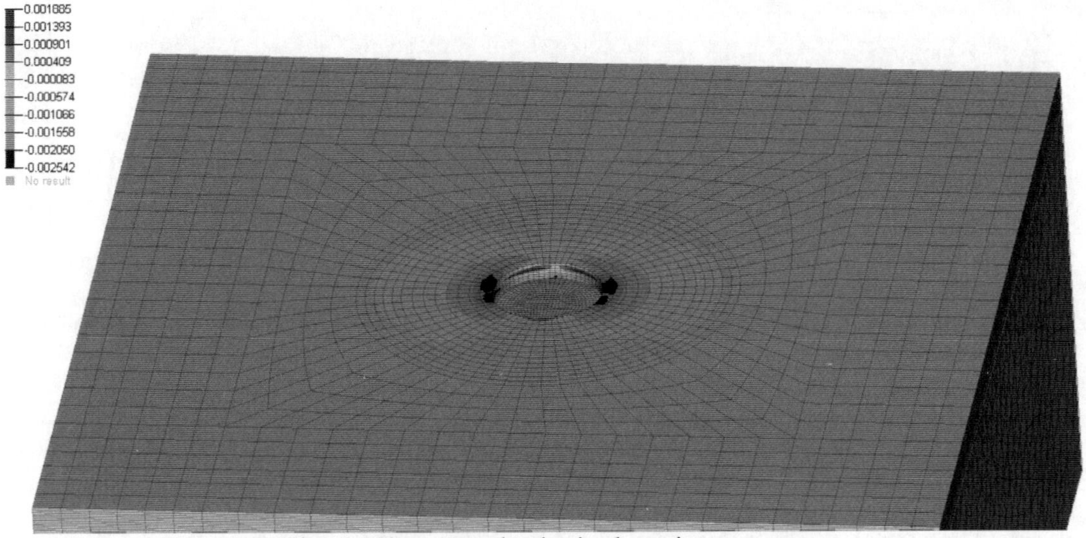

Fig. 6: The second principal strain contours.

Conclusions

Proposed methodology can determine residual stresses, which can be even unevenly distributed along the depth of the drilled hole. In addition to the value of stresses, can be also determined orientation of these stresses. From the numerical and graphical solution of residual stresses can be observed good agreement between evaluation by ASTM E837 [3] and by the proposed approach. Maximum deviation, related to the evaluation by mentioned standard, is 16.1 % and is observed in the second principal stress in the depth of 0.1 mm.

Acknowledgement

The paper has been done in connection with project Innovative and additive manufacturing technology – new technological solutions for 3D printing of metals and composite materials, reg. no. CZ.02.1.01/0.0/17_049/0008407. This work was also supported by Grant Agency of the Czech Republic (GACR) project No. 19-03282S, and by the specific research SP2019/100 project, supported by the Ministry of Education, Youth and Sports of the Czech Republic.

References

[1] ASTM E 837 – 13a: Determining Residual Stresses by the Hole – Drilling Strain-Gage Method.

[2] L. Pečenka, Hybrid Methods for Measuring Residual Stresses, Dissertation Thesis, Department of Applied Mechanics, Faculty of Mechanical Engineering, VŠB-TU Ostrava, 105 Pages, 2019.

[3] Barile, C., C. Casavola, G. Pappalettera and C. Pappalettere. Residual Stress Measurement by Electronic Speckle Pattern Interferometry: A Study of the Influence of Geometrical Parameters. *Structural Integrity and Life*. 2010, Vol. 11, pp.177–182.

[4] K. Kolařik, Z. Pala, N. Ganev, F. Fojtík, Combining XRD with Hole-Drilling Method in Residual Stress Gradient Analysis of Laser Hardened C45 Steel. *Advanced Materials Research*. 2014, Vol.996, pp. 277-282.

EAN 2019
57th conference on
experimental stress analysis

Calculation of the container for municipal waste collection and simulation of welds using FEM

PETR Karel[1,a]

[1]Czech Technical University in Prague, Faculty of Mechanical Engineering, Department Designing and Machine Components, Technická 4, 166 07 Prague 6, Czech Republic

[a]karel.petr@fs.cvut.cz

Keywords: Container, Finite Element Methods, Welds, Stress-strain Analysis, Frame

Abstract. The aim of the article is the design and stress-strain analysis of the container for municipal waste. For stress-strain analysis has been used analytical and FEM (Finite Element Methods) calculations. The container has been designed to be transported on hook-type car and ISO container chassis (the container contains ISO corner elements). For handling, the container was equipped with holes for forklift. The container can be stacked through ISO corner elements.

Introduction

Container for municipal waste designed in this work is shown in Fig. 1. For the design of container was very important implemented following parts (elements) into construction of container, because container will be transported on hook-type car and ISO container chassis.

- Hook – for transported on hook-type car.
- ISO corner elements [1] – for transported on ISO container chassis.
- Two holes for forklift – for transport by forklift.
- ISO corner elements – for stacking the container.

All parts in container were connected with beams (longitudinal, transverse and vertical) by help welds (see Fig. 1.).

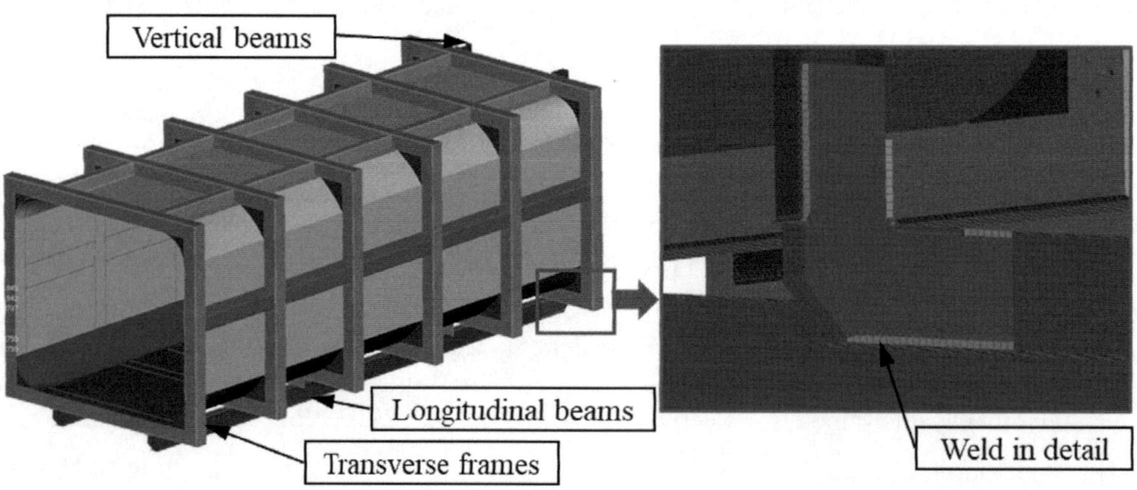

Fig. 1: Model for FEA (Finite Element Analysis) with indication of important parts (left part); Detail on simulation of welds in FEM with adhesive line and FEM mesh (right part)

For stress-strain analysis has been used analytical and FEM (Finite Element Methods) calculations.

Material and Methods

Basic parameters of the container: Outer dimensions of the container (2 438 mm – ISO width [2], 2 000 mm – height, 5 000 mm – length); Total waste weight 12 500 kg; Total container weight 2 354 kg; Volume of waste 13.5 m^3; Material of container – structural steel S355J2H with yield strength Re = 355 Nmm^{-2}; and σweld = 100 Nmm^{-2} [3]; Safety factor k = 1.5.

All profiles of beams used on container are calculated according analytical calculations. Specifically, using Vereščagin's rule of calculating Mohr's integral. According to these moments, profiles were proposed. The Fig. 2 (left part) shown resulting bending moment on the frame (without cylindrical reinforcement plate) – result has been obtained from analytical calculation.

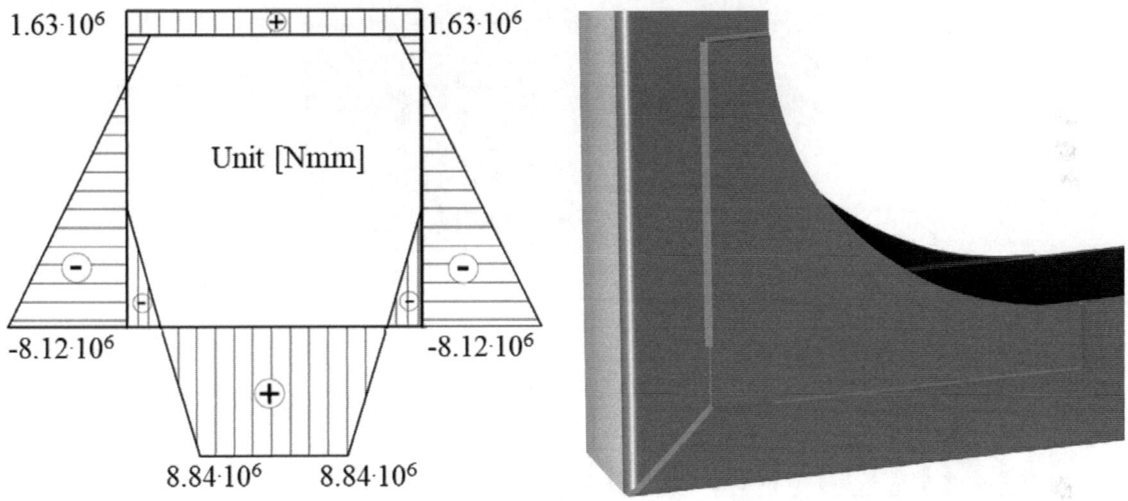

Fig. 2: The resulting bending moment on the frame (without cylindrical reinforcement plate) (left part); Detail of U-profile connection (bent sheet) of frames (right part)

Connected profiles and overall construction was realized as a weldment (see text above). The proposed construction detail (Fig. 2. right side) was made by cutting the profile ends at a 45° angle and then creating a butt weld. 5 mm thick stiffeners are used to reduce butt stress and to reinforcement the inner sheet. These reinforcements are welded with corner welds with sufficient offset from the butt weld to avoid stress concentration.

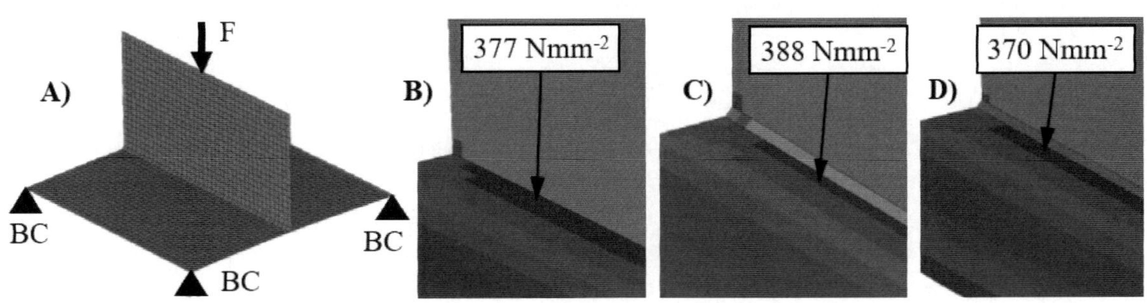

Fig. 3: A) Bearing and load designation; B) Mises stress distribution without weld joint modeling; C) Mises stress distribution for modeling of welds with shell elements; D) Mises stress distribution for modeling of welds with "adhesive line"

Before creating of final calculation FEM model, was tested different types of creation (application) of welds in FEM (see Fig. 3). Sheets of container was realized as shell. Welding modeling was realized in three ways. The first option was without welds - only sheets (Fig. 3 - B). The edge of one sheet had common nodes with the mesh of the other sheet. The second option was to define the weld using a shell layer (Fig. 3 - C). This method was very difficult, because was needed mesh quality. Many elements had to have a given height and angle of 45°. The third option was defined by "adhesive line" (Fig. 3 – C). The weld was created by hexagonal volumetric elements. The volume element nodes were connected to the shell by Distributed coupling – this was the most realistic option and has been used in this work.

The connection between the frame and the longitudinal beams was made using 3D elements – see Fig. 4.

Fig. 4: Modeling of welded joints between frame and longitudinal profiles of container

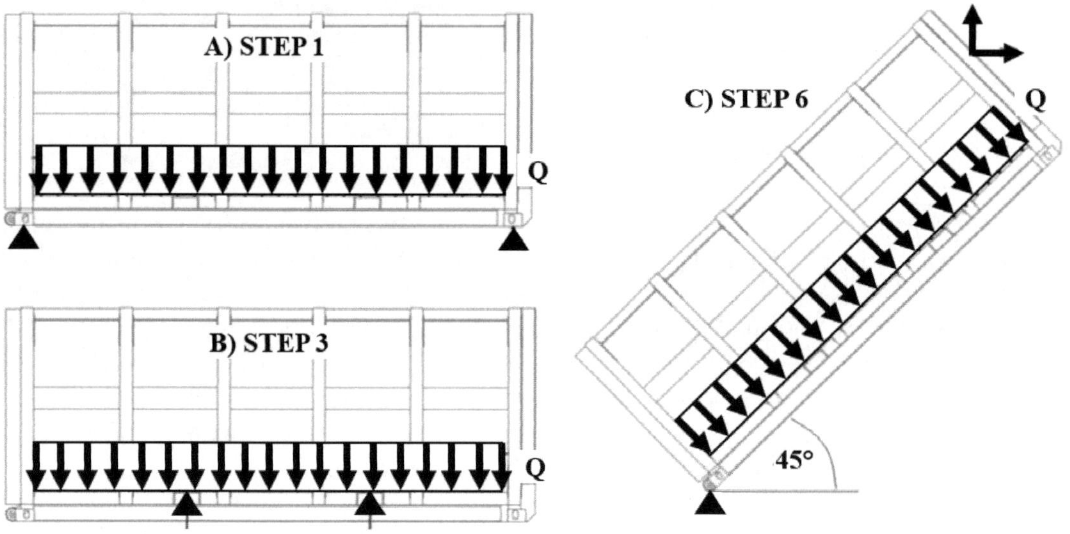

Fig. 5: Load scheme for selected steps – A) STEP 1; B) STEP 3; C) STEP 6

Six steps have been defined for calculations. All steps had same load force and different boundary conditions.

- STEP 1 – Container loaded (Q - total waste weight) in horizontal position and attached to four ISO corner elements (see Fig. 5 – A).
- STEP 2 – Container loaded (Q - total waste weight) in horizontal position and attached to rollers and handling eye.
- STEP 3 – Container loaded (Q - total waste weight) in horizontal position and attached to two holes for forklift (see Fig. 5 – B).
- STEP 4 and 5 – Container loaded (Q - total waste weight) in horizontal position and attached to three ISO corner elements – torsion simulation.

- STEP 6 – Container loaded (Q - total waste weight) in tilting the container at angle 45° and attached to rollers and handling eye (see Fig. 5 – C).

Results

The analytical and FEM stress-strain calculations were performed on the container. The stress should be less than 236 Nmm^{-2} and in welds should be less than 100 Nmm^{-2}. Fig. 6 shown results for STEP 1 – view on whole construction and from bottom. Maximum stress was 228 Nmm^{-2}.

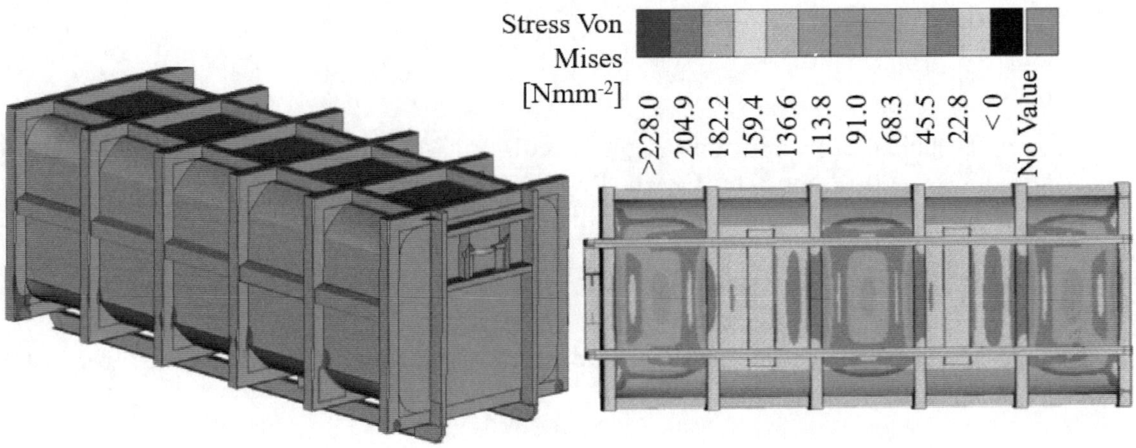

Fig. 6: Example of FEM simulation – Von Mises stress for STEP 1 (3D view and view from bottom)

Fig. 7 shown results for STEP 6 – view on whole construction and detail on handling eye. Maximum stress was 236 Nmm^{-2}.

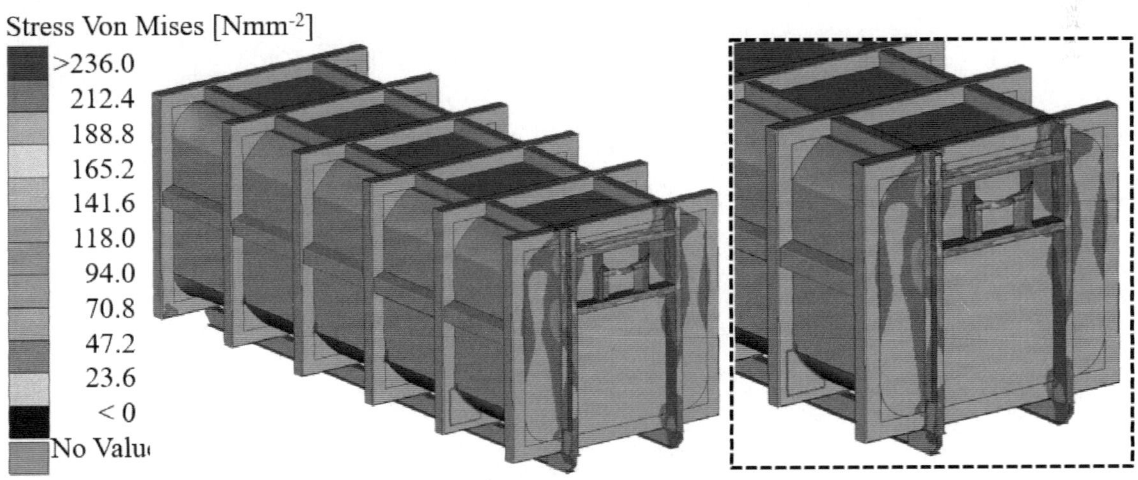

Fig. 7: Example of FEM simulation – Von Mises stress for STEP 6

After FEM calculation was been updated longitudinal beams (height of I-profile) and was been changed connection of hook into container. On the container was used I-profile for longitudinal beams and other profiles were created from bent profiles from sheet metals.

Conclusions

Thanks to the multi-step design, better container parameters have been achieved. The container was calculated in several steps with different stress and boundary conditions. Continuously carried out structural adjustments. This has resulted in a maximum volume and a minimum weight.

Acknowledgement

This work was funded by Operational Programme Prague – Growth Pole of the Czech Republic, project no. CZ.07.1.02/0.0/0.0/16-040/0000363.

References

[1] Czech Technical University in Prague, Faculty of Mechanical Engineering, Technická 1902/4, 160 00 Prague 6, Czech Republic. A shipping container. Inventors: V. DYNYBYL, et al. Czech Republic. Utility Model CZ 32039. 2018-06-22.

[2] ISO 668:2013 Series 1 freight containers – Classification, dimensions and ratings.

[3] HOBBACHER, A. Recommendations for Fatigue Design of Welded Joints and Components. Springer International Publishing. 2016. ISBN 978-3-319-23756-5. DOI 10.1007/978-3-319-23757-2.

EAN 2019
57th conference on experimental stress analysis

Experimental investigation of thermo-mechanical properties of one thermo-responsive shape memory polymer

PETRÍKOVÁ Iva[1,a], MARVALOVÁ Bohdana[1,b], BĚHÁLEK Luboš[1,c] and NĚMEC David[1,d]

[1]Department of Applied Mechanics; Faculty of Mechanical Engineering; Technical University of Liberec; Studentská 2, 460 17 Liberec, Czech Republic

[a]iva.petrikova@tul.cz, [b]bohda.marvalova@tul.cz, [c]lubos.behalek@tul.cz, [d]david.nemec1@tul.cz

Keywords: Shape memory polymers, Thermomechanical behaviour, Dynamical response

Abstract. The thermomechanical behaviour of Estane® ETE 75DT3 was studied experimentally. The material is the aromatic polyether-based thermoplastic polyurethane. It belongs to a group of thermo-responsive shape memory polymers (SMPs) which exhibit the thermally induced shape-memory effect. The thermal properties of the shape memory polymer (SMP) were characterized by Differential Scanning Calorimetry (DSC). The material was investigated by means of the quasi-static tensile tests at the room temperature. The Dynamic Mechanical Analysis (DMA) was carried out at various frequencies and temperatures and the values of dynamical moduli and loss tangent as the functions of the frequency and temperature were determined. The shape memory effect was also studied in the course of the thermo-mechanical loading.

Introduction

The shape memory polymers are smart materials which can change their shape upon application of an external stimulus. The shape memory effect is not an intrinsic property, meaning that polymers do not display this effect by themselves. Shape memory results from a combination of polymer morphology and specific processing and can be understood as a polymer functionalization. [1]. A change in shape caused by a change in temperature is called a thermally induced shape-memory effect. Thermo-responsive SMPs undergo a transition between two different shapes - temporary shape and permanent shape as schematically illustrated in Fig. 1.

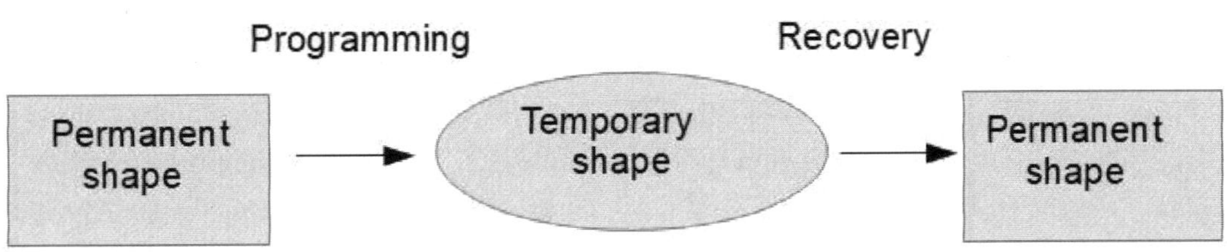

Fig.1: Thermally induced one-way shape-memory effect

First, the material is processed in injection moulding or extrusion where it got its permanent shape. Afterwards, the material undergoes the programming process in which it is heated up above the transition temperature (T_{trans}), deformed and cooled consecutively and a temporary shape is fixed. Heating the sample again above T_{trans} induces the shape memory effect so the recovery of the permanent shape. By further programming, the sample can be brought again into a temporary shape. The transition temperature depends on the type of polymer morphology and it can be either the glass transition temperature T_g or the melting temperature T_m. The shape memory effect can be formed in a number of polymers [2]. It depends on the physical properties of the polymer components, from chemistry and morphology. SMPs are mostly copolymers which consist of at least two types of constituents. One component consisting of hard segments (elastomer) with higher T_g or T_m acts as the physical cross-link and determines the permanent shape. The second soft phase (thermoplastic) enables the fixation of the temporary shape. At room temperature T_r, the hard segments prevent thermoplastic phase from showing rubber-like behaviour. However, when the temperature is increased above the glass temperature T_g, $T_r < T_g < T_m$, the material shows rubber-like behaviour governed by entropy elasticity [3].

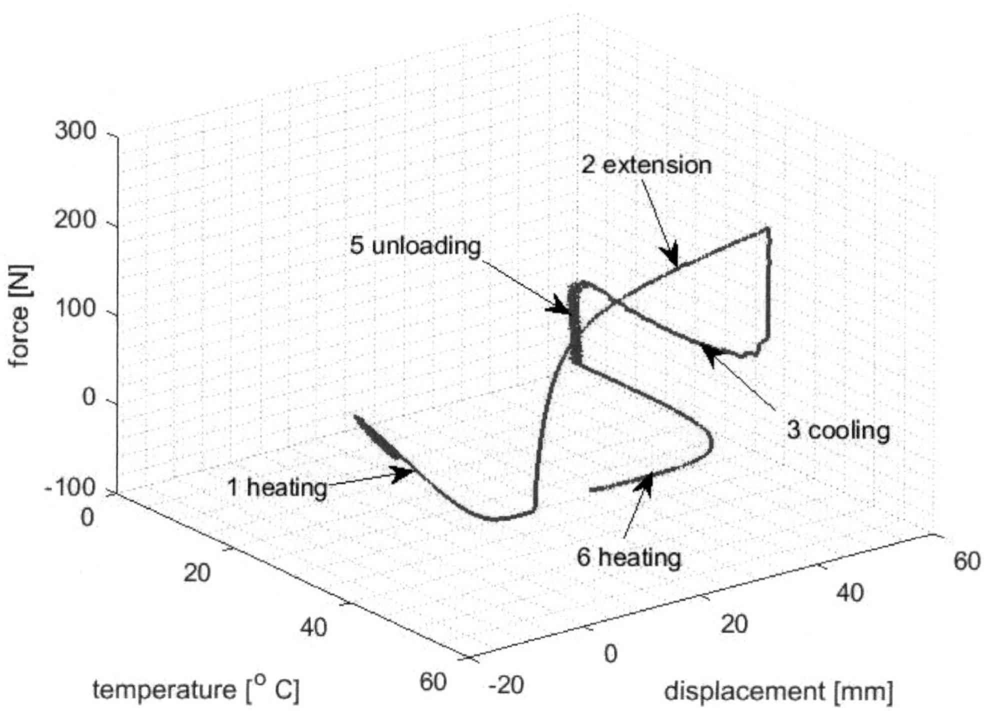

Fig. 2: Cyclic thermo-mechanical loading of thermo-responsive SMP

The thermo-responsive SMP polymer goes through cycles as indicated in Fig. 2. The material is heated in step 1 above the transition temperature ($T_g < T_{trans} < T_{melt}$) then deformed in step 2 to a predetermined shape and fixed into this shape by cooling back below the T_g in step 3 and then unloaded in step 5. The original state can be recovered by reheating the material above its T_g without the mechanical constraint in step 6.

The capability of polymers to be formed to different temporary shapes and subsequently recover their original shape can be used in many applications. Recently a number of biomedical applications have been found for polyurethane-based shape memory polymers. These materials are biocompatible, non-toxic and non-mutagenic in the human body. The advantage of these materials is that their glass transition temperature T_g can be tailored due to self-deployment of different clinical devices and stents inserted in the human body. Textiles offer another way of SMP development. Successful application area may be deployable structures for the aircraft and space industry and in different robots.

Experiment

The thermomechanical behaviour of Estane® ETE 75DT3 was studied by several experimental methods. The material was supplied by the manufacturer Lubrizol in the form of pellets. SMP Estane is an aromatic polyether-based thermoplastic polyurethane. The producer indicates the hardness of 75 Shore D, the tensile strength 41 MPa and the elongation higher than 300% at room temperature.

Differential scanning calorimetry (DSC) was used to determine the response of the material to heating at the rate of 10 K/min according to ISO 11357. Measured values differ significantly from those indicated by the manufacturer, especially the glass transition temperature T_g. However, this temperature is close to the glass transition temperature of Estane $T_g = 54°C$ reported in [3,4]. Results of DSC are in Fig. 3 and in Table 1.

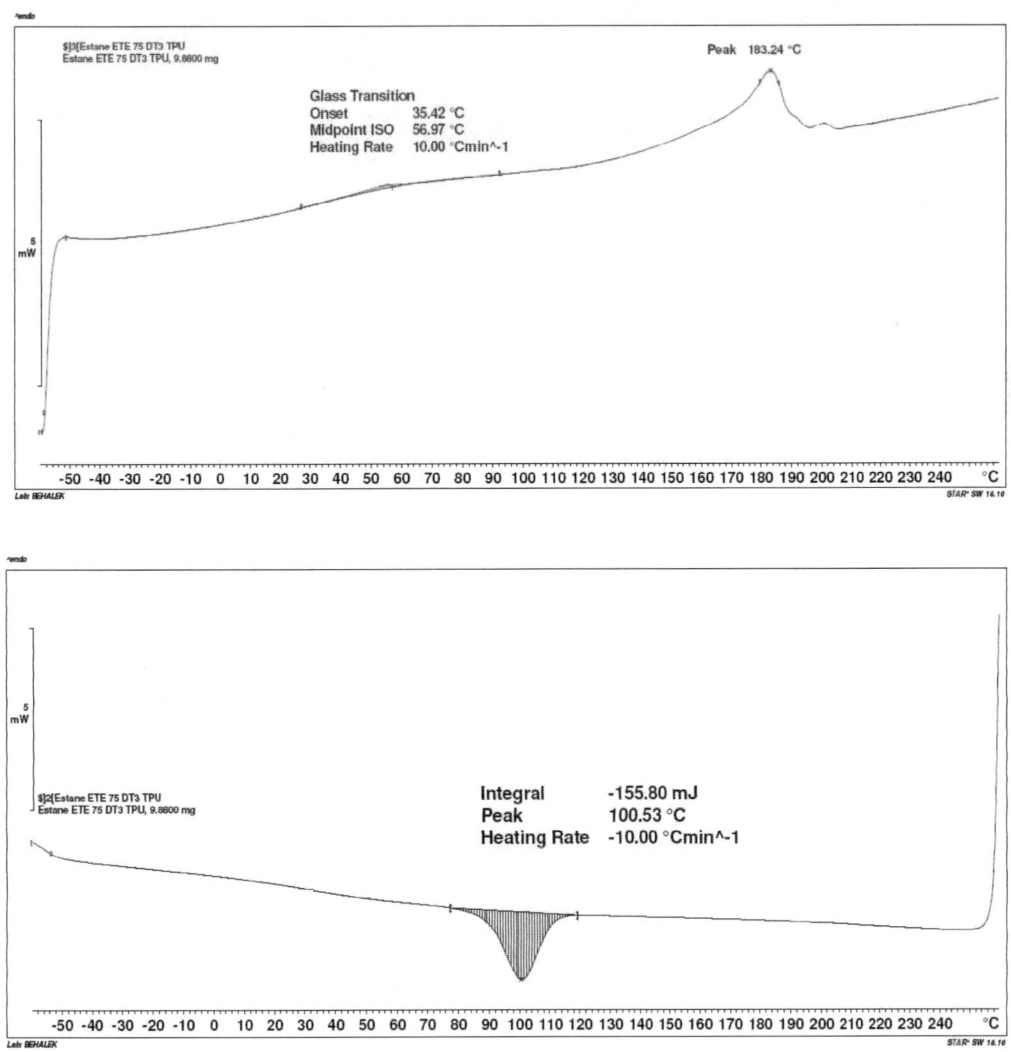

Fig. 3: Results of DSC

Table 1: Thermal properties

Temperature	Measured [°C]	Producer [°C]
T_g	57	-10
T_{melt}	183.2	218
T_{cryst}	100.5	-

The thermo-mechanical measurements were carried out on flat dog bone specimens of length 110 mm and cross section 10 x 4 mm fabricated by the injection moulding. Our measurement of hardness confirmed the value given by the manufacturer 75 Shore D. The quasi-static tensile tests showed the stress dependency on the strain rate so typical for polymers. The results of the tensile tests with rising strain rate are in Fig. 4.

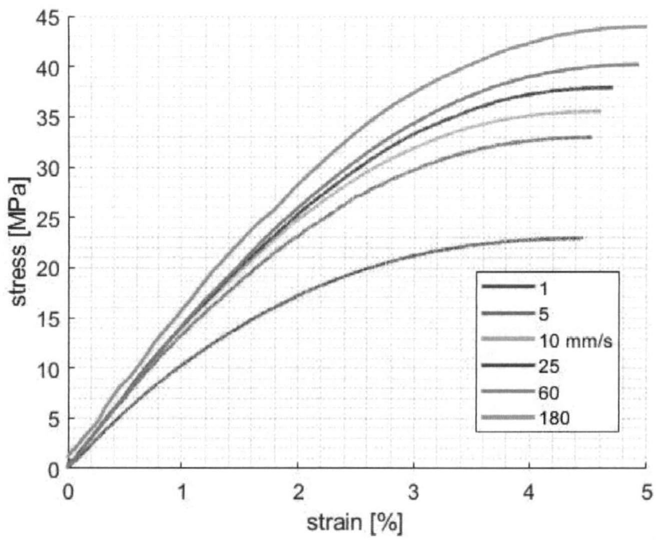

Fig. 4: Tensile tests of Estane with different strain rate

The thermo-mechanical properties were measured by the dynamic thermo-mechanical analysis (DMTA) with different frequencies and the temperature sweep around the glass transition. The dynamic moduli and the loss factor dependency on temperature, frequency and amplitude were determined. DMTA tests in tension deformation mode were performed in the temperature range from 25°C to 45°C using Instron Electropuls 3000 equipped with the temperature chamber and with standard fixtures. The small extension 1mm (around 1%) is applied to the sample in order to maintain it under tension. Then sinus cyclic strain controlled tensile loading was applied with the amplitude 0.01 mm and frequency range from 0.5 Hz to 5 Hz. In order to determine the storage E' and loss modulus E'' of the material, corresponding parameters of the force response and the phase angle δ must be extracted from the recorded raw signals. We suppose that the raw signals i.e. displacement u(t) and force response F(t), are harmonic functions approximately and we use the discrete Fourier transform (DFT) in Matlab in order to determine the phase shift δ at the main excitation frequency [5]. Amplitudes of tensile strain and tensile stress were determined from recorded signals. Dynamic moduli were calculated as

$$E' = \frac{\sigma_a}{\varepsilon_a}\cos(\delta), \quad E'' = \frac{\sigma_a}{\varepsilon_a}\sin(\delta). \tag{1}$$

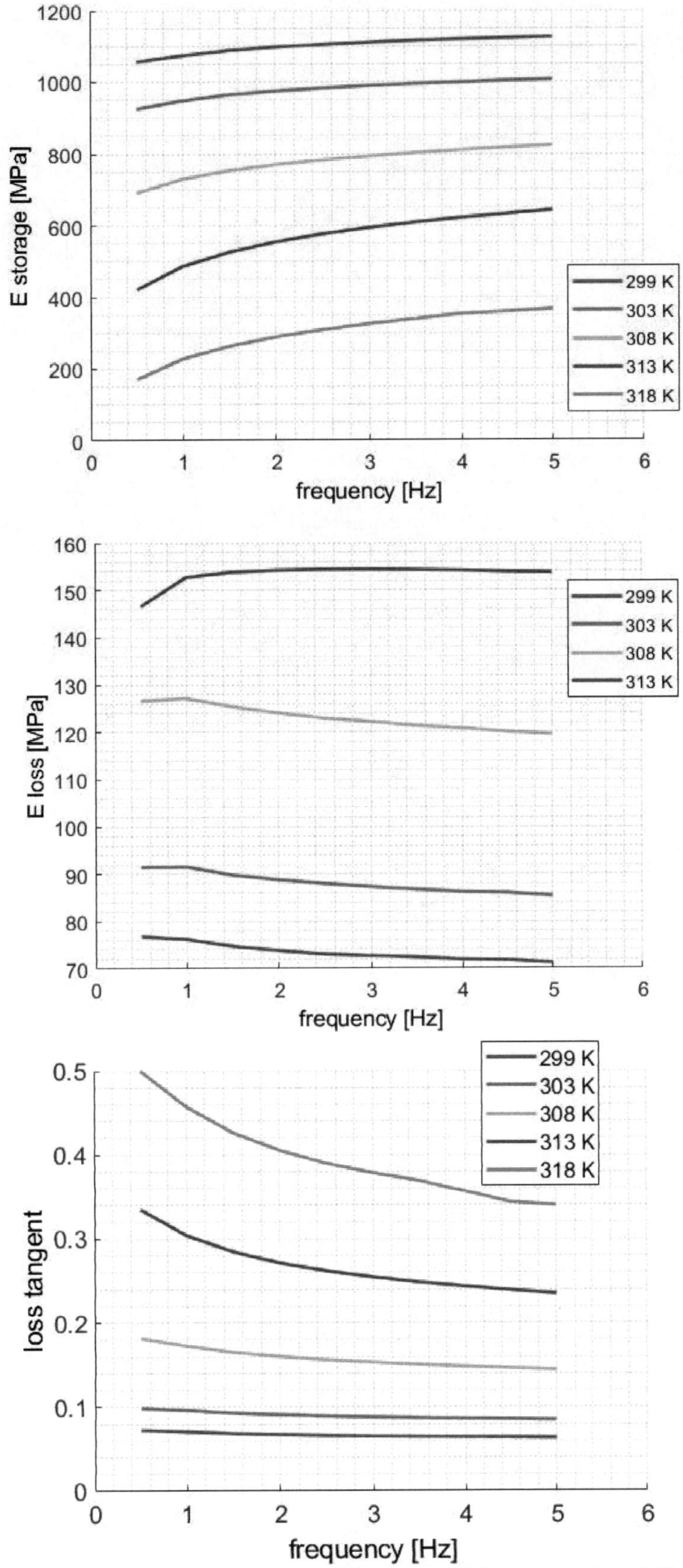

Fig. 5: Dynamic moduli and loss tangent as functions of frequency for different temperatures

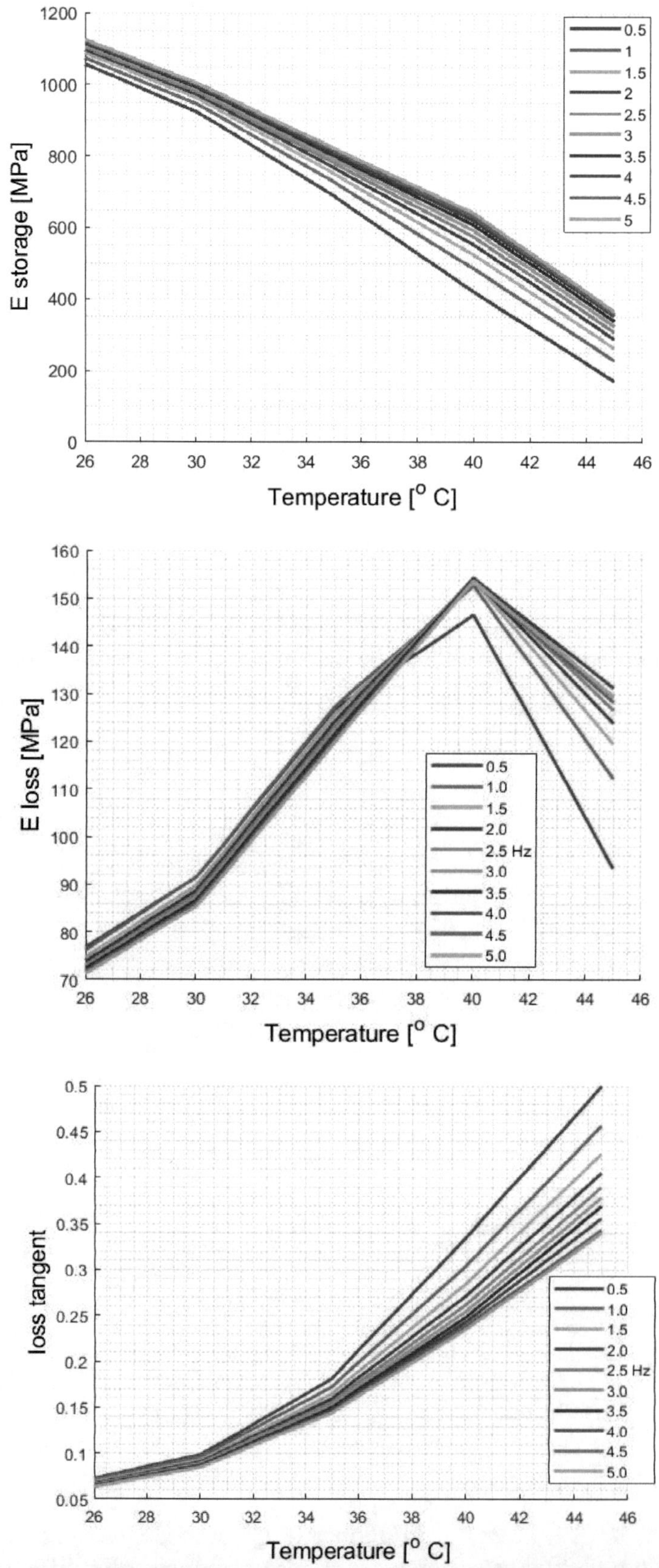

Fig. 6: Temperature dependence of dynamic moduli and loss tangent

The dynamic moduli and loss tangent as the functions of temperature and frequency are depicted in Fig. 5 and 6. It is seen that the storage modulus E' is rising with the rising frequency and decreases gradually with increasing temperature. The loss modulus E'' and loss tangent have the opposite tendency concerning frequency and they show the rising towards the glass transition temperature.

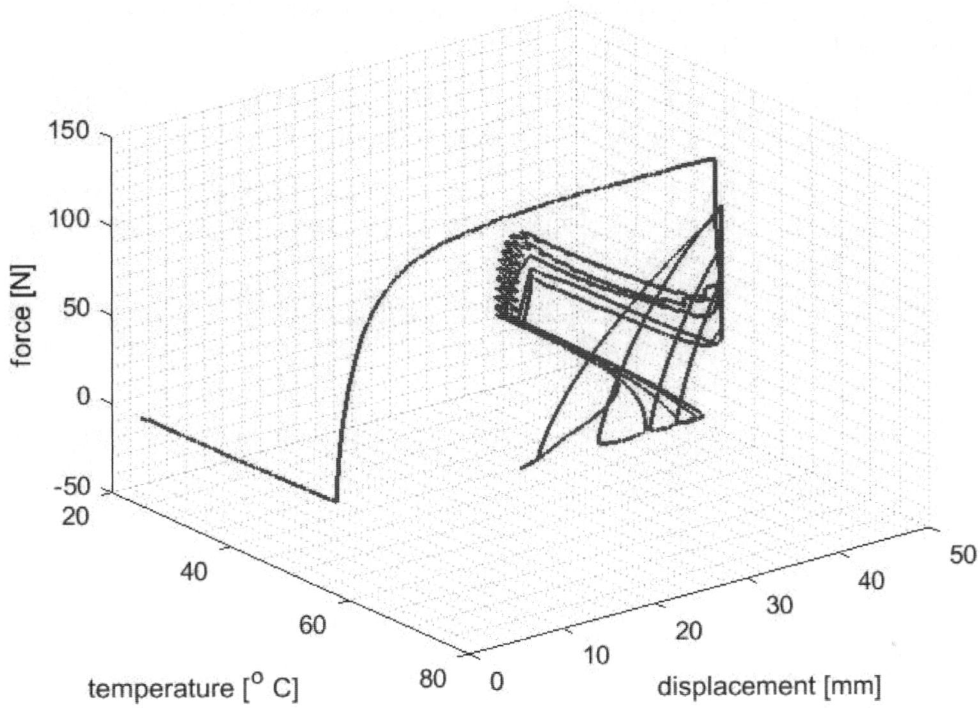

Fig. 7: Result of five cycles Programming - Shape Recovery with the programming temperature T_{max} = 57°C and cooling temperature T_{min} = 25°C

First thermo-mechanic cyclic tests Programming - Shape Recovery were carried out in this research. The result of one experiment is in Fig. 7. We can see the force relaxation at the maximum extension of the sample at the temperature T_{max}. There is also the decrease of the maximum force in subsequent cycles.

Conclusions

We present the first results of experimental research of the behaviour of the shape memory material - aromatic polyether-based thermoplastic polyurethane Estane® ETE 75DT3. The thermal properties of the shape memory polymer (SMP) were characterized by Differential Scanning Calorimetry (DSC). The material was investigated by means of the quasi-static tensile tests at the room temperature. The Dynamic Mechanical Analysis (DMA) was carried out at various frequencies and temperatures and the values of dynamical moduli and loss tangent as the functions of the frequency and temperature were determined. The shape memory effect was also studied in the course of the thermo-mechanical loading.

Acknowledgement

This work was supported by the Ministry of Education, Youth and Sports of the Czech Republic and the European Union - European Structural and Investment Funds in the frames of Operational Programme Research, Development and Education - project Hybrid Materials for Hierarchical Structures (HyHi, Reg. No. CZ.02.1.01/0.0/0.0/16_019/0000843).

References

[1] Lendlein, Andreas, and Steffen Kelch. "Shape memory polymers." Angewandte Chemie International Edition 41.12 (2002): 2034-2057.

[2] Schmidt, Christina, Klaus Neuking, and Gunther Eggeler. "Functional fatigue of shape memory polymers." Advanced Engineering Materials 10.10 (2008): 922-927.

[3] Mogharebi, S., R. Kazakeviciute-Makovska, H. Steeb, G. Eggeler, and K. Neuking. "On the cyclic material stability of shape memory polymer." Materialwissenschaft und Werkstofftechnik 44, no. 6 (2013): 521-526.

[4] Kazakevičiūtė-Makovska, Rasa, Aycan Özlem Özarmut, and Holger Steeb. "Characterization of shape memory polymer estane by means of dynamic mechanical thermal analysis technique." Smart Materials Research 2014 (2014).

[5] Petrikova, I., and B. Marvalova. "Experimental determination of the mechanical properties of naturally aged rubber." In International Conference of Model for Rubber, Dresden, Germany, pp. 143-147. 2009.

Experimental and modelling study of effect of surface treatment on the creep condition of flax fiber reinforced composites

PETRŮ Michal[1,a] and WANG Xiaomeng[1,b]

[1]Institute for Nanomaterials, Advanced Technologies and Innovation, Technical University of Liberec, Studentska 2, Liberec, 461 17, Czech Republic

[a]michal.petru@tul.cz, [b]xiaomeng.wang@tul.cz

Keywords: Key, Word, Paper, Best conference, Contribution template

Abstract. This article was focusing for experimental and mathematical modelling study the creep condition of FFRP (Flax Fiber Reinforced Polymer) under hygrothermal aging condition. In last time FRPs have been adopted as alternatives to traditional structural materials for industrial application due to their high tensile strength, high stiffness to weight ratio and better fatigue resistance. The recent developments of FRPs are towards the growth and usage natural FRP (Fiber Reinforced Polymer) in the field of engineering. FFRP is one of the most commonly used natural FRP. FFRP has the advantages of lightweight, low cost and easy recyclability. In this paper The unidirectional flax fabric was comparison with sample unidirectional flax fabric with chemical treatments of flax fiber are adopted by alkalization. The alkalization group is soaked in 5 wt% NaOH solution for 0.5 h at 25°C. The samples were experimental tested for creep in several days. Results of creep and elastic parameters were different.

Introduction

In present time are researches and developments of FRPs are for growth and use of natural FRP in the field of mechanical engineering, especially for the transport industry [1-2]. FFRP is the most commonly used natural FRP bridge. FFRP has advantages of lightweight, low cost and easy recyclability. However, due to the short history of FFRP in engineering field, the durability and long-term performance of FFRP remains an open question. Therefore we were focusing for experimental and mathematical modelling study the creep condition of FFRP (Flax Fiber Reinforced Polymer) under hygrothermal aging condition.

Experimental study of creep condition

We were using the unidirectional flax fabric which is on the (Fig. 1). We were testing of chemical treatments of flax fiber by alkalization. The alkalization group is soaked in 5 wt% NaOH solution for 0.5 h at 25°C [3]. After the treatment, the flax fabric is washed by distilled water and dried in an oven. The treated flax fabric was used to manufacture FFRP. Epoxy resin were used as the matrix. The FFRP specimen is made of approximately 30% fiber in volume. The dimension of FFRP specimen in rectangular form is 50mm x 10mm x 1.35mm. The creep tests scheme is shown in Table 1. Creep tests were carried out with DMS (Dynamic Mechanical Spectrometer) 6100 (Fig. 2), at a three-point bending mode, with a constant stress level of 20 MPa for a specific time period of 30 min. All experimental values were obtained by an average value of three specimens.

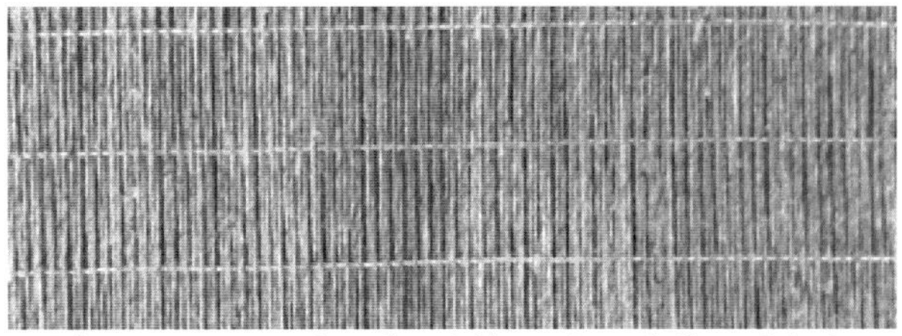

Fig. 1: A sample of unidirectional flax fabric

Table 1: Creep condition test

Surface treatment	Hygrothermal aging time/days				
	0	1	4	9	16
Untreated	U-0	U-1	U-4	U-9	U-16
Alkalization	Al-0	Al-1	Al-4	Al-9	Al-16

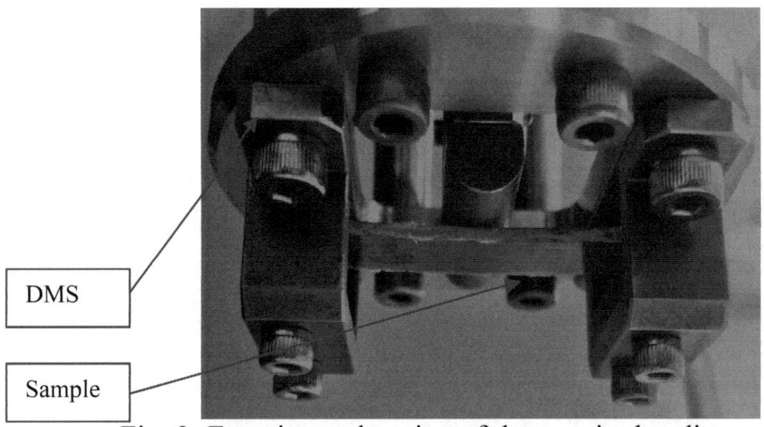

Fig. 2: Experimental testing of three-point bending mode

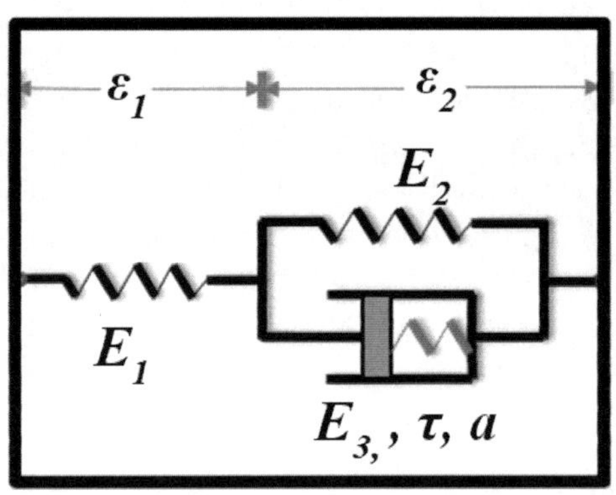

Fig. 3: Reologic model of five-parameter FPT

Mathematical study of creep condition

For mathematical study of creep condition we can used traditional reologic models such as Maxwell model (consists of springs and dashpots connected in series) and Kelvin model (consists of springs and dashpots connected in parallel). For better results instantaneous deformation and stress, we were using a combination of Poynting-Thomson form [4] is adopted in this paper. Fig. 3 shows a five-parameter FPT (Fractional derivative Poynting-Thomson) model, which consists of springs and fractional derivative spring-pot. The constitutive relation of FPT model can be described as follows (1).

$$\sigma + A\tau^{\alpha} D^{\alpha} \sigma = B\varepsilon + C\tau^{\alpha} D^{\alpha} \varepsilon, \tag{1}$$

Where $A = E_3 / (E_1 + E_2)$, $B = E_1 E_2 / (E_1 + E_2)$, $C = E_1 E_3 / (E_1 + E_2)$. By using Laplace transform technique, the creep compliance $J(t)$ of FPT model can be obtained (2).

$$J(t) = \frac{1}{B} + \frac{AB - C}{BC} E_{\alpha,1} \left[-\frac{B}{C} \left(\frac{t}{\tau} \right)^{\alpha} \right], \tag{2}$$

Where Mittag-Leffler function [5] is defined as: $E_{\alpha,\beta}(z) = \sum_{k=0}^{\infty} \frac{z^k}{\Gamma(\alpha k + \beta)}, (\alpha, \beta > 0)$. For simplification of the parameters of the parallel parts in the FPT model, E_2 and E_3 are postulated as the same: $E_3 = E_2 = E^*$. The simplified expression of creep compliance is (3).

$$J(t) = \frac{1}{E_1} + \frac{1}{E^*} \left[1 - E_{\alpha,1} \left(-\left(\frac{t}{\tau} \right)^{\alpha} \right) \right], \tag{3}$$

The material parameters E_1, E^*, τ, α can be obtained by fitting the experimental data as (4).

$$\min f(\mathbf{x}) = \sum_{i=1}^{m} \left[J(\mathbf{x}, t_i) - r_i \right]^2, \tag{4}$$

$$s.t. \begin{cases} x_i \geq 0, i = 1, \cdots c - 1 \\ 0 \leq x_c \leq 1 \end{cases}$$

Where $J(\mathbf{x}, t_i)$ and r_i is the calculated and test result of creep compliance at time t_i, respectively. $\mathbf{x} = (x_1, \cdots x_c)^T$ is the column vector of material parameters. c is the number of parameters. The parameter represents the order of fractional calculus is set as x_c. m is the number of test points, and $m > c$ is required. If the final optimization value of x_c is 1, it indicates that the Newton dashpot is better than the fractional derivative spring-pot, and vice versa.

Results and discussion

The results obtained from the creep test of FFRP are shown in Fig. 4. The creep compliance increases with the hygrothermal ageing time. Under the same hygrothermal ageing time, the chemically treated specimens demonstrate lower creep compliance, which indicates the improvement in creep property of FFRP caused by surface treated. For example, after 16 days of ageing, compared to the untreated group, the creep compliance of alkalization group at the end of the test decreased by 31.54% respectively.

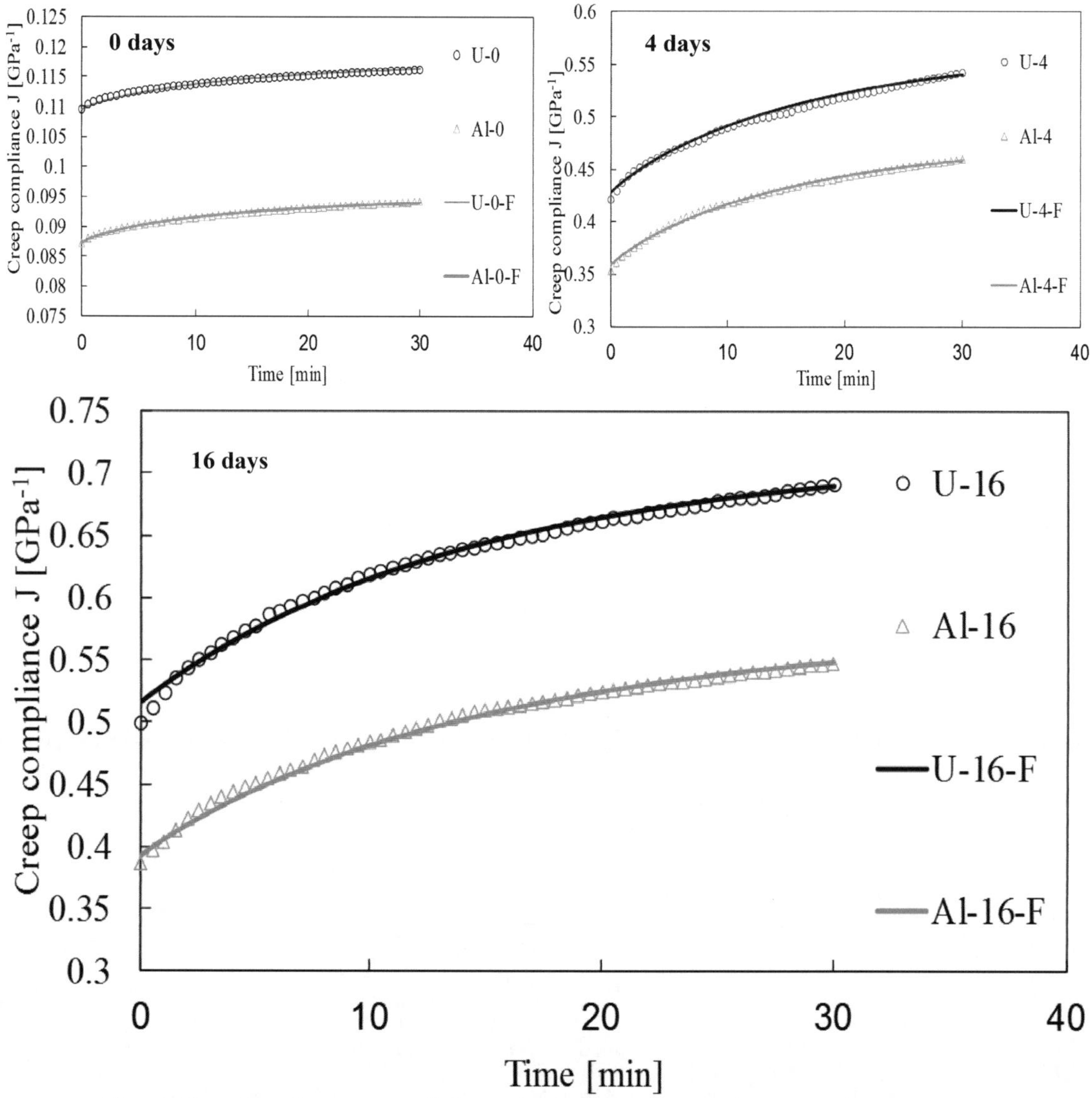

Fig. 4: Results of creep condition of FFRP from experiment and FPT model

Table 1: Parameters of FPT model

Surface treatment	Parameters	Hygrothermal aging time/days				
		0	1	4	9	16
Untreated	E_1	9.133	2.571	2.336	2.225	1.936
	E^*	109.189	8.412	6.953	6.168	4.916
	τ	17.571	17.824	18.129	17.555	15.034
	α	0.820	0.892	0.920	0.959	0.977
Alkalization	E_1	11.464	3.009	2.786	2.569	2.547
	E^*	110.882	8.292	7.981	8.184	5.178
	τ	17.048	16.748	16.953	17.713	16.783
	α	0.786	0.875	0.915	0.927	0.940

The curve-fitting results provided by FPT model presented in **Chyba! Nenalezen zdroj odkazů.** show a fair agreement with test results for both untreated and treated specimens. The influence of hygrothermal ageing time on FPT model parameters is depicted in Table 2. The elastic modulus of the Hook spring in series part E_1 determines the initial strain of FPT model: $\varepsilon_0 = \sigma / E_1$. The value of E_1 decrease after exposure to hygrothermal ageing condition, resulting in increase of ε_0 after 1 day, 4 days, 9 days and 16 days of ageing, respectively. For example, after hygrothermal ageing of 16 days was alkalization group decreased approximately 24%. As demonstrated in Tab.2, hygrothermal ageing does not have a noticeable effect on τ . The value of α increase with hygrothermal ageing time. E.g. after 1 day, 4 days, 9 days and 16 days of ageing, α of untreated group is increased aproximetly 9%, 12%, 17%, 19%, respectively. According to the definition of fractional derivative, the close of α to 1, the nearer the material to the ideal fluid. The trend of α indicates that more viscous character after the hygrothermal aging.

Conclusions

This article was focusing for experimental and modelling study the creep condition of FFRP under hygrothermal aging condition. Comparisons to experimental data reveal that the proposed mathematical FPT model works effectively. The results show that the creep compliance of FFRP significantly decrease with the surface treatment, this is attributed to the enhancement of fiber/matrix interface. In addition, although the modified fractional creep model is established for the FFRP in this paper, this model may have potential applications in creep analysis of other natural fiber reinforced composite materials.

Acknowledgement

This work is partially supported by the Ministry of Education, Youth and Sports of the Czech Republic and the European Union (European Structural and Investment Funds - Operational Programme Research, Development and Education) in the frames of the project "Modular platform for autonomous chassis of specialized electric vehicles for freight and equipment transportation", Reg. No. CZ.02.1.01/0.0/0.0/16_025/0007293.

References

[1] M. Petrů, T. Martinec, J. Mlýnek, Numerical model description of fibres winding process for new technology of winding fibres on the frames, Manufacturing Technology. 16 (2016) 778–785.

[2] M. Petrů, J. Broncek, P. Lepsik, O. Novak, Experimental and numerical analysis of crack propagation in light composite materials under dynamic fracturing, Communications - Scientific Letters of the University of Zilina. 16 (2014) 778–785.

[3] X. Wang, M. Petrů, H. Yu, The effect of surface treatment on the creep behavior of flax fiber reinforced composites under hygrothermal aging conditions, Construction and Building Materials. 208 (2019) 220–227.

[4] R. Hilfer, Applications Of Fractional Calculus In Physics, first ed., World Scientific Publishing Co. Pte. Ltd., Singapore, 2000.

[5] H.J. Haubold, A.M. Mathai, R.K. Saxena, Mittag-Leffler Functions and Their Applications, Journal of Applied Mathematics (2011) 51.

Experimental and FEM modelling for optimization of lightweight steel structures supported by fatigue tests of modern high strength structural steels

NOVÝ František[1,a] and PETRŮ Michal[1,b]

[1]Technical university of Liberec, Faculty of Mechanical Engineering, Studentská 2, 46117 Liberec, Czech Republic

[a]frantisek.novy@fstroj.uniza.sk, [b]michal.petru@tul.cz

Keywords: Optimization of lightweight structures, Fatigue, High strength structural steels

Abstract. Structural optimization has become very popular thanks to modern and powerful software tools. Nowadays a lot of various software for mechanical design and FEM analysis are available. Using those software tools some very nice lightweight structures can be design in fact, but in real practice there can be seen unexpected fatigue failures of those structures very often. This paper shows that optimization of modern lightweight structures using modern software tools can be done relatively easy but it also shows that optimization of lightweight structures which undergo cyclic loading should be in direct connection with fatigue testing of used materials. In this study, the optimization process of real lightweight structure (frame of the semitrailer) was supported by experimental determination of fatigue endurance, mechanical properties, microstructural and fractography analysis of used steels. Connection of structural optimization and detailed materials research resulted in successful and safety design of new lightweight frame of the semitrailer.

Introduction

The mass reduction, while maintaining the same or improving original performances, has become one of the most important goals of newly developed lightweight structures. The principle of lightweight structures design is based on mass reduction using optimization methods. Structural optimization deals with optimal design of load-carrying engineering structures and can be defined as the rational establishment of structural design that is the best of all possible design within a prescribed objective and given set of geometrical and/or behavioural limitations [1].

Various types of structural optimization methods can be used for mass reduction. Sizing, shape and topology optimization methods represent the three different approaches most frequently used in structural optimization. In a sizing optimization problem, the design model is fixed and the objective is to find the optimal dimensions of the structure. For a shape optimization problem, the aim is to optimize the structural shape without modifying the number of borders. Topology optimization allows to design a structure by seeking the optimal material layout within a given design space [2]. There is also fourth approach in optimization, based on application of new materials in design, which seems to be very easy but, it is the most problematic in fact.

In spite of the progress and development in materials science and technology, steel continues to be one of the basic structural materials. This is due to the potential of steel to be widely used in structures because of its low cost, good cold formability, high strength and high modulus of elasticity, as well as weldability. Designers have a great deal of grades of steels with different service properties at their disposal, from which they can select, depending on the demands on the structure being designed. Satisfying all technical and operational requirements is sometimes difficult, and even impossible. In transport machinery and equipment, including semitrailers, the operational conditions significantly affect their service life. The durability of structures operated during driving and carrying cargoes is associated with complex fatigue processes. This gives rise to the issue of not only the appropriate optimization of the part's design, but also the proper selection of its material in order to increase the operational reliability and durability.

Application of modern high strength structural steels allow us to design lightweight structures, but the risk of fatigue failure of modern lightweight structures manufactured from those steels is much higher in comparison with standard structural steels due to the lack of the knowledge about their fatigue behaviour. Due to the lack of the knowledge about fatigue behaviour of modern steels used for cyclically loaded lightweight structures a lot of unexpected failures of those structures can be seen in real practice [3].

Experimental materials

The materials characteristics of steels used in design of the newest lightweight semitrailers should be relatively high. For example, tensile yield strength of steels used for manufacturing of their chassis must not be lower than 600 MPa and the ultimate tensile strength must be above 700 MPa. Therefore two modern high strength structural steels namely STRENX 700MC and HARDOX 450 with improved yield strength were used for optimization in this study. A common S355J2 structural steel was taken in to consideration as a comparative steel.

STRENX 700 MC and HARDOX 450 are advanced high strength steels, designed to meet and exceed the high demands and standards of the vehicle industry. Developed specifically to address the challenges of transport and vehicle manufactures are used to save weight and/or to increase the payload. When using those steels, the transportation vehicles can be made with thinner steel while still retaining their strength properties. This reduces weight, allows for larger payloads, cuts costs and reduces the impact on the environment because less material is needed in production. A well known S355J2 structural steel has a ferritic pearlitic microstructure. The microstructure of STRENX 700MC steel consists of irregular ferrite with approximate average grain size of about 4 - 12 μm, with low amount of pearlite at the grain boundaries of ferritic grains and local occurrence of complex carbonitrides of titanium. The HARDOX 450 steel has microstructure of tempered martensite. Chemical composition and mechanical properties of STRENX and HARDOX steels in delivery state depend on the plate thickness. The detailed chemical composition and mechanical properties of tested steel plates with thickness of 10 mm are given in Tabs. 1, 2.

Table 1: Chemical composition of tested steels in weight %

				S355J2				
C	Si	Mn	P	S	Cr	Ni	Cu	
0.18	0.03	1.47	0.015	0.011	0.03	0.02	0.04	
				STRENX 700MC				
C	Si	Mn	P	S	Al	Nb	V	Ti
0.08	0.35	1.67	0.018	0.0037	0.015	0.06	0.014	0.015
				HARDOX 450				
C	Si	Mn	P	S	Cr	Ni	Mo	B
0.20	0.39	0.80	0. 005	0.005	0.45	0.05	0.01	0.001

Table 2: Experimentally obtained mechanical properties of tested steels; average value from 5 measurements

Steel grade	Re [MPa]	Rm [MPa]	A_5 [%]	Z [%]
S355J2	376 ± 4	564 ± 6	28.1 ± 0.6	41.7 ± 0.9
STRENX 700MC	796 ± 3	849 ± 2	15.4 ± 0.2	36.1 ± 0.3
HARDOX 450	1425 ± 5	1560 ± 7	12.5 ± 0.3	38.0 ± 0.4

There are plenty of producers of high strength structural steels like: SSAB - Sweden, SWEBOR - Sweden, DUFERCO - USA/Brazil, ALGOMA STEEL - USA, ARCELOR MITTAL - USA, NUCOR - USA, HIGH STRENGTH PLATES AND PROFILES - Canada, NLMK - Europe, JFE STEEL - Japan, NSSMC - Japan, KOBE STEEL - Japan, TATA STEEL - India, ESSAR STEEL - India, CHAMPAK INDUSTRIES - India, RUUKKI - Finland, DILLINGER HUTTE - Germany, THYSSEN KRUPP - Germany, ILSENBURGER GROBBLECH - Germany, VOESTALPHINE GROBBLECH - Austria, ISD HUTA - Poland, ACRONI - Slovenia, BISALLOY - Australia, INDUSTEEL – ARCELOR MITTAL group. However none of them can guarantee fatigue properties or give any relevant information about fatigue behaviour of their products. They can guarantee only the chemical composition and basic mechanical properties, but fatigue properties are without any guarantee. Every of above listed producers declares quality of produced steel by newest production technology and everywhere is written that they are selling a high quality steel. What a nice guarantee, isn't it? The real quality of some sheets from S700MC steel on the steel market is depicted in Fig. 1.

Fig. 1: Real (terrible) quality of some sheets from S700MC steel on the steel market; strong heat affection along edges is clearly visible

From Fig. 1 it is clear why fatigue resistance of this steel sheet cannot be guaranteed. There must be very high gradient of internal residual stresses and many problems can arise. What is chemical composition good for a mechanical design? Chemical composition and effect of alloying and microalloying are in the interest of the steelworks. Information about chemical composition of used steel is marginal for mechanical design of structures, because it is important only for the welding and it must be guaranteed. On the other hand, in comparison with the problematics of the influence of microstructural changes in heat affected zones of high strength steels is the problematics of chemical composition guarantee negligible for designers and welding engineers. Chemical composition has to be kept under control. Chemical composition of those steels is known very well and can be checked easy. The secret of success in production of those high strength steels consists in combination of chemical composition (microalloying usually), rolling and heat treatment. Final mechanical properties and behaviour of the steel depends strongly on the rolling and heat treatment. How many designers know that form the steel with chemical composition equal to chemical composition of HARDOX 450 can be produced a soft steel with yield point about 350 MPa or real hard HARDOX with yield point about 1400 MPa? The only difference is in the heat treatment. To prove that, let HARDOX 450 undergo normalizing heat treatment and normalized perlitic-feritic microstructure and yield point about 350 - 400 MPa will be the result.

Experimental methods and results

The aim of this paper is to show extended possibilities of mass reduction of the structural elements using the optimizing methods and introducing new steel grades with improved mechanical properties. FEM simulation were carried out using commercial ANSYS software. In this study about 242274 quadratic tetrahedral elements of C3D10 type and 453358 nodes were used in the mesh of analyzed frame (Fig. 2). For optimization of chosen structural element a mesh consisting of 28632 elements and 51573 nodes was used. In preliminary dimensioning of the structural element are achieved maximal tensile stresses, whose values do not exceed 60-100 MPa. This is due to the geometry of the frame and other structural elements, the material properties and the character of the load.

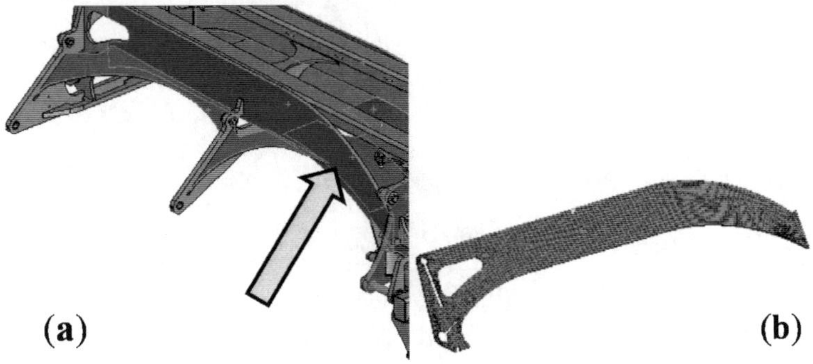

Fig. 2: Example of optimization process; (a) welded frame structure, (b) structural element for optimization

As can be seen from obtained optimization results (Fig. 3a-c) application of modern high strength structural steels can reduce the weight of the structural element from the 140 kg before the optimization to the final weight of 60 kg after optimization, while the relatively high safety coefficient was maintained.

For the comprehensive assessment of the suitability of a change in the thickness of the structural part, the utilization only of optimization calculation is not fully adequate and other experimental tests, such as fatigue tests, are required. This will allow achieve a relevant comprehensive results of the structural part optimizing, which will be able to reliable functionate during their operational lifetime.

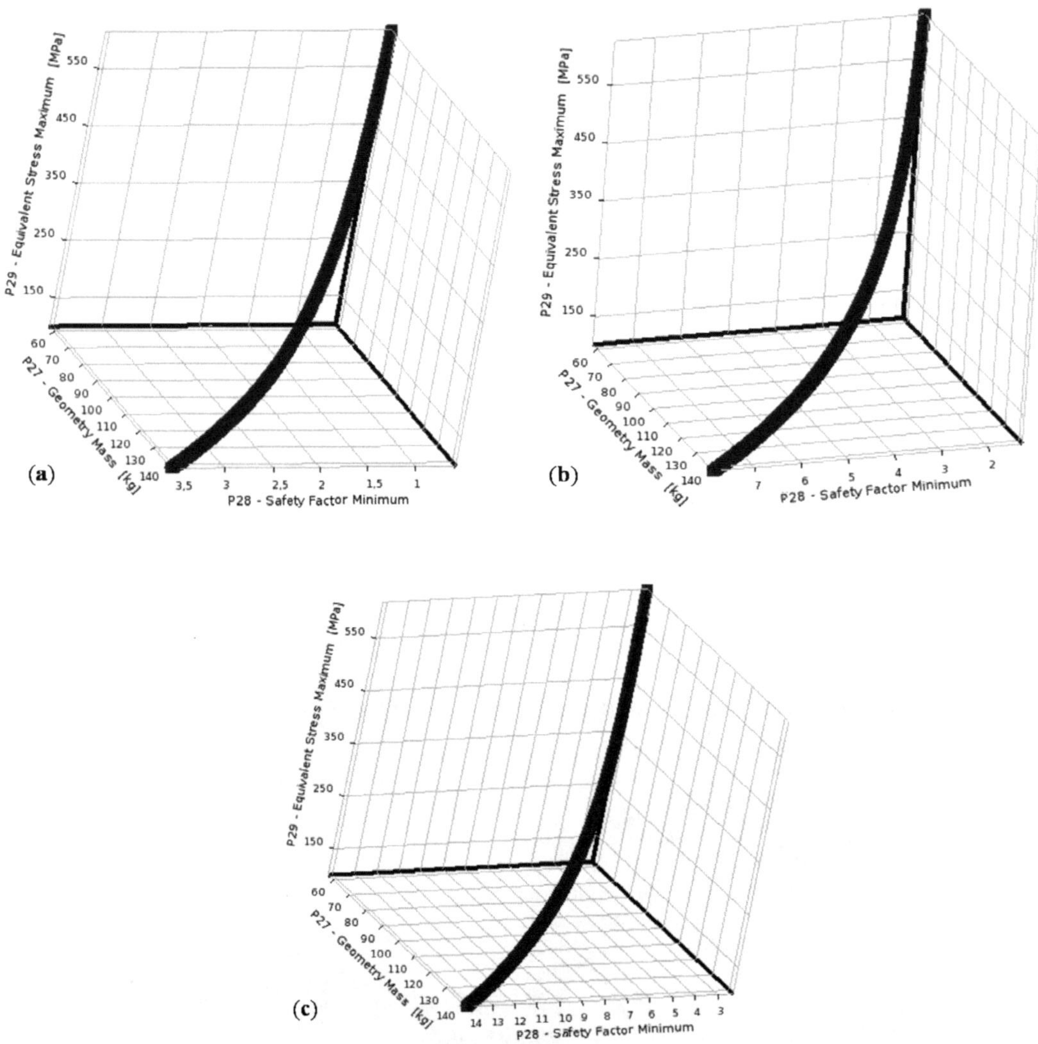

Fig. 3: Example of optimization results for study of stress - safety coefficient - weight of structural part made from: (a) S355 J2, (b) STRENX 700MC, (c) HARDOX 450 steel

Three point bending fatigue tests were carried out at ambient temperature of laboratory air using Amsler 150 HFP 5100 resonant type fatigue testing machine. The fatigue tests with sinusoidal loading were carried out in accordance with the DIN 5010 standard – the S/N curve was determined for assumed stress levels. The experimental set-up illustrating the three point bending loading applied using Zwick/Roel Vibrophore type fatigue testing machine Amsler 150 HFP 5100 can be seen in Fig. 4.

Specimens taken directly from delivered steel sheets were used for the fatigue tests. The rolled plates of S355J2, STRENX 700MC and HARDOX 450 steels with thickness of 10 mm were cut to the required dimensions (15 mm × 55 mm) by water jet abrasive cutting. After that the cutting surfaces were fine grinded to remove possible surface imperfections and influences of cutting process on the material behaviour. Specimens were taken from each steel in the longitudinal orientation to the rolling direction. Finally smooth rectangular (15 × 10 × 55mm) specimens without notches were loaded by sinusoidal cyclic bending using the testing frequency of 100 Hz. Some broken specimens from the fatigue tests of STRENX 700MC and HARDOX 450 steels are documented in Figs. 5 and 6.

Fig. 4: The Amsler 150 HFP 5100 – resonant fatigue testing machine with equipment for three-point bending fatigue tests

Fig. 5: Example of STRENX 700MC broken specimens after three-point bending fatigue tests

Fig. 6: Example of HARDOX 450 broken specimens after three-point bending fatigue tests

Obtained results of three-point bending fatigue tests of S355J2, STRENX 700MC and HARDOX 450 steels are shown in Fig. 7.

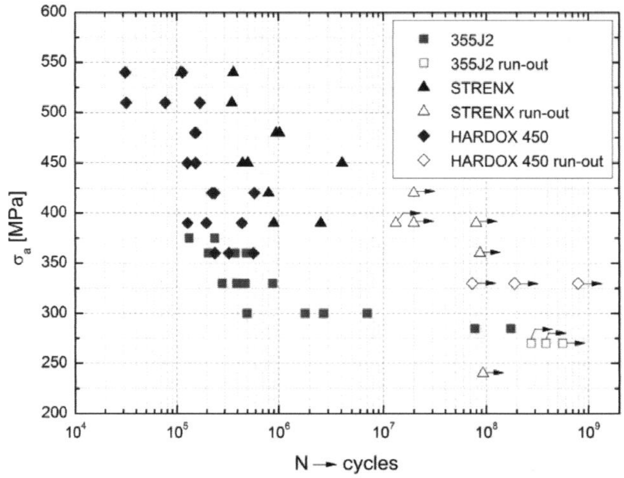

Fig. 7: The fatigue endurance of S355J2, STRENX 700MC and HARDOX 450 steels; the three-point bending fatigue test.

Conclusions

Modern design of the lightweight steel structures using the FEM optimizing methods and introducing new steel grades with improved mechanical properties connected with their fatigue testing used in this study shows extended possibilities of mass reduction and minimization of the risk potentiality of unexpected failures of modern lightweight structures.

Numerous studies have observed that the fatigue endurance of structural steels increases with increasing yield strength and ultimate tensile strength. In this study was found that the increase of tensile strength (see Tab. 2) is not in all cases accompanied by corresponding increase of fatigue properties (see Fig. 7). Some factors influencing this phenomenon were studied and obtained results of microstructural and fractographic analysis gave relevant information about real fatigue behaviour of tested steels. Results obtained from optimization process (Figs. 2, 3) and fatigue tests (Fig. 7) finally resulted in successful and safety design of real lightweight structure. The only way how to reduce the risk potentiality of unexpected failure of modern lightweight structures manufactured from high strength structural steels is to include the knowledge of their fatigue behaviour in to mechanical design and manufacturing processes.

Acknowledgement

This work is partially supported by the Ministry of Education, Youth and Sports of the Czech Republic and the European Union (European Structural and Investment Funds - Operational Programme Research, Development and Education) in the frames of the project "Modular platform for autonomous chassis of specialized electric vehicles for freight and equipment transportation", Reg. No. CZ.02.1.01/0.0/0.0/16_025/0007293.

References

[1] M. Petrů, T. Martinec, J. Mlýnek, Numerical model description of fibres winding process for new technology of winding fibres on the frames, Manufacturing Technology. 16 (2016) 778–785.

[2] H. Eschenauer, N. Olhoff, W. Schnell, Applied Structural Mechanics, Fundamentals of Elasticity, Load-Bearing Structures, Structural Optimization, Springer 1997, pp. 301-306.

[3] R. Ulewicz, F. Nový, Fatigue life of high strength steel for cold forming, Metalurgija, 56 (2017) 115-118.

EAN 2019
57th conference on experimental stress analysis

Unfired earth and diagnostics with methylene blue test

PLAČEK OTCOVSKÁ Tereza[1,a], MUŽÍKOVÁ Barbora[1,b] and PADEVĚT Pavel[1,c]

[1]Department of Mechanics K132; Faculty of Civil Engineering; Czech Technical University in Prague; Thákurova 7/2077, 166 29 Praha 6 Dejvice, Czech Republic

[a]tereza.otcovska@fsv.cvut.cz, [b]barbora.muzikova@fsv.cvut.cz, [c]pavel.padevet@fsv.cvut.cz

Keywords: Unfired earth, Methylene blue dye, Clay, Earth sample, Absorption

Abstract. Unfired earth is one of the oldest building materials. Nowadays, the unfired earth is getting to forefront of interest for its properties that fit into principles of a sustainable building and improving of microclimate in interiors. Final properties of unfired earth are influenced by composition of earth mixtures and this composition can be very diverse. The essential component of earth mixtures is clay because clay fulfils a function of binder and ensures strength of unfired earth. Methylene blue test is a method that helps determine the type and amount of clay in earth mixtures. Significant part of this paper is focused on description of methylene blue test. Research describes in the paper is focused on methylene blue dye absorption by illite-kaolinite clay and dependence of amount of absorbed dye on percentage amount of clay in earth samples.

Introduction

Unfired earth is one of the oldest building materials. There are a lot of historic and modern buildings from unfired earth, e.g. The Great Wall of China or Berlin's Rammed-Earth Reconciliation chapel. Unfired earth had become minority building material in 2nd half of 20th century and it was replace by modern building material like concrete or steel. Nowadays, the unfired earth is getting to forefront of interest for its properties that fit into principles of a sustainable building and improving of microclimate in interiors. The growing interest is evident from a number of scientific papers or quantity of sales of unfired bricks by HELUZ cihlářský průmysl v.o.s [1–5].

Unfired earth is created from earth mixtures. There are three main components of earth: clay, sand, water. The clay fulfils a function of binder, the sand a function of filling agent and the water serves for activation of bonding properties of the clay and for good earth mixture processing. Final properties of unfired earth are influenced by composition of earth mixtures, especially amount and kind of clay and this composition can be very diverse. This is the main reasons why using of unfired earth in construction is complicated [1, 6–8].

The essential component of earth mixtures is clay because clay fulfils a function of binder, similarly like cement in binder, and ensures strength of unfired earth. Ability to determine amount and type of clay in earth mixtures could help solve this problem. Methylene blue test is a method that helps determine the type and amount of clay in earth mixtures [1, 6, 7, 9–11].

Principle of Methylene Blue Test

Basis of methylene blue test is absorption of methylene blue dye (methylthioninium chloride - $C_{16}H_{18}ClN_3S$) by clay. This experimental method is described in norm ČSN EN 933-9 [9].

Methylene blue solution (concentration 20 g/l) is inserted into earth sample (solution of clay and distilled water) by a burette (Fig. 1). Amount of one dose of methylene blue solution is 2 ml. The added dye solution in earth sample is stirred for 1 minute. After mixing, a drop of clay sample is placed on a filter paper.

This process is repeated and each sample drop with increasing amount of methylene blue solution is marked (Fig. 2). Methylene blue test is finished when a blue ring spreads around a drop sample (mark 5 in Fig. 2) because clay has already absorbed the maximum amount of methylene blue.

Fig. 1: Dye dosed into the earth sample

Dye solution consumption increases with increasing amount of clay in clay samples. Amount of dye absorbed also depends on a kind of clay. Amount of dye absorbed is determined by the equations 1. Amount of dye absorbed is defined as the amount of weight of dye in gram that is absorbed by 1 kilogram of test material.

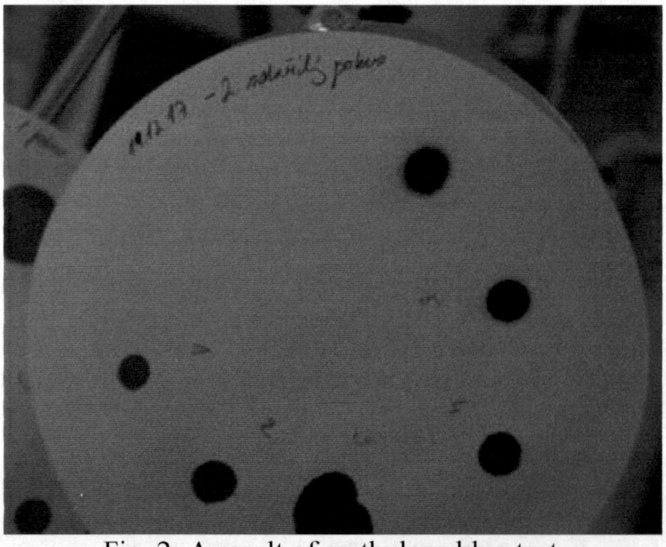

Fig. 2: A result of methylene blue test

$$M_{MB} = \frac{m_S.m_{MB,1ml}.1000}{M_{ES}} \quad (1)$$

$M_{MB} - amount\ of\ dye\ absorbed\ \left[\frac{g}{kg}\right]$
$m_s - amount\ of\ methylene\ blue\ solution\ [g]$
$m_{MB,1ml} - amount\ of\ methylene\ blue\ in\ 1\ g\ of\ methylene\ blue\ solution\ [g]$
$m_{ES} - amounto\ of\ tested\ material\ [g]$

Tested Material and Measured Results

Methylene blue test of illite-kaolinite clay (clay mark KR) is described in this paper. There are 5 types of earth samples. Every type of earth samples includes KR clay, sand and distilled water. Sand and clay ratio is different for these types of earth samples. Methylene blue test is realized 3 times for every type of earth samples.

Table 1: Tested material

Mark of clay	Kind of clay	Clay/sand ratio [-]
KR	Illite-kaolinete	100/0
KR	Illite-kaolinete	75/25
KR	Illite-kaolinete	50/50
KR	Illite-kaolinete	25/75
KR	Illite-kaolinete	0/100

Table 2: Results of experiment

Samples	Number of measure	Amount of dye absorbed [g/kg]	Average amount of dye absorbed [g/kg]	Variance [(g/kg)^2]	Standard deviation [g/kg]
KR 100/0	I. measure	31.68	28.6	5.0	2.2
	II. measure	26.40			
	III. measure	27.72			
KR 75/25	I. measure	25.08	25.52	10.6	3.2
	II. measure	21.78			
	III. measure	29.70			
KR 50/50	I. measure	18.48	18.48	1.2	1.1
	II. measure	17.16			
	III. measure	19.80			
KR 25/75	I. measure	9.24	7.48	1.5	1.2
	II. measure	6.60			
	III. measure	6.60			
KR 0/100	I. measure	0.00	0	0	0
	II. measure	0.00			
	III. measure	0.00			

It was determined that clay-free sand did not absorb any dye. Absorption increases with increasing amount of clay in the sample. Tested earth samples and obtained results are showed in Table 1 and Table 2.

High values of variance, especially for KR 100/0 and KR 75/25 earth samples, is probably because of method of experiment execution.

Methylene blue test was adapted for our experimental measure compared the norm [9]. The main difference is amount of dose dye solution. In the norm is amount of 1 dose 2 ml of dye solution, but we adjusted dosage to reduce experiment time. After added dye solution was earth sample stirred for 1 minute regardless of amount of dye solution added. Example of dye solution dose is showed in Table 3.

We supposed that dependence of amount of absorbed dye on amount of clay would be linear. But, the experimentally obtained data give a nonlinear curve of this dependence (Fig. 3). Reason for obtained these results is probably in method of experiment execution too.

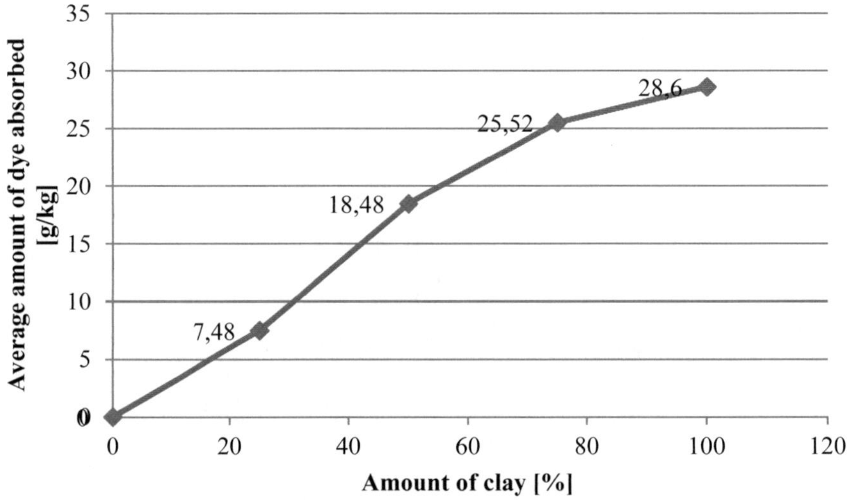

Fig. 3: Dependence of amount of absorbed dye on percentage amount of clay in earth samples

Table 3: Dose of dye solution into earth sample

Earth sample contains clay/sand ratio 50/50		
Number of dose	Amount of dye solution [ml]	Blue ring spreads around
1	20	NO
2	2	NO
3	2	NO
4	2	NO
5	2	NO
6	2	YES

Conclusions

Methylene blue test is good way for facilitating of composition diagnostician of unknown earth. This paper focuses on research of methylene blue dye absorption by illite-kaolinite clay. During this research it was found that the resulting values of absorbed dye are influenced by stirring time of earth sample with dye solution. Method of experiment must be adapted according to these findings for determining more accurate results.

It was determined with certainty that the clay-free material did not absorb methylene blue dye. Furthermore, it was determine that 1 kg illite-kaolinite clay absorbs approximately 29 g methylene blue dye.

Analysis methylene blue dye absorption of other kind of clay is necessary for using of methylene blue test in practical diagnostics of unknown earth.

Acknowledgement

The financial support of this experiment by the Czech Science Foundation (GAČR project NO. 18-10884S) and Faculty of Civil Engineering, Czech Technical University in Prague (SGS project No. SGS16/201/OHK1/3T/11) is gratefully acknowledged.

We would like to express our thanks to LB MINERALS, s.r.o. company for free supply of material necessary for experimental measurement.

References

[1] MINKE, Gernot. *Building With Earth* [online]. B.m.: Ökobuch Verlag, Staufen, 2006 [vid. 2015-08-18]. ISBN 978-3-0346-0822-0. http://archive.org/details/Gernot_Minke-Building_With_Earth

[2] JING, Ai. *A History of the Great Wall of China*. B.m.: World Scientific Publishing Company, 2015. ISBN ISBN: 978-1-938368-32-5.

[3] *Agenda 21 - United Nations Environment Programme (UNEP)* [online]. [vid. 2017-02-14]. http://www.unep.org/Documents.Multilingual/Default.asp?documentid=52

[4] *ČSN 1168-1939 - PODMÍNKY PRO ZEDNICKÉ A PŘIDRUŽENÉ PRÁCE POZEMNÍCH STAVEB*. 1951. bibtex: csn_csn_1951

[5] STEJSKAL, Petr. *e-mail consultation*. April 2017

[6] ŽABIČKOVÁ, Ivana. *Hliněné stavby* [online]. Brno: Era 21, 2002 [vid. 2015-08-18]. ISBN 80-86517-21-7. http://www.kosmas.cz/knihy/107760/hlinene-stavby/

[7] WALKER, Peter. *Rammed Earth: Design and Construction Guidelines*. Watford: IHS BRE Press, 2010. ISBN 978-1-86081-734-2.

[8] NEMECZ, Ernö. *Clay minerals*. 1. ed. B.m.: Akadémiai Kiadó, 1981.

[9] EVROPSKÝ VÝBOR PRO NORMALIZACI. ČSN EN 933-9, Zkoušení geometrických vlastností kameniva - Část 9: Posouzení jemných částic - Zkouška methylenovou modří. 2013.

[10] YELIZ, Yukselen a Kaya ABIDIN. Suitability of the methylene blue test for surface area, cation exchange capacity and swell potential determination of clayey soils. 2008 (102), 38–45.

[11] National Center for Biotechnology. *PubChem* [online]. 2019. https://pubchem.ncbi.nlm.nih.gov/compound/methylene_blue

EAN 2019
57th conference on experimental stress analysis

Non-destructive inspection of L410 NG aircraft structure during full-scale fatigue tests

PODSTAWKA Michal[1,a] and BĚLSKÝ Petr[1,b]

[1]VZLU – Czech Aerospace Research Centre, Strength of Structures Department, Beranovych 130, Prague 9, 199 05, Czech Republic

[a]podstawka@vzlu.cz, [b]belsky@vzlu.cz

Keywords: NDT inspection, Fatigue test, L410 NG

Abstract. A fatigue test is one of the most important steps in the long and demanding way to obtain airworthiness certification of every new aircraft structure. A new type of aircraft structure usually starts with the development and testing a new material, elements, then sub-component and components such as wing or fuselage. A full-scale fatigue test is carried out to prove the design and verify the predicted lifetime and detect damage which can occur during service. Non-destructive inspections are used to detect any damage during the fatigue test. The inspections are divided into the General Visual Inspection, Detailed Visual Inspection or Special Detailed Inspection according to MSG-3 standard. NDT methods such as visual, eddy current and ultrasonic testing are used for the inspections. Also, it is important to determine the schedule of inspections and personnel requirements.

Introduction

Certification of any airframe structure can be performed by numerical or experimental methodology in compliance with airworthiness requirements. Although advanced numerical methods are very sophisticated, experimental verification of any primary airframe structure through full-scale testing is still necessary [1].

The main aim of the full-scale fatigue test is to find out how the aircraft is able to sustain real conditions that will be encountered in-service. A full-scale fatigue testing of airframes and components is carried out to prove the design, to verify the predicted or design life and to reveal structural defects that may appear during service. In generally the full-scale test is an experimental test based on the simulation of flight hours (SFH), which should simulate the real operating condition according to relevant standard [2]. The very important part of any full-scale test is continuous non-destructive inspection (NDI) of the structure. NDI is conducted at different levels and by using different equipment. It is a very time consuming and costly activity. The aim of the paper is to show basic NDI principles and methodologies applied during the fatigue test of L410 NG commuter aircraft.

Full – Scale Fatigue Test of the L410 NG

The structure of L410 NG is designed for 30 000 flight hours (FH). Applied fatigue loading includes all phases of the flight (taxiing, takeoff, cruise including aerodynamic gusts and landing). The set-up of the full-scale test includes many hydraulic cylinders, loads are applied into the structure through the complex whiffle three systems and additionally many gauges and additional measurement systems are installed to the tested structure. The test set-up of the wing together with the middle section of the fuselage is shown in Fig. 1. All the systems are necessary for the test conduction in compliance with the test plan and designer requirements but at the same time all systems make the requested inspections of principal critical parts of the structure more difficult. All these issues must be taking into account during inspection plan development [3, 4].

Fig. 1: Full-scale fatigue test of the wing and fuselage middle section of the L410 NG

Non-Destructive Inspection

The non-destructive inspection plan was created in cooperation with Aircraft Industries, a.s. There are several issues which had to be defined before the start of the experiment:

- inspection areas,
- type of inspection,
- schedule of inspections,
- NDT methods and equipment,
- qualified personnel.

Inspection Areas

Based on the experience and numerical simulations, the critical areas and important parts of the structure were determined and must be inspected. These are some parts of the wing, fuselage, landing gear and tailplane such as spar caps, shear webs, splice and stringer joints, flap hinges, frame segments and bottom skin panels.

Fig. 2: Inspection areas on L410 NG

Type of inspections

NDT inspections performed during full-scale fatigue tests are divided into three types of inspection in accordance with MSG-3 standard:

1. General Visual Inspection (GVI) – this type of visual inspection detects obvious unsatisfactory conditions or discrepancies, including removal of aircraft parts e.g. fairings, access panels, doors etc. due to get close to the inspected area. The inspection is made under normally available lighting condition such as daylight, hangar lighting etc. [5]
2. Detailed Visual Inspection (DET) – this is an intensive visual inspection of a special detail, assembly or installation. Artificial lighting is required if necessary. Aids to inspection such as mirrors, magnifying glasses can be used [5].
3. Special Detailed Inspection (SDET) – the inspection is similar to the DET except that other NDT methods are used, especially eddy current testing or ultrasonic testing depending on the geometry, location, surface condition etc. [5]

The first step is the selection of the inspection method according to MSG-3. The first option is GVI. The general direct visual method enables the inspector to get the overall view of the condition of a structure to be tested. The method focuses mainly on servicing the external areas. If GVI is not able to meet some of the detection requirements, then DET type inspection has to be used.

A DET inspection is a detailed local visual inspection which is able to detect smaller size damage (cracks) than GVI. Both internal and external areas are inspected during the DET.

A SDET inspection is only considered when neither GVI nor DET inspections are able to detect some hardly detectable damage. This more intensive inspection level is achieved through the use of specialized inspection techniques and equipment which go beyond the ability of human sight. A SDET may use other NDT methods to detect small size damage with using special equipment such as endoscopes or flaw detector devices.

Schedule of Inspections

The schedule defines the inspection intervals depending on the simulated flying hours and determines what types of inspection are to be performed [2].

NDT methods

NDT is a wide group of methods, which are used for testing materials, components, parts or the whole assembly such as an aircraft. Basic methods used for the NDT inspections are visual testing, eddy current testing and ultrasonic testing.

Visual Testing (VT)

VT is the basic, oldest and most common NDT method. It is typically the first step of inspection to detect surface discontinuities or defects by human eyes, which use the visible electromagnetic spectrum of radiation. VT is divided according to EN 13018 into direct (uninterrupted optical path of light) or remote (interrupted optical path light) visual testing, which is performed with special equipment (e.g. boroscope, videoscope etc.). The method depends on the following basic conditions:

- visual ability of the performing worker,
- sufficient lighting of the tested surface,

- capability – the knowledge and practical abilities of the worker who is performing VT.

The visual inspection of the L410 NG must be performed precisely either directly by the eyes or with equipment as magnifying glasses, mirrors, etc. eventually remotely with usage of the endoscopes, especially videoscope.

The inspection is performed by ten (full inspection) or six (partial inspection) certified personnel. Each inspector has an assigned section to be inspected. To ensure a high probability of defect detection, inspection of some critical areas is performed by two inspectors.

Some areas are not possible to inspect directly by eyesight, e.g. FWD or AFT spar area. That is why the videoscope must be used for remote visual inspection. For the activities, the videoscope Olympus IPLEX UltraLite is used.

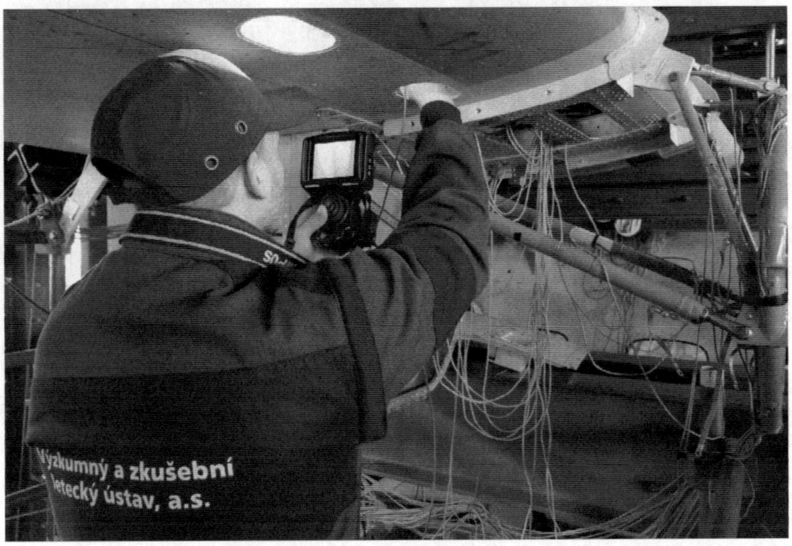

Fig. 3: Remote visual inspection using videoscope

Eddy Current Testing (ECT)

Eddy current testing (ECT) is one of the most extensively used non-destructive techniques for inspecting electrically conductive materials [6]. ECT method employs alternating currents applied to a conducting coil held close to the test piece. The primary alternating magnetic field generated in this way penetrates the inspected material and generates continuous circular eddy currents in it. The induced currents flowing within the test piece generate a secondary magnetic field that tends to oppose the primary magnetic field. Changes in the induced eddy currents may be caused by changes to a material's electromagnetic properties or changes in geometry, including cracks. After the coil passes over a crack, the impedance of the coil changes and this impedance variation is directly illustrated on the screen of the flaw detector in a view called the impedance plane.

The biggest advantage of ECT is that it is a clean NDT method which does not require any couplant. It is extremely sensitive to flaws, very accurate for dimensional analysis, very repeatable and it is able to give an instantaneous response. On the other hand, it has also some disadvantages and limitations. The method is very sensitive to surface variations and therefore requires a good surface. It is applicable only to conducting materials and the depth of penetration is limited.

During a full-scale fatigue test of L410 NG, the ECT methods are used mainly in the frame of SDET inspections of critical areas identified by the manufacturer. For example, they are used for second layer type inspections of the main splice joint in the middle of the wing, inspection of the bottom panel under the engine nacelle cover, bottom spar cap joints inspection, etc.

Along with the conventional ECT techniques, there is used also Eddy Current Array (ECA) technology in some problematic areas. ECA probes use several individual coils grouped together in one probe. Data acquisition is performed by multiplexing the EC sensors. Measured data is referenced to an encoded position and time and represented graphically as a C-scan image. ECA NDT technology has big potential to considerably reduce the duration of the inspections and increase detection probability.

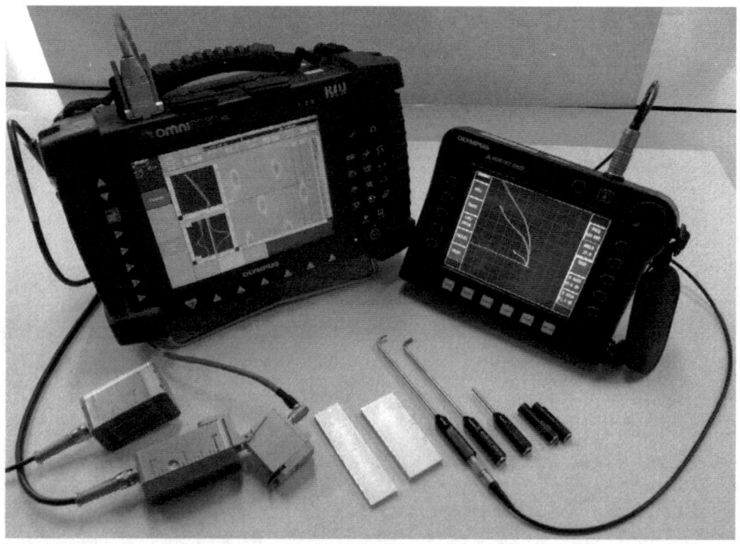

Fig. 4: ECT flaw detectors, probes and calibration standards

Ultrasonic Testing (UT)

Ultrasonic testing is able to detect the presence of internal material discontinuities. Ultrasound, as well as sound and noise, is the mechanical vibration spreading in the environment due to its elastic properties. The frequency range of ultrasonic vibration is out of audible spectrum, i.e. more than 20 kHz. The range of frequencies between 1 MHz and 20 MHz are used for the conventional ultrasonic testing. Ultrasound waves spread into the material by ultrasonic probe transducers which are different due to used angles and frequencies. The wave spread through the part as a sound wave which is reflected at the interface (imperfection) and may be received back by the probe. The received signal appears on the ultrasonic device display in the form of an echo [7].

Ultrasonic testing is performed only in a few locations of the aircraft structure during the fatigue test. There are some areas where there is no access to use either a visual inspection or eddy current testing.

Qualified personnel

The inspections must be performed by qualified personnel. Personnel must be qualified and certified in the appropriate NDT methods according to the relevant standards. Personnel must have satisfactory physical abilities, which means satisfactory vision, especially near vision (Jaeger test no. 1 or Tumbling E in accordance with ISO 18490 from the distance of at least 30,5 cm) and colour vision (Ishihara test). Vision shall be checked at least every 12 months.

Damage

The inspection plan describes specific damage types that may be found on the structure during the fatigue test. Typical damage is e.g. crack, corrosion, loose or broken rivet etc.

Any damage, which is found, has to be written to the summary list of NDI. Each damage is assigned a number. There is also written the location, type of the damage, size, used NDT method and actual number of simulation flight hours. The found damage is monitored due to its propagation.

Some of the damage can be repaired, especially if the damage achieves the critical size or it is a critical point of the structure.

Conclusions

Aviation is an advanced and important field of NDT application. Along with NDT inspections during production or maintenance, NDI is also an integral part of the fatigue test of the new aircraft structure.

The basic principle of NDT inspection during the fatigue test is to find any damage to the structure, where there are always several critical areas. Therefore, the effort to perform the NDI as detailed as possible. Three types of inspection in accordance with MSG-3 standard are used. GVI inspection is used to get the overall view of the structure without using any special equipment. For DET and SDET inspections aids or special equipment to improve damage detection probability are used. The full-scale fatigue test provides also an opportunity to try new NDI techniques such as ECA technology or some developed special equipment for the limited access areas inspections.

Acknowledgement

This project result was developed within the institutional support of the Ministry of Industry and Trade of the CR directed to the development of research organizations.

References

[1] J. Šedek, R. Růžek, J. Raška, J. Běhal, Comparative study of prediction methods for fatigue life evaluation of an integral skin-stringer panel under variable amplitude loading, Procedia Engineering, Volume 114, 2015, 124-131p.

[2] H.J. Morton, B.C. Bishop, G.Clark, NDT of Aircraft Structures During Full-Scale Fatigue Testing, National AINDT Conference, Wollongong 1993, 6p. ISBN 0-86418-265-1.

[3] J. Jironč, Certifikační zkoušky L 410NG, Letectví + kosmonautika, Magnet Press Slovakia, 2019, 28-31p, ISSN 0024-1156.

[4] V. Horak, D. Novotny, Full Scale Fatigue Test and Failure Analysis of Advanced Jet Trainer Wing, 5th International Conference of Engineering Against Failure (ICEAF-V 2018), MATEC Web Conf. Volume 188, 2018, 9p.

[5] ATA MSG-3, Operator/Manufacturer Scheduled Maintenance Development, Air Transport Association of America, Inc., Revision 2007.1, USA, 2007.

[6] García-Martín J, Gomez-Gil J, Vazquez-Sánchez E. Non-destructive techniques based on eddy current testing. Sensors. 2011; 11(3):2525–65

[7] International Atomic Energy Agency, Non-destructive testing for plant life assessment, Training course series no. 26, Vienna 2005, 68p, ISSN 1018-5518.

Experimental verification of analytical model for deflection prediction of hybrid carbon fibre profile under three point bending tests

POŠVÁŘ Zdeněk[2,a], RŮŽIČKA Milan[1,b], KULÍŠEK Viktor[1,c], MAREŠ Tomáš[1,d], UHER Ondrej[2,e] and DOUBRAVA Karel[1,f]

[1]IFaculty of Mechanical Engineering, Czech Technical University in Prague. Technická 4, Praha 6, 166 07, Czech Republic

[2]Compotech Plus s.r.o., Nová 1316, Sušice, 342 01, Czech Republic

[a]zdenek.posvar@compotech.com, [b]milan.ruzicka@fs.cvut.cz, [c]viktor.kulisek@fs.cvut.cz, [d]tomas.mares@fs.cvut.cz, [e]ondrej.uher@compotech.com, [f]karel.doubrava@fs.cvut.cz,

Keywords: Carbon fibre profile, Shear stiffness, Analytical model verification, Hybrid composite, Three point bending test

Abstract

The aim of the article is an experimental verification of new analytical method for prediction of composite profile behaviour under bending and transverse shearing loading when testing the composite profiles for the production machine or automation industry. The article contains the analytic theory for the stiffness prediction and the experimental data from a standard and modified three point bending tests.

Introduction

The composite profiles usually offer a great bending stiffness and reduced mass in comparison with commonly used isotropic structural materials like steel, cast-iron or aluminium, see [1]. These benefits might be limited by a reduced stiffness in out-of plane laminate directions – mainly in transverse shearing.

Fig. 1: Bilsing Automation pressline manipulator based on hybrid carbon profiles

The aim of the article is an experimental verification of new analytical method in revise shear stiffness calculation. This verification was done for composite profile behaviour under bending and transverse shearing loading. The low transverse shearing stiffness might cause issues during the profile application as it might lead to a large global beam deformation or to large local deformation in places, where loading is introduced into the design. The effect of the local compliance, when testing the composite profiles in three-point bending tests and their influence to results is discussed. The experimental results with and without the local compliance effects are compared with developed analytic solution results and with finite element results.

Profile description

A hybrid composite profile of external dimensions 100x100x1000 mm with 8mm wall thickness was selected for the three-point bending test (in Figure 2 first on the left). The profile represents a typical beam used in the production machine, which typical high-stiffness – low-mass application shown in Figure 1. The profile lay-up was of a hybrid composition as it was made from ultra-high modulus carbon fibres (UHMC) and also from high-strength standard carbon fibres (HSC). A resin from epoxy group was used as a matrix system.

Fig. 2: Coupons of structural parts (all hybrid composite profile)

A comparison of material mechanical properties, which are used in production machines, is given in Table1. Properties of the HSC and UHMC laminate are for unidirectional composites with 50% fibre volume fraction in the laminate. Elastic modulus in the direction of fibres E_1, elastic modulus in the transversal direction E_2 and shear elastic modulus G_{12} are important inputs for calculation under bending and transverse shearing loading.

Table 1: Comparison of mechanical properties

Material	ρ [kg·m^{-3}]	E_1 [GPa]	E_2 [GPa]	G_{12} [GPa]
HSC C/E UD	1500	137	9	4
UHMC C/E UD	1750	380	5	3
Epoxy	1200	2.5-4.5	2.5-4.5	1.6
Steel	7800	210	210	80
Cast iron	7200	80-125	80-125	30-55

Analytical approach

Analytical model are really useful for their quick and simple usage. They are mostly used as quick design tools. For isotropic materials, Bernoulli beam theory is often used. However, for anisotropic materials the application of this method is limited. A low shear stiffness of carbon fibre profiles need to be taken into account and in calculation, see [2].

The Timoshenko beam theory works with bending and shear stiffness. Application of the Timoshenko theory to a cantilever beam is shown in Eq. (1),

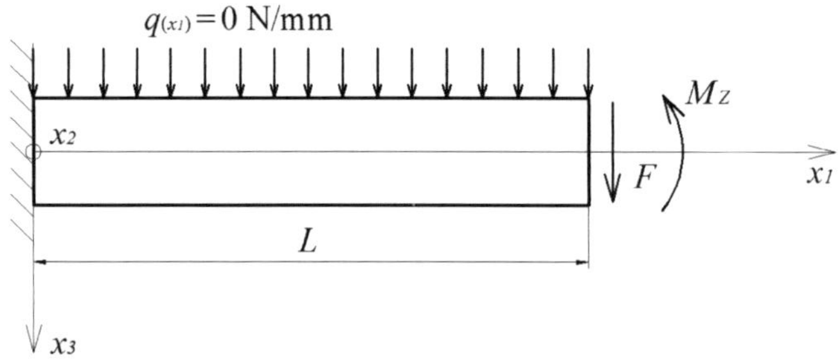

Fig. 3: Beam element

where L is length of a beam, T_{D2} is profile bending stiffness (x_2 is bending neutral axis), T_{A2} is a profile shear stiffness (x_2 is bending neutral axis).

$$w_{(x_1)} = \frac{-F}{6 \cdot T_{D2}} \cdot x_1^3 + \frac{F \cdot L - M_Z}{2 \cdot T_{D2}} \cdot x_1^2 + \frac{F}{\kappa \cdot T_{A2}} \cdot x_1 \tag{1}$$

The bending stiffness T_{D2} and shear stiffness T_{A2} are often evaluated as a sum of every layer stiffness as it is shown in Eq. (2) and Eq. (3), where A is profile cross section area, G_i is shear modulus of i-th layer, E_i is elastic modulus of i-th layer, n is number of layer, A_i is area of i-th layer, J_i is inertia moment of i-th layer and κ is shear revise coefficient.

$$T_{A2} = \int_{(A)} \mathrm{G}\, dA = \sum_i^n G_i \cdot A_i \tag{2}$$

$$T_{D2} = \int_{(A)} \mathrm{E} \cdot {x_3}^2\, dA = \sum_i^n E_i \cdot J_i \tag{3}$$

Direct approach to revise shear stiffness

The big advantage of the direct approach to revise shear stiffness is simple usage, especially for rectangular profiles. Explicit evaluation of the shear revise coefficient is not necessary, because the goal is revise shear stiffness, not the coefficient. Evaluation of shear revise coefficient is difficult for composite materials in general. Raman [3] express the shear revise coefficient as function of many variables (e.g. number of layer, orientation of layer, type of fibres, etc.) but it isn't applicable in analytical methods.

This analytical method is based on energy equivalence between transversal force deformation energy and a shear stress deformation energy as it is shown in Eq. (4).

$$\frac{1}{2} \int_{(l)} \frac{F^2}{\kappa \cdot G \cdot A} dx = \frac{1}{2} \int_{(V)} \frac{\tau^2}{G} dV \tag{4}$$

Where, A is profile cross area, F is transversal force, l is beam length, V is beam volume and τ is shear stress. Revise shear stiffness $T_{A2\,direct}$ is shown in Eq. (5).

$$T_{A2\,direct} = \kappa \cdot G \cdot A \tag{5}$$

The shear stress is determined from Fig.3 by equilibrium equation into Eq. (6).

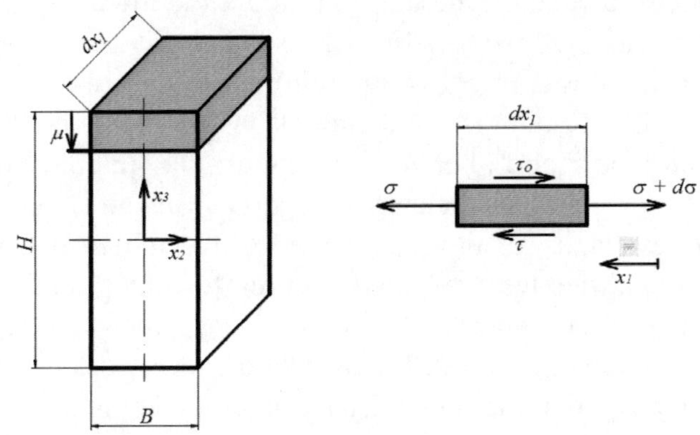

Fig. 4: Element to derive shear stress in cross section

Equilibrium equation is also known as Zhuravskii formulae where σ is axial stress.

$$\tau_{(\mu)} = \frac{d\sigma}{dx_1} \cdot \mu + \tau_o \tag{6}$$

Combine Eq. (6) with Schwedler equation, coordinates transformation and equilibrium equation for the axial stress is obtained Eq. (7) for a prismatic beam with a constant cross sections parameters through the length. Where H is height of web and τ_o is shear stress boundary condition.

$$\tau_{(\mu)} = \frac{\mathrm{F}}{T_{D2}} \cdot E_i \cdot \frac{H-\mu}{2} \cdot \mu + \tau_o \tag{7}$$

The boundary condition on top and bottom flange surface is obviously $\tau = 0$. Because of this fact, we can neglect deformation energy in flange.

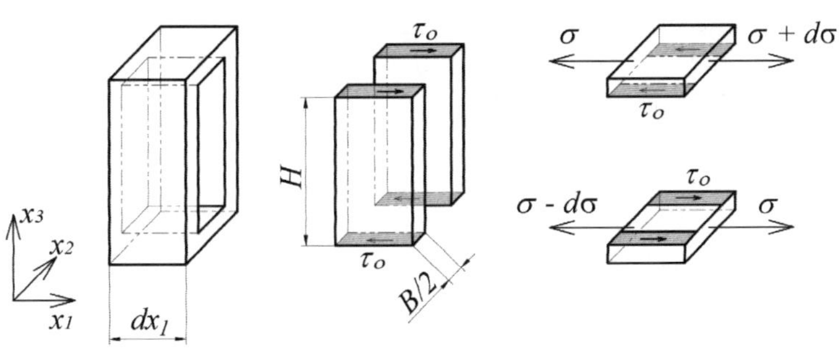

Fig. 5: Element to derive shear stress boundary condition

Shear stress τ_o is the boundary condition between the flange and web, which can be easily evaluated from the axial force in flange F_{axial} , where A_{fl} is flange area, σ_i axial stress of i-th layer, $B/2$ is thickness of web, σ_{mi} average axial stress of i-th layer and A_{if} is area of i-th layer in flange.

$$\tau_o = \frac{F_{axial}}{B \cdot dx_1} = \frac{1}{B \cdot dx_1} \cdot \int_{(A_{fl})} \sigma_i dA = \frac{1}{B \cdot dx_1} \cdot \sum_i^n \sigma_{mi} \cdot A_{if} \tag{6}$$

Three point bending test configurations

The verification of new calculation method required a modified experiment setup, where the beam deformation was captured using a line of displacement sensors at the beam top surface. As the line of displacement was in collision with the actuator, the actuator was moved to the bottom of beam. A modified actuator connection is based on a glued "U" shape clamp and the actuator is placed under the support beam with a roller. The first modified test was done using 600mm span between the supports for easy comparison of the standard 3-point bending test results. But from the results, it was obvious that a huge local deformation in the support areas occurred. The second modified test was done with reinforcement in the areas around the beam supports. Because the coupon length 1000mm, the span between supports was changed to 940mm (no cutting of coupon required by coupon owner). Because of the span change, the results are not directly comparable, but all data are easy comparable with analytical model prediction.

A standard three-point bending configuration was tested using the force actuator on the top surface between the supports of the beam, see Fig. 6 The results look good and compliance 2,53e-02 mm/kN was evaluated.

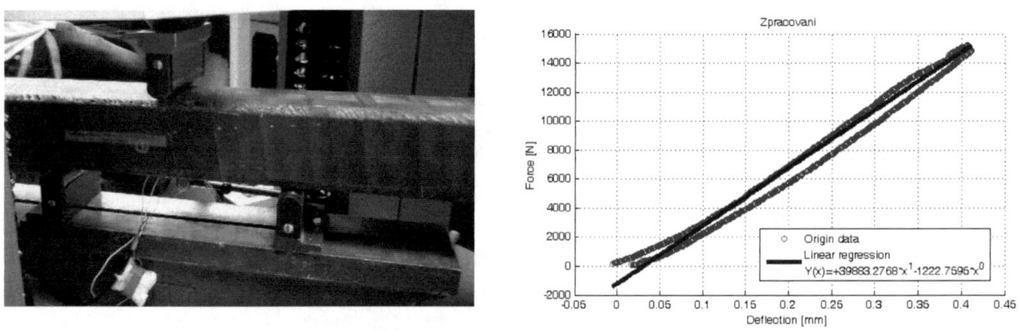

Fig. 6: Standard bending test (600mm between supports)

The three point bending test in the modified configuration was motivated by the need to measure deflection through the whole beam length, including local deformations. The assumption was that the top surface deformation will be the similar as the deformation of the bottom surface or the deformation in the beam middle plane. This assumption could be fulfilled just if the local beam compliance below the loading force actuator and around the supports are negligible. More detailed measurement was done, where the beam deformation was captured using a line of displacement sensors at the beam top surface, see Fig. 7.

Fig. 7: Modified bending test with top surface deflection measurement

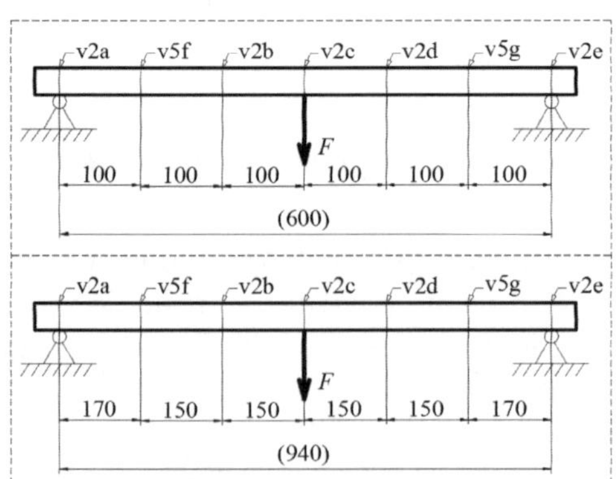

Fig. 8: Test setups

Due to the local beam compliance in supports area, the results of the first modified test were not useable. The effect was studied more deeply by a final element model of the experiment. For the second modification, the coupon support areas were modified, using bonded aluminium inserts, see Fig. 9.

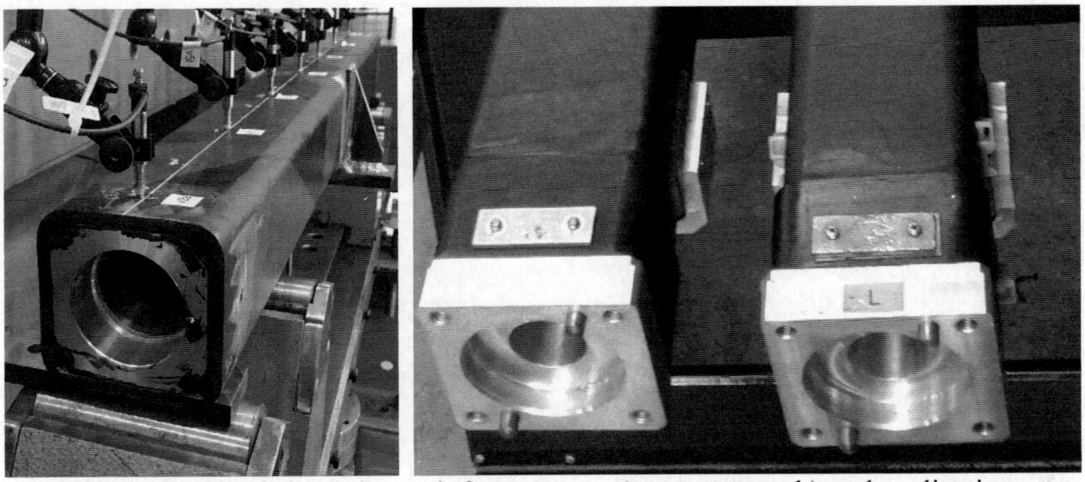

Fig. 9: Detail of aluminium reinforcement – a) test coupon, b) real application

Experimental study was done 3-times for each setup and the captured deformations are shown in Fig. 10. The deformation in supports area was greatly reduced for the second modified test. The assumption is, the captured deformation in supports area for second modified test is deformation of load and measure frame.

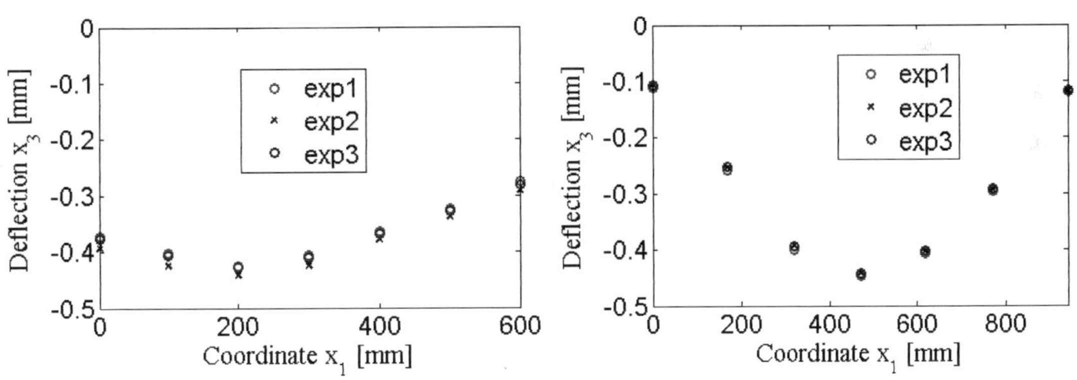

Fig. 10 Raw experimental data for 15kN – a) without end reinforcement, b) with end reinforcement

Results

The comparison of the beam deformation on the top/bottom/middle plane is shown in Fig. 11, both for the modified test without and with ending reinforcements; these data were taken from the finite element analysis, which was performed in ANSYS software.

The effect of local compliance is clearly visible. However, the comparison of the FEA and experimental results demonstrate that the compliance of the real experiment was higher than of the model. During the standard testing, the local deformations were about 40-50% of the total deformation. For the second modified testing, the local deformation was reduced to 5%.

The experimental results were verified by finite element simulations of the tests in the both configurations. The results of the modified test provides reliable for the verification of the developing analytic tools for the stiffness prediction of rigid composite beams, which was the original goal of the testing, see Table 2.

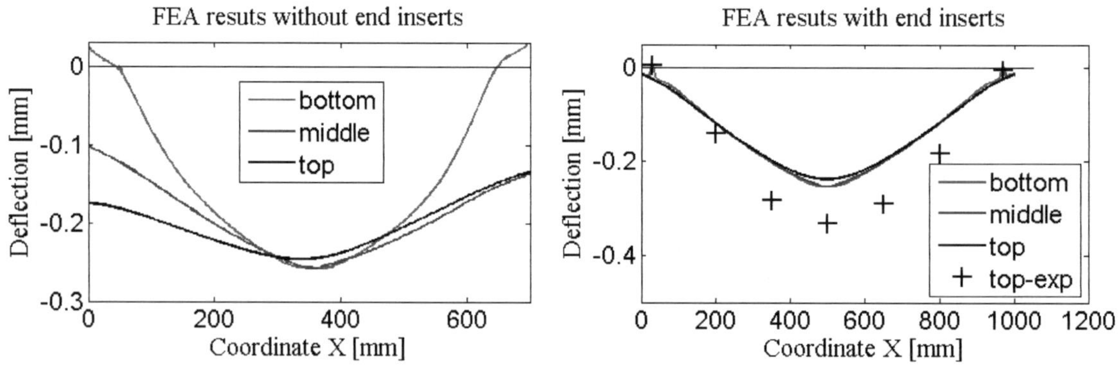

Fig. 11: Bending test FEA results for 15kN – a) without end inserts, b) with end inserts

Direct approach to revise shear stiffness was verified by a modified experiment. This evaluation of shear stiffness is suitable for hybrid carbon profile analytical tool. Deviation -12% in deformation is really good accordance of analytical model with experiment.

Table 2: Comparison of analytical methods with experiment (load 15kN)

Method	Support span [mm]	Deflection [mm]	Compliance [mm/kN]	Δ [%]
Original experiment	600	0.38	2.53e-02	0
Modified experiment	600	0.08	5.33e-03	-79
FEA modified exp. midline	600	0.14	9.33e-03	-63
Bernoulli	600	0.06	4.00e-03	-84
Direct shear model	600	0.10	6.67e-03	-74
Modified experiment v2	940	0.33	2.20e-02	0
FEA modified exp. v2	940	0.25	1.67e-02	-24
Bernoulli	940	0.22	1.47e-02	-33
Direct shear model	940	0.29	1.93e-02	-12

Conclusions

Direct approach to revise shear stiffness has proved as a suitable method for evaluation of shear stiffness of hybrid carbon beam profile. Deviation -12% from modified experiment is evidence of this method usability. The biggest advantage of direct approach to revise shear stiffness is not necessary evaluation of shear revise coefficient, which isn't simple to evaluate for hybrid carbon profiles and coefficient for isotropic material haven't fulfilled derivation condition.

Acknowledgements

The authors would like to thank the Technology Agency of the Czech Republic for supporting this research with project no. TA02010543. Authors would also like to thank the company CompoTech Plus, s.r.o for manufacturing of the specimens.

References

[1] Z. Pošvář, M. Růžička, V. Kulíšek, T. Mareš, K. Doubrava, O. Uher, Design of composite hybrid structures “on request”, Materials Today: Proceedings 3, pages 1129 – 1134, 2015.

[2] T. Mareš, L. Jiran, Odborná zpráva TAČR TA2-0543 V-12-060 Příloha II: Timoshenko beam theory for composite beams, the first approximation of the effect of shear stress. Praha: České vysoké učení technické, 2012.

[3] P. M. Raman, J. F. Davalos, Static shear correction factor for laminated rectangular beams. Elsevier [online]. Great Britain 1996, str. 285-293, ISSN 1359-8368/96

[4] G. R. Cowper, The Shear Coefficient in Timoshenko's Beam Theory. Journal of Applied Mechanics. Červen 1966, roč. 33, čís. 2, s. 335-340. DOI:10.1115/1.3625046

EAN 2019
57th conference on experimental stress analysis

Mechanical properties of cement pastes with fine old concrete, influence of different type of approach mixing

PROŠEK Zdeněk[1,a], TESÁREK Pavel[1,b], TREJBAL Jan[1,c] and PLACHÝ Tomáš[1,d]

[1]Faculty of Civil Engineering, Czech Technical University in Prague, Thákurova 7, 166 29 Prague, Czech Republic

[a]zdenek.prosek@fsv.cvut.cz, [b]tesarek@fsv.cvut.cz, [c]jan.trejbal@fsv.cvut.cz, [d]tomas.plachy@fsv.cvut.cz

Keywords: Recycled concrete, Cement composites, Mechanical properties, Approach mixing

Abstract. The article focuses on the impact of different type of approach mixing on resulting mechanical properties of cement composites. We investigate of influence speed of mixing of fresh mixtures and process of mixing the individual components of the cement composite. The mechanical properties were investigated by using non-destructive resonance method and destructive method on samples with dimensions equal to 40 × 40 × 160 mm. Among the monitored mechanical parameters were dynamic modulus of elasticity, dynamic shear modulus, flexural strength and compressive strength.

Introduction

The amount of waste old concrete is growing year after year and the need for its efficient use with it [1, 2]. Nowadays recycled concrete is used as coarse and fine aggregate in new concrete [3, 4]. The properties of the resulting composite are not only directly dependent on the grain shape which is in the case of recycled concrete is not suitable but also on the size of the transition zone. These effects have a negative effect on the resulting properties [5]. Recently, there have been studies in which the ITZ can be partially eliminated using the correct mixing approach, i.e., the correct procedure of mixing of the components of the cement composite and the mixing speed [6].

Unlike mentioned researches which they focus on recycled coarse aggregate, we focused on mechanical activated old concrete. This is fine aggregate with modified grain shape by using high-speed milling process [7]. The micronizing of the old cement paste into fine particle results in uncovered unhydrated clinker minerals, which can further hydrate upon contact with water [8].

Materials and Samples

Cement pastes composed of 50 wt. % Portland cement CEM I 42.5R from Radotin and 50 wt. % fine recycled old concrete has been prepared for investigating the impact of different mixing approach. Fine recycled old concrete was obtained from an old construction concrete (sleepers and columns) and micronized by high-speed milling process. Water ratio was equal to 0.35. The difference between individual samples was in the approach of mixing fresh mixtures. Two different components were mixed at 132 rpm for 240 seconds in the first step of the mixture preparation. In the second step, the remaining third component was added and the mixtures were mixed for 240 seconds at two different speeds, namely 132 rpm and 421 rpm. Approach of mixing is shown in Table 1.

Table 1: Approach of mixing fresh mixtures

	The first step of mixing		The second step of mixing	
Set	Components	Speed [rpm]	Added component	Speed [rpm]
RW+C_S	Recycled old concrete and Water	132	Cement	132
RW+C_F	Recycled old concrete and Water	132	Cement	421
CR+W_S	Cement and Recycled old concrete	132	Water	132
CR+W_F	Cement and Recycled old concrete	132	Water	421
CW+R_S	Cement and Water	132	Recycled old concrete	132
CW+R_F	Cement and Water	132	Recycled old concrete	421

Experimental Methods

Investigated mechanical parameters were developments of dynamic modulus of elasticity and dynamic shear modulus and values of flexural strength and compressive strength. The development of the dynamic modulus of elasticity and dynamic shear modulus was measured non-destructively in the first 28 days after the start of cement hydration. The Flexural and compressive strength values were measured destructively on 28 day old samples.

Measurement of the dynamic modulus of elasticity was carried out be using resonance method, namely the measurement line from Brüel & Kjær. The measurement line was consisting of a measuring unit type 3560-B-X12 impact hammer type 8206, an acceleration sensor type 4519-003 53422 and software PULSE LabShop. The dynamic modulus of elasticity was achieved by using the basic longitudinal natural frequency and basic flexural natural frequency of the samples. The dynamic shear modulus was achieved by using the basic torsional natural frequency of the samples.

The flexural and compressive strength were determined be using destructive method on the device FHF Strassentest. The flexural strength was determined by a three-point bending test. The testing of flexural strength was displacement controlled at a constant rate of 1 mm/min. The distance between supports for three-point bending test was equal to 100 mm. The compressive strength was determined by using uniaxial compressive test. The uniaxial compressive test was performed on the broken halves of the samples after bending test with effective dimensions of 40 × 40 × 80 mm. The testing of compressive strength was displacement controlled at a constant rate of 5 mm/min.

Results and Discussion

Results of non-destructive testing are shown on Fig. 1. The results show the positive effect of the mixing speed of the fresh mixture on the dynamic modulus of elasticity and the dynamic shear modulus. Differences between the highest and lowest value modules were between samples CR+W_S and samples CR+W_F, namely 0.5 GPa for dynamic modulus of elasticity and 0.3 GPa for dynamic shear module. The highest increase of dynamic modulus of elasticity and dynamic shear modulus is in the first 10 days of sample age. Furthermore, the increase in values is gradual in the next 18 days.

Flexural strength was approximately same value on all testing samples (Fig. 2). Differences were up to size of the value standard deviation. Value of flexural strength was 4.5 MPa. In contrast of it, the compressive strength values (Fig. 2) were different and they have same trend of value of dynamic modulus of elasticity. The greatest value of compressive strength had RW C_F, namely 40.8 ± 2.8 MPa and lowest value of compressive strength had

CW+R_F, namely 28.1 ± 1.1 MPa. Value of standard deviation was approximately 5 % of value compressive strength in all tested case.

The results show that the mixing speed has a positive effect on the resulting compressive strength, with a difference of approximately 5 MPa for slow and fast mixing of RW + C and 4 MPa for slow and fast mixing of CR + W. This is the effect of better homogenization of the fresh mixture by fast mixing. The mixture RW+C (the old concrete was first mixed with water and then with the cement) had the highest compressive strength values, namely 5 MPa higher than the mix where the first old concrete was mixed with the cement. The mixture where the cement was first mixed with water had the lowest compressive strength by about 5 MPa lower than in the case of cement first mixing with the old concrete and 10 MPa lower than the mixture where the recycle was first mixed with water. The process of mixing the individual components of the cement composite affects the ITZ zone between the cement matrix and the recycle grains.

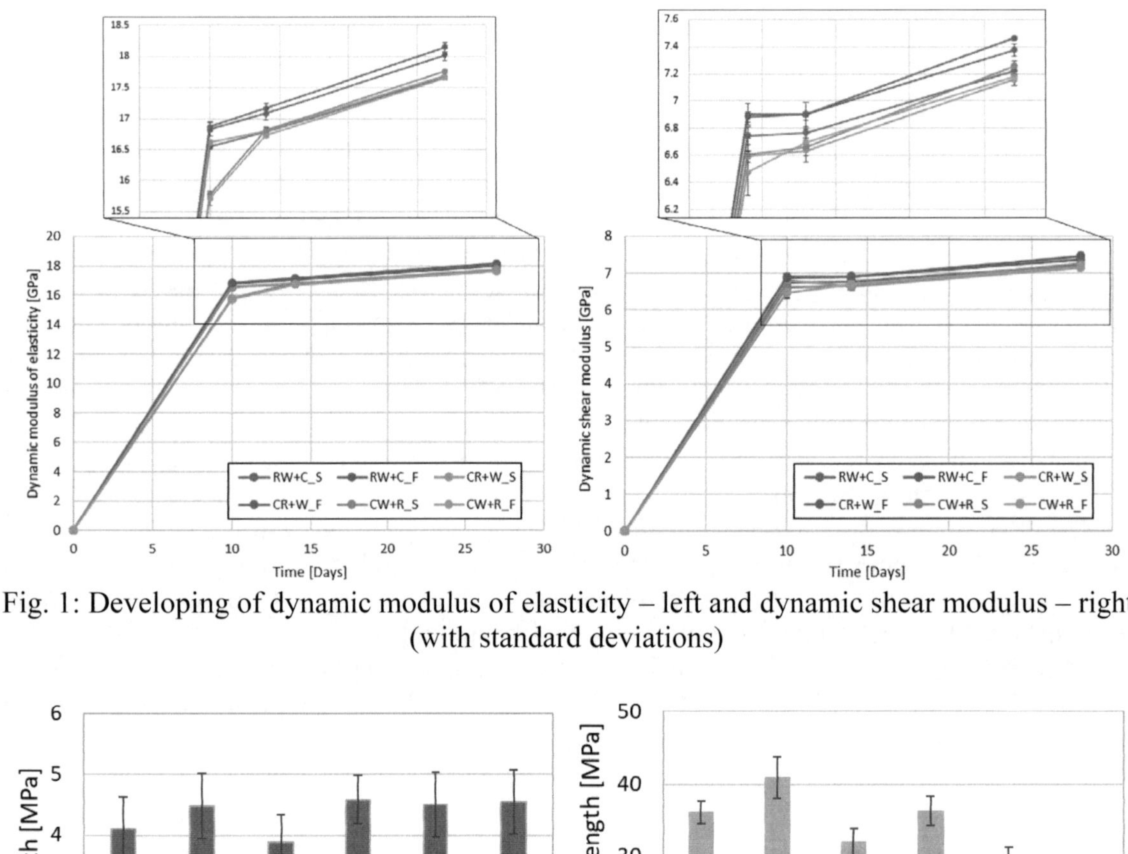

Fig. 1: Developing of dynamic modulus of elasticity – left and dynamic shear modulus – right (with standard deviations)

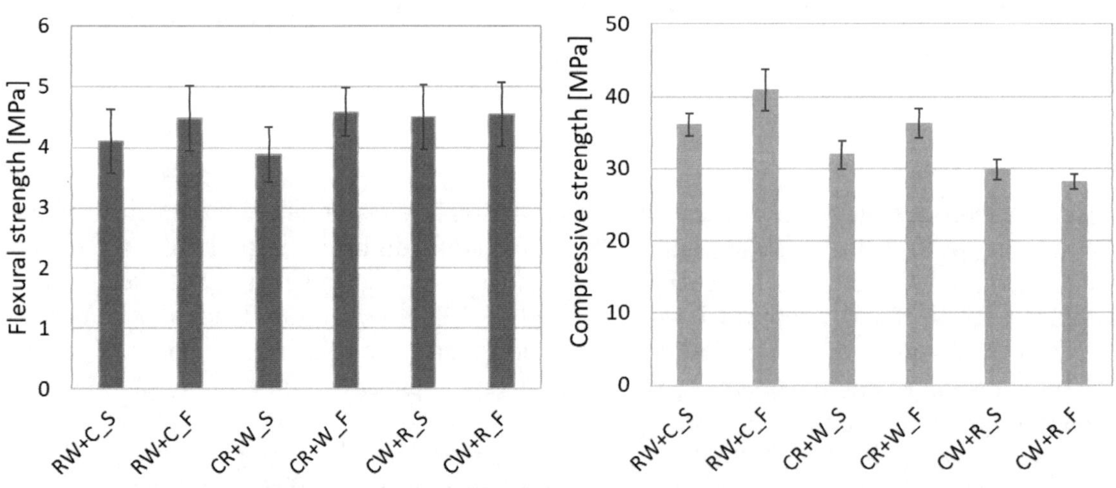

Fig. 2: Value of flexural strength – left and compressive strength – right (with standard deviations)

Conclusions

This work was focused on the influence of different type of approach mixing on mechanical properties of cement composites. The researched samples were composed of Portland cement and micronized recycled old concrete. The difference between individual samples was in the approach of mixing fresh mixtures. Based on the results, it can be concluded that results of the work are in agreement with the statement of V.W.Y. Tam et al. [6]., because the size of ITZ zone depends on the approach of mixing not only in recycled coarse aggregate but also in micronized recycled old concrete.

In the future, the research will focus on confirming influence on approach of mixing on ITZ zone by using scanning electron microscope with X-ray microanalysis, namely an Energy Spectrometer.

Acknowledgement

This paper was financially supported by the GA CR research a project under the number 17-06771S and by Czech Technical University in Prague under No. SGS project SGS19/148/OHK1/3T/11. The authors also thank company LAVARIS Ltd, where a high-speed mill from was used for micronizing of the recycled old concrete.

References

[1] L.P. Chun, D.E. Scorpio, C.J. Kibert, Strategies for successful construction and demolition waste recycling operations, Journal of Construction Management and Economics, 15 (1997) 49–58.

[2] V. Corinaldesi et al., Behaviour of cementious mortars containing different kinds of recycled aggregate, Construction and Building Materials 23 (2009) 289–294.

[3] J. Xiao, W. Li, Y. Fan, X. Huang, Lemaitre, An overview of study on recycled aggregate concrete in China (1996–2011), Construction and Building Materials 31 (2012) 364–383.

[4] K. Šeps, J. Vodička J, Fibre Reinforced Concrete with Recycled Concrete and Insulation Material STERED, Applied Mechanics and Materials 1054 (2014) 262–266.

[5] A.K.H. Kwan, C.F. Mora, H.C. Chan, Particle shape analysis of coarse aggregate using digital image processing, Cement and Concrete Research 29 (1999) 1403–1410.

[6] V.W.Y. Tam, X.F. Gao, C.M. Tam, Microstructural analysis of recycled concrete, produced from two-stage mixing approach, Cement and Concrete Research 35 (2005) 1195–1203.

[7] Z. Prošek, P. Tesárek, V. Hrbek, Characterization of old concrete for new use: Grid nanoindentation and scanning electron microscopy, NANOCON 2018 - Conference Proceedings, 10th Anniversary International Conference on Nanomaterials - Research and Application (2019) 736–741.

[8] J. Hrůza et al., Mechanical properties of the cement paste with fine ground recycled concrete depending on the microstructure, Advanced Materials Research 1144 (2017) 22–27.

Forces caused by vacuum spring bellows during pumping of the quadrupole magnet module

RACHAČ Marek[1,a], DUB Martin[2,b], LOPOT František[2,3,c] ,HOŠEK Jan[1,d] and GALIMOV Artem[4,c]

[1]CTU in Prague, Faculty of mechanical engineering, Department of Instrumentation and Control Engineering, Technická 4, 166 07, Prague, Czech Republic

[2]CTU in Prague, Faculty of mechanical engineering, Department of designing and machine components, Technická 4, 166 07, Prague, Czech Republic

[3]Charles University, Faculty of sports and physical education, Dep. of anatomy and biomechanics, Lab. of extrem loading, José Martího 31, 162 52, Prague, Czech Republic

[4]Joint Institute for Nuclear Research, Veksler and Baldin Laboratory of High Energy Physics Dubna, Russia

[a]marek.rachac@fs.cvut.cz, [b]martin.dub@fs.cvut.cz, [c]frantisek.lopot@fs.cvut.cz, [d]jan.hosek@fs.cvut.cz, [e]galimov@jinr.ru

Keywords: Vacuum spring bellow, NICA booster synchrotron, Beam pipe

Abstract. The paper is focused on effect of forces induced by vacuum bellows due to differential pressure inside and outside of the bellows. This phenomenon can develop stress into rest of the vacuum system. In case of the beam pipe in NICA booster synchrotron bellow forces developed bending moment and there is a risk of the beam pipe destruction. The problem is analysed and FEA analysis is prepared. Due to unknown bellow stiffness between beam pipe modules in FEA analysis the experimental measurement is necessary.

Introduction

Spring bellows are standard parts used in vacuum assemblies. The bellows are often used for vibration decoupling, for example, between vacuum pump and measuring instrument, they can be used as compensators to balance thermal expansion and mounting tolerances and they also can serve as feedthroughs to introduce movements into the vacuum or to separate the vacuum chamber from mechanical parts.

The paper is focused on effect of forces induced by vacuum bellows used for connecting of the beam pipe with the UHV pumping system in quadrupole magnet modules in NICA booster.

NICA (Nuclotron-based Ion Collider fAcility) is a new accelerator complex designed at the Joint Institute for Nuclear Research (JINR) to study properties of dense baryonic matter. The accelerator tube is assembled from similar modules connected inside a big vacuum chamber where fore vacuum helped to reach high vacuum in the beam pipe. All inner parts are at temperature of liquid nitrogen (approximately 70K). The serial assembly and testing of NICA booster magnets started at special facility of the Laboratory of High Energy Physics. It is necessary to assemble and test 48 quadrupole magnets for NICA booster synchrotron. [1]

The quadrupole magnet module design and details of the analysed beam pipe fixed in the lens are shown in the Fig. 1.

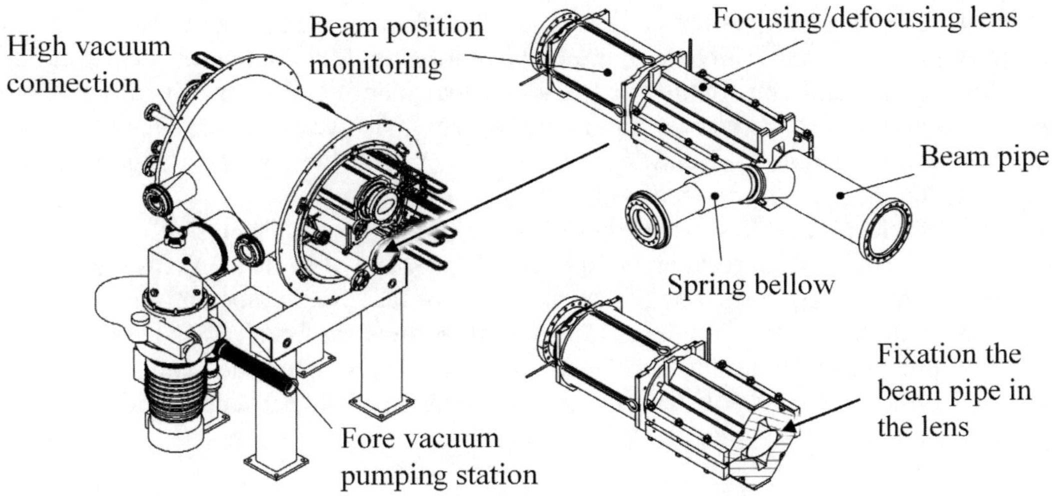

Fig. 1 - The quadrupole magnet module with the beam pipe

Because each accelerator is a source of potentially dangerous ionizing radiation it is necessary to reduce the risk of its failure. Therefore, virtual tests and real tests are applied to all components of the quadrupole magnet module mechanical structure.

Problem - spring bellow force

The important constrain of the vacuum system design is the fact, that the spring bellows are getting shorten during pumping due to presence of external overpressure. This phenomenon can cause stress into rest of the vacuum system.

We took into account two main modes of pressure load during using the quadrupole magnet module in NICA booster. Each part of the module is at the atmospheric pressure – mode 1 or all parts are under certain level of vacuum – mode 2, presented in Fig. 2. In these both cases, there is no problem caused by bellow deformation.

But during commissioning there is used another case of load a mode 3 where a high vacuum is present in the beam pipe and whereas the atmospheric pressure is in the accelerator tube. We want to prevent of extreme force rise from deformed bellow and risk of the beam pipe destruction due to developed bending moment.

To prevent destruction of the beam pipe the FEM analysis of presented load mode is prepared.

Operating mode	p_1	p_2	$F = A\Delta p$
1	p_{vac1}	p_{vac2}	*negligible*
2	p_{atm}	p_{atm}	*0*
3	p_{vac1}	p_{atm}	*critical*

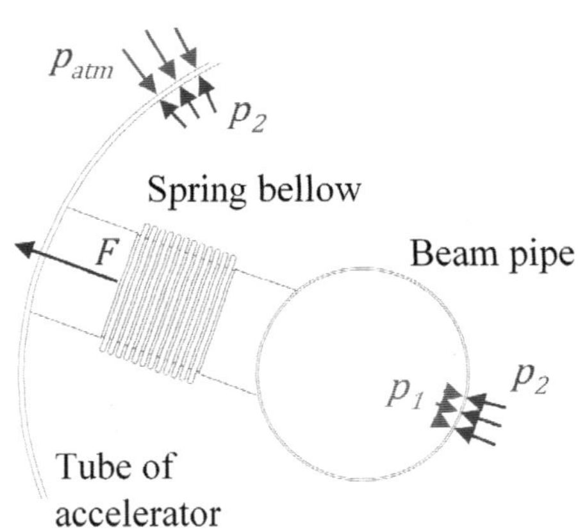

Fig. 2 – Operating modes

Virtual experiment - FEM analysis

The bellow force for the load mode 3 is evaluated from known boundary conditions (dimensions, pressures) and with help of the FEM simulation. Main outcome of calulations are deformations values and constrictions maps with clear identification of critical parts of the design. Calculations are done for two cases. The first case has free end at the site of the circular flange. The second case uses an elastic linkage at the site of the circular flange to simulate connection to another module by means of a bellow.

Used model characteristics. The problem is modeled by means of shell elements with appropriate thicknesses. Both rectangular and triangular shape of finite elements are used.

The model supports are placed on the surface lines of the elliptic tube and modeled as line supports where all the deflections are restrained and rotations are free.

For the second case, the introduced model is supplemented by an elastic linkage with defined stiffness in Z direction to illustrate the impact of connection of the particular modules (Fig. 3).

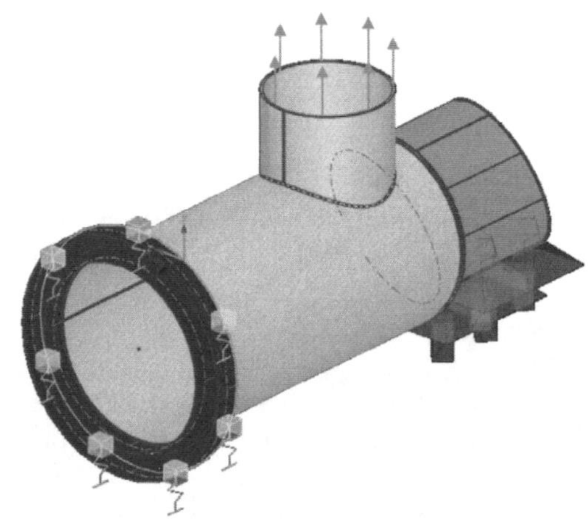

Fig. 3 - Line support with defined stiffness in Z direction

The loading force is distributed to the top circumference of the tube connecting the vacuum pump and modeled as the line load.

The computation was performed with considering geometrically linear analysis. The material is considered linear elastic with modulus of elasticity E = 190000 MPa, Poisson´s ratio ν = 0.27 and yield strength 172 MPa (not considered in computations, but used for results interpretation).

Results. FEM analysis showed the critical stress exceeds shear stress limit of the used material by more than 1.36-times for the first case with free end at the site of the circular flange (Fig. 4).

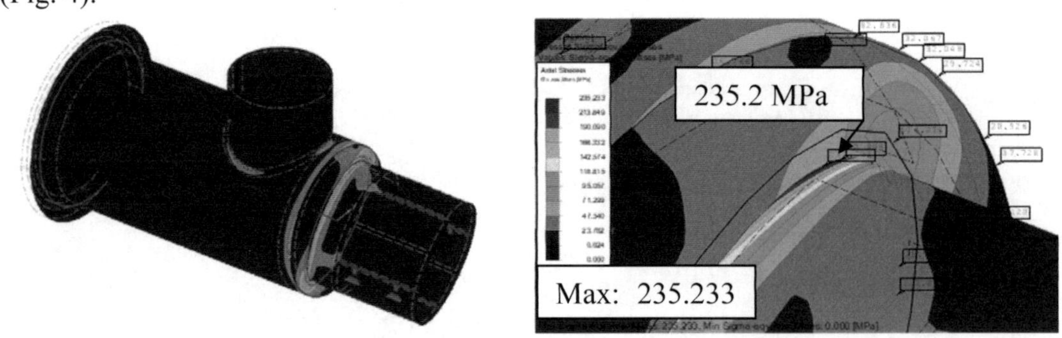

Fig. 4 - von Mises stress - 1st case

For the second case with an elastic linkage simulation indicated that all deformations can be almost neglected and maximum von Mises stress is 63.8 MPa (Fig. 5).

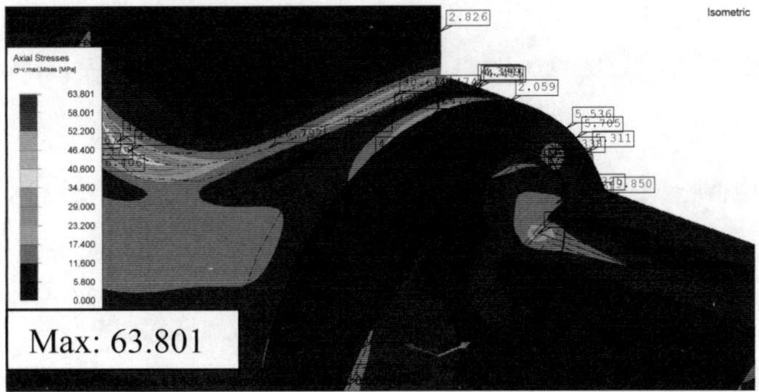

Fig. 5 - von Mises stress - 2nd case

The real stiffness of the elastic linkage used in simulation is unknow paremetr. Due to that it is neccessary to prepare a physical experiment for the result confirmation.

Real experiment

We consider a strain gauge measurement but while we are interested in behaviour just at the level of maximum load we decided to assemble the beam pipe modules together and tested its behaviour under the force effects. It is showed that all assembly totally changed its behaviour compared to a single beam pipe. Influence of connection of the individual modules seems to be crucial. The physical experiment on assembled modules showed that stiffness of comlpete assembly is high enough for safety operation in accelerator.

Decision to test behaviour of comlpete assembly confirmed correctness of the vacuum system design and enable to continue faster in production of magnet modules for NICA booster synchrotron. It confirmed the virtual FEM experiment, allowed to check the beam pipe safety under real conditions, and saved money and time.

Conclusion

FEM analysis as a tool for virtual test can significantly help during vacuum system design phase, but its results are strongly dependent on correctness of boundary conditions. In this project a simple physical experiment helped to verify the beam pipe design safety. Based on the measured data we improved the boundary conditions for next use of the vacuum system FEM model.

Acknowledgement

The project and the paper were supported by grant no. SGS17/176/OHK2/3T/12.

References

[1] Shemchuk, A.V., Borisov, V., Bychkov, A. et al. Phys. Part. Nuclei Lett. (2018) 15: 873. https://doi.org/10.1134/S1547477118070658

Possibilities of airframe structures cost saving through reducing number of fasteners

RŮŽEK Roman[1,a], DOUBRAVA Radek[1,b], KRUSE-STRACK Thomas[2,c] and KOERWIEN Thomas[3,d]

[1]Czech Aerospace Research Centre, Strength of Structure Department; Beranových 130, 199 05 Praha - Letňany, Czech Republic

[2] Airbus Operations GmbH, Airframe Research & Technology, Kreetslag 10 21129 Hamburg Germany

[3] Airbus Defence and Space GmbH, Materials & Processes, Rechliner Str., 85077 Manching, Germany

[a]ruzek@vzlu.cz, [b]doubrava@vzlu.cz, [c]Thomas.Kruse-Strack@airbus.com, [d] Thomas.Koerwien@airbus.com

Keywords: Key, Word, Paper, Best conference, Contribution template (maximum of 5)

Abstract. Reduction of numbers of mechanical fasteners together with bonding technology application can be one of the possibilities how to decrease costs of the structure. In the paper, results based on fatigue testing of novel design approaches for damage tolerant high load transfer (HLT) joints, as e.g. panel joints, are discussed for carbon fibre reinforced plastics (CFRP) adherents. Fatigue experimental results and numerical simulations of Wide Single Lap Shear (WSLS) specimen are presented for different configurations to proof the crack stopping behaviour of "state of the art" fasteners as reference crack arrestor concept.

Introduction

One of the main aims of designers is application of cost-effective strategies. The composite structures can lower costs of an airframe significantly, but there are some obstacles to adopt of advanced technologies like adhesive bonding or welding and prevent to use full strength capabilities of these technologies to real structures. Nowadays, there is no fully certified bonded technology applicable to full-scale airframe structure which comply airworthiness regulations from viewpoint of damage tolerance criteria. Therefore, composite airframes have still a high number of fasteners. Reduction of numbers of mechanical fasteners together with bonding technology application can be one of the possibilities how to decrease costs of the structure. Unfortunately, adhesive bonded joints can lose their capacity without any previous announcement which can be detected by non-destructive inspections. This is the main reason why the application of adhesive bonding technology for primary aerospace structures is limited due to the certification regulations. State of the art is the widely used "chicken rivet" as crack arrestor which is limiting the benefits of bonding technology, particularly in composite bonded joints.

Taking into account airworthiness regulations the only AC20-107B [1] requirement "*The maximum disbonds of each bonded joint consistent with the capability to withstand the loads in paragraph (a)(3) of this section must be determined by analysis, tests, or both. Disbonds of each bonded joint greater than this must be prevented by design features*" can be practically applied for the sizing and certification of bonded joints.

The paper shows the way for decreasing numbers of fasteners in the primary airframe structures and at the same time the way how to keep the required level of reliability of the structure.

Validation methodology

The target of the presented work is to show methodology, how to demonstrate possibility to decrease number of fasters together with bonded joint application, in other word to demonstrate ability a secured crack stopping or arresting in adhesive bonded joint under fatigue loads in case of the presence of a local defect as e.g. a weak bond. In generally the airworthiness regulations allow both experimental and numerical demonstration of a structure. But in practice both methods are necessary.

Damage growth can be classified in the no-growth, slow-growth and arrested-growth criterion [1]. The no-growth criterion allows, as its name already suggests, no crack propagation within the bonded joint. Therefore, the load at which no crack occurs and / or it is stopped is significant for this criterion. On the other hand, the slow-growth criterion allows a certain crack propagation rate, which does not lead to a failure of the joint under the Limit Load (LL). The third criterion states, that a certain crack length is tolerable, which is capable of bearing LL. Yet, the crack growth needs to be arrested using either features or geometry alterations so that it does not fall under LL capability. All three criterions need to be investigated in test campaigns in generally. Figure 1 summarizes all three criterions according to the allowable residual static strength and damage detectability.

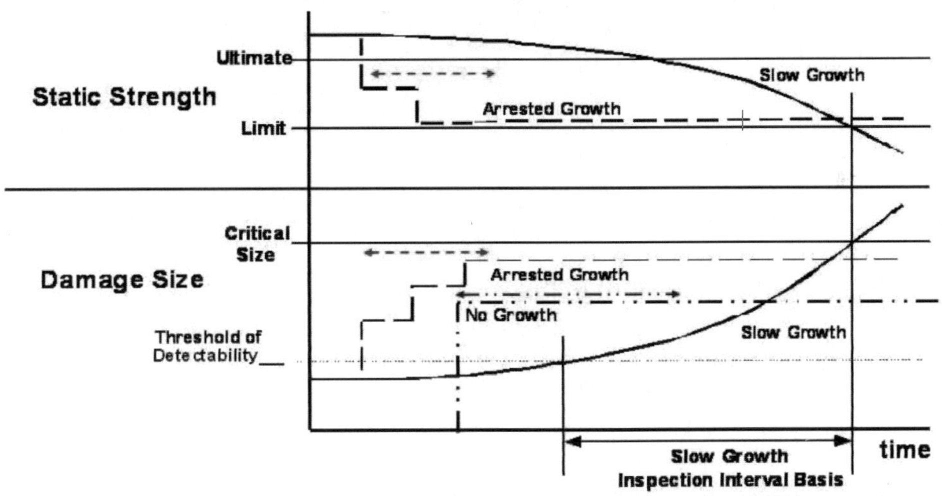

←··—··—··→	Acceptable "No-growth" Inspection Interval for Damage Shown
←- - - - - - -→	Acceptable "Arrested Growth" Inspection Interval for Damage Shown

Fig. 1: Schematic diagram of residual strength and damage size relationship for different approaches to composite structural damage tolerance substantiation [1]

The validation of the crack arresting principle can be divided into the two steps [2]. Firstly, the Cracked Lap Shear (CLS) test can be used for the evaluation of individual crack arresting capability of different design features. Under tension, it features a mixed mode load (in plane shear & peeling) at the bondline interface. Secondly, the wide single lap shear (WSLS) specimen can be used to demonstrate a more realistic application scenario. WSLS specimen has been developed and investigated within the BOPACS project [3]. It represents more realistic a typical high load transfer (HLT) configuration as e.g. a fuselage longitudinal joint.

The implementation of artificial disbonds and different disbond stopping features is part of the validation concept. The main idea is to fulfil the no-growth criteria according the certification requirements. This paper is focused on the crack arrest demonstration by using of WSLS panels.

The evaluation of the damage tolerance and crack arresting behaviour for HLT configurations of bonded joints is based on the expected principle behaviour as sketched in Fig. 2.

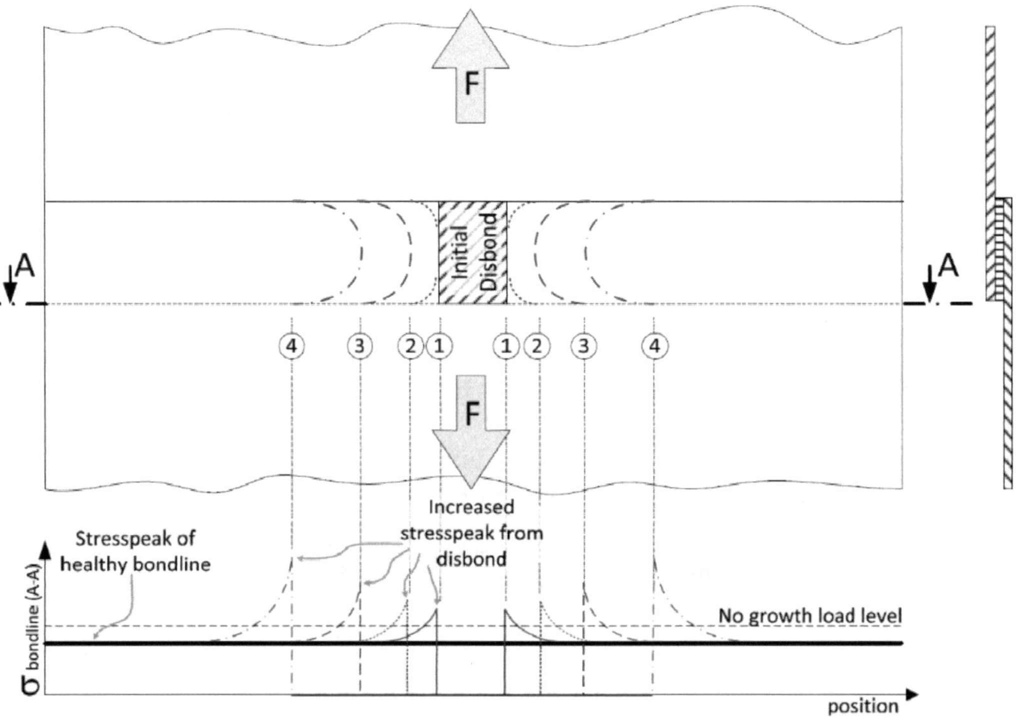

Fig. 2: Structure mechanic principle on high load transfer (HLT) joints

Presence of a local manufacturing defect (weak bond) or in-service damage (impact) in the bondline results in the load transfer interruption. This leads to an increase of the stress peak next to the damage (Figure 2, position 1). Depending on the initial disbond size, the stress peak will exceed the no-growth load level. Therefore, it is assumed that cyclic loading will lead to an accelerated crack growth with increasing stress concentrations at the edges of the disbond (Figure 2, positions 2 to 4). The target of the fasteners is to arrest or guarantee a safe, controlled disbond growth at loads below a limit load (typical fatigue spectrum) similar to the slow-growth criteria for metallic structures.

No-growth criterion

A no-growth criterion and the amount of fatigue cycles needed to represent a design life depends on the structure of the aircraft. Due to the probability of 10^{-5} for a LL event, 100.000 cycles can be representative for one design life of a fuselage seam, which the WSLS specimen stands for. Generally, a fuselage is exposed to loads, which occur due to thermal expansion, differences in pressure and bending moments. Thereby, one flight cycle can be regarded as one fatigue cycle. There are different approaches like the life-factor and load-factor to determine a fatigue life. A composite structure must demonstrate the ability to contain intrinsic manufacturing defects and the maximum allowable service damage(s) in adverse operational conditions and throughout the design life of the structure [4]. For practical reasons, the decision has been made to perform 500.000 cycles, which represent five design lives. This is sufficient to draw the first conclusions out of the testing and apply them to further investigations.

Experimental validation

WSLS specimens were used to investigate the influence of disbond stopping features (DSF) on crack growth. Therefore, it is necessary to have a relatively wide specimen in order to put several DSF in one line. This specimen resembles a longitudinal joint of an aircraft. To realize a homogeneous load distribution the specimen's dimensions were analysed by using a Finite Element Analysis (FEA). A WSLS specimen with a width of 500mm and an overlap of 60mm together with application of ASTM E647 standard clamping conditions has been selected as optimal solution.

The WSLS specimens were made out of Hexcel 8552 resin and IM7 uni-directional fibers, having a quasi-isotropic layout. The adherends were bonded together using an EA9695 film adhesive. Prior bonding, the surface of the adherends was treated with atmospheric plasma in order to activate the surface and to ensure a strong bond. The plasma treatment activation of the surface improves the surface energy leading to higher adhesion [5]. Fig. 3 shows an initial WSLS specimen scheme including its dimensions and components.

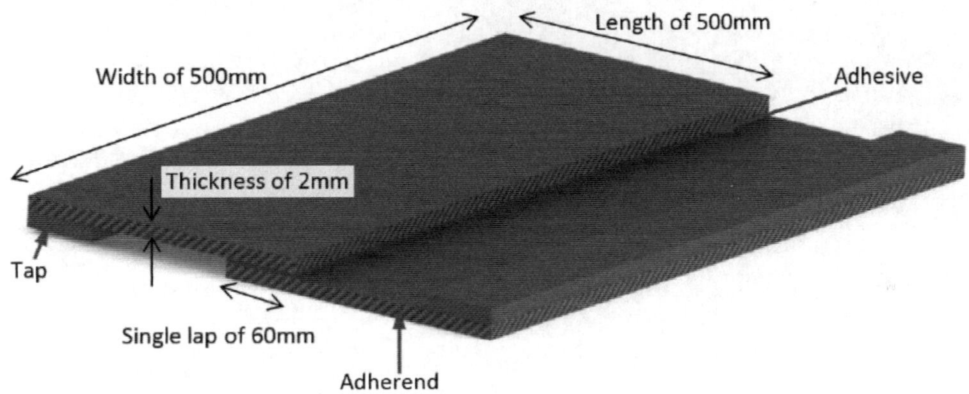

Fig. 3: Dimensions and components of WSLS specimen

The specimen was fixed on both ends, where the taps serve as a fixture for symmetrical clamping. By doing so, an axial misalignment between the direction of force and the adhesive layer is prevented which can result in bending moments. The WSLS were tested for tensile fatigue load. Hence, there occurs a mixed mode behaviour, which is schematically shown in Fig. 4. The specimen is exposed to Mode I, where the ends of the overlap get peeled off due to the occurring moment, as well as to Mode II.

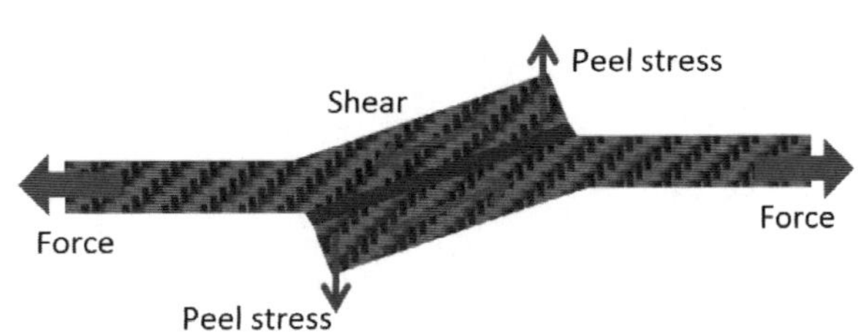

Fig. 4: SLS specimen under tensile load

Uniformity of strain distributions for WSLS specimen in longitudinal and transverse directions were investigated both by experimental and numerical tools. During the experimental investigations the ARAMIS optical contactless digital correlation system together with resistance strain gauges was used. Numerical simulations were performed by using of the NX NASTRAN [6] software. Fig. 5 shows measured uniform load distribution along to perpendicular direction to the applied tension load (see lines 1, 3 and 4 in Fig. 5). The non-uniform strain distribution in longitudinal direction (parallel direction with the applied tension load) in the surrounding of panel (adherends) bonded joint was defined (see line 2 in Fig. 5). The strain measurement and displacement data in longitudinal direction based on strain gauges measurements are shown in Fig. 6. Due to WSLS panel geometric asymmetry additional bending moment of 45% was measured near at the end parts of adherends close to outer sides of bondline. Bending moment induces peel stresses which can have a significant influence on the crack propagation inside of bonded joint of WSLS panel. This self-feature invoked by geometric asymmetry of the panel is not influenced by the gripping concept.

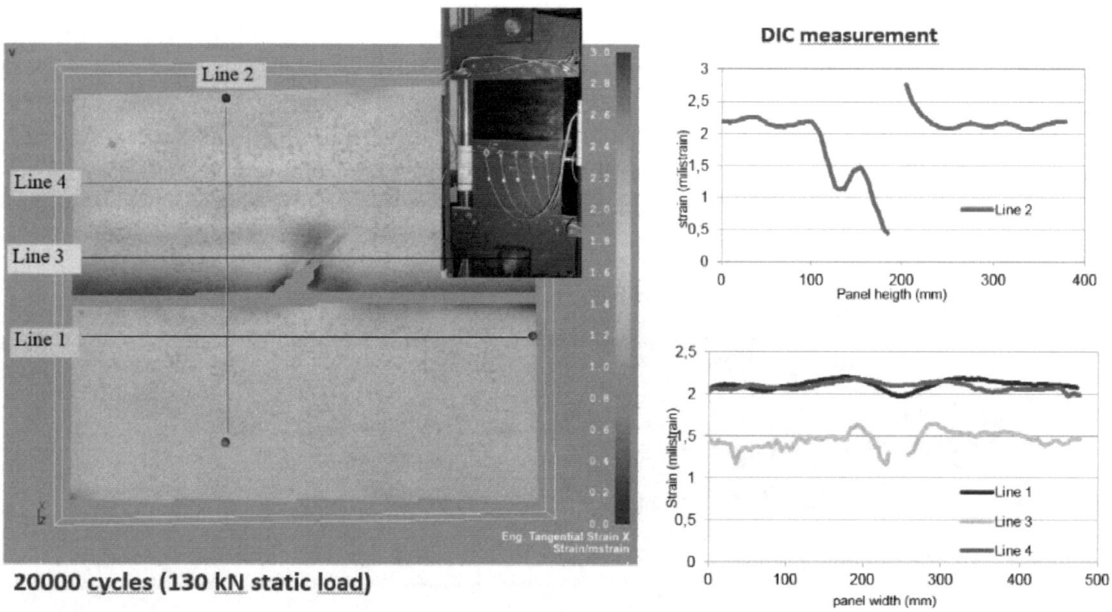

Fig. 5: WSLS panel - Optical measurement result in longitudinal and transverse direction

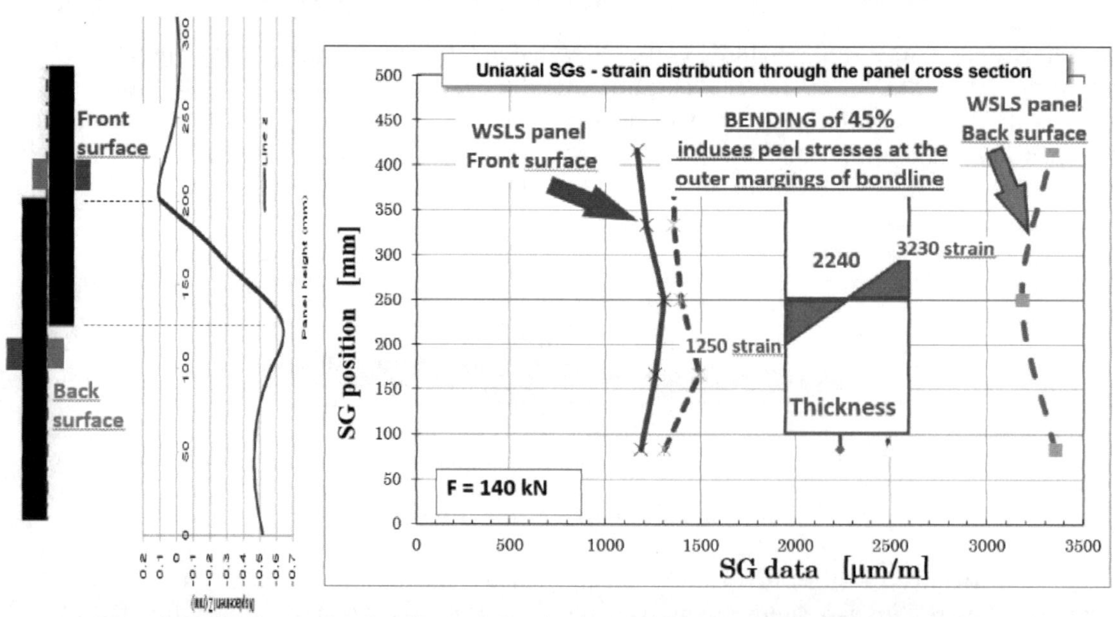

Fig. 6: WSLS panel – Strain and displacement data in longitudinal direction

The WSLS panel behaviour in bending was numerically evaluated for different disbond configurations. The FE results were compared with experimental results based on strain gauges measurement bonded on the panels. Nonlinear analysis (geometry) using the NX NASTRAN solver was used. 2D geometry model for skin and 3D model for adhesive layer was applied [7]. Fig. 7 illustrates the FE results of simulation of geometry eccentricity influence and low bending stiffness behaviour. Both FE analysis and experimental results prove the significant bending effect on strain development together with load level increasing. At the same time, the significant influence of the strain gauge location can be derived. Only a relatively small difference in strain measured (analysed) point can case a high change in strains. Results of peel and shear stress analyses for disbond length of 20 mm and 80 mm are shown in Fig. 8 and Fig.9. Based on additional analyses it was state that increasing disbond length results into decreasing peel and shear stresses. Shear stresses are more uniform with increasing disbond length.

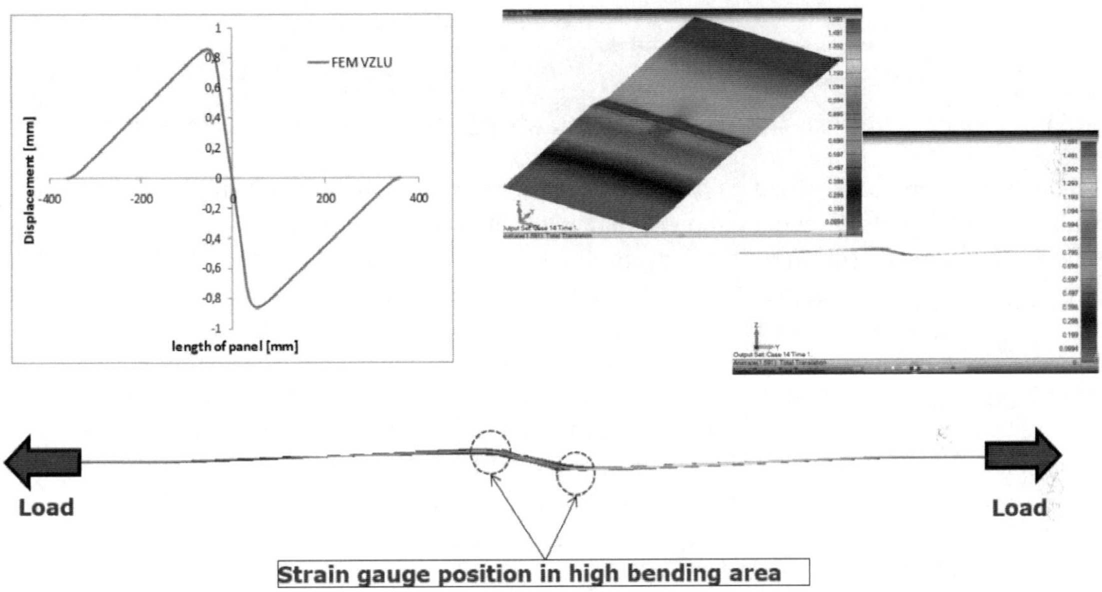

Fig. 7: WSLS numerical simulation – additional bending

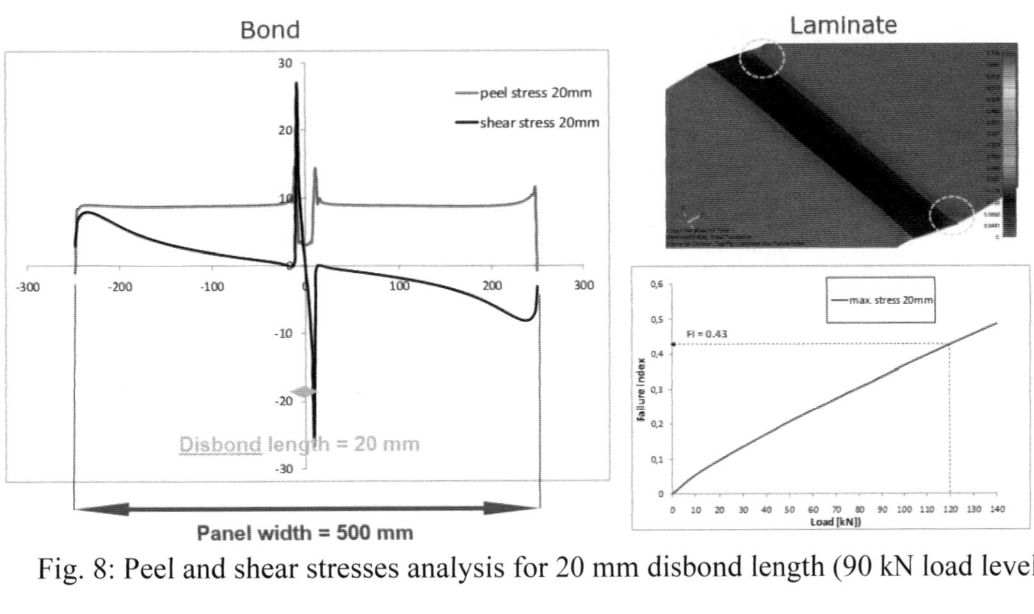

Fig. 8: Peel and shear stresses analysis for 20 mm disbond length (90 kN load level)

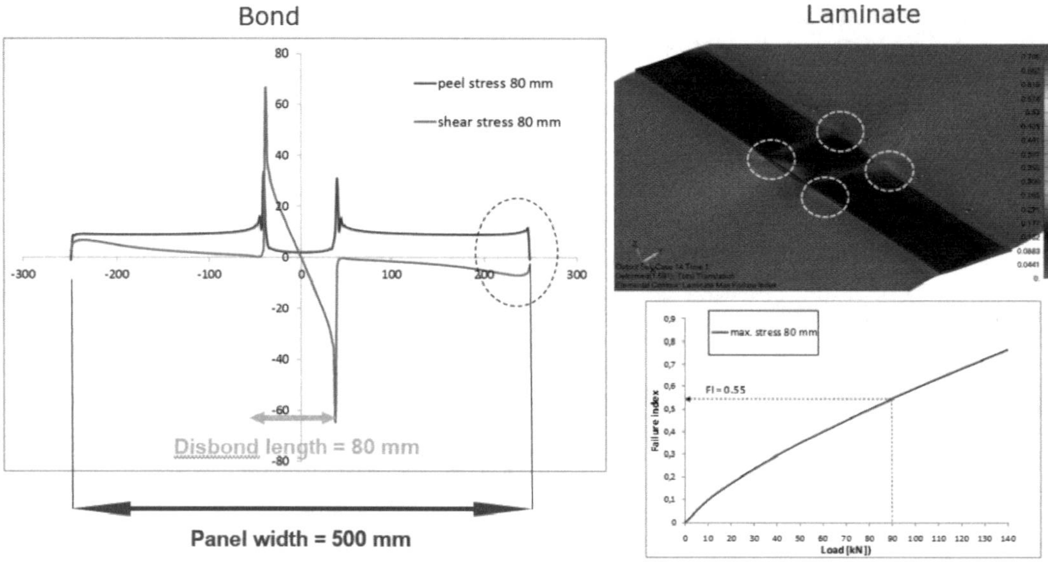

Fig. 9: Peel and shear stresses analysis for 80 mm disbond length (90 kN load level)

The most important phase of work was to identify the behaviour of the WSLS panels containing different disbond, impact damage, laminate lay-up near bondline and crack stopping features. During the verification phase of the disbond crack stopping features (DSF) efficiency different configurations of damage or disbond and DSFs were tested. All the tests were performed under fatigue with a sinusoidal load type. Different influence of DSF can be observed. Fig. 10 shows influence of crack/disbond stopping features (red line) to crack retardation mechanism under fatigue loading for initial disbond length of 80 mm. The DSF increases the lifetime of the adhesively bonded joint with disbond significantly. The lifetime of the joint increases by 10x whereas no additional crack propagation was observed and test was terminated without failure of the bonded joint.

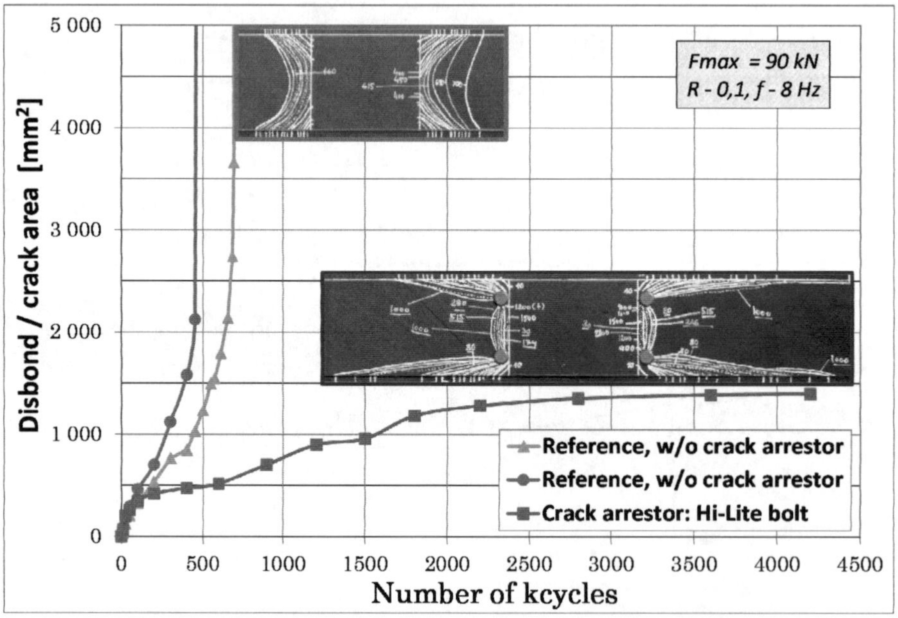

Fig. 10: Disbond area evolution comparison CFRP

Conclusions

FE and SG results confirmed the high sensitivity of the panel geometry (local thickness, disbond length) on stress distribution and accuracy of strain gauges positions on the measured and analysed strain values. The FE simulations also shown load transfer distribution in bond line and mixed mode (peel and shear) disbond propagation. Significant influence of DSF was proved. Multiply lifetime of WSLS panels is achieved compared with reference panels without DSF application. In some cases, full crack stopping is achieved when fasteners are placed close to the disbond area. Slow crack propagation for DSF configurations with bolts positioned close to initial disbond demonstrated no critical spontaneous rupture within a typical lifetime. The position of fasteners has an influence to the crack tip development. Fasteners as a crack arrestor prove high efficiency.

Acknowledgement

This work was funded by the European Union Seventh Framework Programme under grant agreement n° 314180 – Project BOPACS: Boltless Assembling of Primary Aerospace Composite Structures and by Ministry of Industry and Trade of the Czech Republic in the framework of Institutional support of Research organizations.

References

[1] U.S. Department of Transportation - Federal Aviation Administration (FAA), Advisory Circular – Composite Aircraft Structure, AC20-107B, FAA, 2010.

[2] T. Kruse, T. Körwien, R. Růžek, Fatigue behaviour and damage tolerant design of bonded joints for aerospace application, ECCM 2016 - Proceeding of 17th European Conference on Composite Materials, ECCM 2016; Munich; Germany; 26 - 30 June 2016, Scopus 2-s2.0-85018193760, ISBN: 978-300053387-7.

[3] Boltless assembling Of Primary Composite Structures (BOPACS), "Description of Work," EU FP7 Research Project, 2012.

[4] J. Tomblin, W. Seneviratne, Determining the Fatigue Life of Composite Aircraft Structures Using Life and Load-Enhancement Factors, Federal Aviation Administration (FAA), DOT/FAA/AR-10/6 report, 2011.

[5] Composite Materials Handbook (CMH-17), Volume 3 - Polymer Matrix Composites: Materials usage, Design and Analysis, CMH-17-3G, SAE International, 2012.

[6] NX Nastran Quick Reference Guide, Siemens Product Lifecycle Management Software Inc., 2012.

[7] J. Šedek, R. Hron, M. Kadlec, Bond Joint Analysis of Thermoplastic Composite Made from Stacked Tailored Blanks, Appl. Mech. and Mat., Vol. 827 (2016) 161-168, ISBN: 978-3-03835-531-1.

EAN 2019
57th conference on experimental stress analysis

Comparison of mechanical properties of woven carbon reinforcements

RYVOLOVÁ Martina[1,a] and VOSÁHLO Josef[1,b]

[1]Institute for Nanomaterials, Advanced Technology and Innovation, Bendlova 1407/7, Liberec, Czech Republic

[a]martina.ryvolova@tul.cz, [b]josef.vosahlo@tul.cz

Keywords: Carbon fibres, Carbon tow, Spread tow, Woven fabric, Mechanical properties

Abstract Important groups of laminates are long fibre reinforced composites. Carbon reinforced plastics are used in engineering and construction. They have low weight and very good mechanical properties. The properties of the laminate depend on the method of depositing the fibres in the reinforcement and processing the initial longitudinal textiles. Carbon fabrics are the most commonly used type of reinforcement for laminates. The weaving technology provides a variety of interlacing variations. New technology of production and modification of usually used tows bring significant improvement of their mechanical properties. The article described comparison mechanical properties of three woven fabric from carbon fibres according to produce technology original tows. In parallel, mathematical models of fabrics were created and the results were compared with the results of the experiment.

Introduction

The composite material is a combination of two or more materials with different properties. CFRPs are composites made of fibre-reinforced polymer matrix. The properties of the fibres in the composite material have a strong influence on their strength and stiffness. Carbon fibres offer a very good strength/stiffness to weight ratio [1]. Carbon fibre (CF) contains carbon atoms in various modifications. Crystals that are oriented in the direction of the fibre axis form the fibre. The crystal arrangement determines the strength of the fibre. The elemental carbon fibre has a diameter of 5-8μm, the carbon fibre density is 1.75g/m^3. PAN precursor carbon fibre industrial production began in the 1960s, production of carbon fibre from mesophase smol ten years later. CF are technical fibres with high strength and stiffness, but very low elongation, pitch fibres have very good flexural strength [2]. Carbon fibre longitudinal textiles are made by doubling elementary fibrils into tows (1, 3, 6, 12K) or rovings (24, 48, 50K). Carbon fabrics, multiaxial knitted fabrics and braided textiles are produced from carbon longitudinal textiles. Tows, fabrics, knitted textiles and 3D structures serve as a continuous dispersion of fibre reinforced plastics.

Tows 12K are made in two variants: round tow and flat tow format (referred to as WD). The flat tow is an ideal fibre to be used in applications where spreading is required (produce tapes, warp knitting and weaving textiles). Round tow is used, where spreading is not necessarily required (braiding and weaving technology) [3, 4]. The woven fabrics are made of two perpendicularly oriented systems of yarn, weft and warp yarns. The fabrics are produced on weaving loom by repeated weft insertion. The warp and weft material is usually the same, but the properties in the yarns direction are differ. Crossing of warp and weft yarns is called interlacing point. The layout of warp and weft interlacing points is crucial for weave of the

fabric and, above all, for its mechanical properties. For weaving is possible to use standard round tow, WD tow (flat tow format) and spread tow (SPF).
Carbon fabrics are quantify by the following parameters:
- number of elementary fibrils in one tow (number K),
- characteristics of the carbon fibre type used
- area weight,
- sett in warp and weft direction,
- weave [1].

Spread tow fabric (STF) is a tow layout into a wider and ultra-thin tape. Since 1990, there have been several patents that describe the procedure of spread tow production (using acoustic signal and vibration or pneumatic method). Spread tow is basis for produce woven structure. The SPF fabric is ultra-light and ultra-thin. On the market since 2014, Oxeon is the largest producer of this fabric under the trade name TeXtreme®. Two variants are produced; warp and weft orientation 0°/90 ° and warp and weft orientation ± 45° [3]. The tapes in the STF fabric are crimped, minimally deformed and allow maximum tensile load. The STF fabric can replace a conventional low basis weight fabric (made from 1K to 6K), or where radical weight reduction is required.

Experiment

Three carbon fabrics were selected for comparison of mechanical properties. Woven fabrics were produced from carbon tow 12K, tows were produced by different technology. Fabric No.1 was woven from standard tow 12K (round tow). Fabric No.2 was produced from flat tow (WD) – fibres in tow 12K were arranged to 5mm tow width. Fabric No.3 was produced from spread tow (SPF) - fibres in tow 12K were arranged using special spread technology to 15mm tow width. The fabrics were defined by the following parameters: warp and weft material, area weight, square sett and weave. Look of carbon fabrics shows Fig. 1.

Fig. 1: Woven fabric; a) standard tow, b) flat tow, c) spread tow

Six specimens were prepared in the warp direction and in weft direction from each fabric. Specimens were prepared according to ČSN EN ISO 13934-1. Dimensions of specimens were: length 200mm, width 50 mm. Special attentions was given to the selvedge of specimens. The standard prescribes the removal of yarns of 5mm width. Specimens were weighed, fineness of carbon tows – 12K (800tex) was defined. Sett in warp and weft direction and weave were defined according the Standard. The basic fabrics parameters are showed in Table 1.

Table 1: Woven fabric parameters

Fabric	Weave	Weight [gm^{-2}]	Sett wa/we
1 (Fig.1a)	Twill 12K	600	440/440
2 (Fig.1b)	Plain 12K	193	120/120
3 (Fig.1c)	Plain 12K	160	67/67

Thickness of specimens was measured using thickness gauge (SDL M 034/1). Tensile strength test was realized on strength testing machine (Tiratest T2400), speed of test was 2mm/min. Tensile curves, values of maximum break force (F_{max}), maximal strength (R_{max}) and tensile strength modulus (E) were results of test.

Mathematical models

The 2D shell model of carbon fabrics was created to simulate the tensile test in Solidworks. The test sample model was 100 x 50 mm in size. The created model was imported into Ansys. The thickness of the simulated samples varied according to the measured values. The FEM network was introduced into the model. The mash was defined by face mapping with a combination of 1 mm Face Sizing. The mash model created contained 4989 nodes and 4844 elements. Fixing boundary conditions were inserted at the bottom of the model. These boundary conditions prevented the rotation and movement of the sample to all sides. The force boundary condition was inserted into the top of the model [5, 6]. The condition value was changed depending on the properties of specimens. The values used for mathematical models are summarized in Table 1 and Table 2.

Results

Tensile strength test (Fig. 2) was realized for prepared fabrics specimens in warp and weft directions. Table 2 presents the fabric parameters and maximum break force. Tensile curves, values of maximum break force (F_{max}), maximal strength (R_{max}) and tensile strength modulus (E) were results of test. Graph of R_{max} values shows Fig. 3.

Table 2: Results of measurements

Fabric	Sett [m^{-1}]	Thickness [mm]	Number of filaments [m^{-1}]	F_{max} [N]
1 warp	440	0.83	5280K	3955
1 weft	440	0.83	5280K	3448
2 warp	120	0.35	1440K	5206
2 weft	120	0.35	1440K	4552
3 warp	67	0.19	804K	5348
3 weft	67	0.19	804K	5317

Fig. 2: Measurement of fabric

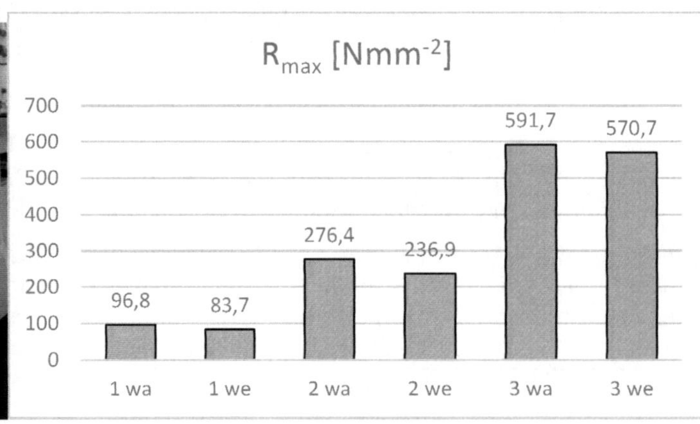

Fig. 3: Graph of maximal strength

The results summarized in Table 2 show significant differences between the thickness of the fabrics, the numbers of elementary fibrils and the strength value R_{max}. The properties of elementary fibrils are comparable, fineness of tow is the same but design and produce technologies of tows are different. The test results indicate that the effective use of the mechanical properties (maximal strength R_{max}) of carbon fibres is relatively low for standard fabrics. Flat tows ensure a higher percentage of fibre utilization. The maximum strength R_{max} of the fabric is three times higher. Spread tow treatment has maximized the mechanical properties; the maximal strength R_{max} of fabric is six time higher. The differences between warp and weft direction is minimal for fabric made of spread tows. Figure 4 shows damage of fibres after test.

Fig. 4: Damage of elementary fibres after test; a) standard tow, b) flat tow, c) spread tow

The simulation results showed that the maximum deflection and stress values are directly dependent on value of force and thickness of the fabric. The results of the mathematical model summarizes Table 3.

Table 3: Results of mathematical model

Fabrics	Deformation [mm]	Stress von-Mises [MPa]
1 warp	2.31	376
1 weft	2.10	341
2 warp	3.14	996
2 weft	2.87	847
3 warp	2.80	2020
3 weft	2.95	2169

Figure 5 shows a comparison of the mathematical model of size and stress distribution in the test specimens in the warp direction during tensile strength test.

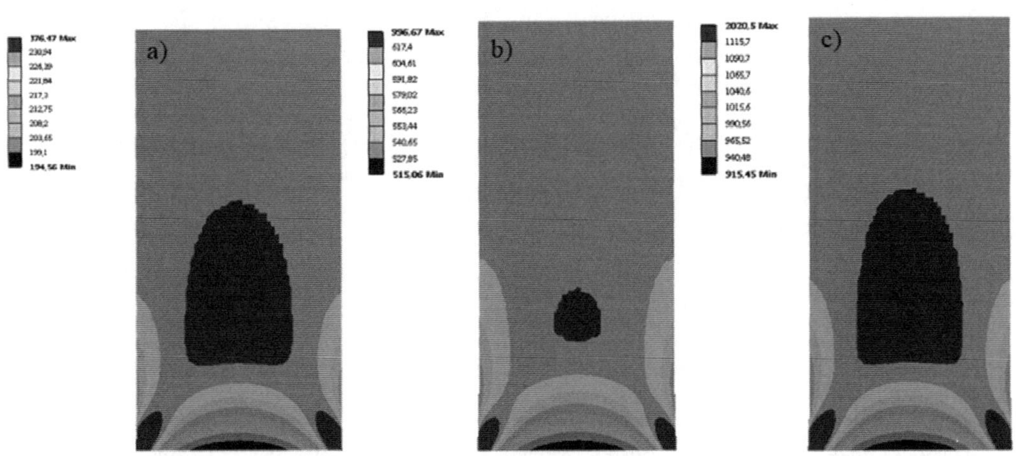

Fig. 5: Stress (von Mises) in a warp direction in a maximum time of 1s for fabrics; a) standard tow, b) flat tow, c) spread tow

Conclusions

The measurement showed that the spread tow treatment significantly affects the mechanical properties of the woven carbon reinforcement. The spread tow fabric contains a significantly lower number of elementary fibrils; but its mechanical properties are significantly higher than those of standard carbon fabric.

A 2D shell model was created into which mechanical properties of individual materials and their thickness was imported. Comparing the results of the tests and calculations of the mathematical model is to show the degree of agreement. The graph of Fig. 6 shows the comparison of mathematical model maximal strength von Mises and measured values of carbon fabric strength. Based on comparison, design elements with mechanical properties for specific applications will be proposed.

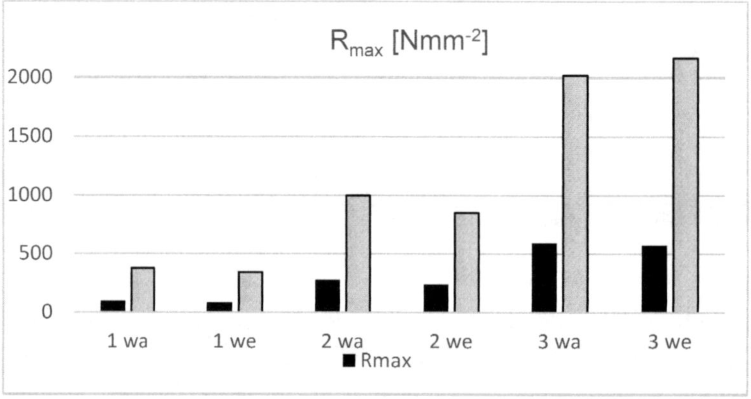

Fig. 6: Comparison of measured values of R_{max} and simulated values of R_{max} (von Mises)

Acknowledgment:

The result was obtained through the financial support of the Ministry of Education, Youth and Sports of the Czech Republic and the European Union (European Structural and Investment Funds - Operational Programme Research, Development and Education) in the frames of the project "Modular platform for autonomous chassis of specialized electric vehicles for freight and equipment transportation", Reg. No. CZ.02.1.01/0.0/0.0/16_025/0007293.

References

[1] I. V. Olsen: Stress Analysis, Lay-up and Production Process for the Tail Section of a Composite Glider Used for Energy Generation. Thesis, Norwegian University of Science and Technology, 2017

[2] J. Grégr: Uhlíková vlákna, 2005; přístupné z: https://www.researchgate.net

[3] H. El-Dessouky: Spread Tow Technology for Ultra Lightweight CFRP Composites: Potential and Possibilities. 10.1515/9783110574, 2017

[4] K. Kawabe, H. Sasayama, S. Tomoda: New Carbon Fiber Tow-Spread Technology and Applications to Advanced Composite Materials, Sampe Journal 45(2):6-17 · March 2009

[5] M. Jirásek, Z. Bažant, Inelastic analysis of structures, Wiley, 2002, ISBN 0-471-98716-6

[6] M. Talia, H. Lankarani, J. E. Talia: New experimental technique for the study and analysis of solid particle erosion mechanisms, Wear pp. 225-229 (1999) pp.1070–1077.

EAN 2019
57th conference on experimental stress analysis

Measurement of residual stresses in the hole of railway axle

ŘEHA Bohuslav[1, a], ZAWADA Josef[1,b], KWACZEK Milan[1,c] and ĎURIŠ Radek[1,d]

[1]BONATRANS GROUP a.s., Revoluční 1234, 73594 Bohumín, Česká republika

[a]bohuslavreha@ghh-bonatrans.com, [b]josefzawada@ghh-bonatrans.com, [c]milankwaczek@ghh-bonatrans.com, [d]radekduris@ghh-bonatrans.com

Keywords: residual stress, X-Ray method, hole drilling method, railway axle

Abstract. Values of the residual stress are very important for fatigue life of machinery components. It is likely to be the same for the railway wheelsets. In the most of the cases the pressure values of stress are required in proximity of the surface e.g. the standart EN 13 261 for railway axles allows the maximum value of stress +100MPa in the distance of 0,1mm under the surface. In many cases the hollow railway axles are used with the advantages of design motor axles. The determination of residual stress is not feasible with the regard to usual diameter of the hole from Ø30mm to Ø110mm. This thesis is focused on the possibility to determin of the values of the residual stress in the bore of railway axle by hole drilling method. The measuring was done with the use of hole drilling device RESTAN 3000. Measured values of residual stress in close proximity of the bore of axle were compared with the measurement of X-ray method.

Introduction

Measurement were done on surface of hardened railway axle from material of A4T quality. With regard to the Ø65mm axle bore diameter, the measurements had to be made in the journal area to allow the installation of the strain gauge RY61-1.5 / 120S.

Measurement by longer milling cutter. A Ø14 mm hole was made in the opposite axle wall against the selected measuring point in the bore. Subsequently, the center of the strain gauge rosette was focused using a laser cross, as shown in Fig. 1. Because of the access during installation, the strain gauge rose was located approximately 45 mm from the end of the hole in the journal of axle.

Fig 1: Measuring middle of the strain gauge rossete

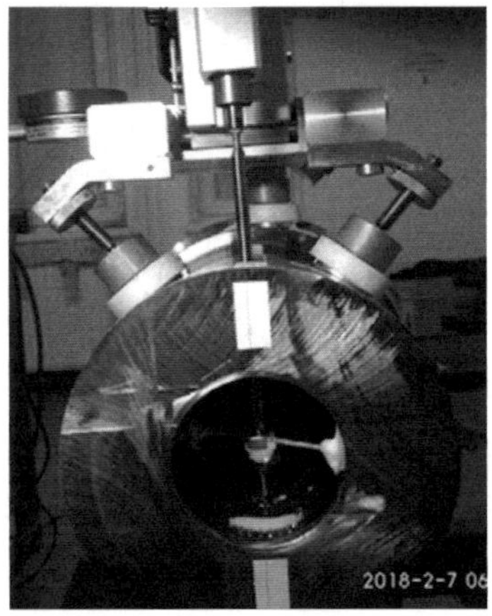

Fig 2: The test arrangement

Fig 3: Drilling detail

After installation and connection of strain gauge type RY61-1,5 / 120S with drilled hole, which was done by drilling cutter. Subsequently, drilling was carried out with an extended length of 160mm, nominal Ø2mm to a depth of 1mm. As can be seen in Figs. 2 and 3, due to the limitation of the milling cutter, this was held inside the hole by a needle bearing. Due to the focal length of the distance microscope, the test arrangement does not allow a direct optical focus of the hole. A special touch probe with a Ø1mm end was used to read the diameter of the hole made by using the micrometric feed table of the RESTAN drill stand. When the probe is attached to the cutter drive with the side of the hole, the electric circuit closes and thus optical and audible signaling occurs. The method of measurement is shown in Fig. 4.

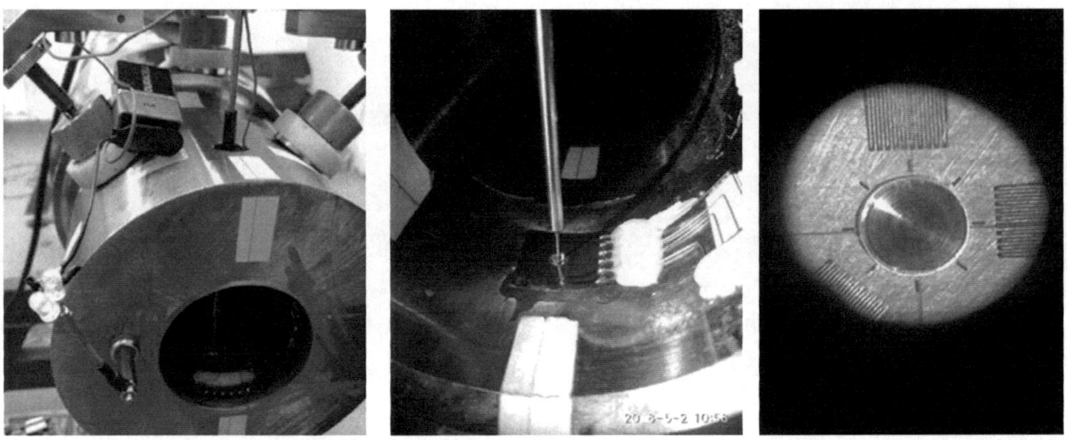

Fig 4: Measuring hole diameter, in the middle is probe in contact with the edge of the hole, on the right view of the drilling hole

Measurement were done by 20 steps of drillng to the 1,0mm depth. We used hole drilling RESTAN stand. We used strain gauge rossete HBM RY61-1.5/120S. Evaluation of the results was performed in the EVAL program using the E837-13 EXT Not Unif method. Depth profiles of directional stresses in the bore of axle are shown in the graph in the following Fig. 5.

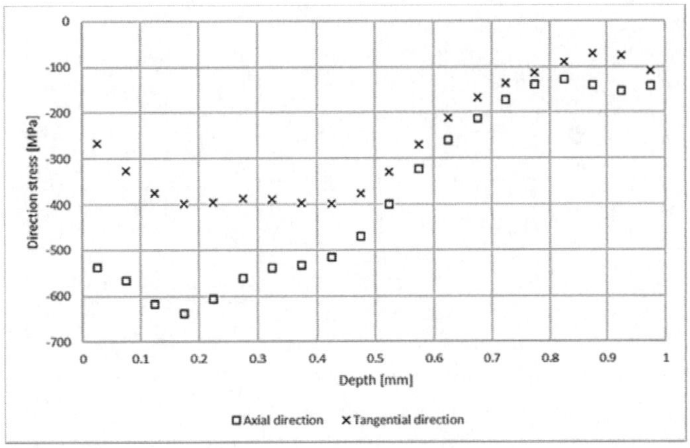

Fig 5: Depth profile of stresses in the bore of axle

Comparasion of results with X-Ray method. Measurement by X –Ray method were done on the small piece of axle. Measurement were done using the Stresstech ROBOT – see fig.6. Electrochemical etching to a depth of 0.2 mm below the surface was performed to obtain a depth stress profile. Calculations of stresses were performed in the XTronic program according to the European standard EN 15305. The determined stress profiles are shown in the graph in FIG.7

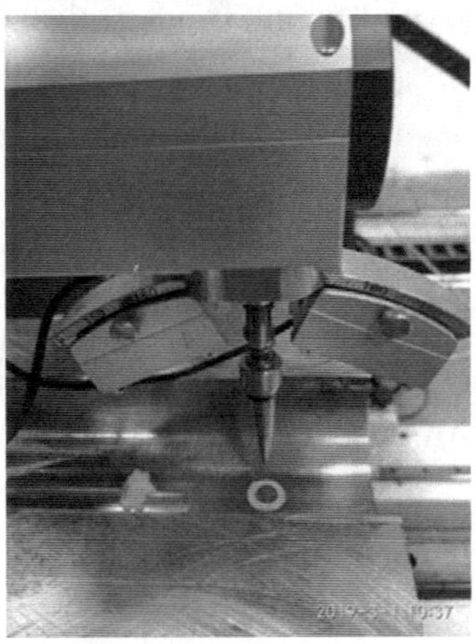

Fig 6: Measurement by X-Ray diffraction

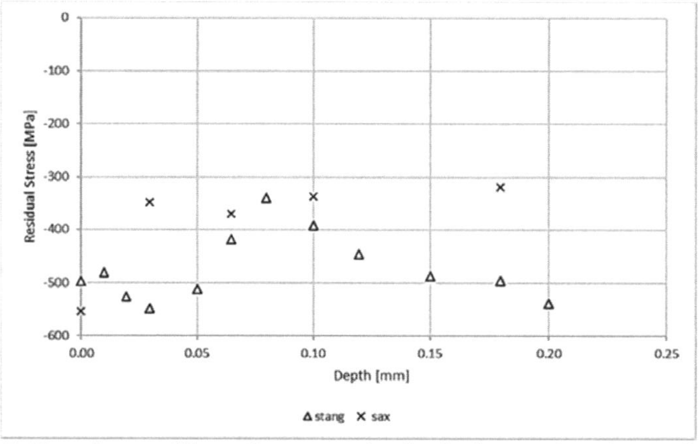

Fig 7: Depths profiles of stresses in the hole on the piece of axle

Conclusions

The use of an longer cutter allowed the determination of residual stress values in the bore of the railway axle without the need to cut the test piece. Strain gauge hole drilling method and X-Ray diffraction confirmed big compressive stresses on the surface of the hole and very near of the hole. During the measurement of longer cutter were residual stress in axial direction in distance 1 mm under the surface from -129 MPa to the-638 MPa, tangential from -71 MPa to the -398 MPa.

Measurement on the piece of axle using by X-Ray method confirmed occurence of compressive stresses. In axial direction were measured values from -319MPa to -554MPa and -339MPa to -549MPa in tangential direction.

References

[1] EN 13261: 2009+A1, „Railway applications Wheelsets and bogies - Axles - Product requirements“ October 2010.

[2] ASTM E837-13a,“Standard Test Method for Determining Residual Stress by the Hole-Drilling Strain-Gage Method“, September 2013

[3] EN 15305, „Non-destructive testing. Test Method for Residual Stress analysis by X-ray Diffraction’’, 2008

Degradation of mechanical properties of magnesium alloy

SEDLÁČEK Radek[1,a], SUCHÝ Tomáš[1,2,b] and PADOVEC Zdeněk[1,c]

[1]Department of Mechanics, Biomechanics and Mechatronics; Faculty of Mechanical Engineering; Czech Technical University in Prague; Technická 4, 160 00 Prague, Czech Republic

[2]Department of Composites and Carbon Materials, Institute of Rock Structure and Mechanics, The Czech Academy of Sciences, V Holešovičkách 41, 182 09 Prague, Czech Republic

[a]radek.sedlacek@fs.cvut.cz, [b]tomas.suchy@fs.cvut.cz, [c]zdenek.padovec@fs.cvut.cz

Keywords: Bioabsorbable metals, Biodegradation, Magnesium, Implant

Abstract. In this study, biodegradation behaviour of WE43 magnesium alloy have been studied and compared during exposure to three different media commonly used to simulate the conditions of human body environment. Magnesium alloys emerged as a new class of bioresorbable implant materials. Their applications reduce certain risks associated with conventional permanent implants. Biodegradation behaviour of the WE43 magnesium alloy was observed under Dulbecco Modified Eagle Medium (DMEM, Sigma Aldrich) supplemented with 5 % fetal bovine serum and gentamicin antibiotic as standard. The samples were stored in the medium at 37° C and in a 5 % CO_2 atmosphere. The second type of medium was Hank's Salt Balanced Solution (HBSS, Sigma Aldrich), which simulates the inorganic composition of blood plasma. HBSS was tempered at 37° C. The last solution was an acidic solution of HCl + NaCl (pH2) with pH ~ 2 (0.01M HCl and 0.14M NaCl) tempered at 37° C. The acidic type of solution was used to simulate a local acidic environment associated with osteoclast activity during bone remodelling. Changes in the mechanical properties of the samples during exposure to simulated body conditions were observed.

Introduction

Magnesium alloys emerged as a new class of bioresorbable implant materials [1-3]. Their applications reduce certain risks associated with conventional permanent implants [1]. The need of reliable *in vitro* models to predict degradation in human body is increasing. In this study, biodegradation behaviour of WE43 magnesium alloy have been studied and compared during exposure to three different media commonly used to simulate the conditions of human body environment. Changes in the mechanical properties and surface area of the samples during exposure to simulated body conditions were observed.

Materials, methods and results

Biodegradation behaviour of the WE43 magnesium alloy was observed under three different simulated body conditions. The first was Dulbecco Modified Eagle Medium (DMEM, Sigma Aldrich) supplemented with 5 % fetal bovine serum and gentamicin antibiotic as standard. The samples were stored in the medium at 37° C and in a 5 % CO_2 atmosphere. The second type of medium was Hank's Salt Balanced Solution (HBSS, Sigma Aldrich), which simulates the inorganic composition of blood plasma. HBSS was tempered at 37° C. Both solutions are physiologically neutral or slightly alkaline. The last solution was an acidic solution of HCl + NaCl (pH2) with pH ~ 2 (0.01M HCl and 0.14M NaCl) tempered at 37° C. The acidic type of solution was used to simulate a local acidic environment associated with osteoclast activity during bone remodelling. Samples (*n* = *5*) in dimensions of 6 mm in diameter and 9 mm in length were exposed to the medium for one and two weeks (see Fig. 1). The solutions were completely replaced every 24 hours. The pH of the solutions was checked throughout the experiment.

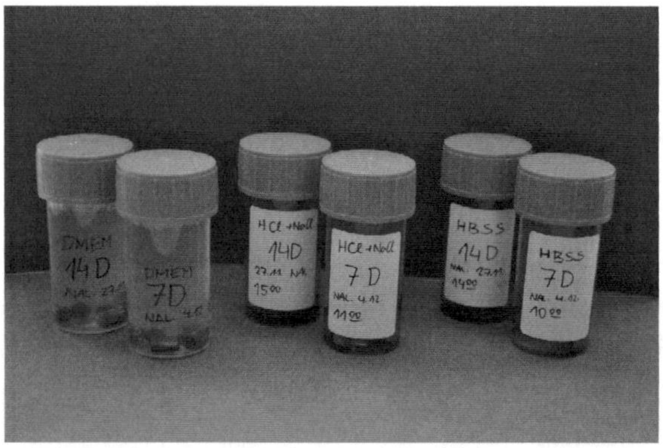

Fig. 1: Cylindrical specimens placed in boxes

Fig. 2: Testing system MTS 858 Mini Bionix

The effect of sample exposure to media was observed by mechanical compression tests (see Fig. 2). Furthermore, the sample surface was scanned on a scanning electron microscope (SEM) and optical microscope. The results of mechanical tests are illustrated in Fig. 3.

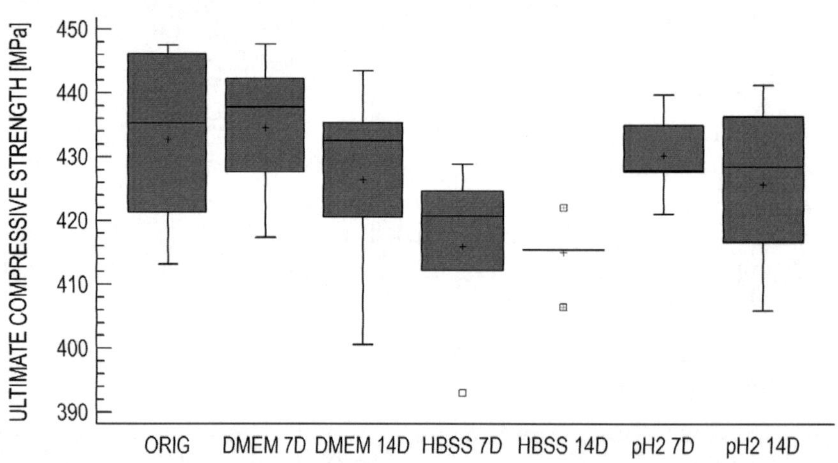

Fig. 3: Compressive strength of magnesium alloys before (ORIG) and seven days (7D) and fourteen days (14D) after exposure to Dulbecco Modified Eagle Medium (DMEM), Hank's Salt Balanced Solution (HBSS) and acidic solution of HCl + NaCl (pH2). There are no statistically significant differences between the samples in the multiple comparison of all groups (Games-Howell test, 0.05)

The results show that the short-term exposure used in all ingested media has no statistically significant effect on the observed mechanical properties ($p = 0.05$). However, the results obtained may also be affected by the sample sizes used. The image analyses showed that the short-term exposure was seen mainly on the surface of the samples, while the reduction of the mechanical properties of the cylindrical samples with a diameter of 6 mm and a height of 9 mm has no effect.

Conclusions

The aim of the biodegradation study was to assess the potential changes in the mechanical properties of cylindrical samples made from magnesium alloy WE43. The need of reliable *in vitro* models to predict degradation in human body is increasing. At present time, there is practically no general consensus on which medium would best and comprehensively simulate events occurring under *in vivo* conditions. Thus, three types of media commonly used to simulate the body environment were selected for the purpose of the study. Two weeks after the exposure, only local surface degradation was observed while no statistically significant effect on mechanical properties was determined. For the short-term degradation-induced loss of mechanical integrity, specimens with smaller dimensions should be used, e.g. specimens with comparable size to that of implants such as cardiovascular stents and orthopaedic wires [1]. The differences among the three kinds of simulated body conditions were not observed within two weeks period.

Acknowledgements

This study was supported by the grant project awarded by the Technology Agency of the Czech Republic No. TACR TH03010418.

References

[1] Galvin E, Jaiswal S, Lally C, MacDonald B, Duffy B, In Vitro Corrosion and Biological Assessment of Bioabsorbable WE43 Mg Alloy Specimens, J. Manuf. Mater. Process. 2017, 1 - 8; doi:10.3390/jmmp1010008

[2] Zhou XH, Jiang L, Wu PP, Sun Y, Yu YD, Wei GY, Ge HL, Effect of Aggressive Ions on Degradation of WE43 Magnesium Alloy in Physiological Environment, Int. J. Electrochem. Sci., 9 (2014) 304 - 314

[3] Jiang L, Xu F, Xu Z, Chen Y, Zhou XH, Wei GY, Ge, HL, Biodegradation of AZ31 and WE43 Magnesium Alloys in Simulated Body Fluid, Int. J. Electrochem. Sci., 10 (2015) 10422 - 10432

Comparison of the published contact configurations for the determination of electrical resistivity of CFRP composite

SCHMIDOVÁ Nikola[1,a], DVOŘÁK Milan[1,b] and RŮŽIČKA Milan[1,c]

[1]Department of Mechanics, Biomechanics and Mechatronics, Faculty of Mechanical Engineering, Czech Technical University in Prague. Technická 4, Praha 6, Czech Republic.

[a]nikola.schmidova@fs.cvut.cz, [b]milan.dvorak@fs.cvut.cz, [c]milan.ruzicka@fs.cvut.cz,

Keywords: Carbon Fibre-Reinforced Polymer (CFRP) composites, Electrical resistance, four-probe method, Delamination, Structural Health Monitoring (SHM)

Abstract. Damage detection methods for electrically conductive composite materials based on the electrical potential or electrical resistance measurement have been widely investigated in the literature. Damage growth inside the material has been also studied using finite element simulation. For the numerical simulations, it is necessary to know nominal resistivity of the material. Several contact configurations have been published in literature for determination of the nominal electrical resistivity in the in-plane and through-thickness directions. For in-plane and through-thickness directions electrical resistivity was experimentally investigated using electrical contact configurations published in literature. Measured electrical resistivity was used for finite element analysis of delamination growth in Carbon Fibre-Reinforced Polymer composites (CFRP).

Introduction

Damage detection of electrically conductive composite materials based on the electrical potential or resistance measurement has been widely investigated in the literature. Damage growth inside the material has been also studied using the numerical simulation. It is effective to simulate various damage scenarios in order to get information about influence of different parameters on the measurement configuration, such as size of the specimen, distance of the electrical contacts or damage size and location. In order to obtain relevant information about the real structure, it is necessary to enter correct material characteristics into the FEA model, primarily values of the nominal resistivity of the examined material.

Several contact configurations for determination of the nominal electrical resistivity in the in-plane and through-thickness direction have been used in recently published works [1- 3].

Overview of such electrical contact configurations is given in the presented article. Three contact configurations were tested in case of determination of in-plane resistivity. The resistance in through-thickness direction was determined using two contact configurations. The obtained values were compared and used subsequently for the numerical simulation of delamination specimen test.

Experimental procedure

Material and Specimen Preparation. Hexply AGP 193PW/8552S RC40 composite material was used for specimen preparation. $[45/0]_{4S}$ lay-up scheme was used. The specimens were prepared from the same material, which was used for manufacturing of specimens in [4]. Electrical contacts (made from a conductive epoxy and a thin copper strips) were manufactured after the curing. Surface of the specimens was slightly sanded at first with sandpaper P120 and P600 and then degreased. Electrical contacts were prepared using a CHO-BOND® 584-29 conductive epoxy and a thin copper strips.

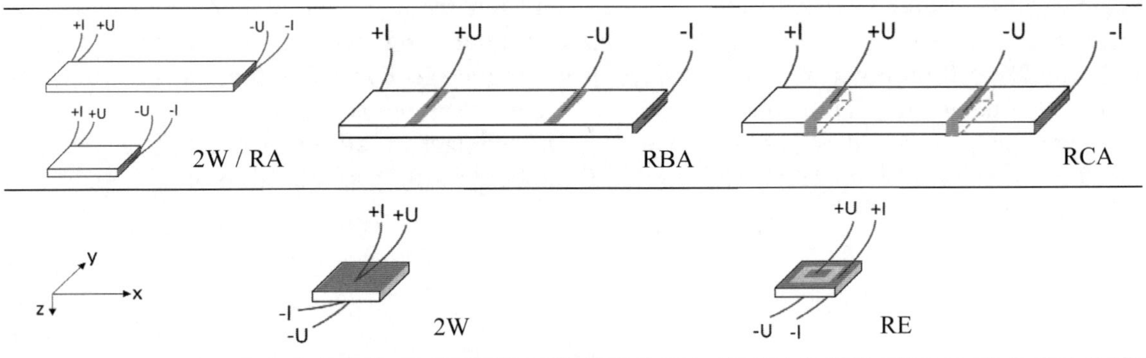

Fig. 1: Configurations of electrical contacts for determination of in-plane and through-thickness resistivity

Measurement procedure. The configurations of the electrical contacts used for electrical resistance measurement are shown in the Fig. 1. When using the RA configuration, measurement has to be performed on specimens with different length. Electrical resistivity of CFRP composite was successfully determined using the RA configuration and electrical contacts prepared by electroplating in [1]. Two-wire (2W) contact configuration is the same as in case of RA, but the contact resistance is neglected. In case when the value of contact resistance is close to the measured values of resistance, it could not be neglected and four probe method must be used. We have performed 2W measurement to demonstrate the influence of using values of electrical resistivity obtained by 2W configuration on the results of finite element analysis. The measurement was performed using the HP E3631A current source and two Agilent 34461A multimeters. The series comparison measurement method was performed.

Table 1: Overview of specimens used for determination of in-plane resistivity and measured electrical resistivity using 2W method

	Length [mm]	Width [mm]	Thickness [mm]	1/Thickness [1/m]	2W - R [Ω]	2W- ρ_x [Ωm]	Temp. t [°C]
L1	88.4	25.2	3.1	0.328	0.293	0.000254	21.4
~~L2~~	~~88.4~~	~~25.2~~	~~3.2~~	~~0.317~~	~~0.792~~	~~0.000711~~	~~21.5~~
L3	87.6	25.2	3.1	0.319	0.316	0.000284	21.8
L4	93.6	25.2	1.6	0.61	0.476	0.000209	21.8
~~L5~~	~~93.1~~	~~25.2~~	~~1.6~~	~~0.63~~	~~1.074~~	~~0.000465~~	~~21.8~~
L6	91.9	25.1	1.6	0.61	0.498	0.000222	21.8
L10	92.5	25.1	1.6	0.63	0.502	0.000218	23.5
L11	92.5	25.1	1.6	0.63	0.412	0.000179	23.7
L12	92.5	25.2	1.6	0.63	0.388	0.000169	23.8
L13	46.3	25.1	1.6	0.63	0.344	0.000298	23.8
L14	46.3	25.1	1.6	0.63	0.270	0.000234	23.7
					Av.	0.000201	
					St.dev.	0.000041	

Experimental Results

In-plane resistivity. Six specimens L1 - L6 were prepared for determination of the in-plane resistivity (ρ_x). Because of the lay-up of the material, we assume that the resistivity in the x and y direction (see Fig. 1) is the same. First 2W and RA method was adopted. Results of 2W measurements are given in Table 1. It was found that measured resistivity of specimen L2 and L4 is outlying from other measured values. For this reason, five more specimens L10 – L14 were prepared and results for specimens L2 and L5 were not considered.

If we assume that it is possible to neglect the contact resistivity, we can determine the electrical resistivity using 2W configuration. Following formula was used for determination of 2W resistivity.

$$R = \rho \cdot \frac{l}{A} = \rho \cdot \frac{l}{w \cdot t} \quad (1)$$

$$\rho = R \cdot \frac{A}{l} \quad (2)$$

Results of 2W measurements are given in Table 1.

Measured electrical resistance of the specimen is dependent on the dimensions of the specimen, see formula (1). When we consider that the contact resistance (R_C) is not negligible, but constant, we can determine the resistivity of the specimen and also the value of the contact resistance. This approach was used successfully for electrical resistivity determination of CFRP specimens with electrical contacts prepared using electroplating, as described in [1]. Here the method is marked as RA. So it is necessary to evaluate 2W measurement on specimens with different dimensions. It is possible to use linear regression for measured data for specimens, which differ in their thickness according to following formula:

$$R = R_C + \rho \cdot \frac{l}{w} \cdot \frac{1}{t} \quad (3)$$

For the linear regression equation of measured data

$$R = 0.6286 \cdot \frac{1}{t} + 0.1012 \quad (4)$$

We can calculate the resistivity for the material as follows

$$\rho \cdot \frac{l}{w} = 0.6286 \rightarrow \rho = 0.6286 \cdot \frac{w}{l} = 0.162\ \Omega mm = 0.000017\ \Omega m \quad (5)$$

First, we evaluated only data for specimens L1, L3, L4, L6 and obtained value of in-plane resistivity $\rho_x = 0.000174\ \Omega m$, see Fig. 2. After preparing specimens L10-L14 we also performed linear regression for measured data of group of specimens L1, L3, L4, L6, L10 - L12, see Fig. 3. According to formula (6), we also proceed linear regression for specimens which differ only in its length (group of results for specimens L10 - L14, see Fig. 3, and group of results for specimens L4, L6, L10 - L14, see Fig. 5.

$$R = R_K + \rho \cdot \frac{1}{w \cdot t} \cdot l \quad (6)$$

Measured electrical resistivity, equation of linear regression and determined values of in-plane resistivity are given in Fig. 2 - 5.The lowest value of in-plane resistivity according to RA configuration was 0.00011 Ωm.

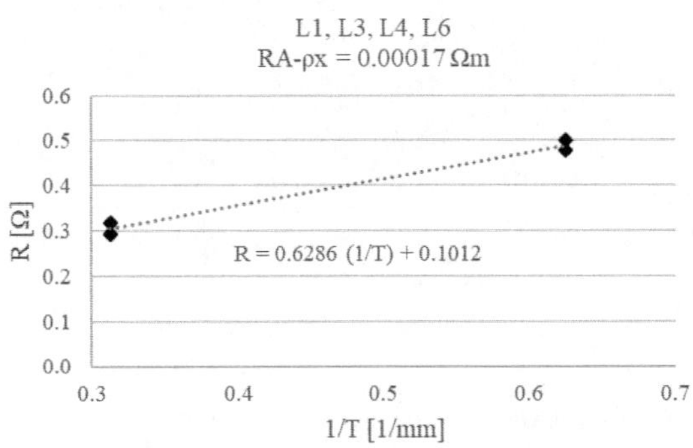

Fig. 2: Evaluation of RA measurement on specimens L1, L3, L4, L6

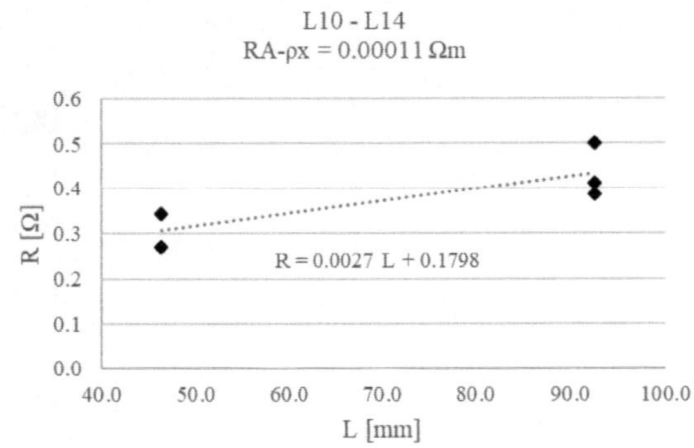

Fig. 3: Evaluation of RA measurement on specimens L10 – L14

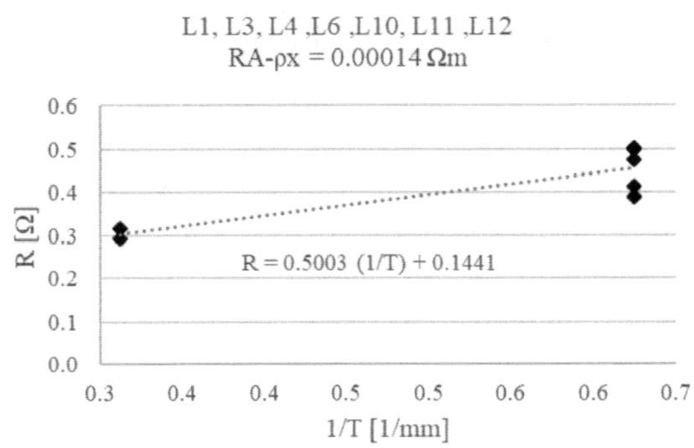

Fig. 4: Evaluation of RA measurement on specimens L1, L3, L4, L6, L10, L11, L12

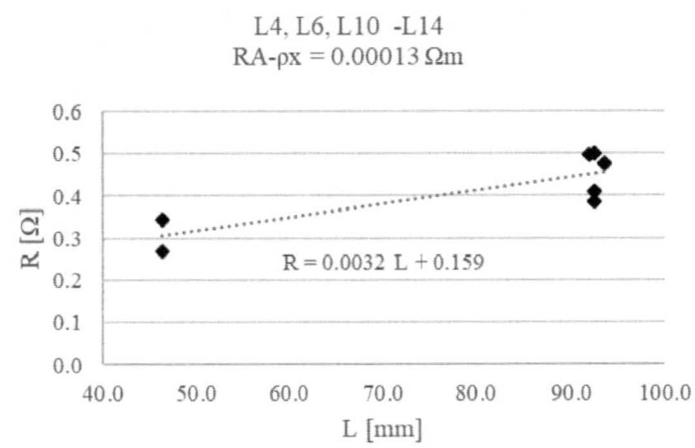

Fig. 5: Evaluation of RA measurement on specimens L4, L6, L10 – L14

The RBA and RCA electrical contact configurations were prepared on specimens L1-L6. The distance between the inner voltage electrodes according to (Fig. 1) was 30 mm for both configurations. For the resistivity determination the same formula as for the 2W configuration (eq. (2)) was used, but the distance between inner electrodes was considered in the formula. Measured data are given in Fig. 6. Although the measurement for the 2W configuration on specimens L2 and L4 were outlying, measurements with RBA and RCA on these two specimens were not outlying.

Through-thickness resistivity. The RA method was not studied for the through-thickness direction. For the 2W and RE measurement configuration four specimens with dimensions (15 x 15 x 1.6 mm) were prepared. For the 2W electrical contact configuration electrical resistivity ρ_z was determined to be 0.12 ± 0.034 Ωm.

The RE electrical configuration was prepared according to Fig.1. The inner voltage electrodes dimensions were 5 x 5 mm. There was distance of 2 mm between the inner voltage and outer current electrodes. The through-thickness electrical resistivity ρ_z according to RE contact configuration was determined to be 0.06 ± 0.005 Ωm.

Table 2: Overview of results of measured electrical resistivity using RBA and RBC method

Specimens	RBA [Ωm]	RBC [Ωm]
L1	0.000075	0.000075
L2	0.000082	0.000085
L3	0.000086	0.000084
L4	0.000068	0.000080
L5	0.000081	0.000078
L6	0.000067	0.000074
Av.	0.000076	0.000079
St.dev.	0.000007	0.000004

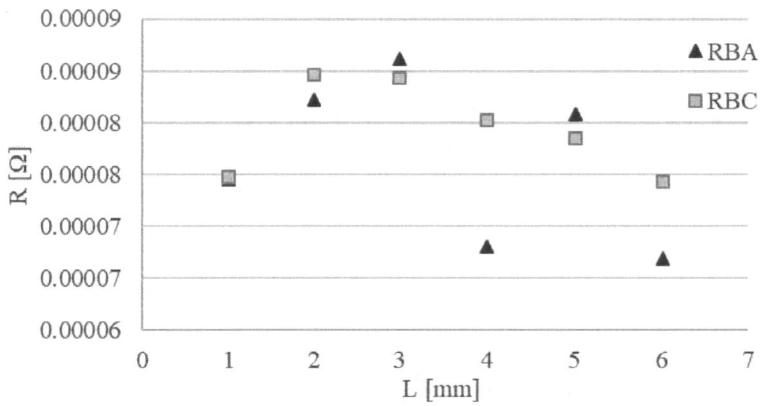

Fig. 6: Results of measured electrical resistance using RBA and RBC method

Numerical Simulation of Delamination Specimen Test

Obtained values of electrical resistivity of the material were used in the finite element analysis of the delamination growth. Simulated data were compared to those obtained during experimental investigations of delamination growth on specimens loaded simultaneously by mode I and mode II loading (Mixed-Mode Bending – MMB Specimens) published in [4]. The dimensions of the specimens were 185 x 25 x 3.3 mm. The numerical analyses were carried out using the commercial code Ansys 17. The MMB specimen was prepared as a 2D model, over 31,000 pane elements were used with element size from 0.008 mm to 0.5 mm. Mechanical loading was not considered during the simulation.

Results of the conducted finite element analysis are given in Fig. 8. and Fig. 9. It can be seen that results obtained with electrical resistivities using 2W method do not correspond to the measured data at all. When using data obtained by RA and RE configuration, the longer the delamination, the bigger the difference between measured and simulated data can be seen.

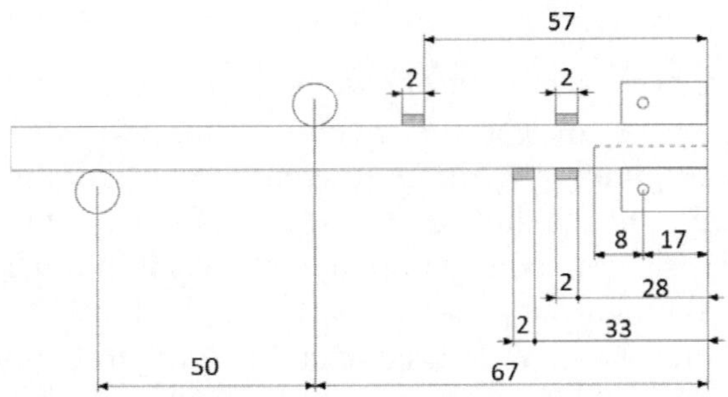

Fig. 7: Dimensions of MMB specimen and electrical contact configuration used for the experimental investigation and numerical simulations

Table 3: Overview of the electrical resistivities used in finite element analysis

	2W – 2W	RBA – 2W	RBA - RE	RA - RE
ρ_x [Ωm]	0.0002	0.000076	0.000076	0.0001
ρ_z [Ωm]	0.12	0.12	0.06	0.06

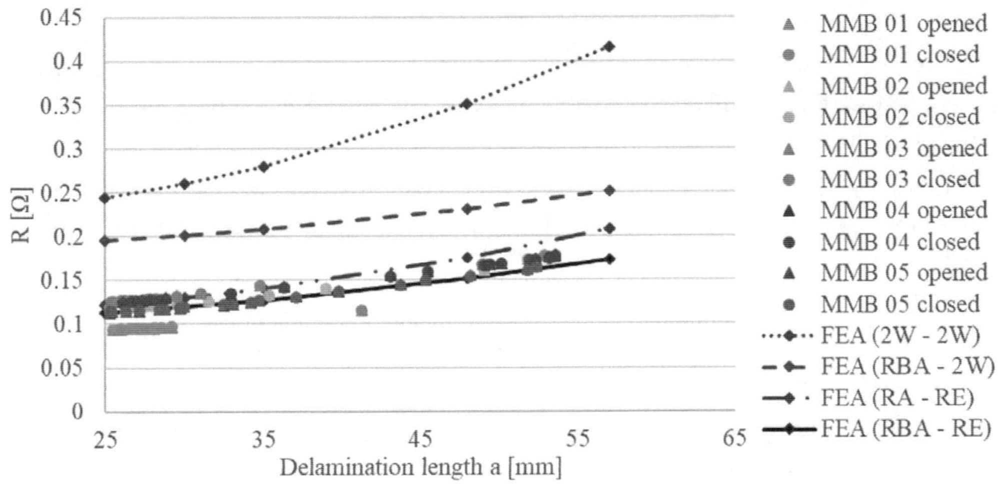

Fig. 8: Results of finite element analysis and measured experimental data

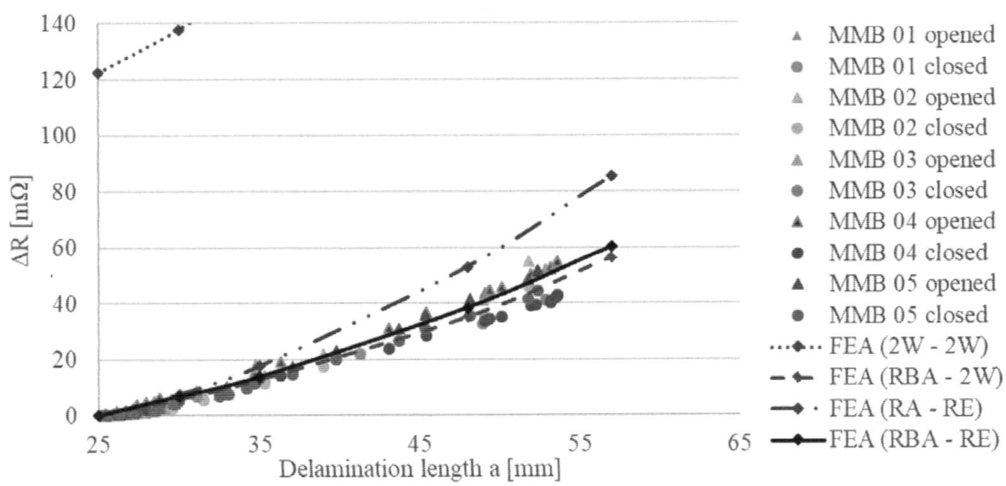

Fig. 9: Results of finite element analysis and measured experimental data

Conclusions

In contrast to the results presented in [1] we prepared electrical contacts for configuration RA using conductive epoxy instead of electroplating. We determined higher electrical resistivity for contact configuration RA than for contact configuration RBA and RCA.

According to obtained values of through-thickness resistivity, it is not possible to neglect the contact resistance and use 2W configuration also when measuring in through-thickness direction, although the through-thickness resistivity is higher by an order than the in-plane resistivity. Negligible difference in measured data for configuration RBA and RCA was found.

According to our investigation it is more convenient to use separated current and voltage electrodes when using conductive epoxy for electrical contact preparation.

The finite element analysis has shown that there is difference in simulated and measured data for electrical resistivities determined according to different contact configurations.

Acknowledgement

The authors would like to thank the Grant Agency of the Czech Technical University in Prague for supporting this research with grant No. SGS18/175/OHK2/3T/12.

References

[1] J. C. Abry, S. Bochard, A. Chateauminois, M. Salvia and Giraud, In situ detection of damage in CFRP laminates by electrical resistance measurement, Composites Science and Technology, no.59, pp.925-935, 1999.

[2] T.Yamane, A.Todoroki, Analysis of electric current densityin carbon fiber reinforced plastic laminated plates with angeled plies, Composites Structures, no. 166, pp.268-276, 2017

[3] M. Zappalorto, F. Panozzo, P.A. Carraro, M. Quaresimin, Electrical response of laminate with a delamination: modelling and experiments, Composite Science and Technology. 143 (2017) 31-45.

[4] N. Schmidová, M. Dvořák, M. Kadlec, M. Růžička, Monitoring of Delamination Growth on the MMB Specimens Using FBG Sensors and Electrical Resistance Measurement Method, in: EAN 2018 56th conference on experimental stress analysis, Conference Proceedings. Praha: Česká společnost pro mechaniku, 2018. p. 367-373. ISBN 978-80-270-4062-9.

Using mathematical analysis for bronchial obstruction detection

SKALICKÝ David[1,a], KOUCKÝ Václav[2,b], VÍTĚZNÍK Martin[3,d] and LOPOT František[1,c]

[1]Department of Anatomy and Biomechanics, Faculty of Physical Education and Sport, Charles University in Prague, José Martího 31, 162 52 Prague 6, Czech Republic

[2]Department of Pneumology, 2nd Faculty of Medicine, Charles University in Prague and Motol University Hospital, V Úvalu 84, 150 06 Prague 5, Czech Republic

[3]Faculty of Biomedical Engineering, Czech Technical University in Prague, nám. Sítná 3105, 272 01 Kladno 2, Czech Republic

[a]skalickydavid@seznam.cz, [b]Vaclav.Koucky@seznam.cz, [c]flopot@seznam.cz, [d]martin.viteznik@gmail.com

Keywords: Asthma, Sound, Wheezing, Harmonic analysis, Fourier transform

Abstract. Around 300 million people all over the world suffer from asthma [1]. This disease affects all ages and patients have primarily difficult breathing with wheezing in respiratory sounds, cough and feeling of constricted chest. Therefore their physical activity is strongly limited [2]. For better treatment and relieving the disease symptoms the early diagnosis of the disease is needed. For asthma (and other pulmonary diseases) diagnosing, we have some reliable methods: spirometry, measuring of peaks of expiratory velocity or measuring of bronchial reactivity. Unfortunately, these methods have their limits: they are not reliable for badly collaboration patients like infants to 3 years old, because these patients can't provide operations required for the diagnosis method (e.g. maximal inhalation and expiration). In this case, the standard diagnosing methods can't be used and therefore it is necessary to develop other diagnosis method without need for difficult cooperation of these patients. One of the most probably working usable principles is observing changes in the breath sound of ill person and detection of wheezing and other sounds which are the typical manifestations of the disease [3]. These typical phenomena can be detected by auscultation or by observing changes in frequency spectra of breath sound recording, which is created by harmonic analysis. This method could be more sensitive and without need patient's collaboration.

Introduction – disease and its manifestation

Asthma is a chronic inflammatory airway disease, which causes a bronchial obstruction [2]. The bronchial obstruction arises as reaction of bronchial tubes on inspiration of foreign substance causing asthma. The bronchial obstruction causes difficult air exchange in lungs and therefore patients with asthma have primarily difficult breathing, caugh and feeling of constricted chest. Patients with asthma have wheezing and crackles (different sound artefacts) in respiratory sound too.

It is important to discover asthma soon, because early diagnosis can help to set up more efficient treatment and therefore patient can be better stabilized [4]. One of the most reliable methods used for asthma diagnosis is spirometry [1]. Spirometry works with patient's breath (maximal inhalation and expiration) and using these data creates a spirometry curve. According to these curves it can be distinguished healthy and ill person. For reliable results of this method it is necessary a patient's cooperation. Unfortunately, small children age up to three years can't provide this operation and reached results are not suitable for reliable diagnosis.

Respiratory sound

Sound is an oscillation of acoustic pressure composed of many sine oscillation characterized by various frequencies. The human ear is able to detect pressure oscillation in frequency range 20 Hz – 20 kHz. The respiratory sound is sound in frequency range from 10 Hz to 1500 Hz [3]. The sound with frequencies up to 600 Hz goes through lung parenchyma better and therefore it is detected the most at lung areas (back or chest). The sound with frequencies over 600 Hz is detected better on jugulum. The normal lung sound spectrum is devoid of discrete peaks and it is not musical [3].

The airway obstruction generates turbulences in air flow, which produces different sounds at higher frequency level. These changes (in breath sound) are represented by wheezing and crackles. For wheezing the frequencies in the range of 300 Hz – 1000 Hz with higher amplitudes in comparison with neighboring areas are typical [4]. The duration of these areas is normally from 0.5 s to 0.75 s [4]. These searched phenomena in the respiratory sound could be emphasized using an intensive breathing e.g. by physical activity [4]. It is important to be able to detect the wheezing in respiratory sound because they accompanying the origin and course of bronchial obstruction. For wheezing detection is used auscultation method. Unfortunately, only quite a large obstruction can be discovered by auscultation, because human ear isn't sufficiently sensitive. Therefore it is necessary to have sufficiently sensitive device for discovering small obstructions, because it is likely, that small bronchial obstruction causes a little loud sound. This little loud wheezing is inaudible during auscultation, but the sound can be decomposed using harmonic analysis (e.g. Fourier transform) and thus these frequencies of wheezes can be discovered - Fig. 1.

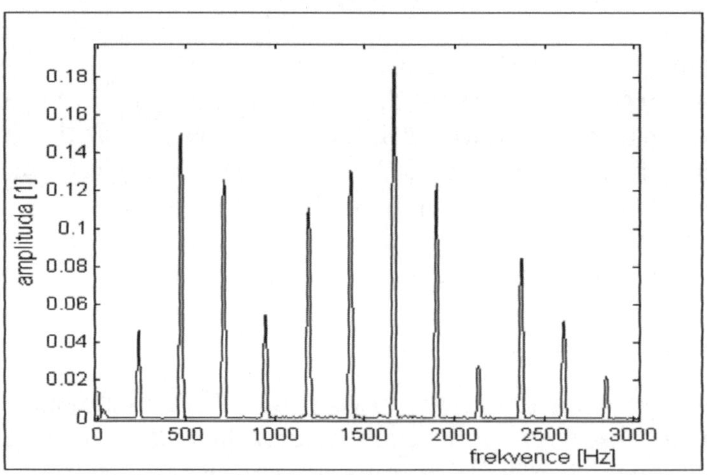

Fig.1: Frequency spectrum of the analyzed and decomposed music sound – the horizontal axis is a frequency scale, the vertical axis indicates an amplitude level of every frequency in the analyzed part of the sound

Data recording and processing

For verification of this hypothesis and for the purposes of our study, software for recording and processing of respiratory sounds was designed. The software operates on the principle of Fast Fourier Transform using Matlab background and creates frequency spectra throughout the length of the analyzed recording working with defined time intervals. The length of the time intervals corresponds to the duration of wheezing approximately. For better clarity of outcomes of performed analysis a specific suitable color scaling for frequencies in obtained frequency spectra was applied - Fig. 2. Finally, the coloured data were rearranged back to the timeline of original recording and so we get a frequency spectrum of the respiratory sound - Fig. 3.

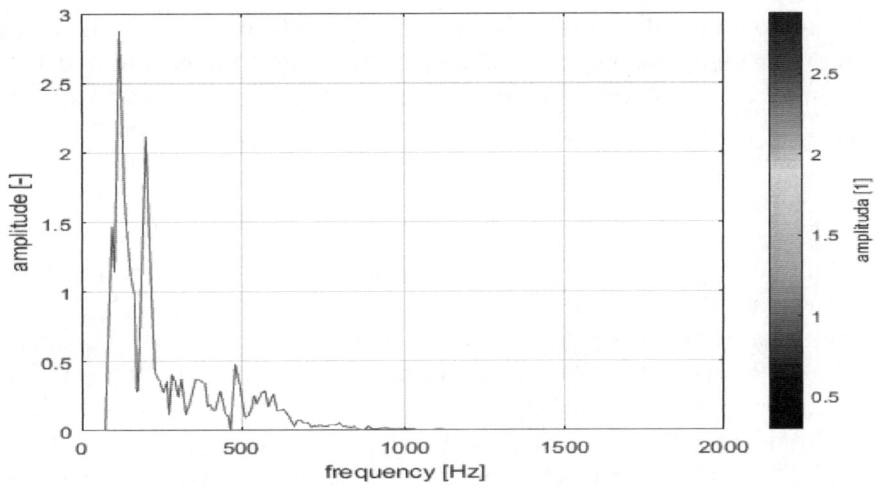

Fig.2: Colour scaling of obtained frequencies

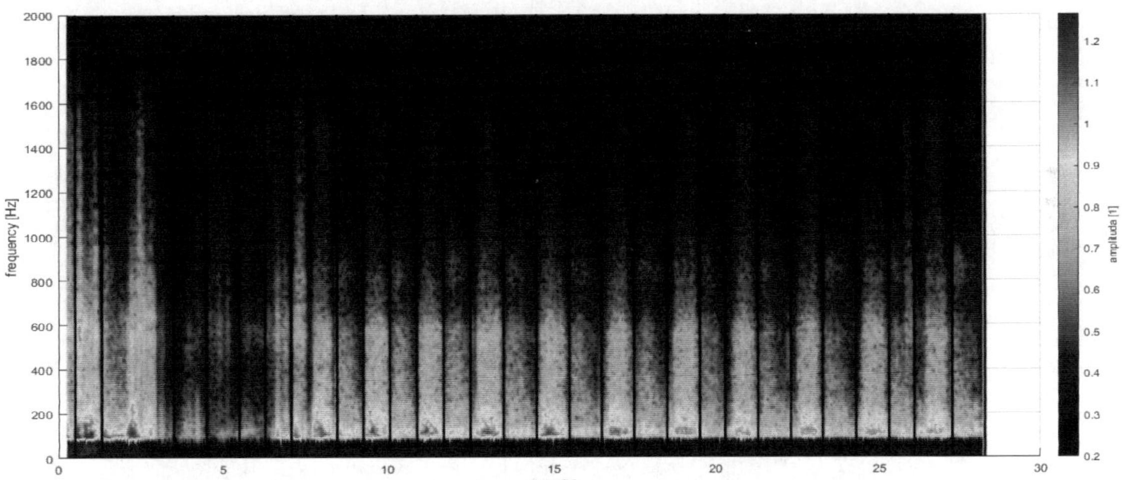

Fig.3: Frequency spectrum of the respiratory sound - displayed as a time course of frequency spectrum and amplitude of each frequency is represented by color. For better orientation the black vertical lines indicate inhalation and expiration phases

For better orientation in the spectrum the ability to detect areas of inhalation and expiration is necessary. Our software can detects these areas using filtration of the sound in specific frequency range and averaging of sound pressure oscillations. Then the software separates these areas in the spectrum using black vertical lines - Fig.3. This approach allows you to compare only inhalation phases and only expiration phases and their levels of amplitudes in various frequency ranges.

All required audio recordings of respiratory sounds were obtained in collaboration with Department of pneumology in UH Motol. In this time, twenty volunteers aged from 5 to 20

years (both healthy men and asthmatic patients) participated on the study. The respiratory sound was recorded before and after the physical activity of the patients and at jugulum areas (previous studies have shown that it is probably the most suitable location for recording).
The commercially available electronic stethoscope recording the heard respiratory sound was utilized in the study. The stethoscope works on 4000 Hz, thus the risk of aliasing and/or other mistakes due to the low sampling frequency were consider as minor ones.
The quality of respiratory sound recording is affected by location, where the sound was recorded and clarity of recording too – the recording should be without ambient noise. The quality of respiratory sound can be affected also by the size of the patient body [3]. Children have a distinct quality of lung sounds, which is generally attributed to acoustic transmission through smaller lungs and thinner chest walls [3].

Results

In the obtained frequency spectra of the respiratory sound recordings we can observe differences between healthy person (Fig. 4) and person with asthma (Fig. 5). It is obvious that sound of patient with asthma (Fig. 5) has higher level of amplitude in the frequency range between 300 Hz – 600 Hz. In this spectrum can be observe wheezing too – it is represented like areas with higher amplitudes in comparison with neighboring (areas highlighted using red ellipses – Fig. 5 and Fig. 6). From this study follows that wheezes in the spectra can be circular or elliptical areas in the frequency range 300 Hz – 800 Hz typically and with length of duration 0.2 – 0.75s, it is depending on the size of the obstruction. We found the areas of wheezing at frequency level around 1200 Hz too (Fig. 6) – it is wider area of occurrence compared to previous studies. It is also probably that it is important to observe new symptoms in the spectra like different amplitude level in the definite frequency range.

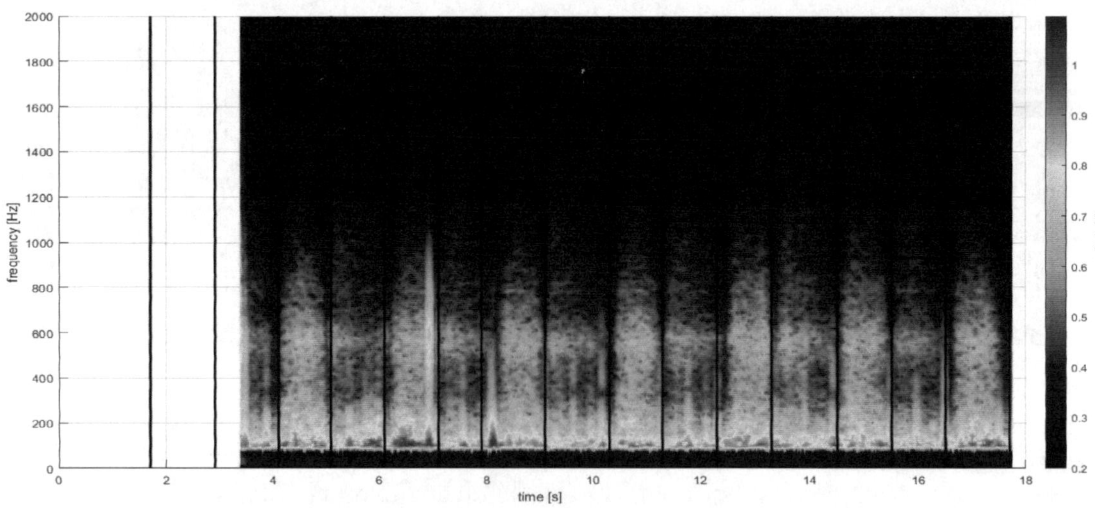

Fig.4: Frequency spectrum of breath sound recording of healthy person – the lower amplitude level of spectrum can be observed in this case

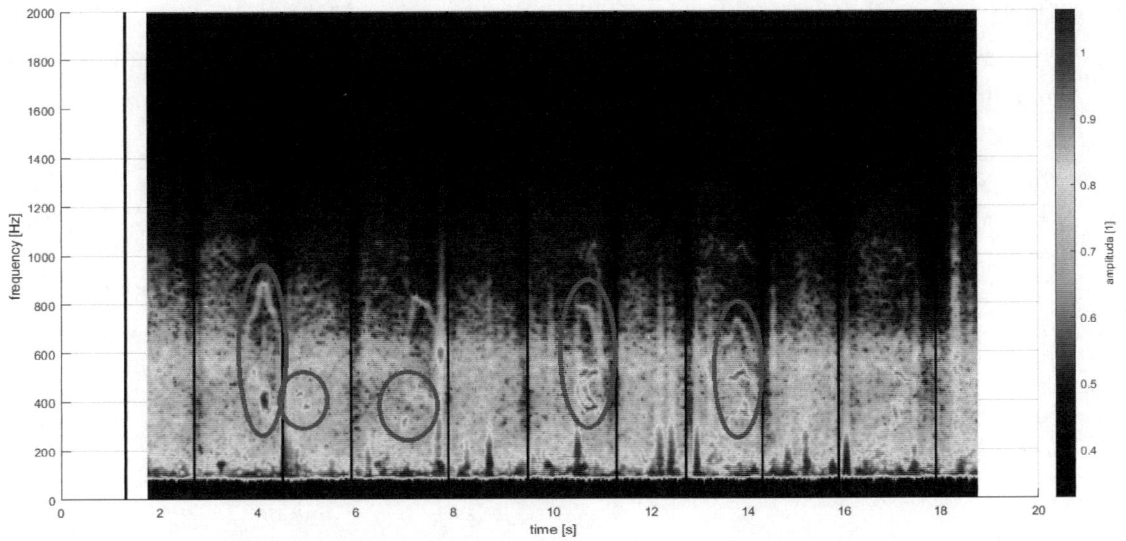

Fig.5: The frequency spectrum of breath of patient with asthma (typical wheezing in red ellipses and their harmonic frequencies)

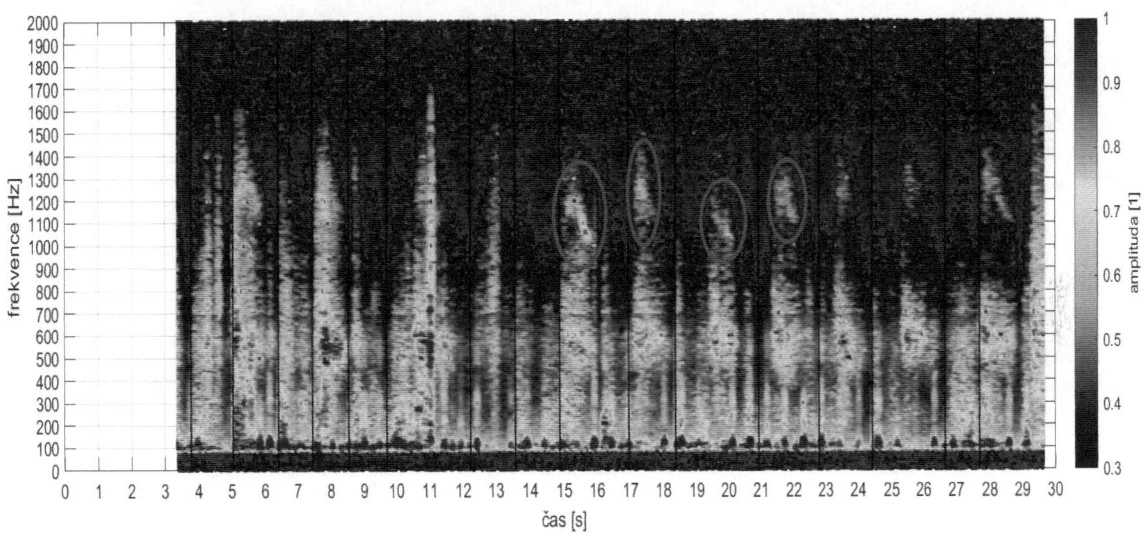

Fig.6: The frequency spectrum of breath of patient with asthma - wheezing around 1200 Hz (red ellipses)

Conclusions

Current results confirm that this method can discover the inaudible obstructions reliably and they also indicate that the method is usable exclusively for intraindividual medical examination. However, it isn't a disadvantage from clinical practice. Now it is necessary to collaborate with physicians to obtain more data and find a clinical significance of discovered phenomena.

Acknowledgements

This text is supported by GAUK grant No. 9022/2018.

References

[1] M. Teřl, P. Pohunek (eds.).: Strategie diagnostiky, prevence a léčby astmatu: uvedení globální strategie do praxe v ČR. Vyd. 1. Praha: Jalna, 2012.

[2] J. G. Ayres. Astma. 1. Vyd. Praha: Grada, 2001. Informace a rady lékaře. ISBN 80-247-0091-3.

[3] H. Pasterkamp, S. S. Kraman, G. R. Wodicka. Respiratory Sounds. American Journal of Respiratory and Critical Care Medicine. 1997, ISSN 1073-449x.

[4] Lekar a technika – Clinician and Technology 2017, vol 47(2), ISSN 2336-5552.

[5] H. S. Fischer, L. C. Puder. Relationship between computerized wheeze detection and lung function parameters in young infants. Pediatric Pulmionology. 2015.

[6] S. S. Kraman. Comparison of lung sound transducers using a bioacoustic transducers testing system. Journal of Applied Physiology. 2006, ISSN 8750-7587.

EAN 2019
57th conference on experimental stress analysis

Experimental FEM validation of the composite recurve bow

SLÁMA David[1,a], NÁVRAT Tomáš[1,b] and KREJČÍ Petr[1,c]

[1]Institute of Solid Mechanics, Mechatronics and Biomechanics; Faculty of Mechanical Engineering; Brno University of Technology; Technická 2896/2, 616 69 Brno, Czech Republic

[a]158230@vutbr.cz, [b]navrat@fme.vutbr.cz, [c]krejci.p@fme.vutbr.cz

Keywords: FEM validation, CFRP, GFRP, recurve bow, orthotropic material, testing, strain gauge, load cell, FFT, modal analysis, large deflections, non-linear static analysis.

Abstract. The main scope of this article is to present the results of several experiments that were performed on the developed home-made composite recurve bow fabricated from ash wood, GFRP and CFRP. The results are compared with validated FEM. The FEM is validated by comparison with: measured draw force, measured strain during the bow draw at two selected locations, measured signals at the same locations of the excited bow without the string from which eigen frequencies are evaluated using FFT. All measured data show good compliance with the validated FEM. The main advantage of those experiments is a possible alternative evaluation of stiffness orthotropic material properties without the need of dedicated test setup generally connected with extra cost.

Introduction

Bows are most commonly described by the maximum draw force (draw weight) and maximum draw displacement (draw length) because those parameters can be different for each person. From technical point of view there is another parameter though that is usually not specified for most of the bows: its energy. This energy can be calculated as an area beneath the curve draw force – displacement. In general, for all types of bows (excluding compound bows) non-linear behaviour between drawing force and deflection is characteristic. Historically, there has been significant progress in this area. Main goal has been to find optimized bow shape which would be able to store maximum energy. Primitive bows had been replaced by long bows and recurve bows. In the end, recurve bow became the winner of this optimization race that took centuries [1].

There still has been a great effort to improve energy of the recurve bows using different types of mathematical models [2], [3], [4]. Several non-linear problems, such as: large deflections, contact between the bow string and limbs, string pretension, complicate the applicability of those models. On top of that, most authors neglect the fact that recurve bows are fabricated from composite materials with orthotropic properties (wood, CFRP, GFRP).

The main scope of this article is to present the results of several experiments that have been performed on developed home-made composite recurve bow fabricated from ash wood, GFRP and CFRP. The results are compared with validated FEM. The main advantage of those experiments is a possible alternative evaluation of stiffness orthotropic properties without the need of dedicated test set-up connected with extra cost.

Materials and methods

Material properties. Thicknesses of glass plies (0.09 mm) and carbon plies (0.25 mm) without epoxy resin were measured. Thicknesses of fabricated recurve bow were measured at three specific locations where the lay-up was known.

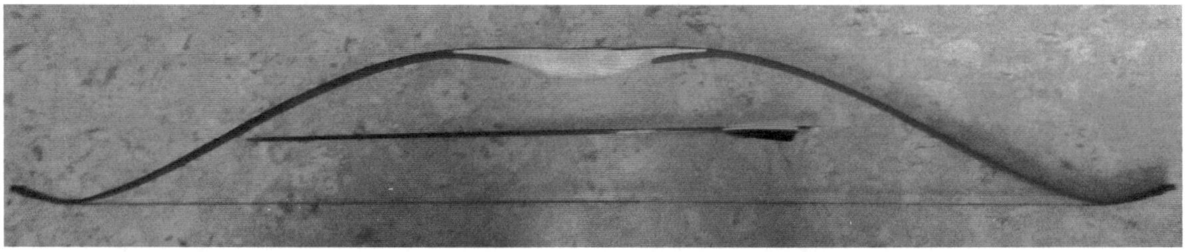

Figure 1: Fabricated recurve bow

Optimisation in terms of minimising sum of relative errors between measured values was performed in software MS Excel while design variables were only CFRP/GFRP fibre volume ratios. Resulting fibre volume ratios were determined as shown in the Table 1. Comparison between calculated thickness values and measured values are herein reported in the Table 2 showing good compliance.

Table 1: Fabricated fibre reinforced polymers volume ratios

CFRP Fibre volume ratio [-]	GFRP Fibre volume ratio [-]
0.53	0.45

Table 2: Comparison of measured and calculated thicknesses at specific locations of the recurve bow

Nr. of CFRP plies [-]	Nr. of GFRP plies [-]	Measured thickness [mm]	Calculated thickness [mm]	Relative error [%]
6	10	4.88	4.83	0.85
6	9	4.70	4.63	1.42
6	7	4.13	4.23	2.63

The resulting fibre volume ratios were applied as inputs for determination of orthotropic stiffness properties for CFRP/GFRP plies considering properties from supplier datasheets applying rules of mixtures [5]. Orthotropic parameters of ash wood [6] which had been used for the riser were also considered as herein reported as shown in the Table 3.

Table 3: Linear orthotropic models of material used in FEM

	CFRP	GFRP	Ash wood
E_1 [GPa]	127.85	15.90	15.00
E_2 [GPa]	5.36	4.93	10.00
E_3 [GPa]	5.36	4.93	10.00
μ_{12} [-]	0.29	0.36	0.20
μ_{23} [-]	0.73	0.68	0.20
μ_{13} [-]	0.29	0.36	0.20
G_{12} [GPa]	2.08	1.72	6.00
G_{23} [GPa]	1.55	1.47	2.80
G_{13} [GPa]	2.08	1.72	6.00

Densities of glass plies, carbon plies and epoxy resin were taken from supplier datasheets. Two GFRP samples masses without 2 mm overall thicknesses were measured.

Table 4: Densities of raw materials

Glass density [kg/m^3]	Carbon density [kg/m^3]	Epoxy density [kg/m^3]
2090	1800	2090

Table 5: Comparison of measured and calculated GFRP sample masses

Measured Sample mass [g]	Measured Sample surface [mm^2]	Calculated mass [g]
22.0	5440	22.7
30.0	7298	30.5

Results showed good compliance between expected and measured masses. Thus, for CFRP plies same relations based on rules of mixtures were applied without additional sample manufacturing. The resulting plie densities reported in Table 6 were implemented into FEM.

Table 6: Linear orthotropic models of material used in FEM

CFRP density [kg/m^3]	GFRP density [kg/m^3]	Ash wood density [kg/m^3]
1936	2090	710

Methods. Finite element modelling was chosen for the problem solution. Analyses were done in the student version of SW Ansys APDL. Created batch file based on parametric input variables was particularly useful during FEM validation loop described below. Bow symmetry was considered in the FEM and symmetric boundary conditions were defined at the bow symmetry axis. Linear SHELL181 elements, recommended for large deflections, were used with lay-up properties described above. Rigid string was modelled with one linear element LINK180 which was set up to be functional in tension only. Connection between LINK and SHELL elements was done by merging two coincident nodes which models the joint transferring only forces while rotations are free. Nodal displacement load in horizontal axis was applied at the second end of the string rigid element. All analyses were done with large displacements turned on which means that the load had been applied step by step and stiffness matrix was updated adequately. For the mode shape extraction Block Lanczos method was used. Same FEM and boundary conditions were applied excluding rigid string element and nodal displacement load.

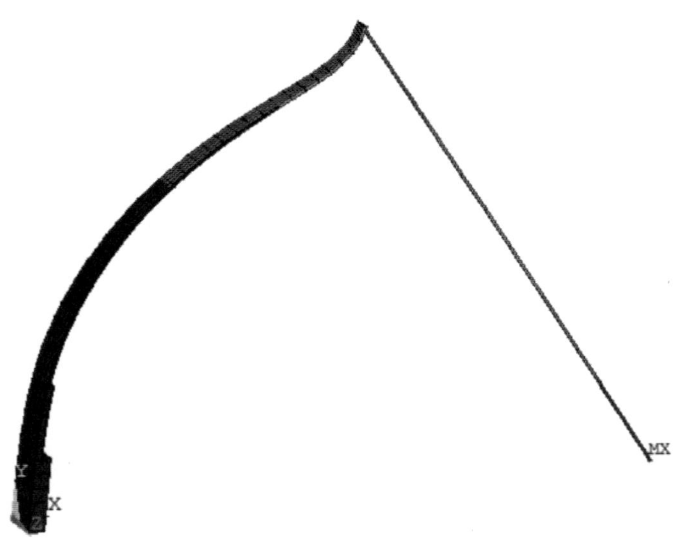

Figure 2: FEM deformed shape at maximum draw length

Figure 3: FEM modal shape of the system without the string at 15.80 Hz

Testing

For the testing of the recurve bow following devices were used:

- strain gauge at height 240 mm from the middle of the bow = “end of the limbs”;
- strain gauge at height 470 mm from the middle of the bow = “middle of the limbs”;
- load cell at the string symmetry axis for monitoring of the Draw Force.

Figure 4: Harness and strain gauge bonded on the recurve bow

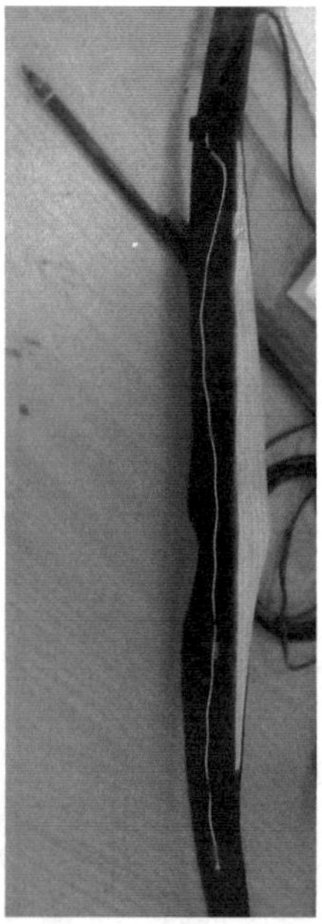

Figure 5: Harness and strain gauges

Figure 6: Detail of the bonded strain gauge 1

Figure 7: Detail of the bonded strain gauge 2

Following tests were performed using devices described above:

- static test - Strain measurements at the outer (external) plane in tension of the bow and Draw Force measurements during drawing of the bow;
- dynamic test - Resonance search of the bow without the string using strain gauges.

Static test was performed as follows:

1. load cell put between the string and the drawing hand;
2. bow supported (hold) at the riser only;
3. drawing of the bow to the predefined displacement values and sustaining at those values for a few seconds for clear data post processing (three measurements done).

Dynamic test was performed as follows:

1. bow supported (hold) at the riser only;
2. excitation of the upper bow limb.

Data processing

All data were processed in software MATLAB.

Static test. Test Data were processed as follows:

1. evaluation of Force-time and Strain-time curves;
2. determination of Force-Displacement and Strain-Displacement curves (relation between time-displacement was known as reported above);
3. averaging Force and Strain values from three measurements done.

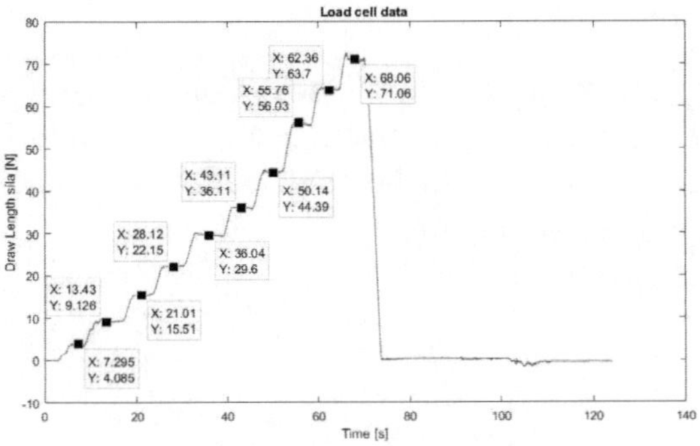

Figure 8: Example of Force evaluation

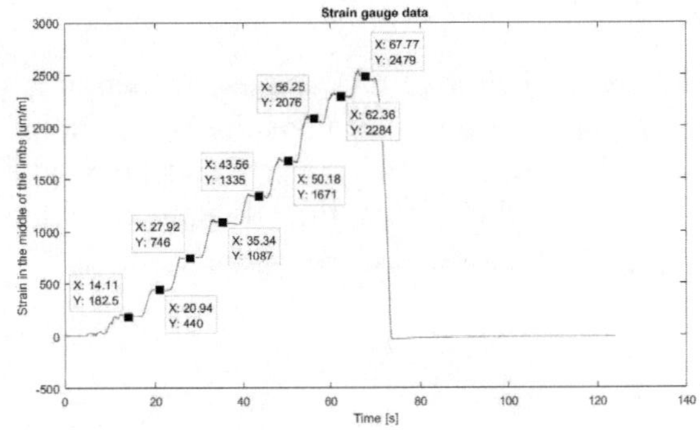

Figure 9: Example of Strain evaluation

Dynamic test. Test Data were processed as follows:

1. signal import and evaluation;
2. selection of only steady-state vibrations for FFT inputs;
3. FFT and resonance evaluation.

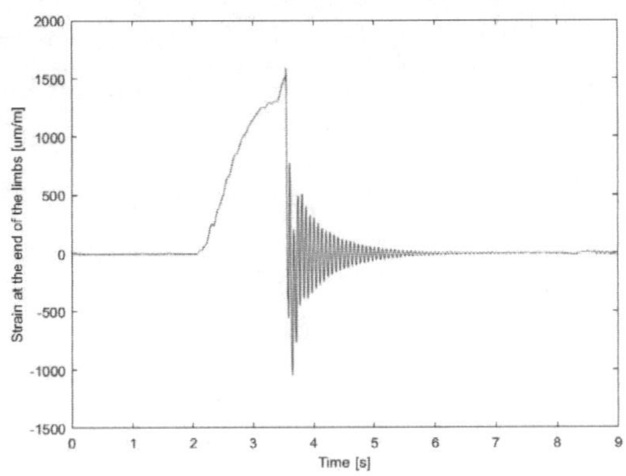

Figure 10: Imported Signal

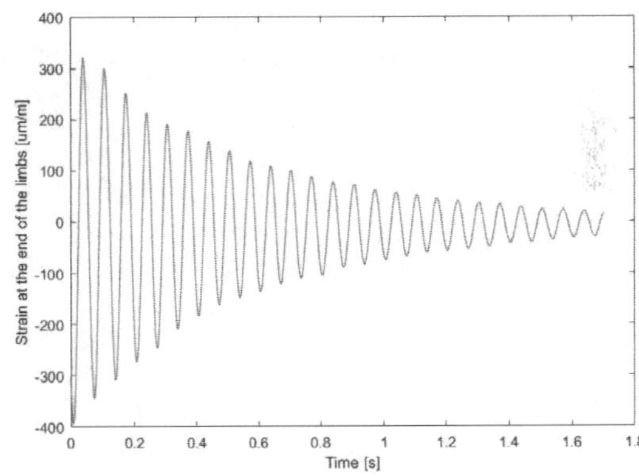

Figure 11: Selected steady-state oscillations for FFT

Experimental FEM validation

The main goal of this activity was to update FEM of the bow to be compliant with the test results. After loop of analyses varying different input parameters, it was found out that external CFRP plies in compression (inner plane closer to the archer) do not have expected stiffness evaluated from rules of mixtures compared to external CFRP plies on the opposite outer plane in tension. The stiffness in compression that well matched with all performed experiments was evaluated as 30 % of expected CFRP stiffness in fibre direction ($E1 = 38.36$ GPa). There was no other need to change other input parameter.

Rationale. Used fabrication process of the recurve bow (hand lay-up moulding excluding vacuum bagging and curing) is not standard fabrication process of FRP in aerospace industry. Thus, it is not surprising that the layer with reduced stiffness properties is the top layer which had been in contact with air during drying. It is possible that this layer had more air bubbles inside and worse properties than the layers beneath which were not in direct contact with air. This effect might be amplified by non-linear effects such as: large deflections and buckling on the fibre level since in FEM reduced stiffness had been applied only to the plies in compression.

Results

The comparison of FEM and experimental results shows good compliance. Considering that modal analysis does not consider the damping of the real system, resonance peaks would be probably shifted to the lower frequencies and compliance would be even better if transient analysis was done in the time domain considering also system damping.

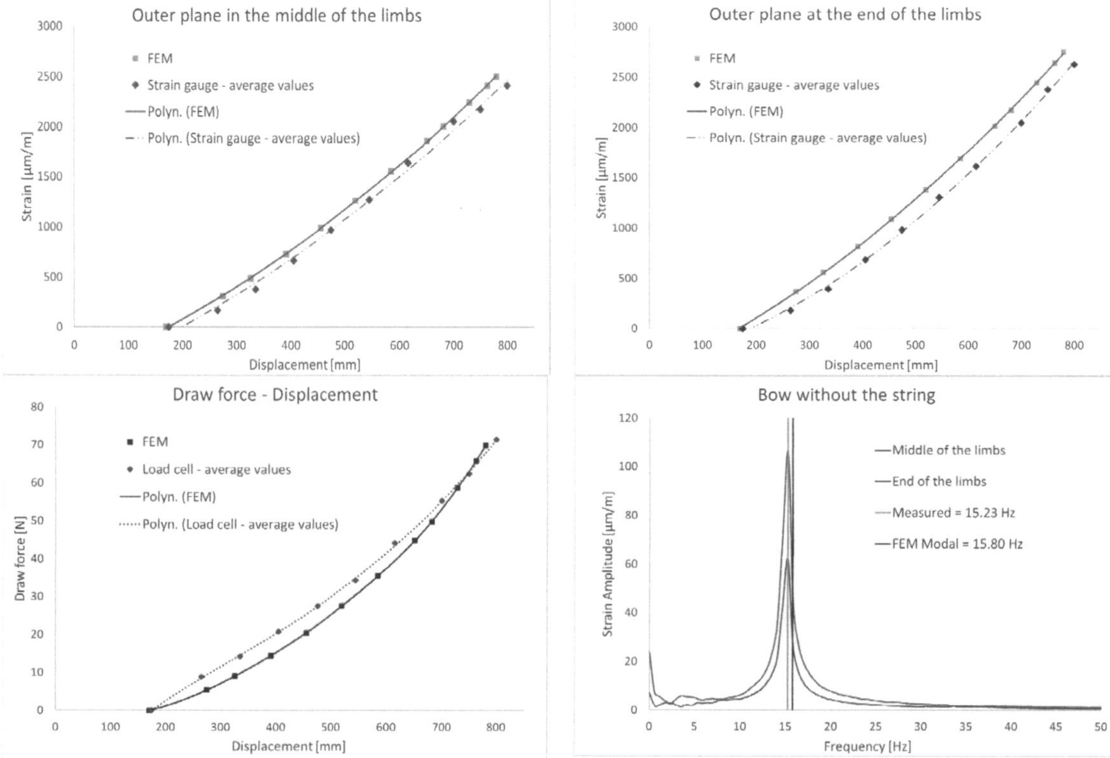

Conclusions

The results show good compliance between FEM prediction and experiment results. Used fabrication process of the recurve bow is characteristic by following material behaviour:

- 100 % of expected stiffness CFRP on the outer plane in tension, E_1 = 127.85 GPa
- 30 % of expected stiffness CFRP on the inner plane in compression, E_1 = 38.36 GPa

Composites in compression generally have lower strength but almost similar stiffness [7]. Thus, it is generally recommended to design composites to be in tension. From the FEM results and experiment results it seems though, that for a hand-layup technology excluding vacuum bagging and curing also stiffness may differ from expected values. This effect might be amplified by non-linear effects such as: large deflections and buckling on the fibre level.

Acknowledgement

This work is an output of successfully finished VUT FSI project: Design and Development of the composite recurve bow, SPP Nr.: RV9080000/11122.

References

[1] LIEU, D.K., Jinho KIM and Ki CHAN, Fundamentals of the Design of Olympic Recurve Bows. University of California, Berkeley, Korea National Sport University.

[2] KOOI, B.W., On the Mechanics of the Modern Working-Recurve Bow. Computational Mechanics 8. 1991, 291-304.

[3] DEMIR, Sermet and Bülent EKICI, 2014. Optimization of design parameters for Turkish Tirkeş (war) bow. Composites Part B. Engineering 66, 147-155. DOI: http://dx.doi.org/10.1016/j.compositesb.2014.04.029.

[4] PAGITZ, M. and K. U. BLETZINGER, Shape optimization of a bow. Struct Multidisc Optim 28. 2004, 73-76. DOI: DOI 10.1007/s00158-004-0413-0.

[5] Agarwal, B. D. and L. J., BROUTMAN, Vláknové kompozity. SNTL Prague, 1987.

[6] PIŠTĚK, A., Soubor podkladů pro pevnostní konstrukce v letectví, 1994, VUT Brno.

[7] WU, Weili, Qingtao WANG and Wei LI, Comparison of Tensile and Compressive Properties of Carbon/Glass Interlayer and Intralayer Hybrid Composites. Materials. 2018(11). DOI: https://doi.org/10.3390/ma11071105.

EAN 2019
57th conference on experimental stress analysis

The analysis of dynamic effects in the frog part of railway switches

SMUTNY Jaroslav[1,a], JANOSTIK Dusan[1,b], NOHAL Viktor[1,c] and PAZDERA Lubos[2,a]

[1]Institute of Railway Structures and Constructions, Faculty of Civil Engineering; Brno University of Technology; Veveri 331/95, 602 69 Brno, Czech Republic

[2]Institute of Physics, Faculty of Civil Engineering; Brno University of Technology; Zizkova 331/17, 602 69 Brno, Czech Republic

[a]smutny.j@fce.vutbr.cz, [b]janostik.d@fce.vutbr.cz, [c]nohal.v@fce.vutbr.cz, [d]pazdera.l@fce.vutbr.cz

Keywords: Railway switch, dynamic effects, vibration analysis in time and frequency domain, non-linear time-frequency transformation, Zhao-Atlas-Marks transformation

Abstract. This paper deals with description and application of the Zhao-Atlas-Marks transformation for dynamic analysis. This transformation belongs to the group of non-linear time-frequency processes. Thanks to its properties it may be successfully used in the area of non-stationary and transitional signals describing various natural processes. The use in the field of railway constructions testing represents a quite interesting application area of the transformation. This paper contains a mathematical description of the transformation; a case study from the area of railway switches dynamic analysis and practical experience obtained and recommendations for its practical use.

Introduction

The fundamental requirements of the particular parts of the railway track are their functional reliability and the safety of the interconnected railway operation. Although the superstructure has been developed nearly to the perfection during more than 150 years of the railway history, it is still possible to find new technical solutions. This especially applies to various types of switches structures, various rail fastenings, various types of rail pads, sleepers etc.

Permanent pressure to increase the transport speed and the operational load on the railway resulted in huge development of new technologies. The decision on modernization of the corridors became the impulse for development of all areas of the railway transport, both in the area of vehicles and infrastructure. In correspondence with the above-mentioned trend, the application of new experimental processes should continue to evaluate the quality and suitability of individual constructional solutions.

The level of dynamic effects in the railway superstructure is influenced by its quality, operational and technical conditions, climatic events and, most importantly by dynamic load on the railway vehicle axle. The fact also relates to construction of the railway switches representing the key points of the railway roads.

The railway switch structures represent one of the key points of the railway transport. Therefore it is necessary to pay a special attention to these structures. The switch structure consists of a number of structural components of different properties which in summary must fulfil the required functions. The most important properties are particularly their reliability and safety within the framework of the railway operation [1].

In accordance to this trend, the development and application of new experimental procedures for assessing the quality and effectiveness of individual design solutions must also proceed.

Zhao-Atlas-Marks transformation

Information about any technical event is represented in the signal by time changes of the instantaneous value of the physical quantity described by the signal. There are a number of methods in the time plane that can be applied to the measured signal. These include the search for global and local highs and lows depending on time, correlation analysis, analysis of signal attenuation, or different statistical methods, etc.

In some cases, important information about the analysed frequency domain event can be obtained. The Fourier transform is the most widely used and best known method to convert from time to frequency domain. Often, it is necessary to analyse non-stationary processes, i.e. events where the frequency content of the signal varies with time.

One of the possible ways to analyse the time incidence of frequency components, especially in transient and non-stationary signals, is to use the so-called time-frequency transformations. These can be divided into two basic classes according to the computational process [2]:

- linear (these include, in particular, Short-Term Fourier Transform and Wavelet Transform)
- nonlinear (include, in particular, quadratic Cohen, affine, and hyperbolic transformations, and other special techniques)

Zhao-Atlas-Marks transform belongs to the class of so-called quadratic, time and frequency invariant transformations. The biggest advantage of this group of transformations is the fact that their resulting resolution in time and frequency is not limited by the so-called Heisenberg uncertainty principle. This is manifested by the high resolution of the transformations in the time-frequency plane. At the same time the so-called core function determines the properties of a particular transformation. It is generally assumed that any quadratic, temporal, and frequency-invariant transformations can be expressed by [3]

$$CT_x(t,\omega) = \int_{-\infty}^{\infty}\int_{-\infty}^{\infty} \Pi(\tau - t, \theta - \omega) \cdot WVT(\tau,\theta) \cdot d\tau \cdot d\theta \,, \tag{1}$$

where

$$\Pi(t,\omega) = \int_{-\infty}^{\infty}\int_{-\infty}^{\infty} \psi(\theta,\tau) e^{-j \cdot 2 \cdot \pi \cdot (\omega \cdot \tau - \theta \cdot t)} \cdot dt \cdot d\omega \,. \tag{2}$$

Note that *WVT* represents the coefficients of the Wigner-Vill transform, Π is the two-dimensional Fourier transform of the core function ψ, the symbol τ represents the time shift, t is the time, ω is the angular frequency, θ the frequency shift and j is complex operator.

The core function for Zhao-Atlas-Marks transformation is defined by [4]

$$\psi(\theta,\tau) = h(\tau) \cdot \frac{\sin a \cdot \theta \cdot \tau}{a \cdot \theta \cdot \tau} \,, \tag{3}$$

where $h(\tau)$ represents a window function, a is a suitably selected parameter for suppressing interference components. In practice, the coefficients of the individual quadratic, temporally and frequency-invariant CTx transformations for discrete signals can be calculated by the fast two-dimensional Fourier transform of the discrete characteristic function $A_x(n, k) \cdot \psi(n, k)$. $Ax(n, k)$ is defined by the equation (4)

$$A_x(\theta,\tau)=\int_{-\infty}^{\infty} x\left(t+\frac{\tau}{2}\right)\cdot x^*\left(t-\frac{\tau}{2}\right)\cdot e^{j\cdot\theta\cdot t}\cdot dt\,. \tag{4}$$

It is a "narrow-band" function that is complex and represents a measure of the time-frequency correlation of a signal, or expresses the degree of similarity between a signal and its shifted version in a time-frequency plane. Thus, it is essentially a time-frequency autocorrelation function.

Case study

The following text presents the use of Zhao-Atlas-Marks transformation in the field of railway engineering. It was an analysis of dynamic effects in the frog part of two switches. Let us note that the frog section formed by wing rails, the frog wedge and the frog rails connected to the frog wedge belong to the most stressed places of the railway line.

The monitored switches no. 3 and no. 4 (Fig. 1) are situated in Třebová station head of the railway station in Ústí nad Orlicí. Both switches are located on a high embankment. They have the same slenderness (1:12 - 500), both are mostly driven by the frog tip and have the same frogs (UIC 60 shape, ZPT frog type - monoblock). Switch no. 4 has a classic fastening with ribbed bases on concrete sleepers with Skl 24 clamps and is located in track no. 2. Switch no. 3 has a new type of rail fastening UNO3718 with ribbed bases on concrete sleepers with Skl 24 clamps and is located in rail no. 1. The main difference between the aforementioned fasteners is primarily in the base under the rail foot. The measurements were performed in March 2019.

Fig. 1: View of rail coupling with switches 3 and 4.

Vibration measurement and analysis methodology was used to test both switches designs. The next part of the paper is devoted to the analysis of vibration transmission passing from the frog through the sleeper to the rail bed. Vibration characteristics were obtained based on the response of tested structures to moving loads, i.e. trainsets. Vibration-accelerating piezoelectric transducers were used to measure the dynamic-effect characteristics. Within this methodology, the sensors are positioned to be able to capture the spreading of vibration energy throughout the system. A three-axis accelerometer located on the foot of the wing rail monitors the magnitude of the dynamic impact that acts on the frog of the switch. In the vertical direction (A_{WRZ}), the size of the vertical component of the dynamic impact is captured on the frog, respectively the part to be transferred to the foot of the wing rail. Depending on the state of the crossing geometry, a larger or smaller part of the dynamic impact is also realized in the longitudinal direction (A_{WRX}). In the transverse direction (A_{WRY}), the wheelset is guided by a retainer. The lateral impact on the frog is very small, but it is interesting to observe this component only from the point of view of the smoothness of the train passing through the switch, therefore, this component is also monitored within the proposed methodology. The other sensors were uniaxial and were positioned to best capture the dynamic impact spreading from the frog crossing through the sleeper to the ballast bed. The A_{SZ} sensor was placed on the sleeper under the frog tip and the A_{BBZ} sensor on the so-called measuring rod (waveguide), which was embedded in the ballast bed near the frog tip.

Note that dynamic parameter measurement was performed after the frog replacement. At both switches, a frog with a high-manganese cast steel central section reinforced by an explosion was installed.

Let us first consider comparing the maximum and minimum vibration acceleration values at the transition from the wing rail to the ballast bed in the time domain.

Tab. 1: The comparison of the maximum values of vibration acceleration

Construction	Train	V [km/h]	a [m/s^2]					
			A_{WRZ}		A_{SZ}		A_{BBZ}	
			Min	Max	Min	Max	Min	Max
3	RailJet	130	-559	719	-80	70	-28	29
4	RailJet	130	-507	515	-141	150	-65	60
3	Pendolino	150	-712	997	-68	65	-79	31
4	Pendolino	150	-477	510	-117	134	-62	60
3	Leo Express	130	-638	839	-70	62	-40	33
4	Leo Express	130	-453	379	-132	145	-59	53
3	RegioJet	130	-729	776	-98	86	-38	43
4	RegioJet	130	-564	497	-132	185	-70	65
3	R 380	130	-696	816	-100	110	-57	32
4	R 380	130	-462	496	-115	157	-73	51

Tab. 2: The comparison of effective vibration acceleration values

Construction	Train	V [km/h]	aef [m/s^2]		
			A_{WRZ}	A_{SZ}	A_{BBZ}
3	RailJet	130	65	10	4
4	RailJet	130	52	18	7
3	Pendolino	150	80	11	3
4	Pendolino	150	70	16	6
3	Leo Express	130	65	9	3
4	Leo Express	130	50	17	6
3	RegioJet	130	67	12	4
4	RegioJet	130	59	20	8
3	380	130	67	11	3
4	380	130	43	14	5

Comparing values indicated in the table Tab. 1 shows that on the wing rail in the vertical direction (A_{WRZ}) higher values were found at the switch no. 3. Values on the sleeper at the crossing frog (A_{SZ}) and at the ballast bed (A_{BBZ}) are already higher at switch no. 4, and the difference is quite significant. Significantly higher vibration acceleration values at the frog sleeper (A_{SZ}) and in the ballast bed (A_{BBZ}) at switch no. 4, mean that more peak values are transmitted to the sleeper and ballast bed. Similar results are shown in Table 2. Note that this table shows comparison of effective acceleration values.

Because of the limited number of pages, the next measurement outputs are presented only for the position of the accelerometer on the sleeper under the frog. The analysis outputs were implemented into three graphs. The upper graph of each image shows the time course of the vibration acceleration. The amplitude spectrum of the vibration response is displayed in the left graph. The middle graph shows a 3D representation of the time-frequency course of the amplitude spectrum of the vibration response. Spectral values are expressed in decibel scale by different colors (or grayscale). Note that the maximum value is black.

It can be seen from the comparison of the amplitude spectra (Fig. 3, Fig. 4) that the calculated spectra are different for both structures. On the wing rail of switch no. 3, the maximum spectrum values at higher frequencies were reached than at switch no. 4. It is already apparent on the sleeper that the shape of the spectra is similar for both structures. The maximum spectral values were reached at switch no. 4.

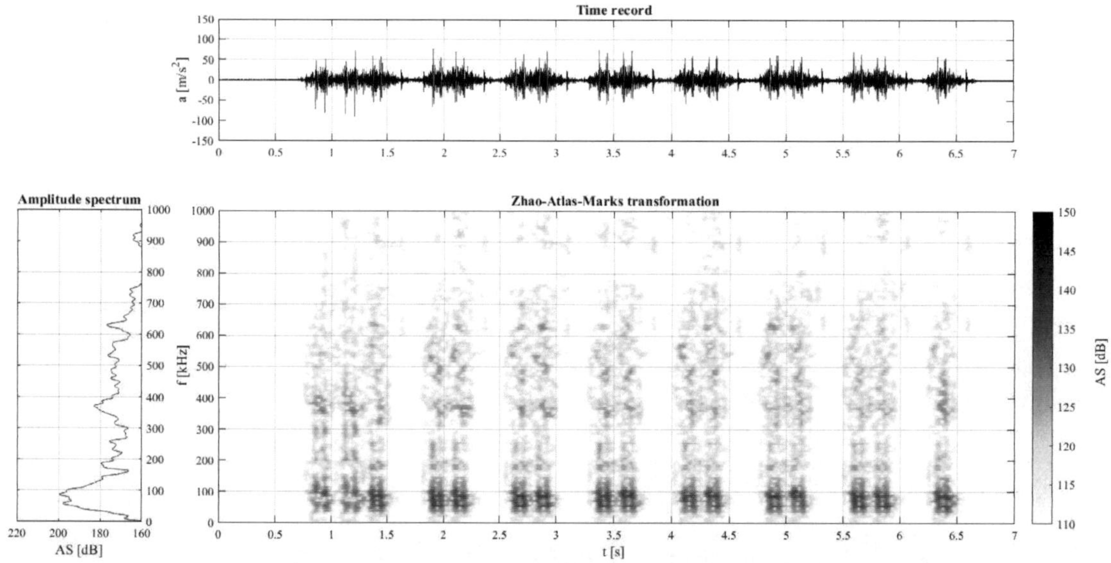

Fig. 2: Time, frequency and time-frequency characteristics of the vibrations, sensor on the sleeper, switch no. 3, vertically direction, RailJet, speed 130 km·h^{-1}

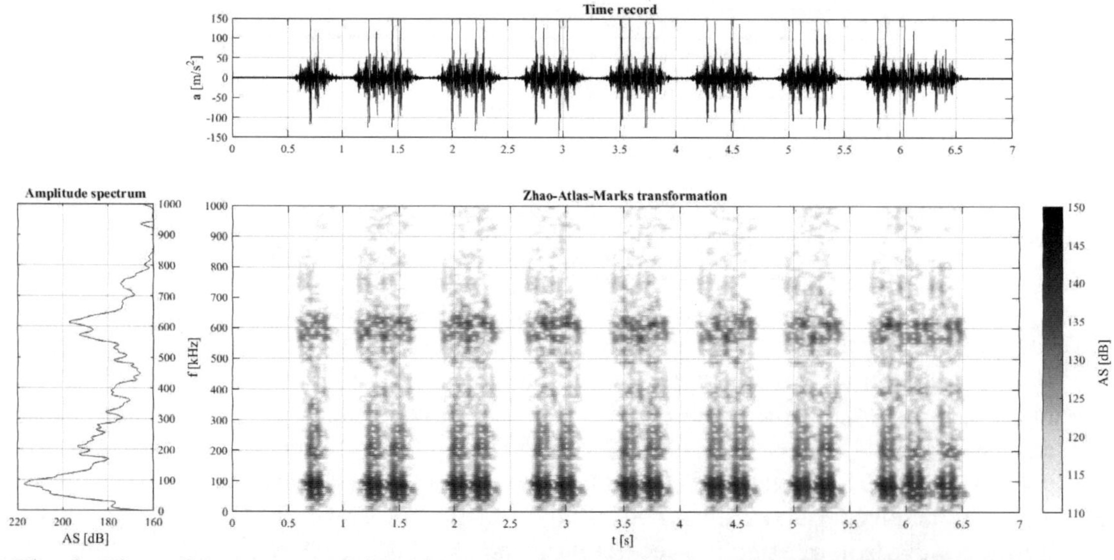

Fig. 3: Time, frequency and time-frequency characteristics of the vibrations, sensor on the sleeper, switch no. 4, vertically direction, RailJet, speed 130 km·h^{-1}

The same conclusions are confirmed by the time-frequency analysis of the Zhao-Atlas-Marks transformation. As can be seen from the middle graphs (Figs. 3 and 4) compared to time and frequency analysis, there is a time localization of significant frequency components.

It is therefore possible to deduce a positive influence of the elasticity in the fastening node from the presented graphs. The new rail fastening node UNO3718 on switch no. 3, even after frog replacement, better dampens peak vibration acceleration values compared to conventional fastening at switch no. 4. In this way the ballast bed is better protected from gradual degradation.

Conclusions

Zhao-Atlas-Marks transformation offers a comprehensive tool for analysing non-stationary signals in particular. Characteristic feature of presented transformation is the fact that its resulting distinguishment in time and frequency in not limited by Heisenberg principle of indefiniteness. This fact includes the high distinguishing ability in the time frequency plane that gives rise to a "precise" localisation of important frequency components in time. This method gives a fast and accurate localisation of frequency components included in the measured signal.

From the vibration measurement analysis in both switches the positive influence of the elasticity in the fastening node can be evaluated. On the sleeper under the frog, there is a smaller vibration acceleration value at the switch no. 3 than at switch no. 4. Elasticized fastening node increases vertical elasticity of superstructure, which ultimately leads to a further reduction of wear and disintegration of aggregate of the ballast bed. In this way, the upper of the switch is protected and the quality of the crossover geometric parameters is maintained. This solution offers railway managers the opportunity to significantly reduce the annual maintenance costs of railway lines.

Acknowledgments

This paper has been worked out under the research project FAST-S-19-5797, The analysis of dynamic response of railway track construction by the affine time frequency transformations method.

References

[1] Moravcik M.: Analysis of vehicle bogie effects on track structure-nonstationary analysis of dynamic response, Communications, Volume 13, 3/2011, EDIS Zilina, ISSN 1335-4205

[2] Smutný, J.; Pazdera, L., The experimental analysis of dynamic processes related to railway transport, monography, ISBN 978-80-7204-827-4, AKADEMICKÉ NAKLADATELSTVÍ CERM, Brno, 2012

[3] Smutný J.: Measurement and Analysis of Dynamic and Acoustic Parameters of Rail Fastening, NDT & E International - Independent Nondestructive Testing and Evaluation, 2004, Volume 37, Issues 8, ELSEVIER, pp. 119-129, ISSN 0963-8695

[4] Smutný J., Pazdera L.: New Techniques in Analysis of Dynamic Parameters Rail Fastening, InSight, The Journal of The British Institute of Non-Destructive Testing. Vol 46. No 10. October. 2004. pp. 612-615. ISSN 13542575

Non-associated flow rule in the prediction of complex stress states and deformations

ŠEBEK František[1,a], KUBÍK Petr[1,b], PETRUŠKA Jindřich[1,c] and ZAPLETAL Josef[2,d]

[1]Institute of Solid Mechanics, Mechatronics and Biomechanics; Faculty of Mechanical Engineering; Brno University of Technology; Technická 2896/2, 616 69 Brno, Czech Republic

[2]Institute of Materials Science and Engineering; Faculty of Mechanical Engineering; Brno University of Technology; Technická 2896/2, 616 69 Brno, Czech Republic

[a]sebek@fme.vutbr.cz, [b]kubik.p@fme.vutbr.cz, [c]petruska@fme.vutbr.cz, [d]zapletal@fme.vutbr.cz

Keywords: Flow rule, Ductile fracture, Large plastic deformations, Lode parameter, Stress triaxiality

Abstract. The paper deals with the application of non-associated flow rule to the non-quadratic yield criterion. The utilization resulted in a significant improvement in the predicted stress states as well as deformations, which matched the theoretical values and experimental observations. The correct calibration of such an advanced plasticity model, and especially its plastic potential, will be subjected to a further research.

Introduction

The correct plasticity modelling is important when large plastic deformations occur. This may be found in many engineering applications [1–4]. The flow rule plays a vital role in the plasticity models. It determines the direction of the plastic strain increment. The flow rule might be either associated or non-associated. The associated one corresponds to equal yield function and plastic potential, which results in the plastic strain increment normal to the yield surface. The non-associated flow rule has a distinct yield function from the plastic potential, so that the plastic strain increment is not normal to the yield surface but to the plastic potential. It might seem as a minor model feature, but it can significantly influence the results [5]. It was shown that the force response can be improved by a non-quadratic plasticity model a lot [6, 7], but it significantly deteriorates the stress state [8–11] as well as the deformations [12].

Experiments

First, standard tensile tests were conducted on aluminium alloy 2024-T351 in order to estimate the elastic constants and flow curve. The flow curve estimation was based on the iterative procedure until the match between the simulation and experiment was satisfying. Then, upsetting tests of notched cylindrical specimens and tensile tests of flat grooved specimens were carried out in order to represent specific stress states [11] and complex deformations [12]. All the loadings in experiments were realized in the rolling direction of the supplied cold-rolled plate. In other words, the specimens' axes were parallel to this rolling direction.

Plasticity modelling

The experiments were modelled using three basic types of plasticity models. The first one was a classical von Mises yield criterion with associated flow rule having the equivalent stress as

$$\bar{\sigma}=\sqrt{\frac{1}{2}\left[\left(\sigma_1-\sigma_2\right)^2+\left(\sigma_2-\sigma_3\right)^2+\left(\sigma_3-\sigma_1\right)^2\right]} \tag{1}$$

where $\sigma_1 \geq \sigma_2 \geq \sigma_3$ are the principal stresses ordered according to their magnitudes.

The second model had an associated flow rule as well, while the yield criterion was close to the Tresca one. This was utilized by a Hershey yield criterion [13] having the equivalent stress defined as

$$\bar{\sigma}=\sqrt[m]{\frac{1}{2}\left[\left|\sigma_1-\sigma_2\right|^m+\left|\sigma_2-\sigma_3\right|^m+\left|\sigma_3-\sigma_1\right|^m\right]} \tag{2}$$

where m is the material constant, which was chosen equal to 100, so that it is close to the Tresca yield criterion. This model was implemented into Abaqus explicit finite element code using VUMAT user subroutine, as it is not available in that commercial software.

The last yield criterion was a Mohr–Coulomb one with non-associated flow rule, which is built in Abaqus. The yield function is [14]

$$f=R_{mc}q-p\tan\phi-c \tag{3}$$

where q is the von Mises equivalent stress according to the Eq. (1), p is the hydrostatic pressure, ϕ is the friction angle, c is the cohesion stress and R_{mc} is a function defined as [14]

$$R_{mc}=\frac{1}{\sqrt{3}\cos\phi}\sin\left(\Theta-\frac{\pi}{3}\right)+\frac{1}{3}\cos\left(\Theta-\frac{\pi}{3}\right)\tan\phi \tag{4}$$

where Θ is the Lode angle. This Lode angle can be normalized as

$$\bar{\Theta}=1-\frac{6}{\pi}\Theta \tag{5}$$

The friction angle was set to zero so that the yield criterion corresponded to Tresca one and was pressure independent. The cohesion stress was half the estimated flow stress with twice the equivalent plastic strain due to its formulation [14]

$$\bar{\varepsilon}_{pl}=\int\frac{1}{c}\boldsymbol{\sigma}:\mathrm{d}\varepsilon_{pl} \tag{6}$$

where $\boldsymbol{\sigma}$ is the Cauchy stress tensor and $\mathrm{d}\varepsilon_{pl}$ is the plastic strain increment. The plastic potential is a hyperbolic function in the meridional stress plane [14] and smooth elliptic function in the deviatoric stress plane, which was proposed by Menétrey and William [15]

$$g=\sqrt{\left(\varepsilon c_0\tan\psi\right)^2+\left(R_{mw}q\right)^2}-p\tan\psi \tag{7}$$

where ε is the meridional eccentricity defining the rate at which the hyperbolic function approaches the asymptote, c_0 is the initial cohesion stress, ψ is the dilation angle and R_{mw} function defined as [14]

$$R_{mw} = \frac{4\left(1-e^2\right)\cos^2\Theta + \left(2e-1\right)^2}{2\left(1-e^2\right)\cos\Theta + \left(2e-1\right)\sqrt{4\left(1-e^2\right)\cos^2\Theta + 5e^2 - 4e}} R_{mc}\left(\frac{\pi}{3},\phi\right) \tag{8}$$

where e is the deviatoric eccentricity and $R_{mc}\left(\frac{\pi}{3},\phi\right)$ from Eq. (4) simplifies into [14]

$$R_{mc}\left(\frac{\pi}{3},\phi\right) = \frac{3-\sin\phi}{6\cos\phi} \tag{9}$$

Meridional eccentricity was set to 0.001 and the dilation angle to zero so the plastic potential was not pressure dependent, while the deviatoric eccentricity was defined as

$$e = \frac{3-\sin\phi}{3+\sin\phi} \tag{10}$$

The plastic potential has the shape according to the von Mises (a circle in a deviatoric plane).

Results and discussion

The results of computational simulations are compared to experiments in Fig. 1.

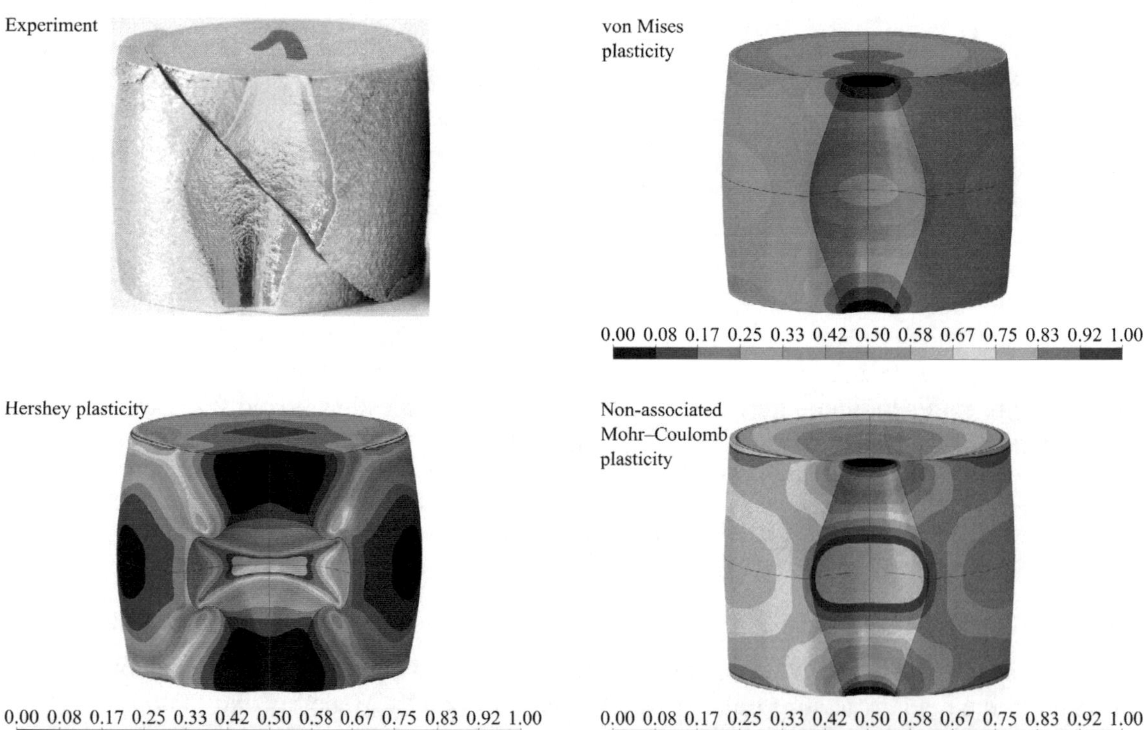

Fig. 1: Comparison of computational (equivalent plastic strain) and experimental results for the compression test

From Fig. 1, it is clear that the non-quadratic yield criterion according to Hershey (close to the Tresca one) predicted incorrectly the deformations. Remaining two plasticity models predicted the deformations in correspondence with experiment, but the Mohr–Coulomb yield criterion with non-associated flow rule resulted in significantly higher equivalent plastic strain.

Computationally obtained stress states for the tensile test of flat grooved specimen are in Fig. 2. Theoretically, the deviatoric stress state measure, normalized Lode angle from Eq. (5) in this case, should approach zero because the specimen underwent a generalized shear. The non-quadratic yield criterion (Hershey) improved the force response, but deteriorated the stress state, when compared to the quadratic one (Mises), both associated. The Mohr–Coulomb yield criterion with non-associated flow rule improved both the force response and stress state. But similar problem as in the case of upsetting test arose for this plasticity model, which predicted twice the equivalent plastic strain at fracture.

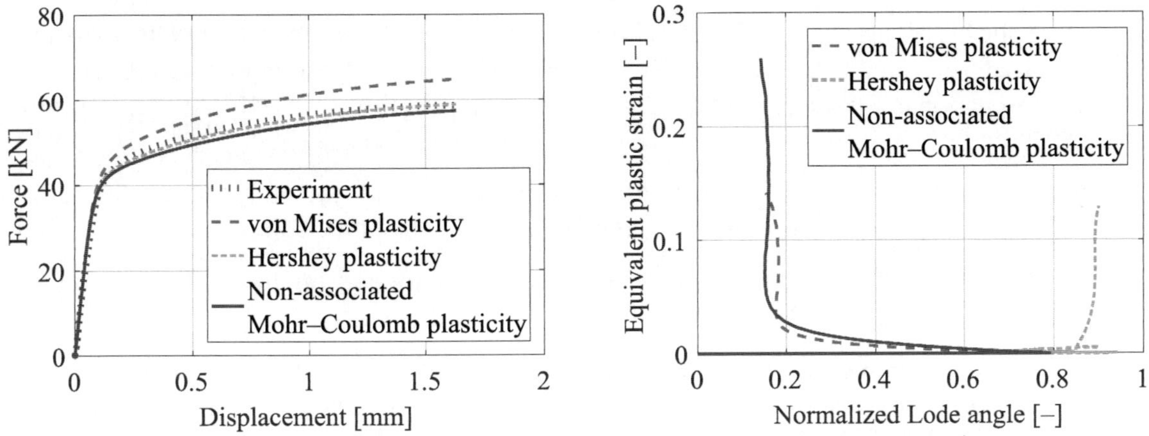

Fig. 2: Computational results for the tensile test of flat grooved specimen

This analysis is important, because correct stress state prediction is important especially for ductile fracture. It may be crucial for accurate simulation of the crack initiation and propagation [16, 17].

Conclusions

The non-associated flow rule has a significant role in the prediction of complex deformations. It has also a decisive role in the predicting of correct stress states, which are crucial in the ductile fracture simulations. This is going to be a hot topic for further research. Especially, the question of higher equivalent plastic strain for the non-associated flow rule when compared to the associated one has to be solved. It is also challenging to calibrate such an advanced plasticity model correctly.

Acknowledgement

This work was supported by the Czech Science Foundation under contract No. 19-20802S "A coupled real-time thermo-mechanical solidification model of steel for crack prediction."

References

[1] F. Šebek, J. Hůlka, P. Kubík, J. Petruška, On the proportionality of damage rule in finite element simulations of the ductile failure, Adv. Mater. Res. 980 (2014) 189–193.

[2] J. Petruška, T. Návrat, F. Šebek, M. Benešovský, Optimal intermeshing of multi roller cross roll straightening machine, AIP Conference Proceedings 1769 (2016) 120002.

[3] M. Peč, P. Vosynek, F. Šebek, T. Návrat, Implementation of American weld connection standards into finite element computations, IOP Conference Series: Materials Science and Engineering 179 (2017) 012056.

[4] M. Španiel, T. Mareš, J. Kuželka, F. Šebek, J. Džugan, Uncoupled material model of ductile fracture with directional plasticity, In: Proceedings of the 14th International Conference on Computational Plasticity - Fundamentals and Applications, COMPLAS (2017) 596–605.

[5] X. Gao, T. Zhang, J. Zhou, S. M. Graham, M. Hayden, Ch. Roe, On stress-state dependent plasticity modeling: Significance of the hydrostatic stress, the third invariant of stress deviator and the non-associated flow rule, Int. J. Plast. 27 (2011) 217–231.

[6] F. Šebek, J. Petruška, P. Kubík, The performance and prediction ability of advanced approach to ductile fracture, In: Proceedings of the 14th International Conference on Computational Plasticity - Fundamentals and Applications, COMPLAS (2017) 588–595.

[7] F. Šebek, P. Kubík, J. Petruška, Application of coupled ductile fracture criterion to plane strain plates, Defect Diffus. Forum 382 (2018) 186–190.

[8] Y. Bai, T. Wierzbicki, A new model of metal plasticity and fracture with pressure and Lode dependence, Int. J. Plast. 24 (2008) 1071–1096.

[9] M. Kroon, J. Faleskog, Numerical implementation of a J_2- and J_3-dependent plasticity model based on a spectral decomposition of the stress deviator, Comput. Mech. 52 (2013) 1059–1070.

[10] M. Algarni, Y. Bai, Y. Choi, A study of Inconel 718 dependency on stress triaxiality and Lode angle in plastic deformation and ductile fracture, Eng. Fract. Mech. 147 (2015) 140–157.

[11] F. Šebek, J. Petruška, P. Kubík, Behavior of lode dependent plasticity at plane strain condition and its implication to ductile fracture, Solid State Phenom. 258 (2017) 213–216.

[12] P. Kubík, F. Šebek, J. Petruška, Notched specimen under compression for ductile failure criteria, Mech. Mater. 125 (2018) 94–109.

[13] A. V. Hershey, The plasticity of an isotropic aggregate of anisotropic face-centered cubic crystals. J. Appl. Mech. 53 (1954) 241–249.

[14] Abaqus. SIMULIA User Assistance 2017. Dassaul Systèmes Simulia Corporation, Providence, Rhode Island.

[15] P. Menétrey, K. J. William, Triaxial failure criterion for concrete and its generalization, ACI Struct. J. 92 (1995) 311–318.

[16] J. Petruška, P. Kubík, J. Hůlka, F. Šebek, Ductile fracture criteria in prediction of chevron cracks, Adv. Mater. Res. 716 (2013) 653–658.

[17] F. Šebek, P. Kubík, J. Petruška, Chevron crack prediction using the extremely low stress triaxiality test, MM Sci. J. (2015) 617–621.

High velocity testing of concrete composites

ŠEBÍK Marek[1,a], POPOVIČ Miloslav[1,b] and DRDLOVÁ Martina[2,c]

[1]SVSFEM s.r.o., Škrochova 42; 615 00, Brno; CZ

[2]Výzkumný ústav stavebních hmot, a. s., Hněvkovského 65, 617 00 Brno; CZ

[a] msebik@svsfem.cz, [b] mpopovic@svsfem.cz, mdrdlova@vustah.cz

Keywords: Hopkinson test, explicit simulations, concrete composites, Tri-Split Hopkinson Bar

Abstract. The aim of this study is a new test method focused on dynamic behaviour of brittle materials especially concretes and their composites. Modified Hopkinson testing device was developed. The dynamic test conditions were simulated with advance of explicit numerical solution. The tensile test was performed as the so-called Brazilian test and standard tensile test as well.

Introduction

Besides quasi-static loads, many concrete structures are subjected to short duration loads, such as the impacts from missiles and projectiles, wind gusts, earthquakes etc. The aim of this paper consists in an experimental study of the influence of non-metallic fibers both on quasi static and dynamic mechanical properties. The concrete specimens with carbon, aramide, glass, polyprophylen and wollastonite fibers have been tested.

The response of the given materials to the quasi-static compressive and flexural loading has been studied. The dynamic loading was realized using of the Split Hopkinson Pressure Bar technique (SHPB). Three kinds of dynamic tests were carried out: a classical compressive test, tensile – splitting test (Brazilian test) and classical tensile test.

Material and Experimental Technique.

Five material types with reinforcing fibers were prepared (including SIFCON matrix):

- Aramid fibers with length of 1 mm. Their volume content is 1.5%.
- Carbon fibers with length of 3 mm. Their volume content is 1.5%.
- Recycled glass laminate fibers length < 0.3 mm. Their volume content is 3%.
- Polypropylen fibers, length 3 mm. Their volume content is 1.5%.
- Wollastonite fibers, length < 0.3 mm. Their volume content is 2%.

The dynamic mechanical properties were obtained using of the Split Hopkinson Pressure Bar technique (SHPB). The used specimens had the form of the cylinders, whose diameter and thickness were 30 and 15 mm respectively. The position of the specimen is shown in the next picture.

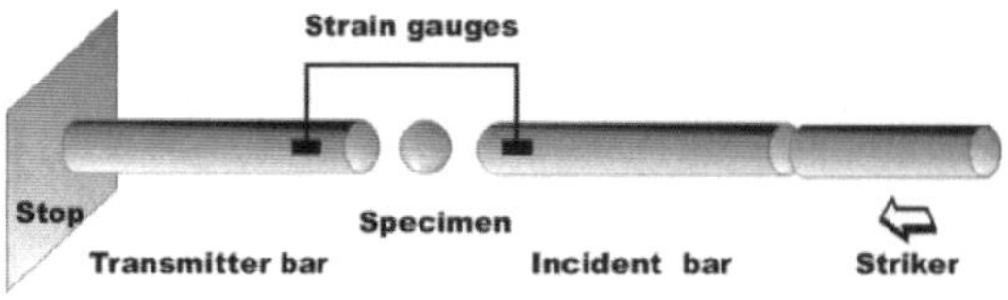

Fig.1. Experimental arrangement of the dynamic tensile – split test (Brazilian test).

This arrangement, which is known as the Brazilian test, or diametric compression test, is a typical indirect tensile test, which was developed to determine the tensile strength of brittle materials such as rock, ceramics and concrete [1-2]. In this case the diametral loading generates tension perpendicular to the load plane, which eventually causes the specimen to split. To obtain a detailed dynamic deformation and fracture process of the brittle specimens in the dynamic Brazilian test, using a high-speed camera, requires a high frame rate. For a high-speed camera, there is a compromise between the frame rate and image resolution. The fracture of the specimens during the Brazilian test was recorded using the high-speed photography (160000 frames per second).

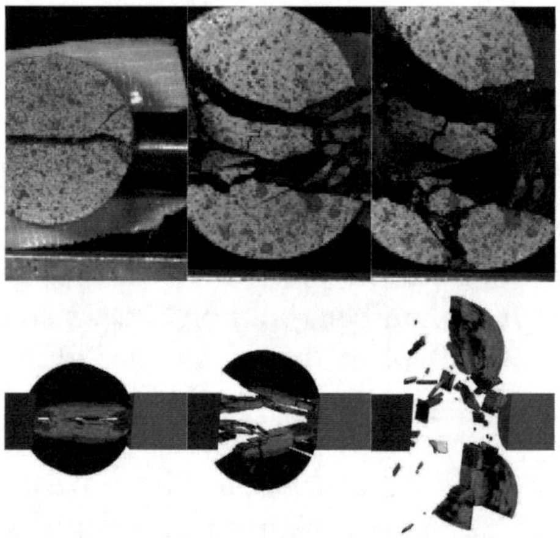

Fig.2. Result comparison

A relatively good agreement between experiment and numerical simulation was found at the beginning of the specimen fracture. The next development of the specimen fracture differs from the results of the numerical simulation. Experimental records show that the specimen fracture is not plane symmetric. This is probably a consequence of the material inhomogeneity

Numerical simulation of the tests on the indirect tensile strength has been performed using LS DYNA 3D finite element code. The concrete composite material has been described in terms of Winfrith concrete model. Its main advantage consists in the fact that this model does not need much knowledge on the concrete.

New experimental device was developed. It enables classical tensile test with sample with diameter about 50mm. The longitudinal pressure wave from the punch impact is transferred to the material sample and the loading sensors.

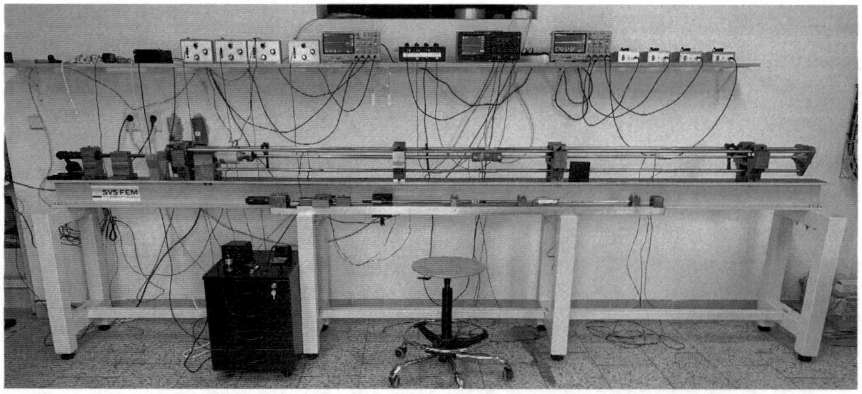

Fig. 3: Tensile Split Hopkinson Bar device

The specimen is connected to measuring bars by bonding joint. It was developed during set of quasi static tests.

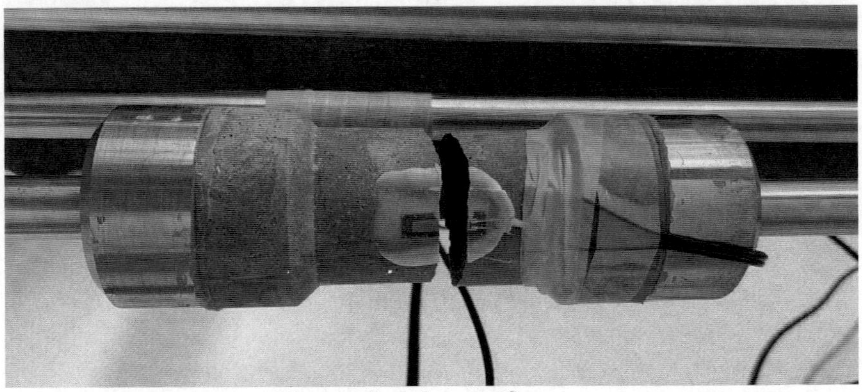

Fig. 4: Specimen after test

The stress history from the specimen is evaluated by four independent methods.

1. directly from a strain gauge installed on the specimen
2. indirectly from strain gauges installed on the measuring bars (standard evaluation from Hopkinson test principle
3. indirectly from a force transducer installed on the back measurement bar
4. optically with high speed extensometer

Conclusions

Static and dynamic properties of concrete reinforced by different fibers were evaluated. The fiber reinforcement of the concrete leads to the improve of its mechanical properties like the compressive strength and tensile strength.

References

[1] Yan,F.;X.T.Feng,,X.T.;Chen,P.R. Rock Mech. Rock. Eng. 2012,45, 341.

[2] ChenJ.J.;Guo,B.Q.;H.B.Lin,H.B.;Chen,W.; Strain 2014,50,563. DOI: 10.1111/str.12118

Fatigue crack growth reconstitution in the specimen representing longitudinal joint of an integral panel

ŠEDEK Jakub[1,a] and MICHALCOVÁ Lenka[1,b]

[1]Czech Aerospace Research Centre (VZLÚ), Beranových 130, 19905, Prague, Czech Republic

[a]sedek@vzlu.cz, [b]michalcova@vzlu.cz

Keywords: Aluminium alloy, Crack growth, Fatigue, Reconstitution, SEM

Abstract. The fatigue crack growth reconstitution is presented on the crack developed in the specimen representing longitudinal joint of an integral panel. The fractographic observation was carried out using optical and scanning electron microscopy (SEM). The major crack was investigated in detail. The striation spacing was determined from several locations and processed to obtain its average values. The approach assuming the relation between the striation spacing and the crack growth rate was employed. The integral method to obtain the crack growth curve was realized. The crack growth determined by fractographic reconstitution was faster by 22% of the fatigue life. The result is on the conservative side from the point of view of safety.

Introduction

The joint is the critical point of the structure from the aspect of strength and fatigue. The life of the component then depends on the crack initiation and growth periods. Fatigue cracks can initiate not only from the connecting elements, but also from other locations. The contact of connected parts is sometimes serious from the point of crack initiation by fretting.

The joint of the integral panel and the flange of the spar in the wing structure of a commuter airplane was tested in Czech Aerospace Research Centre in order to determine S-N curves and to monitor the crack growth. The cracks were expected to initiate at the riveted joint but in some cases the crack initiated elsewhere and determined the fatigue life of the specimen. Apparently, the fretting damage was the reason for such crack initiation as shown in Fig. 1. Due to unexpected crack initiation position, the crack growth was not monitored as the cracks initiating from the rivet hole. Therefore, the crack reconstitution by means of fractography was carried out on one specimen to show possibilities of the method.

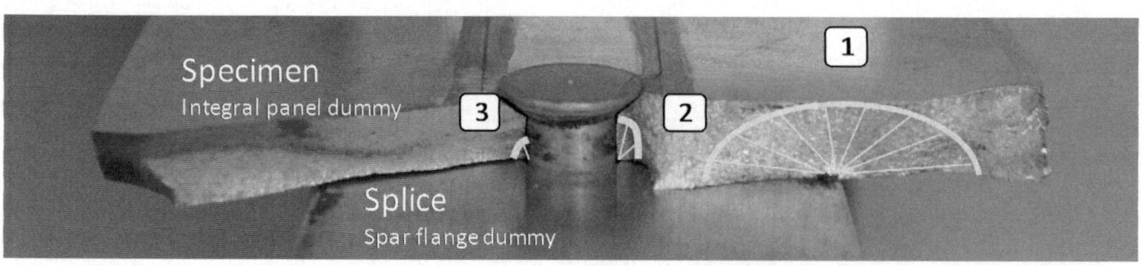

Fig. 1: Fatigue cracks in the specimen; yellow colour marks the visually visible crack fronts

Specimen and test description. Specimens represented the longitudinal joint of the integral wing panel and the spar flange of a commuter airplane. The specimens were made from two components as shown in Fig. 2. One of 4.5 mm thick flat part representing the integral panel made of aluminium alloy 7475-T7351 and the splice representing the flange of 7.5 mm thick made of aluminium alloy 2124-T851. The joint was realized by rivets 5 mm in diameter with countersunk head.

Specimens were subjected to constant amplitude cyclic loading and with cycle asymmetry $R = 0.05$. Analysed specimen was loaded at the cycle upper level of 67 MPa for 5 000 000 cycles and then the load level was increased up to 220 MPa. The failure subsequently occurred after 5 878 cycles.

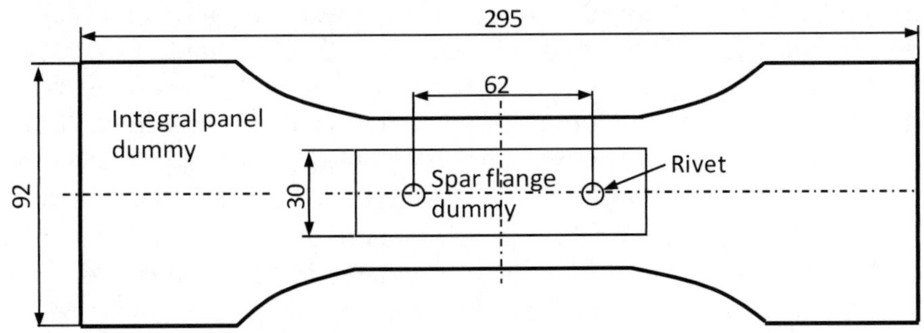

Fig. 2: Specimen representing the longitudinal joint of integral wing panel and the spar flange; dimensions in mm

Fractographic observation of crack surfaces. The visual inspection of fracture surfaces was carried out primarily using optical microscopy. The major crack was observed on the cracked surfaces and the initiation point was easily identified due to clear radial steps emanating from one point and due to the clearly visible border between two regions, obviously different from the point of view of micromorphology. The initiation spot was located at the contact with the splice. Due to the different stiffness of connected parts, relative displacements were present and fretting occurred. At the crack initiation, the flaw was 0.36 × 1 mm in size. The crack development was regular to all directions according to beach mark on the surface.

Minor cracks were located opposite each other at the edge of the rivet hole adjacent to the splice. The initiation of the minor cracks marked as no. 2 and no. 3 started from the grooves made by the drill tool and from the rough surface inside the hole. The microcracks linked together formed the corner crack. Their development was slower than the major crack and so they remained only quarter-circle shaped at the time of the final failure.

The surface of major crack was analysed in more detail by means of scanning electron microscopy (SEM) using TESCAN VEGA apparatus. The analysed crack is marked as crack no. 1 in Fig. 1.

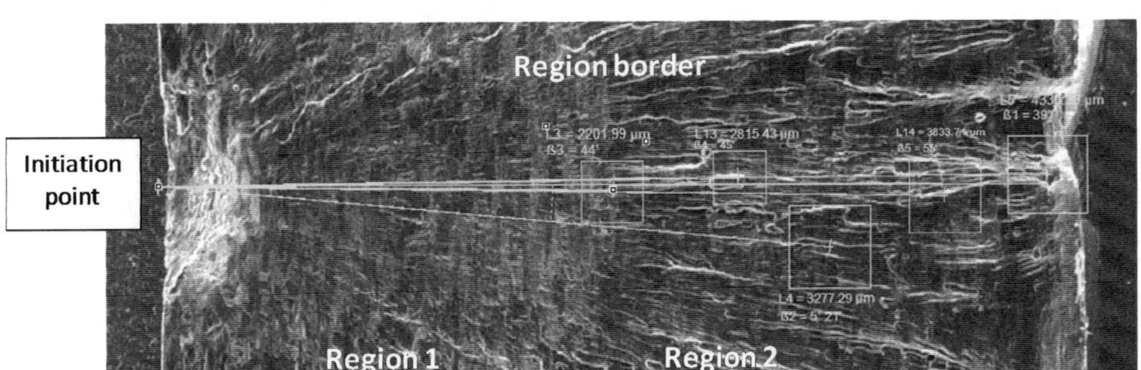

Fig. 3: SEM picture of the major crack; rectangles mark inspected locations in more detail

The beach mark distinguishing fracture surface into two regions is the significant attribute of the crack no. 1. The regions are marked in Fig. 3. The region border is the half circle shaped reaching up to half of the thickness. Different micromorphology in the regions can be attributed to the crack growth under different loading conditions before and after the change of the load level. The focus for SEM was given to the region 2, where the striations were expected. In the region 1, the striations were very difficult do observe.

Fractographic reconstitution of crack growth. Based on the work [1] the fractographic reconstitution of crack growth in region 2 was realized. The quantitative fractographic data based on striation spacing are necessary to determine from several locations by means of SEM. Five locations were chosen for a more detailed analysis (see Fig. 3). In each location, several measurements of the series of striations were carried out. The average striation spacing s_{ij} in the direction of macrocrack growth and the average angle ϑ_{ij} of local and global crack growth is the result from each location. Micromorphology pictures were evaluated at the crack lengths of 2.20, 2.82, 3.27, 3.83 and 4.43 mm. The results are stated in Tab. 1 and the example of the local processing at length of 4.43mm is shown in Fig. 4.

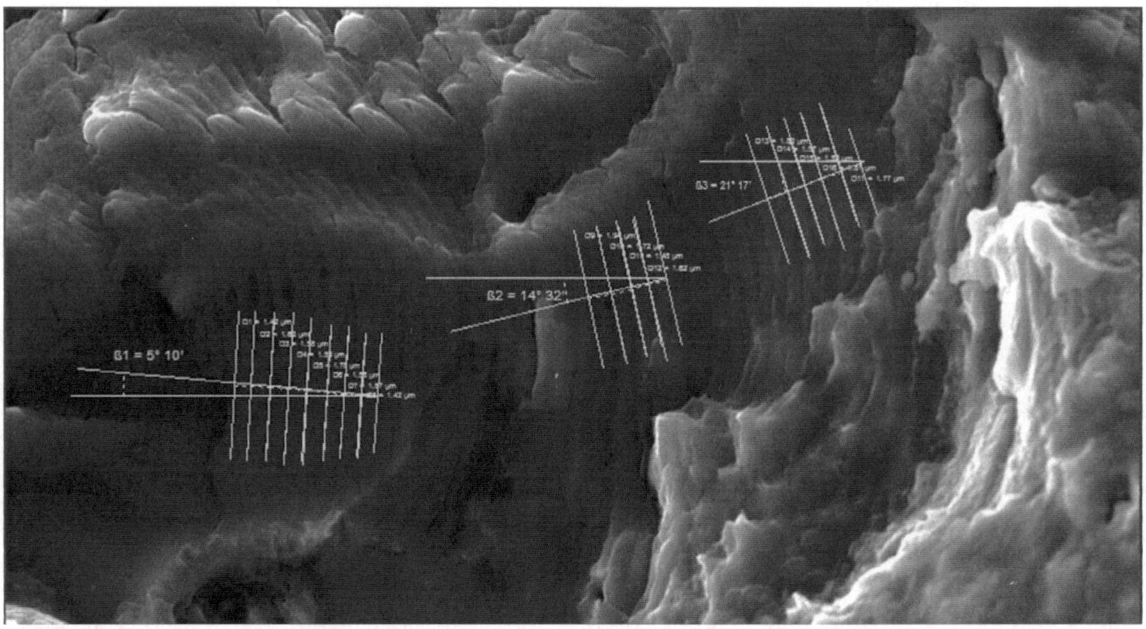

Fig. 4: SEM picture of the major crack detail at distance of 4.43 mm from the initiation point; Parallel lines mark the striation fronts and wedges represents the inclination of the global and local crack growth.

The SEM| pictures *ij* do not lie on the same line and the crack front is also curved, so the striation spacing in the global direction of crack growth $s_{ij}{}^{g}$ was also determined.

Tab. 1: Basic micromorphological characteristics from analysed locations in region 2 of the crack n. 1.

Distance from the initiation point (mm)	2.201	2.815	3.269	3.834	4.320
s_{ij} (μm)	0.37	0.32	0.58	0.62	1.64
ϑ_{ij} (rad)	0	-0.04	0.07	0.49	0.12
$s_{ij}{}^{g}$ (μm)	0.36	0.33	0.58	0.55	1.61

The average striation spacing $s_{ij}{}^{g}$ (thereinafter denoted as s) in dependence on the crack length a was fitted by equation

$$s = 1.41\,E - 06 \cdot exp(a/0.637) + 0.234\,E - 03 \tag{1}$$

The relationship $s=s(a)$ is used for the reconstitution of crack kinetics utilizing the dependency of crack length a on the elapsed cycles N. The number of elapsed cycles N_x at corresponding crack length a_x was determined according to

$$N_x = \int_{a_1}^{a_x} \frac{da}{D(s) \cdot s(a)} + N_i \tag{2}$$

where the parameter D denotes the relation between striation spacing and the macroscopic crack growth $v = da/dN$ and is dependent on several variables; e.g. material, its heat treatment, or loading conditions. Generally, D requires to be determined experimentally. N_i is the initial cycle number and a_1 denotes initial crack length. Parameter D can be expressed for the correlation of macroscopic and microscopic characteristics as $D=v/s$.
Influence of loading conditions is being described by the dependency on the stress intensity factor K_{eff}.. For the aluminium alloy 7475-T7351, the match of crack growth rate da/dN and the striation spacing s was found for the cycle asymmetry $R = 0.1$ and v in the range 0.18 -2.02 μm/cycle [2]. The coefficient $D = 1$ was assumed for presented work also for $R = 0.05$ and the whole range of ΔK.
Integrating Eq. (2) from a_i up to a_x the number of cycles can be obtained. Substituting the Eq. (1) into Eq. (2) the direct formula was achieved according to [3]

$$N_x = \frac{a_x - a_i}{C} + \frac{1}{BC} \ln \frac{|C + A\exp(Ba_i)|}{|C + A\exp(Ba_x)|} + N_i \tag{3}$$

where using constants $A = 1.41$ E-06, $B = 1/0.637$ and $C = 0.234$ E-03. The result of crack growth curve is shown in Fig. 6. In comparison with experimental data, the reconstituted number of cycles up to the failure in crack growth region 2 is less by 22%.

Reconstutition of the crack growth by methods of fracture mechanics. The crack growth curve was independently determined also by the simulation of crack growth, because the crack growth data was not available from the test. The elapsed cycles at load level change up to 220 MPa and the specimen failure at 5878 cycles were only available data. The beach mark distinguishing between region 1 and region 2 was attributed to the load level change. From the fractographic analysis the shape of the crack front at this point is known and with material crack growth data the simulation could be performed.

The AFGROW code [4] was utilized for the crack growth simulation. The shape function of the surface crack implemented in the code was used for the calculation of stress intensity factor K. The model represented only one half of the specimen due to lack of the model including the hole. Representative solution can be assumed up to reaching the whole thickness of the specimen by the crack. Substitutive model dimensions are shown in Fig. 5.

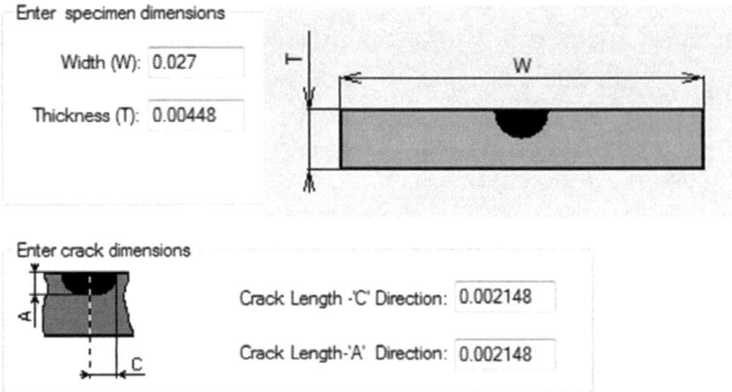

Fig. 5: Crack growth curve obtained by fractographic reconstitution in region 2 of fracture surface.

Material characteristics are used according to Ref. [5] in the form of the relationship of *da/dN* versus ΔK for load assymmetry $R = 0.02$ described by the linear formula. Crack growth rate was computed according to Paris law [6]

$$\frac{da}{dN} = C\Delta K^{m} \quad (4)$$

where $C = 1.816$ E-10 and $m = 2.814$. Characteristics determined at load ratio $R = 0.02$ are assumed not to be significantly different from those determined at $R = 0.05$ and are used for the simulation. Crack growth increments are integrated from the crack length of 2.15 mm up to 4.5 mm. The results of simulation in comparison with the fractographic reconstitution are shown in Fig. 6.

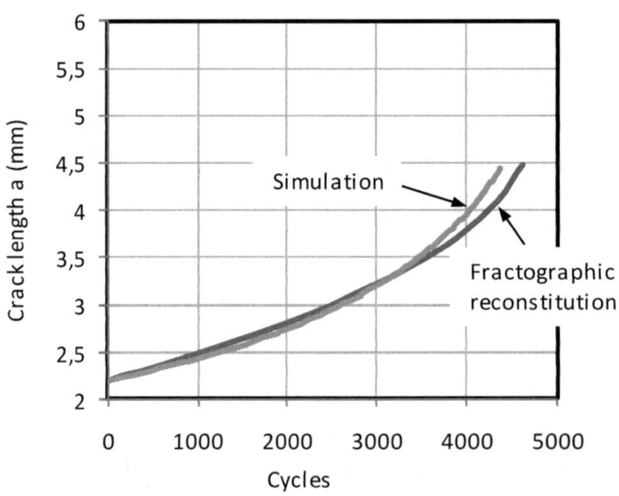

Fig. 6: Crack growth curve determined by the simulation in comparison with the curve determined by fractographic reconstitution in region 2 of fracture surface.

Conclusions

Fatigue crack growth reconstitution in the specimen representing the longitudinal joint of integral panel was performed. Fractographic analysis was employed to determine the striation spacing in several locations of the cracked surface. The integral method to obtain the crack growth curve was realized. The crack growth determined by the fractographic reconstitution is faster by 22% of final life. The results were also compared with the simulated crack growth curve showing good match of both curves with slight deviation at the last 20% of the specimen life. The result is on the conservative side from the point of view of safety.

Acknowledgments

This project result was developed within the institutional support of the Ministry of Industry and Trade of the CR directed to the development of research organizations.

References

[1] Nedbal J. et al. Fractographic Reconstitution of Fatigue Crack History – Part I. Fatigue and Fracture of Engineering Materials and Structures, 31 (2), 2008. pp.164-176. doi: 10.1111/j.1460-2695.2007.01211.x

[2] Ruckert C.O.F.T. et al. On the Relation Between Micro- and Macroscopic Fatigue Crack Growth Rates in Aluminum Alloy AMS 7475-T7351

[3] Rektorys K et al. Přehled užité matematiky. 3. vydání, SNTL, Praha, 1973.

[4] Harter J.A. AFGROW Users' Guide and Technical Manual, Air Force Research Laboratory, 1999. AFRL-VA-WP-1999-3016

[5] Homola et al. Materiálové charakteristiky desky tl. 76 mm vyrobené ze slitiny 7475-T7351 ve směru L-T (Alcoa), VZLÚ, 2013. R-5650

[6] P.C. Paris and F. Erdogan. A critical Analysis of Crack Propagation Laws, Journal of Basic Engineering, 85 (4), 1963. pp. 528-533, doi:10.1115/1.3656900

EAN 2019
57th conference on experimental stress analysis

Mechanical structures made of high-strength sheets or composites for electric vehicles

ŠEVČÍK Ladislav[1,a] and LUFINKA Aleš[1,b]

[1]Department of machine design; Faculty of Mechanical Engineering; Technical university of Liberec; Studentská 2, 461 17 Liberec, Czech Republic

[a]ladislav.sevcik@tul.cz, [b]ales.lufinka@tul.cz

Keywords: High-strength steel, Welding, Tensile test, car protection

Abstract. The paper deals with the problem of calculations of high-strength sheet metal structures. These sheets are mainly used for parts that deform in non-standard situations. These parts protect the crew in this case of a electro truck. The success of the calculation depends mainly on the technological execution of the welds of these sheets. The temperature affecting the structure of the high-strength sheet affects its strength and flexibility. Welding technology varies considerably in companies. It is therefore very difficult to estimate the strength properties and so at least the tensile test must always be performed on these sheets.

Introduction

Nowadays designer exhibited question if key parts of an electric vehicle to propose a high-strength steel, or fiberglass or carbon fiber. Both types of materials are commonly available on the market today and companies have technologies to process them the battery. Meanwhile, it is not possible to accumulate electricity so mass and the density was comparable to other types of fuels such as diesel, petrol, gas. Due to this current state of knowledge, it is necessary to reduce the weight as much can be from many point of view and to increase the strength of the supporting parts of the electric vehicle as much as possible.
The paper deals with the problem of connecting individual parts of the electric car frame. In the case of composites, an insert is used to attach parts to the frame, and in the case of high-strength steels, mostly different type of welds. Welding technology for high-strength sheets is the manufacturing secrets of the company. The aim is to minimize recrystallization of the structure of high-strength sheets by heat. Since the calculations affect the elastoplastic area, it is necessary to measure the characteristics of the welded high strength parts.

Measurement devices

The standard tensile testing device FU250 was used to measure the properties of welded parts. The test device is equipped with hydraulically operated jaws for specimens anchoring. The result of the tensile test is the dependence of normal tension σ on the relative elongation ε. So, two quantities - loading force and specimen elongation - need to be measured during the test. The force sensor is a standard equipment of the test device but the external extensometer needs to be used for the elongation measurement. An extensometer based on the incremental length sensor Renishaw was used for this measurement. This sensor measuring accuracy is 0.001 mm. A wide choice of initial length L_0 and the sensor destruction resistance during the specimen breaking are the two biggest advantages of this solution. The extensometer digital output cannot be connected to the test device FU250 control system, so the DEWE5000 universal measurement device was used for data record. The extensometer and the force signal from the FU250 control system were connected to this device. The extensometer principal and block scheme is shown in Fig. 1.

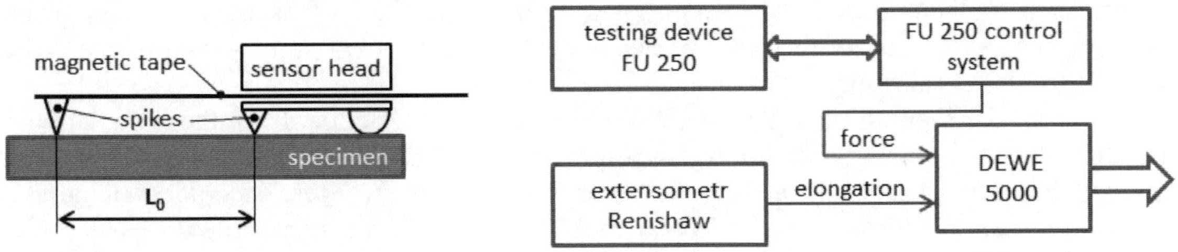

Fig. 1: The extensometer Renishaw principle (left) and measurement block scheme (right)

Solution procedure

Two 10 mm thick Simaxx 690QL sheets were welded together and cut into square bars samples approximately 10x10mm [2]. Broken samples are in Fig.2. Shredder was equipped with extensometer of own production. It was attached to the rod by elastic elements. Its advantages are in the possibility of attachment to the given square sample and resistance to mechanical damage. See Fig. 3.

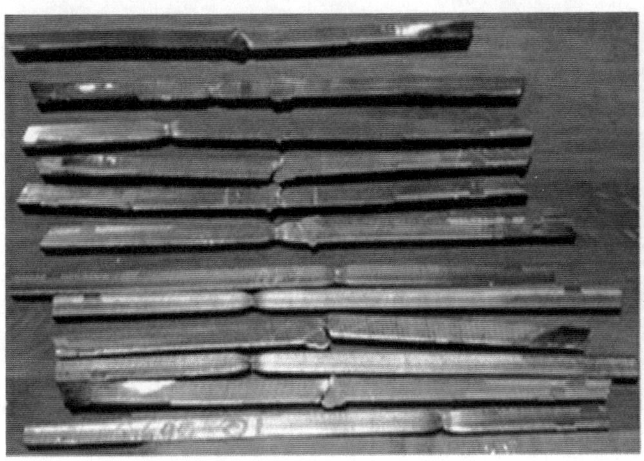

Fig. 2: Samples of welded cut sheets to the measured after the test

Fig. 3: Fitting of the extensometer on the specimen

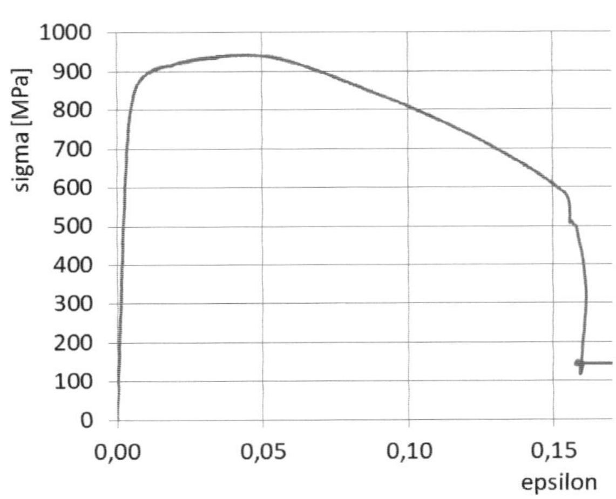

Fig. 4: Material 690, welded, broken outside the weld - weld made OK

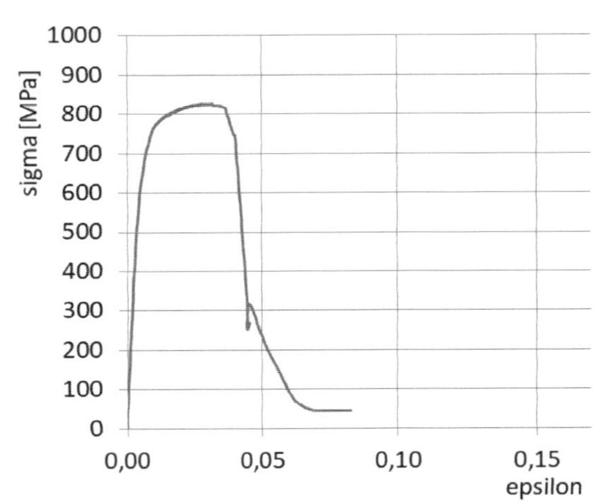

Fig. 5: Material 690 welded, broken in weld

The FEM calculation was based on the tensile test data of the material, Fig. 3, 4 by [1]. The device is tested by loading forces in two directions. Forces operate gradually over time. Load first in the direction of travel. The device deforms plastically. It is then unloaded but partially deformed. It is then loaded with a horizontal force see Fig.5. This again causes plastic deformations. As a result of the positive evaluation, the permanent deformation of the device does not reach a certain space.
Furthermore, the loading procedure (tests) according to the standard for the calculation of operator protection during overturning and the size of welds, in particular at welding points, was entered. The plastic and elastic deformations and stresses were calculated in gradual loading by individual tests in Ansys Workbench 17.2, module transient structure. The project scheme is shown in Fig. 6. In linear static structure were counted deformation and stress separately for each load. In transient structural the loading was linked to the previous one.

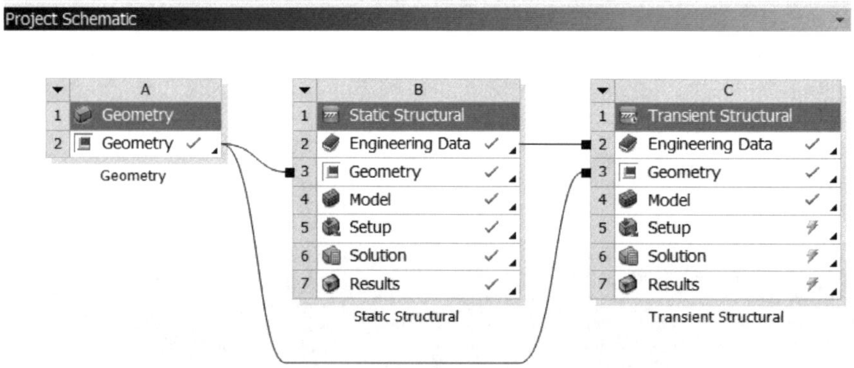

Fig. 6: Project scheme

In transient structural the loading was linked to the previous one. In the graph, Fig.7 the load intervals are plotted on the horizontal axis, and load values in [N] are plotted on the vertical axis.

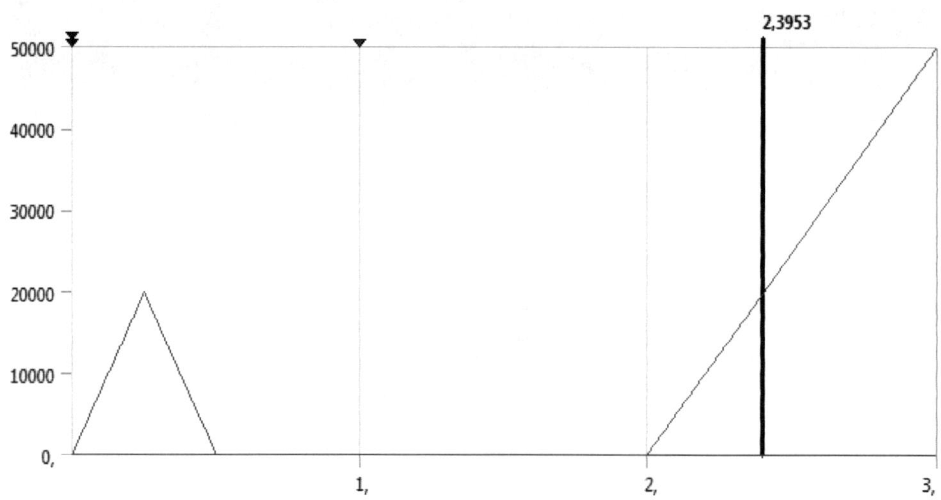

Fig. 7: Loading process

Results

Taking into account the post-test deformation and stress of transient loading, see Fig. 8, 9. Result of equivalent plastic deformation and elastoplastic deformation are in Fig. 10, 11.

Fig. 8: Deformation after transient loading

Fig. 9: Stress after transient loading

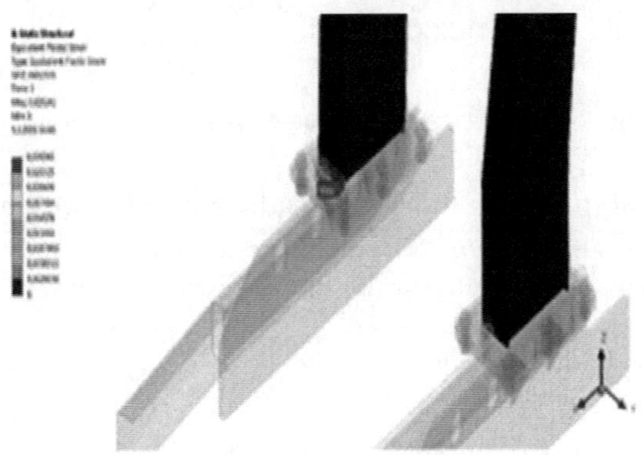

Fig. 10: Equivalent plastic deformation is 2%

Fig. 11: Elastic and plastic deformation is 3%

Conclusions

As predicted, the resulting load capacity of the electric car frame made of welded high strength plates will depend on the technology of the welds. All measured samples reached the reported strength even in welds. Particularly the different was in the percentage of elastoplastic deformation. Poorly made welds are brittle and elongation is less to 2%. In executed welds, the elongation was more than 10%. In the FEM calculations was used the variant of the incorrectly brittle weld.

Acknowledgment

The result was obtained through the financial support of the Ministry of Education, Youth and Sports of the Czech Republic and the European Union (European Structural and Investment Funds - Operational Programme Research, Development and Education) in the frames of the project "Modular platform for autonomous chassis of specialized electric vehicles for freight and equipment transportation", Reg. No. CZ.02.1.01/0.0/0.0/16_025/0007293 and "Innovation of the products, equipments and processes in engineering practice" with the support of the Specific University Research Grant.

References

[1] Xiaolin Chen and Yijun Liu, Finite Element Modeling and Simulation with ANSYS Workbench, CRC Press 2019, ISBN-13: 978-1138486294

[2] Parrish A. Mechanical Engineer's Reference Book, 11th Edition, Elsevier, 19. 5. 2014 – pages 1612

EAN 2019
57th conference on experimental stress analysis

Prediction of the tensile breaking force and extension at break of woven fabrics in an arbitrary load direction

ŠIMIĆ PENAVA Diana[1,a*], PENAVA Željko[2,b] and KNEZIĆ Željko[3,c]

[1]Faculty of Civil Engineering, University of Zagreb, Kačićeva 26, 10000 Zagreb, Croatia

[2,3]Faculty of Textile Technology, University of Zagreb, Pr. b. Filipovića 28a, 10000 Zagreb, Croatia

[a]dianas@grad.hr, [b]zpenava@ttf.hr, [c]zknezic@ttf.hr

Keywords: Anisotropy, Tensile breaking force, Extension at break, Woven fabric

Abstract. Anisotropy is the characteristic which is typical for most materials, specially woven fabrics. Influence of direction of tensile force acting on the properties of the fabric is big and frequently tested. The woven fabric can be defined as orthogonal elastomer. This paper presents an experimental testing and determination of tensile breaking forces, extension at break and ultimate strength of woven fabric for different angles of force direction. Cotton fabric samples in plain weave were tested under tensile forces at the angles of 0°, 15°, 30°, 45°, 60°, 75° and 90° with respect to the weft direction. Based on the experimentally obtained values, theoretically calculated breaking forces, extension at break and ultimate strength for arbitrarily chosen fabric directions was calculated. This paper compares experimental and theoretical values of breaking forces, extension at break and ultimate strength. Experimental results show the applicability of theoretical equations for calculating the breaking force and extension at break for arbitrarily chosen direction of cutting the fabric specimen. Therefore, the measurements need to be implemented when the tensile force acting on the fabric only in the warp (90°), weft (0°) and at angle of 45°.

Introduction

Woven fabrics are generally non homogeneous, anisotropic and discontinuous objects. Special types of anisotropic materials having two mutually perpendicular planes of elastic symmetry are called orthotropic materials. These planes of elastic symmetry are planes of orthotropy, and their cross sections are axes of orthotropy (x and y) [1]. So the woven fabric is elastic orthotropic material. In a biaxial woven structure two main directions are defined: longitudinal (warp, y-axes) and transverse (weft, x-axes). Theoretical analysis of fabric behaviour is often very complex, so that the experimental verification of theoretical predictions for them is more important than for other materials. The method for measuring anisotropic tensile properties of fabrics is called "uniaxial test method" or the method of force action in only one direction [2]. When the angle of an external load (tensile force) changes, tensile breaking forces and extension at break change too. The aim of this work is to analyse the impact of anisotropy and different weaves on the tensile breaking forces and extension at break of woven fabrics in various directions and to determine the degree of agreement between experimental results and calculated obtained values.

Anisotropy of breaking force and extension at break

When the tensile force F is applied to the woven fabric in arbitrary directions, the normal stresses are varied depending on the direction of the force acting. The highest values of these tensile forces which cause breakage of the fabric are breaking tensile force F_φ (N). Eq. (1a) and (1b) calculate the values of extension at break for any combination of reversed axes in the plane of orthotropy [3, 4].

$$\varepsilon_\varphi = \frac{\varepsilon_0}{\cos^4\varphi + b\cdot\sin^2(2\cdot\varphi) + c\cdot\sin^4\varphi} \tag{1a}$$

where are: $\varepsilon_\varphi = \varepsilon_k$, $\varepsilon_y = \varepsilon_{90}$, $\varepsilon_x = \varepsilon_0$, $b = \dfrac{\varepsilon_0}{\varepsilon_{45}} - \dfrac{1+\varepsilon_0/\varepsilon_{90}}{4}$, $c = \dfrac{\varepsilon_0}{\varepsilon_{90}}$. (1b)

The angle φ is the angle between the axis x (weft direction) and direction of tensile force acting [4, 5]. ε_φ is the value of extension at break for the arbitrary direction of force acting relative to the weft of the fabric, ε_0, ε_{90} are the extension at break in the orthotropy axes (weft and warp direction), and ε_{45} is the extension at break at angle φ=45 °.

Theoretical breaking tensile forces F_φ (N) of orthotropic materials on the directions of all angles φ can be determined by measured values of breaking tensile forces in the directions at an angle of 0 °, 45 °, 90 ° by using the expressions (2a) and (2b) [3, 4]:

$$F_\varphi = \frac{F_0}{\cos^4\varphi + b\cdot\sin^2(2\cdot\varphi) + c\cdot\sin^4\varphi} \tag{2a}$$

where are: $F_\varphi = F_k$, $F_y = F_{90}$, $F_x = F_0$, $b = \dfrac{F_0}{F_{45}} - \dfrac{1+F_0/F_{90}}{4}$, $c = \dfrac{F_0}{F_{90}}$. (2b)

The values of breaking forces F_0, F_{45}, F_{90} are obtained in laboratory tests under uniaxial stretching of the sample when the tensile forces acting in the directions φ = 0°, 45°, 90°.

The second way of calculating the values of breaking forces is that on the basis of the experimental values of measured breaking forces of fabric samples tested at different angles, sets the relationship between force F and angle φ. The relationship is shown in Eq. (3) [5]:

$$F(\varphi) = a\varphi_i^2 + b\varphi_i + c \tag{3}$$

The mathematically most acceptable approach is to customize the experimental data to the quadratic equation-the polynomial second order equation (3) [5]. The condition must be met:

$$\sum_{i=0}^{n}(a\varphi_i^2 + b\varphi_i + c - F_i)^2 = \min \tag{4}$$

where *a*, *b* i *c* - the second degree polynomial parameters. Partial derivation of the Eq. (4) by *a*, *b*, and *c* and by solving the resulting system, the second degree polynomial parameters are obtained. From the Eq. (3) the theoretical values of the breaking forces are obtained.

Based on the experimental values of extension at break, the relationship between extension and the angle of force action is set by Eq. (5):

$$\varepsilon(\varphi) = a\varphi_i^2 + b\varphi_i + c \tag{5}$$

Again can be used quadratic function to fit the results.

$$\sum_{i=0}^{n}(a\varphi_i^2 + b\varphi_i + c - \varepsilon_i)^2 = \min \tag{6}$$

Partial derivation of the Eq. (6) by *a, b,* and *c* and by solving the resulting system, the second degree polynomial parameters are obtained. From the Eq. (5) the theoretical values of the extension at break are obtained.

Experimental determination of breaking force and extension at break

The experiment was carried out by measuring fabric spatial deformation under the action of tensile force till rupture. Two different cotton fabrics of different weaves (plain and twill) were used. Their structural properties are shown in Table 1. Width of each specimen is b=50 mm.

Table 1: Description of fabrics

	Warp direction			Weft direction				
Fabric structure	Yarn fibres	Yarn count (tex)	Density (cm^{-1})	Yarn fibres	Yarn count (tex)	Density (cm^{-1})	Weight (g/m^2)	Thickness fabric (mm)
Plain	Cotton	36	29,3	Cotton	36	20,0	193,7	0,48
Twill 1/3	Cotton	36	29,3	Cotton	36	20,0	191,6	0,53

Table 2: Mean experimental, theoretical and fitted values of breaking force and extension at break for plain weave fabric

φ (°)	ε (%)	ε_{calc} (%)	ε_{poly} (%)	F (N)	F_{calc} (N)	F_{poly} (N)
0	18,24	18,24	12,87	145,3	145,3	97,0
15	22,12	20,13	23,49	198,4	161,0	213,3
30	26,01	26,53	32,17	251,5	216,4	308,7
45	38,29	38,29	38,91	330,9	330,9	383,0
60	42,30	49,72	43,72	444,6	478,9	436,3
75	46,57	51,97	46,59	497,5	548,1	468,6
90	50,84	50,84	47,53	550,4	550,4	479,9
105	48,16	51,97	46,54	501,1	548,1	470,1
120	45,47	49,72	43,60	451,8	478,9	439,4
135	38,29	38,29	38,74	339,2	330,9	387,6
150	32,15	26,53	31,94	261,7	216,4	314,8
165	26,01	20,13	23,20	203,0	161,0	221,1
180	10,41	18,24	12,53	151,3	145,3	106,2

Yarn linear density was determined by the gravimetric method according to standard ISO 2060:1994. Number of threads per unit length was determined according to standard ISO 7211-2:1984. Standard ISO 5084:1996 describes a method for the determination of the thickness of fabric. Before testing all specimens were conditioned under the conditions of standard atmosphere (relative air humidity 65 ± 2%, at a temperature of 20 ± 2°C). For the purposes of this testing standard specimens with dimensions 300 x 50 mm were clamped in clamps of the tensile tester at a distance of 200 mm and pulling speed: 100 mm/min and subjected to uniaxial tensile load till rupture. The specimens were cut in seven different directions: warp direction (φ=90°), weft direction (φ=0°), and at angles 15°, 30°, 45°, 60°, 75° to the weft. Three tests were done for each mentioned direction of force action on the fabric specimen. Tensile properties of all specimens were tested according with standard ISO 13934-1:2008 using the strip method for measuring fabric strength on a tensile strength tester Textechno Statimat M [6]. The measurement results were collected and stored on the hard disk by a computer program of tensile strength tester. The experimental mean values of breaking force F_φ (N) and the extension at break ε_φ (%) are shown in Tab. 2 and Tab. 3

According to Eq. (2) and based on the experimental data $F_{0°}$, $F_{90°}$, $F_{45°}$, in Table 2 and Table 3, theoretical values of breaking force F_{calc} were calculated for various angles of tensile force action. According to Eq. (3), the fitted values of the breaking force Poly. (F (N)) was obtained. According to Eq. (1) and based on the experimental data $\varepsilon_{0°}$, $\varepsilon_{90°}$, $\varepsilon_{45°}$, in Table 2 and Table 3, theoretical values of extension at break ε_{calc} were calculated According to Eq. (5), the fitted values of extension at break Poly. (ε (%)) was obtained. The experimental and calculated values for plain weave fabric are shown in Tab. 2, and for twill weave fabric are in Tab. 3.

Table 3: Mean experimental, theoretical and fitted values of breaking force and extension at break for twill weave fabric

φ (°)	ε (%)	ε_{calc} (%)	ε_{poly} (%)	F (N)	F_{calc} (N)	F_{poly} (N)
0	16,10	16,10	13,08	131,2	131,2	90,7
15	21,45	17,81	22,52	179,8	145,3	195,5
30	26,90	23,69	30,25	240,7	195,0	281,6
45	34,68	34,68	36,27	298,1	298,1	348,8
60	41,60	45,47	40,58	405,4	434,5	397,4
75	42,59	47,34	43,17	457,4	503,1	427,1
90	46,12	46,12	44,06	507,8	507,8	438,1
105	43,56	47,34	43,24	436,5	503,1	430,3
120	38,40	45,47	40,71	412,2	434,5	403,7
135	35,48	34,68	36,47	308,1	298,1	358,4
150	29,11	23,69	30,51	267,5	195,0	294,4
165	24,70	17,81	22,85	200,9	145,3	211,5
180	9,52	16,10	13,48	138,7	131,2	109,9

The diagram of experimental and calculated values of breaking forces for plain weave fabric for each 15° is shown in Fig 1a. The diagram of experimental and calculated values of extension at break for plain weave fabric for each 15° is shown in Fig 1b.

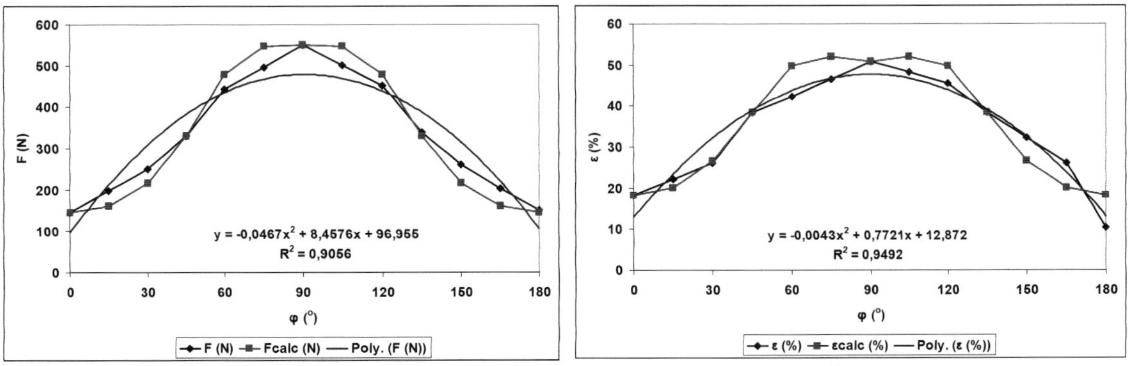

a. b.

Fig. 1: Diagram of experimental and calculated values for plain weave fabric: a) breaking forces, b) extension at break

The diagram of experimental and calculated values of breaking forces for twill weave fabric for each 15° is shown in Fig 2a. The diagram of experimental and calculated values of extension at break for twill weave fabric for each 15° is shown in Fig 2b.

In Fig. 1 and Fig. 2 is evident that woven fabrics of both weaves have the lowest value of breaking forces and extension at break in the transverse or weft direction ($\varphi = 0°$, $\varphi = 180°$). The values of breaking forces and the values of extension at break gradually increase from the weft direction ($\varphi=0°$) and the highest values are achieved in the longitudinal or warp direction ($\varphi = 90°$). Diagrams in Fig. 1 and Fig. 2 have symmetric values in relation to bias angle $\varphi = 90°$. The values of breaking force for plane weave fabric samples in the longitudinal direction ($\varphi = 90°$) is higher by about 3.8 times in relation to the transverse direction ($\varphi = 0°$), and in relation to the diagonal direction ($\varphi = 45°$) is higher by 1.7 times. The values of extension at break for plane weave fabric samples in the longitudinal direction ($\varphi = 90°$) is higher by about 2.8 times in relation to the transverse direction ($\varphi = 0°$), and in relation to the diagonal direction ($\varphi = 45°$) is higher by 1.3 times.

The values of breaking force for twill weave fabric samples in the longitudinal direction ($\varphi = 90°$) is higher by about 3.9 times in relation to the transverse direction ($\varphi = 0°$), and in relation to the diagonal direction ($\varphi = 45°$) is higher by 1.7 times. The values of extension at break for twill weave fabric samples in the longitudinal direction ($\varphi = 90°$) is higher by about 2.9 times in relation to the transverse direction ($\varphi = 0°$), and in relation to the diagonal direction ($\varphi = 45°$) is higher by 1.3 times.

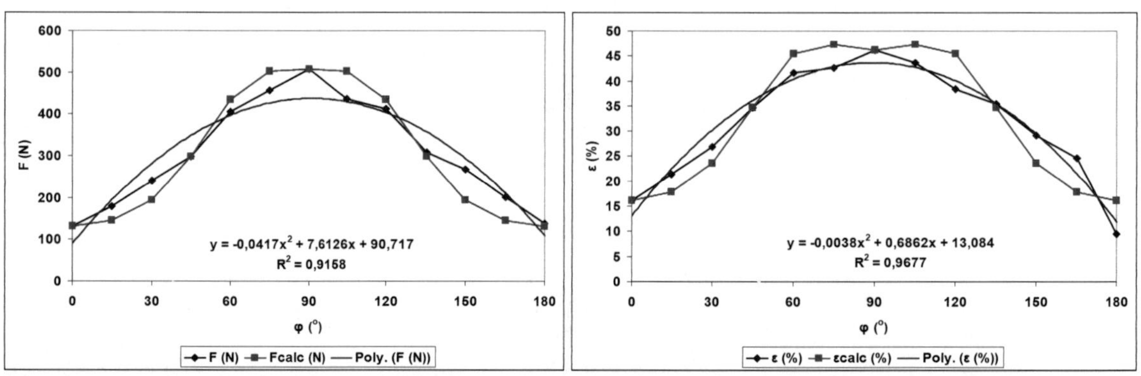

a. b.

Fig. 2: Diagram of experimental and calculated values for twill weave fabric: a) breaking forces, b) extension at break

Conclusions

The values of breaking force and extension at break are changed depending on the direction of force acting. Prediction of breaking forces and extension at break can be carried out in two ways. Theoretical equation with high accuracy can be used to calculate values of breaking forces and extension at break of fabrics for an arbitrarily chosen direction of action of tensile force. The values of breaking force and extension, also can be obtained by fitting the experimental values with the quadratic equation. A good agreement between experimental results and the calculated obtained values of breaking forces and extension at break was shown. Therefore, the measurements need to be implemented when the tensile force acting on the fabric only in the warp, weft and at an angle of 45°.

In plain weave fabric, a very high coefficient of determination R^2 ($R^2 = 0.9056$) is obtained, which means that there is almost an exact dependence between the value of breaking force F obtained by quadratic function and angle φ. For the dependence between extension at break ε obtained by quadratic function and angle φ, the coefficient of determination is $R^2 = 0.9492$.

In twill weave fabric, a very high coefficient of determination R^2 ($R^2 = 0.9158$) is obtained, which means that there is almost an exact dependence between the value of breaking force F obtained by quadratic function and angle φ. For the dependence between extension at break ε obtained by quadratic function and angle φ, the coefficient of determination is $R^2 = 0.9677$.

The values of breaking force and extension of plain weave fabric are about 10% higher than the values of breaking force and extension of twill.

References

[1] R. Kovar, B. S. Gupta, Study of the Anisotropic Nature of the Rupture Properties of a Plain Woven Fabric, Textile Research Journal, 79 (2009) 506-516.

[2] Ž. Penava, D. Šimić Penava, Ž. Knezić, Determination of the Elastic Constants of a Plain Woven Fabrics by Tensile Test in Various Directions, Fibres & Textiles in Eastern Europe 22 (2014) 57-63.

[3] E. K. Aškenazi, E. V. Ganov, Anizotropija konstrukcionih materialov, Mašinostroenie, Lenjingrad, 1972.

[4] I. Alfirević, Uvod u tenzore i mehaniku kontinuuma, Golden marketing, Zagreb, 2003.

[5] B. Ram, Numerical Methods, Pearson India, New Delhi, 2010.

[6] Ž. Penava, D. Šimić, Analysis of the elastic constants of woven fabrics for at random chosen extension directions, Tekstil, 61 (2012) 169–179.

EAN 2019
57th conference on experimental stress analysis

Stress analysis of a tank for bulk materials

ŠTÁDLER Marek[1,a] and DUB Martin[1,b]

[1]Czech Technical University in Prague, Faculty of Mechanical Engineering, Department of Designing and Machine Components, Technická 4, 166 07, Prague 6, Czech Republic

[a]marek.stadler@fs.cvut.cz, [b]martin.dub@fs.cvut.cz

Keywords: Bulk materials, Tank trailer, Stress analysis, Janssen´s theory, Optimization

Abstract. The paper deals with a stress analysis of the existing tank trailer for bulk materials as well as with the effects of bulk materials on the tank trailer wall. Technical parameters: Volume 62m^3, Max. working pressure 2 bar, Max. load weight 32 000 kg, Material EN AW-5083 0/H111. Designed according to AD 2000 Merkblatt. Analysis is based on extreme load conditions of the structure identified from the trailer working cycle. Based on its results, design optimization is made so as to eliminate the places with increased stress and, if possible to reduce the weight of the structure.

Introduction

Cargo transport is nowadays the most widespread and growing type of transport of goods. With its development, the amount of conveyed bulk materials is also growing, with increasing number of tank trailers being used. With increasing demands on ecology, traffic economy and safety, there is a growing demand for weight reduction and increased safety and lifespan of tank structures.

Due to the number and complex shapes of tank trailers and the specific types of transported materials, classical analytical computational procedures are inadequate. For newly designed tanks and optimization of the existing ones, the accuracy of the design is often verified by FEM analysis.

The aim of this article is design optimization based on a stress analysis (FEM) of the tank trailer for bulk materials (see Fig.1) made of EN AW-5083 0/H111.

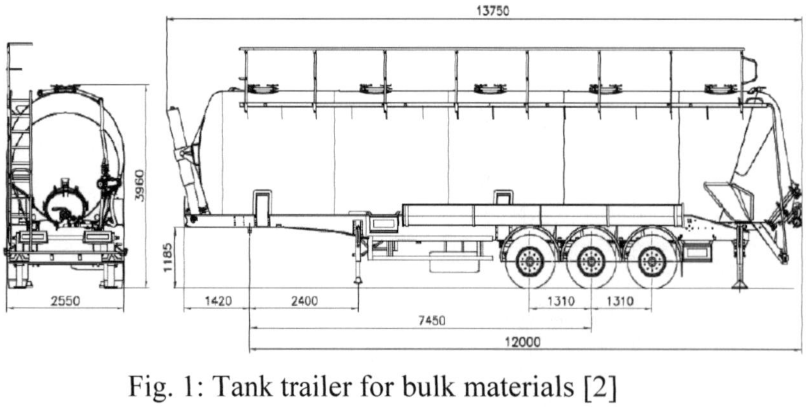

Fig. 1: Tank trailer for bulk materials [2]

Analysis and evaluation

From the tank working cycle three extreme load conditions were identified (see Table 1) of the structure, it was proved to be crucial to define the effects of bulk material on the tank trailer wall.

Table 1: Extreme load conditions

Condition:	Media pressure [MPa]:	Media density [kg/m3]:	Tank tilt angle: [deg]	Material type:
Start of tilting	0,2	516	0	bulk
End of tilting	0,2	516	45	bulk
Pressure test	0,3	1000	0	fluid

The effect of bulk material was simulated by commonly used Janssen´s theory [1] for the shallow bins (see Fig.2). The chosen theory was applied to the wall of the tank and was transformed for the case where the gravity is acting in reversed y-direction (1,2).

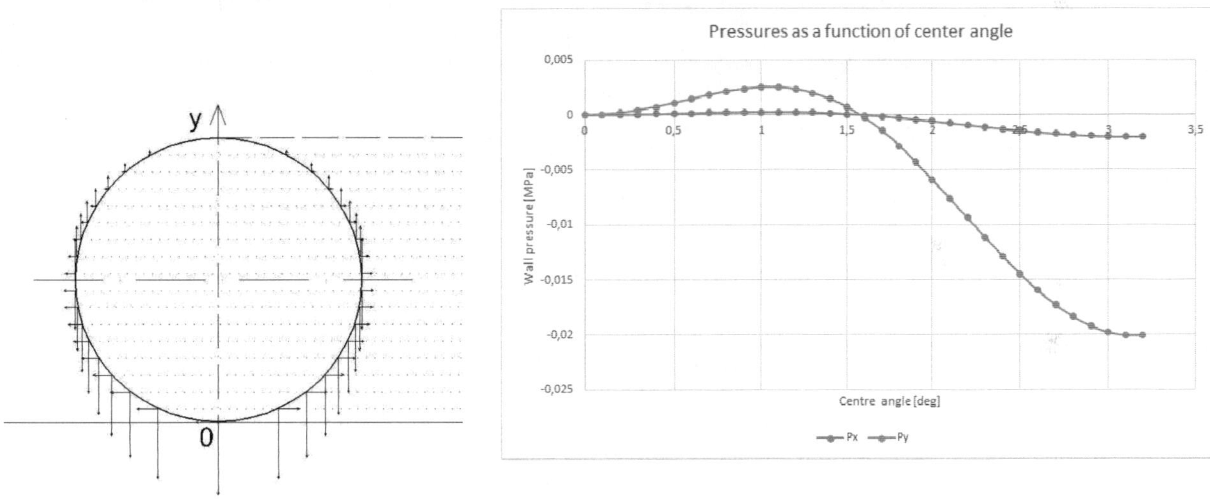

Fig. 2: Simulated pressure distribution on a circular wall

$$p_y = p_H \cdot \cos\varphi = R{\cdot}\rho \cdot g \cdot (1 - \cos\varphi) \cdot \cos\varphi \qquad (1)$$

$$p_x = p_y \cdot tg^2 \left(\frac{\pi}{4} - \frac{\rho}{2}\right) \qquad (2)$$

Verification of correctness of transformation and the application of Janssen's theory to the FEM program environment was performed by two examples for which it was possible to determine the results using FEM analysis as well as by classic analytical procedures, and these results were compared.

The first verification example was to determine total weight of the bulk material in cylindrical wall as the resulting force of simulated wall pressures (1,2)
The second example was comparison of the Mises stress (see Fig.3) on a half circular curved beam (0;π) with one end built-in and the other end free.

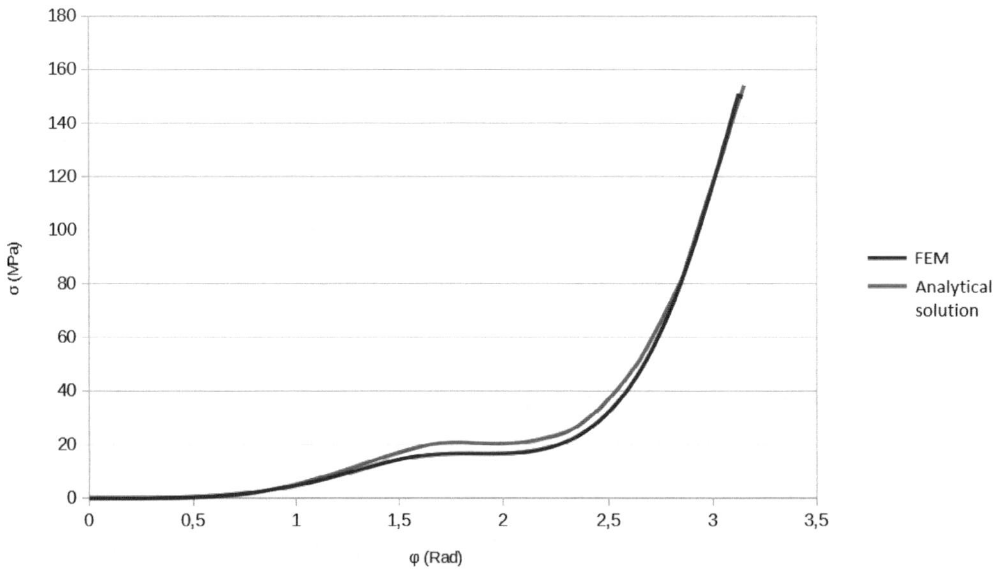

Fig. 3: Comparison of Mises stress on a curved beam as a function of center angle

After verification of application of Janssen's theory, FEM analysis was made, which shown that several places can be found, where minimum yield stress of material was exceeded (see Fig.4). (EN AW-5083 0/H111 - $MinRp_{0,2}$ =125 N/mm^2)

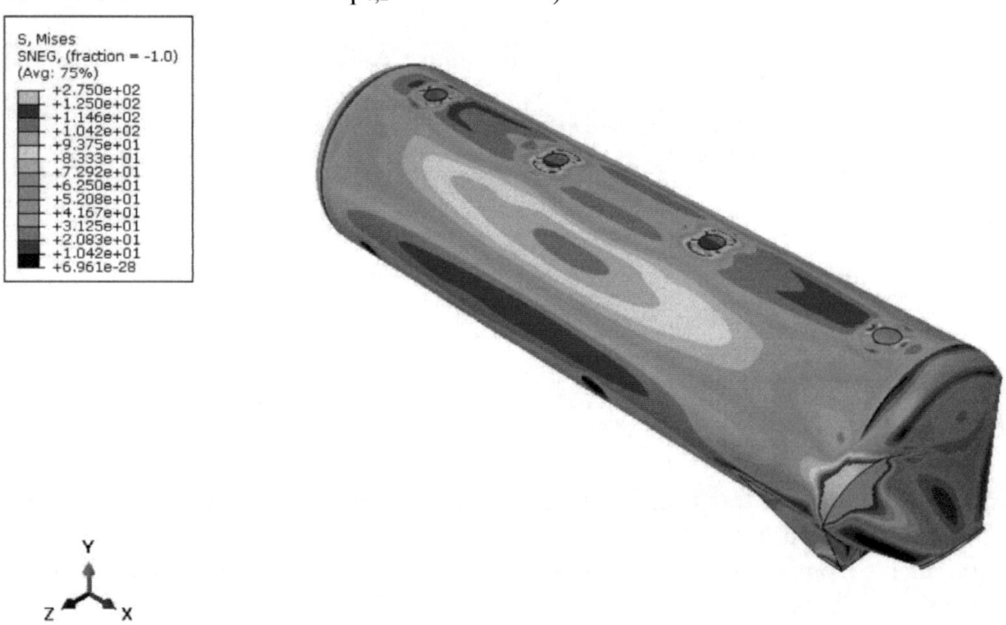

Fig. 4: Structural stress - start of tilting (general)

The first place was in the circular mid-section of tank, symmetrically on both sides the stress increased up to 125 N/mm^2. This stress was caused by combination of boundary conditions of tank and pressure fields of bulk material acting along tank wall.
The second place was connection between circular section and conical end section (see Fig.4), where the increased stress was caused by incorrect geometry of joint, as well as low tank toughness in the ZY plane caused by reduced shell thickness.

Design optimization

As a result of several analyses, design optimization was made (see Fig.5). After consultation with manufacturer only joint between mid and end section was optimized. Material used for tanks has significantly higher yield stress $ReaIRp_{0,2}$ =164 N/mm^2, so design modification of circular mid-section wasn´t necessary.

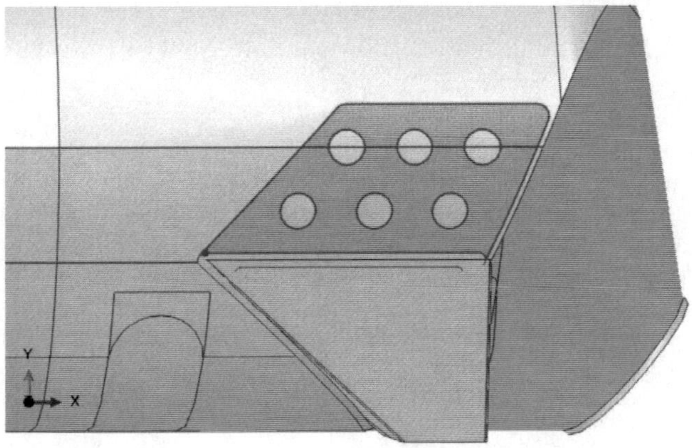

Fig. 5: Structural stress comparison – strengthening of joint

Optimization consisted of strengthening tank toughness in the ZY plane by extension of reinforcing plates from back support legs and rounding of cone section in joint for smooth transition. The designed modifications were compared with the existing design. As it can be seen (see Fig.6;7;8) modifications resulted in a considerable reduction of stress in tank wall.

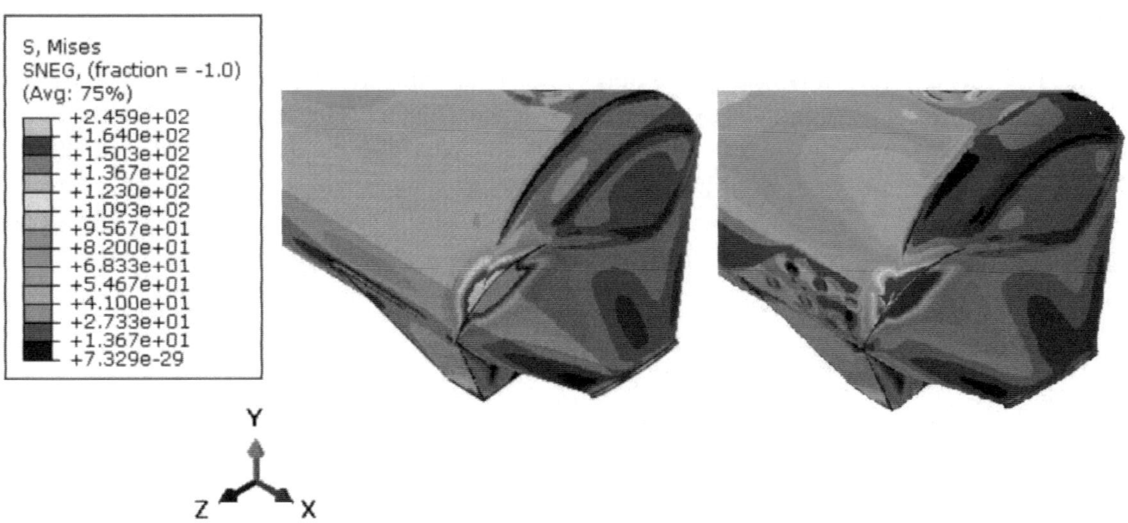

Fig. 6: Structural stress comparison - start of tilting

Fig. 7: Structural stress comparison - end of tilting

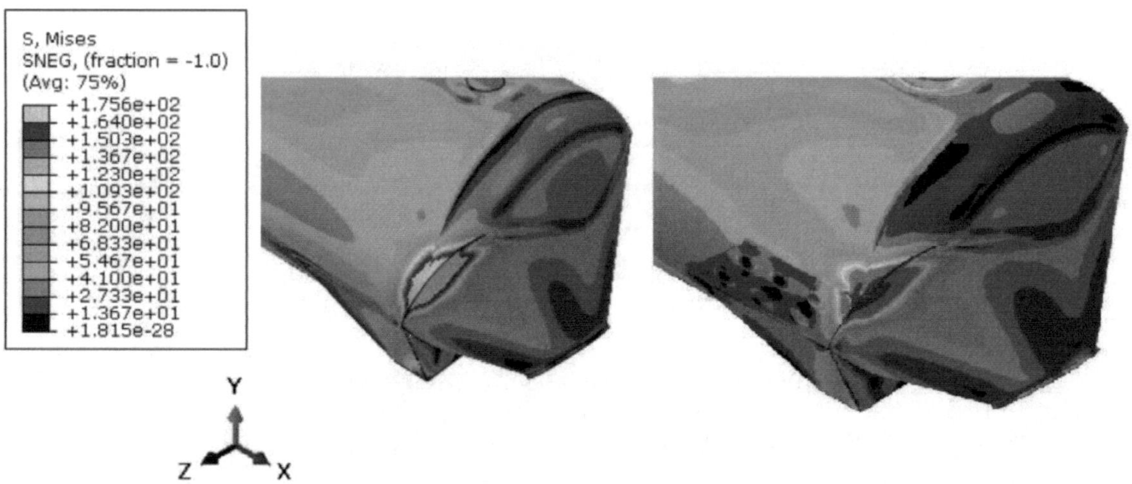

Fig. 8: Structural stress comparison - pressure test

Conclusion

Series of FEM analysis of tank construction showed that there can be found locations where the yield strength of material was exceeded (see Fig.4). Based on the analysis, the locations were selected and the possible causes were discussed. After that the design of the structural modifications were made, according to the manufacturer's technical requirements.

References

[1] Cowin SC. The Pressure Ratio in the Theory of Bin Pressures. ASME. J. Appl. Mech. 1979;46(3):524-528. doi:10.1115/1.3424600

[2] Přepravníky | ZVVZ a.s. - dodavatel zařízení pro ekologii. ZVVZ a.s. - dodavatel zařízení pro ekologii [online]. Copyright © 2017, ZVVZ a.s. [cit. 21.06.2017]. Dostupné z: http://www.zvvz.cz/zvvz-machinery/vyrobky/cisternove-navesy.html.

EAN 2019
57th conference on experimental stress analysis

Use of durometer to determine trapezius muscle relaxation

ŠTOČEK Ondřej[1,a], HAVLÍČEK Martin[1,b], LOPOT František[1,c] and KUBOVÝ Petr[2,d]

[1]CTU in Prague, Faculty of Mechanical Engineering, Department od Designing and Machine Components Technická 4, 166 07 Prague 6, Czech Republic

[2]Charles University in Prague, Faculty of Physical Education and Sport, Department of Anatomy and Biomechanics, Jose Martího 31, 162 52, Prague 6, Czech Republic

[a]ondrej.stocek@fs.cvut.cz, [b]martin.havlicek@ fs.cvut.cz, [c]frantisek.lopot@ fs.cvut.cz, [e]kubovy.petr@seznam.cz

Keywords: Durometer, Palopation test, Myotonometer

Abstract:
Measurement of the muscle tension are normally done with specialized apparatus, a.k.a. myotonometers. These devices need specialized personnel, sophisticated electric drive, feedback control and mostly static construction. Hence the high purchase and operating costs of these devices. For these reasons, the requirement for our device arose. The basic requirements are the manual size of the apparatus and the possibility of measuring various muscles all over the body. The apparatus works on the principle of indentor injection into the tissue. The measured quantity is the resistance that the tissue puts. The maximum indentation depth is adjustable and constant during all partial measurements.
Another use of apparatus is to objectify the Palpation Test. The evaluation of this test is highly subjective and depends on the experience of the implementing specialist. With help of this apparatus it is possible to visualize result while facilitating the transfer of experience among clinical experts.
The apparatus is designed to by attached to the respective muscles and the indentor is manually pushed into the muscle. A compression spring is used to reverse movement. The threaded rod is used as a beam between the measuring sensor and the handle. This makes it possible to use nuts as an adjustable depth of indentation. The measuring sensor is one-axis pressure gauge from Kisler.

Introduction

This paper deals with the device for measurement of the muscle tension in various modes of human musculoskeletal activity. The demand for its development was initiated by the Department of Physiotherapy and Anatomy and Biomechanics, Faculty of Physical Education and Sport at Charles University (FPES) to enable the palpation test with a simple portable apparatus. On the market there are available various types of specialized apparatus, a.k.a myotonometers. The absolute majority of them are quite sophisticated with the electric drive and feedback control and they are thus relatively expensive and often non-mobile apparatus. Practically without the exception is the principle of all instruments inspired by the durometric principle, where the indentor is at a constant speed pushed into the superficies of the muscle. The device measure two parameters: the indentation depth and the reaction force of loaded tissues.

Palopation tests are an indispensable part of soft tissue examination methods. Based on this, is a diagnosis of the patient problems, source localization or area of clinical manifestation of the problem and the success of the intervention is monitored. Because is performance and evaluation of the palpation test heavily dependent of the experience of the practitioner, there is an understandable attempt to objectify of such findings and activities so that can by information about them better transferred between clinical expert within the scope learning process and minimize risk of potential negative human impact in clinical application.

Requirements for the properties of our mynotonometer – assignment

Requirements for the properties of our apparatus arose from the fact that it should primarily serve the needs of teaching physiotherapists and possible the research of muscle tone within the scope of diploma and dissertation theses of FPES students.

- The apparatus must by design for manual control manual (max. weight 0,5 kg).
- The apparatus must enable measuring different muscles on the body.
- The apparatus must record measured data.
- The apparatus must represent measuring data online on a display.
- The apparatus must by easy to use.
- The apparatus may not be able to measuring force and depths of the indentor over time – the maximum values achieved are important.
- The apparatus must be producible at minimum cost:
    - Utilize the production possibilities of the technological hinterland of the laboratory BEZ, Department of Anatomy and Biomechanics.
    - Utilize the sensors and measuring technology of the laboratory BEZ, Department of Anatomy and Biomechanics.

Similar to commercially available apparatus, we have used the durometric principle of indentor

Properties of our mynotonometer – compliance assignment

Base on the given assignment, we designed an apparatus with the following properties and the following design:

- Drive: Manual.
- Depth indent of indentor: Non-measured, constant and configuratable for goven measurements.
- Indentor force: Measured throughout pushing.
- Sensor: Kisler 9317
- Measuring station: Dewesoft Sirius
- Control software: Dewesoft X3
- Sampling frequency: 1000 Hz
- 3D printing (string): body of the apparatus
- 3D printing (lithography): indentor, push handle
- Other components: standard machine components

Constructional solutions our mynotonometer:

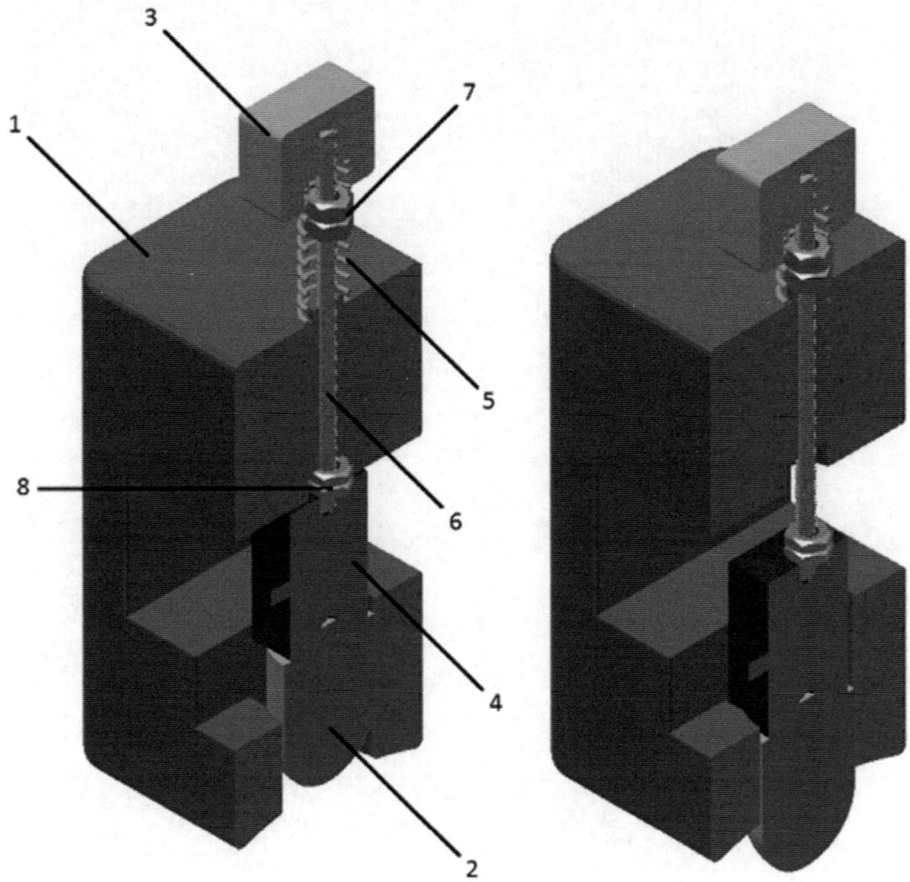

Fig. 1: 1 – body of the apparatus, 2 – indentor, 3 – push handle, 4 – sensor, 5 – spring, 6 – threaded rod, 7-8 – nuts

Fig. 1 provides an overall view of the myotonometer.

The measuring apparatus consists of apparatus body (Fig. 1 pos. 1), which is a complete supporting construction and a frame for linear reversible movement of the measuring apparatus, the indentor (Fig. 1 pos. 2) and the handle (Fig. 1 pos. 3). To produce these parts, we used the 3D printing method. The measuring apparatus includes a one-axis pressure sensor by Kisler 9317 (Fig. 1 pos. 4), threaded rod M6 (Fig. 1 pos. 6), nut for end-stops (Fig. 1 pos. 7 and 8) and the compression spring (Fig. 1 pos. 5) for reversible movement. To transfer the measured data to PC, the device is equipped with the connector, which is in the opening. The Sensor (Fig. 1 pos. 4) is located between the indentor and the threaded rod on which end is the handle (Fig. 1 pos. 3). Two pairs of nuts are placed on the threaded rod to define the start and stop position for the movement of the measuring system.

It can be seen from Fig. 2 that the apparatus has a surface applied to the measured muscle to by rounded to make better and truly large area have contact with muscle ant not compress locally examined tissue.

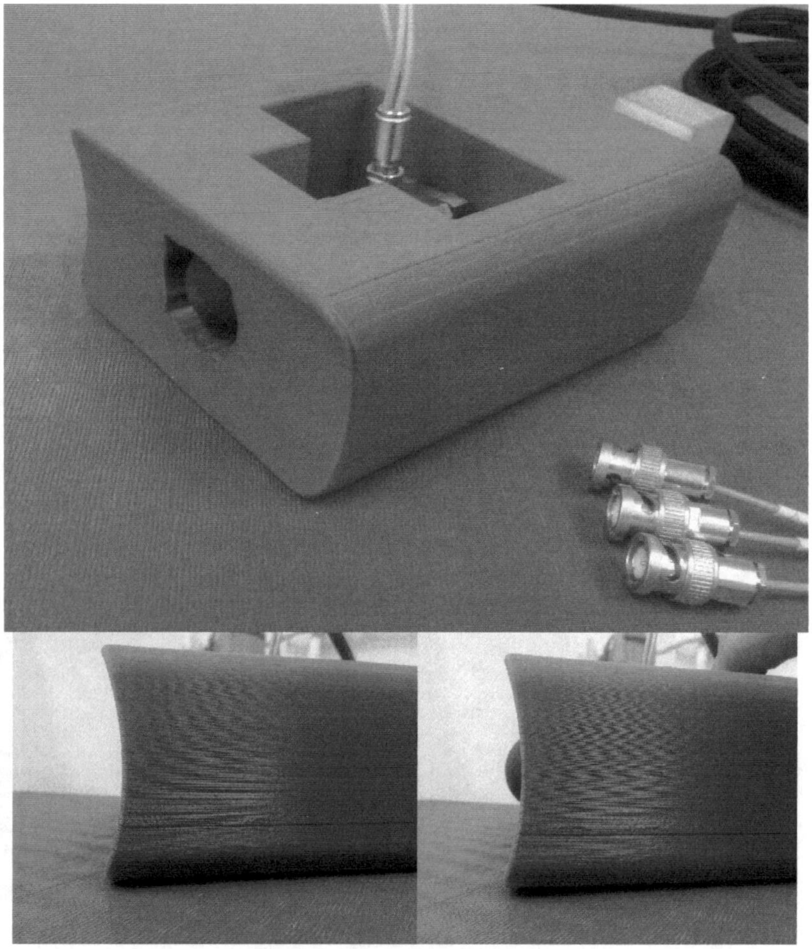

Fig. 2: Myotonometer

Actual used of our myotonometer:

The apparatus is currently used in the master thesis Bc. Miroslava Patschová, who examines the changes in tone of selected muscles after applying Foam roller immediately after physical performance. Fig. 3 is a control software environment showing a contact force measurement record.

Conclusion:

The purpose of the article was to show how beneficial the "engineering" approach can be when it is applied with respect to the industry of its immediate application. Within the scope of this article we presented apparatus, that despite its simplicity, it provides useful data. At the same time, the design provides practically unlimited possibilities for its expansion to measure other phenomena that are subject to more detailed research in muscle tissue. Currently, next generation of the device is under the development. It will offer a more ergonomic profile and a more comfortable setting of the depths for the indentation.

EAN 2019
57th conference on experimental stress analysis

Substituting a conventional ram frame with a variants of sandwich structure

TANČIN Milan
University of West Bohemia, Univerzitní 22, 306 14, Plzeň, Czech Republic
milax@rti.zcu.cz

Keywords: Frame, ram, sandwich, simulation, modal analysis, static analysis, frequency response

Abstract

The main goal is to determine if a sandwich structure (with steel sheets as outer plates and aluminium foam as filler) is suitable for a ram in an overhead gantry. The main expected features of the new design are fewer deformations, lower weight and equal or higher natural frequencies than the current solution. A flat semi-circular sandwich structure, wrinkled (horizontal and vertical) semi-circular sandwich structure and angular-shaped sandwich structure will be implemented into the adjusted frame. It is also necessary to keep in mind the possibility of mounting the sandwich structure into the adjusted frame. Comparative analysis were done to determine if and how the new design is better than the current solution. The primary tools used for the comparison are modal analysis, static analysis and frequency response for several states.

Introduction

Strong competition between machine tool manufacturers is the driving force behind innovations. Therefore it is necessary to develop and improve current solutions used in machine tools. One of the main component groups of machine tools are rams. The main feature of a ram is to hold the machining head which is connected to the ram by bolts on a circular flange and afterwards connected with other parts of whole machine tools. The slide is connected to the ram and performs horizontal transverse movements. Longitudinal movements are performed by a crossbeam which is placed on two supports which transfers all the loads to the base. Sleeve bearings connect the basic parts of the machine tool. Basic outer dimensions of the whole machine tool are (l x w x h) 10100 mm x 5280 mm x 7112 mm. Dimensions of the fixing desk are 6200 mm x 2670 mm. The ram performs vertical movements in a defined range. Furthermore, the ram is another part of a machine tool for which it is possible to use the advantages of sandwich materials, such as high bending stiffness, low weight and vibration damping. Several approaches for implementing sandwich materials will be shown. The approaches will be verified by simulations, which will be performed on a conventional frame and on the newly designed frame.

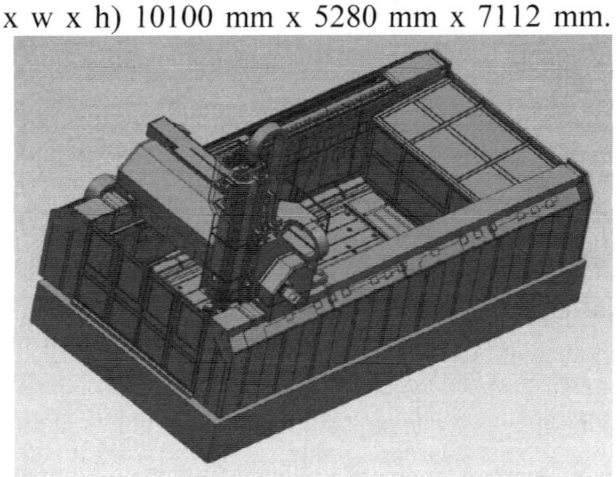

Figure 1 Overhead gantry with ram (in yellow)

Material selection for designing a sandwich structure

A sandwich material composed from two steel sheets as outer plates and aluminium foam filling was selected for the design. Steel sheets were selected considering the connection with the rest of the frame by various methods including welding. The main advantages of aluminium foam are high stiffness, low density, high toughness, corrosion resistance and very good vibration damping. Aluminium foam can be used not only for sandwich structures but also for filling more complex products such as cast engine brackets. Foam can be prepared in several ways. It is a very porous material and its structure is similar to natural structures such as bone and coral. Cellular structures have relatively high stiffness. In comparison with other metal foams aluminium foam has many advantages, as mentioned above. The principle is to create bubbles in the melt and then cool the melt. There are two main types of alloys for manufacturing the foam – alloys that are suitable for casting and alloys that are suitable for forming. The first have smaller, irregular shaped pores, the walls are thin and the structure is non-uniform. Alloys suitable for forming create foams with spherical pores and the walls are thicker. Bubbles can be created as a result of melting the semiproduct, which contains a frother. Another way is to blast gas into the melt from an external gas reservoir. Thermal decomposition of added frother in the melt is also possible. When the melt has low viscosity, the resultant foam contains bigger spherical pores, which rise to the surface. The tendency is to create smaller pores. Manufacturers of aluminium foams include LKR, Cymat and Shinko Wire Company.

Figure 2 Sandwich material composed from two steel sheets and aluminium foam [2]

Principle of substitution

The objective is to use sandwich material in the area shown in Figure 3. The area is composed from inner and outer steel sheet envelopes which are connected to each other by ribs with various thickness. The weight of the existing frame is 2640 kg. The dimensions of the frame are (l x w x h) 705 mm x 620 mm x 3000 mm. The new design of the frame should preserve these dimensions to ensure the simple replacement of the conventional frame with the new design, which is without the struts and instead has a sandwich panel. The new design is adjusted in the modified area for placing the sandwich structure in grooves. The sandwich panel is fixed above by a steel element with the same groove. Grooves are adjusted to each design. The adjusted frame, sandwich panel and upper element are connected to each other by welding. This is shown in yellow in Figure 4. The sandwich panel is principally composed from two bent steel sheets as outer plates and filled with aluminium foam. The series of new designs are slightly different from the current state and its profile are semi-circular, semi-circular with wrinkles (horizontal and vertical) and angular. Semi-circular flat profile and angular profile are the best for manufacturing because of the simplest shape. The sandwich

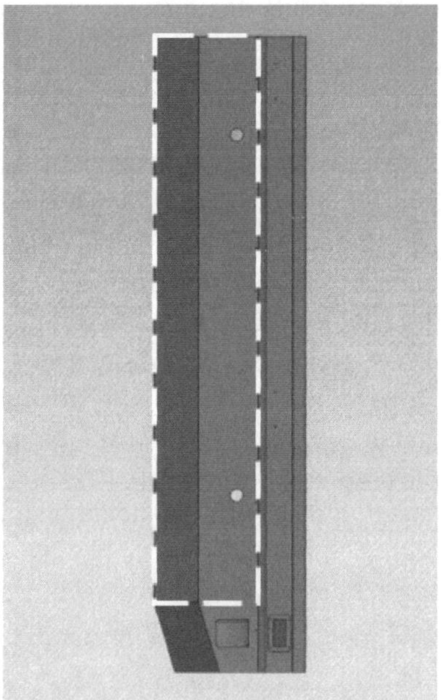

Figure 3 Conventional ram with area to be replaced by sandwich structure (white dashed line)

plate is completely implemented as one part. Small holes and hollows are not taken into consideration during the following simulations due to their small influence on the final results. Four versions of the new design are tested – V1, V2, V3 and V4. These variants have different shape and design but the same layer thicknesses. V1 has flat semi-circular profile, V2 has horizontally wrinkled semi-circular profile, V3 has angular shape profile and V4 has vertically wrinkled semi-circular profile. Total wall thickness in the current state is 85mm, newly designed composition is 50mm thick so wall thicknesses are about 41% thinner.

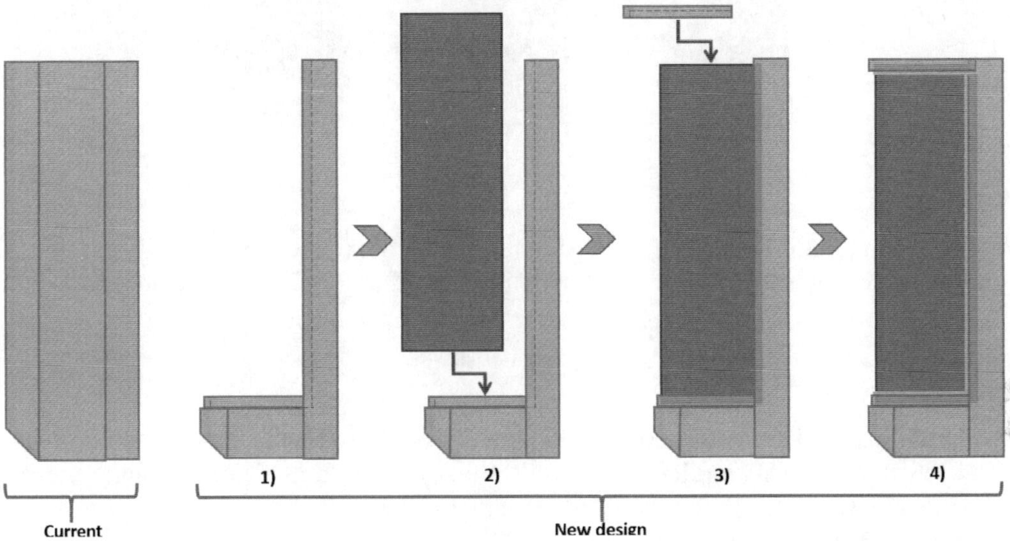

Figure 4 Diagram of current solution and new design including manufacturing steps

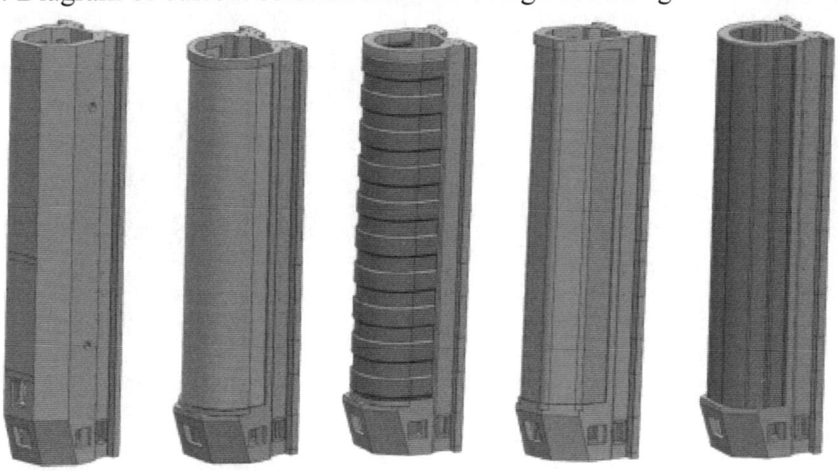

Figure 5 Comparison of current and new frames (from left: current, V1, V2, V3, V4)

Preparing the models of conventional and new frames

Conventional frame

First, modal analysis, static and frequency response were performed and these results are compared with the new designs. CTETRA4 with element size 25mm with the possibility of local refinement was selected as the mesh. For modal analysis, the model was fixed in maximal overhang and not loaded. Static simulation was carried out for minimum and maximum overhang and with a load (1000N) on two axes in the horizontal plane. Frequency response was carried out for maximal overhang in whole range of frequencies of modal analysis for most suitable variant. Modal analysis was used as basis for frequencies range determination.

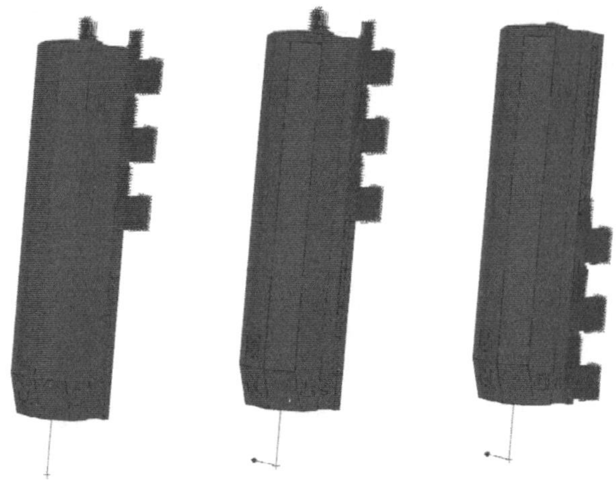

Figure 6 Computational models of conventional frame. (Left) modal analysis; (middle) static analysis and frequency response - high overhang; (right) static analysis - low overhang

New design – V1, V2, V3, V4

The sandwich material was defined by a laminate modeller (Siemens NX), which means that the sandwich structure was modelled as two 2D faces. First, both 2D faces ware meshed by a 2D dependent mesh with 25mm elements to ensure the defined layer composition between those two faces is filled. Thicknesses of layers were modelled the same for all variants. It is

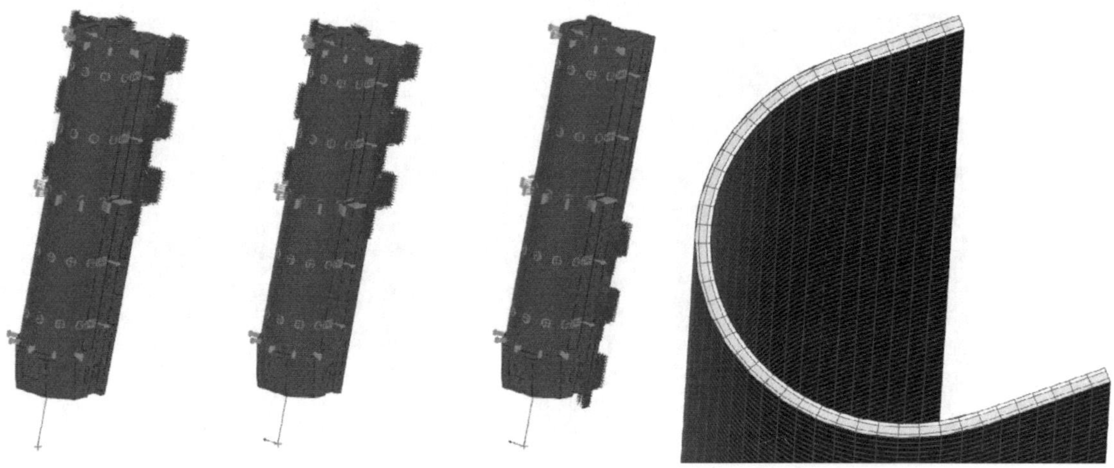

Figure 7 Left: Computational models of new designed frame – from left: modal analysis, static and harmonic analysis – high overhang, static analysis - low overhang; Right: Layer composition of sandwich panel

composed of 12 mm steel sheets for the outer layers and 26 mm aluminium foam (Density 400 kg/m^3, Young´s Modulus 2374 MPa, Poisson´s Ratio 0.29) as filler. Total thickness is 50mm. The rest of the frame was adjusted by cutting and removing the conventional struts in the modified area. The grooves and the upper element, mentioned above, were modelled in this area. All these adjustments create one integral part. The outer steel sheets of the resultant sandwich panel are connected to the adjusted frame by Surface-To-Surface Gluing. This is a simplifying element which substitutes for the connection by welding.

Results

The results of the modal, static and frequency response are used as reference results for the following evaluation of the new designs. Figure 8 shows the first four natural frequencies and Figure 11 shows the results of static analysis. The same results are also shown for the modified variant V4 - Figure 9 shows natural frequencies and Figure 10 shows static analysis. Only V4 was selected because its results are comprehensively most similar to the current solution. The first 10 natural frequencies for each design (current frame, V1, V2, V3, V4) were simulated and four loading states were prepared for each design. To verify how big displacements occur during modal analysis, frequency response for the most suitable was performed. In general, the resultant natural frequencies of modal analysis for V1 have lowest deviations in compare with the conventional frame. In general all new variants have lower deformations than conventional frame. The deformations of V4 are the lowest in compare with all another variants (average improvement is about 16% across all load states). This is main advantage of V4. It is also about 10% lighter. In absolute terms, this means a saving of 274 kg. Results of frequency response shows, that V4 has lower deformation in the most critical mode. It is the same mode as in conventional frame – mode 1. It is about 20% lower deformation than for conventional frame.

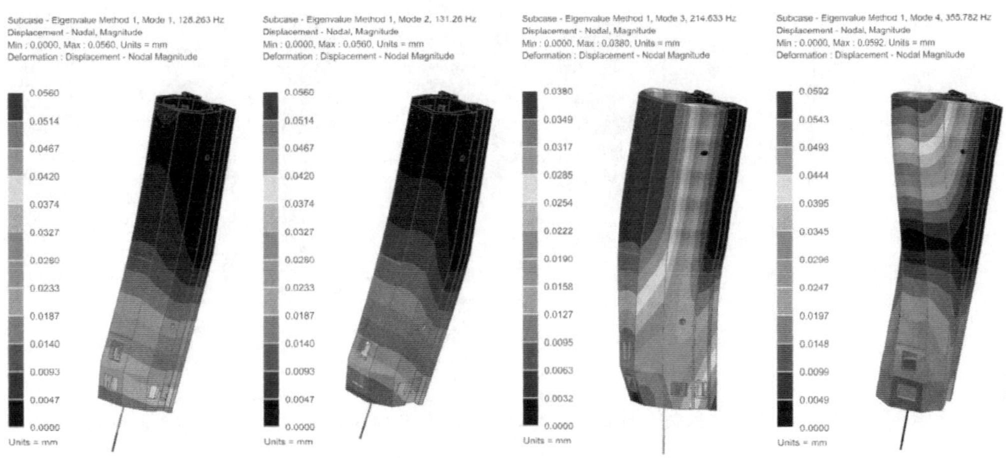

Figure 8 Results: Modal analysis of conventional frame with directions of first 4 natural frequencies

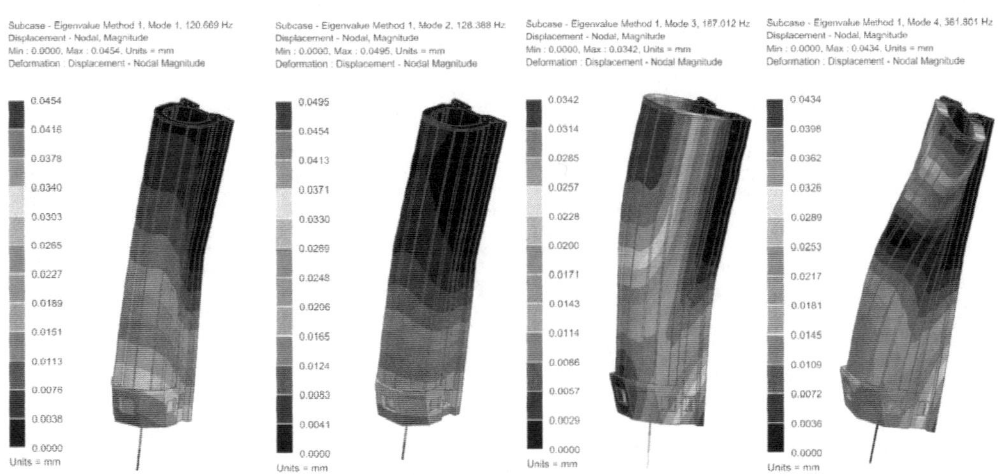

Figure 9 Results: Modal analysis of new design V4 with directions of first 4 natural frequencies

	CURRENT SOLUTION	V1		V2		V3		V4	
MODE	VALUE (HZ)	VALUE (HZ)	DIFFERENCE (%) (V1 VS. CURRENT)	VALUE (HZ)	DIFFERENCE (%) (V2 VS. CURRENT)	VALUE (HZ)	DIFFERENCE (%) (V3 VS. CURRENT)	VALUE (HZ)	DIFFERENCE (%) (V3 VS. CURRENT)
1	128.3	123.8	-3.5	116.7	-9.0	122.4	-4.6	120.7	-5.9
2	131.3	130.8	-0.4	117.5	-10.5	132	0.5	128.4	-2.2
3	214.6	196.5	-8.4	192.3	-10.4	177.6	-17.2	187	-12.9
4	355.8	364.7	2.5	364	2.3	335.5	-5.7	361.8	1.7
5	477.7	474	-0.8	445.5	-6.7	444.7	-6.9	459	-3.9
6	505.8	474.4	-6.2	463.5	-8.4	462.3	-8.6	471	-6.9
7	576.7	505.7	-12.3	489.3	-15.2	471.8	-18.2	472	-18.2
8	657.4	563.7	-14.3	559.7	-14.9	524.9	-20.2	535	-18.6
9	669.7	597.7	-10.8	582.8	-13.0	574.7	-14.2	584.5	-12.7
10	683.3	616.4	-9.8	588.9	-13.8	589.4	-13.7	603.7	-11.6

Table 1 Comparison of results – natural frequencies

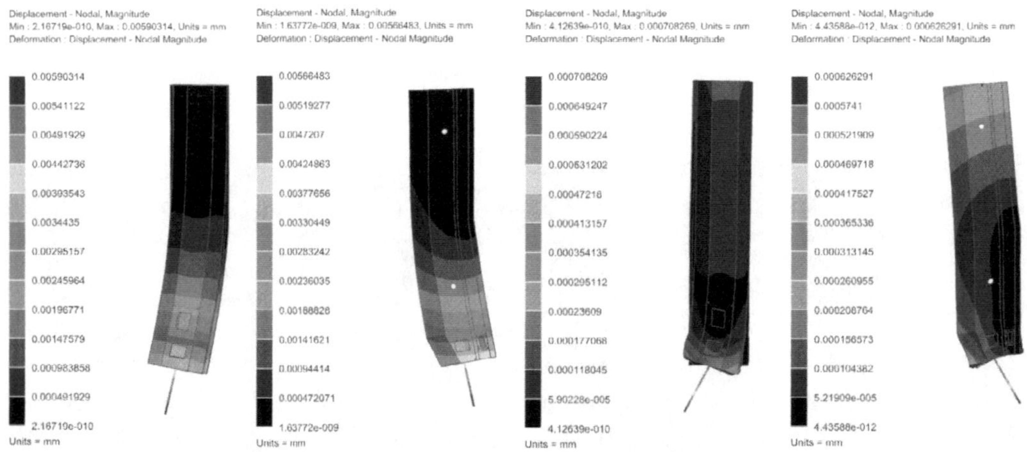

Figure 11 Results: Static analysis of conventional frame. From left: 1) low overhang, load in X direction, 2) low overhang, load in Z direction; 3) high overhang, load in X direction; 4) high overhang, load in Z direction

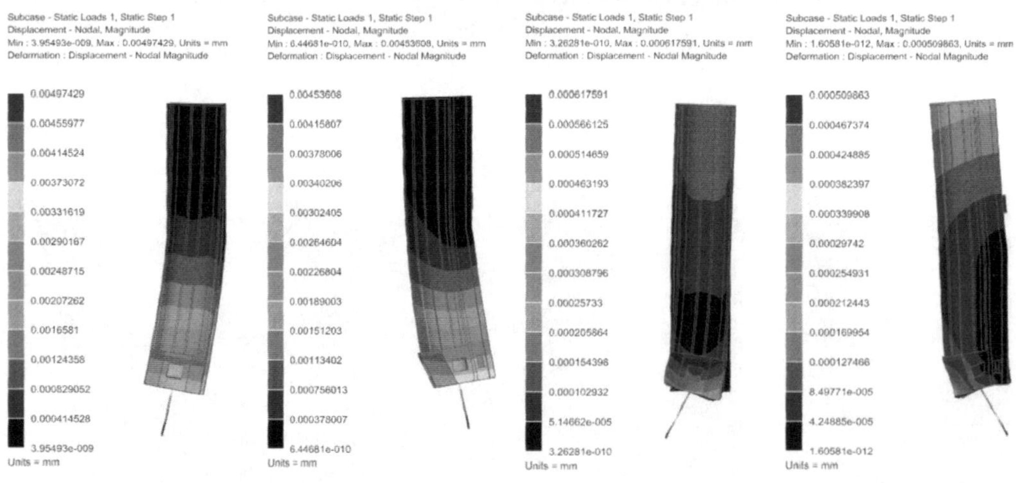

Figure 10 Results: Static analysis of new design V2. From left: 1) low overhang, load in X direction, 2) low overhang, load in Z direction; 3) high overhang, load in X direction; 4) high overhang, load in Z direction

	MINIMUM OVERHANG (mm) Max Deformation X	MINIMUM OVERHANG (mm) Max Deformation Y	MAXIMUM OVERHANG (mm) Max Deformation X	MAXIMUM OVERHANG (mm) Max Deformation Y	Total weight (kg)
Conventional frame	0.000708	0.000626	0.005903	0.005665	2641.1
V1	0.000642	0.000540	0.005711	0.005630	2356,3
V2	0.000645	0.00053	0.006017	0.005851	2343,2
V3	0.000685	0.000590	0.005591	0.005432	2336,3
V4	0.000620	0.000510	0.004970	0.004540	2367,4
DIFFERENCE (%) Conventional vs. V1	-9.3	-13.7	-3.3	-0.6	-10,8
DIFFERENCE (%) Conventional vs. V2	-8,9	-16,1	1,9	3,3	-11,3
DIFFERENCE (%) Conventional vs. V3	-3,2	-5,8	-5,3	-4,1	-11,5
DIFFERENCE (%) Conventional vs. V4	-12,4	-18,5	-15,8	-19,9	-10,4

Table 2 Comparison of results - static analysis and weight

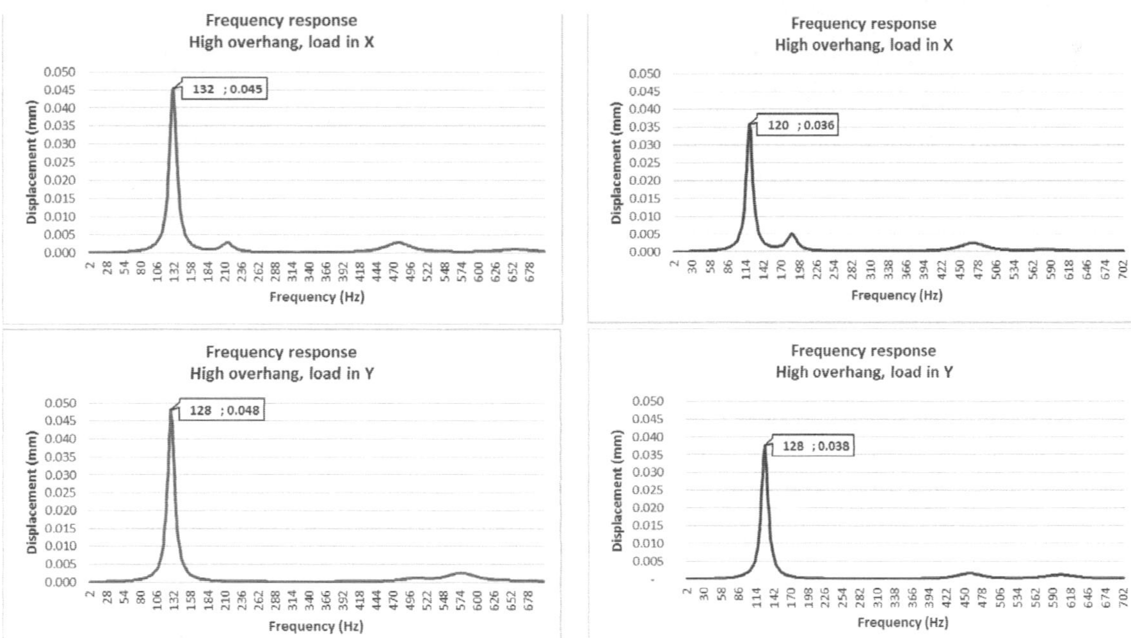

Figure 12 Results - frequency response; left top: Conventional frame, high overhang, load in X; right top: V4, high overhang, load in X; bottom left: Conventional frame, high overhang load in Y; bottom right: Conventional frame, high overhang, load in Y

Conclusion

This article deals with the comparison of a conventional ram frame and an alternative construction in several variants. They are composed of sandwich material. Four variants of the new design with aluminium foam filler were simulated. Simulations confirmed that this material is useable for similar operations as the conventional ram. Substitution of the conventional design in the ram with similar natural frequencies and lower weight and mainly lower deformations is possible by using variant V4. The method is also applicable for previously manufactured conventional frames which can be also adjusted as mentioned above. In general terms, the method cannot be prepared without simulations of a virtual frame. Substitution cannot be based only on estimation. There are still opportunities for optimization. The frame could be reinforced by several struts designed around the sandwich panel which would mean that the steel panels of the sandwich panel would be thinner, ensuring lower final weight. It would be the next step to use the struts to reinforce the structure and investigate these influences on the results. Another way is to replace the material of the outer plates. Aluminium sheet can be used instead of steel sheet. In this case, connection by bolts or glue is better. Investigations will mainly focus on the reduction of weight, resultant deformation, preservation of current natural frequencies and manufacturing costs. The optimized ram can be also tested on a whole machine tool. Other parts of the machine tool can be the same. These results will be also important because the new design should be usable mainly for a complete machine tool and its cutting conditions.

Acknowledgments

I would like to thank mainly my supervisor for her valuable advice, and also my colleagues.

References

[1] Nová, Iva a Machuta, Jiří. Výrobky z hliníkové pěny. *MM Spektürm.* Inovace, 18. 12 2007, 12, str. 58.

[2] Banhart, John. Manufacture, characterisation and application ofcellular metals and metal foams. *Progress in Material Science.* 2001, 46, str. 578.

[3] *Materials engineer / Materiálový inžinier.* místo neznámé : Peter Oslanec - Entos, 2009. ISSN 1337 - 8953.

[4] Tančin, M., & Limberg, L. (2018). Substituting a Conventional Ram Frame with a Sandwich Structure. *EXPERIMENTAL STRESS ANALYSIS - Conference Proceedings* (stránky 407-415). Liberec: Czech Society for Mechanics.

Optimization of cement composites composition with milled recycled concrete and admixture of slag and lime hydrate

TESÁREK Pavel[1,a], ZOBAL Ondřej[1,b], PROŠEK Zdeněk[1,c] and PLACHÝ Tomáš[1,d]

[1]Czech Technical University in Prague, Faculty of Civil Engineering, Department of Mechanics, Thákurova 7, 166 29 Prague 6, Czech Republic

[a]pavel.tesarek@fsv.cvut.cz, [b]ondrej.zobal@fsv.cvut.cz, [c]zdenek.prosek@fsv.cvut.cz, [d]plachy@fsv.cvut.cz

Keywords: Cement, Recycled concrete, Slag, Lime Hydrate, Optimization

Abstract. Sustainable development and construction is a very topical topic. This is related to waste management, where construction waste is a significant part and waste concrete is one of the most voluminous components. It turns out that after the necessary modifications (such as milling) waste concrete can be returned back into the building process, not only in the form of aggregates, but also as part of the binder. This paper deals with optimization of the composition of cement composites using recycled ground concrete, lime hydrate and other waste material – slag. Selected parameters such as workability, modulus of elasticity and strength are monitored.

Introduction

This paper builds on previous research and composition of the mixtures has been optimized and designed on the basis of several phases of experiments [1-3]. Milled recycled concrete is only composed of a combination of structural concretes, because concretes as used in other applications (e.g. drain channel) as a potential component of the binder not worked. However, it was confirmed that treatment of recycled concrete by milling works as a mechanical activation. The selected admixtures slag and lime hydrate and their amounts should help to alkaline activate the unhydrated grains of clinkers of milling recycled concrete, i.e. chemical activation. The paper presents the effect of composition optimization on wet mixture workability and on the mechanical properties of the hardened mixture.

Used materials and composition of the mixtures

It has been proposed and solved a total of 7 mixtures with designation CR-LS I-VII. The mixture CR-LS I was reference and consisted only of Portland cement CEM I 42.5 R from Radotín and milled recycled concrete in ratio 50:50. The mixtures CR-LS II-VII consisted of 50 % by weight milled recycled concrete and a various ratio of cement (35.0; 37.5 and 40.0 %), slag (10 %) and lime hydrate (2.5 and 5.0 %). Milled recycled concrete was obtained from structural concrete. Slag was used in two variants, slag I is blast furnace slag and slag II is steel slag. The slag is used same as a fly ash for its hydraulic properties and the ability to reduce the hydration temperatures in the building industry in mixed cement and concrete [4]. All mixtures had the same water coefficient of 0.35 and workability was tested. Table 1 summarizes the designation and composition of the mixtures and their workability.

Table 1: Composition of the mixtures and their workability.

Designation of the mixture	Composition of the mixture [%]					w/b [-]	Spillage – avg. [mm]
	Cement	Rec. concrete	Lime	Slag I	Slag II		
CR-LS I	50.0	50.0	0.0	0.0	0.0	0.35	201.0
CR-LS II	40.0	50.0	0.0	10.0	0.0	0.35	200.0
CR-LS III	37.5	50.0	2.5	10.0	0.0	0.35	201.5
CR-LS IV	35.0	50.0	5.0	10.0	0.0	0.35	201.5
CR-LS V	40.0	50.0	0.0	0.0	10.0	0.35	204.0
CR-LS VI	37.5	50.0	2.5	0.0	10.0	0.35	203.5
CR-LS VII	35.0	50.0	5.0	0.0	10.0	0.35	197.0

Test specimens and methodology and evaluation of experiments

Test specimens had the standard shape of beam with dimensions 40 × 40 × 160 mm. For all fresh mixtures, workability resp. the fluidity of the mixture was verified by a standard cone spill test on fresh mixture. On all test samples, the dynamic and shear modulus of elasticity was evaluated using a non-destructive resonance method after 7 and 28 days [5]. Thereafter, 28 days compressive strength and tensile strength in bend were measured with help destructive tests according ČSN EN 196-1 [6].

Workability. The workability was verified with help spillage of cone. This methodology is described in the standard ČSN EN 12350-5 [7]. The flow table test is displayed on the Fig. 1. Each cone-shaped test specimen was tapped 15 times after mixing and the diameter and height were measured in two perpendicular directions.

The Table 1 shows evaluation of workability. Average spillage of reference mixture CR-LS I was 201 cm. This value is only about 8 % lower than spillage of mixture composited from only cement with same water ratio [3]. This fact means good workability of mixtures with milled recycled concrete. The used admixtures of slag and lime hydrate and the proposed substitution size had no big effects on consistency and workability. This is another positive fact.

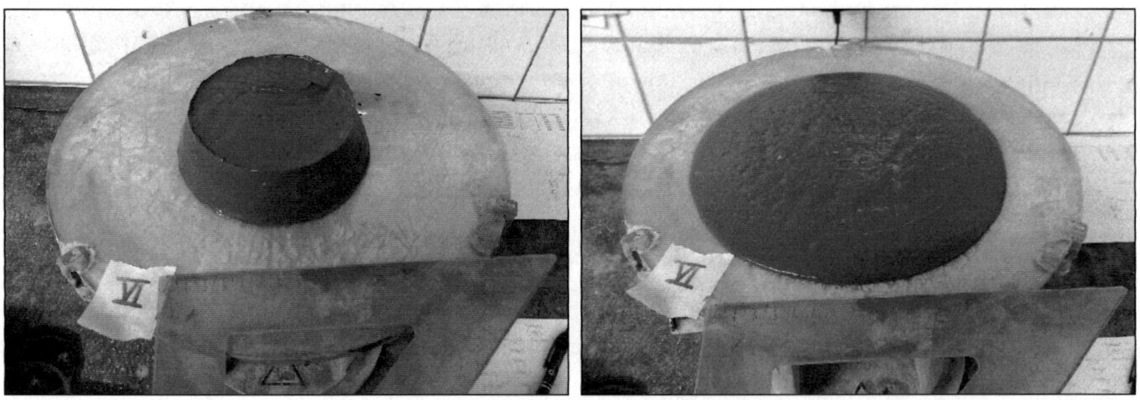

Fig. 1 The flow table test – mixture CR-LS VI.

Dynamic and shear modulus of elasticity. All test specimens were stored in the water bath at temperature 21 °C throughout the hardening period. Test beams were dried, measured and weighed after taking out from the water bath. Required parameters (dimensions and volume density) for evaluation of the modules of elasticity were obtained. Average volume density of reference mixture CR-LS I was 1956.6 kg/m^3. This value is only about 7 % lower than average value of volume density of mixture composited from only cement with same water ratio [3].

The influence of admixtures slag and lime hydrate to the volume density is minimal (maximal 2.2 %), as well as the 50% substitution of cement by the recycled concrete (difference maximal 11 %, [3]). The transverse frequency for each test body was measured. The measurement itself was non-destructive and was made after 7 and 28 days to observe the impact of the additives. Tables 2 and 3 show the average values of the volume density ρ_{avg}, the dynamic modulus of elasticity E_{dyn} and the shear modulus of elasticity G and their standard deviations (by volume density for all mixtures maximal 1 %, by modulus for all mixtures maximal 3 %).

Table 2: Average values of the volume density and dynamic modulus of elasticity after 7 and 28 days.

Designation of the mixture	ρ_{avg} [kg/m^3]	$E_{p,7}$ [GPa]	$E_{p,28}$ [GPa]
CR-LS I	1956.6 ± 6.0	17.040 ± 0.043	19.038 ± 0.093
CR-LS II	1967.1 ± 20.1	16.029 ± 0.332	18.316 ± 0.454
CR-LS III	1924.1 ± 7.5	14.966 ± 0.412	17.016 ± 0.410
CR-LS IV	1914.1 ± 6.5	14.727 ± 0.214	16.912 ± 0.296
CR-LS V	1922.4 ± 6.4	15.229 ± 0.119	17.498 ± 0.138
CR-LS VI	1918.1 ± 3.2	14.773 ± 0.082	16.929 ± 0.196
CR-LS VII	1923.5 ± 9.4	14.459 ± 0.081	16.553 ± 0.057

Table 3: Average values of the volume density and shear modulus of elasticity after 7 and 28 days.

Designation of the mixture	ρ_{avg} [kg/m^3]	$G_{p,7}$ [GPa]	$G_{p,28}$ [GPa]
CR-LS I	1956.6 ± 6.0	6.839 ± 0.037	7.635 ± 0.039
CR-LS II	1967.1 ± 20.1	6.327 ± 0.038	7.391 ± 0.189
CR-LS III	1924.1 ± 7.5	6.021 ± 0.090	6.916 ± 0.151
CR-LS IV	1914.1 ± 6.5	5.829 ± 0.105	6.961 ± 0.113
CR-LS V	1922.4 ± 6.4	6.114 ± 0.059	7.119 ± 0.065
CR-LS VI	1918.1 ± 3.2	5.783 ± 0.039	6.928 ± 0.100
CR-LS VII	1923.5 ± 9.4	5.675 ± 0.028	6.786 ± 0.016

The average value of the dynamic modulus of elasticity of reference mixture CR-LS I was 17.0 GPa after 7 days and 19.0 GPa after 28 days (increase by about 12 %). This value is about 25 % lower after 7 days and about 28 % lower after 28 days (at an increase of about 16 %) than the dynamic modulus of elasticity of mixture composited from only cement with same water ratio [3]. The used admixtures of slag and lime hydrate and the proposed size of substitution caused a further reduction of the dynamic modulus of elasticity by up to 15 % (increase between 7 and 28 days by about 15 %).

The average value of the shear modulus of elasticity of reference mixture CR-LS I was 6.8 GPa after 7 days and 7.6 GPa after 28 days (increase by about 11.6 %). This value is about 24 % lower after 7 days and even about 28 % lower after 28 days (at an increase of about 18 %) than the dynamic modulus of elasticity of mixture composited from only cement with same water ratio [3]. The trend in the development of the values of the both monitored modulus of elasticity for the reference mixtures is thus identical. The used admixtures of slag and lime hydrate and the proposed size of substitution caused a further reduction of the shear modulus of elasticity by up to 7 % after 7 days and only up to 2 % after 28 days (increase between 7 and 28 days between 16 to 19 %).

Compressive and tensile strength. All test beams with dimensions 40 × 40 × 160 mm (ie 3 pieces for each mixture) were used for destructive measurement of tensile strength after 28 days (after non-destructive test). On the broken test samples after the tensile test, the compressive strength was tested (ie 6 pieces for each mixture). The measurement of tensile strength was carried out as a classic three point test. Distance of supports was 100 mm and test was calculated by help of the maximum force. The area under compressive load had size 160 mm2 and has been demarcated by the loading device. Table 4 summarizes the average values of compressive and tensile strengths and their standard deviation after 28 days. Fig. 2 and 3 show graphically the development of the strengths of the individual mixtures in %.

The value of tensile strength of the reference mixture CR-LS I was 6.3 MPa. This value is about 10.4 % lower than the tensile strength of mixture composited from only cement with same water ratio [3]. The used admixtures of slag and lime hydrate and the proposed size of substitution caused reduced the tensile strength between 5.5 to 31.8 %. Standard deviations fluctuated between 9.6 to 48.2 %. This large variance is probably due to shrinkage and follow-up microcracks. However this isn't so a big problem, because the test specimens aren't reinforced in any way.

The value of compressive strength of the reference mixture CR-LS I was 46.1 MPa. This value is about 60.0 % lower than the compressive strength of mixture composited from only cement with same water ratio [3]. The used admixtures of slag and lime hydrate and the proposed size of substitution caused reduced the compressive strength between 4.6 to 33.7 %. The increase in compressive strength over time is assumed for mixtures with slag. [8; 9]. Standard deviations fluctuated between 2.5 and 14.5 % (so high only for mixture CR-LS VI).

Table 4: Average values of the tensile and compressive strength after 28 days.

Designation of the mixture	$f_{t,28,p}$ [MPa]	$f_{c,28,p}$ [MPa]
CR-LS I	6.391 ± 2.125	46.1 ± 2.0
CR-LS II	6.043 ± 0.969	44.0 ± 3.7
CR-LS III	4.528 ± 2.184	42.9 ± 1.1
CR-LS IV	5.985 ± 1.172	35.2 ± 2.5
CR-LS V	5.741 ± 0.553	40.3 ± 3.4
CR-LS VI	4.361 ± 0.561	42.0 ± 6.1
CR-LS VII	4.704 ± 1.489	40.1 ± 3.2

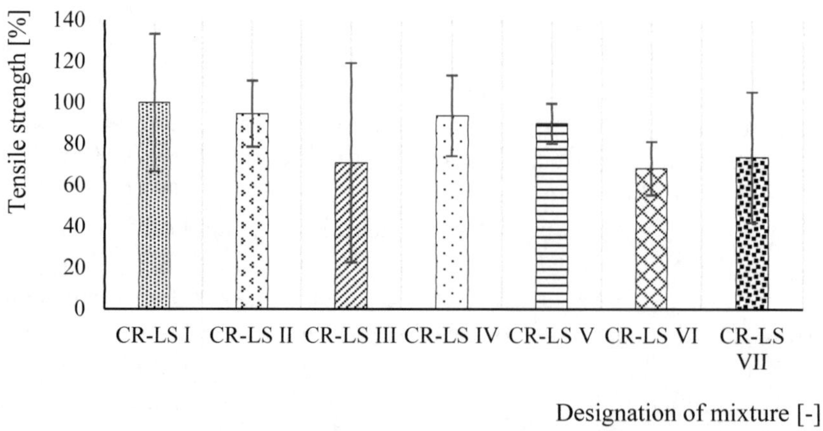

Fig. 2 Average values of tensile strength after 28 days.

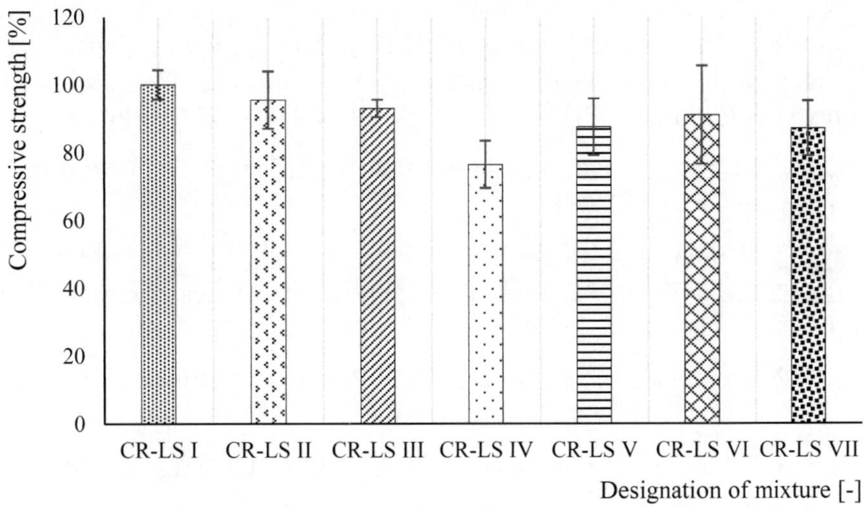

Fig. 3 Average values of compressive strength after 28 days.

Conclusions

The proposed cement substitution by milled recycled concrete in the matrix was 50 % by weight for all mixtures. Furthermore, admixtures of slag and lime hydrate were used. Selected properties were compared with the reference mixture. The influence on volume density was almost zero. The substitution also had minimal effect on the workability of the mixture. The change of dynamic modulus of elasticity resp. the shear modulus between proposed mixtures was maximal up to 15 % resp. only up to 7 % after 28 days.

The change of tensile strength resp. the compressive between proposed mixtures was maximal up to 31.8 % resp. only up to 33.7 % after 28 days. The difference of one of the most important parameters of compressive strength compared to a compound composed of only cement is high 60 %, but with a strength about 40 MPa after 28 days, some special applications can be easily considered (e.g. foundation constructions). Thus, optimized designed blends with recycled concrete and admixture of slag and lime hydrate and the values of the monitored parameters show the correct direction of research with the possibility of use in building practice.

Acknowledgement

This paper has been supported by the Grant Agency of the Czech Republic GA ČR, project No. 17-06771S and CTU in Prague a student project under number SGS19/148/OHK1/3T/11.

References

[1] Y. J. Kim et al., Utilization of waste concrete powder as a substitution material for cement. Construction and Building Materials 30 (2012) 500–504.

[2] M. Lidmila et al., Utilization of Recycled Fine-ground Concrete from Railway Sleepers for Production of Cement-based Binder. 51nd International Conference on Experimental Stress Analysis 2014, 323–326.

[3] V. Hrbek et al., P. Effect of Activation Recycled Concrete Using Different Admixtures on the Strength and Modulus of Elasticity of the Resulting Cement Composite, In: EAN 2018 56th conference on experimental stress analysis, Conference Proceedings. Praha: Česká společnost pro mechaniku, 2018. p. 118-125. ISBN 978-80-270-4062-9

[4] ČSN EN 197-1 Cement – Part 1: Composition, specifications and conformity criteria for common cements, Prague, 2011

[5] T. Plachý et al. Nondestructive Determination of Young's Modulus of Gypsum Specimens Using Impulse Excitation Method In: Experimental Stress Analysis 2010. Olomouc: Univerzita Palackého, 2010. pp. 339-344. ISBN 978-80-244-2533-7.

[6] ČSN EN 196-1 Methods of testing cement – Part 1: Determination of strength, Prague, 2005.

[7] ČSN EN 12350-5 Testing fresh concrete – Part 5: Flow table test, Prague, 2009.

[8] O. Zobal, V. Šmilauer, L. Kopecký a kol., Analýza betonu z přehrady Orlík po 50 letech, Beton TKS 14, 2014

[9] A. Neville, Properties of concrete, 2009

Residual stresses of laser-welded pressure vessel steel determined by X-ray and neutron diffraction

TROJAN Karel[1,a], VLK Aleš[1,b], ČAPEK Jiří[1,c], HERVOCHES Charles[2,d], and GANEV Nikolaj[1,e]

[1]Department of Solid State Engineering, Faculty of Nuclear Sciences and Physical Engineering, Czech Technical University in Prague, Trojanova 13, Prague, Czech Republic

[2]Nuclear Physics Institute, ASCR, v.v.i., Department of Neutron Physics, Řež, Czech Republic

[a]karel.trojan@fjfi.cvut.cz, [b]vlkales2@fjfi.cvut.cz, [c]jiri.capek@fjfi.cvut.cz, [d]hervoches@ujf.cas.cz, [e]nikolaj.ganev@fjfi.cvut.cz

Keywords: Laser welding, Pressure vessel steel, Residual stresses, X-ray diffraction, Neutron diffraction

Abstract. To gain insight into the quality of the laser weld, information about the residual stress state across the weld is very useful. In this contribution, the residual stress profiles for low-alloy carbon steel plates of P355NL1 grade, which were laser welded from both sides, are presented. To separate the effects of the production of plates from the welding process, the samples were annealed for stress relieving. Using X-ray and neutron diffraction, the surface and bulk RS profiles were obtained. From the obtained data, first, the difference in the character of the longitudinal (parallel to the weld axis) and the transversal component of the RS tensor can be seen. The longitudinal component has a tensile character with a maximum value approximately 500 MPa. Values and character of the transversal component vary strongly with depth and distance from the weld axis.

Introduction

Welding is a widely used method in fabrication industries producing ships, trains, steel bridges, pressure vessels, and more. Therefore, high demands are laid on mechanical properties and durability of welds used to connect two or more components together. For the superior properties, such as high welding speeds, low thermal load of the surrounding material, precision and strength of the weld, or possibility of joining components of a wide range of thickness (0.01 to 50 mm), the laser welding has become one of the suitable welding techniques [1,2].

However, despite the advantages of the laser welding, due to a heterogeneous application of energy and localized fusion which occur during the welding process, great residual stresses (RSs) can be present in a region near and in the weld itself. These RSs are a superposition of thermal and transformation processes and can reach high values and subsequently cause fatigue or in combination with cracklike defects promoting of brittle fractures [3]. Therefore, it is necessary to reduce the values of the RSs and size of the area affected by the welding process. These properties of the weld can be influenced by the used method, parameters used in the process and by the treatment of the material before and after welding.

Using X-ray and neutron diffraction, the residual stress state across the laser weld were examined. Specifically, double-side square butt welds performed on P355NL1 steel plates for high-pressure vessels were analysed. This fine-grained low-alloy carbon steel is widely used for fabrication of high-pressure vessels, steam boiler parts, pressure piping, compressors, etc.

Experiment

The P355NL1 hot-rolled steel plates of dimensions of $150 \times 300 \times 10$ mm^3 and $150 \times 300 \times 8$ mm^3 were studied. The composition of this steel determined by glow-discharge optical emission spectroscopy (GDOES) is given in Table 1. The error at low concentrations is up to 10%. The determined composition is in agreement with the European Standard EN 10028-3:2017 [4] with one exception, manganese, whose weight fraction is below the minimum value 1.10 wt% stated in the norm.

In order to separate the effect of the production of the plates and the laser welding, the samples were separated into two groups, where the first group was laser welded with no extra treatment. The second group was firstly annealed for stress relieving in inert gas atmosphere. Samples were held on the temperature of 560 °C for one hour and subsequently cooled down to 350 °C in the inert gas atmosphere. Cooling to room temperature was performed in air. The double-sided square butt weld (the "Up" side was welded first and "Down" as second) was performed under the same conditions for the first and second group. The most important welding parameters, such as laser power P, welding speed v, welding mode and laser wavelength λ are given in Table 2.

Table 1: Chemical composition of the P355NL1 steel determined for 10 and 8 mm thicknesses using glow-discharge optical emission spectroscopy

Element	Fe	C	Mn	Si	Ni	Cr	Cu	V	Al
Weight fraction [wt%], 10 mm	98.92	0.067	0.80	0.004	0.005	0.021	0.018	0.002	0.037
Weight fraction [wt%], 8 mm	99.10	0.071	0.65	0.002	0.004	0.022	0.021	0.000	0.038

Table 2: Welding parameters

P [W]	v [mm·s^{-1}]	mode	λ [nm]
3000	5.5	Continuous	900–1080

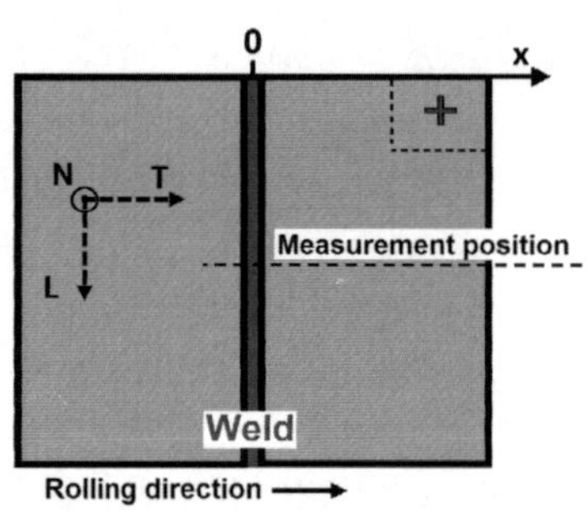

Fig. 1: Scheme of the sample. The cross in the right upper corner denotes the place where the depth profile was measured

RS measurements were performed on PANalytical X'Pert PRO MPD diffractometer with Cr$K\alpha$ radiation ($\lambda = 0.229$ nm). The surface residual stress state across the weld and the depth dependence up to approximately 300 μm was determined by X-ray diffraction on {211} α-Fe lattice planes. The position of the measurements is depicted in Fig. 1. The data from different depths under the surface were obtained using electrolytic polishing to remove the material. In order to prevent redistribution of the RSs due to the polishing, the size of the electrolytically polished area was chosen to be 150 mm^2, which was approximately 1/40 of the total area of the sample.

To non-destructively determine the RS profile across the weld in the different depths in the material, the neutron diffraction was used. Measurements were performed in the Institute of Nuclear Physics of Czech Academy of Sciences using the HK4-Strain scanner. For this purpose, the wavelength of the radiation used for the diffraction on {110} α-Fe lattice planes was $\lambda = 0.213$ nm and $2\Theta = 65°$.

Results and discussion

First of all, to analyse the effect of annealing, the depth profiles of RSs up to 300 μm were analysed. From this comparison depicted in Fig. 2, a significant difference between annealed and non-annealed plate can be seen. With respect to the results for the annealed plate, where the values of RSs insignificant, the next presented RS profiles for annealed plates can be related only to the welding process and not to the preparation (hot-rolling) of the plates.

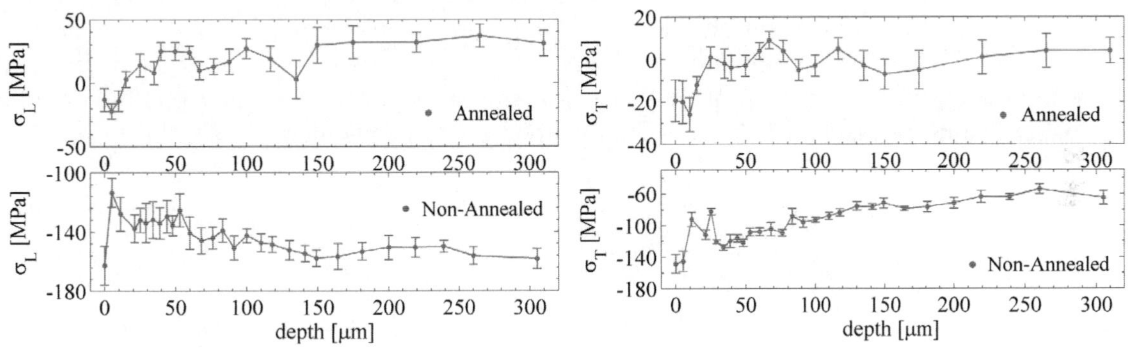

Fig. 2: Longitudinal L (parallel to the weld axis) and transversal T (perpendicular to the weld axis) RSs with respect to depth under the surface

A comparison of longitudinal RSs across the weld for annealed and non-annealed plates of both thicknesses is presented in Fig. 3. From this comparison, the differences between annealed and non-annealed plates are small. The main difference is the smaller stress affected zone (SAZ) of annealed plates. The approximate beginning of this zone is depicted by the lines of a relevant colour. For non-annealed plates, the SAZ extends approximately 3–15 mm further from the weld axis than for annealed plates.

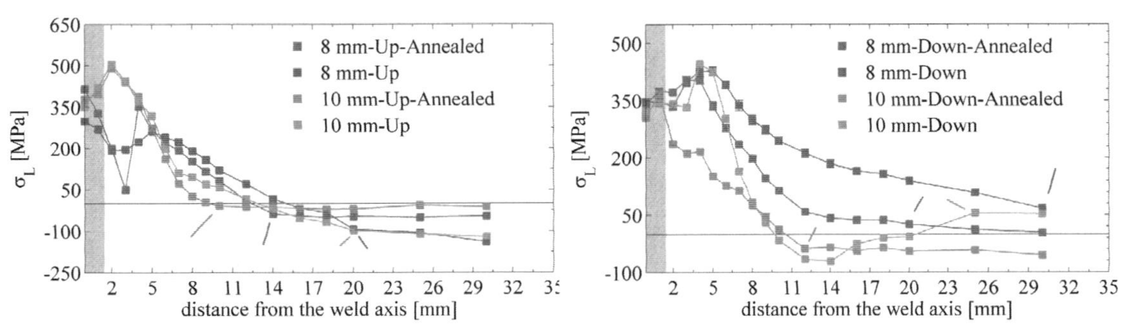

Fig. 3: Longitudinal RSs on "Up" and "Down" side determined for annealed and non-annealed plates

From the obtained depth profiles, it can be seen, that the RSs in longitudinal direction have mainly tensile character, on the other hand, the character of the RSs in the transversal direction is compressive, see Fig. 4 and 5. A value of 0 mm corresponds to the "Down" side. This is in agreement with the mechanisms giving rise the RSs during the welding process, wherein the longitudinal component of stress tensor is mainly given by thermal stresses caused by shrinkage of the material along the weld. In the case of the transversal component, the contribution of stresses given by phase transformation, which material undergoes during cooling, is more significant and causes smaller tensile RSs or even high compressive RSs. For the detailed description of the origin of the RSs see reference [5].

RSs for both the thicknesses have similar pattern. In the longitudinal direction, RSs reach the maximum values. Only for 8 mm thick sample, there is a certain asymmetry for the 6.5 mm line. On the contrary, in the transversal direction, the compressive RS reach about 100 MPa higher values for 8 mm thick sample. Conversely, for 10 mm thick sample on the "Up" side, tensile RSs are approximately 200 MPa higher.

The comparison of both the methods (X-ray and neutron diffraction) showed that longitudinal components of the surface RS obtained by X-ray diffraction are comparable with the longitudinal components of the residual stresses in the bulk material determined by neutron diffraction. On the other hand, in the case of transversal components, great differences across the thickness were observed. These differences could be explained due to the fact that length of heated material and subsequent shrinkage is many times shorter than along the weld and residual stresses due to transformation could have occurred. This volume change has different impact in the surface layers and inside the weld. Another significant influence is the double-side welding and the associated warping in the direction perpendicular to the weld.

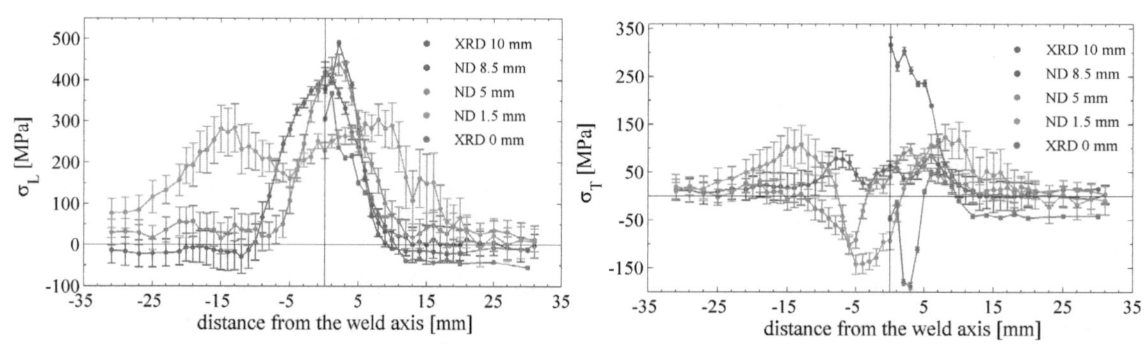

Fig. 4: Longitudinal and transversal RSs for 10 mm thick sample with respect to distance determined across the weld for different depths

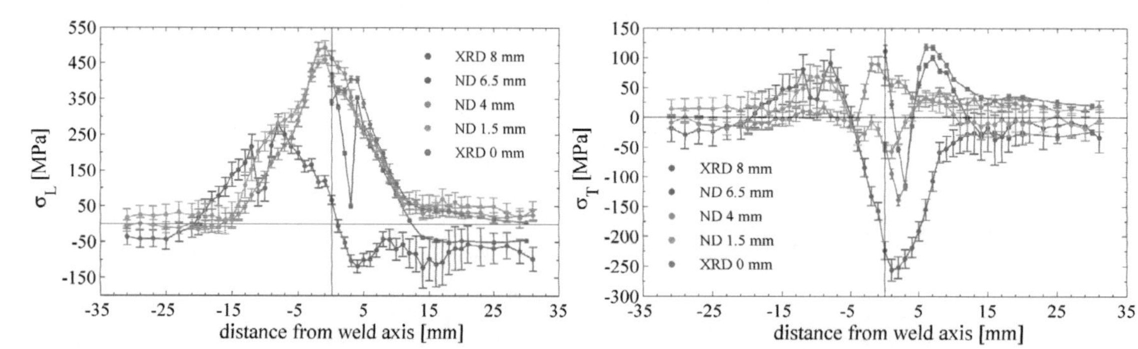

Fig. 5: Longitudinal and transversal RSs for 8 mm thick sample with respect to distance determined across the weld for different depths

Conclusions

From this observation, it can be said that the longitudinal surface RSs obtained by X-ray diffraction are, with respect to measurement errors and differences between samples, representative for the longitudinal state of residual stresses across the weld. On the other hand, the measurements show that the transversal RSs change significantly with depth and the nature of the residual stresses below the surface cannot be assessed on the basis of the surface state of stresses. Other non-destructive methods, e.g. neutron diffraction or ultrasonic velocity measurement [6], should be used to describe subsurface gradient of residual stresses. Alternatively, destructive approach using X-ray diffraction together with electrochemical polishing (removal) of surface layers can be applied.

Acknowledgements

Measurements were supported by the project TH02010664 of the Technology Agency of the Czech Republic. Neutron diffraction measurements were carried out at the CANAM infrastructure of the NPI ASCR Rez supported through MEYS project No. LM2015056. Presented neutron diffraction results were obtained with the use of infrastructure Reactors LVR-15 and LR-0, which is financially supported by the Ministry of Education, Youth and Sports - project LM2015074. This work was supported by the Grant Agency of the Czech Technical University in Prague, grant No. SGS19/190/OHK4/3T/14. GDOES experiments were performed by RNDr. Zdeněk Weiss, CSc. at Department of Material Analysis at Institute of Physics, ASCR, v.v.i..

References

[1] R. Oyyaravelu, P. Kuppan, N. Arivazhagan, Metallurgical and mechanical properties of laser welded high strength low alloy steel, J. Adv. Res. 7.3 (2016) 463-472.

[2] S. Karayama, Introduction: fundamentals of laser welding. In: Handbook of laser welding technologies. Woodhead Publishing, 2013.

[3] D. Radaj, Heat effects of welding: temperature field, residual stress, distortion, Springer Science & Business Media, 2012.

[4] EN 10028-3:2017. Flat products made of steels for pressure purposes - Part 3: Weldable fine grain steels, normalized. European Committee for Standardization, 2017, 22p. 77.140.50.

[5] T. Nitschke-Pagel, K. Dilger, Sources and consequences of residual stresses due to welding, Mater. Sci. Forum. 783 (2014) 2777–2785.

[6] C.O. Ruud, A review of selected non-destructive methods for residual stress measurement, NDT Inter. 15.1 (1982) 15–23.

EAN 2019
57th conference on experimental stress analysis

Groove grinding method for near the surface residual stress measurement

VÁCLAVÍK Jaroslav[1,a], HEJMAN Marek[2,b] and WEINBERG Otakar[1,c]

[1]Výzkumný a zkušební ústav Plzeň s.r.o., Tylova 1581/46, 301 00 Plzeň, Czech Republic

[2]Fatigue Analysis RI s.r.o., V Zátiší 423, 330 08 Zruč-Senec, Czech Republic

[a]vaclavik@vzuplzen.cz, [b]hejman@fatari.cz, [c]weinberg@vzuplzen.cz

Keywords: Residual stresses, Semi-destructive method, Groove grinding method, Slitting method, Strain gauge

Abstract. The grove grinding method for the semi-destructive measurement of residual stresses near the surface is presented in the article. The principle of the method is that an annular grove is ground perpendicularly to the surface and the stresses are calculated from the relieved strains at the strain gauge positioned near the groove. The calibration coefficients for one-dimensional distribution of residual stresses for homogenous stress profile and for stresses changing with the depth using integral method were derived and a handy tool for grinding the grove was developed. Initial measurements on rectangular bar are presented and compared with hole-drilling method.

Introduction

Residual stress is an important parameter in respect of the failure of a structure or mechanical component. Positive near the surface residual stresses can significantly lower the fatigue strength of the structure. Several technics are available for subsurface residual stress measurement as hole-drilling, ring-core, slitting and X-Ray or ultrasonic methods [1]. Here proposed groove grinding method is a modification of the slitting and hole-drilling methods. It can be applied for one dimensional distribution of residual stresses normal to the surface plane near to the surface.

Theoretical considerations

The principle of the method is that an annular grove is ground perpendicularly to the surface and the stresses are calculated from the relieved strains at the strain gauge glued perpendicularly on given distance from the groove edge with the help of calibration constants. The real dimensions of used grinding disc and strain gauge position are visible in Fig. 1.

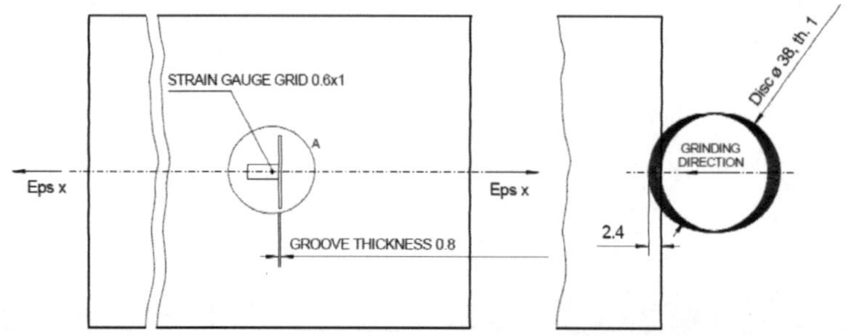

Fig. 1: Principle of the groove grinding method

Theory. A shallow annular grove is incrementally ground to the object surface at the depth z_i and the residual stress σ_{xk} at the depth z_j is estimated using linear elasticity theory from relieved strains ε_j, measured at one or more linear strain gauges, glued near to the edge of the groove.

$$\varepsilon_i = \frac{1}{E}\sum_{k=1}^{j} a_{jk} \cdot \sigma_{xk} \tag{1}$$

The calibration constants a_{jk} indicate the relieved strain in a groove j steps deep, due to unit stresses within groove step k.

Using the integral method, the residual stresses within each groove depth step can be computed from measured strains solving the matrix equation

$$\bar{\boldsymbol{a}}\boldsymbol{\sigma} = E\boldsymbol{\varepsilon} \tag{2}$$

To reduce the high dispersion of evaluated stress due to ill-conditioned matrix $\bar{\boldsymbol{a}}$, the tri-diagonal matrix **c** is used, implementing Tichonov second derivative regularization. The factor α controls the amount of regularization

$$(\bar{\boldsymbol{a}}^T\boldsymbol{a} + \alpha \boldsymbol{c}^T\boldsymbol{c})\boldsymbol{\sigma} = E\bar{\boldsymbol{a}}^T\boldsymbol{\varepsilon} \tag{3}$$

FEM analysis. Numerical values of calibration constants were calculated for given grinding wheel diameter, for strain gauge HBM 0.6LY11 grid dimensions and for given distance of the strain gauge from the groove edge. FEM analysis was made inside system ANSYS v 19.1. Used type of elements is quadratic TETRA. Specimen was cut on 1/8 of real test specimen to reduce number of elements and to simplify of boundary conditions. The loading was considered as stress 1 MPa in each layer of thickness 0.2 mm for each of 12 depth of grinding. Average stresses in place of strain gauge are evaluated in to triangular matrix $\bar{\boldsymbol{a}}$. Here corresponding member a_{jk} of matrix $\bar{\boldsymbol{a}}$ is average stress on area of strain gauge $S_{StrainGaugeArea}$.

$$a_{jk} = \{\int \sigma_{x,jk} dS\}/\{S_{StrainGaugeArea}\} \tag{4}$$

The FEM model was made with fine mesh in relevant surrounding of strain gauge and grinding disk, see Fig. 2 The calibration constants were derived as for homogenous as for non-homogenous stress profile.

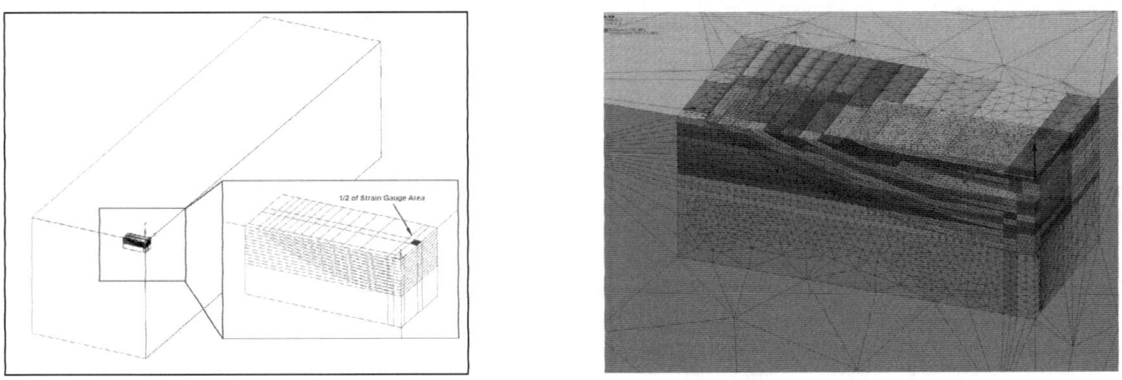

a) b)

Fig. 2: Geometry of 1/8 test specimen, position of strain gauge area (a) and FEM model with forth drilled depth and introduced stress in the second upper removed layer (b)

The relieved strain related to the homogenous nominal strain ε_n for dimensionless grove depth ξ, given as the ratio of the depth and the distance of the strain gauge centre from the groove edge, is presented in the Fig. 3a. The triangular matrix of calibration constants for non-homogenous stress profile is schematically given in Fig. 3b expressed as the curves for partial depths.

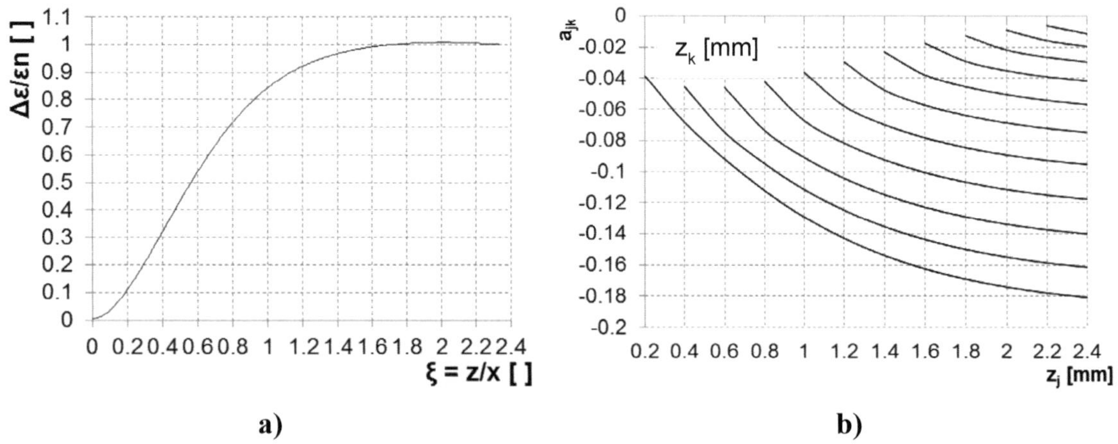

Fig. 3: Calibration coefficients for homogenous and non-homogenous residual stress profile

Implementation

Calibration constants for homogenous stress profile were verified experimentally by tension and bending tests. The bending test was performed on the rectangular beam on tension upper side (Fig. 4a). The resulting strain relaxation obtained experimentally (x = 2) was compared with the relaxation curves, obtained using FEM analysis for various distances of the strain gauge centre x from the groove edge. It is obvious, that the experiment fits to the calculated curves very well.

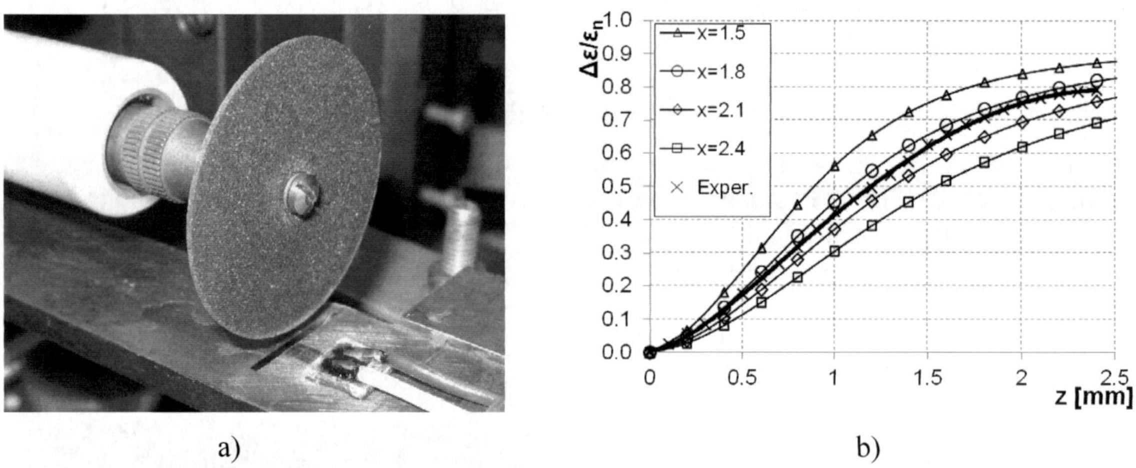

Fig. 4: Verification of calculated calibration constants with bending test

A handy grinding tool has been developed for in-situ tests (Fig. 5). The measurement of longitudinal residual stress in forged bar, used for the manufacture of LP turbine blades is demonstrated in next text. Residual stress is presented in forgings of some producers, which causes the deformations of blades after their milling. Two SG were used, 2 mm and 4 mm from the groove edge, considering that the change of released strains in larger depths is greater at the second strain gauge. Evaluated stresses are presented in Fig. 6b. Two bottom lines represent the stress, evaluated for homogenous stress field. Unfortunately using this access, the evaluated stresses at larger depths are influenced from high stresses at the surface and are evaluated not properly (the residual stresses are compressive for all depths). The real residual stresses are given from the upper diamond filled curve, which represents the stress, calculated with integral method from the first strain gauge positioned just at the grooved edge. It is obvious, that below 0.3 mm depth the residual stresses are tensile. The crosshair curve represents the stresses, evaluated using integral method from released strain at the second strain gauge, positioned 4 mm from the groove edge. It is visible, that the evaluated stresses are not representative near the surface, but the dispersion of stresses is lower in the depths, approaching the depth of 2 mm. The last upper curve represents the longitudinal stress, measured using hole-drilling method and evaluated also using integral method. This curve fit very well with the stresses, obtained with grinding groove method, evaluated with integral method.

Fig. 5: Tool for measurement residual stresses using groove grinding method (a) and detail of the positions of the relieved strain measuring two strain gauges (right 2 mm and the left 4 mm from the groove edge)

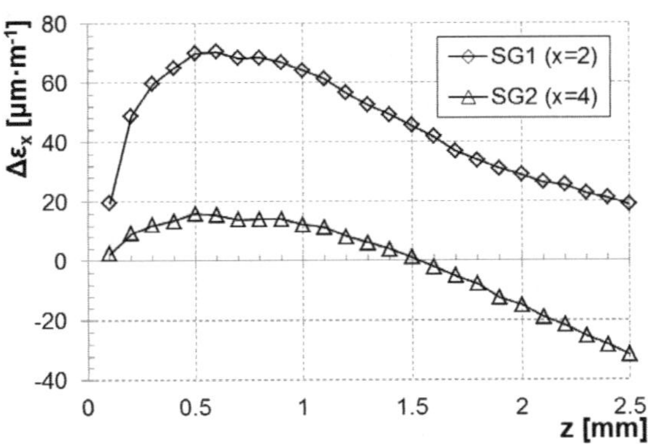

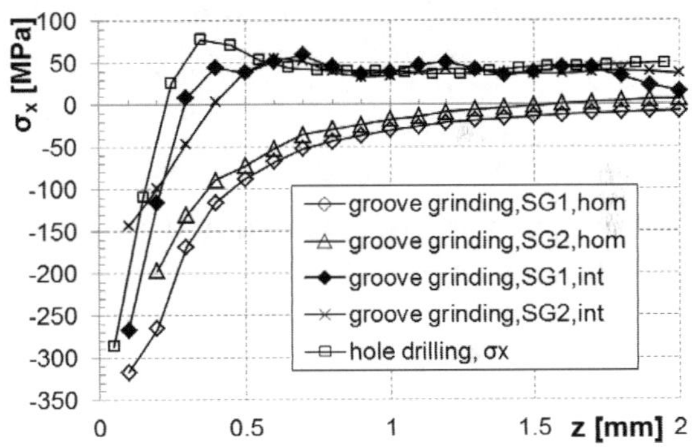

Fig. 6: Measured relieved strains at strain gauges (a) and evaluated residual stress profile (b)

Conclusions

Proposed method is very well suited for stress profile measurement of single stress component, is easier in comparison with hole-drilling method, portable and has good repeatability. Multi-cuts can be performed for multi stress components.

The article has originated in the framework of the institutional support for the long-term conceptual development of the research organization.

References

[1] G.S. Schajer, Practical residual stress measurement methods. first ed., John Wiley & Sons Ltd, United Kingdom, 2013.

EAN 2019
57th conference on experimental stress analysis

Numerical analyses of the gyroid structure for use in dental and orthopaedic implants

VAKRČKA Petr[1,a], JÍRA Aleš[1,b] and ŘEHOUNEK Luboš[1,c]

[1]Czech Technical University in Prague, Faculty of Civil Engineering, Department of Mechanics, Thákurova 7, 166 29, Prague, Czech Republic

[a]petr.vakrcka@fsv.cvut.cz, [b]jira@fsv.cvut.cz, [c]lubos.rehounek@fsv.cvut.cz

Keywords: Numerical, Gyroid, Trabecular, Implant, FEM

Abstract. The proposed work deals with the numerical analyses of different variants of the gyroid structure. The analyses were carried out to determine the optimal geometry (wall thickness, spacing) of the gyroid structure for use in dental and orthopaedic implants. Since the human body represents an environment of (compared to metals) soft tissues, it is not always desirable to introduce a structure with high stiffness as an implant. The gyroid structure (on the macroscopical level) represents a porous structure that reduces the overall stiffness of the implant if applied on the outer layer of the implant. What is more, this kind of mesh provides interesting features for tissue scaffolding due to high porosity which gives an opportunity to design a structure that can also meet the requirement for the optimal pore size with regard to the most effective bone ingrowth. Furthermore, we can minimize the stress shield effect which can lead to the bone resorption after implant insertion.

Introduction

Implants made of titanium and its alloys have a long history of good success rates [1]. Nevertheless, they still share a common shortcoming, which lies in the difference between material properties of the metal and human tissue [2, 3]. This shortcoming can be addressed by reducing the overall stiffness of the implant. This reduction serves to better the conditions for osseointegration and reduce the stress-shielding effect, which is often associated with implant loosening. Aseptic loosening is, in fact, one of the main reasons of implant failure [4]. The proposed solution to this problem is shown in Fig. 1. An outer porous gyroid layer was introduced into the implant body to lower the global modulus of the implant and also to make the conditions for osseointegrations better [5] by allowing bone ingrowth.

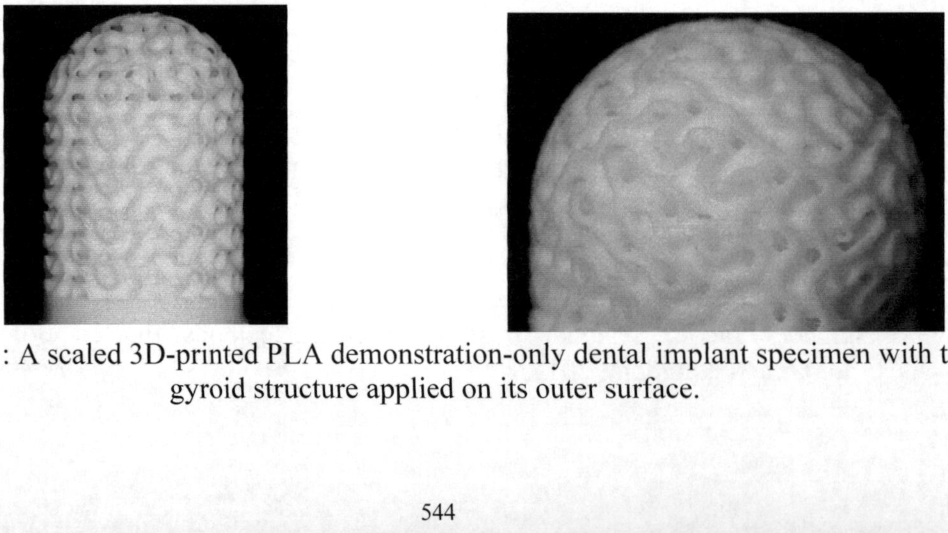

Fig. 1: A scaled 3D-printed PLA demonstration-only dental implant specimen with the gyroid structure applied on its outer surface.

Furthermore, the gyroid structure has other benefits which result from its geometry. Due to the absence of sharp edges, it is much better for osseointegration. Also, since it is not formed by beams, but rather by continuous walls of material, it is much less prone to chipping and local crack defects, which have been observed in the past by the research team. Due to the fluent change of curvature, local stress concentrations are also expected to be lower. Also, wall-like behavior is superior to beam structures in preventing local failure and transferrance of potential implant debris into the body of the patient.

Numerical Simulations

The numerical analyses were performed in ANSYS Workbench 19.0. The aim of the analyses was to determine the trend of increase in stiffness and prepare for a more complex analysis comparing global moduli in relation to that of human bone ($E_{c,bone} \approx$ 10-25 GPa [6], $E_{Ti\text{-}6\text{-}Al\text{-}4V} \approx$ 120 GPa [7]). The material of both the model and future specimens is the Ti-6Al-4V alloy, the most common alloy in the implant industry [8]. It is a bio-inert material. In general, an implant from the Ti-6Al-4V alloy needs to have a surface modification to be effectively used for osseointegration. A different approach of dealing with the osseointegration problem is using the gyroid morphology which can effectively substitute the surface modification. Based on previous research the authors conducted [9], it was assumed that a structure with a global modulus in the range of units of GPa can be considered low stiffness and viable. The final assigned structure can then be manufactured and uniaxial mechanical tests can verify the precision of the model. After model verification, numerical simulations can serve to design specific structures without prior mechanical tests to conserve the costly material. Images showing three of the analysed structures are shown in Fig. 2.

The loading of individual structures was simulated using the mode of controlled displacement. The speed of displacement was 1 mm/min, according to the speed prescribed in the "ISO 13314:2011 – Mechanical testing of metals – Ductility testing – Compression test for porous and cellular metals" standard [10]. This was done to prepare the models for comparison with data from future mechanical tests of individual structures, which will be done in accordance with this standard.

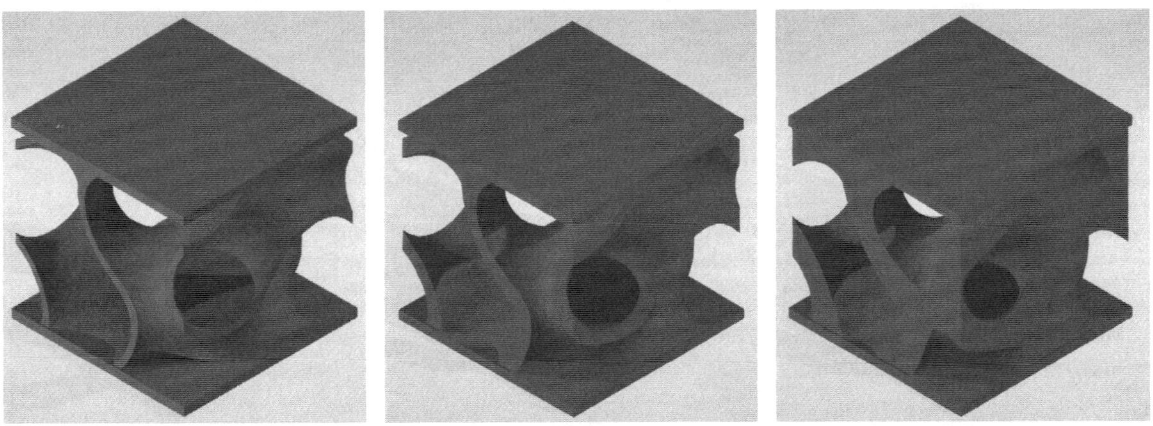

a) Spec. 0.2 mm b) Spec. 0.4 mm c) Spec. 0.8 mm

Fig. 2: Three basic cells of the gyroid structure varying in wall thickness. Upper homogeneous part designed to meet the criteria for future uniaxial mechanical tests, which the model will be compared against.

Results

The specimens with varying wall thickness (0.2, 0.4, 0.6, and 0.8 mm wide) were numerically analysed in order to determine the trend in increase of stiffness as well as to prepare the environment for future analyses after successful manufacturing of real specimens for mechanical tests. Results from the analyses are shown in Fig. 4. The graph demonstrates increase of the vertical force reaction with increasing wall thickness. The force was measured after 1 minute of controlled displacement loading according to the "ISO 13314:2011 – Mechanical testing of metals – Ductility testing – Compression test for porous and cellular metals" standard, which prescribes a loading speed of 1 mm/min. In Fig. 3, the displacement loading is shown, along with the FEM mesh generated and used for the numerical analyses in Ansys Workbench 19.0.

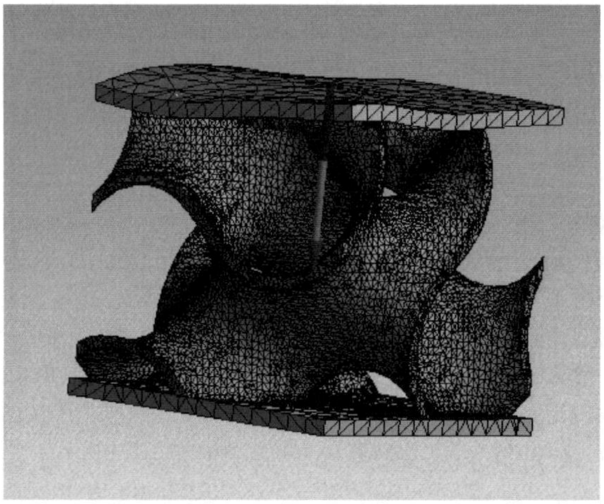

Fig. 3: A figure showing the FEM mesh used to perform the simulation and the deformation of the unit cell of the 0.2mm specimen after 1 minute of loading prescribed by ISO 13314:2011 standard.

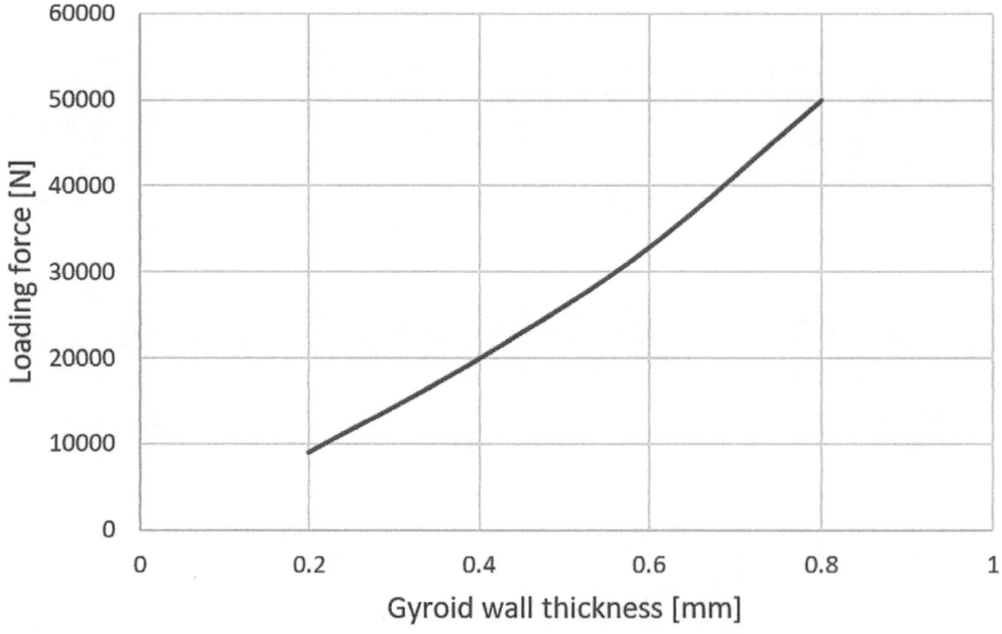

Fig. 4: A graph showing the final loading force against the thickness of an individual gyroid structure after 1 minute of displacement loading prescribed by ISO 13314:2011 standard.

Conclusions

A total of 5 different variants of the gyroid structure were numerically analysed to determine the trend of increase in stiffness and wall thickness in regard to the potential application in dental and orthopaedic implants. The values obtained from the analyses are also subject to further investigation as other factors, such as bone osteon ingrowth and manufacturability of the 3D-printed parts come into play. Manufacturing fine details and thin walls might prove to be a challenge in the future as the resolution in the xy direction varies from printer to printer. Authors have encountered problems with SLS metal printing in the past and expect the technology to improve and enable manufacture of functional, fine parts, considering the rate of improvement in the field of 3D printing.

The gyroid structure has many benefits as it is unique in its wall-like mechanical behavior that prevents local strain-induced failure and subsequent transfer of the implant material into the body of the patient. Another benefit of the gyroid structure is absence of sharp corners and edges, making it a fit candidate for the implant industry, where predictability, ease of operation and viability for osseointegration make for a successful, long-lasting and practical implant.

References

[1] T. Hasegawa, S. Kawabata, D. Takeda, et al., Survival of Brånemark System Mk III implants and analysis of risk factors associated with implant failure. International journal of oral and maxillofacial surgery 46(2) (2016) 267–273.

[2] Ridzwan, M. I. Z., et al., Problem of stress shielding and improvement to the hip implant designs: a review. J. Med. Sci7.3 (2007) 460-467.

[3] M. Long, H. J. Rack, Titanium alloys in total joint replacement–a materials science perspective. Biomaterials 19(18) (1998) 1621–1639.

[4] Le Guéhennec, L., Soueidan, A., Layrolle, P. and Amouriq, Y., Surface treatments of titanium dental implants for rapid osseointegration. Dental materials, 23(7) (2007) 844-854.

[5] Řehounek, Luboš, et al., Geometry and mechanical properties of a 3D-printed titanium microstructure. Acta Polytechnica, 57(3) (2018) 104-108.

[6] Rho, Jae Young, Richard B. Ashman, and Charles H. Turner., Young's modulus of trabecular and cortical bone material: ultrasonic and microtensile measurements. Journal of biomechanics 25(2) (1993) 111-119.

[7] Niinomi, M., Mechanical properties of biomedical titanium alloys. Materials Science and Engineering: A, 243(1) (1998) 231-236.

[8] Niinomi, M., Recent metallic materials for biomedical applications. Metallurgical and materials transactions A, 33(3) (2002) 477-486.

[9] L. Řehounek, “Mechanical and numerical analyses of titanium trabecular structures of dental implants formed by 3d printing,” Acta Polytechnica, vol. 57, no. 3, pp. 218–228, 2017.

[10] Standard, I. "ISO 13314: 2011 (E) (2011) Mechanical testing of metals—ductility testing—compression test for porous and cellular metals." Ref Number ISO 13314.13314: 1-7.

EAN 2019
57th conference on experimental stress analysis

Mechanical joining of thermoplastic composites - bearing behaviour simulation and validation

ZDĚNEK Pavel[1,a], RŮŽEK Roman[1,b], KUCHARSKÝ Pavel[1,c] and KŘENA Josef[2,d]

[1]VZLU Czech Aerospace Research Centre, Beranových 130, 199 05 Prague 9 – Letňany, Czech Republic

[2]Latecoere LATECOERE Czech Republic s.r.o., Beranových 65, 199 02 Prague 9 – Letňany, Czech Republic

[a]pavel.zdenek@vzlu.cz, [b]ruzek@vzlu.cz, [c]kucharsky@vzlu.cz, [d]Josef.Krena@latecoere.aero

Keywords: Bearing, FEA (Finite Element Analysis), Composite, Joining, Thermoplastic

Abstract. The paper discusses issues associated with failure mechanism of the mechanical fastened joint – bearing strength and clamp-up conditions. Work is focused on carbon fibres with polyphenylenesulfide thermoplastic resin composite material. The bearing strength behaviour of the two different layup alternatives, hole positions, plate thicknesses and tightening conditions are investigated. Numerical simulations were checked with experimental measurements. Propositions to improve simulation results has been made.

Introduction

Linear polyphenylenesulfide (PPS) thermoplastic is widely used in carbon fiber reinforced composite structures in aerospace parts and structures. Increasing interest in carbon fiber/PPS thermoplastic composite materials (C/PPS) is driven by its excellent mechanical properties. It means mainly resistance to impact damage or better formability, especially in comparison with conventional epoxy-based laminates. Widespread application of C/PPS thermoplastic composites is dependent on the manufacturing costs, ability to utilize all the advantages of thermoforming process and optimization of the lay-up. Different studies relating to mechanical behaviour and damage tolerance evaluation of C/PPS thermoplastics were published in the past, for example experimental investigation of factors considered for the strength evaluation [1], investigation of fatigue loading rate to lifetime of the composite with variable thickness [2,3], bearing strength [4] and in-plane shear behaviour [5] etc.

Different methods can be used for thermoplastics joining like bonding [6,7], welding [8] and mechanical fastening. Joining by using of mechanical fasteners is still one of the most effective, widespread and reliable methods. This is caused by the fact that bonding and/or welding methods are not yet fully certificated for airspace applications in primary structures. There are several issues which must be solved in joints with fasteners: bearing strength, preparation of fasteners hole, fastener installation, corrosion, fastener interaction (pull-through, bending), clamp-up, stress concentrations, multiple row limitations, ply configuration effects and off-axis loading.

The paper discusses two main issues associated with failure mechanism of the joint – bearing strength and clamp-up. Typically, the bearing strength of composite materials is relatively lower compared with metals, and additionally the bearing strength is a function of the ply orientation and composition. At the same time tightening of joint could increase the bearing strength of a composite structure. This problem associates with the composite through the thickness feature. Despite several numerical approaches published in the past, widespread applications of carbon/PPS require huge experimental program for fastened joint behaviour verification.

The paper is focused on the numerical simulation and experimental verification of the bearing strength behaviour of two alternatives of the lay-up, positions of the hole, plate thicknesses and tightening conditions. Presented work associates with the design, simulation and optimization [9] of the thermoplastic rib [10], and possibilities to take into account correctly the boundary conditions of the real rib joining into the door structure.

Theory

Major failure modes of single-hole bolted composite joints [11] are shown on Fig. 1. The occurrence of a particular failure mode depends on joint geometry (bolt diameter, laminate width, edge distance and thickness) and laminate lay-up. Tension failure (see a) in Fig. 1) typically occurs, when the ratio of bolt diameter to strip width is sufficiently high. Cleavage failure is prone to occur if the bolt is in the proximity of end of specimen. However, for any given geometry may failure mode vary as a function of lay-up, stacking sequence, geometry or clamping boundary conditions.

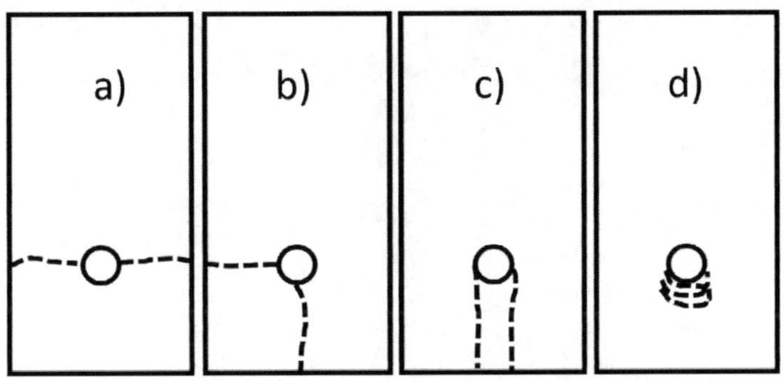

Fig. 1: Failure modes of composite bolted joint. a) Tension failure, b) Cleavage failure, c) Shear-out failure, d) Bearing failure.

Mechanical fastened joint in fibrous composite structures at first behaves like one-phase homogenous engineering material on which it is usually modelled. Together with increasing load the nonlinear behaviour prevails, and final failure occurs long after the nonlinearities appeared. Both fibers and matrix are essentially linear until failure; however, microcracks and delamination appear around bolt hole and cause the internal load redistributions, not accounted for in common mathematical models. More realistic simulation can be achieved, when each lamina is modelled separately.

Woven composites [12] present an interesting alternative to fiber-reinforced composites. In comparison to unidirectional (UD) lamina composites, woven composites have lower in-plane stiffness and strength due to the tow waviness. On the contrary, woven composites are more resistant to delamination and have superior impact response.

Continuum damage models are mostly developed for UD lamina composites. Typical models are Hashin's Failure Criteria [13], LaRC03, LaRC04, Maximum Stress, Tsai Wu or Puck Criteria for initiation of failure and Matzenmiller model [14] for constitutive material with accounted damage progression. These models, however, cannot be directly extrapolated to woven composites, due to the differences resulting from their dissimilar morphologies.

Materials and Methods

Paper is focused on AIMS05-09-002 [15] carbon fibre reinforced thermoplastic material with satin weave and PPS matrix. The two nominal thicknesses of 4 mm and 5.6 mm were investigated. Two variations of layups were used for specimen production. Two boundary conditions of the joint were considered - material clamped with clearance of 0.5 mm and tightened joint with torque of 20 Nm. The bearing strength was experimentally investigated based on the procedures defined in ASTM D5961 standard [16]. Basic test configuration is shown in Fig. 2.

Ultimate bearing strength σ_u and bearing strain were evaluated using following equations:

$$\sigma_u = \frac{P_u}{dt_n}, \tag{2}$$

$$\varepsilon_{br} = \frac{\delta}{d}, \tag{3}$$

where P_u is the maximum load sustained by specimen during the test, d is the nominal bolt diameter, t_n is the laminate thickness and δ is the extensometer displacement.

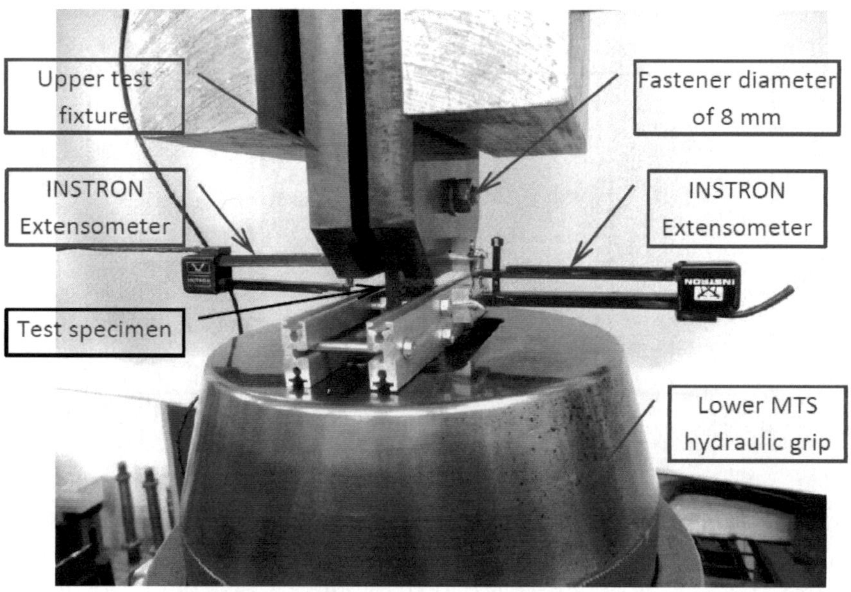

Fig. 2: Bearing test configuration

The ABAQUS software [17] for numerical simulation was used. Finite element model consists of the three parts – specimen, cylindrical fastener and clamping rigs. The fastener and clamping rigs were considered as rigid, undeformable parts. This assumption was made for the purpose of contact convergence; however, real stiffness of deformable part (composite specimen) is only about four times lower then real stiffness of presumably rigid parts (steel bolt and clamping rigs). For this reason, contact stresses cannot be assumed as strictly realistic. Idealization of bearing test in FEM is shown in Fig. 4.

Each composite ply was represented by a separate shell layer meshed with continuum shell elements SC8R [17]. Layers were connected by the cohesive contact with interlaminar stiffness characterization. Initiation of delamination damage was controlled by the Quadratic traction criterion; damage propagation was based on the mixed-mode Benzeggagh-Kenane law [18]. Hashin's Failure criteria for carbon fibre-reinforced polymer [13] were applied to account criterion for damage of matrix and fibres. Degradation process was estimated as quasistatic, hence the Abaqus implicit solver was used.

Simulations were made for each alternative of the joint configuration and boundary condition. Results obtained experimentally and numerically were compared in terms of bearing stress – bearing strain curves and failure modes.

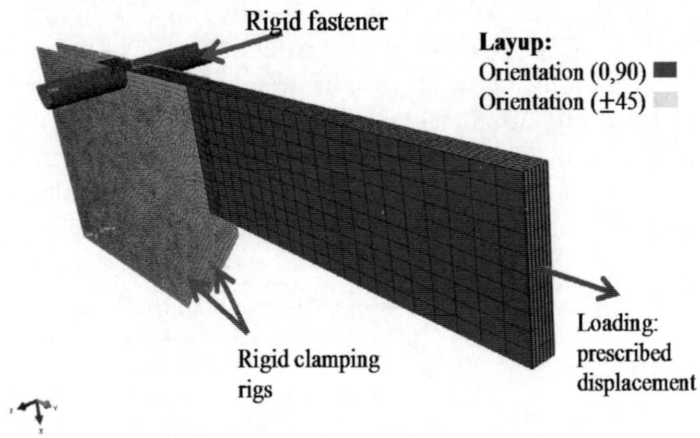

Fig 3: Idealization of bearing test facility in finite element model

Results

Influence of clamping boundary conditions and hole distances from the edge of specimen in the load direction on the ultimate bearing strength is shown in Fig. 4.

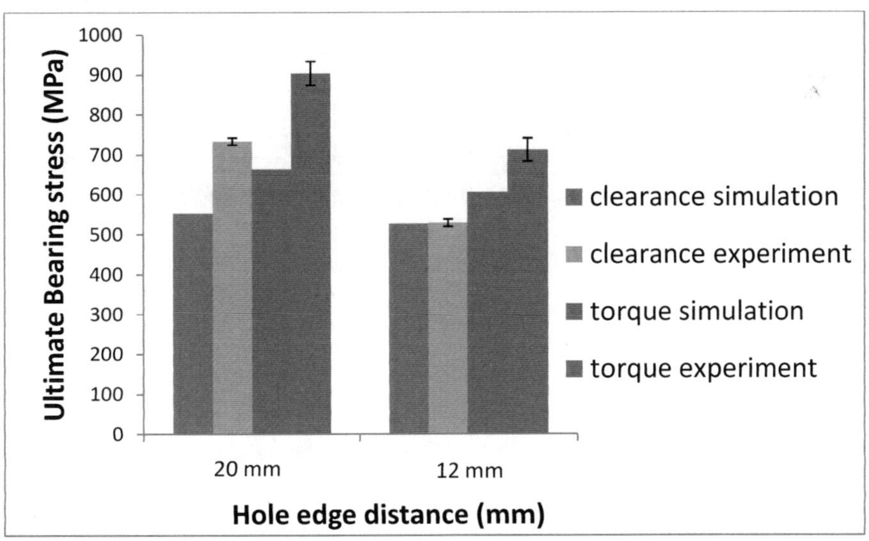

Fig. 4: The influence of boundary conditions of clamping with torque or with clearance and hole edge distance on Ultimate bearing stress determined numerically and experimentally.

Comparisons of bearing stress – bearing strain curves for different boundary conditions are shown in Fig. 5-6. Damage progression is illustrated on the example of Hashin Matrix Failure Initiation Criterion shown in Fig. 7. Results in Fig. 7 correspond to 4 mm thick specimen clamped with clearance and to the hole distance of 12 mm from the edge of the specimen (ED). Specimen is loaded in tension in horizontal direction. Results are shown for two representative plies with orientation (±45) and (0,90) and for three stages of damage progression. Scale value of 1 in Fig. 7 indicates the state in which the failure initiation criterion has been met and element undergoes stiffness degradation according to the associated evolution law.

Experimentally, significant increase of bearing strength was observed for clamping with torque in comparison to clamping with clearance. Similarly, bearing strength depends on the hole distance from the edge. Larger hole distance from the edge of the material led to higher bearing strength. Influence of used composite thickness, resp. layup variations, was unneglectable. Numerical simulations showed similar trends; however, positive effect of boundary conditions was significantly smaller in comparison with experimental data. It could be caused by using of continuum shell elements SC8R instead of solid elements for ply by ply calculation. Solid elements should ensure more precise distribution of contact and through-the-thickness stresses in comparison to continuum shell elements.

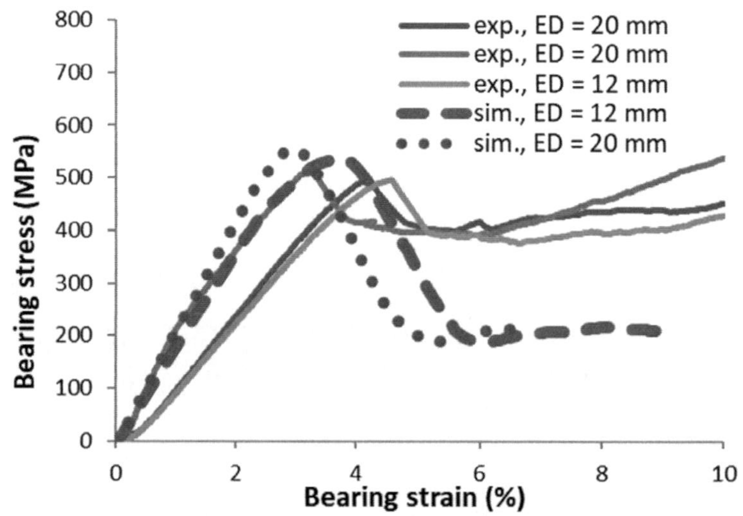

Fig. 5: Comparison of bearing stress – bearing strain curves obtained experimentally (exp.) and from numerical simulations (sim.) for specimens 5.6 mm thick clamped with clearance. ED stands for distance from the hole to the edge of the specimen.

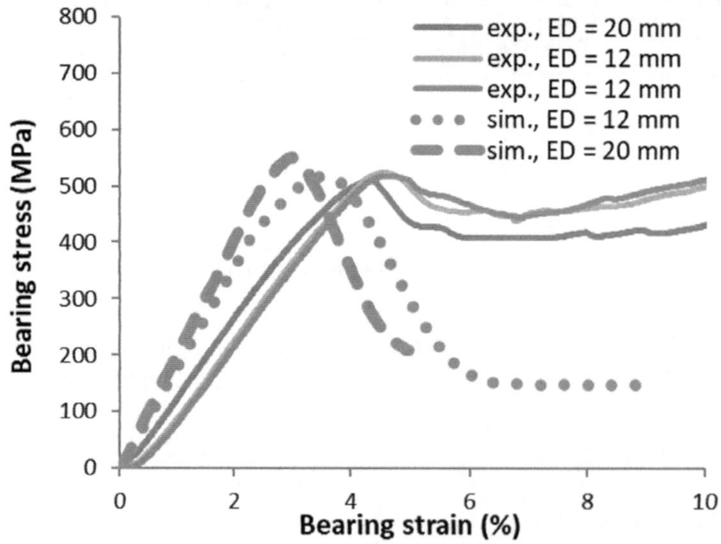

Fig. 6: Comparison of bearing stress – bearing strain curves obtained experimentally (exp.) and from numerical simulations (sim.) for specimens 5.6 mm thick clamped with clearance. ED stands for distance from the hole to the edge of the specimen.

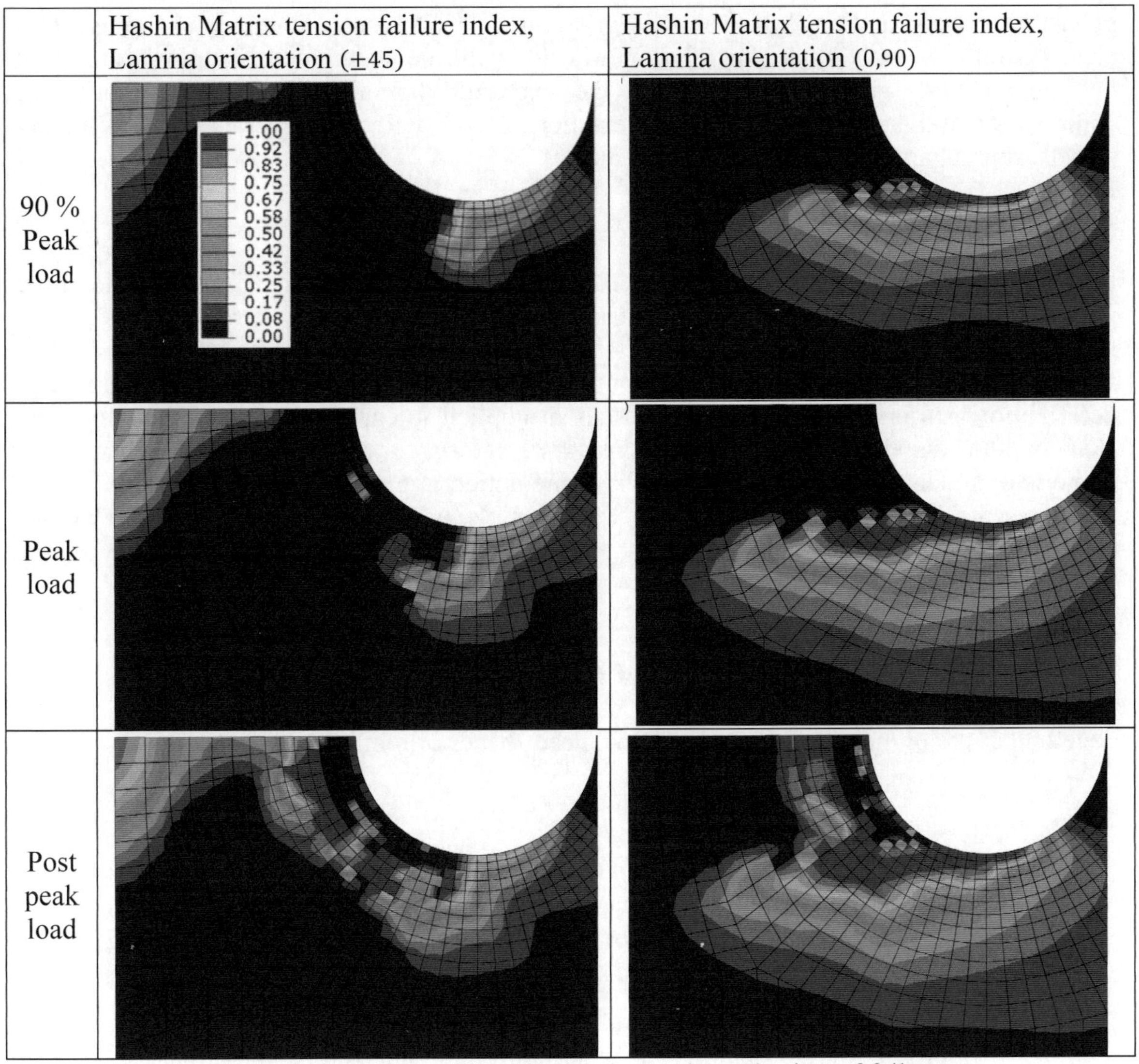

Fig. 7: Evolution of Hashin matrix tension criterion of failure.

The reasons for the other discrepancies between experiments and simulations may have resulted from multiple model idealizations. Firstly, the assumption of rigid fastener is incorrect and leads to more severe damage of specimen. Secondly, the continuum damage model based on Hashin failure criteria for fiber reinforced polymers, underestimates the influence of fiber reinforcement in transverse direction for woven reinforced composites. Moreover, real pressure distribution for boundary condition of clamping with torque is probably concentrated in the vicinity of the hole, while in simulation, the pressure was assumed uniform on the whole clamping surface.

Chosen modelling method was based on several assumptions and simplifications such as rigid fastener and clamping rigs, continuum shell elements for modelling of composite layer, mesh sizing or implicit integration scheme. The desired outcome consisted of relatively low time consumption of both the computation and evaluation of the analysis, easy accessibility of progressive continuum damage model, sufficient contact convergence and portability of the method to the global model of part of real construction.

Alternative choices are joined with considerable pros and cons. The explicit integration scheme should lead to better contact convergence for the price of higher time consumption of the analysis and higher numeric accumulation of errors. Softer meshing would lead to more precise stress distribution, however, with a significant increase of computation time. Solid elements for ply by ply modelling of composite and deformable bolt and clamping would lead to better contact stress distribution in the exchange for higher time consumption of computation and the necessity of implementation of user defined progressive damage model (e.g. Hashin's Failure Criteria for solid elements).

Conclusions

Influence of boundary conditions, such as clamping force and hole distance to the edge of the part, to bearing strength were experimentally verified. Similar trends were qualitatively but not quantitatively observed also numerically. Modelling of composite failure proved to be very complex task. Progressive damage modelling in combination with contact and large deformations suffers with convergence issues in implicit integration scheme. Using of rigid elements for dummy of load introduction system lead to convergence improvement of numerical models, but by contrast the lack of prediction is much higher. Explicit integration together with ply by ply methodology with flexible parts may to be more suitable for simulation.

Acknowledgement

This work was funded by the Ministry of Industry and Trade of the Czech Republic in the framework of TRIO platform, project No. FV30033 - Design and manufacturing process development of primary aircraft parts of advanced shapes of reinforced thermoplastics.

References

[1] M. Kadlec, L. Nováková, R. Růžek, An Experimental Investigation of Factors Considered for the Short Beam Shear Strength Evaluation of Carbon Fiber–reinforced Thermoplastic Laminates, Journal of Testing and Evaluation, 42(3) (2014) 580-592, http://dx.doi.org/10.1520/JTE20120043.

[2] R. Růžek, M. Kadlec, L. Petrusová, Effect of fatigue loading rate on lifespan and temperature of tailored blank C/PPS thermoplastic composite. International Journal of Fatigue, 113 (2018) 253-263. https://doi.org/10.1016/j.ijfatigue.2018.04.023.

[3] P. Homola, M. Kadlec, R. Růžek, J. Šedek, Fatigue behaviour of tailored blank thermoplastic composites with internal ply-drops, Procedia Struct. Integr. 5 (2017) 1342–1348, https://doi.org/10.1016/j.prostr.2017.07.144.

[4] A. Zeole, Experimental Investigation and Analysis for Bearing Strength Behavior of Composite Laminates, Wichita State University, Dept. of Mech. Engineering, 2006.

[5] Z. Chen, T. Phung, R. Paton, P. de Bruijn, Characterisation of a Reinforced PPS Thermoplastic Laminate For Forming Simulations. Composite Techologies for 2020: Proceedings of the Fourth Asian-Australasian Conference on Composite Materials (Accm 4), Elsevier, 2004, ISBN: 1845690621.

[6] J. Šedek, R. Hron, M. Kadlec, Bond Joint Analysis of Thermoplastic Composite Made from Stacked Tailored Blanks. Applied Mechanics and Materials, 827 (2015) 161-168, ISSN 1662-7482. DOI: 10.4028/www.scientific.net/AMM.827.161.

[7] J. Šedek, R. Hron, M. Kadlec, Evaluation of a bond joint in a carbon weave reinforced thermoplastic manufactured from stacked tailored blanks. In EAN 2015 - 53rd

Conference on Experimental Stress Analysis (EAN2015). Český krumlov (CZ), 1 – 4 June 2015, 399-402.

[8] A. Yousefpour, M., Hojjati, J.P. Immarigeon, Fusion Bonding/Welding of Thermoplastic Composites, Journal of Thermoplastic Composite Materials, 17(4) (2004) 303–341. https://doi.org/10.1177/0892705704045187.

[9] J. Šedek, Optimization of composite airframe rib using tailoring blank technology. In: EAN 2016 - 54th International Conference on Experimental Stress Analysis. Srní (CZ), 30 May – 2 June 2016.

[10] R. Růžek, J. Šedek, M. Kadlec, P. Kucharský, Mechanical behavior of thermoplastic rib under loading representing real structure conditions. In: EAN 2016 - 54th International Conference on Experimental Stress Analysis. Srní (CZ), 30 May – 2 June 2016.

[11] ASM Handbook Volume 21: Composites, ASM International, Editor: D.B. Miracle and S.L. Donaldson, 2001, ISBN: 978-0-87170-703-1.

[12] C.G. Huguet, A continuum damage mechanics model for woven composites. Masters Thesis. Delft University of Technology, 2017, http://resolver.tudelft.nl/uuid:63f1527a-b2f2-4474-9a63-20234a6057ae.

[13] Z. Hashin, A. Rotem, A fatigue criterion for fiber-reinforced materials. Journal of Composite Materials, 7 (1973) 448–464.

[14] Matzenmiller A., Lubliner J., Taylor R.L., „A constitutive model for anisotropic damage in fiber-composites“, Mechanics of Materials, 20 (1995) 125-152.

[15] AIMS Airbus Material Specification, AIMS05-09-002, 2007.

[16] ASTM D5961 Standard Test Method for Bearing Response of Polymer Matrix Composite Laminates. ASTM International, 100 Barr Harbor Drive, PO Box C700, West Conshohocken, PA 19428-2959, United States.

[17] Abaqus 2017 Documentation. Dassault Systemes Simulia Corporation; 2017.

[18] M.L. Benzeggagh, M. Kenane, Measurement of Mixed-Mode Delamination Fracture Toughness of Unidirectional Glass/Epoxy Composites with Mixed-Mode Bending Apparatus, Composites Science and Technology, 56 (1996) 439–449.

Index of Authors